Emergence, Complexity and Computation

Volume 40

The Emergence, Complexity and Computation (ECC) series publishes new developments, advancements and selected topics in the fields of complexity, computation and emergence. The series focuses on all aspects of reality-based computation approaches from an interdisciplinary point of view especially from applied sciences, biology, physics, or chemistry. It presents new ideas and interdisciplinary insight on the mutual intersection of subareas of computation, complexity and emergence and its impact and limits to any computing based on physical limits (thermodynamic and quantum limits, Bremermann's limit, Seth Lloyd limits…) as well as algorithmic limits (Gödel's proof and its impact on calculation, algorithmic complexity, the Chaitin's Omega number and Kolmogorov complexity, non-traditional calculations like Turing machine process and its consequences,…) and limitations arising in artificial intelligence. The topics are (but not limited to) membrane computing, DNA computing, immune computing, quantum computing, swarm computing, analogic computing, chaos computing and computing on the edge of chaos, computational aspects of dynamics of complex systems (systems with self-organization, multiagent systems, cellular automata, artificial life,…), emergence of complex systems and its computational aspects, and agent based computation. The main aim of this series is to discuss the above mentioned topics from an interdisciplinary point of view and present new ideas coming from mutual intersection of classical as well as modern methods of computation. Within the scope of the series are monographs, lecture notes, selected contributions from specialized conferences and workshops, special contribution from international experts.

Indexed by zbMATH.

More information about this series at https://link.springer.com/bookseries/10624

Xiong Wang · Nikolay V. Kuznetsov ·
Guanrong Chen
Editors

Chaotic Systems with Multistability and Hidden Attractors

Springer

Editors
Xiong Wang
Institute for Advanced Studies
Shenzhen University
Shenzhen, China

Guanrong Chen
Department of Electrical Engineering
City University of Hong Kong
Hong Kong, China

Nikolay V. Kuznetsov
Faculty of Information Technology
The University of Jyväskylä
Jyväskylä, Finland

Faculty of Mathematics and Mechanics
Saint Petersburg State University
Saint Petersburg, Russia

ISSN 2194-7287 ISSN 2194-7295 (electronic)
Emergence, Complexity and Computation
ISBN 978-3-030-75820-2 ISBN 978-3-030-75821-9 (eBook)
https://doi.org/10.1007/978-3-030-75821-9

This Springer imprint is published by the registered company Springer Nature Switzerland AG
The registered company address is: Gewerbestrasse 11, 6330 Cham, Switzerland

Dedicated to the Memory of
Professor Gennady Alexeyevich Leonov
(2 February 1947–23 April 2018)

Preface

Chaos theory for three-dimensional autonomous systems has been intensively and extensively studied since the time of Edward N. Lorenz in the 1960s, and the theory has become quite mature today. However, some of the recent findings reveal that the complexity and richness of this subject are far beyond our wildest imagination.

Recently, a great deal of interest in a new classification of attractors, proposed by Leonov and Kuznetsov, has emerged from various scientific and engineering communities. According to this new classification, there is a new kind of attractors named "hidden attractors" besides the conventional self-excited attractors. The term "hidden attractor" refers to the fact that the attractor is not associated with any unstable equilibrium and thus is invisible in simulations, because it may occur in a tiny region of the parameter space and has a small basin of attraction in the phase space. From a computational standpoint, it requires specific numerical methods to find the presence of such hidden attractors.

Hidden attractors have significant importance in both theory and practice. For example, there is a connection between multistability and hidden attractors; systems with only one stable equilibrium point can generate hidden chaotic attractors; hidden attractors allow unexpected and potentially disastrous responses to perturbations in a certain structure like a bridge or an airplane wing, and so on. Studying multistability and hidden attractors, therefore, is of extreme importance, which may reveal novel phenomena in nonlinear dynamical systems, such as non-Shilnikov type chaos, coexistence of different types of attractors, and spontaneous symmetry breaking in chaotic systems.

The aim of this book is to collect new articles from experts and active researchers to present their recent findings and progress in this new area of scientific research, including both theoretical advances and potential applications.

The Editors

Shenzhen, China — Xiong Wang
Jyväskylä, Finland/Saint Petersburg, Russia — Nikolay V. Kuznetsov
Hong Kong, China — Guanrong Chen
October 2020

Contents

Part I

Introduction

Xiong Wang, Nikolay V. Kuznetsov, and Guanrong Chen

1 Classical Chaotic Systems

1.1 Lorenz System

Chaos theory for three-dimensional autonomous systems has been intensively and extensively studied since the time of Edward N. Lorenz in the 1960s. When studying the atmosphere convection phenomenon, Lorenz found a chaotic attractor in the shape of a butterfly from a three-dimensional quadratic autonomous system described by

$$\begin{cases} \dot{x} = \rho(y - z) \\ \dot{y} = rx - y - xz \\ \dot{z} = -bz + xy \end{cases} \tag{1}$$

which is chaotic when $\rho = 10,\ r = 28,\ b = 8/3$.

This work was done within the Russian Science Foundation (project 19-41-02002).

X. Wang (✉)
Institute for Advanced Study, Shenzhen University, Shenzhen, Guangdong 518060, People's Republic of China
e-mail: wangxiong8686@szu.edu.cn

N. V. Kuznetsov
Faculty of Mathematics and Mechanics, Saint Petersburg State University, Saint Petersburg 198504, Russia; Faculty of Information Technology at the University of Jyväskylä, Box 35, Jyväskylä FIN-40014, Finland
e-mail: nkuznetsov239@gmail.com

G. Chen
Department of Electrical Engineering, City University of Hong Kong, Hong Kong SAR 999077, People's Republic of China
e-mail: eegchen@cityu.edu.hk

X. Wang et al. (eds.), *Chaotic Systems with Multistability and Hidden Attractors*, Emergence, Complexity and Computation 40,
https://doi.org/10.1007/978-3-030-75821-9_1

Ever since its discovery in 1963, the Lorenz system has been a paradigm of chaos and the Lorenz attractor has become an emblem of chaos. Lorenz himself thus has been marked by history as an icon of chaos theory.

System (1) is the most typical chaotic system with only two quadratic terms in its nonlinearity, which is very simple in the algebraic structure and yet is fairly complex in the dynamical behavior. The observation of these two seemingly contradictory aspects of the Lorenz system thereby triggered a great deal of interest from the scientific community to seek closely-related Lorenz-like systems, spurred by a variety of motivations from different perspectives.

1.2 Rössler System

Otto Rössler [30, 31] designed his chaotic system in 1976. The chaotic Rössler system was found to be quite useful in modeling equilibrium in chemical reactions. The system is described by

$$\begin{cases} \dot{x} = -y - z \\ \dot{y} = x + ay \\ \dot{z} = -b + z(x - c) \end{cases} \tag{2}$$

which is chaotic when $a = 0.20,\ b = 0.20,\ c = 5.7$.

Two types chaotic attractors were found from system (2): one is a single lobe chaotic attractor (spiral-type), and the other is a more complicated chaotic attractor (screw-type) [9, 36].

1.3 Chua's Circuit

The genesis of Chua's circuit was reported by Leon O. Chua and his colleagues right after Lorenz and Rössler, as the first circuit-implementable electronic system, known as Chua's circuit [26, 33]. Chua's circuit is considered an ideal paradigm for research on chaos by means of laboratory experiments and computer simulations, because it admits an adequate model for circuit implementation. In the simplest case, Chua system read as

$$\begin{cases} \dot{x} = \alpha(-x + y - h(x)) \\ \dot{y} = x - y + z \\ \dot{z} = -\beta y \end{cases} \tag{3}$$

where the nonlinear function $h(x)$ has the form:

$$h(x) = m_1 x + \frac{1}{2}(m_0 - m_1)(|x + 1| - |x - 1|)\,,$$

which is chaotic when $\alpha = 10.0,\ \beta = 14.87,\ m_0 = -1.27,\ m_1 = -0.68$.

Numerical analysis has verified that Chua's circuit could exhibit several distinct routes to chaos with a double-scroll chaotic attractor. There are three visible equilibria of the saddle-focus type, which is in sharp distinction with most known attractors of three-dimensional systems.

1.4 Chen System

To investigate whether the discovery of Lorenz system was just a lucky incident or there actually exist other closely related systems around it, many researchers had made further endeavor in the pursuit. In 1999, Chen [4, 37] gave a definite answer to this question by applying an engineering feedback anti-control approach, which could intentionally yield many chaotic systems. The coined chaotic system was lately referred to as the Chen system by others. It is in the following form:

$$\begin{cases} \dot{x} = a(y - z) \\ \dot{y} = (c - a)x + cy - xz \\ \dot{z} = -bz + xy \end{cases} \tag{4}$$

which is chaotic when $a = 35,\ b = 3,\ c = 28$.

Thereafter, it has been verified that the attractor in this system (4) exists [46], and it displays even more sophisticated dynamical behaviors than the Lorenz system [45].

Connection between Lorenz and Chen Systems

It is also interesting to recall a unified chaotic system [24], which was constructed to connect both the Lorenz system and the Chen system. This unified chaotic system is by nature a convex combination of the two systems, and is described by

$$\begin{cases} \dot{x} = (25\alpha + 10)(y - z) \\ \dot{y} = (28 - 35\alpha)x + (29\alpha - 1)y - xz \\ \dot{z} = -\frac{\alpha+8}{3}z + xy \end{cases} \tag{5}$$

Note that

- When $\alpha = 0$, the system (5) is the Lorenz system.
- When $\alpha = 1$, the system (5) is the Chen system.
- For any value of α in the interval (0,1), the system remains to be chaotic.

Sprott Systems

Besides the above classical chaotic systems, there are many other similar or related chaotic systems reported [1, 6, 7, 23, 25, 27, 39, 42–44]. It is worthwhile to refer to the nineteen Sprott systems, listed in Table 1 [34], some of which will be repeatedly referred to in the sequel and all of which are algebraically simpler than any of the systems that were previously known.

Beyond the Classical Chaotic Systems

To this end, despite all these exciting discoveries and fruitful results, the very essential and important question for chaos theory remains open:

What is the fundamental generating mechanism of chaos?

2 Šil'nikov Theory

To understand what types of systems can exhibit chaotic behavior, while for what types of systems this may not be impossible, a theorem referred to as the "*Šil'nikov theorem*" [32] is widely used. This theorem gives a proof to the situation where a three-dimensional hyperbolic system can generate self-excited chaos. It states that *If a three-dimensional hyperbolic system has a homoclinic (or heteroclinic) orbit connecting its equilibriums, and if its characteristic eigenvalues satisfy certain inequalities, then the system will have a Smale horseshoe attractor and in this sense it is chaotic*. For the present concern of chaos, a local C^1-linearization theorem of Hartman-Grobman [10] and a hyperbolicity criterion [11, 12] are applicable to the non-wandering part of the Smale horseshoe [3].

Since it is important to make this general question approachable, our first step is to focus on the dynamical systems described by a set of simple, mostly three-dimensional, autonomous ordinary differential equations with only one or two quadratic terms. Notice that, although limited to such simplest cases, the complexity and richness of the subject are still beyond our wildest imagination in general. This book attempts to "see a world in a grain of sand," and discusses chaotic systems from algebraic, geometric and analytical perspectives.

In the next chapter, the Šil'nikov theorem will be introduced and described in detail, its limitations will be discussed.

Table 1 Sprott systems [34]

Case	Equations	Equilibrium points	LEs	Dimension
A	$\begin{cases}\dot{x}=y\\ \dot{y}=-x+yz\\ \dot{z}=1-y^2\end{cases}$	*None*	0.014 0 −0.014	3.000
B	$\begin{cases}\dot{x}=yz\\ \dot{y}=x-y\\ \dot{z}=1-xy\end{cases}$	$(1,1,0)$ $(-1,-1,0)$	0.210 0 −1.210	2.174
C	$\begin{cases}\dot{x}=yz\\ \dot{y}=x-y\\ \dot{z}=1-x^2\end{cases}$	$(1,1,0)$ $(-1,-1,0)$	0.163 0 −0.163	2.140
D	$\begin{cases}\dot{x}=-y\\ \dot{y}=x+z\\ \dot{z}=xz+3y^2\end{cases}$	$(0,0,0)$	0.103 0 −1.320	2.078
E	$\begin{cases}\dot{x}=yz\\ \dot{y}=x^2-y\\ \dot{z}=1-4x\end{cases}$	$(0.25,0.063,0)$	0.078 0 −1.078	2.072
F	$\begin{cases}\dot{x}=y+z\\ \dot{y}=-x+0.5y\\ \dot{z}=x^2-z\end{cases}$	$(0,0,0)$ $(-2,-4,4)$	0.117 0 −0.617	2.190
G	$\begin{cases}\dot{x}=0.4x+z\\ \dot{y}=xz-y\\ \dot{z}=-x+y\end{cases}$	$(0,0,0)$ $(-2.5,-2.5,1)$	0.034 0 −0.634	2.054
H	$\begin{cases}\dot{x}=-y+z^2\\ \dot{y}=x+0.5y\\ \dot{z}=x-z\end{cases}$	$(0,0,0)$ $(-2,4,-2)$	0.117 0 −0.617	2.190
I	$\begin{cases}\dot{x}=0.2y\\ \dot{y}=x+z\\ \dot{z}=x+y^2-z\end{cases}$	$(0,0,0)$	0.012 0 −1.012	0.012
J	$\begin{cases}\dot{x}=2z\\ \dot{y}=-2y+z\\ \dot{z}=-x+y+y^2\end{cases}$	$(0,0,0)$	0.076 0 −2.076	2.037
K	$\begin{cases}\dot{x}=xy-z\\ \dot{y}=x-y\\ \dot{z}=x+0.3z\end{cases}$	$(0,0,0)$ $(-3.333,-3.333,11.111)$	0.038 0 −0.890	2.042
L	$\begin{cases}\dot{x}=y+3.9z\\ \dot{y}=0.9x^2-y\\ \dot{z}=1-x\end{cases}$	$(1,1.111,-0.231)$	0.061 0 −1.061	2.057
M	$\begin{cases}\dot{x}=-z\\ \dot{y}=-x^2-y\\ \dot{z}=1.7+1.7x+y\end{cases}$	$(2.406,-5.791,0)$ $(-0.706,-0.499,0)$	0.044 0 −1.044	2.042

(continued)

Table 1 (continued)

Case	Equations	Equilibrium points	LEs	Dimension
N	$\begin{cases} \dot{x} = -2y \\ \dot{y} = x + z^2 \\ \dot{z} = 1 + y - 2z \end{cases}$	$(-0.25, 0, 0.5)$	0.076 0 −2.076	2.037
O	$\begin{cases} \dot{x} = y \\ \dot{y} = x - z \\ \dot{z} = x + xz + 2.7y \end{cases}$	$(0, 0, 0)$ $(-1, 0, -1)$	0.049 0 −0.319	2.154
P	$\begin{cases} \dot{x} = 2.7y + z \\ \dot{y} = -x + y^2 \\ \dot{z} = x + y \end{cases}$	$(0, 0, 0)$ $(1, -1, 2.7)$	0.087 0 −0.481	2.181
Q	$\begin{cases} \dot{x} = -z \\ \dot{y} = x - y \\ \dot{z} = 3.1x + y^2 + 0.5z \end{cases}$	$(0, 0, 0)$ $(-3.1, -3.1, 0)$	0.109 0 −0.609	2.179
R	$\begin{cases} \dot{x} = 0.9 - y \\ \dot{y} = 0.4 + z \\ \dot{z} = xy - z \end{cases}$	$(-0.444, 1.111, -0.4)$	0.062 0 −1.062	2.058
S	$\begin{cases} \dot{x} = x - 4y \\ \dot{y} = x + z^2 \\ \dot{z} = 1 + x \end{cases}$	$(-1, 0.25, 1)$ $(-1, 0.25, -1)$	0.188 0 −1.188	2.151

3 Chaos beyond Šil'nikov

As mentioned above, for the three-dimensional hyperbolic type of chaotic systems, a commonly-used criterion for proving the existence of chaos is the Šil'nikov theory. However, this theorem offers only a sufficient condition, therefore it does not rule out the possibility for chaos to exist in many other systems.

Moreover, according to the Hartman-Grobman theorem, mathematically the local behaviors of a dynamical system in the neighborhood of a hyperbolic equilibrium point is quantitatively the same as the behavior of its linearized version near this equilibrium point. No eigenvalues of the linearized system with real parts equal to zero are considered by the theorem. Therefore, likewise it does not imply any dynamical behaviors in the domains far from the equilibria. As such, chaotic behavior in such domains still needs to be further investigated.

After getting rid of the shackles of the Šil'nikov criterion, one can immediately open up a new horizon, where a variety of new chaotic systems beyond Šil'nikov might exist and be found.

The journey of exploring various new chaotic systems, beyond Šil'nikov, began from 2012 when Wang and Chen [38] found a new interesting system that has one and only one stable equilibrium, by modifying the Sprott E system. A systematic search for similar chaotic systems with one stable equilibrium was pursued by Sprott and his colleagues later [25]. Thereafter, non-Šil'nikov type chaotic systems have been extensively studied by many others [13, 28, 29, 40, 41].

A variety of non-Šil'nikov type chaotic systems with no equilibria, a line of equilibria, a surface of equilibria, and even any desired number of equilibria, will be introduced and studied in the following chapters. By means of Lyapunov exponents, fractional dimensions, positive entropy, continuous broad frequency spectrums and bifurcation diagrams, the dynamical properties of these systems with hidden attractors will be analyzed in various details.

4 Hidden Attractors and Multi-Stability

4.1 Hidden Attractors

The attractors of the above-mentioned chaotic systems are of great interest and a new category of attractors has recently been characterized, as "hidden attractors" besides the conventional self-excited attractors. The term *hidden attractor* refers to the scenario where the attractor is not associated with any unstable equilibrium point and is often invisible in simulations. Because it may occur in a tiny region of the parameter space and have a small basin of attraction in the phase space. The concept and mathematical definition of hidden attractors were first introduced by Leonov and Kuznetsov [21, 22]. From a computational standpoint, it requires specific numerical methods to identify the presence of such hidden attractors [2, 5, 19, 20, 35]. Hidden attractors have significant importance in both theory and practice. For example, systems with only one stable equilibrium point can also generate hidden attractors, quite surprisingly from a theoretical point of view, and hidden attractors permit unexpected and potentially disastrous responses to perturbations in a certain structure like a bridge or an airplane wing, from a practical point of view [8, 17]. The classification of attractors as self-exited or hidden ones was a fundamental premise for the emergence of the theory of hidden oscillations, which represents the modern development of Andronov's theory of oscillations [14, 18]. The analysis of hidden oscillations, regarding their absence or presence as well as location, requires developing special analytical and numerical methods. It is key to determining the exact boundaries of the global stability, parts of which can be classified as trivial (i.e., determined by local bifurcations) or as hidden (i.e., determined by non-local bifurcations and by the birth of hidden oscillations) [18]–[15].

4.2 Multi-Stability

Referring to the properties of hidden attractors, it is easy to notice that there is a connection between multi-stability and hidden attractors. Multi-stability exists in almost all scientific and technological fields, such as physics, chemistry, mechanics, lasers, biological systems, lactose utilization networks, and even the Greenland ice sheet. Thus, it is of great importance since it poses potential threats and damages to practical engineering applications.

Most higher-dimensional systems, and even some lower-dimensional ones with multiple nonlinearities, often have coexisting attractors within their separate attracting basins. Therefore, studying multi-stability and hidden attractors is of extreme importance, which may reveal novel phenomena in nonlinear dynamical systems, such as Non-Šil'nikov type of chaos, coexistence of different types of chaotic attractors, and spontaneous symmetry breaking in chaotic systems.

Encouraged by the recent research progress, a variety of new discoveries of chaotic systems with hidden attractors will be introduced in this book. The multi-stability of chaotic systems is analyzed from the perspectives of observation, designs and detections. Along with the introduction of such new and interesting chaotic systems, chaos control, identification and symmetry breaking, circuit implementation as well as some engineering applications will be presented and discussed.

5 Organization of the Book

This book can roughly be divided into the following five parts.

5.1 Classical Šil'nikov Chaos

The first part is about the background and motivation of this book, and an introduction of the classical Šil'nikov chaos theory.

Chapter 2 discusses a useful analytical tool—the "Šil'nikov theorem". This chapter analyzes the complex dynamics in the neighborhood of a homoclinic loop to a saddle-focus, or the dynamics in the neighborhood of a heteroclinic loop of the simple type. The simplest form of the Šil'nikov theorem for three-dimensional autonomous hyperbolic systems, as well as the basic concepts and properties related to the Šil'nikov theory, will be discussed. This chapter provides an analytical background of the basic chaos theory, which will be useful throughout the whole book.

The rest of the book is organized as follows.

5.2 Chaotic Systems with Various Equilibria

The second part of the book mainly investigates the relationships between chaos and equilibrium points, which includes five chapters.

These chapters study a fundamental problem in the chaos theory—the relationship between the global dynamic behavior of the system and the number and type of its equilibrium points—what kind of equilibrium characteristics of a dynamical system may produce chaos? The traditional Šil'nikov type of chaos relies on the unstable equilibrium point(s) of the system. Homoclinic orbits and heteroclinic orbits based on the unstable equilibrium point(s) are necessary for the existence of chaos in the system. This part reviews most of the recently found new chaotic systems with structures beyond the traditional Šil'nikov framework. The focus is on the relatively simple three-dimensional autonomous quadratic systems.

Chapter 3 discusses a few simple three-dimensional autonomous quadratic systems with only one, or more than one stable equilibrium point. These new systems are of a non-hyperbolic type, for which the Šil'nikov homoclinic criterion is inapplicable. Having positive Lyapunov exponents, fractional dimensions, continuous broad frequency, and period-doubling routes, these systems are chaotic, which is a striking discovery.

Chapter 4 discusses chaotic flows without equilibria. Neither homoclinic nor heteroclinic orbits exist in such systems, thus it is also inapplicable by the Šil'nikov theorem. Since chaotic behavior can still be observed in the phase space for such systems, hidden attractors are investigated mainly via numerical simulations.

Chapters 5 and 6 introduce an attractive topic of finding three-dimensional chaotic systems with an infinite number of equilibria. Chapter 6 illustrates chaotic systems with a curve of equilibria, and the relation between its properties and the complex dynamics. Despite the fact that the basin of attraction in such a system has an intersection with the equilibria in some portions, an infinite number of the remaining equilibria are still located outside the basin of attraction, due to the essential "hidden" feature. This Chapter contains three parts: chaos with lines of equilibria, chaos with closed-curves of equilibria, and chaos with open-curves of equilibria. Chapter 6 describes some rare three-dimensional chaotic flows with surfaces of equilibria and presents a systematic method for finding them.

Chapter 7 shows how to construct a chaotic system that can have any preassigned number of equilibria and how the number of equilibria determines the shapes of the chaotic attractors. Two effective approaches are introduced and discussed.

5.3 Chaotic Systems with Various Components

After reviewing and discussing the aforementioned new chaotic systems, in simple three-dimensional autonomous quadratic forms with various types of equilibria, the next part of the book studies new chaos in more general and more complex systems both theoretically and numerically, with potential applications discussed. These various types of systems include hyperchaotic systems (from three dimensions to higher dimensions), fractional systems (from integer order to fractional order), jerk systems (from lower-order to higher-order ordinary differential equation), and memristor systems (from mathematical equations to physical circuit systems).

These different types of systems are more complex than the simple three-dimensional autonomous systems discussed above. They are of great importance in both theory and application. It will be shown that these systems have hidden attractors and multi-stabilities in various forms.

Chapter 8 summarizes some four-dimensional hyperchaotic systems with hidden attractors, while the basin of attraction of a hidden attractor does not intersect with a sufficiently small domain of any equilibrium. Various such hyperchaotic systems will be introduced, with no equilibria, limited equilibria, a curve of equilibria, or a surface of equilibria, respectively.

Chapter 9 applies the fractional calculus to chaotic systems and obtains fractional-order systems that still exhibit chaos. The Riemann-Liouville derivative and the Caputo derivative are most widely used. Through some manipulations, both three-dimensional and four-dimensional chaotic systems can be recast to fractional-order forms. The sample systems with hidden attractors are also of great importance as the fractional-order systems without equilibria or with other cases of the equilibria. It is shown that, by utilizing a new control scheme consisting of a single state variable, a fractional-order system with a stable equilibrium point can be stabilized.

Chapter 10 combines chaotic systems with circuit implementation. A memristive system with multi-stability and a hidden attractor is presented and discussed.

Chapter 11 discusses the simplest explicit third-order ordinary differential equation in the "jerk" form, which exhibits chaos. The term "jerk" represents the time derivative of acceleration in physics. Such a system is considered as being in the simplest form of three-dimensional dissipative systems with quadratic nonlinearities that can generate chaos. A systematic method for constructing such a jerk system is introduced. Then, a hyper-jerk system, coexistence of multiple attractors, and a jerk system with hidden attractors are investigated.

5.4 Multi-Stability in Various Systems with Different Characteristics

The fourth part mainly studies multi-stability.

Hidden attractors often coexist in more than one type. Even in the first-found example of chaotic system with only one stable equilibrium point, there are several attractors of different types. The equilibrium point is stable, therefore it is a point attractor of dimension zero. Chaotic attractors are a kind of strange attractors typically having dimensions between 2 and 3. In this system, a periodic solution is also found to be attractive, thus it is a one-dimensional attractor. So, including the chaotic one, there are three coexisting attractors of different dimensions in this system.

In the study of multi-stability properties, the multi-stability of symmetric systems is particularly interesting, because the symmetry of the system equations makes the system dynamics have some kinds of natural symmetry. If there is only one chaotic attractor in this kind of system, its attraction domain is symmetrical and its attractor is also symmetrical, such as the familiar Lorenz and Chen systems. But, if such a system has multi-stability, e.g., with multiple attractors, they often appear in symmetric pairs, but each attractor itself is not symmetrical.

Hidden attractors are closely related to multi-stability. The following chapters discuss multi-stability, including multi-stability in symmetric systems, multi-stability in asymmetric systems, multi-stability in conditionally symmetric systems, multi-stability in self-reproducing systems, and investigate the issue of multi-stability detection.

Chapter 12 describes different regimes of multi-stability in symmetric chaotic systems, including coexisting symmetric and/or symmetric pairs of strange attractors in three-dimensional chaotic systems, and coexisting symmetric pairs of strange attractors with tori in a four-dimensional hyperchaotic system.

Chapter 13 describes multi-stability in asymmetric dynamical systems. As a typical example, coexisting asymmetric strange attractors or limit cycles or even a mixture of chaotic and periodic orbits are found from the Rössler system. Additional feedback or extra dimension may destroy the symmetry of the dynamical structure and, therefore, hatch the coexisting asymmetric attractors.

Chapter 14 describes multi-stability in conditional symmetric dynamical systems. Conditional symmetry represents a special structure, which stands in the line of asymmetry but provides a new possibility for hosting coexisting symmetric attractors. In this case, the polarity balance is retored by the offset boosting of some of dimensions. Typically, the absolute value function or a trigonometric function can provide a polarity reversal under the offset boosting of the corresponding variable. A couple of new cases with conditional symmetry are found by an exhaustive computer search. Mismatched offset boosting results in asymmetric coexisting attractors because of the unbalanced polarity. From this point of view, conditional symmetry as a new structure may bridge symmetry and asymmetry together, reflecting the polarity balance of the variables in a dynamical system.

Chapter 15 describes multi-stability in self-reproducing systems. Offset boosting is a direct approach for moving an attractor in the phase space. When a periodic function is introduced into any dimension, the corresponding system becomes a self-reproducing one exhibiting infinitely distributed attractors. One-dimensional, two-dimensional and three-dimensional infinitely many chaotic or hyperchaotic attractors are obtained by periodically offset boosting from the initial conditions when a trigonometric function is introduced and a suitable period is set to the underlying system. Self-reproducing systems, therefore, represent general multi-stability for hosting countably infinitely many similar attractors with the same Lyapunov exponent spectrum.

Chapter 16 discusses the detection of multi-stability. Two basic methods for multi-stability detection are described, based on amplitude control and offset boosting, respectively. For a multi-stable system, basins of attraction are consequently linearly scaled or offset boosted, leaving a chance for a fixed initial condition to visit various basins of attraction. Furthermore, the method for extracting independent amplitude control is demonstrated based on the Sprott B system. Offset boostable chaotic systems are found via exhaustive computer search. As a result, when the initial condition goes through different basins of attraction triggering corresponding oscillations or dropping into a fixed point, the Lyapunov exponent spectrum shows random hopping for the fractional structure of the attraction region, which gives a flag for detecting multi-stable states.

5.5 Various Theoretical Advances and Potential Applications

This part is a collection of chapters presenting very recent achievements and progress in the concerned area of scientific research, including both theoretical advances and potential applications.

Chapter 17 describes a special class of delay impulsive differential equations with the coefficient of the delayed term given by a nilpotent matrix of index 2.

Chapter 18 discusses the possibility of mutual fusion of evolutionary algorithms, and hidden attractors identification and their synthesis. This chapter presents numerical demonstration of identification of the basin of attraction for hidden attractors (via evolutionary algorithms) as well as the possibility of its synthesis and design. In this chapter, there are also discussions on a few research questions, joined with biologically inspired algorithms, chaos theory, and hidden attractors.

Chapter 19 presents a new three-dimensional autonomous chaotic system with a sine trigonometric function, which can display hidden attractors. The fundamental dynamical properties of the system are discovered by calculating its equilibrium points as well as using phase portraits, Poincaré map, bifurcation diagram, Lyapunov exponents, and Kaplan–Yorke fractional dimension.

Chapter 20 discusses the finding of a new four-dimensional hyperchaotic system with infinitely many isolated equilibria, or without equilibria, which is obtained by adding a linear feedback controller to the first equation of a three-dimensional chaotic system.

Chapter 21 studies piecewise-affine systems and investigates the existence of homoclinic orbits and heteroclinic cycles, and extends the Šil'nikov theory to the chaos theory for a class of piecewise-smooth systems.

Chapter 22 proposes a new three-dimensional chaotic system with infinitely many equilibria located on a line. Investigation of dynamical properties of the system shows its potential to have various complex behaviors. Circuit implementation declares the feasibility of the system in real applications.

Chapter 23 analyzes a simple chaotic system. In some ranges of its parameters, this system has no equilibria, which means that it has hidden attractors. It is found that this system has two unstable equilibria located in other ranges of its parameter values.

Chapter 24 introduces a new three-dimensional chaotic system. The nonlinear terms of the system are quadratic, while there is no any linear term. Interestingly, the system has eight symmetric equilibrium points, which are located at the vertexes of a cube. Bifurcation analysis and circuit implementation of the system are also reported.

Chapter 25 introduces the *globally* attracting hidden attractors. Since not only is every initial condition in their basin of attraction, but every initial condition lies on the attractor, and thus they could hardly be less hidden. Such attractors have been known and studied long before the recent hoopla about hidden attractors, and they have other remarkable properties to be recounted here.

5.6 Discussions and Perspectives

The final part of the book presents some discussions of the origin of multi-stability in nonlinear dynamic systems, and the physics implication.

Chapter 26 discusses spontaneous symmetry breaking (SSB) which, in contrast to explicit symmetry breaking, is a spontaneous process in a system governed by a symmetrical dynamic procedure, ending up in an asymmetrical state. Thus, the symmetry of the equations is not reflected by the individual solutions, but is reflected by the symmetrical coexistence of asymmetrical solutions. The SSB provides a way of understanding the complexity of nature without renouncing the fundamental symmetry, which leads to the belief or preference of symmetric-to-asymmetric fundamental physical laws. Many illustrations of SSB are discussed from daily life scenario to mathematical nonlinear dynamic systems.

Chapter 27 discusses multi-ability of chaotic systems in reference to symmetry, diversity and practical theorems. The premise of multi-stability is that there is a unified kinematic equation governing the whole state space, so multi-stability does not emphasize on the diversity caused by different laws. In addition, the Noether and Goldstone theorems are introduced to promote a better understanding of the vacuum and the ground state of a multi-scroll system, as well as the mathematical representation of the Higgs mechanism.

References

1. S.K. Agrawal, M. Srivastava, S. Das, Synchronization of fractional order chaotic systems using active control method. Chaos, Solitons Fractals **45**(6), 737–752 (2012)
2. V.O. Bragin, V.I. Vagaitsev, N.V. Kuznetsov, G.A. Leonov, Algorithms for Finding Hidden Oscillations in Nonlinear Systems. The Aizerman and Kalman Conjectures and Chua's Circuits. Int. J. Comput. Syst. Sci. **50**(4), 511–543 (2011)
3. S.S. Cairns, *Differential and Combinatorial Topology* (Princeton University Press, Princeton, 1965)
4. G. Chen, T. Ueta, Yet another chaotic attractor. Int. J. Bifurc. Chaos **9**(07), 1465–1466 (1999)
5. G. Chen, N.V. Kuznetsov, G.A. Leonov, T.N. Mokaev, Hidden attractors on one path: Glukhovsky-Dolzhansky, Lorenz, and Rabinovich systems. Int. J. Bifurc. Chaos **27**(8), 1750115 (2017)
6. V. Daftardar-Gejji, S. Bhalekar, Chaos in fractional ordered Liu system. Comput. Math. Appl. **59**(3), 1117–1127 (2010)
7. W.H. Deng, C.P. Li, Chaos synchronization of the fractional lü system. Phys. A **353**(1), 61–72 (2005)

8. D. Dudkowski, S. Jafari, T. Kapitaniak, N.V. Kuznetsov, G.A. Leonov, A. Prasad, Hidden attractors in dynamical systems. Phys. Rep. **637**, 1–50 (2016)
9. R. Genesio, G. Innocenti, F. Gualdani, A global qualitative view of bifurcations and dynamics in the Rössler system. Phys. Lett. A **372**(11), 1799–1809 (2008)
10. P. Hartman, On local homeomorphisms of Euclidean space. Bol. Soc. Mat. Mexicana **5**(2), 220–241 (1960)
11. M. Hirsch, C. Pugh, Stable manifolds and hyperbolic sets. Proc. Symp. Pure Math. **14**, 133–163 (1970)
12. P.J. Holmes, A strange family of three-dimensional vector fields near a degenerate singularity. J. Differ. Equ. **37**(3), 382–403 (1980)
13. S. Kingni, S. Jafari, H. Simo, P. Woafo, Three-dimensional chaotic autonomous system with only one stable equilibrium: Analysis, circuit design, parameter estimation, control, synchronization and its fractional-order form. Eur. Phys. J. Plus **129**(5), 76 (2014)
14. N.V. Kuznetsov, M.Y. Lobachev, M.V. Yuldashev, R.V. Yuldashev, E.V. Kudryashova, O.A. Kuznetsova, E.N. Rosenwasser, S. M Abramovich, The birth of the global stability theory and the theory of hidden oscillations. 2020 European Control Conference Proceedings (European Control Association, St. Petersburg, 2020), pp. 769–774
15. N.V. Kuznetsov, T.N. Mokaev, O.A. Kuznetsova, E.V. Kudryashova, The Lorenz system: Hidden boundary of practical stability and the Lyapunov dimension. Nonlinear Dyn. (2020). https://doi.org/10.1007/11071-020-05856-4
16. N.V. Kuznetsov, V. Reitmann, *Attractor Dimension Estimates for Dynamical Systems: Theory and Computation (Dedicated to Gennady Leonov)* (Springer, Cham, 2021)
17. N.V. Kuznetsov, Hidden attractors in fundamental problems and engineering models. A short survey. Lect. Notes Electrical Eng. **371**, 13–25 (2016)
18. N.V. Kuznetsov, Theory of hidden oscillations and stability of control systems. Int. J. Comput. Syst. Sci. **59**(5), 647–668 (2020)
19. N.V. Kuznetsov, G.A. Leonov, M.V. Yuldashev, R.V. Yuldashev, Hidden attractors in dynamical models of phase-locked loop circuits: limitations of simulation in MATLAB and SPICE. Commun. Nonlinear Sci. Numer. Simul. **51**, 39–49 (2017)
20. N.V. Kuznetsov, G.A. Leonov, T.N. Mokaev, A. Prasad, M.D. Shrimali, Finite-time Lyapunov dimension and hidden attractor of the Rabinovich system. Nonlinear Dyn. **92**(2), 267–285 (2018)
21. G.A. Leonov, N.V. Kuznetsov, "Hidden attractors in dynamical systems. From hidden oscillations in Hilbert-Kolmogorov, Aizerman, and Kalman problems to hidden chaotic attractor in Chua circuits. Int. J. Bifurc. Chaos, **23**(1), 1330002 (2013)
22. G.A. Leonov, N.V. Kuznetsov, T. Mokaev, Homoclinic orbits, and self-excited and hidden attractors in a Lorenz-like system describing convective fluid motion. Eur. Phys. J.: Special Topics **224**(8), 1421–1458 (2015)
23. C. Li, G. Chen, Chaos and hyperchaos in the fractional-order rössler equations. Phys. A **341**(1–4), 55–61 (2004)
24. J. Lü, G. Chen, S. Zhang, The compound structure of a new chaotic attractor. Chaos, Solitons Fractals **14**(5), 669–672 (2002)
25. M. Molaie, S. Jafari, Simple chaotic flows with one stable equilibrium. Int. J. Bifurc. Chaos **23**(11), 699 (2013)
26. I. Petráš, A note on the fractional-order Chua's system. Chaos, Solitons Fractals **38**(1), 140–147 (2008)
27. I. Petráš, D. Bednárová, *Fractional-Order Chaotic Systems* (Springer, Berlin, 2011)
28. V.-T. Pham, C. Volos, T. Kapitaniak, Systems with stable equilibria, in *Systems with Hidden Attractors* (Springer, Berlin, 2017), pp. 21–35
29. V.T. Pham, X. Wang, S. Jafari, C. Volos, T. Kapitaniak, From Wang-Chen system with only one stable equilibrium to a new chaotic system without equilibrium. Int. J. Bifurc. Chaos **27**(6), 1750097 (2017)
30. O.E. Rössler, Continuons chaos: Four prototype equation. Ann. N. Y. Acad. Sci. **316**, 376–392 (1979)

31. O.E. Rössler, An equation for hyperchaos. Phys. Lett. A **71**(2–3), 155–157 (1979)
32. L.P. Shilnikov, Acontribution to the problem of the structure of an extended neighborhood of Arough equilibrium state of saddle-focus type. Math. USSR-Sbornik **10**(1), 92–103 (1970)
33. L.P. Shil'nikov, Chua's circuit: Rigorous results and future problems. Int. J. Bifurc. Chaos **04**(03), 9400037 (2014)
34. J.C. Sprott, Some simple chaotic flows. Phys. Rev. E **50**(2), R647 (1994)
35. N.V. Stankevich, N.V. Kuznetsov, G.A. Leonov, L.O. Chua, Scenario of the birth of hidden attractors in the Chua circuit. Int. J. Bifurc. Chaos **27**(12), 1730038 (2017)
36. M.S. Tavazoei, M. Haeri, Unreliability of frequency-domain approximation in recognising chaos in fractional-order systems. IET Signal Proc. **1**(4), 171–181 (2007)
37. T. Ueta, G. Chen, Bifurcation and chaos of Chen's equation. Proceedings of the IEEE International Symposium on Circuits and Systems, vol. 5 (IEEE, New York, 2000), pp. 505–508
38. X. Wang, G. Chen, A chaotic system with only one stable equilibrium. Commun. Nonlinear Sci. Numer. Simul. **17**(3), 1264–1272 (2012)
39. X.Y. Wang, M.J. Wang, Dynamic analysis of the fractional-order Liu system and its synchronization. Chaos **17**(3), 304–311 (2007)
40. Z. Wang, W. Sun, Z. Wei, S. Zhang, Dynamics and delayed feedback control for a 3d jerk system with hidden attractor. Nonlinear Dyn. **82**(1–2), 577–588 (2015)
41. Z. Wei, W. Zhang, Hidden hyperchaotic attractors in a modified Lorenz-Stenflo system with only one stable equilibrium. Int. J. Bifurc. Chaos **24**(10), 1450127 (2014)
42. X. Wu, H. Wang, A new chaotic system with fractional order and its projective synchronization. Nonlinear Dyn. **61**(3), 407–417 (2010)
43. Q. Yang, G. Chen, A chaotic system with one saddle and two stable node-foci. Int. J. Bifurc. Chaos **18**(05), 1393–1414 (2008)
44. Q. Yu, B.C. Bao, F.W. Hu, Q. Xu, M. Chen, J. Wang, Wien-bridge chaotic oscillator based on first-order generalized memristor. Acta Physica Sinica, **63**(24), 240505 (2014)
45. T. Zhou, Y. Tang, G. Chen, Complex dynamical behaviors of the chaotic Chen's system. Int. J. Bifurc. Chaos **13**(09), 2561–2574 (2003)
46. T. Zhou, Y. Tang, G. Chen, Chen's attractor exists. Int. J. Bifurc. Chaos **14**(09), 3167–3177 (2004)

Šil'nikov Theorem

Yu-Ming Chen

1 Dynamics in the Neighborhood of a Homoclinic Loop to a Saddle-Focus

The possibility of the existence of transitive invariant compact sets, which contain a countable set of rough periodic motions, is one of the fundamental singularities of multi-dimensional dynamical systems. Although some examples of such systems have been shown by Poincare, Birkhoff, Morse, and some other researchers, the conceptual foundation of systems with a countable set of rough periodic motions was laid in the works of Smale, based on the idea of roughness presented by Andronov and Pontrjagin. In 1961, Smale [1] first introduced an example of a rough diffeomorphism on a two-dimensional sphere, having an invariant subset that is homeomorphic to a topological Bernoulli process with two symbols. On the basis of the construction of this diffeomorphism, Smale developed a construction, which is now known as the "Smale horseshoe".

In Ref. [2], Šil'nikov approached a homoclinic loop to a saddle-focus with a positive saddle value, and found a countable set of periodic motions of the saddle type. He discovered that a map T, to which the problem reduces, has a countable number of "Smale horseshoes".

The simplest case corresponds to a three-dimensional autonomous system that has a homoclinic loop to an equilibrium, with one positive eigenvalue and two complex conjugate eigenvalues with negative real parts. By placing such an equilibrium at the origin $O(0, 0, 0)$, the three-dimensional system can be written in the form

Y.-M. Chen (✉)
Department of Mathematics and Computer Science, Gannan Normal University, Ganzhou 341000, P.R. China
e-mail: blkhpz@126.com

X. Wang et al. (eds.), *Chaotic Systems with Multistability and Hidden Attractors*, Emergence, Complexity and Computation 40,
https://doi.org/10.1007/978-3-030-75821-9_2

$$\dot{x} = \rho x - \omega y + P(x, y, z),$$
$$\dot{y} = \omega x + \rho y + Q(x, y, z),$$
$$\dot{z} = \lambda z + R(x, y, z),$$

where P, Q, R are smooth functions, along with their first-order partial derivatives, vanishing at the origin O. The eigenvalues of the equilibrium at the origin of this system are $\lambda > 0$ and $\lambda_{1,2} = \rho \pm i\omega$, where $\rho < 0$ and $\omega > 0$. The unstable manifold W^u is a curve, which is tangent to the z-axis. The stable manifold W^s is a two-dimensional surface, which is tangent to the plane $z = 0$. If one restricts the system to its stable manifold W^s, then the equilibrium O will be a stable focus. This is why the equilibrium O is called a saddle-focus in the full system.

Under the assumption that there is a homoclinic orbit to the saddle-focus O of the above system, the following theorem was established in Ref. [2].

Theorem 2.1 *Reference [2] Let the saddle value $\lambda + \rho$ be positive. Then, in any arbitrarily small neighborhood of the homoclinic loop to the saddle-focus, there exist infinitely many saddle periodic orbits.*

A countable set of three-dimensional "Smale horseshoes" was demonstrated for a four-dimensional system with a saddle-focus equilibrium in Ref. [3].

In Ref. [4], which employs a countable set of symbols to describe the structure of a neighborhood of a homoclinic loop to a saddle-focus, it presents an extension of the results in Refs. [2] and [3] to systems of arbitrary dimensions.

Consider a system of $m + n$-dimensional differential equations,

$$\frac{dx}{dt} = Ax + f(x, y), \quad \frac{dy}{dt} = By + g(x, y), \tag{1}$$

where A is an m-dimensional matrix with eigenvalues $\lambda_1, \cdots, \lambda_m$ lying to the left of the imaginary axis, and B is an n-dimensional matrix with eigenvalues $\gamma_1, \cdots, \gamma_n$ lying to the right of the imaginary axis. Here, $f(x, y)$ and $g(x, y)$ are analytic functions in some regions of $\mathbb{R}^{m+n}$, and these functions and their first derivatives vanish at $x = y = \mathbf{0}$. Without loss of generality, assume that

$$\mathrm{Re}\gamma_1 \leq \mathrm{Re}\gamma_i, \ i = 2, \ldots, n, \quad \mathrm{Re}\gamma_1 \leq -\mathrm{Re}\lambda_j, \ j = 1, \ldots, m.$$

Let W^s be an m-dimensional stable manifold with a saddle point O, and let W^u be an n-dimensional unstable manifold.

Assuming that, for the system (1), there exists a homoclinic loop Γ to the saddle equilibrium O, and the intersection of the manifolds W^s and W^u along Γ satisfies

$$\dim(W^s_M \bigcap W^u_M) = 1,$$

where W^s_M and W^u_M are the tangent spaces to manifolds W^s and W^u, respectively, at the point $M \in \Gamma$. Furthermore, assume that

1. The eigenvalues γ_1 and γ_2 are complex conjugate, and $\mathrm{Re}\gamma_1 < -\mathrm{Re}\lambda_j$, $j = 1, \ldots, m$.
2. $\mathrm{Re}\gamma_1 < \mathrm{Re}\gamma_i$, $i = 3, \ldots, n$. Under this assumption, almost all trajectories which lie in W^u will be improper O^--trajectories, and as $t \to -\infty$ they will be tangent to a two-dimensional conducting plane.
3. Γ converges to the saddle point O as $t \to -\infty$, being tangent to the conducting plane.
4. There is a certain quantity $\delta \neq 0$ (see Lemma 3.2 of [4]).

Consider the set of doubly infinite sequences,

$$\Omega(\rho) = (\ldots, j_i, j_{i+1}, \cdots), \tag{2}$$

consisting of the symbols 0, 1, 2, ..., and for arbitrary adjacent indices in (2) satisfying the condition that $j_{i+1} < \rho j_i$ for some $\rho > 1$. Under the above assumptions, the following theorem can be obtained.

Theorem 2.2 *Reference [4] In any extended neighborhood of the saddle-focus of the system (1), which satisfies conditions 1, 2, 3 and 4, there exists a subset of trajectories that is in one-to-one correspondence with the set $\Omega(\rho)$, where ρ does not exceed $(-Re\lambda_1)Re\gamma_1^{-1}$.*

Remark 2.1 Under other general assumptions, the condition that the complex eigenvalues with real parts being the smallest in absolute value is necessary for the existence of periodic motions. Otherwise, under the assumption that γ_1 is a real root, as shown in Ref. [5], generally an arbitrary neighborhood of Γ does not contain periodic motions, and by small perturbations to Γ only one periodic motion can be obtained.

Remark 2.2 For the case of $n = 2$, conditions 2 and 3 are automatically satisfied.

The method for proving the Theorem 2.2 is analogous to the method for proving the Poincaré-Birkhoff problem [6], which constructs a map T whose domain of definition is representable as a countable union of mutually disjoint domains where T is of saddle type.

With the help of a local C^1-linearization theorem by Hartman [7] and the hyperbolicity criteria [8, 9], which apply in particular to the non-wandering set of a horseshoe [10], Tresser [11] improved (at least in dimension 3) Theorems 2.1 and 2.2 obtained by Šil'nikov for certain real analytic vector fields, as further discussed below.

Theorem 2.3 *Reference [11] Consider the system*

$$\begin{aligned}
\dot{x} &= \rho x - \omega y + P(x, y, z),\\
\dot{y} &= \omega x + \rho y + Q(x, y, z),\\
\dot{z} &= \lambda z + R(x, y, z),
\end{aligned} \tag{3}$$

where P, Q, R are $C^{1,1}$ functions which, as well as their first order partial derivatives, vanish at the origin $O = (0, 0, 0)$. Suppose that there exists a homoclinic orbit

Γ, bi-asymptotic to O, which remains at a finite distance from any other singularity. Suppose, at least, that

$$\lambda > -\rho > 0. \tag{4}$$

Suppose also that θ is a first return map correctly defined on a well-chosen piece of surface π_0, and

$$\Sigma^m = \{1, \cdots, m\}^{\mathbb{Z}},$$
$$\Sigma^{*,\alpha} = \left\{ \underline{s} = \{s_i\}_{i=-\infty}^{+\infty} | s_i \in \mathbb{Z} - \{0\} \text{ and } |s_{i+1}| \geq \frac{|s_i|}{\alpha} \right\}.$$

Then, the following conclusions hold:

(a) for each positive integer m, there exists a map,

$$h_m : \Sigma^m \to \pi_0,$$

which is a homeomorphism of Σ^m onto $O_m = h_m(\Sigma^m)$, such that

$$\theta|_{O_m} = h_m \circ \sigma \circ h_m^{-1};$$

(b) the set O_m is hyperbolic;
(c) for each real α with $1 \leq \alpha \leq \lambda/|\rho|$, there exists a map,

$$h_{*,\alpha} : \Sigma^{*,\alpha} \to \pi_0,$$

which is a homeomorphism of $\Sigma^{*,\alpha}$ onto $O_{*,\alpha} = h_{*,\alpha}(\Sigma^{*,\alpha})$, such that

$$\theta|_{O_{*,\alpha}} = h_{*,\alpha} \circ \sigma \circ h_{*,\alpha}^{-1}.$$

Remark 2.3 In Theorem 2.3, conclusion (c) implies (a). Conclusion (a) is isolated since (a) and (b) together describe the well-known hyperbolic sets.

Remark 2.4 One of the consequences of (a) and (c) is that any neighborhood of Γ contains infinitely many periodic orbits of the saddle type, which was the main conclusion in Refs. [2] and [4], as shown by Theorems 2.1 and 2.2.

Remark 2.5 The most direct consequence of Theorem 2.3 is that it applies as well to system (3), but with

$$-\lambda > \rho > 0 \tag{5}$$

to replace the eigenvalues condition (4), since one only has to reverse time to get the hypotheses of Theorem 2.3.

Remark 2.6 On the contrary, when

$$-\rho > \lambda > 0 \quad (\text{or } \rho > -\lambda > 0), \tag{6}$$

under the hypotheses of Theorem 2.3, except that the eigenvalues condition (6) is used to replace (4), there is no periodic orbit in any neighborhood of Γ if it is chosen small enough [5].

2 Dynamics in the Neighborhood of a Heteroclinic Loop of the Simple Type

Let $X \in \mathbb{R}^3$ be a $C^{1,1}$ vector field and $O_i, i = 1, \ldots, n$, be hyperbolic equilibria of X. All equilibria O_i are supposed to possess a one-dimensional unstable manifold $W^u_{O_i}$ and a two-dimensional stable manifold $W^s_{O_i}$, with

$$\Gamma_i = W^u_{O_i} \cap W^s_{O_{i+1}} \neq \emptyset, \quad i = 1, \ldots, n,$$

where the simplified notation $O_{n+i} = O_i$ has been used.

The intersections $\Gamma_i, i \in \{1, \ldots, n\}$ are called heteroclinic connections and the union

$$\Gamma_0 = \bigcup_{i=1}^{n} (\Gamma_i \cup O_i)$$

is called a "heteroclinic loop".

In the remainder of this section, the following notations and hypotheses will be used and assumed:

1. $W_i^{u^+} = \Gamma_i \cup O_i$, the other part of $W^u_{O_i}$ being denoted by $W_i^{u^-}$;
2. each Γ_i is bounded away from all equilibria, except O_i and O_{i+1};
3. the vector field X has m saddles, with $0 \leq m < n$, and for each O_j of the saddle type, the eigenvalues of the linearized flow at O_j are

$$\lambda_{j,1} > 0 > \lambda_{j,2} > \lambda_{j,3};$$

4. for each saddle-focus O_i, the eigenvalues of the linearized flow at O_i are

$$\lambda_i > 0, \quad \rho_i \pm \mathrm{i}\omega_i \text{ with } \rho_i < 0;$$

5. the non-resonance hypotheses of saddles for flows are satisfied, namely,

$$\forall k: \quad |e^{\lambda_k}| \neq |e^{\lambda_i}| \cdot |e^{\lambda_j}|, \quad |e^{\lambda_i}| \leq 1 \leq |e^{\lambda_j}|. \tag{7}$$

Definition 2.1 *(Saddle of the simple type)* Reference [4] Let Γ_0 be a heteroclinic loop joining the equilibria O_i, $1 \leq i \leq n$ and let O_j be a saddle for some $j \in \{1, \cdots, n\}$.

Equilibrium O_j is said to be of the simple type if $W^s_{O_{j+1}} \cup O_j$ contains a disk, which in turns contains $W^{u^+}_{O_j}$, and the local part of the strongly stable manifold corresponding to $\lambda_{j,3}$. In the sequel, such a disk, say D_j, is assumed to contain O_j and O_{j+1} in its interior. Each D_j generates two half-tubes $\mathfrak{T}^{\pm}_j$ and $\mathfrak{T}^{+}_j$ is referred to the one that contains the local part of Γ_{j-1} near O_j, which is tangent to the eigenvector corresponding to $\lambda_{j,2}$.

Definition 2.2 *(Heteroclinic loop of the simple type)* Reference [4] A heteroclinic loop involving at least one saddle-focus is said to be of the simple type if the following conditions are satisfied:

1. all saddles are of the simple type;
2. for each of them, $\mathfrak{T}^{+}_j$ contains $W^{u^+}_{j+1}$; and
3. Γ_0 always follows the leading direction in $W^s_{O_j}$ near the saddles.

With the above notations and hypotheses, the following theorem about the dynamics in the neighborhood of heteroclinic loop Γ_0 of the simple type can be obtained.

Theorem 2.4 *Let X be a $C^{1,1}$ vector field in $\mathbb{R}^3$, Γ_0 be a heteroclinic loop of the simple type such that all saddles in Γ_0 verify the non-resonance conditions (7) [4]s, and let*

$$p = \left(\prod_{O_i \text{ is a saddle-focus}} \left(-\frac{\lambda_i}{\rho_i}\right)\right) \times \left(\prod_{O_j \text{ is a saddle}} \left(-\frac{\lambda_{j,1}}{\lambda_{j,2}}\right)\right).$$

Then,

1. if $p > 1$, one has the conclusions of Theorem 2.3, with $\Sigma^{*,\alpha}$ replaced by $\Sigma^{*,\alpha}_{2(n-m-1)}$ and $1 \leq \alpha \leq p$; and
2. if $p < 1$, one has the conclusion of Remark 2.6.

3 Simplest Form of the Šil'nikov Theorem

In this section, based on the above theorems, which are about the dynamics in the neighborhoods of a homoclinic loop to a saddle-focus and of a heteroclinic loop of the simple type, in Ref. [12] it presents a tutorial look at the simplest form of the Šil'nikov Theorem and its various extensions.

Consider the three-dimensional autonomous dynamical system

$$\frac{dx}{dt} = f(x),\ t \in \mathbb{R},\ x \in \mathbb{R}^3, \tag{8}$$

where the vector field $f(x) : \mathbb{R}^3 \to \mathbb{R}^3$ is a C^p $(p \geq 1)$ function. Let x_e be a saddle-focus equilibrium of system (8), and the eigenvalues of x_e be

$$\lambda, \quad \rho \pm \mathrm{i}\omega, \quad \rho\lambda < 0, \omega \neq 0.$$

Theorem 2.5 *(Homoclinic Šil'nikov method) Reference [12] Given the three-dimensional autonomous system (8), where $f(x)$ is a C^2 vector field on $\mathbb{R}^3$. Let x_e be an equilibrium of system (8). Suppose that*

1. *the equilibrium x_e is a saddle-focus with characteristic eigenvalues satisfying the Šil'nikov inequality*

$$|\lambda| > |\rho| > 0; \tag{9}$$

2. *there exists a homoclinic orbit Γ based at x_e.*

Then,

1. *the Šil'nikov map defined in a neighborhood of Γ possesses a countable number of Smale horseshoes in its discrete dynamics;*
2. *for any sufficiently small C^1-perturbation $g(x)$ of $f(x)$, the perturbed system*

$$\frac{dx}{dt} = g(x), \ x \in \mathbb{R}^3 \tag{10}$$

   has at least a finite number of Smale horseshoes in the discrete dynamics of the Šil'nikov map defined near Γ;
3. *both the original system (8) and the perturbed system (10) exhibit horseshoe chaos (also known as homoclinic chaos).*

Remark 2.7 Function $g(x)$ is a C^1-perturbation of $f(x)$, which means that the norm of the difference $g(x) - f(x)$ and its first derivative are sufficiently small in a neighborhood containing Γ.

Remark 2.8 Under the condition (9), one has the eigenvalues condition (4) in Theorem 2.3, or eigenvalues condition (5) in Remark 2.5. Thus, the conclusions of Theorem 2.3 hold.

Theorem 2.6 *(Heteroclinic Šil'nikov method) [12] Given the three-dimensional autonomous system (8), where $f(x)$ is as in Theorem 2.5. Let x_{e_1} and x_{e_2} be two distinct equilibria of (8). Suppose that*

1. *both x_{e_1} and x_{e_2} are saddle-foci that satisfy the Šil'nikov inequality*

$$|\lambda_i| > |\rho_i| > 0 \quad (i = 1, 2) \tag{11}$$

   with the constraints

$$\rho_1\rho_2 > 0 \quad or \quad \lambda_1\lambda_2 > 0; \tag{12}$$

2. *there is a heteroclinic loop* Γ_0 *joining* x_{e_1} *to* x_{e_2} *that is made up of two heteroclinic orbits* Γ_i $(i = 1, 2)$.

Then, the conclusion $1 - 3$ *of Theorem 2.5 hold again, with* Γ *being replaced by* Γ_0*, equilibrium* x_e *by* x_{e_i} $(i = 1, 2)$*, and homoclinic chaos by the corresponding term heteroclinic chaos.*

Remark 2.9 Under the eigenvalues conditions (11) and (12), the number p in Theorem 2.4 is bigger than one. Thus, the first conclusion of Theorem 2.4 holds.

References

1. S. Smale, A structurally stable differentiable homeomorphism with an infinite number of periodic points. Nelin. Dinam. **ii**, 445–446 (1961)
2. L.P. Šil'nikov, A case of the existence of a denumerable set of periodic motions. Sov. Math. Doklady, **160**(3), 558–561 (1965)
3. L.P. Šil'nikov, Existence of a countable set of periodic motions in a four-dimensional space in an extended neighborhood of a saddle-focus. Dokl. Akad: Nauk USSR, **172**(1), 54–57 (1967)
4. L.P. Šil'nikov, A contribution to the problem of the structure of an extended neighborhood of a rough equilibrium state of saddle-focus type. Math. USSR-Sbornik **10**(1), 92–103 (1970)
5. L.P. Šil'nikov, On the generation of a periodic motion from trajectories doubly asymptotic to an equilibrium state of saddle type. Math. USSR-Sbornik **6**(3), 461–472 (1968)
6. L.P. Šil'nikov, On a poincarÉ-birkhoff problem. Math. USSR-Sbornik, **3**(3), 353–371 (1967)
7. P. Hartman, On local homeomorphisms of Euclidean space. Bol. Soc. Mat. Mexicana **5**, 220–241 (1960)
8. M. Hirsch, C. Pugh, Stable manifolds and hyperbolic sets. Proc. Symp. Pure Math. **14**, 133–163 (1970)
9. P.J. Holmes, A strange family of three-dimensional vector fields near a degenerate singularity. J. Differ. Equ. **37**(3), 382–403 (1980)
10. S.S. Cairns et al., *Differential and Combinatorial Topology: A Symposium in Honor of Marston Morse* (Princeton University Press, Princeton, 1965)
11. C. Tresser, *About some theorems by L. P. Šil'nikov* (Physique théorique, Annales De L'I. H. P., 1984)
12. C.P. Silva, Šil'nikov's theorem: A tutorial. IEEE Trans. Circuits Syst. I: Fundam. Theory Appl. **40**(10), 675–682 (1993)

Part II

Chaotic Systems with Stable Equilibria

Xiong Wang and Guanrong Chen

1 Introduction

In this chapter, some three-dimensional autonomous quadratic chaotic systems with stable equilibria are introduced. The new systems are of non-hyperbolic type, therefore the homoclinic Šil'nikov method is inapplicable. Having positive largest Lyapunov exponents, fractional dimensions, continuous broad frequency spectra and period-doubling routes, the new systems are indeed chaotic. Several systems proposed by Sprott [1–3], Wang [4], Wei [5–9], Yang [6, 7, 10, 11], and some others will be introduced as well, providing an overview of such new chaotic flows with stable equilibria.

2 Motivation

Many early examples of three-dimensional chaotic flows occur in system with one or more unstable saddle points, which are justified by the Šil'nikov theorem. A self-excited attractor defined by Leonov and Kuznetsov [12] is involved in such a system having a basin of attraction associated with an unstable equilibrium. These systems are expected to be chaotic in the sense of Šil'nikov [13], allowing homoclinic and heteroclinic orbits [14].

X. Wang (✉)
Institute for Advanced Study, Shenzhen University, Shenzhen 518060, Guangdong, People's Republic of China
e-mail: wangxiong8686@szu.edu.cn

G. Chen
Department of Electrical Engineering, City University of Hong Kong, Hong Kong SAR 999077, People's Republic of China
e-mail: eegchen@cityu.edu.hk

X. Wang et al. (eds.), *Chaotic Systems with Multistability and Hidden Attractors*, Emergence, Complexity and Computation 40,
https://doi.org/10.1007/978-3-030-75821-9_3

Although the Šil'nikov theorem ensures horseshoe chaos to exist with a homoclinic orbit if its characteristic eigenvalues with negative real parts at the equilibria satisfy some specific conditions, it does not rule out the possibility of encountering chaos in systems with stable equilibria.

According to the Hartman–Grobman theorem, mathematically the local behaviors of a dynamical system in the neighborhood of a hyperbolic equilibrium point is qualitatively the same as the behavior of its linearized system near this equilibrium point. However, it does not cover the domain far from the equilibrium, thus chaotic behavior in such a domain is still possible, hence needs further investigation.

In the last few years, the concept of hidden attractor was introduced, which is different from self-excited attractors and is considered to have a basin of attraction that does not intersect with small neighborhoods of any equilibrium point. In 2008, Yang and Chen [10] found a three-dimensional system with three fixed points: one saddle and two stable equilibria. Moreover, a few new chaotic systems with stable equilibria were discovered by Wei and Yang [6, 11, 15], which can generate strange attractors not satisfying the Šil'nikov theorem. In 2012, Wang and Chen [4] coined a new system based on the Sprott E system (see Table 1), referred to as the Wang-Chen system, which has one and only one stable equilibrium point.

All the above had motivated a great deal of research effort in studying hidden attractors and chaos in systems with stable equilibria.

3 First Example on Chaos with One Stable Equilibrium

Sprott searched and found 19 distinct simple examples of chaotic flows with quadratic nonlinearities [1]. Some systems with only one equilibrium are listed in Table 1. In this table, systems I, J, L, N and R all have only one saddle-focus equilibrium, while systems D and E both have a degenerate equilibrium.

It is clear that the equilibria of systems D and E are not stable. Nevertheless, it is quite natural to imagine that a small perturbation to the system may be able to change such a degenerate equilibrium to a stable one. Therefore, to generate such a new system, Wang and Chen [4] started from the Sprott E system (see Table 1), having a degenerate equilibrium $E(\frac{1}{4}, \frac{1}{16}, 0)$, with eigenvalues $\lambda_1 = -1$ and $\lambda_{2,3} = \pm 0.5i$, described by

$$\begin{cases} \dot{x} = yz \\ \dot{y} = x^2 - y \\ \dot{z} = 1 - 4x. \end{cases} \tag{1}$$

In order to change the stability of its single equilibrium to a stable one, while preserving its chaotic dynamics, Wang and Chen added a constant control parameter a to the Sprott E system (see Table 1), so as to form a new system as follows:

Table 1 Equilibria and eigenvalues of the Sprott systems [4]

Systems	Equations	Equilibria	Eigenvalues
Case D	$\dot{x} = -y$ $\dot{y} = x + z$ $\dot{z} = xz + 3y^2$	(0, 0, 0)	$0, \pm i$
Case E	$\dot{x} = yz$ $\dot{y} = x^2 - y$ $\dot{z} = 1 - 4x$	(0.25, 0.0625, 0)	$-1, \pm 0.5i$
Case I	$\dot{x} = -0.2y$ $\dot{y} = x + z$ $\dot{z} = x + y^2 - z$	(0, 0, 0)	$-1.13449, 0.06725 \pm 0.58996i$
Case J	$\dot{x} = 2z$ $\dot{y} = -2y + z$ $\dot{z} = -x + y + y^2$	(0, 0, 0)	$-2.31462, -0.15730 \pm 1.30515i$
Case L	$\dot{x} = y + 3.9z$ $\dot{y} = 0.9x^2 - y$ $\dot{z} = 1 - x$	(1, 0.9, −0.23077)	$-1.43329, -0.21664 \pm 1.63526i$
Case N	$\dot{x} = -2y$ $\dot{y} = x + z^2$ $\dot{z} = 1 + y - 2z$	(−0.25, 0, 0.5)	$-2.31460, -0.15730 \pm 1.30515i$
Case R	$\dot{x} = 0.9 - y$ $\dot{y} = 0.4 + z$ $\dot{z} = xy - z$	(−0.44444, 0.9, −0.4)	$-1.23212, -0.11606 \pm 0.84674i$

$$\begin{cases} \dot{x} = yz + a \\ \dot{y} = x^2 - y \\ \dot{z} = 1 - 4x. \end{cases} \tag{2}$$

With appropriate values of a, system (2) has only one equilibrium point $E(\frac{1}{4}, \frac{1}{16}, -16a)$, and the stability of this single equilibrium point is fundamentally different from that of system (1), as can be verified and compared from the results showed in Tables 1 and 2, respectively. Particularly, when $a = 0$, the system (2) is the Sprott E system shown in Table 1. System (2) is the revised Sprott E system when $a \neq 0$. The homoclinic Šhi'lnikov method or criterion might be applied to this system to show the existence of chaos; however, it involves somewhat subtle mathematical analysis and computations.

By linearizing system (2) at the equilibrium point E, the Jacobian matrix is obtained as

$$J = \begin{bmatrix} 0 & z & y \\ 2x & -1 & 0 \\ -4 & 0 & 0 \end{bmatrix} = \begin{bmatrix} 0 & -16a & \frac{1}{16} \\ \frac{1}{2} & -1 & 0 \\ -4 & 0 & 0 \end{bmatrix}. \tag{3}$$

The characteristic equation of system (2) is

Table 2 Equilibria and eigenvalues of the Wang-Chen system [4]

Cases	Equations	Equilibria	Eigenvalues
$a = -0.005$	$\dot{x} = yz + a$ $\dot{y} = x^2 - y$ $\dot{z} = 1 - 4x$	$(0.25, 0.0625, 0.08)$	$-1.03140, 0.01570 \pm$ $0.49208i$
$a = 0.006$	$\dot{x} = yz + a$ $\dot{y} = x^2 - y$ $\dot{z} = 1 - 4x$	$(0.25, 0.0625, -0.096)$	$-0.96069, -0.01966\pm$ $0.50975i$
$a = 0.022$	$\dot{x} = yz + a$ $\dot{y} = x^2 - y$ $\dot{z} = 1 - 4x$	$(0.25, 0.0625, -0.352)$	$-0.84580, -0.07710\pm$ $0.53818i$
$a = 0.030$	$\dot{x} = yz + a$ $\dot{y} = x^2 - y$ $\dot{z} = 1 - 4x$	$(0.25, 0.0625, -0.48)$	$-0.78217, -0.10891\pm$ $0.55476i$
$a = 0.050$	$\dot{x} = yz + a$ $\dot{y} = x^2 - y$ $\dot{z} = 1 - 4x$	$(0.25, 0.0625, -0.8)$	$-0.60746, -0.19627\pm$ $0.61076i$

$$\lambda^3 + \lambda^2 + \left(\frac{1}{4} + 8a\right)\lambda + \frac{1}{4} = 0 \tag{4}$$

By solving the characteristic equation (4), one obtains the Jacobian eigenvalues, as shown in Table 2, for some chosen values of the parameter a.

In addition, to verify the chaoticity of system (2), its Lyapunov exponents and Lyapunov dimension are calculated in [4].

The Lyapunov exponents are denoted by L_i, $i = 1, 2, 3$, and ordered as $L_1 > L_2 > L_3$. Recall that a system is considered chaotic if $L_1 > 0,\ L_2 = 0,\ L_3 < 0$ with $|L_1| < |L_3|$. It is noted that for this system, the largest Lyapunov exponent decreases as the parameter a increases from -0.01 to 0.05.

The Lyapunov dimension is defined by

$$D_L = j + \frac{1}{|L_{j+1}|}\sum_{i=1}^{j} L_i, \tag{5}$$

where j is the largest integer satisfying $\sum_{i=1}^{j} L_i \geqslant 0$ and $\sum_{i=1}^{j+1} L_i < 0$.

According to the Routh–Hurwitz criterion, the equilibrium of the system

$$\begin{cases} \dot{x} = yz + a \\ \dot{y} = x^2 - y \\ \dot{z} = 1 - 4x \end{cases} \tag{6}$$

is stable.

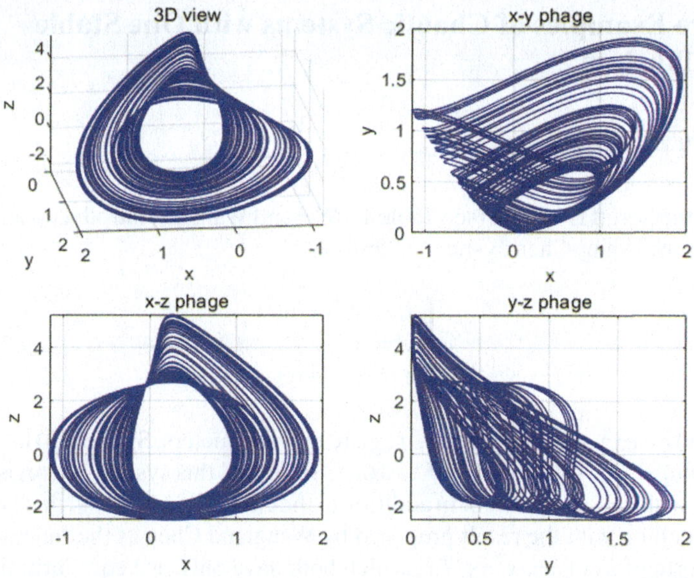

Fig. 1 Chaotic attractor with $a = 0.006$ in Wang-Chen system [4], including 3D views on the x-y plane, x-z plane and y-z plane

Through the analysis of the largest Lyapunov exponent (positive), and the fact that the chaotic waveform of $y(t)$ is a continuous broadband spectrum, Wang and Chen verified that the new system is indeed chaotic.

Take the case of $a = 0.006$ as an example. As shown in Table 2, the equilibrium of the new system becomes a node-focus when $a > 0$, therefore the Šhi'lnikov homoclinic criterion is not applicable to this case. Remarkably, when $a = 0.06$, three characteristic values of the Jacobian of the linearized equation, evaluated at the equilibria point E_0, are: $\lambda_1 = -0.96069$, $\lambda_{2,3} = -0.01966 \pm 0.50975i$. Furthermore, numerical calculation of the Lyapunov exponents yields

$$L_1 = 0.0489, \quad L_2 = 0, \quad L_3 = -1.0485, \tag{7}$$

indicating the existence of chaos in this system with one and only one stable equilibrium, as shown in Fig. 1.

Moreover, the coexistence of point, periodic and strange attractors in the Wang-Chen system was found by Sprott et al. [2].

As the first example of chaos from a system with one stable equilibrium, the Wang-Chen system is a milestone for finding or constructing new chaotic systems with mysterious features of chaos. It was a striking discovery, which motivated considerable subsequent research and studies.

4 More Examples of Chaotic Systems with One Stable Equilibrium

4.1 Wei System

Based on the Sprott E system (see Table 1), Wei and Wang [9] introduced an extended version of the Wang-Chen system, as follows:

$$\begin{cases} \dot{x} = yz + h(x) \\ \dot{y} = x^2 - y \\ \dot{z} = 1 - 4x. \end{cases} \tag{8}$$

where $h(x) = ex^2 + fx + g$ and e, f, g are real parameters. System (8) has only one equilibrium, $E = (\frac{1}{4}, \frac{1}{16}, -e - 4f - 16g)$. Similarly, this system is a revised Sprott E system, with $e = f = g = 0$. In addition to the choice of $h(x) = g$, this Wei system is also matching with the result proposed by Wang and Chen as the function has the specific form of $h(x) = ex^2$ or fx, which both have only one equilibrium:

$$\begin{cases} \dot{x} = yz + ex^2 \\ \dot{y} = x^2 - y \\ \dot{z} = 1 - 4x. \end{cases} \tag{9}$$

$$\begin{cases} \dot{x} = yz + fx \\ \dot{y} = x^2 - y \\ \dot{z} = 1 - 4x. \end{cases} \tag{10}$$

By linearizing the system (8) at the equilibrium point E_0, the Jacobian matrix is obtained, as

$$J = \begin{bmatrix} \frac{e}{2} + f & -e - 4f - 16g & \frac{1}{16} \\ \frac{1}{2} & -1 & 0 \\ -4 & 0 & 0 \end{bmatrix}.$$

Then, the characteristic equation of system (8) is given by

$$|\lambda I - J| = 0 \tag{11}$$

According to the Routh–Hurwitz criterion, the real parts of all the roots λ are negative if and only if

$$\delta_1 = 1 - f - \frac{e}{2} > 0, \delta_2 = \left(1 - f - \frac{e}{2}\right)\left(\frac{1}{4} + f + 8g\right) - \frac{1}{4} > 0.$$

By solving these inequalities, one obtains

$$f < 1 - \frac{e}{2}, e < \frac{2(3f - 4f^2 + 32g - 32fg)}{1 + 4f + 32g}. \tag{12}$$

In addition, the negative Lyapunov exponent spectra of system (8), when $f = -0.1$, $g = 0.02$ and $e \in [-0.4, -0.303)$, clearly illustrates that $-0.015 \leq e < 0.08$ is a periodic window; $-0.015 \leq e < 0.029$ is a period-2 orbit region; $0.029 < e < 0.048$ is a period-4 orbit region, and as e increases in the range of $0.048 < e < 0.08$, the orbit of system (8) is attracted into a sink. Hence, the equilibrium point E of system (8) is asymptotically stable when the above conditions are met.

In addition, a modified projective synchronization between the extended system (8) and the Sprott E system (see Table 1) was reported in [9]. Degenerate Hopf bifurcation, adaptive control, and Lyapunov stability in the extended system (8) were also studied [16].

4.2 Multiple-delayed Wang-Chen System

Wei and Pham [17] created a model of the Wang-Chen system with multiple delays:

$$\begin{cases} \dot{x} = yz + a + k_1 x(t - \tau_1) + k_2 x(t - \tau_2) \\ \dot{y} = x^2 - y \\ \dot{z} = 1 - 4x. \end{cases} \tag{13}$$

where τ_i (i = 1, 2) are time delays, and k_i (k = 1, 2) are the gains of the time delays τ_i (i = 1, 2), respectively.

Clearly, system (13) has one and only one equilibrium

$$E(x_0, y_0, z_0) = \left(\frac{1}{4}, \frac{1}{16}, -16a - 4(k_1 + k_2)\right). \tag{14}$$

By linearizing system (13) at the equilibrium point E_0, the characteristic equation with appropriate constraints on the eigenvalues is given by

$$\lambda^3 + a_2\lambda^2 + a_1\lambda + a_0 - k_1(\lambda^2 + a_2\lambda)e^{-\lambda\tau_1} - k_2(\lambda^2 + a_2\lambda)e^{-\lambda\tau_2} = 0, \tag{15}$$

where $a_2 = 1$ and $a_1 = (\frac{1}{4} + 8a + 2k_1 + 2k_2)$, $a_0 = \frac{1}{4}$.

When $\tau_1 = \tau_2 = 0$, Eq. (15) becomes

$$\lambda^3 + (1 - k_1 - k_2)\lambda^2 + \left(\frac{1}{4} + 8a + k_1 + k_2\right)\lambda + \frac{1}{4} = 0. \tag{16}$$

According to the Routh–Hurwitz criterion, Eq. (16) has three roots with negative real parts under the following condition:

$$1 - k_1 - k_2 > 0, (1 - k_1 - k_2)\left(\frac{1}{4} + 8a + k_1 + k_2\right) - \frac{1}{4} > 0. \tag{17}$$

Therefore, equilibrium E is an asymptotically stable node or foci-node, when

$$\frac{3 - 32a - \varphi(a)}{8} < k_1 + k_2 < \frac{3 - 32a - \varphi(a)}{8}, \tag{18}$$

where $\varphi(a) = \sqrt{1024a^2 + 320a + 9}$.

For system (13), its stability and the existence of Hopf bifurcation were determined, and the direction, stability as well as the period of the bifurcating periodic solution were analyzed using the normal form method and center manifold theorem in [17].

The theoretical results and numerical simulations of [17] show that the chaotic dynamics of system (13) can be controlled by manipulating the time delays. Theoretically, when only one equilibrium is asymptotically stable, the chaotic attractor is converted into a stable state, an unstable periodic orbit or another chaotic attractor again, if the delay passes through a certain value. Indeed, as the delay increases further, numerical simulations show that the periodic solution disappears and the chaotic attractors appear again.

4.3 Lao System

To overcome some limitations of the conventional methods [18–20] for parameter estimation in various models of chaos, which are caused by the sensitivity of chaos to initial conditions, Lao [21] used a geometry-based cost function to build a statistical model on the distribution of the real system attractors in the state space, whereby a new chaotic system with only one equilibrium was also introduced:

$$\begin{cases} \dot{x} = -z \\ \dot{y} = -x - z \\ \dot{z} = 2x - 1.3y - 2z + x^2 + z^2 - xz. \end{cases} \tag{19}$$

There is one equilibrium point at the origin $E(0, 0, 0)$ in the Lao system (19), with three corresponding eigenvalues:

$$\lambda_1 = -1.9783, \quad \lambda_2 = -0.0108 + 0.8106i, \quad \lambda_3 = -0.0108 - 0.8106i. \tag{20}$$

Since the three eigenvalues of the characteristic equation contain one negative root and two imaginary roots with negative real parts, the equilibrium point E is a stable focus.

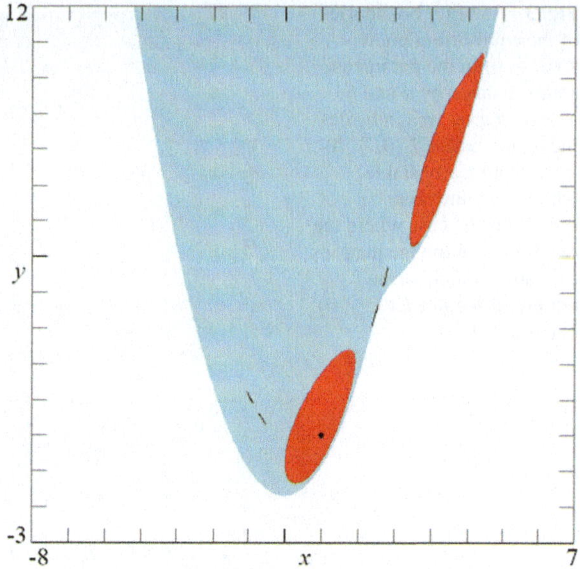

Fig. 2 Cross-section of the basins of attraction of the two attractors in the x-y plane at $z = 0$. Initial conditions in the white region lead to unbounded orbits, those in the red region lead to the point attractor, shown by a black dot, and those in the light blue region lead to the strange attractor, shown in cross-section as a pair of black lines [21]

In addition, the Lyapunov spectrum for the strange attractor was calculated, yielding $LE_1 = 0.018$, $LE_2 = 0$, $LE_3 = -2.018$. Thus, the Lao system (19) is chaotic with only one stable equilibrium.

Moreover, there exists a point attractor in the basin of attraction of the strange attractors. The coexistence of these two attractors leads to a cross-section, as shown in Fig. 2, where the cross-section of the strange attractor nearly touches its basin boundary as is typical for lower-dimensional chaotic flows. Lao also discovered an interesting phenomenon that, besides the area around the origin, there is another separate area that converges to the stable equilibrium.

4.4 *Kingni System*

Kingni [22] introduced a new three-dimensional autonomous system with only one equilibrium:

$$\begin{cases} \dot{x} = -z \\ \dot{y} = -x - z \\ \dot{z} = 3x - ay + x^2 - z^2 - yz + b, \end{cases} \tag{21}$$

where a and b are real parameters, assuming that $a \neq 0$. The system (21) possesses only one equilibrium point, $E(0, \frac{b}{a}, 0)$ if $a \neq 0$.

The characteristic equation of the system Jacobian at E_0 is

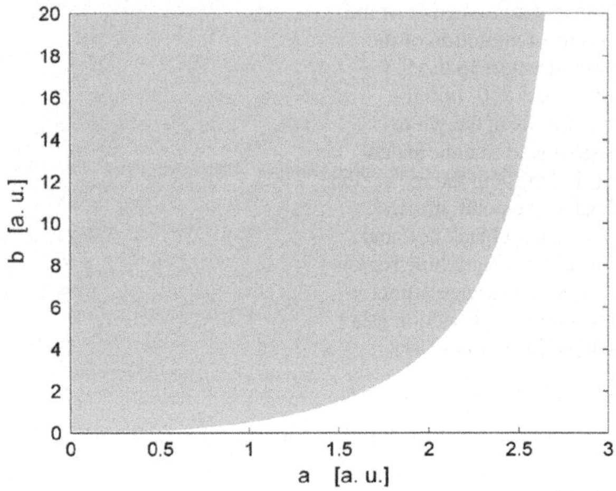

Fig. 3 Stability boundaries of the equilibrium point $E(0, \frac{b}{a}, 0)$ in the parameter space spanned by a and b. The light grey area indicates the region where $E(0, \frac{b}{a}, 0)$ is asymptotically stable, while the white area represents the case where the conditions on the inequations (23) are not met, so the equilibrium point $E(0, \frac{b}{a}, 0)$ is unstable [22]

$$\lambda^3 + \frac{b}{a}\lambda^2 - (-3 + a)\lambda + a = 0. \tag{22}$$

According to the Routh–Hurwitz conditions, this equation has all roots with negative real parts if and only if

$$A > 0, C > 0, AB - C > 0, \tag{23}$$

where $A = \frac{b}{a}$, $B = 3 - a$, $C = a$. By varying a from 0.002 to 3.0, and b from 0 to 20.0, one can obtain the stability boundaries of the equilibrium point $E(0, \frac{b}{a}, 0)$, as shown by the light grey area in Fig. 3. By choosing $a = 1.3$ and $b = 1.01$, Kingni solved the characteristic equation (22) and obtained three eigenvalues:

$$\lambda_1 = -0.7678519459, \lambda_{2,3} = -0.004535565486 \pm 1.301158769i\,.$$

Therefore, the equilibrium point $E(0, \frac{1.01}{1.30}, 0)$ becomes a node-focus, therefore the Kingni system is stable.

4.5 *From an Infinite Number of Equilibria to Only One Stable Equilibrium*

As shown by Pham [14], it is possible to construct a new system with one stable equilibrium from another system with an infinite number of equilibria.

Start with a chaotic flow, LE_1, introduced by Jafari and Sprott [23],

$$\begin{cases} \dot{x} = y \\ \dot{y} = -x + yz \\ \dot{z} = -x - axy - bxz. \end{cases} \tag{24}$$

in which a and b are two positive parameters. It is clear that the LE_1 system has infinitely many equilibria:

$$E(0, 0, z).$$

Now, by adding a positive control parameter c to the LE_1 system, one obtains a new system

$$\begin{cases} \dot{x} = y \\ \dot{y} = -x + yz + c \\ \dot{z} = -x - axy - bxz. \end{cases} \tag{25}$$

This system has only one equilibrium,

$$E\left(c, 0, -\frac{1}{b}\right).$$

The Jacobian matrix at the equilibrium E is

$$J_E = \begin{bmatrix} 0 & 1 & 0 \\ -1 & -\frac{1}{b} & 0 \\ 0 & -ac & -bc \end{bmatrix} \tag{26}$$

and its characteristic equation is

$$\lambda^3 + \left(bc + \frac{1}{b}\right)\lambda^2 + bc = 0. \tag{27}$$

According to the Routh–Hurwitz stability criterion, the equilibrium is stable when

$$\begin{cases} bc + \frac{1}{b} > 0 \\ c + 1 > 0 \\ bc > 0 \\ (bc + \frac{1}{b})(c + 1) > bc. \end{cases} \tag{28}$$

From Eq. (27), one obtains the eigenvalues

$$\lambda_1 = -bc, \ \lambda_{2,3} = -0.5 \pm 0.866i. \tag{29}$$

Hence, one can see that the system (25) has a chaotic attractor as shown in Fig. 4, although it has only one equilibrium.

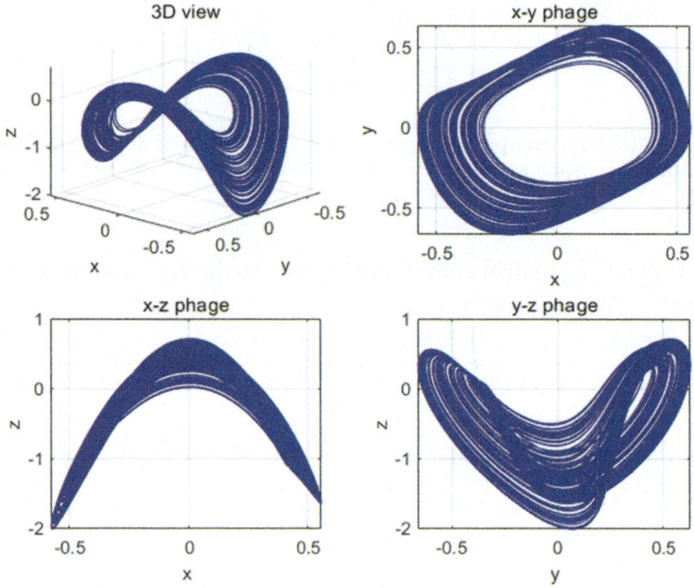

Fig. 4 Phase portrait on the x-y plane of the system with only one stable equilibrium (25) with $a = 15, b = 1, c = 0.001$, and initial conditions $(x(0), y(0), z(0)) = (0, 0.5, 0.5)$ [14]

5 Systematic Search for Chaotic Systems with One Stable Equilibrium

5.1 *Jerk System*

Consider a jerk system with one stable equilibrium for generating chaotic flows, described by

$$\begin{cases} \dot{x} = y \\ \dot{y} = z \\ \dot{z} = f(x, y, z) \\ f = a_1 x + a_2 y + a_3 z + a_4 x^2 + a_5 y^2 + a_6 z^2 + a_7 xy + a_8 xz + a_9 yz + a_{10}. \end{cases} \tag{30}$$

By solving Eq. (30), one obtains $y^* = z^* = 0$ as the equilibrium. The eigenvalues λ must satisfy

$$\lambda^3 - f_z \lambda^2 - f_y \lambda - f_x = 0, \tag{31}$$

in which $f_x = a_1 + 2a_4 x^*$, $f_y = a_2 + a_7 x^*$, and $f_z = a_3 + a_8 x^*$.

Using the Routh–Hurwitz stability criterion, to have a stable equilibrium point it is required that

$$f_z < 0, \tag{32}$$

$$f_y f_z + f_x > 0\,,$$
$$f_x < 0\,.$$

By setting $y^* = z^*$, one obtains

$$a_1 x + a_4 x^2 + a_{10} = 0\,, \tag{33}$$

and, for $a_4 \neq 0$,

$$x^*_{1,2} = (-a_1 \pm \sqrt{\Delta})/2a_4\,, \tag{34}$$

where $\Delta = a_1^2 - 4a_4 a_{10}$.

To obtain an equilibrium, Δ should not be less than 0, while the stability condition $f_x < 0$ at x_1^* requires $\sqrt{\Delta} < 0$, which is impossible.

In conclusion, a quadratic jerk system cannot have two stable equilibria. Therefore, the general case (30) can be modified to

$$\begin{cases} \dot{x} = y \\ \dot{y} = z \\ \dot{z} = a_1 x + a_2 y + a_3 z + a_4 y^2 + a_5 z^2 + a_6 xy + a_7 xz + a_8 yz + a_9\,, \end{cases}$$

where there is no x^2 term in the $\dot{z}$ equation to ensure one and only one equilibrium to exist.

Such a system has only one equilibrium at $(-a_9/a_1, 0, 0)$, whose stability requires

$$a_1 < 0,\ \left(a_3 - \frac{a_7 a_9}{a_1}\right) < 0,\ \left(a_2 - \frac{a_6 a_9}{a_1}\right) < \frac{-a_1}{\left(\frac{a_7 a_9}{a_1}\right)}\,. \tag{35}$$

Considering several thousands of combinations of the coefficients a_1 through a_9 and the initial conditions subject to the constraints (35), an exhaustive computer search was carried out to seek cases with the largest Lyapunov exponent greater than 0.001. For each case found, the space of coefficients was searched over, while as many coefficients as possible are set to zero with the others set to ± 1 if possible, or otherwise to a small integer or decimal fraction with the lowest possible digits.

Cases $SE_1 - SE_6$ in Table 3 are six simple examples found in this way.

5.2 17 Simple Chaotic Flows

By applying similar calculations of the jerk system method, Jafari and Sprott [3] investigated 17 simple structures of chaotic flows with only one equilibrium, as listed in Table 4, from SE_7 to SE_{23}. The Lyapunov spectra and Kaplan–Yorke dimensions are also shown in Table 4, along with initial conditions that are close to the attractors.

Table 3 (Part I) 23 simple chaotic flows with one stable equilibrium [3]

Model	Equations	Equilibrium	Eigenvalues	LEs	D_{KY}	(x_0, y_0, z_0)
SE_1	$\begin{cases} \dot{x} = y \\ \dot{y} = z \\ \dot{z} = -x - 0.6y - 2z + z^2 - 0.4xy \end{cases}$	0 0 0	−1.9548 −0.0226 ±0.7149i	0.0377 0 −2.0377	2.0185	4 −2 0
SE_2	$\begin{cases} \dot{x} = y \\ \dot{y} = z \\ \dot{z} = -0.5x - y - 0.55z - 1.2z^2 - xz - yz \end{cases}$	0 0 0	−0.5103 −0.0198 ±0.9896i	0.0804 0 −0.4889	2.1644	−1 0 1
SE_3	$\begin{cases} \dot{x} = y \\ \dot{y} = z \\ \dot{z} = -3.4x - y - 4z + y^2 + xy \end{cases}$	0 0 0	−3.9641 −0.0179 ±0.9259i	0.0711 0 −4.0711	2.0175	−2 0 2.4
SE_4	$\begin{cases} \dot{x} = y \\ \dot{y} = z \\ \dot{z} = -x - 1.7z + y^2 + 0.6xy - 1 \end{cases}$	−1 0 0	−1.6942 −0.0029 ±0.7683i	0.0434 0 −1.7434	2.0249	0.5 1 0
SE_5	$\begin{cases} \dot{x} = y \\ \dot{y} = z \\ \dot{z} = -x - z - z^2 + 0.4xy - 2.7 \end{cases}$	−2.7 0 0	−0.9600 −0.0200 ±1.0204i	0.0136 0 −1.0136	2.0134	−6.1 1 1
SE_6	$\begin{cases} \dot{x} = y \\ \dot{y} = z \\ \dot{z} = -x - 2.9z^2 + xy + 1.1xz - 1 \end{cases}$	−1 0 0	−1.0526 −0.0237 ±0.9744i	0.0638 0 −1.0638	2.0600	−2.2 0.6 0

It is constructive to find most of the elementary forms of chaotic flows with one stable equilibrium, as these 17 additional cases, either equivalent to one of the cases listed by some linear transformations of variables or they were extensions of these cases with additional terms.

A three-dimensional view of the state-space diagram of the cases shown in Tables 3, 4 are dissipative, with attractors as shown in Fig. 5. From a common point of view, these 23 systems have the attractor dimensions only slightly above 2.0, whereas the largest one is SE_{20} with $D_{KY} = 2.1753$.

They were analyzed as in the previous sections. However, it should be noted that all the equilibria calculated in [3] are spiral nodes with one pair of complex conjugate eigenvalues, rather than simple nodes with all real eigenvalues. Therefore another common feature of these systems is the small negative real part in the complex pair of eigenvalues, compared to the real eigenvalue case.

In conclusion, simple chaotic systems with one and only one stable equilibrium are common, rather than unusual, as they were once thought to be. Some more constructive results will be illustrated in the following chapters.

6 Chaos with Stable Equilibria

6.1 Yang-Chen System

A chaotic system with more than one equilibrium was reported earlier than a chaotic system with only one equilibrium.

Connecting the original Lorenz system and the original Chen system, Yang and Chen [10] first introduced a chaotic system with stable equilibrium points in the following three-dimensional autonomous system:

$$\begin{cases} \dot{x} = a(y - z) \\ \dot{y} = cx - xz \\ \dot{z} = -bz + xy. \end{cases} \tag{36}$$

where a, b and c are parameters, referred to as the Yang-Chen system. This system has one saddle and two stable node-foci, which is topologically different from all the well-known systems such as the Lorenz system [24], Chen system [25], and Lü system [26].

The Šil'nikov criterion is applicable to the Yang-Chen system. If one chooses a condition with $a = c = 35$, $b = 3$, then system (36) becomes

$$\begin{cases} \dot{x} = 35(y - x) \\ \dot{y} = 35x - xz \\ \dot{z} = -3z + xy. \end{cases} \tag{37}$$

Table 4 (Part II) 23 simple chaotic flows with one stable equilibrium [3]

Model	Equations	Equilibrium	Eigenvalues	LEs	D_{KY}	(x_0, y_0, z_0)
SE_7	$\begin{cases} \dot{x} = y \\ \dot{y} = -x + yz \\ \dot{z} = -2z - 8xy + xz - 1 \end{cases}$	0 0 −0.5	−2.0000 −0.2500 ±0.9682i	0.0360 0 −25.6798	2.0014	1 −0.7 0
SE_8	$\begin{cases} \dot{x} = y \\ \dot{y} = -x + yz \\ \dot{z} = -z - 0.7x^2 + y^2 - 0.1 \end{cases}$	0 0 −0.1	−1.0000 −0.0500 ±0.9987i	0.1412 0 −1.3649	2.1034	0 0.9 0
SE_9	$\begin{cases} \dot{x} = y \\ \dot{y} = -x + yz \\ \dot{z} = 2x - 2z + y^2 - 0.3 \end{cases}$	0 0 −0.15	−2.0000 −0.0750 ±0.9972i	0.0203 0 −2.4751	2.0082	0 0.8 −0.2
SE_{10}	$\begin{cases} \dot{x} = y \\ \dot{y} = -x + yz \\ \dot{z} = x - 0.3y - 2z + xz - 0.1 \end{cases}$	0 0 −0.05	−2.0000 −0.0250 ±0.9997i	0.0963 0 −15.7010	2.0061	3.9 0 1
SE_{11}	$\begin{cases} \dot{x} = y \\ \dot{y} = -x + yz \\ \dot{z} = -y - 12z + x^2 + 9xz - 1 \end{cases}$	0 0 −1/12	−12.0000 −0.0417 ±0.9991i	0.0801 0 −14.1917	2.0056	−2 0 0.1
SE_{12}	$\begin{cases} \dot{x} = y \\ \dot{y} = -x + yz \\ \dot{z} = -66z + y^2 + 35xz - 1 \end{cases}$	0 0 −1/66	−66.0000 −0.0076 ±1.0000i	0.0259 0 −61.6130	2.0004	2 0.6 0

(continued)

Table 4 (continued)

Model	Equations	Equilibrium	Eigenvalues	LEs	D_{KY}	(x_0, y_0, z_0)
SE_{13}	$\begin{cases} \dot{x} = y \\ \dot{y} = -x + yz \\ \dot{z} = -4.9z + 0.4y^2 + xy - 1 \end{cases}$	0 0 −1/4.9	−4.9000 −0.1020 ±0.9948i	0.0540 0 −4.8228	2.0112	0 −2.2 0
SE_{14}	$\begin{cases} \dot{x} = z \\ \dot{y} = x + z \\ \dot{z} = -y - 3z^2 + xy + yz - 0.7 \end{cases}$	0 −0.7 0	−0.6082 −0.0459 ±1.2814i	0.0657 0 −1.6407	2.0401	0.5 −1 0
SE_{15}	$\begin{cases} \dot{x} = -z \\ \dot{y} = x - z \\ \dot{z} = 0.9y + 0.2x^2 + xz + yz + 1 \end{cases}$	0 −10/9 0	−1.0618 −0.0247 ±0.9203i	0.0414 0 −6.6641	2.0062	3 2 0
SE_{16}	$\begin{cases} \dot{x} = -z \\ \dot{y} = -x + z \\ \dot{z} = -7y - 1.4z + x^2 + xz - yz \end{cases}$	0 0 0	−1.0549 −0.1726 ±2.5702i	0.0775 0 −6.7190	2.0115	1 6 −6
SE_{17}	$\begin{cases} \dot{x} = z \\ \dot{y} = x - y \\ \dot{z} = -3.1x - 0.3xz + 0.2yz + 0.57 \end{cases}$	0.57/3.1 0.57/3.1 0	−1.0000 −0.0092 ±1.7607i	0.0832 0 −0.6549	2.1262	7.5 0 −5
SE_{18}	$\begin{cases} \dot{x} = z \\ \dot{y} = -y + z \\ \dot{z} = -2.1x - 0.1z - y^2 + 0.11xz + 0.5yz \end{cases}$	0 0 0	−1.0000 −0.0500 ±1.4483i	0.1469 0 −3.8348	2.0383	−28 0 0
SE_{19}	$\begin{cases} \dot{x} = z \\ \dot{y} = -y + z \\ \dot{z} = -x - 2xy + 1.7xz - 0.3 \end{cases}$	−0.3 0 0	−1.3766 −0.0667 ±0.8497i	0.0241 0 −49.8730	2.0005	0.2 6 7

(continued)

Table 4 (continued)

Model	Equations	Equilibrium	Eigenvalues	LEs	D_{KY}	(x_0, y_0, z_0)
SE_{20}	$\begin{cases} \dot{x} = z \\ \dot{y} = -y - z \\ \dot{z} = -11x + 2y - 2y^2 - z^2 - yz \end{cases}$	0 0 0	−0.8543 −0.0728 $\pm 3.5875i$	0.2125 0 −1.2125	2.1753	−2.1 0.1 5
SE_{21}	$\begin{cases} \dot{x} = z \\ \dot{y} = -y - z \\ \dot{z} = -7.1x + y - 2y^2 + xz - yz \end{cases}$	0 0 0	−0.8875 −0.0563 $\pm 2.8279i$	0.0484 0 −2.8617	2.0169	0 −3 8.2
SE_{22}	$\begin{cases} \dot{x} = z \\ \dot{y} = -y - z \\ \dot{z} = -6x - 2y^2 + xz - yz - 0.9 \end{cases}$	−0.15 0 0	−1.0000 −0.0750 $\pm 2.4483i$	0.0557 0 −2.8695	2.0194	−6 3.8 0
SE_{23}	$\begin{cases} \dot{x} = -z \\ \dot{y} = -y - z \\ \dot{z} = 4x - 0.2z^2 + xy - 2 \end{cases}$	0.5 0 0	−0.9060 −0.0470 $\pm 2.1006i$	0.0159 0 −1.0159	2.0156	−0.4 1 −9

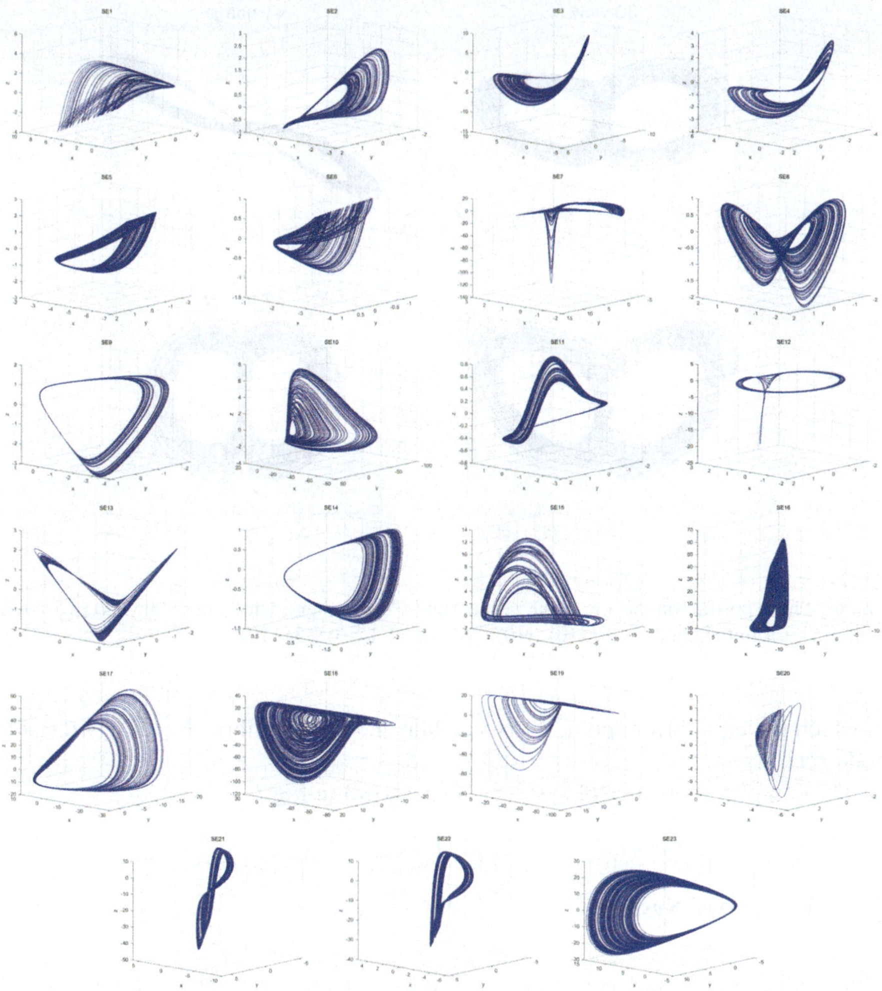

Fig. 5 State-space diagrams of the cases in Tables 3 and 4 [3]

It has three equilibria:

$$\begin{cases} E_1(0,0,0), \\ E_2(\sqrt{105}, \sqrt{105}, 35), \\ E_3(-\sqrt{105}, -\sqrt{105}, 35). \end{cases} \tag{38}$$

By solving the characteristic equation, the eigenvalues at the three equilibria are obtained, as

$$\begin{cases} E_1 : \lambda_1 = -3,\ \lambda_2 = -\frac{35(\sqrt{5}+1)}{2},\ \lambda_3 = \frac{35(\sqrt{5}-1)}{2} \\ E_{2,3} : \lambda_1 = -37.6122,\ \lambda_{2,3} = -0.1939 \pm 13.9778i\ , \end{cases} \tag{39}$$

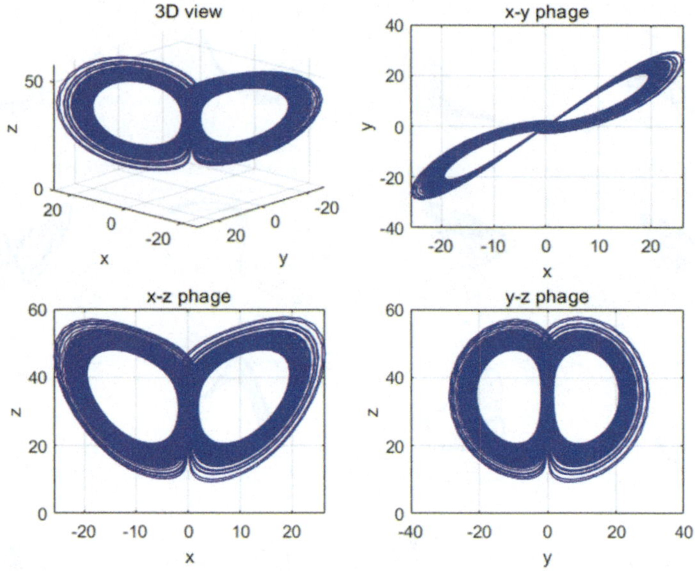

Fig. 6 Phase portrait on the x-y plane of the Yang-Chen system, with one saddle and two stable node-foci, for initial conditions $(x(0), y(0), z(0)) = (1.15, 3.5, 3)$ [10]

in which the equilibrium point E_1 is a saddle and the equilibria $E_{2,3}$ are two stable node-foci.

The complex dynamics of system (37) is shown in Fig. 6.

6.2 *Yang-Wei System*

Motivated by the Yang-Chen system, it is interesting to ask whether or not there are similar three-dimensional autonomous chaotic systems with only stable node-foci. Yang and Wei [11] gave a positive answer to this question. They introduced and analyzed a chaotic system with two stable node-foci as its only equilibria, which enhance or even modify the current theory and characterization of the structural properties of chaotic attractors in general.

The Yang-Wei system is described by

$$\begin{cases} \dot{x} = a(y - x) \\ \dot{y} = -cy - xz \\ \dot{z} = -b + xy, \end{cases} \tag{40}$$

where a, b and c are real parameters. If and only if the conditions of $a \neq 0$ and $b > 0$ are satisfied, the Yang-Wei system has two equilibria:

$$E_1(x_0, y_0, z_0) = \left(\sqrt{b}, \sqrt{b}, -c\right)$$

and

$$E_2(-x_0, -y_0, z_0) = \left(-\sqrt{b}, -\sqrt{b}, -c\right).$$

For $a = 10$, $b = 100$ and $c = 11.2$, the system (40) becomes

$$\begin{cases} \dot{x} = 10(y - x) \\ \dot{y} = -11.2y - xz \\ \dot{z} = -100 + xy, \end{cases} \tag{41}$$

with two equilibria

$$\begin{cases} E_1(10, 10, -11.2) \\ E_2(-10, -10, -11.2). \end{cases} \tag{42}$$

Three characteristic values of the Jacobian of its linearized system, evaluated at the equilibria $E_{1,2}$, are given by

$$\lambda_1 = -20.9778, \ \lambda_{2,3} = -0.1111 \pm 9.7635i\,, \tag{43}$$

which indicates that $E_{1,2}$ are two stable node-foci.

It is interesting to note that the system (40) contains the well-known diffusionless Lorenz system, the Burke–Shaw system, and some others, as special cases.

Hopf bifurcation and singularly degenerate heteroclinic and homoclinic orbits were analyzed in detail in [11].

6.3 Delayed Feedback of Yang-Wei System

In order to investigate the effect of delayed feedback on the three-dimensional chaotic system with only two stable node-foci, found by Yang et al. [8], the system controlled by the time-delay controlling forces is designed in [8], as

$$\begin{cases} \dot{x} = a(y - x) \\ \dot{y} = -cy - xz + k[y(t - \tau) - y] \\ \dot{z} = -b + xy, \end{cases} \tag{44}$$

where τ is the time delay, k is the gain of the time delay feedback.

The associated characteristic equation of the linearized system is

$$\lambda + (a + c + k)\lambda^2 + (b + ak)\lambda + 2ab - (\lambda^2 + a\lambda)ke^{-\lambda\tau} = 0. \tag{45}$$

When $\tau = 0$, Eq. (45) becomes

$$\lambda + (a+c)\lambda^2 + b\lambda + 2ab = 0. \tag{46}$$

According to the Routh–Hurwitz criterion, Eq. (46) has three roots with negative real parts under the following condition:

$$b > 0, \ c > a\,. \tag{47}$$

Therefore, the two equilibria E_1 and E_2 are both stable nodes or node-foci.

6.4 More Examples

To design a control law, such as the Yang-Wei feedback nonlinear control system [6], can be understood as adding a perturbation to the chaotic system such that the feedback controlled system undergoes a controllable Hopf bifurcation. In so doing, Wei and Yang introduced a feedback controller

$$u = m(x-y) + n(x-y)^3\,, \tag{48}$$

to control the system, yielding

$$\begin{cases} \dot{x} = a(y-x) \\ \dot{y} = -cy - xz \\ \dot{z} = -b + xy + u. \end{cases} \tag{49}$$

The associated Jacobian matrix at the equilibrium point $O(0,0,0)$ is

$$J_O = \begin{bmatrix} -a & a & 0 \\ -z_0 & -c & -x_0 \\ y_0 + m & x_0 - m & 0 \end{bmatrix}. \tag{50}$$

The corresponding characteristic equation is

$$\lambda^3 + (a+c)\lambda^2 + (b - m\sqrt{b})\lambda + 2ab = 0, \tag{51}$$

which indicates that the control gain u has no influence on the eigenvalues of the matrix J_O.

It is clear that the controlled system (49) does not change the divergence of the Yang-Wei system (40) and the controller (48) still keeps the equilibrium structure of the system (40).

Based on numerical analysis, parameter conditions can be found under which the system (49) presents Hopf bifurcation at the equilibria, some degenerate cases, Lyapunov coefficients and the criticality of Hopf bifurcation. The unknown dynamical behaviors of chaotic attractors with stable equilibria are also investigated.

Wei and Yang [27] also introduced a new system with six terms, having only one nonlinear term of an exponential function, as follows:

$$\begin{cases} \dot{x} = a(y - x) \\ \dot{y} = -by + mxz \\ \dot{z} = n - e^{xy}, \end{cases} \tag{52}$$

where a, b, m and n are real parameters.

When $n > 1$ and $m \neq 0$, system (52) possesses two equilibria:

$$E_1\left(\sqrt{\ln n}, \sqrt{\ln n}, \frac{b}{m}\right)$$

and

$$E_2\left(-\sqrt{\ln n}, -\sqrt{\ln n}, \frac{b}{m}\right).$$

In particular, for parameters $a = 0.8696$, $b = 2.1$, $m = 0.756144$ and $n = 10.5$, the equilibria $E_{1,2}$ are both stable, with characteristic eigenvalues

$$\lambda_1 = -1.94684, \ \lambda_{2,3} = -0.511378 \pm 4.05168i\,. \tag{53}$$

Therefore, system (52) has neither homoclinic orbits nor heteroclinic orbits joining $E_{1,2}$, but it can generate a double-scroll chaotic attractor. Clearly, it does not satisfy the Šil'nikov conditions.

By generalizing the Sprott C system (see Table 1), another interesting system was constructed in [15], as

$$\begin{cases} \dot{x} = a(y - x) \\ \dot{y} = -cy - xz \\ \dot{z} = y^2 - b, \end{cases} \tag{54}$$

where a, b, c are real parameters.

When $c = 0$ and $a = b^2 > 0$, system (54) is topologically equivalent to the original Sprott C system (see Table 1).

If and only if $b > 0$ and $a \neq 0$, according to the form for the generalized Lorenz systems, system (54) has two fixed equilibria:

$$E_1\left(\sqrt{b}, \sqrt{b}, -c\right), \quad E_2\left(-\sqrt{b}, -\sqrt{b}, -c\right). \tag{55}$$

In particular, for parameters $a = 10$, $b = 100$, $c = 0.4$, a chaotic attractor can be observed, which however does not satisfy the Šil'nikov conditions either. The eigenvalues of such cases are

$$\lambda_1 = -10.1357, \ \lambda_{2,3} = -0.1321 \pm 1.0465i\,. \tag{56}$$

When $c > 0$, system (54) generates very complex and abundant dynamics in a very wide parameters domain, with stable equilibria.

References

1. J.C. Sprott, Some simple chaotic flows. Phys. Rev. E **50**(2), R647 (1994)
2. J.C. Sprott, X. Wang, G. Chen, Coexistence of point, periodic and strange attractors. Int. J. Bifurc. Chaos **23**(05), 1350093 (2013)
3. M. Molaie, S. Jafari, S.M.R.H. Golpayegani, Simple chaotic flows with one stable equilibrium. Int. J. Bifurc. Chaos **23**(11), 699–283 (2013)
4. X. Wang, G. Chen, A chaotic system with only one stable equilibrium. Commun. Nonlinear Sci. Numer. Simul. **17**(3), 1264–1272 (2012)
5. Z. Wei, Dynamical behaviors of a chaotic system with no equilibria. Phys. Lett. A **376**(2), 102–108 (2011)
6. Z. Wei, Q. Yang, Anti-control of Hopf bifurcation in the new chaotic system with two stable node-foci. Appl. Math. Comput. **217**(1), 422–429 (2010)
7. Z. Wei, Q. Yang, Controlling the diffusionless Lorenz equations with periodic parametric perturbation. Comput. Math. Appl. **58**(10), 1979–1987 (2009)
8. Z. Wei, Delayed feedback on the 3D chaotic system only with two stable node-foci. Comput. Math. Appl. **63**(3), 728–738 (2012)
9. Z. Wei, Chaotic behavior and modified function projective synchronization of a simple system with one stable equilibrium. Kybernetika **2**(2), 359–374 (2013)
10. Q. Yang, G. Chen, A chaotic system with one saddle and two stable node-foci. Int. J. Bifurc. Chaos **18**(05), 1393–1414 (2008)
11. Q. Yang, Z. Wei, G. Chen, An unusual 3D autonomous quadratic chaotic system with two stable node-foci. Int. J. Bifurc. Chaos **20**(04), 1061–1083 (2010)
12. N. Kuznetsov, G. Leonov, Hidden attractors in dynamical systems: systems with no equilibria, multistability and coexisting attractors. IFAC Proc. **47**(3), 5445–5454 (2014)
13. T. Zhou, G. Chen, S. Celikovsky, Si'lnikov chaos in the generalized Lorenz canonical form of dynamical systems. Nonlinear Dyn. **39**(4), 319–334 (2005)
14. V.-T. Pham, C. Volos, T. Kapitaniak, Systems with stable equilibria, *Systems with Hidden Attractors* (Springer, Berlin, 2017), pp. 21–35
15. Z. Wei, Q. Yang, Dynamical analysis of the generalized Sprott C system with only two stable equilibria. Nonlinear Dyn. **68**(4), 543–554 (2012)
16. Z. Wei, I. Moroz, A. Liu, Degenerate Hopf bifurcations, hidden attractors, and control in the extended Sprott E system with only one stable equilibrium. Turk. J. Math. **38**(4), 672–687 (2014)
17. Z. Wei, V.-T. Pham, T. Kapitaniak, Z. Wang, Bifurcation analysis and circuit realization for multiple-delayed Wang-Chen system with hidden chaotic attractors. Nonlinear Dyn. **85**(3), 1635–1650 (2016)
18. J.F. Chang, Y.S. Yang, T.L. Liao, J.J. Yan, Parameter identification of chaotic systems using evolutionary programming approach. Expert Syst. Appl. **35**(4), 2074–2079 (2008)
19. Y. Tang, X. Guan, Parameter estimation of chaotic system with time-delay: a differential evolution approach. Chaos, Solitons Fractals **42**(5), 3132–3139 (2009)
20. Y. Tang, X. Zhang, C. Hua, L. Li, Y. Yang, Parameter identification of commensurate fractional-order chaotic system via differential evolution. Phys. Lett. A **376**(4), 457–464 (2012)
21. S.-K. Lao, Y. Shekofteh, S. Jafari, J.C. Sprott, Cost function based on Gaussian mixture model for parameter estimation of a chaotic circuit with a hidden attractor. Int. J. Bifurc. Chaos **24**(01), 1450010 (2014)
22. S. Kingni, S. Jafari, H. Simo, P. Woafo, Three-dimensional chaotic autonomous system with only one stable equilibrium: analysis, circuit design, parameter estimation, control, synchronization and its fractional-order form. Eur. Phys. J. Plus **129**(5), 76 (2014)

23. S. Jafari, J.C. Sprott, Simple chaotic flows with a line equilibrium. Chaos, Solitons Fractals **57**(4), 79–84 (2013)
24. E.N. Lorenz, Deterministic nonperiodic flow. J. Atmos. Sci. **20**(2), 130–141 (1963)
25. G. Chen, T. Ueta, Yet another chaotic attractor. Int. J. Bifurc. Chaos **9**(7), 1465–1466 (1999)
26. J. Lu, G. Chen, S. Zhang, The compound structure of a new chaotic attractor. Chaos, Solitons Fractals **14**(5), 669–672 (2002)
27. Z. Wei, Q. Yang, Dynamical analysis of a new autonomous 3-D chaotic system only with stable equilibria. Nonlinear Anal.: Real World Appl. **12**(1), 106–118 (2011)

Chaotic Systems Without Equilibria

Xiong Wang and Guanrong Chen

1 Introduction

Although a large number of chaotic systems with a certain number of equilibria have been known, in the last decade special chaotic systems without equilibrium received considerable attention. Dissipative systems without equilibria can also be considered as systems with hidden attractors. Chaotic systems with hidden attractors do not satisfy the Šil'nikov criterion. Thus, they have neither homoclinic nor heteroclinic orbits [1]. From a computational point of view, numerical localization of the hidden attractors in such systems is interesting but challenging, due to the fact that hidden attractors have basins of attraction not intersecting small neighborhood of any equilibria, which leads to unpredictable chaotic behaviors. The perhaps oldest such example is the Sprott A system (see Table 1), which is a special case of the Nose–Hoover oscillator [2, 3]. By modifying the Sprott D system (see Table 1), a no-equilibrium chaotic system with six terms was proposed by Wei [4]. Then, Wang and Chen [5] found a chaotic system without equilibria when constructing systems with any number of equilibria, a subject to be studied in detail in a following chapter. Inspired by these works, Jafari and Sprott [6] carried out a systematic computer search to find 17 additional three-dimensional chaotic systems without equilibria. Then, Leonov and Kuznetsov [7] pointed out that these systems belong to the category of systems with hidden attractors.

X. Wang (✉)
Institute for Advanced Study, Shenzhen University,
Shenzhen 518060, Guangdong, People's Republic of China
e-mail: wangxiong8686@szu.edu.cn

G. Chen
Department of Electrical Engineering, City University of Hong Kong,
Hong Kong SAR 999077, People's Republic of China
e-mail: eegchen@cityu.edu.hk

X. Wang et al. (eds.), *Chaotic Systems with Multistability and Hidden Attractors*,
Emergence, Complexity and Computation 40,
https://doi.org/10.1007/978-3-030-75821-9_4

Motivated by the initial Sprott A model, another no-equilibrium system with cubic nonlinearity was constructed by Maaita [8]. In addition, Akgul et al. [9] proposed a three-dimensional no-equilibrium chaotic system based on a chaos-based random number generation design. Moreover, along the same line of investigation on multi-scroll chaotic systems [10], a multi-scroll chaotic system without equilibria was investigated [11–13]. More recently, Pham and Zhen, and some others, found several new chaotic systems of this kind [14–18]. The dynamics and circuit realization with a boostable variable was studied [16].

2 Examples That Have Been Discovered

2.1 *Sprott A System*

Sprott A system (see Table 1) is the oldest example of a chaotic system without equilibria, which is described by

$$\begin{cases} \dot{x} = y \\ \dot{y} = -x + yz \\ \dot{z} = 1 - y^2, \end{cases} \tag{1}$$

where x, y, z are stable variables.

It is easy to verify that this Sprott A system has no equilibria by solving $\dot{x} = 0$, $\dot{y} = 0$ and $\dot{z} = 0$. Its Kaplan–Yorke dimension is 3.0 for most initial conditions such as (0, 5, 0) and Lyapunov exponents are (0.0139, 0, −0.0139), with

$$\sum_{i=1}^{3} L_i = L_1 + L_2 + L_3 = 0.0139 + 0 - 0.0139 = 0. \tag{2}$$

This implies that the system (1) is nonuniformly conservative; therefore it has no self-excited attractors but a chaotic sea coexisting with a set of nested tori, as shown in Fig. 1.

Note that this Sprott A system is a special case of the Nose–Hoover oscillator [2], with variables that can be boosted. This means that the DC offset of the variable x can be changed to any level [20–22].

The following two variants are introduced by Pham [14]. When replacing the variable x with a new term $x + k_x$, which has no effect on the dynamics, the system becomes

$$\begin{cases} \dot{x} = y \\ \dot{y} = -x - k_x + yz \\ \dot{z} = 1 - y^2, \end{cases} \tag{3}$$

in which k_x is a control constant.

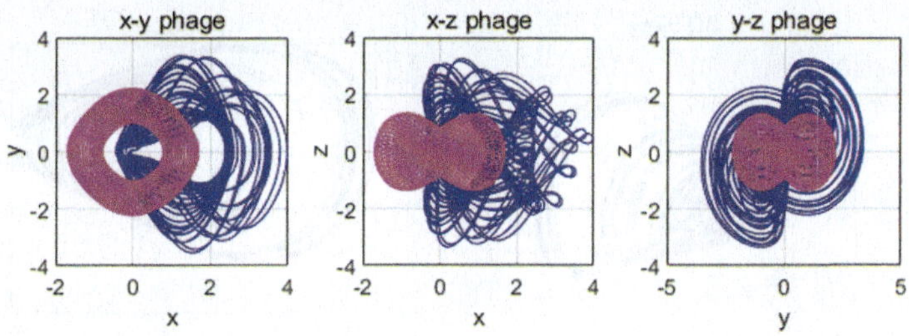

Fig. 1 The Sprott A system with initial condition (0, 5, 0) resulting in a chaotic set (blue), and (0, 1, 0) resulting in a conservative torus (pink) [19]

Also, a chaotic sea can be observed when $a \geq 0$, by replacing y^2 with $\|y\|^a$ in system (1), which becomes a variant of the above Sprott A system:

$$\begin{cases} \dot{x} = y \\ \dot{y} = -x - yz \\ \dot{z} = 1 - |y|^a. \end{cases} \tag{4}$$

2.2 Wei System

Wei [4] added a constant controller to the original Sprott D system to form a new three-dimensional system without equilibria, which is the first known dissipative system in this category.

The Sprott D system (see Table 1):

$$\begin{cases} \dot{x} = -y \\ \dot{y} = x + z \\ \dot{z} = 3y^2 + xz. \end{cases} \tag{5}$$

The Wei system [4]:

$$\begin{cases} \dot{x} = -y \\ \dot{y} = cx + z \\ \dot{z} = ay^2 + xz - d. \end{cases} \tag{6}$$

It is easy to see that system (5) has only one non-hyperbolic equilibrium, $E(0, 0, 0)$, with eigenvalues $\lambda_1 = 0$, $\lambda_{2,3} = \pm i$, coexisting with a chaotic attractor. System (6) actually performs a tiny perturbation on the system (5), so as to have no equilibria but preserving its chaotic dynamics.

In the Wei system, a, c, d are real parameters. It is the Sprott D system (5) when $a = 3, c = 1, d = 0$ and is the Falkner–Skan system [23] when $a = d, c = 0$.

It has been verified that the perturbation d can change the type and the number of equilibria and that the system (6) contains no equilibria when $d > 0$.

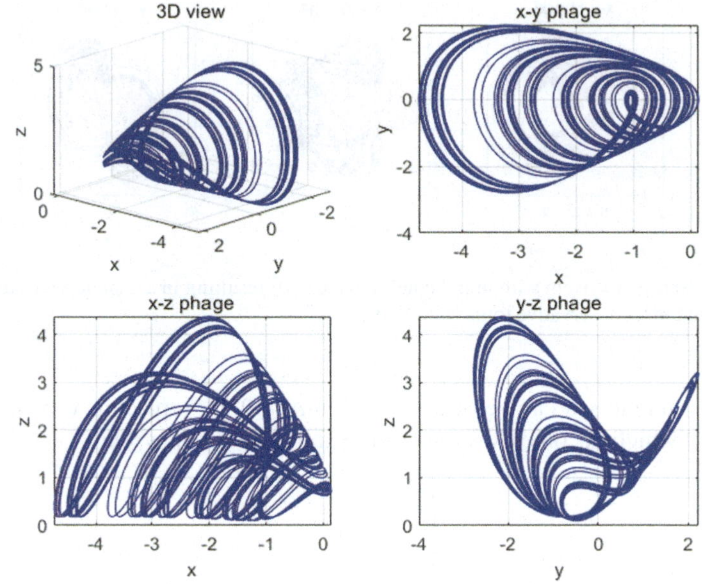

Fig. 2 Chaotic phase portrait of system (6) in the three-dimensional space, when parameters $(a, c, d) = (2, 1, 0.35)$ and initial value $(-1.6, 0.82, 1.9)$ [14]

With $a = 2, c = 1, d = 0.35$, and initial value $(-1.6, 0.82, 1.9)$, the Wei system becomes

$$\begin{cases} \dot{x} = -y \\ \dot{y} = x + z \\ \dot{z} = 2y^2 + xz - 0.35 \, , \end{cases} \tag{7}$$

with Lyapunov exponents $\lambda_{L_1} = 0.0793$, $\lambda_{L_2} = 0$, $\lambda_{L_3} = -1.5034$ and the Kaplan–Yorke dimension $D_L = 2.0528$. Therefore, system (7) displays a single-scroll chaotic attractor without equilibria, as shown in Fig. 2.

2.3 Wang-Chen System

When constructing a chaotic system with any number of equilibria, Wang and Chen [5] introduced another three-dimensional autonomous system,

$$\begin{cases} \dot{x} = y \\ \dot{y} = z \\ \dot{z} = -y + 3y^2 - x^2 - xz + a \, , \end{cases} \tag{8}$$

where a is a real parameter.

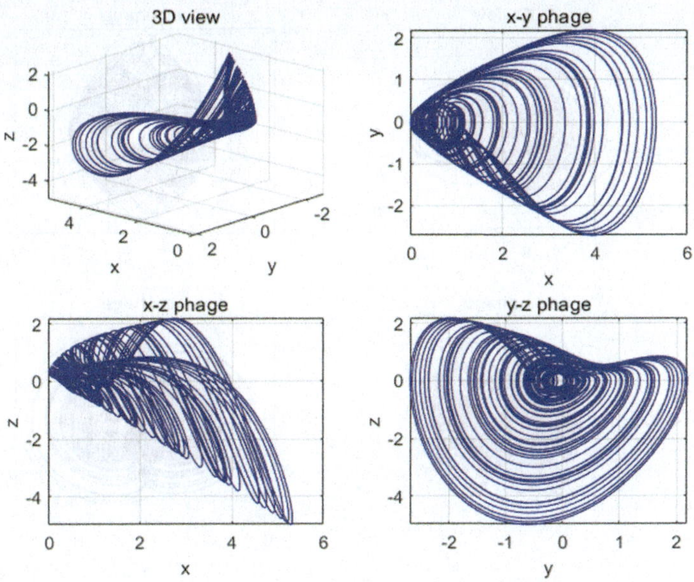

Fig. 3 Three-dimensional view of the chaotic behavior of system (8) when $a = -0.05$ and initial conditions (0.5, 0.5, 0.5)

By determining the plus-minus signs of the real parameter a, system (8) can be a no-equilibrium system when $a < 0$, but still can generate a chaotic attractor, as shown in Fig. 3. The Lyapunov exponents with respect to parameter a were also calculated and analyzed.

2.4 Maaita System

By modifying the original Sprott A system (see Table 1) with a cubic nonlinearity, Maaita et al. [8] proposed a three-dimensional autonomous system without equilibria:

$$\begin{cases} \dot{x} = y \\ \dot{y} = -x^3 - zy \\ \dot{z} = y^2 - a\,, \end{cases} \tag{9}$$

where a is a real parameter.

When $a > 0$, the system has no equilibria, which was analyzed through numerical stimulations on Lyapunov exponents, bifurcations diagrams and Poincaré maps [24, 25].

For the case of $a = 5.16$ with initial condition $(x_0, y_0, z_0) = (-0.8, 0, 1.0)$, system (9) has the Poincaré map with both chaotic behavior and regular orbits, as well

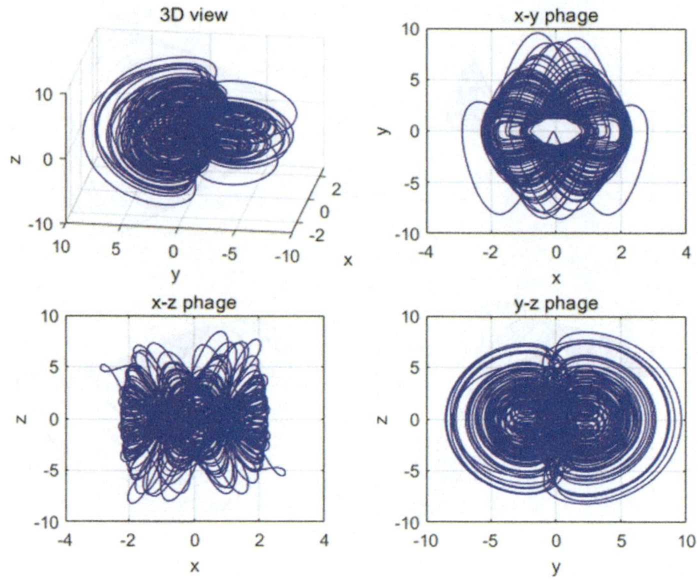

Fig. 4 Trajectory in the three-dimensional space with $a = 5.16$ and initial condition $(x_0, y_0, z_0) = (-0.8, 0, 1.0)$ [8]

as a detectable 3-tori periodicity. Also, the system has a positive Lyapunov exponent, a negative one and a zero one, which confirms the chaotic behavior.

For the case of $a = 0.6$ with initial condition $(x_0, y_0, z_0) = (-1.2, 0, 0)$, system (9) has three Lyapunov exponents that are equal to zero, which implies that the system has regular orbits coexisting with the chaotic ones instead of chaotic behavior.

Moreover, based on Kirchhoff's circuit laws, the circuit realization of the system was designed by using an Op Am with three integrators and one inverting amplifier, as well as four signal multipliers. The trajectory of the system (9), for the first case mentioned above, is shown in Fig. 4.

2.5 *Akgul System*

Akgul et al. [9] implemented a new three-dimensional chaotic system without equilibria, described by

$$\begin{cases} \dot{x} = ay - x + zy \\ \dot{y} = -bxz - cx + zy + d \\ \dot{z} = e - fxy - x^2 , \end{cases} \tag{10}$$

where a, b, c, d, e and f are real parameters.

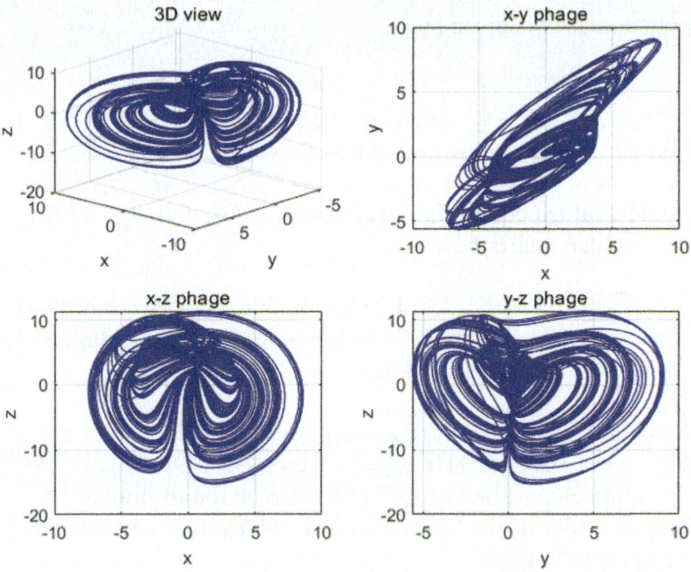

Fig. 5 Trajectory of system (11) in the three-dimensional space with initial value $(x_0, y_0, z_0) = (0, 0, 0)$ [9]

With initial condition $(x_0, y_0, z_0) = (0, 0, 0)$, its chaotic dynamics can be easily observed. But, when $(x_0, y_0, z_0) \neq (0, 0, 0)$, different parameters can generate different chaotic behaviors.

With $a = 2.8$, $b = 0.2$, $c = 1.4$, $d = 1$, $e = 10$ and $f = 2$, the system becomes

$$\begin{cases} \dot{x} = 2.8y - x + zy \\ \dot{y} = -0.2xz - 1.4x + zy + 1 \\ \dot{z} = 10 - 2xy - x^2 , \end{cases} \tag{11}$$

which exhibits chaotic dynamics as well.

Although Akgul et al. [9] aimed to design a new FPGA-based chaotic oscillator and focused on encryption studies, the electronic circuit realization, Lyapunov exponent spectrum, fractal dimension and bifurcation analysis were computed by Akgul and Pehlivan [17].

On the other hand, four complex-valued equilibria of the chaotic system with initial condition $(x_0, y_0, z_0) = (0, 0, 0)$ were found:

$$E_{1,2}(2.707 \pm 0.573i,\ 0.413 \pm 0.661i,\ -1.584 \pm 3.332i)$$

and

$$E_{3,4}(-2.962 \pm 0.739i,\ -0.107 \pm 0.766i,\ -3.215 \pm 3.924i).$$

The Jacobian matrix is given by

$$J = \begin{bmatrix} -1 & a+z & y \\ -bz-c & z & -bx+y \\ -fy-2x & -fx & 0 \end{bmatrix}. \tag{12}$$

For the complex-valued equilibria $E_1(2.707 + 0.573i,\ 0.413 - 0.661i,\ -1.584 + 3.332i)$, the Jacobian matrix becomes

$$J = \begin{bmatrix} -1 & 1.215 + 3.332i & 0.413 - 0.661i \\ -1.083 - 0.666i & -1.584 + 3.332i & -0.128 - 0.776i \\ -6.242 + 0.175i & -5.415 - 1.147i & 0 \end{bmatrix}. \tag{13}$$

By solving the characteristic equation $\Delta(\lambda I - J(E_1)) = 0$, the eigenvalues are obtained: $\lambda_1 = 1.020 + 2.941i, \lambda_2 = -1.068 - 2.149i, \lambda_3 = -2.537 + 2.540i$. The corresponding eigenvalues of $E_{2,3,4}$ can also be found similarly.

The phase portraits of the chaotic system with initial condition $(x_0, y_0, z_0) = (0, 0, 0)$ is displayed in Fig. 5.

2.6 Pham System

Based on the Jafari LE_i model [26], Pham [14] promoted the development of chaotic systems without equilibria.

Given the following two Jafari LE systems:

Jafari LE_5 system:

$$\begin{cases} \dot{x} = y \\ \dot{y} = -1.5x + zy \\ \dot{z} = -x^2 + y^2 - 5xy. \end{cases} \tag{14}$$

Jafari LE_6 system:

$$\begin{cases} \dot{x} = y \\ \dot{y} = -x + zy \\ \dot{z} = 0.04y^2 - xy - 0.1xz. \end{cases} \tag{15}$$

By adding a constant parameter a ($a \neq 0$) to system (14) and system (15), respectively, two new modified system are obtained, as follows:

Modified Jafari LE_5 system:

$$\begin{cases} \dot{x} = y \\ \dot{y} = -1.5x + zy \\ \dot{z} = -x^2 + y^2 - 5xy + a. \end{cases} \tag{16}$$

Modified Jafari LE_6 system:

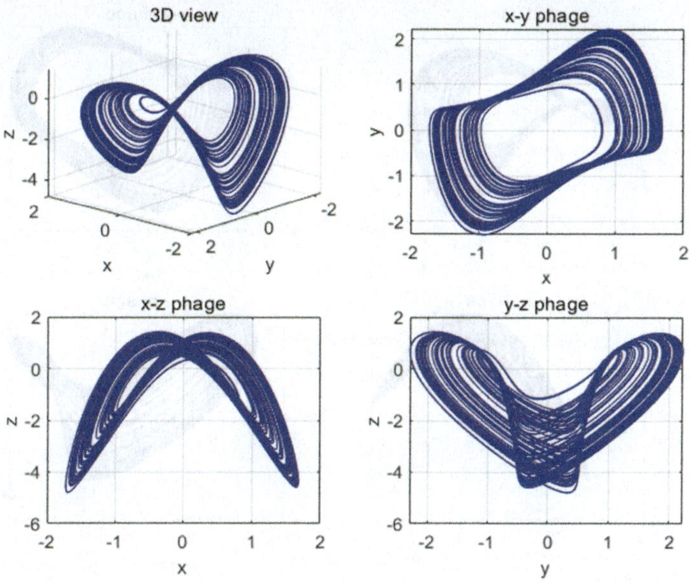

Fig. 6 Phase portrait of system (16) in the three-dimensional space with $a = 0.001$ and initial condition $(x_0, y_0, z_0) = (0.7, 1, 0)$ [14]

$$\begin{cases} \dot{x} = y \\ \dot{y} = -x + zy \\ \dot{z} = 0.04y^2 - xy - 0.1xz + a. \end{cases} \tag{17}$$

Both of the two systems, (16) and (17), are systems without equilibria but are chaotic when $a = 0.001$ with different initial conditions, as shown in Figs. 6 and 7.

In addition, Pham [14] constructed a special system with hidden attractors:

$$\begin{cases} \dot{x} = y \\ \dot{y} = 0.4xz - a \\ \dot{z} = 0.3y - 0.1z - 1.4y^2 - bxy - c, \end{cases} \tag{18}$$

where a, b, c are positive parameters.

Computational results show that the chaotic system (18) is dissipative. By setting $\dot{x} = \dot{y} = \dot{z} = 0$ and choosing appropriate values of parameters, the equilibria of the system can be found.

When $a = c = 0$, system (18) has an infinite number of equilibria, $E(x, 0, 0)$; when $a \neq 0$ and $c \neq 0$, system (18) has one equilibrium point, $E(-\frac{a}{4c}, 0, -10c)$. To determine the stability of the equilibrium by the eigenvalues of the Jacobian matrix at the equilibrium point E, one has

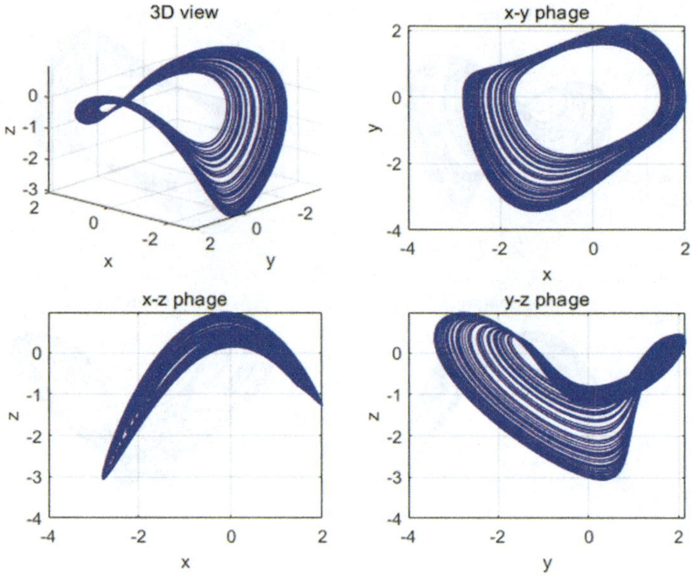

Fig. 7 Phase portrait of system (17) in the three-dimensional space with $a = 0.001$ and initial condition $(x_0, y_0, z_0) = (1, 2, 0)$ [14]

$$J = \begin{bmatrix} 0 & 1 & 0 \\ -4c & 0 & -\frac{0.1a}{c} \\ 0 & 0.3 + \frac{ab}{4c} & -0.1 \end{bmatrix}. \tag{19}$$

The corresponding characteristic equation is given by

$$\lambda^3 + 0.1\lambda^2 + \left(\frac{0.12ac + 0.1a^2b}{4c^2} + 4c\right)\lambda + 0.4c = 0 \tag{20}$$

Using the Routh–Hurwitz criterion, $0.1(\frac{0.12ac+0.1a^2b}{4c^2} + 4c) > 0.4c$, the system has one and only one equilibrium.

In particular, when $a \neq 0$ and $c = 0$, system (18) has no equilibria.

A phase portrait of system (18) without equilibrium is shown in Fig. 8.

Recently, Pham and Akgul [16] investigated another chaotic system without equilibria, which is given by

$$\begin{cases} \dot{x} = y + a \\ \dot{y} = -x + z \\ \dot{z} = -bx^2 + z^2 + c, \end{cases} \tag{21}$$

where a, b and c are positive parameters.

By solving $\dot{x} = \dot{y} = \dot{z} = 0$, the result gives

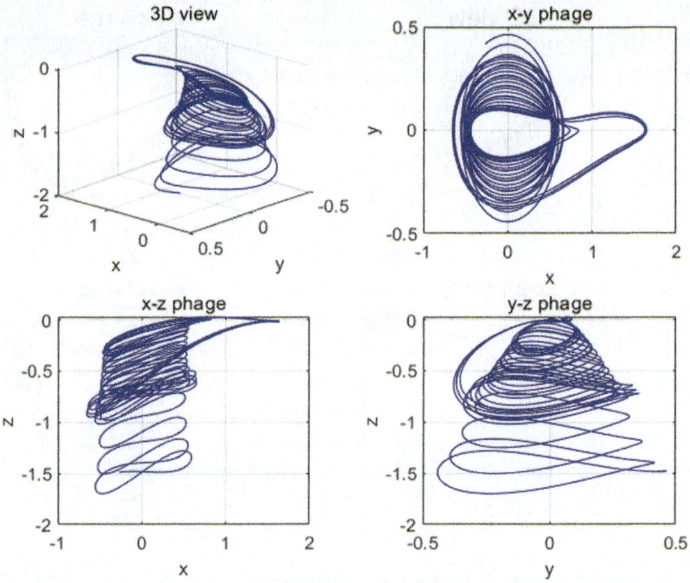

Fig. 8 Phase portrait of system (18) in the three-dimensional space, with $a = 0.005$, $b = 0.2$, $c = 0$ and initial condition $(x_0, y_0, z_0) = (-1.53, 0.33, 0.39)$ [14]

$$x^2 = \frac{c}{b-1}, \tag{22}$$

for $b \neq 1$. Thus, when $b < 1$, Eq. (22) is inconsistent, so system (21) has no equilibria. It is a system with hidden attractors.

For $a = 1, b = 0.8, c = 2$ and initial condition $(x_0, y_0, z_0) = (0, 3, 0)$, Lyapunov exponents are calculated, as $L_1 = 0.026$, $L_2 = 0$, $L_3 = -6.8624$, and the Kaplan–Yorke dimension is $D_{KY} = 2.0038$. The Phase portrait for this case is shown in Fig. 9.

2.7 *Wang System*

By applying a systematic approach for investigating new no-equilibrium systems introduced by Jafari et al. [6], which is to be discussed in more detail in the next section, Wang et al. [18] proposed a new chaotic system without equilibria:

$$\begin{cases} \dot{x} = y \\ \dot{y} = z \\ \dot{z} = k_1xy + k_2xz + k_3yz + k_4y^3 + k_5z_3 + k_6, \end{cases} \tag{23}$$

where k_i are adjusting parameters and particularly $k_6 \neq 0$.

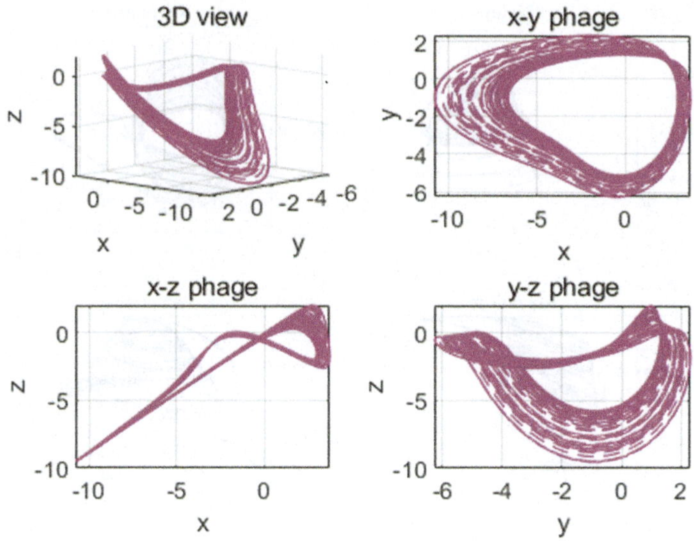

Fig. 9 Phase portrait of system (21) in the three-dimensional space with $a = 1$, $b = 0.8$, $c = 2$ and initial condition $(x_0, y_0, z_0) = (0, 3, 0)$ [16]

By solving $\dot{x} = \dot{y} = \dot{z} = 0$, one gets $k_6 = 0$, which is inconsistent because it has conflict with the assumption of $k_6 \neq 0$. Thus, system (23) has no equilibria.

With parameters $a = 0.49$, $b = 0.75$ and initial condition $(x_0, y_0, z_0) = (0, 3, 0)$, the phase portrait of this system is shown in Fig. 10. Also, Lyapunov exponents are $L_1 = 0.034$, $L_2 = 0$, $L_3 = -0.173$. Other dynamical behaviors of this system were analyzed in [18].

3 Systematic Approach for Finding Chaotic Systems Without Equilibria

Systematic approach is a useful method for finding rare systems without equilibria or with hidden attractors. In the book *Systems with Stable Equilibrium* [14], Pham shows an interesting example of using systematic approach by considering a general equation having quadratic nonlinearities with eight coefficients $(a_1 - a_8)$, as follows:

$$\begin{cases} \dot{x} = y \\ \dot{y} = z \\ \dot{z} = a_1 y + a_2 z + a_3 y^2 + a_4 z^2 + a_5 x y_3 + a_6 x z + a_7 y z + a_8, \end{cases} \tag{24}$$

where parameters $a_8 \neq 0$.

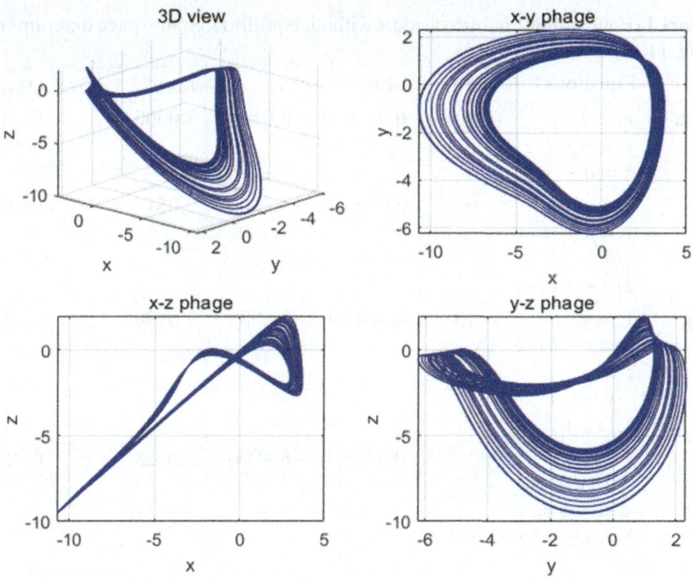

Fig. 10 Phase portrait of system (21) with $a = 1$, $b = 0.8$, $c = 2$ and initial condition $(x_0, y_0, z_0) = (0, 3, 0)$ [18]

By solving $\dot{x} = \dot{y} = \dot{z} = 0$, one finds that $a_8 = 0$, which is inconsistent with the assumption of $a_8 \neq 0$. Thus, no equilibria exist in the system (24).

Pham's approach suggests an effective approach to finding new chaotic systems with hidden attractors. As mentioned in Sect. 2.7, Jafari and Sprott [6] performed a systematic search to find the simplest three-dimensional chaotic systems without equilibria, based on the method proposed in [3], right after the reports of the Wei system and Wang system discussed above.

Method 1 By adding a constant parameter a to other non-hyperbolic systems, one can find, for example, a new system like

$$\begin{cases} \dot{x} = y \\ \dot{y} = -x + z \\ \dot{z} = k_1 x^2 + k_2 y^2 + k_3 z^2 + a\,. \end{cases} \tag{25}$$

With $a = 0$, system (25) has an equilibrium at $O(0, 0, 0)$, whose eigenvalues are zero and a pair of pure imaginary numbers. By adjusting and simplifying the terms k_1, k_2, k_3 and a, it generates chaotic systems listed as NE_7 and NE_{11} in Tables 1 and 2, respectively.

Method 2 There are cases where the equilibria are imaginary. Jafari et al. [14] added ten terms into a chaotic system, so as to obtain

Table 1 **(Part 1)** Four simple chaotic systems without equilibria. State-space diagrams of the cases shown in Fig. 11 [6]

Model	Equations	a	LEs	D_{KY}	(x_0, y_0, z_0)
NE_1 (Sprott A)	$\dot{x} = y$ $\dot{y} = -x - zy$ $\dot{z} = a - y^2$	1.0	0.0138, 0, −0.0138	3.0000	(0, 5, 0)
NE_2(Wei)	$\dot{x} = -y$ $\dot{y} = x + z$ $\dot{z} =$ $2y^2 + xz - a$	0.35	0.0776, 0, −1.5008	2.0517	(0, 0.4, 1)
NE_3(Wang-Chen)	$\dot{x} = y$ $\dot{y} = z$ $\dot{z} =$ $-y + 0.1x^2 +$ $1.1xz + a$	1.0	0.0522, 0, −2.6585	2.0196	(1, 1, −1)
NE_4	$\dot{x} =$ $-0.1y + a$ $\dot{y} = x + z$ $\dot{z} = xz - 3y$	1.0	0.0235, 0, −8.4800	2.0028	(−8.2, 0, −5)

$$\begin{cases} \dot{x} = y \\ \dot{y} = z \\ \dot{z} = k_1x + k_2y + k_3z + k_4x^2 + k_5y^2 + k_6z^2 + k_7xy + k_8xz + k_9zy + a \end{cases} \tag{26}$$

where $k_1^2 - 4k_4a \leq 0$.

By adjusting and simplifying the terms k_i $(i = 1, 2, 3, \ldots, 9)$ and a, one obtains the systems listed as NE_3 and NE_6 in Tables 1 and 2, respectively.

In particular, when $a = 1.0$ with initial condition $(x_0, y_0, z_0) = (1, 1, -1)$, system NE_3 is a simplified Wang-Chen system.

In addition, the Jafari system NE_6 is one of the most elegant systems [14], where the state variable x is conveniently controllable by replacing it with $x + k_x$. For $k_1 = k_3 = k_4 = k_5 = k_6 = k_7 = 0$, $k_2 = k_8 = k_9 = -1$ and $a = -0.75$, the NE_6 system becomes

$$\begin{cases} \dot{x} = y \\ \dot{y} = z \\ \dot{z} = -y - xz - yz - 0.75\,. \end{cases} \tag{27}$$

By replacing x with $x + k_x$, system (27) is rewritten as

$$\begin{cases} \dot{x} = y \\ \dot{y} = z \\ \dot{z} = -y - (x + k_x)z - yz - 0.75\,. \end{cases} \tag{28}$$

Table 2 **(Part 2)** The rest of the seventeen simple chaotic systems without equilibria. State-space diagrams of the cases shown in Fig. 11 [6]

Model	Equations	a	LEs	D_{KY}	(x_0, y_0, z_0)
NE_5	$\dot{x} = 2y$ $\dot{y} = -2x - z$ $\dot{z} = -y^2 + z^2 + a$	2.0	0.0168, 0, −0.3622	2.0465	(0.98, 1.8, −0.7)
NE_6	$\dot{x} = y$ $\dot{y} = z$ $\dot{z} = -y - xz - yz - a$	0.75	0.0280, 0, −3.4341	2.0082	(0, 3, −0.7)
NE_7	$\dot{x} = y$ $\dot{y} = -x + z$ $\dot{z} = -0.8x^2 + z^2 + a$	2.0	0.0252, 0, −6.8524	2.0037	(0, 2.3, 0)
NE_8	$\dot{x} = y$ $\dot{y} = -x - zy$ $\dot{z} = xy + 0.5x^2 - a$	1.3	0.0314, 0, −10.2108	2.0031	(0, 0.1, 0)
NE_9	$\dot{x} = y$ $\dot{y} = -x - zy$ $\dot{z} = -xz + 7x^2 - a$	0.55	0.0504, 0, −0.3264	2.1544	(0.5, 0, 0)
NE_{10}	$\dot{x} = z$ $\dot{y} = z - y$ $\dot{z} = -0.9y - xy + xz + a$	0.6	0.0061, 0, −1.3002	2.0047	(1, 0.7, 0.8)
NE_{11}	$\dot{x} = y$ $\dot{y} = -x + z$ $\dot{z} = z - 2xy - 1.8xz - a$	1.0	0.0706, 0, −0.6456	2.1094	(0, 1.6, 3)
NE_{12}	$\dot{x} = z$ $\dot{y} = x - y$ $\dot{z} = -4x^2 + 8xy + yz + a$	0.1	0.0654, 0, −2.0398	2.0321	(0.5, 0, −1)
NE_{13}	$\dot{x} = -y$ $\dot{y} = x + z$ $\dot{z} = xy + xz + 0.2yz - a$	0.4	0.1028, 0, −2.1282	2.0483	(2.5, 0, 0)
NE_{14}	$\dot{x} = y$ $\dot{y} = z$ $\dot{z} = x^2 - y^2 + 2xz + yz + a$	1.0	0.0532, 0, −11.8580	2.0045	(1, 0, −4)
NE_{15}	$\dot{x} = y$ $\dot{y} = z$ $\dot{z} = x^2 - y^2 + xy + 0.4xz + a$	1.0	0.1101, 0, −1.3879	2.0793	(0, 1, −4.9)
NE_{16}	$\dot{x} = -0.8x - 0.5y^2 + xz + a$ $\dot{y} = -0.8y - 0.5z^2 + yx + a$ $\dot{z} = -0.8z - 0.5x^2 + zy + a$	1.0	0.0607, 0, −0.1883	2.3224	(0, 1, −1)
NE_{17}	$\dot{x} = -y - z^2 + 2.3xy + a$ $\dot{y} = -z - x^2 + 2.3yz + a$ $\dot{z} = -x - y^2 + 2.3xz + a$	2.0	0.2257, 0, −1.7477	2.1292	(1, −1, 0)

This system (28) can amend the level of amplitude by changing the value of the control term k_x, which is considered as "*elegant*". Another similar discussion is illustrated in book [14].

Method 3 By adding a constant term to each of the derivatives in a known chaotic system, a new method for investigating chaos showing that numerically calculated equilibria do not exist, was proposed as follows:

$$\begin{cases} \dot{x} = k_1 y + a_1 \\ \dot{y} = k_2 x + k_3 z + a_2 \\ \dot{z} = k_4 x + k_5 xz + k_6 y + a_3 \,, \end{cases} \tag{29}$$

where $a_{1,2,3}$ are added constants. This method finds the systems listed as NE_4 in Tables 1 and 2.

4 Multi-scroll Attractors in Chaotic Systems Without Equilibria

4.1 Jafari System

As discussed in the above sections, a considerable number of chaotic systems without equilibria were found and studied in the literature [4–9, 27]. Several attempts on finding or constructing special chaotic systems, particularly systems with multi-scroll chaotic attractors [11–13] were reported, as introduced below.

By modifying the original Sprott A system (see Table 1), Jafari et al. [11] constructed the following system:

$$\begin{cases} \dot{x} = y \\ \dot{y} = -x + ayz + by \cdot sin(z) \\ \dot{z} = 1 - y^2 \,, \end{cases} \tag{30}$$

where a, b are positive parameters.

System (30) can generate 2×3-grid scroll chaotic flows, although it has no equilibria. For specific case with $a = 0.1$, $b = 2.9$, and initial condition $(x_0, y_0, z_0) = (0, 1, 0)$, Fig. 12 shows a torus coexisting with a chaotic sea. Moreover, dynamical behaviors of multi-scroll attractors for parameters $a = 0.1$, $b = 2.9$ and initial condition $(x_0, y_0, z_0) = (0, 5, 0)$ are shown in Fig. 13.

4.2 Hu System

By introducing nonlinear functions into the Sprott A system (see Table 1), Hu et al. [13] constructed two simple chaotic systems with multi-scroll hidden attractors, based on a sine function and a sign function, respectively.

Consider a chaotic system,

$$\begin{cases} \dot{x} = y \\ \dot{y} = -x + yz - af(x) \\ \dot{z} = 1 - y^2 \end{cases} \tag{31}$$

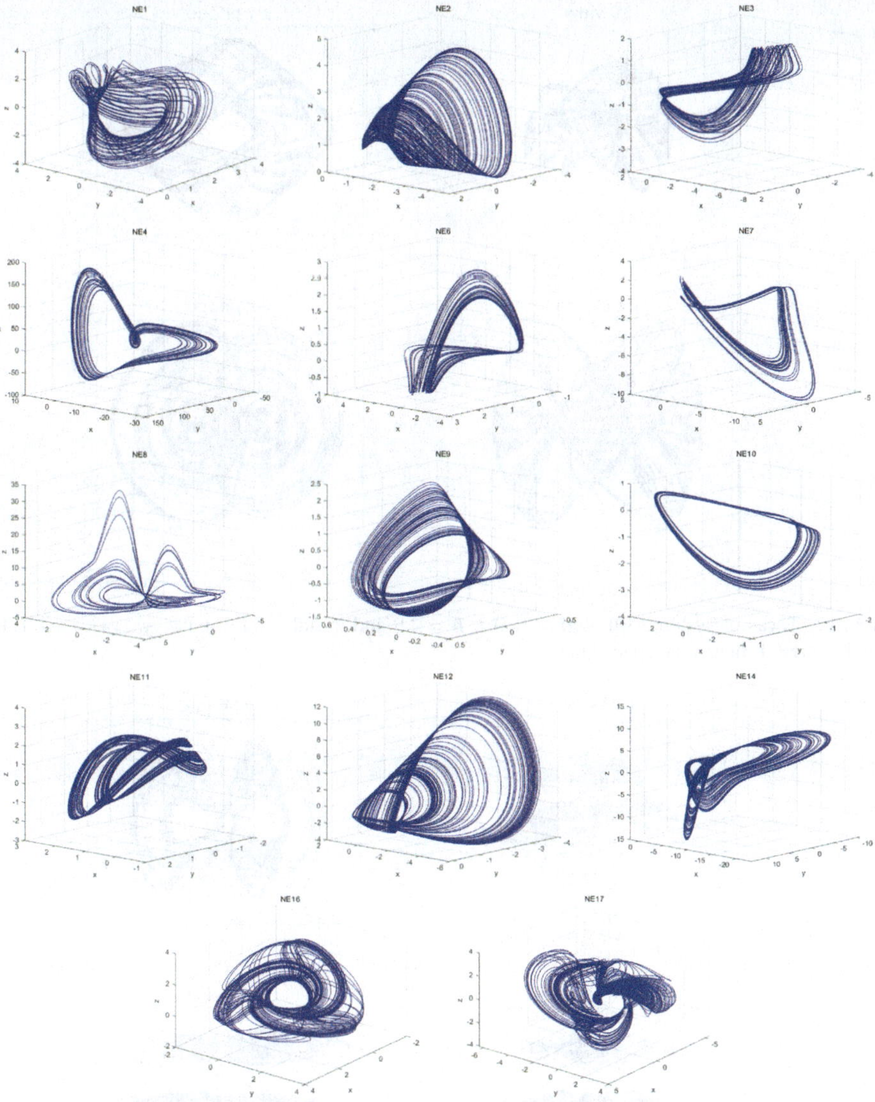

Fig. 11 State-space diagrams of the cases in Tables 1 and 2, projected onto the x-y plane [6]

where a is real parameter and $f(x)$ is nonlinear function.

When $a = 0$, system (31) is the Sprott A system (see Table 1). Theoretically, by solving

$$\dot{x} = \dot{y} = \dot{z} = 0, \tag{32}$$

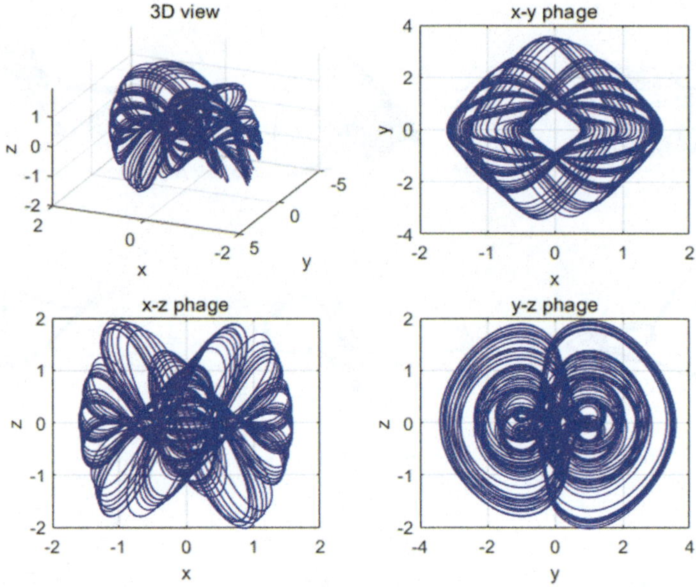

Fig. 12 Torus of system (30), with $a = 0.1$, $b = 2.9$ and initial condition $(x_0, y_0, z_0) = (0, 1, 0)$ in the three-dimensional space [11]

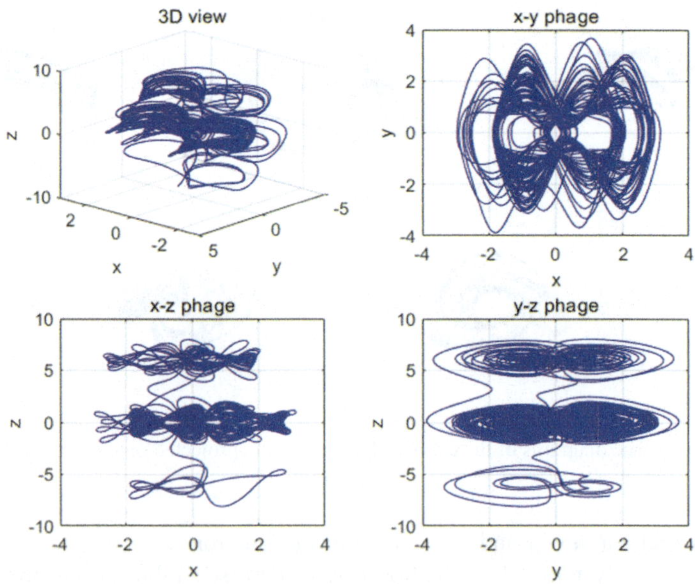

Fig. 13 A 2×3-grid scroll chaotic flow of system (30), with $a = 0.1$, $b = 2.9$ and initial condition $(x_0, y_0, z_0) = (0, 5, 0)$ in the three-dimensional space [11]

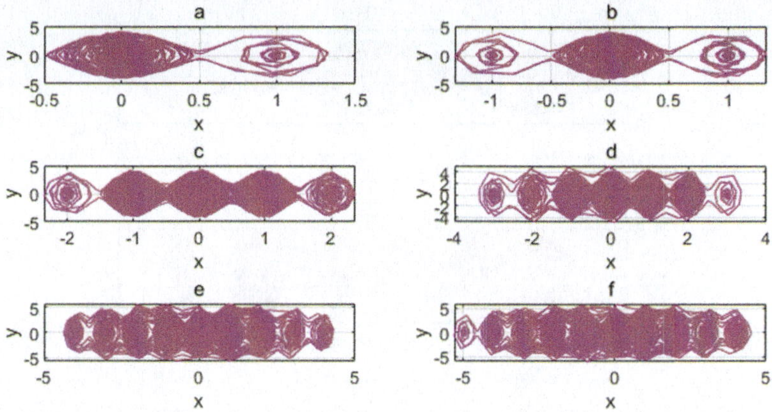

Fig. 14 Multi-scroll hidden attractors (2, 3, 5, 7, 9 and 10) obtained by setting different transient simulation time t (time units = 0.1). **a** $t = 455$, **b** $t = 500$, **c** $t = 1900$, **d** $t = 2200$, **e** $t = 3200$, **f** $t = 3500$ [13]

the equilibria of system (31) can be found. However, there is no solution for Eq. (32) for any value of the parameter a.

Consequently, Hu et al. [13] proposed two improved Sprott A system, as follows:

Hu System I

$$\begin{cases} \dot{x} = y \\ \dot{y} = -x + yz - a \cdot sin(2\pi bx) \\ \dot{z} = 1 - y^2, \end{cases} \tag{33}$$

where b is a real parameter. For $a = 25$, $b = 1$ with initial condition $(x_0, y_0, z_0) = (0, 0.1, 0)$, multi-scroll hidden attractors can be observed and the number of multi-scroll hidden attractors depends on the transient time of the system self-oscillations, as shown in Fig. 14.

Hu System II

$$\begin{cases} \dot{x} = y \\ \dot{y} = -x + yz - \frac{1}{2}\, a \cdot sin(2\pi bx)[sgn(x-c) - sgn(x-d)] + x[2 - sgn(x-c) + sgn(x-d)] \\ \dot{z} = 1 - y^2, \end{cases} \tag{34}$$

where a, b, c and d are real parameters and $sgn(x)$ is the sign function defined by

$$sgn(x) = \begin{cases} 1, x > 0 \\ 0, x = 0 \\ -1, x < 0 \end{cases} \tag{35}$$

Equation (34) is composed of two parts: a nonlinear part and a linear part. The nonlinear part is the product of a *sine* function and a *sign* function, which is bounded with dynamical range being determined by the terms c and d. While the linear part is

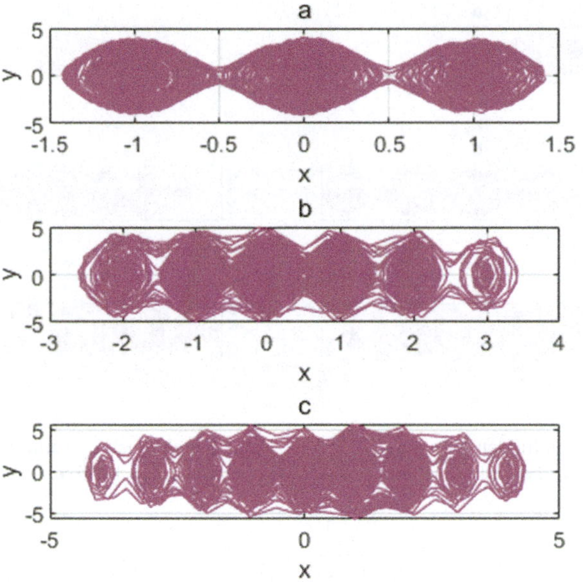

Fig. 15 Multi-scroll hidden attractors (3, 6 and 9), where the simulation time is set to $t = 5000$ units. **a** 3-scroll hidden attractors with $c = -1.5$ and $d = 1.5$; **b** 6-scroll hidden attractors with $c = -2.5$ and $d = 3.5$; **c** 9-scroll hidden attractors with $c = -4.5$ and $d = 4.5$ [13]

the product of the state variable x and the $sign$ function, which means that one can obtain different numbers of multi-scroll hidden attractors by appropriately choosing the system parameters c and d. The phase trajectory with different numbers of multi-scroll hidden attractors within a finite range of simulation time is illustrated in Fig. 15.

References

1. W. Wang, Q.C. Zhang, R.L. Tian, Shilnikov sense chaos in a simple three-dimensional system. Chin. Phys. B **19**(3), 207–216 (2010)
2. H.A. Posch, W.G. Hoover, F.J. Vesely, Canonical dynamics of the Nosé oscillator: stability, order, and chaos. Phys. Rev. A **33**(6), 4253 (1986)
3. J.C. Sprott, Some simple chaotic flows. Phys. Rev. E **50**(2), R647 (1994)
4. Z. Wei, Dynamical behaviors of a chaotic system with no equilibria. Phys. Lett. A **376**(2), 102–108 (2011)
5. X. Wang, G. Chen, Constructing a chaotic system with any number of equilibria. Nonlinear Dyn. **71**(3), 429–436 (2013)
6. S. Jafari, J. Sprott, S.M.R.H. Golpayegani, Elementary quadratic chaotic flows with no equilibria. Phys. Lett. A **377**(9), 699–702 (2013)
7. N.V. Kuznetsov, G.A. Leonov, Hidden attractors in dynamical systems: systems with no equilibria, multistability and coexisting attractors. IFAC Proc. **47**(3), 5445–5454 (2014)
8. J. Maaita, C.K. Volos, I. Kyprianidis, I. Stouboulos, The dynamics of a cubic nonlinear system with no equilibrium point. Nonlinear Dyn. **2015**, Article ID 257923 (2015)

9. A. Akgul, H. Calgan, I. Koyuncu, I. Pehlivan, A. Istanbullu, Chaos-based engineering applications with a 3D chaotic system without equilibrium points. Nonlinear Dyn. **84**(2), 481–495 (2016)
10. J. Lü, G. Chen, Generating multiscroll chaotic attractors: theories, methods and applications. Int. J. Bifurc. Chaos **16**(4), 775–858 (2006)
11. S. Jafari, V.T. Pham, T. Kapitaniak, Multiscroll chaotic sea obtained from a simple 3D system without equilibrium. Int. J. Bifurc. Chaos **26**(02), 1650031 (2016)
12. F. Yu, P. Li, K. Gu, B. Yin, Research progress of multi-scroll chaotic oscillators based on current-mode devices. Opt.: Int. J. Light Electron Opt. **127**(13), 5486–5490 (2016)
13. X. Hu, C. Liu, L. Liu, J. Ni, S. Li, Multi-scroll hidden attractors in improved Sprott a system. Nonlinear Dyn. **86**(3), 1725–1734 (2016)
14. V.-T. Pham, C. Volos, T. Kapitaniak, Systems with stable equilibria, *Systems with Hidden Attractors* (Springer, Berlin, 2017), pp. 21–35
15. V.-T. Pham, S. Jafari, C. Volos, T. Gotthans, X. Wang, D.V. Hoang, A chaotic system with rounded square equilibrium and with no-equilibrium. Opt.: Int. J. Light Electron Opt. **130**, 365–371 (2017)
16. V.-T. Pham, A. Akgul, C. Volos, S. Jafari, T. Kapitaniak, Dynamics and circuit realization of a no-equilibrium chaotic system with a boostable variable. AEU Int. J. Electron. Commun. **78**, 134–140 (2017)
17. A. Akgul, I. Pehlivan, A new three-dimensional chaotic system without equilibrium points, its dynamical analyses and electronic circuit application. Tech. Gaz. **23**(1), 209–214 (2016)
18. Z. Wang, A. Akgul, V.-T. Pham, S. Jafari, Chaos-based application of a novel no-equilibrium chaotic system with coexisting attractors. Nonlinear Dyn. **89**(3), 1877–1887 (2017)
19. S. Jafari, J. Sprott, F. Nazarimehr, Recent new examples of hidden attractors. Eur. Phys. J.: Spec. Top. **224**(8), 1469–1476 (2015)
20. C. Li, J. Sprott, Amplitude control approach for chaotic signals. Nonlinear Dyn. **73**(3), 1335–1341 (2013)
21. C. Li, J. Sprott, Finding coexisting attractors using amplitude control. Nonlinear Dyn. **78**(3), 2059–2064 (2014)
22. C. Li, J.C. Sprott, Variable-boostable chaotic flows. Opt.: Int. J. Light Electron Opt. **127**(22), 10 389–10 398 (2016)
23. V.M. Falkneb, S.W. Skan, Lxxxv. Solutions of the boundary-layer equations. Lond. Edinb. Dublin Philos. Mag. J. Sci. **12**(80), 865–896 (1931)
24. M. Sandri, Numerical calculation of Lyapunov exponents. Math. J. (1996)
25. C. Skokos, The Lyapunov characteristic exponents and their computation. Lect. Notes Phys. **790**(790), 63 (2008)
26. S. Jafari, J. Sprott, Simple chaotic flows with a line equilibrium. Chaos, Solitons Fractals **57**, 79–84 (2013)
27. V.-T. Pham, C. Volos, S. Jafari, Z. Wei, X. Wang, Constructing a novel no-equilibrium chaotic system. Int. J. Bifurc. Chaos **24**(05), 1450073 (2014)

Chaotic Systems with Curves of Equilibria

Xiong Wang and Guanrong Chen

1 Introduction

In the recent years, finding three-dimensional chaotic systems with an infinite number of equilibria has become an attractive topic for research as reported in the literature, since there are some new mysterious features of such chaotic systems with important applications in engineering [1]. The presence of such systems provides some new insights in the relationships between the local properties of a line or curve of equilibria and the complex dynamical behaviors of the underlying chaotic systems [2].

Despite the fact that the basin of attraction in such a chaotic system intersects with the equilibria in some regions, an infinite number of the other equilibria are located outside the basin of attraction [3]. Therefore, the conventional methods for identifying chaotic attractors associated with unstable equilibria do not help in such special systems, illustrating that the "hidden" feature of the uncountable number of attractors actually leads to some unexpected interesting phenomena.

Chaotic systems with an infinite number of equilibria may be classified into three categories, as shown in Fig. 1. Different dynamical models with lines of equilibria, closed-curves of equilibria and open-curves of equilibria have been discovered recently. One of the most impressive discoveries is the first chaotic model with a line of equilibria, containing generalized mathematical functions, coined by Jafari and Sprott [4]. After that, chaotic systems with parallel or perpendicular lines of

X. Wang (✉)
Institute for Advanced Study, Shenzhen University,
Shenzhen 518060, Guangdong, People's Republic of China
e-mail: wangxiong8686@szu.edu.cn

G. Chen
Department of Electrical Engineering, City University of Hong Kong, Hong Kong SAR 999077, People's Republic of China
e-mail: eegchen@cityu.edu.hk

X. Wang et al. (eds.), *Chaotic Systems with Multistability and Hidden Attractors*,
Emergence, Complexity and Computation 40,
https://doi.org/10.1007/978-3-030-75821-9_5

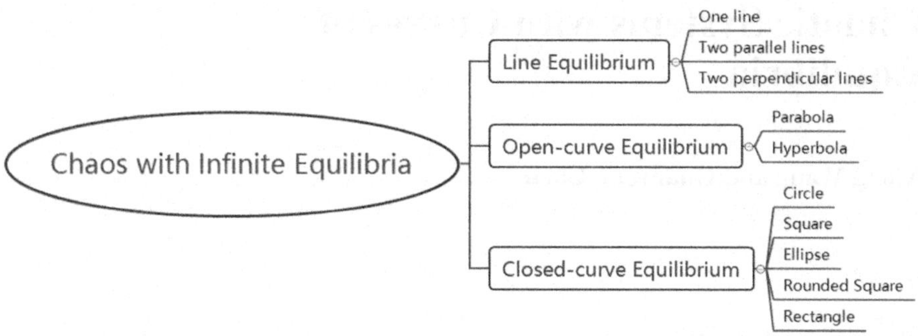

Fig. 1 Chaotic systems with an infinite number of equilibria located on different lines and curves [13]

equilibria [5, 6] and systems with hyper-dimensional chaos [7–9] were discovered. Then, a symmetric time-reversible (STR) model was reported in [10].

Recent studies of chaotic systems with curves of equilibria, including open-curves and closed-curves, started form Gotthans and Petrzela's discovery [11]. There are different shapes of curves of equilibria founded, such as parabola [3], hyperbola [12, 13], circle [14], [2], square [15], ellipse [2], rectangle [2], and some other special shapes [16].

This chapter will provide an overview of chaos generated by systems with lines and curves of equilibria, mostly founded in the last few years. Their dynamical behaviors, circuit design and implementation will be briefly discussed.

2 Constructing a Chaotic System with Infinite Equilibria

Recall the original Sprott A system (see Table 1):

$$\begin{cases} \dot{x} = y \\ \dot{y} = -x + yz \\ \dot{z} = 1 - y^2 , \end{cases} \tag{1}$$

which is a nonuniformly conservative chaotic system without equilibria. This system is an example of the oldest and best-known interesting Nose–Hoover oscillator [17], which describes many phenomena in nature.

By modifying system (1), Jafari and Sprott [4] proposed a general parametric system with quadratic nonlinearities,

$$\begin{cases} \dot{x} = y \\ \dot{y} = a_1 x + a_2 yz \\ \dot{z} = a_3 x + a_4 y + a_5 x^2 + a_6 y^2 + a_7 xy + a_8 xz + a_9 yz , \end{cases} \tag{2}$$

Table 1 LE system: six simple chaotic flows with a line of equilibria [4]

Case	Equations	(a, b)	Equilibrium	Eigenvalues	LEs	D_{KY}	(x_0, y_0, z_0)
LE_1	$\begin{cases} \dot{x} = y \\ \dot{y} = -x + yz \\ \dot{z} = -x - axy - bxz \end{cases}$	$a = 15$ $b = 1$	0 0 z	$\frac{z \pm \sqrt{z^2-4}}{2}$ 0	0.0717 0 −0.5232	2.1371	0 0.5 0.5
LE_2	$\begin{cases} \dot{x} = y \\ \dot{y} = -x + yz \\ \dot{z} = -x - axy - bxz \end{cases}$	$a = 17$ $b = 1$	0 0 z	$\frac{z \pm \sqrt{z^2-4}}{2}$ 0	0.0564 0 −0.2927	2.1927	0 0.4 0
LE_3	$\begin{cases} \dot{x} = y \\ \dot{y} = -x + yz \\ \dot{z} = x^2 - axy - bxz \end{cases}$	$a = 18$ $b = 1$	0 0 z	$\frac{z \pm \sqrt{z^2-4}}{2}$ 0	0.0556 0 −0.3245	2.1714	0 −0.4 0.5
LE_4	$\begin{cases} \dot{x} = y \\ \dot{y} = -x + yz \\ \dot{z} = -axy - bxz - yz \end{cases}$	$a = 4$ $b = 0.6$	0 0 z	$\frac{z \pm \sqrt{z^2-4}}{2}$ 0	0.0539 0 −0.3147	2.1712	0.2 0.7 0
LE_5	$\begin{cases} \dot{x} = y \\ \dot{y} = -ax + yz \\ \dot{z} = -x^2 - y^2 - bxz \end{cases}$	$a = 1.5$ $b = 5$	0 0 z	$\frac{z \pm \sqrt{z^2-4}}{2}$ 0	0.1386 0 −1.3764	2.1007	0.7 1 0
LE_6	$\begin{cases} \dot{x} = y \\ \dot{y} = -x + yz \\ \dot{z} = -ay^2 - xy - bxz \end{cases}$	$a = 0.04$ $b = 0.1$	0 0 z	$\frac{z \pm \sqrt{z^2-4}}{2}$ 0	0.0543 0 −0.6314	2.0860	12 2 0

(continued)

Table 1 (continued)

Case	Equations	(a, b)	Equilibrium	Eigenvalues	LEs	D_{KY}	(x_0, y_0, z_0)
LE_7	$\begin{cases} \dot{x} = z \\ \dot{y} = x + yz \\ \dot{z} = -ax^2 - xy - byz \end{cases}$	$a = 1.85$ $b = 0.3$	0 y 0	$\frac{-0.3y \pm \sqrt{0.09y^2 - 4y}}{2}$ 0	0.1144 0 −1.0270	2.0140	5.1 7 0
LE_8	$\begin{cases} \dot{x} = z \\ \dot{y} = -x - yz \\ \dot{z} = ax^2 - xy - bxz \end{cases}$	$a = 3 \;\; b = 1$	0 y 0	$\pm\sqrt{y}$ 0	0.0521 0 −0.8053	2.0647	0 −0.3 −1
LE_9	$\begin{cases} \dot{x} = z \\ \dot{y} = -ay + xz \\ \dot{z} = z - bz^2 + xy \end{cases}$	$a = 1.62$ $b = 0.2$	x 0 0	$\frac{-0.62 \pm \sqrt{6.8644 - 4x^2}}{2}$ 0	0.0642 0 −0.6842	2.0939	0 1 0.8

where $a_i (i = 1, 2, 3, \ldots, 9)$ are real parameters.

Equation (2) has a line of equilibria, $(0, 0, z)$, which means an infinite number of equilibria, without any other equilibria. With the consideration of millions of combinations of the coefficients a_1 to a_9, and also initial conditions, system (2) is considered "elegant" [18], since many coefficients have been set to zero, while the rests are set to ± 1 or otherwise to a small integer or decimal fraction with the fewest possible digits.

To that end, Pham [19] expanded the model and proposed a more general one with 12 coefficients $(a_1 - a_{12})$:

$$\begin{cases} \dot{x} = z \\ \dot{y} = a_1 x + a_2 z + a_3 z^2 + a_4 xz + a_5 yz \\ \dot{z} = a_6 x + a_7 z + a_8 x^2 + a_9 z^2 + a_{10} xy + a_{11} xz + a_{12} yz \, , \end{cases} \tag{3}$$

which has a line of equilibria, $E(0, y, 0)$.

Other two general parametric systems generating chaotic systems with an infinite number of equilibria were reported [3], in the following form:

$$\begin{cases} \dot{x} = y \\ \dot{y} = a_1 x + a_2 yz \\ \dot{z} = a_3|x| + a_4|y| + a_5 x + a_6 y + a_7 xy + a_8 xz + a_9 yz + a_{10} x^2 + a_{11} y^2 \, , \end{cases} \tag{4}$$

and

$$\begin{cases} \dot{x} = a_1 z, \\ \dot{y} = z f_1(x, y, z, |x|, |y|, |z|), \\ \dot{z} = f_2(x, y, |x|, |y|) + z f_3(x, y, z) \, , \end{cases} \tag{5}$$

where

$$\begin{aligned} f_1(x, y, z, |x|, |y|, |z|) =& a_2|x| + a_3|y| + a_4|z| + a_5 x + a_6 y + a_7 z \\ &+ a_8 xy + a_9 xz + a_{10} yz \\ &+ a_{11} x^2 + a_{12} y^2 + a_{13} z^2 + a_{14} \, , \end{aligned}$$

$$\begin{aligned} f_3(x, y, z) =& a_{15} x + a_{16} y + a_{17} z \\ &+ a_{18} xy + a_{19} xz + a_{20} yz \\ &+ a_{21} x^2 + a_{22} y^2 + a_{23} z^2 + a_{24} \, , \end{aligned}$$

in which a_i are real parameters.

It is noted that system (4) is also an expanded form of system (2), with a line of equilibria, $E(0, 0, z)$. While in system (5), modified from the Gotthans–Petrzela system [11], $f_2(x, y, |x|, |y|)$ is a predefined nonlinear function. An infinite number of equilibria are found, located on the curve $f_2(x, y, |x|, |y|) = 0$ in the plane $z = 0$.

3 Chaotic Systems with Lines of Equilibria

3.1 LE System and a General Equation

There is a new general model in order to find chaotic systems with a line of equilibria, mentioned in Sect. 2, system (2), where $a_1 - a_9$ are real parameters.

As shown in Table 1, cases $LE_1 - LE_6$ are six simple ones founded in the system with six terms, and $LE_7 - LE_9$ are three other similar cases found via a similar procedure. Based on the cases in Table 1, dozens of equivalently additional cases have also been found.

As shown in Fig. 2, where attractors of the above-mentioned cases are plotted, these cases have dissipative properties. In addition, cases listed in Table 1 all have dimensions slightly greater than 2.0, typical for generating strange attractors in three-dimensional autonomous systems.

For the LE_8 system, let

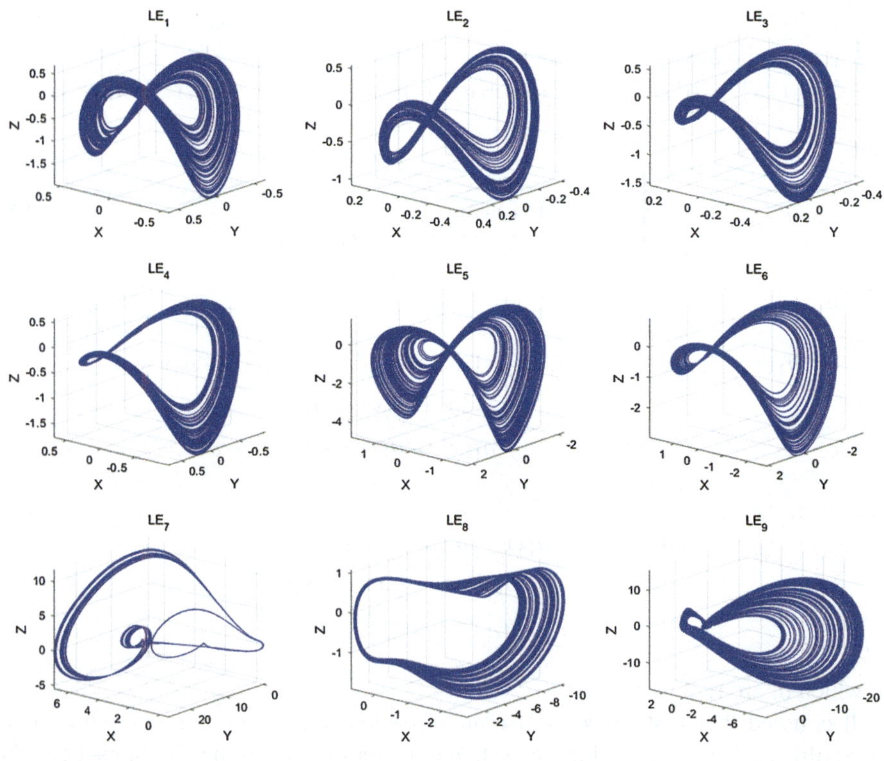

Fig. 2 State-space plots of the cases listed in Table 1

$$\begin{cases} \dot{x} = z \\ \dot{y} = x - yz \\ \dot{z} = -ax^2 + xy + xz , \end{cases} \tag{6}$$

for which the Jacobian matrix at the equilibrium point is

$$J = \begin{bmatrix} 0 & 0 & 1 \\ 0 & 0 & -y^* \\ y^* & 0 & 0 \end{bmatrix} . \tag{7}$$

Then, one can obtain its eigenvalues by solving the characteristic equation $\lambda(\lambda^2 - y^*) = 0$: $\lambda_1 = 0$, $\lambda_{2,3} = \pm\sqrt{y^*}$.

Table 1 lists several analytical features of the systems, including the equilibria, eigenvalues, Lyapunov exponent spectra, and Kaplan–Yorke dimensions, along with initial conditions. Among the examples listed in this table, LE_2 has the largest $D_{KY} = 2.1927$, for which no effort to tune the parameters was needed for generating chaos. All the cases approach chaotic behaviors through successive period-doubling limit cycles, while a typical example with a decreasing a and fixed $b = 1$ shows that the attractor is destroyed through a boundary crisis as a decreases further.

More importantly, it was observed that LE_{1-8} systems generate strange attractors around a line of equilibria, while the LE_9 system has a line of equilibria lying outside the strange attractor. All these cases do not have equilibrium intersecting the attractors, so no homoclinic orbits are expected to exist.

3.2 SL System

Chunbiao Li and Sprott [5] introduced another chaotic system with a line of equilibria, which is more practical because the amplitude and frequency control knob of this system can provide a well-suitable secure key for secret communications. One can also replace the extra amplifier and attenuator in radars and other communication systems, if such a chaotic system is used as the signal source.

This chaotic system has the following form:

$$\begin{cases} \dot{x} = f(x, y) + a_1x^2 + a_2y^2 + a_3z^2 + a_4xy + a_5xz + a_6yz \\ \dot{y} = a_7x^2 + a_8y^2 + a_9z^2 + a_{10}xy + a_{11}xz + a_{12}yz \\ \dot{z} = a_{13}x^2 + a_{14}y^2 + a_{15}z^2 + a_{16}xy + a_{17}xz + a_{18}yz , \end{cases} \tag{8}$$

where a_i are real parameters and $f(x, y)$ can be a single non-quadratic term.

In the following three cases, the multiplicative coefficient of $f(x, y)$ can be set to generate chaos, thereby yielding several chaotic systems with lines of equilibria.

For the case with $f(x, y) = 0$, system (8) has nullclines that are planes or lines passing through the origin with a limit cycle or a strange attractor. The simplest example of such a case is

$$\begin{cases} \dot{x} = yz \\ \dot{y} = xz \\ \dot{z} = -xy\,. \end{cases} \tag{9}$$

Such a system has each variable depending only on the other two, which can generate three perpendicular lines of equilibria:

$$(x, 0, 0) \quad , \quad (0, y, 0) \quad , \quad (0, 0, z)\,,$$

with the corresponding eigenvalues

$$\lambda_1 = 0, \quad \lambda_{2,3} = \pm xi \quad for \quad (x, 0, 0);$$

$$\lambda_1 = 0 \quad , \quad \lambda_{2,3} = \pm yi \quad for \quad (0, y, 0);$$

$$\lambda_1 = 0 \quad , \quad \lambda_{2,3} = \pm z \quad for \quad (0, 0, z).$$

Therefore, the chaotic dynamics of system (9) depends only on the initial conditions, and are constrained to a one-dimensional manifold, which is a hyperbola on the x-y plane.

For the case with $f(x, y) = \pm x$, the simplest chaotic systems were found with six terms in which one is a single linear function of x.

Among the five models illustrated in Table 2, SL_4 and SL_5 have lines of equilibria. While considering quadratic nonlinearities, the SL_4 system is unstable with a negative x, with zero being the largest real part of the eigenvalues, but the SL_5 system has a line of equilibria that is stable with a growing oscillation on the y-z-plane, followed by a relaxation oscillation in the x-direction. The corresponding state-space diagram of the $SL_{4,5}$ system is shown in Fig. 3.

For the case with $f(x, y) = y$, the stability is determined by the nonlinearities and there are five non-quadratic terms. There are different numbers and types of equilibria with a largest eigenvalue that is either zero or negative for the chosen parameters. Among them, SL_9, SL_{10}, SL_{11} and SL_{12} have lines of equilibria. On the other hand, all the isolated equilibria (points that are not part of a line of equilibria) are neutrally stable. Also, SL_{12} generates two parallel lines of equilibria with a single isolated equilibrium point. Moreover, systems $SL_{9,10,11,12}$ are all asymmetric, though symmetric cases exist sometimes.

3.3 AB System

In engineering applications, amplitude control of a chaotic signal is of high importance. From a computational point of view, the amplitude control parameter can be identified from the retained isolated equilibria or the additional line of equilibria.

A general method for controlling the amplitudes of variables in a chaotic system is introduced by Li and Sprott [6] via modifying the Sprott B system. They proposed four models of linearization, as follows:

Table 2 SL system: chaotic flows with a single linearity in x [5]

Model	Equations	(a, b)	Equilibria	Eigenvalues	LEs	D_{KY}	(x_0, y_0, z_0)
SL_1	$\begin{cases} \dot{x} = -x + ay^2 - xy \\ \dot{y} = xz \\ \dot{z} = z^2 - bxy \end{cases}$	$a = 2 \quad b = 1$	$(0, 0, 0)$	$(0, 0, -1)$	0.1889 0 −1.6864	2.1120	3.4 3 0
SL_2	$\begin{cases} \dot{x} = -ax + xy \\ \dot{y} = z^2 + xz \\ \dot{z} = y^2 - byz \end{cases}$	$a = 2 \quad b = 1$	$(0, 0, 0)$ $(-\frac{a}{b}, a, a)$	$(0, 0, -2)$ $(-2.9311, 0.4656 \pm 1.5851i)$	0.2191 0 −2.2191	2.0987	−2 0 3
SL_3	$\begin{cases} \dot{x} = x + ay^2 - z^2 \\ \dot{y} = x^2 - by^2 \\ \dot{z} = xz \end{cases}$	$a = 2.4 \quad b = 1$	$(0, 0, 0)$ $(-\frac{b}{a}, \pm\frac{\sqrt{b}}{a}, 0)$	$(1, 0, 0)$ $(-0.4167, 0.0833 \pm 0.9091i)$ $(2.2103, -0.3770 - 0.4167i)$	0.0734 0 −1.5866	2.0453	0 0.9 ±0.6
SL_4	$\begin{cases} \dot{x} = -x + by^2 + xz \\ \dot{y} = xz \\ \dot{z} = -axy + yz \end{cases}$	$a = 0.1 \quad b = 1$	$(0, 0, z)$	$(0, 0, z - 1)$	0.2390 0 −1.3956	2.1713	8 0 0.7
SL_5	$\begin{cases} \dot{x} = -x + az^2 \\ \dot{y} = z^2 - bxz \\ \dot{z} = xy - yz \end{cases}$	$a = 1 \quad b = 2$	$(0, y, 0)$ $(\frac{1}{ab^2}, 0, \frac{1}{ab})$	$(0, -1, -y)$ $(0.2685, -0.6342 \pm 0.2516i)$	0.0748 0 −0.8856	2.0845	2 3 0
SL_6	$\begin{cases} \dot{x} = y - z^2 \\ \dot{y} = -axz \\ \dot{z} = x^2 - yz \end{cases}$	$a = 0.9$	$(0, 0, 0)$	$(0, 0, 0)$	0.1304 0 −3.9246	2.0332	0 0 1.4

(continued)

Table 2 (continued)

Model	Equations	(a, b)	Equilibria	Eigenvalues	LEs	D_{KY}	(x_0, y_0, z_0)
SL_7	$\begin{cases} \dot{x} = -y - yz \\ \dot{y} = x^2 + axz \\ \dot{z} = z^2 + byz \end{cases}$	$a = 14 \quad b = 1$	$(0, 0, 0)$ $(0, \frac{1}{b}, -1)$ $(a, \frac{1}{b}, -1)$	$(0, 0, 0)$ $(-2.7937, 0.8969 \pm 2.0511i)$ $(0.0714, -0.5357 \pm 13.9925i)$	0.1340 0 −0.5489	2.2441	0 1 −0.7
SL_8	$\begin{cases} \dot{x} = -y - y^2 \\ \dot{y} = az^2 + xy \\ \dot{z} = -x^2 - bxy \end{cases}$	$a = 0.3 \quad b = 1$	$(0, 0, 0)$ $(0, 1, 0)$ $(-1, 1, \pm\frac{1}{\sqrt{a}})$	$(0, 0, 0)$ $(0, \pm i)$ $(-1.5302, 0.2651 \pm 0.8035i)$ $(0.4098, -0.7049 \pm 1.4752i)$	0.0337 0 −0.2544	2.1324	0 1.3 −1
SL_9	$\begin{cases} \dot{x} = y \\ \dot{y} = ay2 - xz \\ \dot{z} = x^2 + xy - bxz \end{cases}$	$a = 0.4 \quad b = 1$	$(0, 0, z)$	$(0, \pm\sqrt{z})$	0.0749 0 −0.7391	2.1014	0 4 5
SL_{10}	$\begin{cases} \dot{x} = y + axz \\ \dot{y} = yx - xz \\ \dot{z} = x^2 + bxy \end{cases}$	$a = 0.2 \quad b = 3$	$(0, 0, z)$ $(-\frac{1}{a}, \frac{1}{ab}, \frac{1}{ab})$	$(0, \frac{z \pm \sqrt{z^2 - 100z}}{10})$ $(0.3565, -2.5116 \pm 7.9885i)$	0.0280 0 −0.2397	2.1167	−0.2 0 0
SL_{11}	$\begin{cases} \dot{x} = y + y^2 - ayz \\ \dot{y} = -z^2 + byz \\ \dot{z} = xy \end{cases}$	$a = 0.9 \quad b = 1$	$(x, 0, 0)$ $(0, -1, 0)$ $(0, \frac{1}{ab-1}, \frac{ab}{ab-1})$	$(0, 0, 0)$ $(-0.7113, 0.3556 \pm 1.1311i)$ $(0.9911, -5.4956 \pm 8.4079i)$	0.1401 0 −0.8573	2.1634	0.8 −2 0

(continued)

Table 2 (continued)

Model	Equations	(a, b)	Equilibria	Eigenvalues	LEs	D_{KY}	(x_0, y_0, z_0)
SL_{12}	$\begin{cases} \dot{x} = -y + x^2 - y^2 \\ \dot{y} = -xz \\ \dot{z} = ax^2 + bxy \end{cases}$	$a = 0.3 \;\; b = 1$	$(0, 0, z)$ $(0, -1, z)$ $(\frac{ab}{a^2-b^2}, \frac{a^2}{b^2-a^2}, 0)$	$(0, \pm\sqrt{z})$ $(0, \pm\sqrt{-z})$ $(-5690, -0.0452 \pm 0.2351i)$	0.0096 0 −0.2660	2.0362	0 −0.3 2

Table 3 AB system: amplitude-controllable Sprott B systems [6]

Model	Equations	Parameters	Equilibria	Eigenvalues	LEs	D_{KY}
AB_2	$\begin{cases}\dot{x} = zsgn(y)\\ \dot{y} = x - y\\ \dot{z} = a\|x\| - mxy\end{cases}$	$a = 1 \quad m = 1$	$(\frac{a}{m}, \frac{a}{m}, 0)$ $(0, 0, z)$ $(-\frac{a}{m}, -\frac{a}{m}, 0)$	$(-1.5448, 0.2724 \pm 0.8760i)$ $(0, 0, -1)$ $(-1.5448, 0.2724 \pm 0.8760i)$	0.0906 0 −1.0906	2.1294
AB_3	$\begin{cases}\dot{x} = zsgn(y)\\ \dot{y} = x - y\\ \dot{z} = a\|x\| - mxy\|y\|\end{cases}$	$a = 1.5 \quad m = 1$	$(\sqrt{\frac{a}{m}}, \sqrt{\frac{a}{m}}, 0)$ $(0, 0, z)$ $(-\sqrt{\frac{a}{m}}, -\sqrt{\frac{a}{m}}, 0)$	$(-1.8637, 0.4319 \pm 1.1930i)$ $(0, 0, -1)$ $(-1.8637, 0.4319 \pm 1.1930i)$	0.1486 0 −1.1486	2.1294
AB_5	$\begin{cases}\dot{x} = yz\\ \dot{y} = x\|x\| - y\|x\|\\ \dot{z} = m\|x\| - axy\end{cases}$	$a = 1 \quad m = 1$	$(\frac{m}{a}, \frac{m}{a}, 0)$ $(0, y, 0)$ $(0, 0, z)$ $(-\frac{m}{a}, -\frac{m}{a}, 0)$	$(-1.4656, 0.2328 \pm 0.7926i)$ $(\sqrt{a}\|y\|, 0, -\sqrt{a}\|y\|)$ $(0, 0, 0)$ $(-1.4656, 0.2328 \pm 0.7926i)$	0.0993 0 −1.1783	2.0843
AB_6	$\begin{cases}\dot{x} = yz\\ \dot{y} = x\|x\| - y\|x\|\\ \dot{z} = a\|xy\| - mxy\|y\|\end{cases}$	$a = 1.4 \quad m = 1$	$(\frac{a}{m}, \frac{a}{m}, 0)$ $(0, y, -0)$ $(0, 0, -z)$ $(-\frac{a}{m}, -\frac{a}{m}, 0)$	$(-2.1964, 0.3982 \pm 1.2612i)$ $(\sqrt{a}\|y\|, 0, -\sqrt{a}\|y\|)$ $(0, 0, 0)$ $(-2.1964, 0.3982 \pm 1.2612i)$	0.1006 0 −1.2861	2.0782

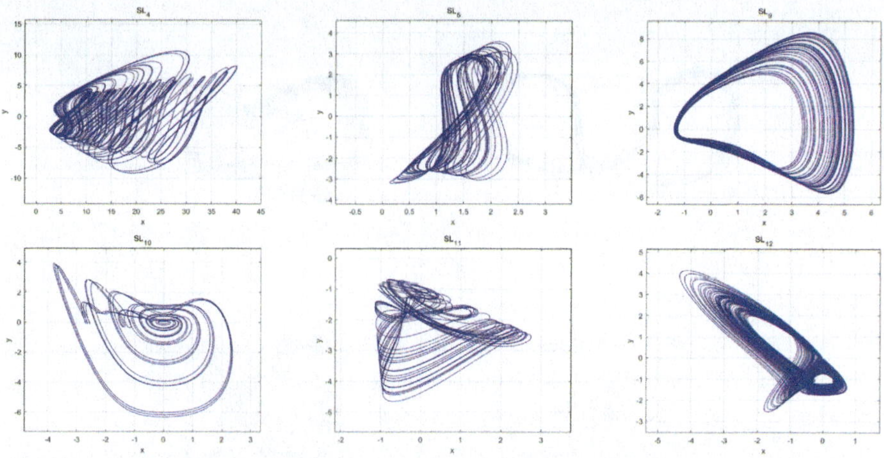

Fig. 3 State-space diagrams for the cases in Table 2, projected onto the x-y plane [5]

$$\begin{cases} \dot{x} = z\, sgn(y) \\ \dot{y} = x - y \\ \dot{z} = m - ay\, sgn(x)\,; \end{cases} \tag{10}$$

$$\begin{cases} \dot{x} = z\, sgn(y) \\ \dot{y} = x - y \\ \dot{z} = m - ax\, sgn(y)\,; \end{cases} \tag{11}$$

$$\begin{cases} \dot{x} = y\, sgn(z) \\ \dot{y} = x - y \\ \dot{z} = m - ay\, sgn(x)\,; \end{cases} \tag{12}$$

$$\begin{cases} \dot{x} = y\, sgn(x) \\ \dot{y} = x - y \\ \dot{z} = m - ax\, sgn(y)\,, \end{cases} \tag{13}$$

where a is the bifurcation parameter and m is the amplitude parameter.

Four examples of the above AB systems, derived from Eqs. (10)–(13), respectively, are listed in Table 3.

The preserved isolated equilibrium coordinates are reversely proportional to the amplitude control parameters, while lines of equilibria result from the additional absolute-value terms, which will have some influence on the dynamics, along with the alteration of the nonlinearities in the system.

As listed in Table 3 and Fig. 4, the AB_2 and AB_3 systems have lines of equilibria, while the AB_5 and AB_6 systems have two perpendicular lines of equilibria. By observing the rearrangement of the basins of attraction with multi-stability, equilibria

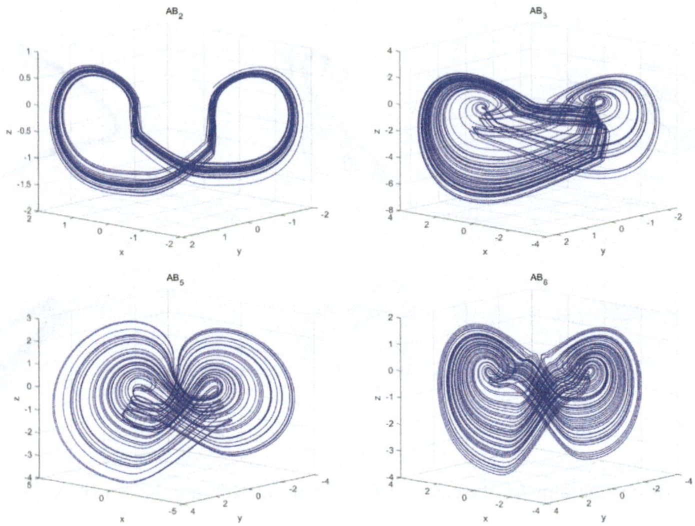

Fig. 4 AB system: strange attractors observed from the amplitude-controllable Sprott B systems in Table 3, for $m = 1$, projected in the three-dimensional space [6]. Green color indicates the value of the local largest LEs with positive values, while red color indicates those LEs with negative values

of all these systems are considered "safe" [6], which are either unstable or neutrally stable in general.

By computing the non-full ranks of the new Jacobian matrices, eigenvalues of such systems are obtained: $AB_{2,3}$ $(0, 0, 1)$ and AB_5 $(0, 0, 0)$ for line $(0, 0, z)$ and $(\sqrt{a}|y|, 0, \sqrt{a}|y|)$ for line $(0, y, 0)$, which indicates that the line of $AB_{2,3}$ is stable in one direction, while one of the lines of AB_5 is neutrally stable and the other one has unstable saddle nodes.

In addition, there are symmetric limit cycles coexisting with strange attractors in the above AB systems, including also those systems without lines of equilibria. Moreover, the line of equilibria threads the attractors with multi-stability for different segments of lines, therefore the corresponding attractors are hidden.

The amplitude control is decreased by the existence of multi-stability, when initial conditions switch the basin of attraction, thereby initiating the state-shift of the coexisting attractors.

3.4 STR System

To investigate the chaotic systems with strange attractors in both forward and reversed time, with the same parameters, Sprott [10] introduced a three-dimensional autonomous chaotic system with polynomial nonlinearities, which generate strange attractors, in the following form:

$$
\begin{cases}
\dot{x} = a_1 + a_2 x + a_3 y + a_4 z + a_5 x^2 + a_6 y^2 + a_7 z^2 + a_8 xy + a_9 xz + a_{10} yz \\
\dot{y} = a_{11} + a_{12} x + a_{13} y + a_{14} z + a_{15} x^2 + a_{16} y^2 + a_{17} z^2 + a_{18} xy + a_{19} xz + a_{20} yz \\
\dot{z} = a_{21} + a_{22} x + a_{23} y + a_{24} z + a_{25} x^2 + a_{26} y^2 + a_{27} z^2 + a_{28} xy + a_{29} xz + a_{30} yz \,,
\end{cases} \tag{14}
$$

where $a_1 - a_{30}$ are real parameters.

System (14) is time-reversal invariant, whereas the condition $(x, y, z, t) \rightarrow (x, y, z, -t)$ reverses the sign of a_i. To obtain chaotic solutions, a simple case was considered:

$$
\begin{cases}
\dot{x} = -yz \\
\dot{y} = (ax + y + z^2)z \\
\dot{z} = x - x^3.
\end{cases} \tag{15}
$$

This system has two equilibria, $(-1, 0, \pm\sqrt{2})$, coexisting with three lines of equilibria, $(0, 0, z)$ and $(\pm 1, y, 0)$. With $a = 2$, eigenvalues of the equilibrium $(-1, 0, -\sqrt{2})$ are obtained as $(-2.090161, 0.337974 \pm 2.301872i)$. On the other hand, the eigenvalues of the system on the equilibrium line $(0, 0, z)$ can be obtained by solving the characteristic equation

$$
\lambda^3 - z\lambda^2 + az^2\lambda - 3z^3 = 0 \,.
$$

Besides, one can get eigenvalues $(0, \pm\sqrt{2y})$ for the system with a line of equilibria, $(\pm 1, y, 0)$. Figure 5 shows a dense set of symmetric invariant periodic orbits of system (15), coexisting with non-asymmetric strange attractors. The periodic orbit, with a Lyapunov exponent $(0, 0, 0)$ and initial condition $(-1, -2, 1)$, will return to that value within an error of 1×10^{-4}, thus repeatedly yielding millions of cycles.

In addition, a hidden attractor is found in such a system, which has Lyapunov exponents $(0.0892, 0, -1.2270)$ and the Kaplan–Yorke dimension 2.0727.

3.5 *IE System*

Two general parametric forms for finding chaotic systems with infinitely many equilibria were proposed by Pham [3], as shown in Eqs. (4) and (5). It was found that there is a line of equilibria, $(0, 0, z)$, in system (3) and curves of equilibria in system (4).

Eight elegant chaotic systems with an infinite number of equilibria were introduced, where five systems ($LE_1 - LE_5$) generate chaotic dynamics with a line of equilibria as listed in Table 4. By applying the Wolf algorithm and with computational analysis, the equilibria, parameter values, Lyapunov exponents, Kaplan–Yorke

Table 4 IE system: five chaotic systems with a line of equilibria [3]

Case	Equations	Parameters	Equilibria	ICs	LEs	D_{KY}
IE_1	$\begin{cases}\dot{x}=y\\ \dot{y}=-x+yz\\ \dot{z}=a\lvert x\rvert-bxy-xz\end{cases}$	$a=0.25$ $b=19$	Line $(0,0,z)$	0 −0.35 0.45	0.0363 0 −0.2786	2.1303
IE_2	$\begin{cases}\dot{x}=y\\ \dot{y}=-ax+yz\\ \dot{z}=-x^2+b\lvert y\rvert-cxy\end{cases}$	$a=1.5$ $b=1.11$ $c=5$	Line $(0,0,z)$	0.7 1 0	0.1082 0 −1.4490	2.0747
IE_3	$\begin{cases}\dot{x}=y\\ \dot{y}=-x+yz\\ \dot{z}=a\lvert y\rvert-xy-bxz\end{cases}$	$a=0.065$ $b=0.1$	Line $(0,0,z)$	1 2 0	0.0584 0 −0.6026	2.0969
IE_4	$\begin{cases}\dot{x}=az\\ \dot{y}=z(by^2+cxz)\\ \dot{z}=d\lvert y\rvert-1-xyz\end{cases}$	$a=0.6$ $b=0.3$ $c=0.5$ $d=1.22$	Two parallel lines $y=\pm\frac{1}{d}$ $z=0$	0 0.7 −1.3	0.0258 0 −0.8077	2.0319
IE_5	$\begin{cases}\dot{x}=-z\\ \dot{y}=xz^2\\ \dot{z}=x-a\lvert y\rvert+z(by^2-z^2)\end{cases}$	$a=0.4$ $b=3$	Piece-wise lines $x=a\lvert y\rvert$ $z=0$	0.5 0.5 0.5	0.0756 0 −0.3276	2.2308

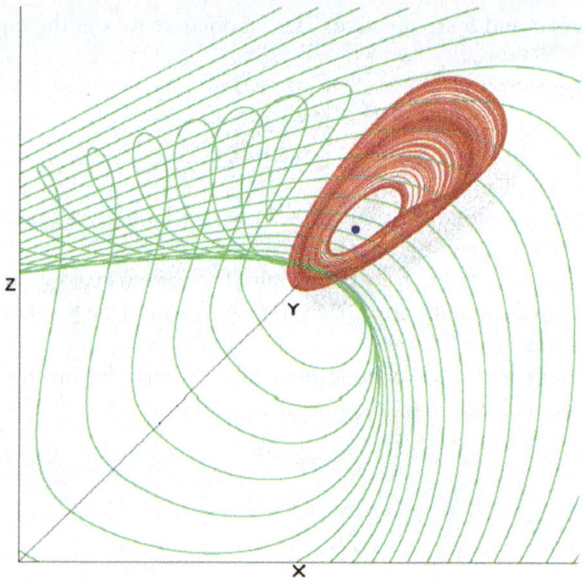

Fig. 5 A perspective view of strange attractors (red) along with selected conservative periodic orbits (green) for Eq. (14) with $a = 2$ [10]

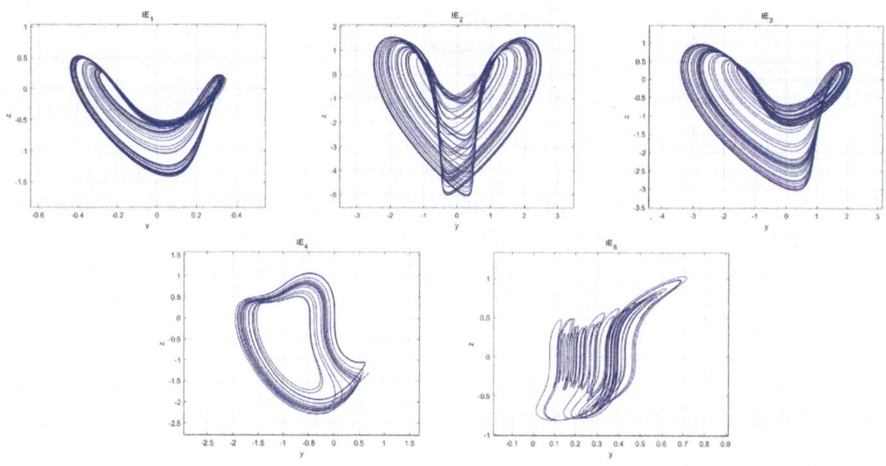

Fig. 6 Phase portraits of five elegant systems ($IE_1 - IE_5$) on the y-z plane [3]

dimension, and ICs are shown in Table 4. Phase portraits of IE_{1-5} are illustrated in Fig. 6.

The IE_1 system with six terms is given as follows:

$$\begin{cases} \dot{x} = y \\ \dot{y} = -x + yz \\ \dot{z} = a|x| - bxy - xz\,, \end{cases} \tag{16}$$

where parameters a and b are positive. The Jacobian matrix at the equilibrium is

$$J = \begin{bmatrix} 0 & 1 & 0 \\ -1 & z & 0 \\ -z & 0 & 0 \end{bmatrix}. \tag{17}$$

The characteristic equation is $\lambda(\lambda^2 - z\lambda + 1) = 0$, which gives one zero eigenvalue and two eigenvalues depending on z ($z > 0$).

For the case with $a = 0.25$, $b = 19$ and initial condition $(x_0, y_0, z_0) = (0, -0.35, 0.45)$, the Lyapunov exponents are 0.0363, 0, −0.2786 and the Kaplan–Yorke dimension $D_{KY} = 2.1303$.

The IE_1 system generates chaotic behaviors, where the basin of attraction of strange attractors intersects with the line of equilibria.

3.6 CE System

Inspired by the Gotthans–Petrzela system [11], Barati [12] introduced a simple chaotic system with two parallel lines of equilibria, in the following form:

$$\begin{cases} \dot{x} = az \\ \dot{y} = z(by^2 + cxa) \\ \dot{z} = y^2 - 1 - xyz\,, \end{cases} \tag{18}$$

where a, b and c are real parameters.

For the case with $a = 0.6$, $b = 0.3$, $c = 0.5$ and initial condition $(x_0, y_0, z_0) = (0, 0.7, -1.3)$, the equilibria of system (18) are obtained as $y = \pm 1$ and $z = 0$, indicating that it is a chaotic flow with two parallel lines of equilibria.

The Lyapunov exponents are $(0.0345, 0, -0.8817)$, while the Kaplan–Yorke dimension $D_{KY} = 2.0391$. State-space plots of this CE_3 system are shown in Fig. 7.

3.7 Petrzela–Gotthans System

While investigating the current-mode network structures dedicated to simulation, Petrzela and Gotthans [1] introduced five dynamical systems, each with a line of equilibria, characterized by a more general expression. They also investigated a chaotic flow with multiple lines of equilibria.

The generalized dynamical system is given in the following form:

$$\begin{cases} \dot{x} = y \\ \dot{y} = -x + yf_1(x) \\ \dot{z} = -xf_2(x) - yf_3(x)\,, \end{cases} \tag{19}$$

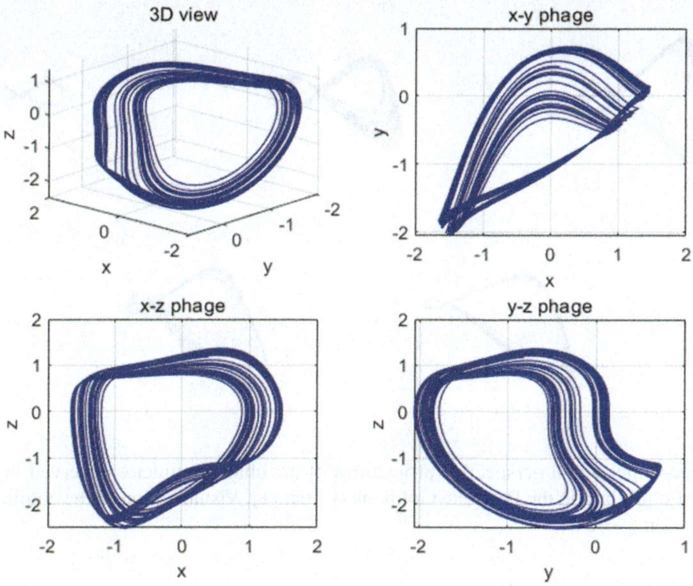

Fig. 7 State-space plots of the CE_3 system [12]

where the line of equilibria is given implicitly by $X_e = (0\ 0\ z)^T$. Based on system (19), the first characterized system can be generalized by setting a certain condition for $f_i(x)$, as follows:

$$\begin{cases} f_1(x) = z \\ f_2(x) = 1 + ay + bz \\ f_3(x) = 0\,, \end{cases} \tag{20}$$

where a and b are real parameters.

For the case with $a = 15,\ b = 1$ and initial condition $X_e = (0.2\ 0\ 0)^T$, the three-dimensional projection is shown in Fig. 8a.

The Second characterized system with a line of equilibria is described by

$$\begin{cases} f_1(x) = z \\ f_2(x) = ay + bz \\ f_3(x) = 1\,. \end{cases} \tag{21}$$

Figure 8b shows the corresponding three-dimensional projection of system (21), with $a = 17,\ b = 1$ and $X_e = (0\ 0.4\ 0)^T$.

The third characteristic system with a single line of equilibria is obtained as

$$\begin{cases} f_1(x) = z \\ f_2(x) = -x + ay + bz \\ f_3(x) = 1\,. \end{cases} \tag{22}$$

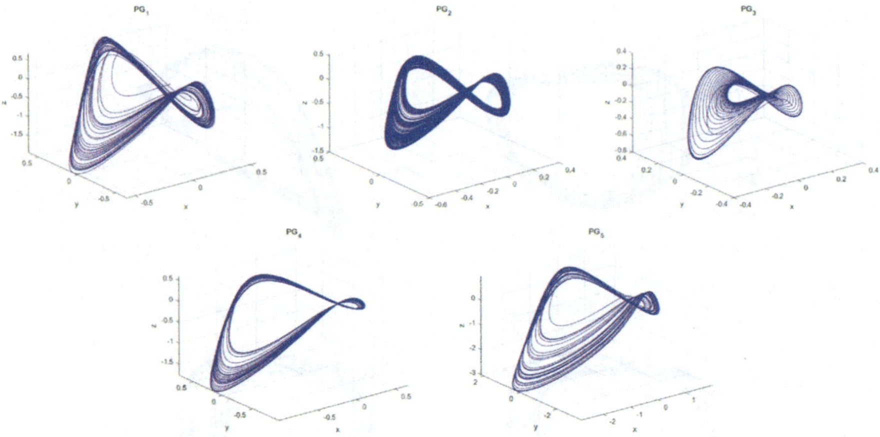

Fig. 8 Three-dimensional perspective projections of the chaotic attractor observed in the system with a line of equilibria in the Petrzela–Gotthans system [1]. Visualization of the equilibria and the Poincaré section $z = 0$

Figure 8c shows the corresponding three-dimensional projection of system (22), with $a = 18,\ b = 1$ and $X_e = (0\ \ -0.4\ 0.5)^T$.

Similarly, the fourth characteristic system with a single line of equilibria is given by

$$\begin{cases} f_1(x) = z \\ f_2(x) = ay + bz \\ f_3(x) = z\,, \end{cases} \tag{23}$$

Figure 8d shows the corresponding three-dimensional projection of system (23), with $a = 4,\ b = 0.6$ and $X_e = (0.2\quad 0.7\quad 0)^T$.

Lastly, the fifth system has the form

$$\begin{cases} f_1(x) = z \\ f_2(x) = y + bz \\ f_3(x) = -ay\,. \end{cases} \tag{24}$$

The corresponding attractor is visualized in Fig. 8e for system (24) with $a = 0.04$, $b = 0.1$ and $X_e = (0.8\quad 0.8\quad 0)^T$.

As illustrated in Fig. 8, there are similar geometrical structures of the Petrzela–Gotthans systems (20)–(24). These chaotic systems with a line of equilibria are mathematically related, because they were discovered by the same method with the same starting model (19) having many quadratic terms. Generally, a large number of chaotic systems proposed by Jafari and Sprott [4] satisfy this general dynamical model introduced by Petrzela and Gotthans [1], and preserve lines of equilibria on the x-y plane with Poincaré section $z = 0$.

The above five chaotic systems are all dissipative and are not time-reversible.

In addition, circuit implementation of these systems have the same number of active elements and network complexity.

Furthermore, Petrzela and Gotthans [1] introduced a new chaotic system with multiple lines of equilibria, which is described by

$$\begin{cases} \dot{x} = az \\ \dot{y} = z(by^2 + cxz) \\ \dot{z} = -xyz + y^2 - 1 , \end{cases} \tag{25}$$

where a, b and c are real parameters.

For the case with $a = 0.6,\ b = 0.3,\ c = 0.5$ and initial condition $X_e = (1\ 0\ 0)^T$, system (25) generates a couple of parallel lines of equilibria. Only a small number of such parallel lines is responsible for the formation of the strange attractors in system (25), which are bounded in a certain range of volume elements.

4 Chaotic Systems with Closed-Curves of Equilibria

4.1 Circular Curve of Equilibria

Inspired by Jafari and Sprott's work [4, 20], Gotthans and Petrzela [11] proposed a three-dimensional system with circular equilibria.

The new model is in the following form:

$$\begin{cases} \dot{x} = az \\ \dot{y} = zf_1(x, y, z) \\ \dot{z} = x^2 + y^2 - r^2 + zf_2(x, y, z) , \end{cases} \tag{26}$$

where r is the radius of the circular equilibria, a is a free parameter, and f_1 and f_2 are nonlinear functions with several quadratic terms, given by

$$\begin{cases} f_1(x, y, z) = bx + cz^2 , \\ f_2(x, y, z) = dx , \end{cases}$$

in which b, c, d are constants.

For a particular case with

$$a = -0.1 \ , \ b = 3 \ , \ c = -2.2 \ , \ d = -0.1 \ , \ r = 1 ,$$

and initial condition $X_0 = (0 \quad 0 \quad 0)^T$, the Jacobian matrix is state-dependent:

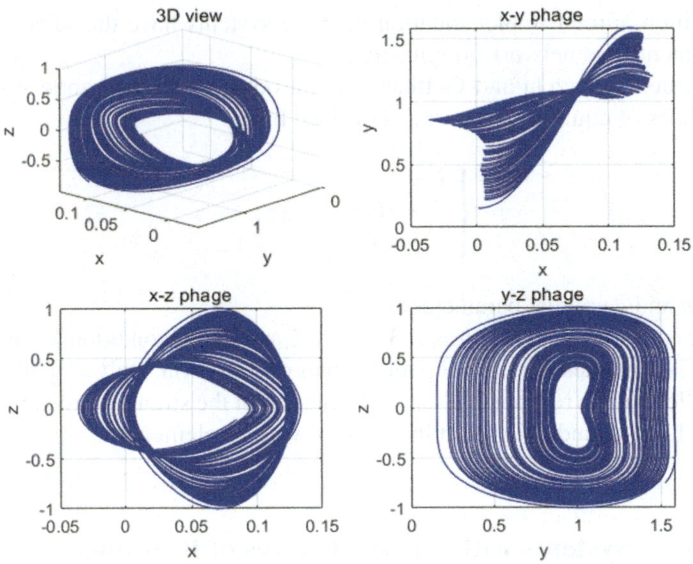

Fig. 9 Chaotic motion with initial condition near the equilibrium circle [11]

$$J(x) = \begin{bmatrix} 0 & 0 & a \\ bz & 0 & bx + 3cz^2 \\ 2x + dz & 2y & dx \end{bmatrix}. \tag{27}$$

Thus, the characteristic equation is

$$\lambda^3 - dx\lambda^2 - 2x(a \pm b\sqrt{r^2 - x^2})\lambda = 0. \tag{28}$$

One eigenvalue is found to be zero, while the other two depend on the positions of the equilibrium circle.

Figure 9 illustrates the chaotic motion in a neighborhood of the circle, with various configurations of the two-dimensional subspace of system (26). The Lyapunov exponents and other properties can also be determined.

Following the above logical consideration of constructing systems with circular equilibria, the so-called IE systems and CE systems are proposed in [11]. In particular, the IE_8 system in [3] and the CE_1 system in [12] also have a circle of equilibria. The corresponding mathematical functions, Lyapunov exponents, and Kaplan–Yorke dimensions are shown in Table 5.

Motivated by the above chaotic systems with circular equilibria (26), a new three-dimensional autonomous chaotic system with circular equilibria was proposed in [11]:

$$\begin{cases} \dot{x} = z \\ \dot{y} = z^3 + z^2 + 3xz \\ \dot{z} = x^2 + y^2 - r^2 - 4yz^2, \end{cases} \tag{29}$$

Table 5 IE & CE systems: two chaotic systems with a circle of equilibria [3, 12]

Case	Equations	Parameters	Equilibria	ICs	LEs	D_{KY}
IE_8	$\begin{cases} \dot{x} = z \\ \dot{y} = -z(a\lvert y\rvert + xz) \\ \dot{z} = x^2 + y^2 - 1 + z(y^2 - bz^2 + x) \end{cases}$	$a = 0.98$ $b = 2$	$x^2 + y^2 = 1$ $z = 0$	0 0.8 0.6	0.0337 0 −0.8441	2.0399
CE_1	$\begin{cases} \dot{x} = z \\ \dot{y} = -z(y^2 + xz) \\ \dot{z} = x^2 + y^2 - 1 + z(y^2 - az^2 + x) \end{cases}$	$a = 2$	$x^2 + y^2 = 1$ $z = 0$	0 0.8 0.61	0.0653 0 −0.8227	2.0794

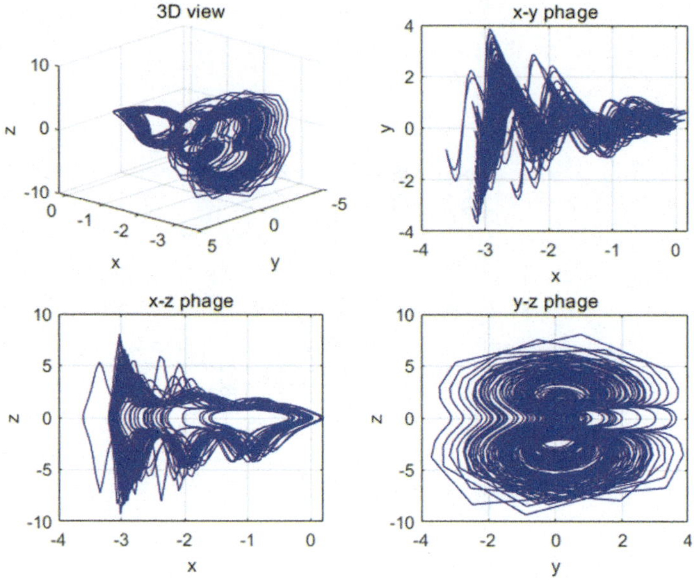

Fig. 10 Phase portrait on the *y-z* plane, with $r = 0.992$ and initial condition $(x_0, y_0, z_0) = (-3.14, -2.2, -6.91)$ [14]

where r is the only parameter. The circular equilibria of system (29) are

$$E = \{(x, y, z) \in R^3 \mid x = x^*, y = \pm\sqrt{r^2 - (x^*)^2}, z = 0\}, \tag{30}$$

in which $r^2 - (x^*)^2 \geq 0$ and the Jacobian matrix at the equilibrium $E = (x^*, \pm\sqrt{r^2 - (x^*)^2}, 0)$ is given by

$$J = \begin{bmatrix} 0 & 0 & 1 \\ 0 & 0 & 3x^* \\ 2x^* & \pm 2\sqrt{r^2 - (x^*)^2} & 0 \end{bmatrix}. \tag{31}$$

The characteristic equation of system (29) is

$$\lambda[\lambda^2 - 2x^*(1 \pm 3\sqrt{r^2 - (x^*)^2})] = 0\,. \tag{32}$$

Supposing $1 \pm 3\sqrt{r^2 - (x^*)^2} > 0$, the roots of Eq. (29) are obtained as $\lambda_1 = 0$, $\lambda_{2,3} = \pm\sqrt{2x^*(1 \pm 3\sqrt{r^2 - (x^*)^2})}$ if $x^* > 0$. Under this condition, the eigenvalues have at least one positive real root associated with an unstable equilibrium. Otherwise, $\lambda_1 = 0$, $\lambda_{2,3} = \pm i\sqrt{2x^*(1 \pm 3\sqrt{r^2 - (x^*)^2})}$.

Figure 10 shows the phase portrait on the y-z plane, with $r = 0.992$ and initial condition $(x_0, y_0, z_0) = (-3.14, -2.2, -6.91)$, where Lyapunov exponents are $(0.05996, 0, -0.2513)$ and the Lyapunov dimension is $D_L \approx 2.2386$.

Recently, Pham [2] introduced a new general form based on system (26), to investigate systems with an infinite number of equilibria, which is given by

$$\begin{cases} \dot{x} = az \\ \dot{y} = zf_1(x, y, z) \\ \dot{z} = f_3(x, y) + zf_2(x, y, z) , \end{cases} \tag{33}$$

where $f_{1,2}(x, y, z)$ and $f_3(x, y)$ are three nonlinear functions.

Note that in [2], the descriptions of the functions $f_{1,2}(x, y, z)$ are chosen to be exactly the same as they are in [11], namely,

$$f_1(x, y, z) = bx + cz^2$$

and

$$f_2(x, y, z) = dx .$$

The third nonlinear function $f_3(x, y)$ in system (33) is chosen as

$$f_3(x, y) = \left(\frac{x}{m}\right)^k + \left(\frac{y}{n}\right)^k - r^2 ,$$

where m, n and k are real parameters.

By selecting $f_3(x, y) = x^2 + y^2 - 1$, one can obtain the very dynamical system introduced by Petrzela and Gotthans (26). This chaotic system also has circular equilibria.

4.2 *Square Curve of Equilibria*

The new general model (33) proposed by Pham [2] can also fit for the case that the equilibria are in an approximate square shape, by setting $f_3(x, y) = x^{12} + y^{12} - 1$. In this case, the characteristic model becomes

$$\begin{cases} \dot{x} = az \\ \dot{y} = z\,(bx + cz^2) \\ \dot{z} = x^{12} + y^{12} - 1 + z\,dx , \end{cases} \tag{34}$$

where $d = -0.2$, $k = 12$ and $m = n = 1$.

As shown in Fig. 11, system (34) generates chaos, and has a square-shaped curve of equilibria, with Lyapunov exponents $(0.0135, 0, -0.1387)$.

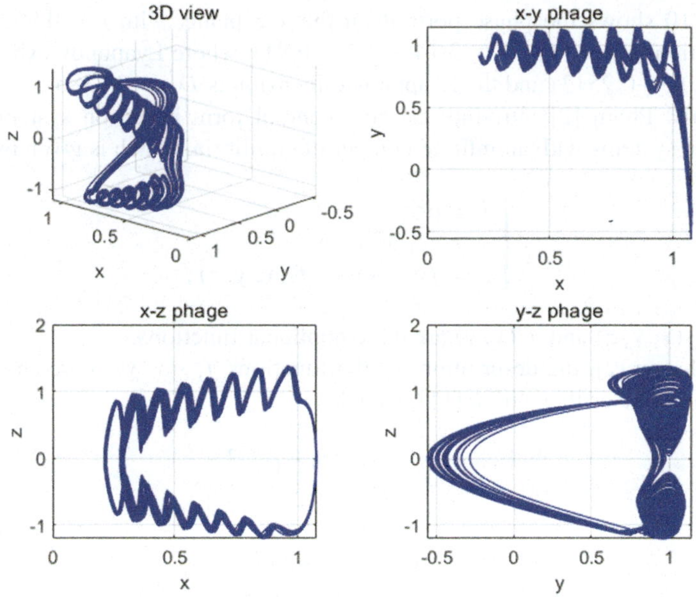

Fig. 11 Phase portrait of system (34), with $a = -0.1,\ b = 3,\ c = -2.2$, and initial condition (1, 1, 1) [2]

In addition, by linearizing system (33), Gotthans et al. [15] introduced a piecewise-linear chaotic system, with a square-shaped curve of equilibria, given by

$$\begin{cases} \dot{x} = z \\ \dot{y} = -z(ay + by^2 + xz) \\ \dot{z} = x^2 + y^2 - 1. \end{cases} \tag{35}$$

The corresponding Jacobian matrix is

$$J = \begin{bmatrix} 0 & 0 & 1 \\ -|z| & -az - bz * sgn(y) & -ay - b|y| - x * sgn(z) \\ sgn(x) & sgn(y) & 0 \end{bmatrix}. \tag{36}$$

whereas the characteristic equation is

$$\begin{aligned} &-\lambda^2(az + \lambda) + sgn(x)[az + \lambda + bz * sgn(y)] \\ &-sgn(y)\{b\lambda|y| + |z| + \lambda[ay + bz\lambda + x * sgn(x)]\} = 0\,. \end{aligned}$$

Therefore, the eigenvalues are obtained, as

$$\lambda_1 = 0 \quad , \quad \lambda_{2,3} = \pm\sqrt{sgn(x) - ay * sgn(y) - b|y| * sgn(y)}. \tag{37}$$

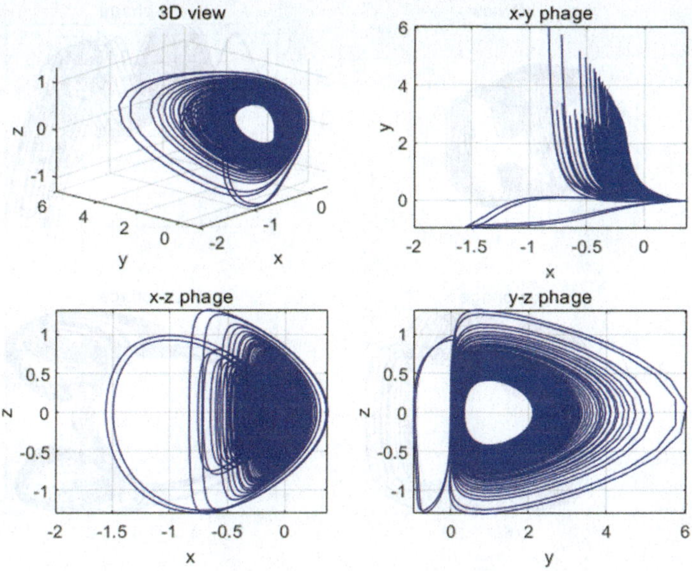

Fig. 12 Three-dimensional state portrait, with $a = 5$, $b = 3$ and initial condition (0, 0, 0). The black quadrangle is the equilibrium square [15]

The state projection in the neighborhood of the square curve of unstable equilibria is obtained, as shown in Fig. 12, which also shows the square curve of equilibria surrounded by some concentric periodic orbits.

4.3 Ellipse Curves of Equilibria

Other interesting shapes of equilibrium curves, such as ellipse and rectangle, can be formed from the same chaotic model of Pham et al. [2].

For the case with $d = -0.1$, $k = 2$, $m = 0.95$, $n = 0.9$ in system (33), one has

$$f_3(x, y) = \left(\frac{x}{m}\right)^2 + \left(\frac{y}{n}\right)^2 - 1\,. \tag{38}$$

This chaotic system has an ellipse curve of equilibria, with Lyapunov exponents $(0.0181, 0, -0.0793)$, as shown in Fig. 13.

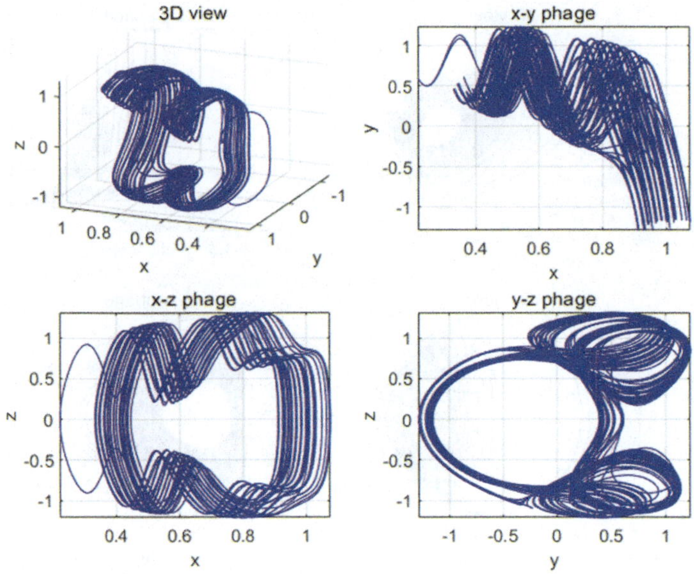

Fig. 13 Phase portrait of the dynamical system with an ellipse curve of equilibria and initial condition (0.2, 0.2, 0.2) [2]

4.4 Rectangle Shape

By selecting another function for $f_3(x, y)$ in system (33), a model with an approximate rectangular curve of equilibria is obtained:

$$\begin{cases} \dot{x} = az \\ \dot{y} = z(bx + cz^2) \\ \dot{z} = \left(\frac{x}{m}\right)^{12} + \left(\frac{y}{n}\right)^{12} - 1 + dxz\,, \end{cases} \tag{39}$$

where $d = -0.2, k = 12, m = 0.95, n = 0.5$ and the corresponding Lyapunov exponents are $(0.0150, 0, -0.1314)$. The three-dimensional projection of the chaotic attractor, along with a rectangular curve of equilibria, are illustrated in Fig. 14.

4.5 Rounded-Square Curves of Equilibria

Motivated by the notable features of the Gotthans–Petrzela model (26) [15], Pham and Jafari [21] introduced a new dynamical system with hidden attractors and an infinite number of rounded-square curves of equilibria. The new system has the following form:

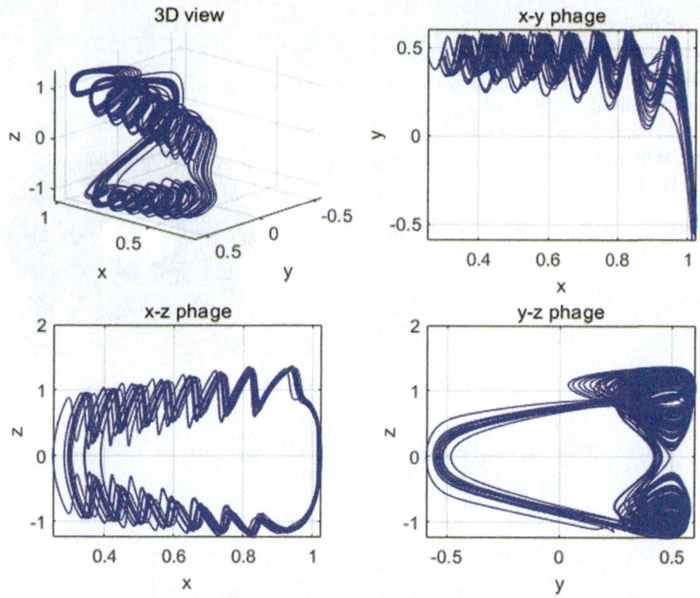

Fig. 14 Three-dimensional projection of the dynamical system (39), with a rectangular curve of equilibria and initial condition (1, 0.5, 1) [2]

$$\begin{cases} \dot{x} = z \\ \dot{y} = -zf_1(x, y, z) - c \\ \dot{z} = f_2(x, y)\,, \end{cases} \tag{40}$$

where c is a positive parameter and $f_1(x, y, z), f_2(x, y)$ are nonlinear functions.

For $c = 0$, there is an infinite number of equilibria, while the equilibrium curve of the system is given by $f_2(x, y) = 0$. Else, the system has no equilibria.

While considering the case of $c = 0$, different nonlinear functions can be chosen for $f_1(x, y, z)$ and $f_2(x, y)$, such as

$$f_1(x, y, z) = ay + by^2 + xz \quad , \quad f_2(x, y) = x^4 + y^4 - 1\,. \tag{41}$$

In this case, the new chaotic system has rounded-square curves of equilibria. The system is described by

$$\begin{cases} \dot{x} = z \\ \dot{y} = -z(ay + by^2 + xz) \\ \dot{z} = x^4 + y^4 - 1\,. \end{cases} \tag{42}$$

where a and b are real parameters.

The equilibria of system (42) are $E_0(x^*, y^*, 0)$, and the Jacobian matrix is

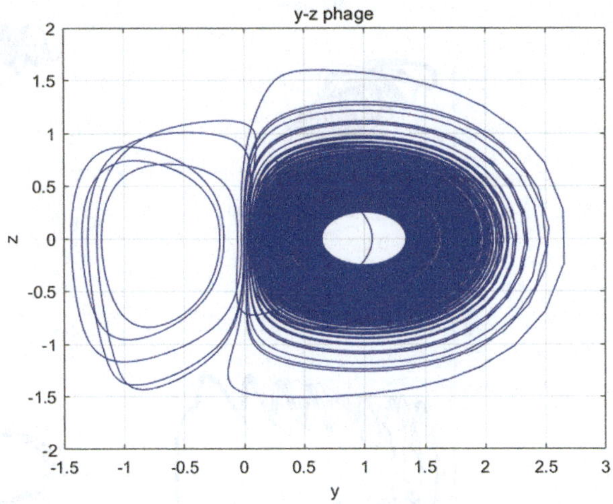

Fig. 15 Phase projection of the dynamical system (42), with a rounded-square curve of equilibria and initial condition (0.5, 0.5, 0.5) on the y-z plane, with $a = 5$, $b = 3$, $c = 0$ [21]

$$J = \begin{bmatrix} 0 & 0 & 1 \\ 0 & 0 & -ay^* - b(y^*)^2 \\ 4(x^*)^3 & 4(y^*)^2 & 0 \end{bmatrix}. \tag{43}$$

The corresponding characteristic equation is

$$\lambda(\lambda^2 + 4(x^*)^3\lambda + 4a(y^*)^4 + 4b(y^*)^5 \quad = \quad 0\,.$$

So, there is one zero eigenvalue but two eigenvalues that depend on the location of E_0.

Figure 15 shows the projection of the system state, with rounded-square curves of equilibria, for the case of $a = 5$, $b = 3$, $c = 0$. The Lyapunov exponents are $(0.0231, 0, -0.0593)$, while the Kaplan–Yorke dimension is $D_{KY} = 2.3895$.

Additionally, in [22], based on the same system (33) and a systematic search [11, 20], another chaotic model is introduced, with rounded-square curves of equilibria, described by

$$\begin{cases} \dot{x} = -az \\ \dot{y} = bxz + cz^3 \\ \dot{z} = x^4 + y^4 - k - dxz\,, \end{cases} \tag{44}$$

where a, b, c, d, k are positive parameters.

By solving $\dot{x} = 0$, $\dot{y} = 0$ and $\dot{z} = 0$, one obtains $z = 0$, which leads to $x^4 + y^4 - k = 0$. This gives a curve of infinitely many equilibria in the system. Moreover, the equilibria of the system (44) are given by $E_0(x^*, \pm\sqrt[4]{k - (x^*)^4})$ $(|x| \leq k^{\frac{1}{4}})$ and the shape of the equilibrium curve can be described by a particular parameter-depending function:

$$(x^*)^4 + (y^*)^4 = k\,.$$

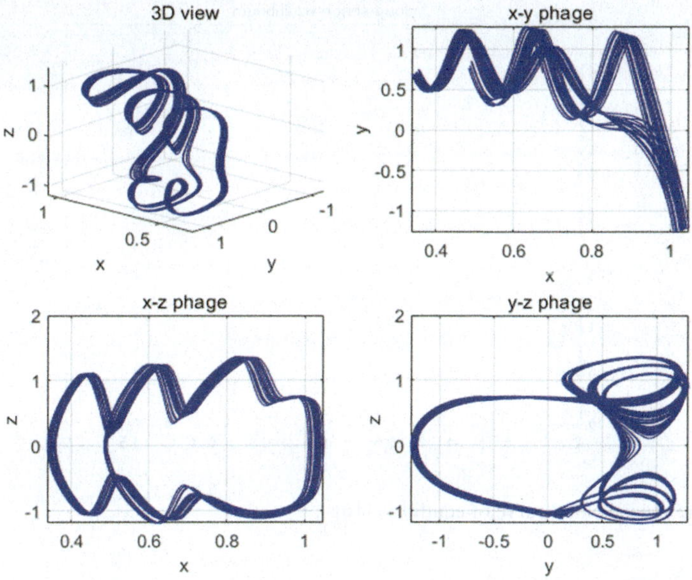

Fig. 16 Phase projection of the dynamical system (44), with rounded-square curves of equilibria, in the case of $a = 0.1,\ b = 3,\ c = 2.2,\ d = 0.2,\ k = 0.81$, with initial condition (0.5, 0.5, 0.5) [22]

Note that the above function makes sense if and only if $k > 0$; otherwise, there will be no equilibria or will be one single equilibrium point located at the origin.

Figure 16 illustrates the chaotic attractors, with the rounded-square curves of equilibria of the system.

4.6 Cloud Curves of equilibria

An interesting example of chaotic systems with a closed curve of equilibria is proposed by Wang [16].

The special chaotic model is in the following form:

$$\begin{cases} \dot{x} = z \\ \dot{y} = -z(ay + by^2 + xz) \\ \dot{z} = x^2 + y^2 - |xy| - 1\,, \end{cases} \tag{45}$$

where a and b are real parameters. This system is formed based on system (33), but it has a special curve of equilibria, different from those typical shapes (circle, square, rectangle, rounded-square or ellipse).

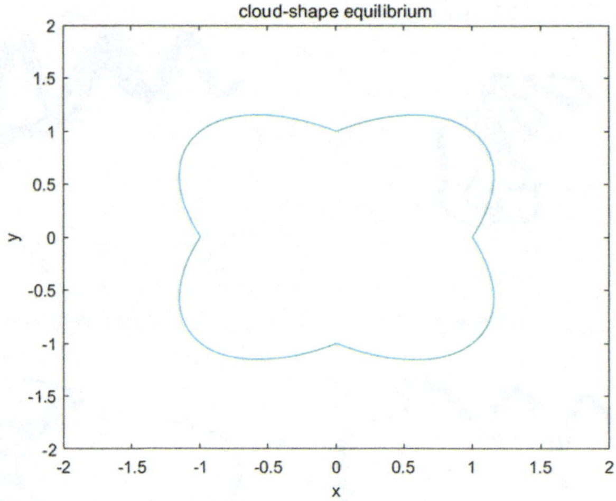

Fig. 17 The special cloud curve of equilibria [16]

By solving $\dot{x} = 0$, $\dot{y} = 0$ and $\dot{z} = 0$, one obtains $z = 0$. Therefore, one has $x^2 + y^2 - |xy| - 1 = 0$, which means that system (45) has a cloud curve of equilibria, as shown in Fig. 17. For the case of $a = 4$, $b = 2.5$, with initial condition $(x_0, y_0, z_0) = (0.01, 0.02, 0.01)$, the Lyapunov exponents are calculated by using the Wolf method for 10,000 iterations, yielding $(0.13, 0, -0.6853)$. The Kaplan–Yorke dimension is $D_{KY} = 2.1897$. The projection of the system state is displayed in Fig. 18.

5 Open Curves of Equilibria

Chaotic systems with open curves of equilibria were first found by Pham and Jafari [3, 23], in the following form:

$$\begin{cases} \dot{x} = -z \\ \dot{y} = xz^2 \\ \dot{z} = x - a|y| + bzy^2 - z^3 , \end{cases} \tag{46}$$

where the real parameters a and b are positive.

The set of equilibria in system (46) is $E_0(a|y^*|, y^*, 0)$, which are located on a piecewise linear curve. By solving the corresponding characteristic equation $\lambda(\lambda^2 - b(y^*)^2\lambda + 1) = 0$, where $\Delta = b^2b(y^*)^4 - 4$, the eigenvalues of the system can be determined by the sign of Δ. Generally,

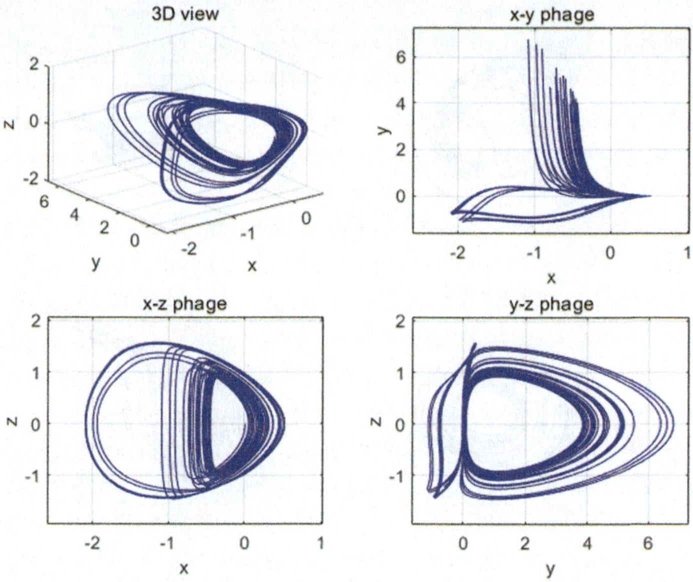

Fig. 18 Projections of the chaotic attractor, with a cloud curve of equilibria, when $a = 4$, $b = 2.5$ [16]

$$when \quad \Delta > 0, \quad \lambda_1 = 0 \ , \ \lambda_{2,3} = \frac{b(y^*)^2 \pm \sqrt{\Delta}}{2}$$

$$when \quad \Delta < 0, \quad \lambda_1 = 0 \ , \ \lambda_{2,3} = \frac{b(y^*)^2 \pm i\sqrt{\Delta}}{2}$$

From a computational point of view, this equilibrium of system (46) is unstable since $b > 0$.

Chaotic behaviors generated by the system are shown in Fig. 19.

For the case of $a = 0.4$ and $b = 3$, with initial condition $(x_0, y_0, z_0) = (0.5, 0.5, 0.5)$, the Lyapunov exponents are (0.0756, 0, −0.3276) while the Kaplan–Yorke dimension is $D_{KY} = 2.2308$.

Four more mathematical models are introduced in Table 6. All of them have similar properties and generate chaotic behaviors with an open curve of equilibria. The equilibria of the IE_6 system and the CE_4 system have the shape of a parabolic curve, while those of the IE_7 system and the CE_2 system have the shape of a hyperbolic curve. Their Lyapunov exponents, D_{KY} and ICs are shown in Table 6. Chaotic motions of the models listed in Table 6 are illustrated in Fig. 20.

Recently, based on the statistical results of existing research reports on the microcontroller implementation [24–26], Giakoumis and Volos [27] offered a microcontroller-implementational perspective for chaotic systems with open curves of equilibria. The dynamical model is described by

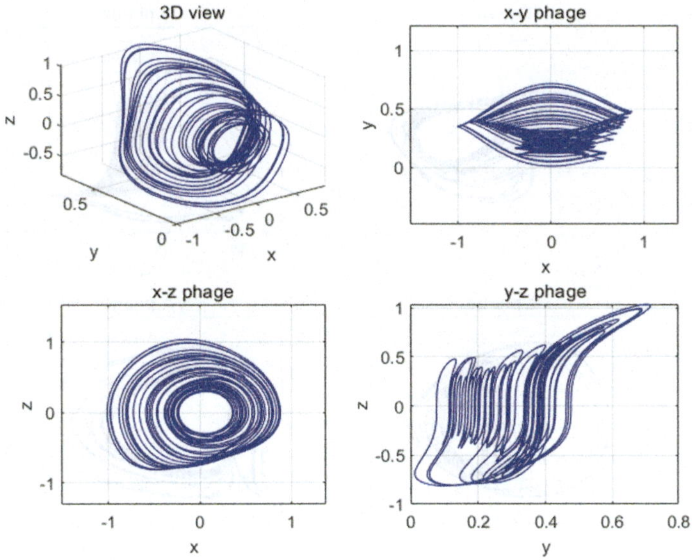

Fig. 19 Phase portraits of different systems with open curves of equilibria in the three-dimensional space [3, 23]

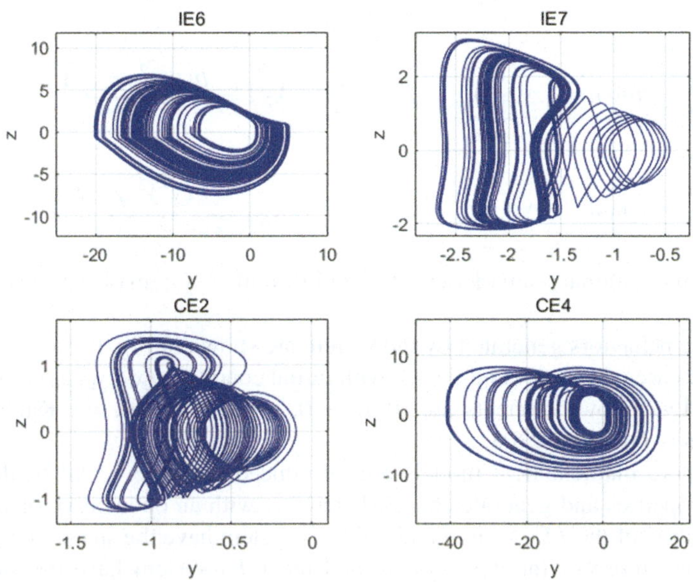

Fig. 20 State-space plots of systems in Table 6 [12]

Table 6 IE & CE systems: four chaotic systems with an open curve of equilibria [3, 12]

Case	Equations	Parameters	Equilibria	ICs	LEs	D_{KY}
IE_6	$\begin{cases} \dot{x} = -az \\ \dot{y} = -bz\|z\| \\ \dot{z} = x^2 + y + z(z - xy) \end{cases}$	$a = 2b = 3$	Parabola $y = -x^2$ $z = 0$	0.2 3.8 2	0.1243 0 −5.7488	2.0216
IE_7	$\begin{cases} \dot{x} = -z \\ \dot{y} = z(a\|z\| - 1) \\ \dot{z} = x^2 - y^2 - 1 + z(y^2 - z^2) \end{cases}$	$a = 0.63$	Hyperbola $x^2 - y^2 = 1$	0.7 −1 0	0.0363 0 −0.2786	2.1303
CE_2	$\begin{cases} \dot{x} = -z \\ \dot{y} = z(z^2 - 1) \\ \dot{z} = x^2 - y^2 - 1 + z(y^2 - z^2) \end{cases}$		Hyperbola $x^2 - y^2 = 1$	0.75 −0.9 0	0.1250 0 −0.7538	2.1658
CE_4	$\begin{cases} \dot{x} = -az \\ \dot{y} = -z^3 \\ \dot{z} = x^2 + y + z(z - xy) \end{cases}$	$a = 2$	Parabola $y = -x^2$ $z = 0$	0.23 3.89 2	0.1433 0 −8.7143	2.0164

$$\begin{cases} \dot{x} = -z \\ \dot{y} = xz^2 \\ \dot{z} = x - a\ sinh(y) + zy^2 - z^3\,, \end{cases} \tag{47}$$

where the parameter a is positive.

To find the equilibria of the system, one can solve $\dot{x} = 0$, $\dot{y} = 0$ and $\dot{z} = 0$, thus obtaining the equilibrium point $E(a\ sinh(y), y^*, 0)$. The corresponding Jacobian matrix at the equilibrium point is

$$J_E = \begin{bmatrix} 0 & 0 & -1 \\ 0 & 0 & 0 \\ 0 & -a\ cosh(y^*) & (y^*)^2 \end{bmatrix}. \tag{48}$$

The characteristic equation is $\lambda(\lambda^2 - (y^*)^2\lambda + 1) = 0$, with the discriminant $\Delta = (y^*) - 4$. Generally,

$$\begin{aligned} &when \quad \Delta > 0, \quad \lambda_1 = 0\ ,\ \lambda_{2,3} = \frac{(y^*)^2 \pm \sqrt{\Delta}}{2} \\ &when \quad \Delta = 0, \quad \lambda_1 = 0\ ,\ \lambda_{2,3} = \frac{(y^*)^2}{2} = 2 \\ &when \quad \Delta < 0, \quad \lambda_1 = 0\ ,\ \lambda_{2,3} = \frac{(y^*)^2 \pm i\sqrt{\Delta}}{2}\,. \end{aligned}$$

For the case with $a = 0.25$ and initial condition $(x_0, y_0, z_0) = (0.4, 0.2, 0.1)$, system (47) generates chaotic behaviors with only one positive Lyapunov exponent, as shown in Fig. 21a.

Another chaotic flow with an open curve of equilibria is studied by Pham and Volos [13]. Based on the typical model with an infinite number of equilibria, $E(x^*, y^*, 0)$, under the condition of $z = 0$, as proposed by Gotthans and Petrzela [11], the new system is given by

$$\begin{cases} \dot{x} = -z \\ \dot{y} = xz^2 + a\ tanh(bz) \\ \dot{z} = x + cy - c\ tanh(by) + dzy^2 - z^3\,, \end{cases} \tag{49}$$

where parameters a, b, c, d are all positive.

By linearizing system (49) at the equilibrium point, the Jacobian matrix is obtained as

$$J_E = \begin{bmatrix} 0 & 0 & -1 \\ 0 & 0 & ab \\ 1 & c - bc(1 - tanh^2(by^*)) & d(y^*)^2 \end{bmatrix}. \tag{50}$$

The corresponding characteristic equation is

$$\lambda[\lambda^2 - d(y^*)^2\lambda + 1 - abc + ab^2c - ab^2\ tanh^2(by^*)] = 0\,,$$

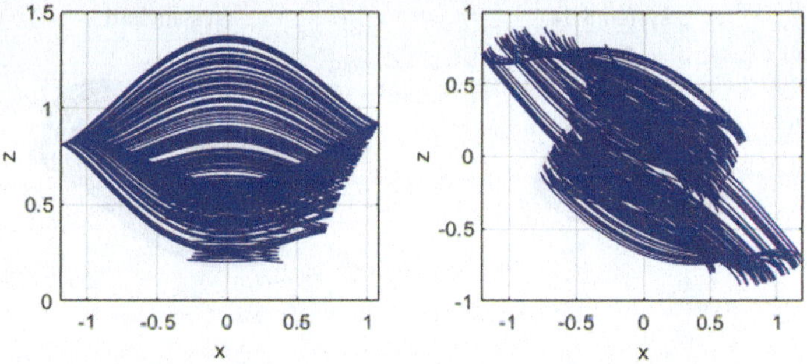

Fig. 21 Projection of chaotic attractors of system (47)–(49) on the x-y plane [13, 27]

and the discriminant is

$$\Delta = d^2(y^*)^4 - 4(1 - abc + ab^2c - ab^2\, tanh^2(b\, y^*))\,. \tag{51}$$

The eigenvalues of the system depend on the sign of the discriminant (51). Therefore,

$$when \quad \Delta > 0, \quad \lambda_1 = 0\ , \quad \lambda_{2,3} = \frac{d(y^*)^2 \pm \sqrt{\Delta}}{2}$$

$$when \quad \Delta = 0, \quad \lambda_1 = 0\ , \quad \lambda_{2,3} = \frac{d(y^*)^2}{2}$$

$$when \quad \Delta < 0, \quad \lambda_1 = 0\ , \quad \lambda_{2,3} = \frac{d(y^*)^2 \pm i\sqrt{\Delta}}{2}\,.$$

In all the above cases, the equilibrium point of system (49) is unstable for $d > 0$ and $y^* \neq 0$. It is interesting to observe the chaotic behaviors generated by such a system. Figure 21b shows the projection of the chaotic attractor on the x-y plane for the case of $a = 0.28$, $b = 10$, $c = 0.1$, $d = 2.2$, with initial condition $(x_0, y_0, z_0) = (0.1, 0.1, 0.1)$. The Lyapunov exponents and Kaplan–Yorke dimension are also calculated, for Fig. 21b, as $L = (0.1042, 0, -0.2855)$ and $D_{KY} = 2.3650$.

Two additional mathematical models, with equilibrium curves located on the horizontal plane $z = 0$, are introduced by Petrzela and Gotthans [1].

The first model is described by

$$\begin{cases} \dot{x} = az \\ \dot{y} = z^3 - z \\ \dot{z} = x^2 + (z-1)y^2 - z^3 - 1\,, \end{cases} \tag{52}$$

which generates typical chaotic behaviors, with a hyperbolic equilibrium curve, as shown in Fig. 22a ($a = -1$ and IC is (0, –0.6, 0)).

The second model is described by

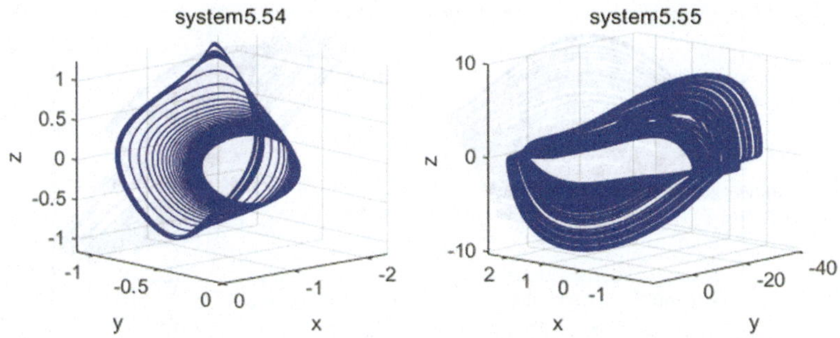

Fig. 22 Three-dimensional rainbow-scaled projections of system (52)–(53), visualizing the equilibrium curve and Poincaré section defined by plane $z = 0$ [1]

$$\begin{cases} \dot{x} = az \\ \dot{y} = -z^3 \\ \dot{z} = x^2 + z^2 - xyz + y\,, \end{cases} \tag{53}$$

which generates chaotic behaviors, with a parabolic equilibrium curve, as shown in Fig. 22b ($a = -2$ and IC is (0, 10, 0)).

References

1. J. Petrzela, T. Gotthans, M. Guzan, Current-mode network structures dedicated for simulation of dynamical systems with plane continuum of equilibrium. J. Circuits Syst. Comput. **27**(09), 1830004 (2018)
2. V.T. Pham, S. Jafari, X. Wang, J. Ma, A chaotic system with different shapes of equilibria. Int. J. Bifurc. Chaos **26**(04), 1650069 (2016)
3. V. Pham, S. Jafari, C. Volos, T. Kapitaniak, A gallery of chaotic systems with an infinite number of equilibrium points. Chaos, Solitons Fractals **93**, 58–63 (2016)
4. S. Jafari, J. Sprott, Simple chaotic flows with a line equilibrium. Chaos, Solitons Fractals **57**, 79–84 (2013)
5. C. Li, J.C. Sprott, Chaotic flows with a single nonquadratic term. Phys. Lett. A **378**(3), 178–183 (2014)
6. C. Li, J.C. Sprott, Z. Yuan, H. Li, Constructing chaotic systems with total amplitude control. Int. J. Bifurc. Chaos **25**(10), 1530025 (2015)
7. C. Li, J.C. Sprott, W. Thio, Bistability in a hyperchaotic system with a line equilibrium. J. Exp. Theor. Phys. **118**(3), 494–500 (2014)
8. J. Ma, Z. Chen, Z. Wang, Q. Zhang, A four-wing hyper-chaotic attractor generated from a 4D memristive system with a line equilibrium. Nonlinear Dyn. **81**(3), 1275–1288 (2015)
9. J.P. Singh, B.K. Roy, A novel hyperchaotic system with stable and unstable line of equilibria and sigma shaped Poincare map. IFAC PapersOnLine **49**(1), 526–531 (2016)
10. J.C. Sprott, Symmetric time-reversible flows with a strange attractor. Int. J. Bifurc. Chaos **25**(05), 759– (2015)
11. T. Gotthans, J. PetrŽela, New class of chaotic systems with circular equilibrium. Nonlinear Dyn. **81**(3), 1–7 (2015)
12. K. Barati, S. Jafari, J.C. Sprott, V.T. Pham, Simple chaotic flows with a curve of equilibria. Int. J. Bifurc. Chaos **26**(12), 511–543 (2016)

13. V.-T. Pham, A. Akgul, C. Volos, S. Jafari, T. Kapitaniak, Dynamics and circuit realization of a no-equilibrium chaotic system with a boostable variable. AEU: Int. J. Electron. Commun. **78**, 134–140 (2017)
14. S.T. Kingni, V.T. Pham, S. Jafari, G.R. Kol, P. Woafo, Three-dimensional chaotic autonomous system with a circular equilibrium: analysis, circuit implementation and its fractional-order form. Circuits Syst. Signal Process. **35**(6), 1933–1948 (2016)
15. T. Gotthans, J.C. Sprott, J. Petrzela, Simple chaotic flow with circle and square equilibrium. Int. J. Bifurc. Chaos **26**(08), 1650137 (2016)
16. X. Wang, V.T. Pham, C. Volos, Dynamics, circuit design, and synchronization of a new chaotic system with closed curve equilibrium. Complexity **2017**, 1–9 (2017)
17. T. Chawanya, Coexistence of infinitely many attractors in a simple flow. Phys. D: Nonlinear Phenom. **109**(3–4), 201–241 (1997)
18. J.C. Sprott, *Elegant Chaos: Algebraically Simple Chaotic Flows* (World Scientific, Singapore, 2010)
19. V.-T. Pham, C. Volos, T. Kapitaniak, Systems with stable equilibria, *Systems with Hidden Attractors* (Springer, Berlin, 2017), pp. 21–35
20. S. Jafari, J.C. Sprott, S.M.R.H. Golpayegani, Elementary quadratic chaotic flows with no equilibria. Phys. Lett. A **377**(9), 699–702 (2013)
21. V.-T. Pham, S. Jafari, C. Volos, T. Gotthans, X. Wang, D.V. Hoang, A chaotic system with rounded square equilibrium and with no-equilibrium. Opt.: Int. J. Light Electron Opt. **130**, 365–371 (2017)
22. V.T. Pham, S. Jafari, C. Volos, A. Giakoumis, S. Vaidyanathan, T. Kapitaniak, A chaotic system with equilibria located on the rounded square loop and its circuit implementation. IEEE Trans. Circuits Syst. II: Express Briefs **63**(9), 878–882 (2017)
23. V.-T. Pham, S. Jafari, C. Volos, S. Vaidyanathan, T. Kapitaniak, A chaotic system with infinite equilibria located on a piecewise linear curve. Opt.: Int. J. Light Electron Opt. **127**(20), 9111–9117 (2016)
24. J.D. Serna, A. Joshi, Visualizing the logistic map with a microcontroller. IOP Sci.: Phys. Educ. **47**(6), 736 (2011)
25. C.K. Volos, I.M. Kyprianidis, I.N. Stouboulos, Experimental investigation on coverage performance of a chaotic autonomous mobile robot. Robot. Auton. Syst. **61**(12), 1314–1322 (2013)
26. R. Chiu, M. Mora-Gonzalez, D. Lopez-Mancilla, Implementation of a chaotic oscillator into a simple microcontroller. IERI Procedia **4**(2013), 247–252 (2013)
27. A.E. Giakoumis, C.K. Volos, I.N. Stouboulos, I.K. Kyprianidis, V.T. Pham, A chaotic system with equilibria located on an open curve and its microcontroller implementation, in *International Conference on Modern Circuits and Systems Technologies* (2017), pp. 1–4

Chaotic Systems with Surfaces of Equilibria

Xiong Wang and Guanrong Chen

1 Introduction

Recently, finding a new catalog of chaotic systems with surfaces of equilibria has become an attractive focal topic of research. As the advent of computers becomes faster and faster, one can now use it to explore mathematical models in order to find chaotic features and the desired properties of complex dynamical systems [1]. A particular complex category of chaotic systems are those with hidden attractors, some of which have an infinite number of equilibria on a surface lying outside the basin of attraction that may possibly intersect the equilibrium surface in some sections. As a result, conventional methods of finding chaotic attractors cannot be applied to such systems.

Notice that there was no study on three-dimensional chaotic systems with surfaces of equilibria, an important step in finding such chaotic systems, to be introduced in this chapter, includes a new systematic method and the ES_i systems proposed by Jafari et al. in Ref. [1]. Some hyper-dimensional chaos with surfaces or planes of equilibria [1–3] will be discussed in some following chapters.

A new three-dimensional chaotic system with a surface of equilibria was introduced in Ref. [1], including different types of surfaces, such as multiple planes [4, 5], spheres [6, 7], circular cylinders [8, 9], hyperbolic cylinders [10, 11], paraboloid [12, 13] and saddles. Such systems satisfy at least one of the three necessary conditions that any newly proposed system should have, which was established by Sprott in Ref. [14].

X. Wang (✉)
Institute for Advanced Study, Shenzhen University,
Shenzhen 518060, Guangdong, People's Republic of China
e-mail: wangxiong8686@szu.edu.cn

G. Chen
Department of Electrical Engineering, City University of Hong Kong, Hong Kong SAR 999077,
People's Republic of China
e-mail: eegchen@cityu.edu.hk

X. Wang et al. (eds.), *Chaotic Systems with Multistability and Hidden Attractors*,
Emergence, Complexity and Computation 40,
https://doi.org/10.1007/978-3-030-75821-9_6

2 Systematic Method for Finding Chaotic Systems with Surfaces of Equilibria

To construct a three-dimensional chaotic system with a surface of equilibria, Jafari et al. [1] applied a systematic computer method by constructing a specific mathematical function in the following form:

$$\begin{cases} \dot{x} = Q_1(x, y, z) \\ \dot{y} = Q_2(x, y, z) \\ \dot{z} = Q_3(x, y, z)\,, \end{cases} \tag{1}$$

where

$$\begin{aligned} Q_1 =& a_1x + a_2y + a_3z + a_4x^2 + a_5y^2 + a_6z^2 \\ &+ a_7xy + a_8xz + a_9yz + a_{10} \\ Q_2 =& a_{11}x + a_{12}y + a_{13}z + a_{14}x^2 + a_{15}y^2 + a_{16}z^2 \\ &+ a_{17}xy + a_{18}xz + a_{19}yz + a_{20} \\ Q_3 =& a_{21}x + a_{22}y + a_{23}z + a_{24}x^2 + a_{25}y^2 + a_{26}z^2 \\ &+ a_{27}xy + a_{28}xz + a_{29}yz + a_{30}\,, \end{aligned}$$

with a_{ij} being real parameters.

By multiplying a function $f(x, y, z)$ into system (1), it is possible to obtain a surface on which all points are equilibria. Notice that such an equilibrium surface can always be there, even with $f(x, y, z) = 1$ in the following model:

$$\begin{cases} \dot{x} = f(x, y, z)Q_1(x, y, z) \\ \dot{y} = f(x, y, z)Q_2(x, y, z) \\ \dot{z} = f(x, y, z)Q_3(x, y, z)\,. \end{cases} \tag{2}$$

Therefore, the candidates for such surface functions $f(x, y, z)$ can be quite simple, e.g., a plane $f = x$, two orthogonal planes $f = xy$, or three orthogonal planes $f = xyz$. Also, some standard quadrics may be properly chosen, such as paraboloid, ellipsoids, and hyperboloid.

3 Twelve Cases: ES Systems

After a systematic search was finished, Jafari and Sprott [15] obtained twelve ES systems with dissipative properties.

These twelve cases are shown in Table 1. Notice that all the ES_i $(i = 1, 2, 3, ..., 12)$ systems have the largest Lyapunov exponents greater than 0.001, thus implying chaos in the system, more precisely with chaotic attractors on the x-y plane. Also, the equi-

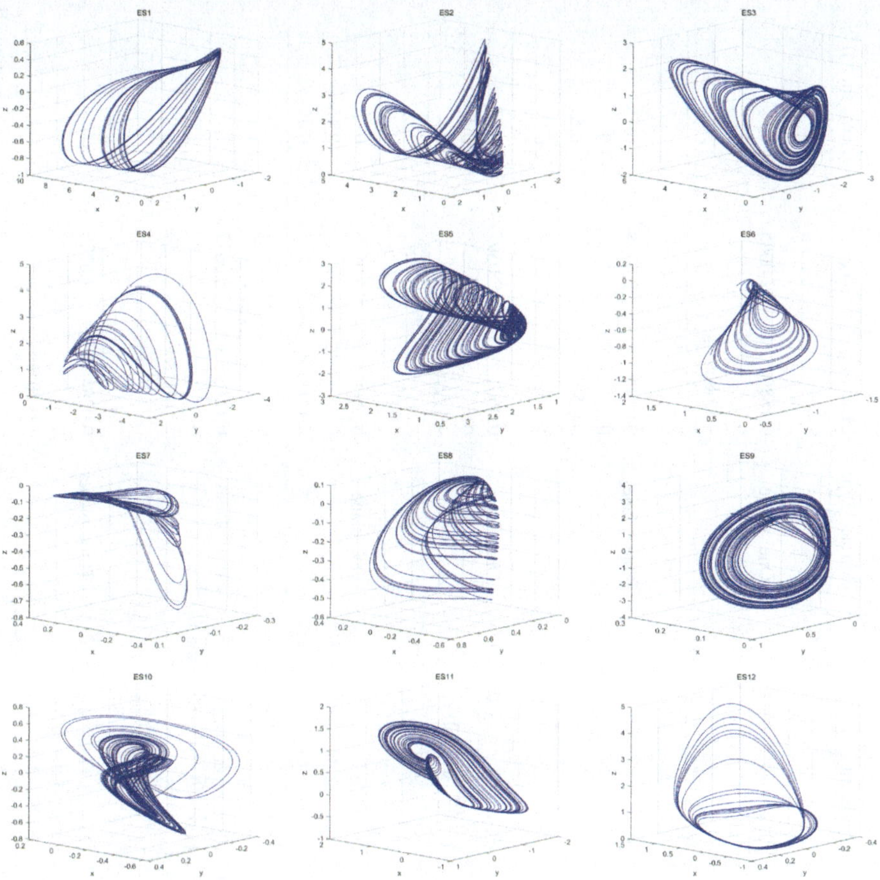

Fig. 1 State-space diagrams of the cases in Table 1 [1]

libria, Kaplan-Yorke dimensions and initial conditions are given in Table 1, along with the surface type for each ES system. For example, with $a = 0.1$, $b = 1$ and initial condition $(0, -0.08, 0)$, the ES_{10} system generates a hyperbolic cylinder of equilibria, with Lyapunov exponents $(0.0420, 0, -0.2230)$ and Kaplan-Yorke dimension $D_{KY} = 2.1883$. On the other hand, the ES systems in Table 1 approach chaos through a successive period-doubling route of limit cycles as parameter a increases. Yet, as a further increases in the ES_1 system, the strange attractor will be destroyed at a boundary crisis.

Figure 1 is a three-dimensional visualization of the state-space diagrams of the new systems. The strange attractors in such cases are considered as hidden attractors since an uncountable number of equilibria are found on the plane, on which a tiny fraction intersect the basin of attraction of the chaotic attractor.

Recall the cases with lines of equilibria discussed in the last chapter, the strange attractor there is surrounding the lines of equilibria [16–20]. However, in the present

Table 1 Twelve simple chaotic flows with surfaces of equilibria [1]

Case	Equations	(a, b, c)	Equilibria	Surface Type	LEs	D_{KY}	(x_0, y_0, z_0)
ES_1	$\begin{cases} \dot{x} = fy \\ \dot{y} = fz \\ \dot{z} = f\,(x + ay^2 - xz) \\ f = x \end{cases}$	$a = 1.54$	$(0, y, z)$	One plane	0.1889 0 −1.0869	2.0065	6 0 −1
ES_2	$\begin{cases} \dot{x} = yf \\ \dot{y} = -x + az \\ \dot{z} = (by^2 - xz)f \\ f = x \end{cases}$	$a = 1\ b = 3$	$(0, y, z)$	One plane	0.0644 0 −0.8279	2.0778	0.15 0 0.8
ES_3	$\begin{cases} \dot{x} = (y^2 + axy)f \\ \dot{y} = -zf \\ \dot{z} = f(b + xy) \\ f = x \end{cases}$	$a = 2\ \ b = 1$	$(0, y, z)$ $(\sqrt{\frac{b}{a}}, -\sqrt{ab}, 0)$ $(-\sqrt{\frac{b}{a}}, \sqrt{ab}, 0)$	One plane	0.0661 0 −1.664	2.0397	0.87 0.4 0
ES_4	$\begin{cases} \dot{x} = -yf \\ \dot{y} = (x + z)f \\ \dot{z} = f(ay^2 + xz - b) \\ f = x \end{cases}$	$a = 2$ $b = 0.35$	$(x, y, 0)$	One plane	0.0560 0 −1.0855	2.0516	0 0.46 0.7
ES_5	$\begin{cases} \dot{x} = -azf \\ \dot{y} = (b + z^2 - xy)f \\ \dot{z} = (x^2 - xy)f \\ f = xy \end{cases}$	$a = 0.4\ \ b = 1$	$(0, y, z)$ $(x, 0, z)$ $(\sqrt{b}, \sqrt{b}, 0)(-\sqrt{b}, -\sqrt{b}, 0)$	Two planes	0.1242 0 −1.8356	2.0677	1 1.44 0
ES_6	$\begin{cases} \dot{x} = (y + ayz)f \\ \dot{y} = (bz + y^2 + cz^2)f \\ \dot{z} = (x^2 - y^2)f \\ f = xyz \end{cases}$	$a = 2\ b = 8$ $c = 7$	$(0, y, z)$ $(x, 0, z)$ $(x, y, 0)$ $(\pm 1.5, 1.5, -0.5)$ $(\pm 1.5, -1.5, -1.5)$	Three planes	0.0294 0 −0.4051	2.0725	1 −1.3 −1

(continued)

Table 1 (continued)

Case	Equations	(a, b, c)	Equilibria	Surface Type	LEs	D_{KY}	(x_0, y_0, z_0)
ES_7	$\begin{cases} \dot{x} = ayf \\ \dot{y} = xzf \\ \dot{z} = (-z - x^2 - byz)f \\ f = 1 - x^2 - y^2 - z^2 \end{cases}$	$a = 0.4$ $b = 6$	$x^2 + y^2 + z^2 = 1$ $(0, 0, 0)$	Sphere	0.0113 0 −0.9501	2.0119	0 0.1 0
ES_8	$\begin{cases} \dot{x} = (az + y^2)f \\ \dot{y} = (-y + bx^2)f \\ \dot{z} = -xyf \\ f = 1 - x^2 - y^2 - z^2 \end{cases}$	$a = 1$ $b = 5$	$x^2 + y^2 + z^2 = 1$ $(0, 0, 0)$	Sphere	0.0232 0 −0.9552	2.0338	0.24 0.2 0
ES_9	$\begin{cases} \dot{x} = (y^2 - axy)f \\ \dot{y} = xzf \\ \dot{z} = (1 - by^2)f \\ f = 1 - x^2 - y^2 \end{cases}$	$a = 5$ $b = 7$	$x^2 + y^2 = 1$ $(-0.0756, -0.3781, 0)$ $(0.0756, 0.3781), 0$	Circular cylinder	0.0388 0 −1.2078	2.0321	0.06 0 1
ES_{10}	$\begin{cases} \dot{x} = (a - z^2)f \\ \dot{y} = xzf \\ \dot{z} = (y + bxz)f \\ f = 1 + x^2 - y^2 \end{cases}$	$a = 0.1$ $b = 1$	$y^2 - x^2 = 1$ $(0, 0, -0.3162)$ $(0, 0, 0.3162)$	Hyperbolic cylinder	0.0420 0 −0.2230	2.1883	0 −0.08 0
ES_{11}	$\begin{cases} \dot{x} = yzf \\ \dot{y} = (x - axz)f \\ \dot{z} = (x - bz^2)f \\ f = z + x^2 + y^2 \end{cases}$	$a = 1$ $b = 0.6$	$x^2 + y^2 + z = 0$ $(0, y, 0)$ $(0.6, 0, 1)$	Paraboloid	0.0283 0 −0.6171	2.0458	0.46 0 0.8
ES_{12}	$\begin{cases} \dot{x} = yzf \\ \dot{y} = -axf \\ \dot{z} = (-z + by^2 + xz)f \\ f = z + x^2 + y^2 \end{cases}$	$a = 0.1$ $b = 6$	$-x^2 + y^2 - z = 0$ $(0, 0, 0)$	Saddle	0.0068 0 −0.4998	2.0135	1 0 1

cases with surfaces of equilibria, the scenario is quite different since the normal component of the flow is zero on the surfaces. Therefore, the equilibrium surface will divide the three-dimensional space into two, four, or eight regions where strange attractors cannot span.

Additionally, these systems can be useful in practice for such as secure communications. Take a sphere of equilibria as an example. No perturbation smaller than the radius of the sphere can produce unbounded solutions, so it can ensure no entrance and no exit of system orbits. Thus, it is a perfect protection shield of the strange attractor inside or a protective cavity that any exterior chaotic orbit cannot penetrate.

References

1. S. Jafari, J.C. Sprott, V.T. Pham, C. Volos, C. Li, Simple chaotic 3D flows with surfaces of equilibria. Nonlinear Dyn. **86**(2), 1–10 (2016)
2. S. Jafari, J.C. Sprott, M. Molaie, A simple chaotic flow with a plane of equilibria. Int. J. Bifurc. Chaos **26**(06), 1650098 (2016)
3. B. Bao, T. Jiang, G. Wang, P. Jin, H. Bao, M. Chen, Two-memristor-based Chua's hyperchaotic circuit with plane equilibrium and its extreme multistability. Nonlinear Dyn. **60**, 1–15 (2017)
4. A.M. Macbeath, Geometrical realization of isomorphisms between plane groups. Bull. Am. Math. Soc. **71**(1965), 629–630 (1965)
5. S. Fukushima, T. Okumura, Estimating the three-dimensional shape from a silhouette by geometrical division of the plane. Proceedings of International Conference on Industrial Electronics, Control and Instrumentation, vol. 3 (IEEE, New York, 1991), pp. 1773–1778
6. J.C. Hart, Sphere tracing: a geometric method for the antialiased ray tracing of implicit surfaces. Visual Comput. **12**(10), 527–545 (1996)
7. P. Richard, L. Oger, J.P. Troadec, A. Gervois, Geometrical characterization of hard-sphere systems. Phys. Rev. E, **60**(4 Pt B), 4551–4558 (1999)
8. H. Tamura, M. Kiya, M. Arie, Numerical study on viscous shear flow past a circular cylinder. JSME Int. J. **23**(186), 1952–1958 (2008)
9. A. Ekmekci, D. Rockwell, Effects of a geometrical surface disturbance on flow past a circular cylinder: a large-scale spanwise wire. J. Fluid Mech. **665**(12), 120–157 (2010)
10. C.O. Bloom, Diffraction by a hyperbolic cylinder. Bull. Am. Math. Soc. **74**(1968), 587–589 (1968)
11. M. Gidea, J.P. Marco, Diffusion along chains of normally hyperbolic cylinders (2017). arXiv:1708.08314
12. M. Reinert, Geometric form of the thorax: An elliptic paraboloid. Praxis Und Klinik Der Pneumologie **40**(12), 470 (1986)
13. T. Fukusumi, S. Fujiwara, Geometrical nonlinear analysis of hyperbolic-paraboloid structure with pin joints. Memoirs Faculty Eng. Kobe Univ. **46**, 23–32 (1999)
14. J.C. Sprott, A proposed standard for the publication of new chaotic systems. Int. J. Bifurc. Chaos **21**(09), 2391–2394 (2011)
15. J.C. Sprott, *Elegant Chaos: Algebraically Simple Chaotic Flows* (World Scientific, Singapore, 2010)
16. S. Jafari, J. Sprott, Simple chaotic flows with a line equilibrium. Chaos Solitons Fractals **57**, 79–84 (2013)
17. C. Li, J.C. Sprott, Chaotic flows with a single nonquadratic term. Phys. Lett. A **378**(3), 178–183 (2014)
18. C. Li, J.C. Sprott, Z. Yuan, H. Li, Constructing chaotic systems with total amplitude control. Int. J. Bifurc. Chaos **25**(10), 1530025 (2015)

19. V. Pham, S. Jafari, C. Volos, T. Kapitaniak, A gallery of chaotic systems with an infinite number of equilibrium points. Chaos Solitons Fractals **93**, 58–63 (2016)
20. K. Barati, S. Jafari, J.C. Sprott, V.T. Pham, Simple chaotic flows with a curve of equilibria. Int. J. Bifurc. Chaos **26**(12), 511–543 (2016)

Chaotic Systems with Any Number and Various Types of Equilibria

Xiong Wang, Guanrong Chen, and Julien Clinton Sprott

1 Introduction

For a hyperbolic system, generally the Šil'nikov criterion [1–6] is applicable for proving the existence of chaos.

In the last few decades, it has been observed that although most dynamical systems are of hyperbolic type, there are still many others that are not so. For non-hyperbolic chaotic systems, the saddle-focus equilibrium point is usually not involved, such as those proposed by Sprott [7–10].

Recently, many chaotic systems, which can display three-wing, four-wing and even multi-wing chaotic attractors, are observed. For example, the Chen system [11, 12], Lorenz system [13, 14] and other Lorenz-like systems [15–17, 17–19] all can generate a two-wing attractor with two unstable saddle-foci and one unstable node. There lies an unstable saddle-focus equilibrium near the center of each butterfly-shaped wing. Additionally, it was found that there lies an unstable saddle-focus near the center of each wing of a multi-wing chaotic attractor. This common feature may indicate that the number of equilibria can determine the shape of a multi-wing chaotic attractor. Therefore, it is interesting to ask if it is possible to construct a chaotic system

X. Wang (✉)
Institute for Advanced Study, Shenzhen University,
Shenzhen, Guangdong 518060, People's Republic of China
e-mail: wangxiong8686@szu.edu.cn

G. Chen
Department of Electrical Engineering, City University of Hong Kong, Hong Kong, SAR 999077, People's Republic of China
e-mail: eegchen@cityu.edu.hk

J. Clinton Sprott
University of Wisconsin-Madison, Madison, WI, USA
e-mail: sprott@physics.wisc.edu

X. Wang et al. (eds.), *Chaotic Systems with Multistability and Hidden Attractors*,
Emergence, Complexity and Computation 40,
https://doi.org/10.1007/978-3-030-75821-9_7

with an arbitrary number of equilibria, and whether the number of equilibria also determines the shape of the chaotic attractor.

On the other hand, the stability of the equilibrium of an autonomous dynamical system is of high importance in the study of chaotic theory. Recall that Yang and Chen [20] introduced a dynamical system with one saddle and two stable node-foci, and they [21] also introduced an unusual three-dimensional autonomous quadratic Lorenz-like system with only two stable node-foci. Moreover, Wang and Chen [22] found a chaotic system with only one stable node-focus. These examples motivate another interesting question as if it is possible for a chaotic system to generate chaos with two, or three, or even an arbitrarily large number of stable or unstable equilibria.

This chapter introduces several examples of chaotic systems, proposed by Wang [23], with an arbitrarily preassigned number of equilibria, and another method, proposed by Zhang et al. [24], for a chaotic system to have any preassigned number of equilibria.

Recall that almost all the typical examples of chaotic attractors in three-dimensional autonomous systems of ordinary differential equations have one or more spiral saddle points, which satisfy the Šil'nikov condition [25]. These attractors are "self-excited" and they can be found by choosing initial conditions on the unstable manifold in the vicinity of one of these equilibria [26, 27].

For systems that admit a wide variety of chaotic solutions like the generalized version of the Nose-Hoover oscillator [28, 29], described by the following equations, it is natural to ask whether a chaotic attractor can exist if the system has only one equilibrium:

$$\begin{cases} \dot{x} = y \\ \dot{y} = -x + yz \\ \dot{z} = f(x, y, z), \end{cases} \tag{1}$$

where

$$f(x, y, z) = a_1x + a_2y + a_3z + a_4x^2 + a_5y^2 + a_6xy + a_7xz + a_8yz + a_9 \tag{2}$$

is determined to guarantee the system to have one single equilibrium, at $(x, y, z) = (0, 0, -a_9/a_3)$ with $a_3 \neq 0$ or $a_9 = 0$.

Note that it is also possible to include the term z^2 in Eq. (2) and adjust the constant a_9 so that the two roots of $f(x, y, z) = 0$ coincide, but this turned out to be unnecessary for the present purpose.

The simplest chaotic system in the form of (1) has $f = 1 - y^2$, which has been extensively studied in the Sprott A system. Nevertheless, such a system is conservative without any equilibria. However, in a weakly dissipative form of (1), with $f = 1 + \epsilon tanhx - y^2$, it has no equilibria since $a_3 = 0$ and $a_9 \neq 0$, but nevertheless can produce a chaotic attractor. All systems in the form of (1) have a nonlinear damping whose magnitude and sign both depend on the time average of z along the orbit due to the presence of the $y - z$ term.

As shown in Ref. [30, 31], systems (1) with various forms of $f(x, y, z)$ can have chaotic attractors for two types of the equilibria. By selecting appropriate constraints

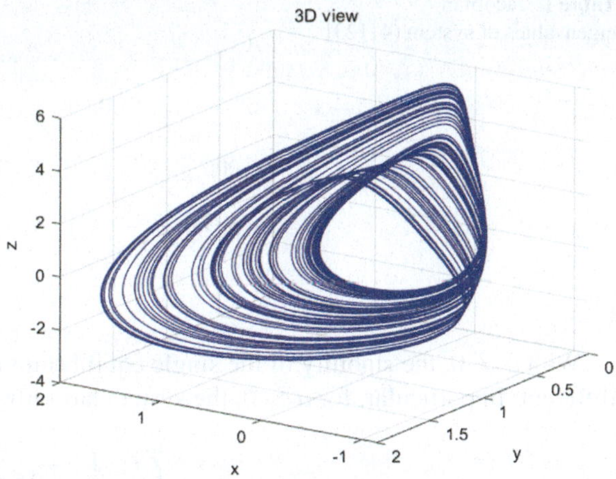

Fig. 1 Three-dimensional view of the chaotic attractor of system (4), with $a = 0.006$ [23]

on the eigenvalues λ given by

$$\lambda^3 + \left(\frac{a_9}{a_3} - a_3\right)\lambda^2 + (1 - a_9)\lambda - a_3 = 0 \tag{3}$$

and searching over the nine-dimensional parameter space and three-dimensional space of initial conditions, it is possible to find bounded and dissipative solutions with a positive Lyapunov exponent. Consequently, by simplifying the resulting system by removing unnecessary terms and setting some parameters to unity, it is possible to find some desirable chaotic systems as further discussed in more details below.

2 Chaotic Systems with Any Desired Number of Equilibria

2.1 A Modified Sprott E System with One Stable Equilibrium

By modifying the Sprott E system (see Table 1), a new dynamical system with only one equilibrium was introduced by Wang and Chen [22].

The new system has the following form:

$$\begin{cases} \dot{x} = yz + a \\ \dot{y} = x^2 - y \\ \dot{z} = 1 - 4x. \end{cases} \tag{4}$$

where a is a constant parameter and the equilibrium of the system is a stable node-focus.

With $a = 0$, the system reduces to the original Sprott E system (see Table 1).

Table 1 Jacobian eigenvalues of system (4) [23]

a	*Jacobian eigenvalues*
0	$-1, \pm 0.5i$
0.006	$-0.9607, -0.0197 \pm 0.5098i$
0.022	$-0.8458, -0.0771 \pm 0.5382i$

With $a \neq 0$, the stability of the single equilibrium turns out to be fundamentally different. In particular, for $a > 0$, the system has only one stable equilibrium:

$$E(x, y, z) = \left(\frac{1}{4}, \frac{1}{16}, -16a\right).$$

Through numerical simulation, the eigenvalues of the corresponding Jacobian matrix, with $a = 0$ and $a > 0$ respectively, are found as listed in Table 1. Interestingly, system (4) can display a one-scroll chaotic attractor, as illustrated in Fig. 1.

Moreover, with different numbers of stable equilibria, the system can remain to be chaotic, as further discussed in the following.

2.2 Chaotic System with Two Equilibria

To obtain a chaotic system with two equilibria, the system (4) is rewritten in terms of u, v, w, as follows:

$$\begin{cases} \dot{u} = vw + a \\ \dot{v} = u^2 - v \\ \dot{w} = 1 - 4u. \end{cases} \tag{5}$$

By selecting the following coordinate transformation:

$$\begin{cases} \dot{u} = x^2 - z^2 \\ \dot{v} = y \\ \dot{w} = 2xz, \end{cases}$$

which is a rotation symmetry about the y axis, denoted $\mathbb{R}_y(\pi)$, the system (5) is transformed to be in (u, v, w) with two original points $(\pm x, \pm y, \pm z)$ corresponding to the new variables (u, v, w). After the transformation, the new system reads

$$\begin{cases} \dot{x} = \frac{1}{2}\frac{z+2x^2yz+ax-4x^2z+4z^3}{x^2+z^2} \\ \dot{y} = (x^2 - z^2)^2 - y \\ \dot{z} = -\frac{1}{2}\frac{-x+2xyz^2+az-4xz^2+4x^3}{x^2+z^2}. \end{cases} \tag{6}$$

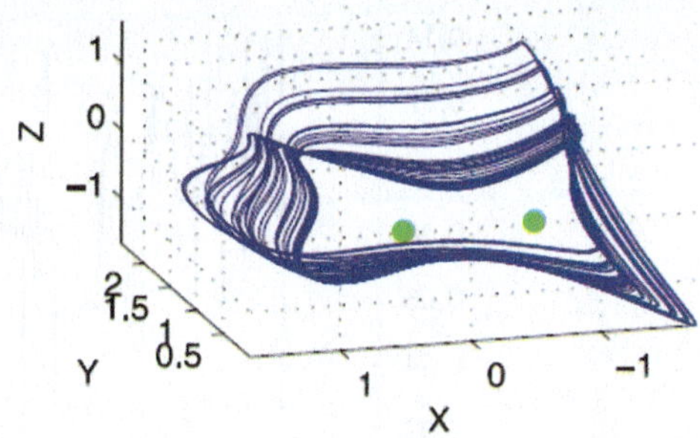

Fig. 2 The three-dimensional view of the chaotic attractor of system (6), with $a = 0.003$. The new two-petal chaotic attractor with stable equilibria is shown [23]

The system (6) is not globally but only locally topologically equivalent to system (4) since it has a singularity at the origin. The new system (6) has two equilibria, which are symmetrical and independent of the parameter a, as follows:

$$E_1\left(\frac{1}{2}, \frac{1}{16}, 0\right) \quad , \quad E_2\left(-\frac{1}{2}, \frac{1}{16}, 0\right) .$$

By linearizing system (6) at equilibrium $E_1(\frac{1}{2}, \frac{1}{16}, 0)$, the Jacobian matrix is obtained as

$$J = \begin{bmatrix} -2a & 0 & \frac{1}{16} \\ \frac{1}{2} & -1 & 0 \\ -1 & 0 & -2a \end{bmatrix} .$$

and the corresponding characteristic equation is

$$\lambda^3 + (1 + 4a)\lambda^2 + \left(4a^2 + 4a + \frac{1}{4}\right)\lambda + 4a^2 + \frac{1}{4} = 0 .$$

Hence, the eigenvalues of system (6) at the equilibrium $E_1(\frac{1}{2}, \frac{1}{16}, 0)$ are obtained as

$$\begin{aligned} \lambda_1 &= -1 \quad < 0 \\ \lambda_2 &= -2a + 0.5i \\ \lambda_3 &= -2a - 0.5i \,. \end{aligned}$$

It is clear that one can control the stability of these equilibria by adjusting the parameter a. Figure 2 shows a symmetrical two-petal chaotic attractor of the system (6). Especially, for $a < 0$ these equilibria are unstable and for $a > 0$ they are stable. The largest Lyapunov exponent of system (6) is illustrated in Fig. 3. Therefore, the existence of chaotic behavior for certain values of the parameter a is verified.

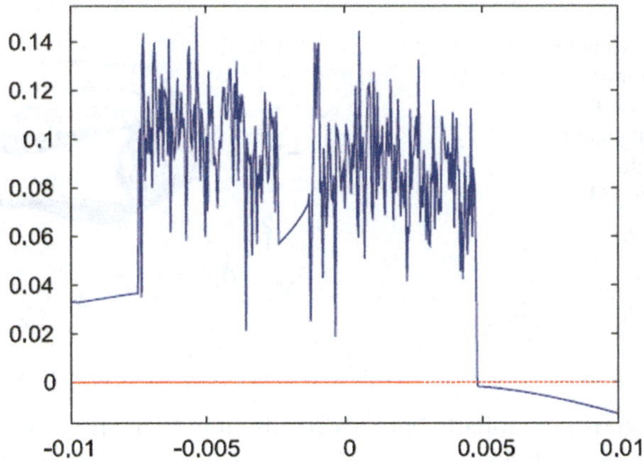

Fig. 3 The largest Lyapunov exponent of system (6) with respect to parameter a (Color figure online) [23]

2.3 *Chaotic System with Three Equilibria*

To obtain a chaotic system with three equilibria, the following transformation is applied:

$$\begin{cases} \dot{u} = x^3 - 3xz^2 \\ \dot{v} = y \\ \dot{w} = 3x^2z - z^3 , \end{cases}$$

which is rotation about the y-axis with symmetry, denoted as $\mathbb{R}_y(\frac{2}{3}\pi)$. It changes system (5) to be

$$\begin{cases} \dot{x} = \frac{1}{3}\frac{3x^4yz - 4x^2yz^4 + ax^2 - 8x^4z + 2xz + 24x^2z^3 + yz^5 - az^2}{2x^2z^2 + x^4 + z^4} \\ \dot{y} = (x^3 - 3xz^2)^2 - y \\ \dot{z} = -\frac{1}{3}\frac{6x^3yz^2 - 2xyz^4 + 2axz + 4x^5 - x^2 - 16x^3z^2 + z^2 + 12xz^4}{2x^2z^2 + x^4 + z^4} . \end{cases} \tag{7}$$

The system has a symmetrical three-petal equilibria, which are dependent on the parameter a. For $a < 0$, these equilibria are unstable, and for $a > 0$ they are stable.

Analysis on the equilibria and the corresponding Jacobian eigenvalues for two particular values of parameter, $a = 0.01$ and $a = -0.01$, are shown in Table 2.

The existence of chaos in system (7) is also supported by numerical calculation. The three-dimensional view of the new three-peral equilibria, when $a = 0.01$, is shown in Fig. 4.

Table 2 Equilibria and Jacobian eigenvalues of system (7) for certain values of parameter a [23]

a	Equilibria	Jacobian eigenvalues
0.01	$E_1 = (0.6550, 0.0625, -0.1258)$ $E_2 = (-0.2186, 0.0625, 0.6300)$ $E_3 = (-0.4365, 0.0625, -0.5044)$	$-1.0617, 0.0308 \pm 0.4843i$
−0.01	$E_1 = (0.6550, 0.0625, 0.1258)$ $E_2 = (-0.2186, 0.0625, -0.6300)$ $E_3 = (-0.4365, 0.0625, 0.5044)$	$-9334, 0.0333 \pm 0.5165i$

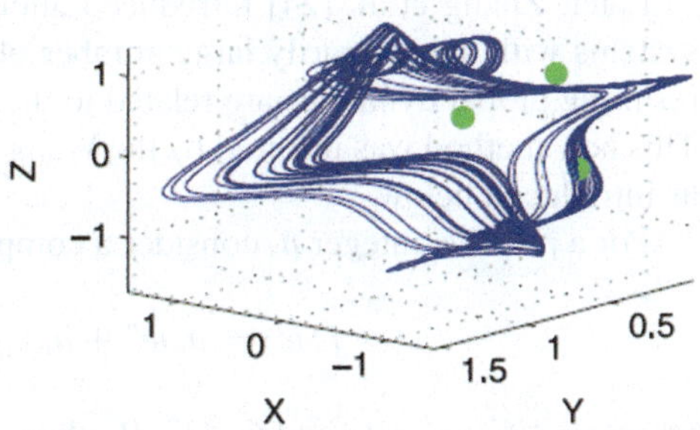

Fig. 4 The three-dimensional view of the chaotic attractor of system (7), with $a = 0.01$. The new three-petal chaotic attractor with symmetrical stable equilibria is shown [23]

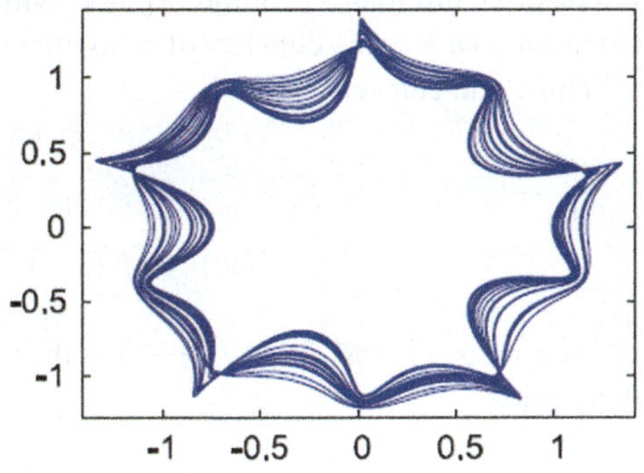

Fig. 5 Chaotic attractor of system (8), with $a = 0$ [23]

2.4 Constructing a Chaotic System with Any Number of Equilibria

To continue, a dynamical system with five equilibria is obtained via the following transformation:

$$\begin{cases} \dot{u} = x^5 - 10x^3z^2 + 5xz^4 \\ \dot{v} = y \\ \dot{w} = 5x^4z - 10x^2z^3 + z^5 , \end{cases} \tag{8}$$

which is a rotation about the y-axis with symmetry, denoted as $\mathbb{R}_y(\frac{2}{5}\pi)$. Applying this transformation to system (5), a new chaotic system is obtained, with five symmetrical equilibria, as shown in Fig. 5.

In general, the above Wang's method is to apply the transformation $(x + iz)^n = (u + iw)$ to obtain a new chaotic system with n equilibria, in theory. Note that the new systems so obtained all have global chaotic attractors and space the chaotic flows locally converge to the stable equilibrium.

Wang [23] also pointed out that the above systematic method for constructing new chaotic systems with any number of equilibria only provides a theoretical framework, but it had been noticed the chaotic attractor destines to flatten out as the value of the number n increases. Theoretically, there may still exist a possibly very narrow range of parameter values of a for chaos to present.

Later, Zhang et al. [24] introduced another method for constructing dynamical systems with an arbitrarily large number of chaotic attractors, and showed that the resulting global dynamics are related to the equilibria and the discontinuous points. This new method was inspired by the Wang's method [23] and some others [32, 33], as introduced below.

For a positive integer n, consider a complex polynomial function:

$$f(w) = a_n w^n + a_{n-1} w^{n-1} + \cdots + a_0 ,$$

where $a_i \in \mathbb{C}$, $a_n \neq 0$, and $w \in \mathbb{C}$. By the classical theory of complex analysis, there is a positive number N_1 for any $w_0 \in \mathbb{C}$ with $|w_0| \geq N_1$, such that the solution of the equation $f(w) = w_0$ consists of n distinct complex numbers.

Thus, one can set

$$w := x = +iy$$

and let

$$f(w) := f_1(x_1, x_2) + i f_2(x_1, x_2) ,$$

where $i = \sqrt{-1}$, and $f_1 : \mathbb{R}^2 \to \mathbb{R}$ is the real part while $f_2 : \mathbb{R}^2 \to \mathbb{R}$ is the imaginary part.

Then:

1. Take an autonomous system, denoted by $\dot{\mathbf{p}} = \Phi(\mathbf{p})$ with $\mathbf{p} \in \mathbb{R}^3$. A chaotic attractor and the basin of attraction will be contained in the following region:

$$\{(p_1, p_2, p_3) : |p_1| \leq M_1, |p_2| \leq M_2\}$$

2. Let the geometric structure of the attractor not be affected by the coordinate transformation below, so that one can "move" the basin of attraction to a "safe place". Theoretically, one can set $N := max\{N_1, M_1, M_2\}$. The coordinate transformation in terms of $\mathbf{u} = (u_1, u_2, u_3)$ reads

$$\begin{cases} p_1 &= u_1 - 2N \\ p_2 &= u_2 - 2N \\ p_3 &= u_3 . \end{cases}$$

3. For the new dynamical system $\dot{\mathbf{u}} = \Phi(\mathbf{u})$, apply the complex function f to generate a system with an arbitrarily large number of chaotic attractors. In so doing, consider the coordinate transformation in terms of $\mathbf{x} = (x_1, x_2, x_3)$, as follows:

$$\begin{cases} u_1 &= f_1(x_1, x_2) \\ u_2 &= f_2(x_1, x_2) \\ u_3 &= x_3 \,. \end{cases}$$

Following the above discussion, it can be verified that the new chaotic system

$$\dot{\mathbf{x}} = \hat{\Phi}(\dot{\mathbf{x}}) \tag{9}$$

has n chaotic attractors.

Theoretically, one might observe chaotic attractors in similar shapes. The geometric structures of the attractors will not be affected by the above transformation. Also, numerical experiments should be used to discover the proper system parameters and initial conditions.

An example using this method to construct a chaotic system with any number of equilibria is the following system, obtained from the Lorenz system:

$$\begin{cases} \dot{x_1} = 10(x_2 - x_1) \\ \dot{x_2} = 28x_1 - x_1x_3 - x_2 \\ \dot{x_3} = x_1x_2 - \frac{8}{3}x_3 \,. \end{cases} \tag{10}$$

Following the new method, a function is constructed:

$$f(w) = w^n = (x_1 + ix_2)^n = \sum_{k=0}^{n} C_n^k x_z^{n-k} (ix_2)^k = f_1(x_1, x_2) + if(x_1, x_2)\,,$$

where

$$C_n^k = \frac{n!}{k!(n-k)!}, \quad 0 \le k \le n.$$

Choose, for example. $N_1 = 0.1$. Then, the region of the attractor is obtained:

$$\{(p_1, p_2, p_3): \quad |p_1| < R_1 := 40, |p_2| < R_2 := 40, -2 < p_3 < R_3 := 80\}\,.$$

The coordinate transformation is

$$\begin{cases} p_1 &= u_1 - 2N &= u_1 - 80 \\ p_2 &= u_2 - 2N &= u_2 - 80 \\ p_3 &= u_3 \,, \end{cases}$$

where $N = max\{N_1, R_1, R_2\} = 40$, with the basin of attraction contained in the set

$$\{(u_1, u_2, u_3) \,:\, 40 < u_1 < 120, 40 < u_2 < 120, -2 < u_3 < 80\},$$

for which the corresponding system is given by

$$\begin{cases} \dot{u_1} = 10(u_2 - u_1) \\ \dot{u_2} = 28u_1 - u_1u_3 - u_2 - 2160 \\ \dot{u_3} = (u_1 - 80)(u_2 - 80) - \frac{8}{3}u_3 \,. \end{cases} \tag{11}$$

Considering the simplicity in numerical simulation, the value of the integer n is set to be $n = 2$. Thus,

$$f_1(x_1, x_2) = x_1^2 - x_2^2 \quad , and \quad f_2(x_1, x_2) = 2x_1x_2 \,,$$

hence the coordinate transformation gives

$$\begin{cases} u_1 &= x_1^2 - x_2^2 \\ u_2 &= 2x_1x_2 \\ u_3 &= x_3 \,. \end{cases}$$

Finally, the new chaotic system is obtained, as

$$\begin{cases} \dot{x_1} = \frac{48x_1^2x_2 - 10x_1^3 + 8x_2^2x_1 - 28x_2^3 - x_1^2x_2x_3 + x_2^3x_3 + 80x_2x - 3 - 2160x_2}{2x_1^2 + 2x_2^2} \\ \dot{x_2} = \frac{48x_2^2x_1 - 10x_2^3 + 8x_1^2x_2 - 28x_1^3 - x_2^2x_1x_3 + x_1^3x_3 + 80x_2x - 3 - 2160x_1}{2x_1^2 + 2x_2^2} \\ \dot{x_3} = (x_1^2 - x_2^2 - 80)(2x_1x_2 - 80) - \frac{8}{3}x_3 \,. \end{cases} \tag{12}$$

The chaotic behavior of this system, with initial condition (10, 10, 10), is shown in Fig. 6, which has a similar geometric structure with the classical Lorenz attractor.

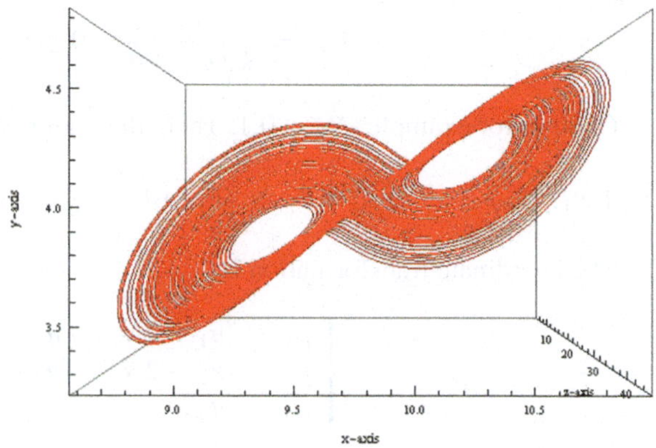

Fig. 6 The phase portrait of system (12), with initial condition (10, 10, 10) [24]

3 Chaotic Systems with Any Type of Equilibria

3.1 System with No Equilibria

A simple example of a dissipative chaotic system, in the form of Eq. (1) without equilibria, is given by

$$\begin{cases} \dot{x} = y \\ \dot{y} = -x + yz \\ \dot{z} = x^2 - 4y^2 + 1 \, . \end{cases} \tag{13}$$

For initial condition (0, 2, 0), it has a chaotic attractor, with Lyapunov exponents $(0.0131, 0, -0.0155)$ and a relatively large KaplanâŁ"Yorke dimension, 2.8455, as shown in Fig. 7.

This chaotic attractor is "hidden", which cannot be found by using an initial condition in the vicinity of an equilibrium, since simply no such points exist [26, 27].

The system (13) is time-reversal invariant under the transformation $(x, y, z, t) \rightarrow (x, -y, -z, -t)$. Its chaotic attractor is slightly asymmetric and is accompanied by a repellor that is symmetric with it under a 180° rotation about the x-axis and intertwined with it. The attractor and repellor exchange their roles when time is reversed. The asymmetry in z is the source of the weak nonlinear damping, since its time-averaged value along the chaotic orbit is $< z > \approx -0.0024$.

This system is unusual, because the chaotic attractor is intertwined with a set of nested conservative tori, which are symmetric under rotation about the x-axis and $< z > = 0$.

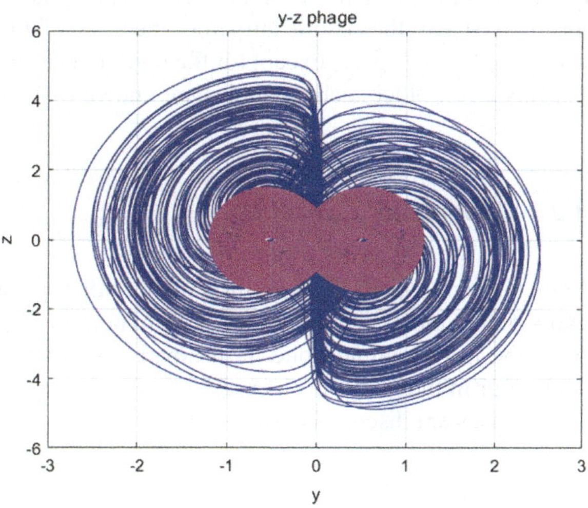

Fig. 7 Chaotic attractor (in pink) of System (13), coexisting with a conservative invariant torus (in green), projected onto the $x - y$ plane

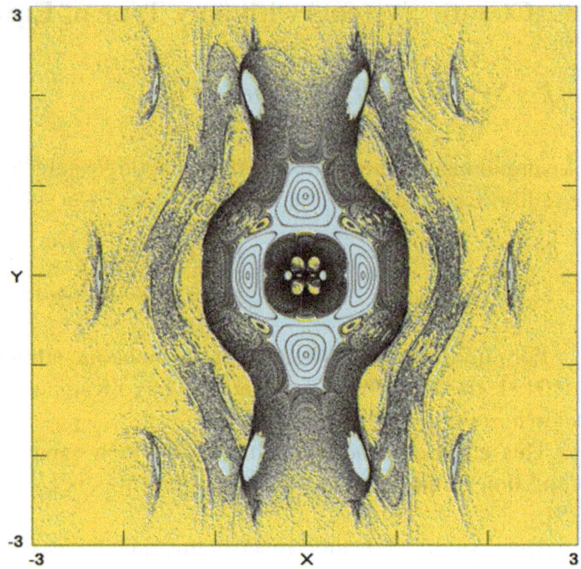

Fig. 8 Cross section on the $z = 0$ plane of the nested tori, surrounded by a chaotic attractor of System $ET0$. The blue background shows the initial conditions that give conservative orbits (tori), and the yellow background is the basin of attraction of the chaotic attractor [39]

Figure 7 shows one such torus with initial condition (0, 1.2, 0), for which the Lyapunov exponents are (0, 0, 0).

Figure 8 shows a cross-section on the $z = 0$ plane of the nested tori, surrounded by what looks like a chaotic sea but is actually a weakly dissipative chaotic attractor.

Sixty-three initial conditions were spaced uniformly over the range $-2.9625 \leq x \leq -0.6522$ with $y = z = 0$. The blue background shows the initial conditions that generate conservative orbits (tori), and the yellow background is the basin of attraction for the chaotic attractor. It appears that there are additional thin tori toward the outer edge of the chaotic attractor. The basin of attraction for the chaotic attractor is the whole state space, except for the region of a finite volume occupied by the tori.

Only a few other such examples are known today [34–36].

3.2 Hyperbolic Examples

Eight types of hyperbolic equilibria in three-dimensional flows are shown in Table 3 and Fig. 9. Among the eight cases with hyperbolic equilibria, Type 7 is overwhelmingly the most common one in dissipative chaotic systems, while other examples also occur in the other seven cases.

All cases are discussed below.

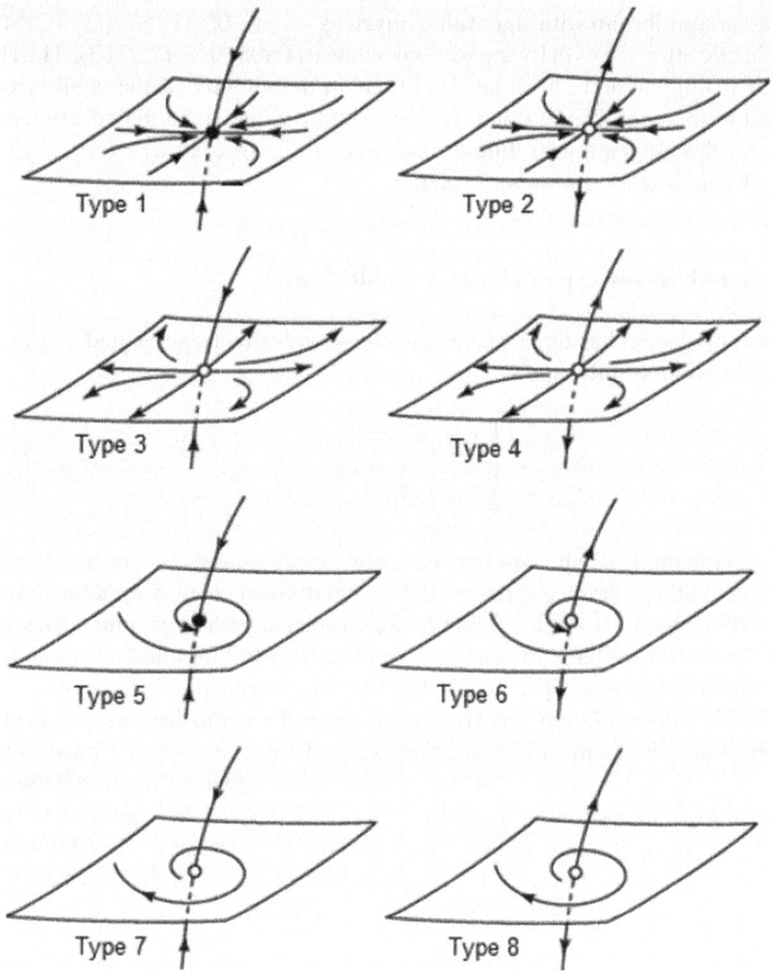

Fig. 9 Types of hyperbolic equilibria in three-dimensional flows [39]

3.2.1 Equilibrium Type 1 (Index-0 Node)

This type of system has a single equilibrium with three real eigenvalues, all negative. Hence, it is a stable node with index 0, where the index is the number of eigenvalues with a positive real part, or equivalently, the dimension of the unstable manifold.

A typical system of this type is

$$\begin{cases} \dot{x} = y \\ \dot{y} = -x + yz \\ \dot{z} = -z - 8xy + 0.3xz - 3\,. \end{cases} \tag{14}$$

It has an equilibrium with eigenvalues given by $\lambda = (-0.381966, -1, -2.618034)$ and a chaotic attractor with Lyapunov exponents (0.0505, 0, −17.2283). The chaotic attractor is hidden, and all initial conditions in the vicinity of the equilibrium are attracted to the equilibrium point. It is also multi-stable since the chaotic attractor coexists with a point attractor. Initial conditions close to the attractor are (−2, 1, 0.7), and the basin of attraction is very small.

3.2.2 Equilibrium Type 2 (Index-1 Saddle Node)

This type of system has three real eigenvalues, with two negative and one positive. A typical system of this type is

$$\begin{cases} \dot{x} = y \\ \dot{y} = -x + yz \\ \dot{z} = 0.5z - y^2 + 5\,. \end{cases} \tag{15}$$

This system has an equilibrium with eigenvalues $\lambda = (0.5, -0.101021, -9.898980)$ and a symmetric pair of tightly intertwined chaotic attractors with Lyapunov exponents (0.0141, 0, −0.3030), which coexist with three limit cycles, a symmetric one with Lyapunov exponents (0, −0.1239, −0.2336) and a symmetric pair with Lyapunov exponents (0, −0.0264, −0.0264). This is possible, because System (14), like the Lorenz system, has a rotational symmetry about the z-axis, as evidenced by its invariance under the transformation $(x, y, z) \to (-x, -y, z)$. Hence, either the solutions share that symmetry, or there is a symmetric pair of them. All the attractors are hidden, and all initial conditions in the vicinity of the equilibrium produce unbounded orbits. Initial conditions that give the five attractors are $(\pm 0.9, 0, -2)$, (0.43, 2, 0.18), and $(\pm 0.4, \pm 3, 1)$. The basins of attraction of the chaotic attractors are relatively small and bounded (with a finite volume).

3.2.3 Equilibrium Type 3 (Index-2 Saddle Node)

This type of system has three real eigenvalues, with two positive and one negative. A typical system of this type is

$$\begin{cases} \dot{x} = y \\ \dot{y} = -x + yz \\ \dot{z} = -x - 0.1z - y^2 + 0.3\,. \end{cases} \tag{16}$$

It has an equilibrium with eigenvalues $\lambda = (2.618034, 0.381966, -0.1)$ and a self-excited chaotic attractor with Lyapunov exponents (0.02191, 0, −0.3181). Initial conditions close to the attractor are (0, 0.1, 0), and the basin of attraction is relatively large.

3.2.4 Equilibrium Type 4 (Index-3 Repellor)

This type of system has three real and positive eigenvalues, and hence it is an unstable node, referred to as a "repellor". A system of this type [31] has the following form:

$$\begin{cases} \dot{x} = y \\ \dot{y} = -x + yz \\ \dot{z} = z + 8.894x^2 - y^2 - 4. \end{cases} \tag{17}$$

It has an equilibrium with eigenvalues $\lambda = (3.732051, 1, 0.267949)$ and a chaotic attractor with Lyapunov exponents $(0.1767, 0, -0.9158)$. The equations have a rotational symmetry since they are invariant under the transformation $(x, y, z) \rightarrow (-x, -y, z)$, and the system has a symmetric pair of solutions for some parameters, but not for the ones given above. The symmetric chaotic attractor of this system is hidden and all initial conditions chosen in the vicinity of the equilibrium lead to unbounded solutions. Initial conditions close to the attractor are $(0, 3.8, 0.7)$, and the basin of attraction is very small.

3.2.5 Equilibrium Type 5 (Index-0 Spiral Node)

This type of system has one real negative eigenvalue and a complex conjugate pair with negative real parts. Twenty-three chaotic examples of this type have been reported in Ref. [30]. One typical system is

$$\begin{cases} \dot{x} = y \\ \dot{y} = -x + yz \\ \dot{z} = 2x - 2z + y^2 - 0.3\,. \end{cases} \tag{18}$$

It has an equilibrium with eigenvalues $\lambda = (-2, -0.075 \pm 0.997184i)$ and a chaotic attractor with Lyapunov exponents $(0.0203, 0, -2.4751)$. All chaotic systems of this type are multi-stable since the chaotic attractor coexists with a stable equilibrium, and the chaotic attractor is hidden, which cannot be found by using any initial condition in the vicinity of the equilibrium. Initial conditions close to the chaotic attractor are $(0.9, 0, 0.7)$, and the basin of attraction is very large.

3.2.6 Equilibrium Type 6 (Index-1 Spiral Saddle)

This type of system has one real positive eigenvalue and a complex conjugate pair with negative real parts. A typical system of this type is

$$\begin{cases} \dot{x} = y \\ \dot{y} = -x + yz \\ \dot{z} = 0.28z - xy + 0.48\,. \end{cases} \tag{19}$$

It has an equilibrium with eigenvalues $\lambda = (0.28, -0.857143 \pm 0.515079i)$ and a symmetric pair of chaotic attractors with Lyapunov exponents $(0.0677,\ 0, -1.5020)$. The equations have a rotational symmetry since they are invariant under the transformation $(x, y, z) \to (-x, -y, z)$. The chaotic attractors are hidden, and all initial conditions in the vicinity of the equilibrium lead to unbounded orbits. Initial conditions close to the attractors are $(0, \pm 4, 2)$, and the basins of attraction are very small.

3.2.7 Equilibrium Type 7 (Index-2 Spiral Saddle)

This type of system has one real negative eigenvalue and a complex conjugate pair with positive real parts. This is overwhelmingly the most common type with abundant examples, including the familiar Lorenz system [13] and Rössler system [37], although they have multiple equilibria. The simplest such system with a single equilibrium is the jerk system [9]: $\dddot{x} + 2.017\ddot{x} - \dot{x}^2 + x = 0$. Other simple examples are the Sprott I, J, L, N, and R systems [8]. There are also some systems in the form of Eq. (1), which have not been carefully studied, including the following model:

$$\begin{cases} \dot{x} = y \\ \dot{y} = -x + yz \\ \dot{z} = -z + xy + 0.39\,, \end{cases} \tag{20}$$

which is functionally the same as System $ET6$ with rotational symmetry but having different parameters.

This system has an equilibrium with eigenvalues $\lambda = (-1, 0.195 \pm 0.980803i)$ and a symmetric pair of chaotic attractors with Lyapunov exponents $(0.0820, 0, -0.6920)$. The chaotic attractors are self-excited, although the equilibrium lies on their basin boundary. Its attractor can be found, depending on where in the vicinity of the equilibrium the initial condition is chosen. Initial conditions close to the attractors are $(\pm 1.4, \pm 1, 1)$, and the basins of attraction are relatively large.

3.2.8 Equilibrium Type 8 (Index-3 Spiral Repellor)

This type of system has one real positive eigenvalue and a complex conjugate pair with positive real parts. A typical system of this type is

$$\begin{cases} \dot{x} = y \\ \dot{y} = -x + yz \\ \dot{z} = 0.2z + 0.1y^2 - xy - 0.08\,. \end{cases} \tag{21}$$

This system has an equilibrium with eigenvalues $\lambda = (0.2, 0.2 \pm 0.979796i)$ and a chaotic attractor with Lyapunov exponents $(0.1083, 0, -3.2555)$. The equations have a rotational symmetry since they are invariant under the transformation

$(x, y, z) \rightarrow (-x, -y, z)$. For the given parameters, the chaotic attractor is symmetric and self-excited. Initial conditions close to the attractor are $(1, -2, 0.4)$, and the basin of attraction is very large.

3.3 *Non-Hyperbolic Systems*

A non-hyperbolic equilibrium has one or more eigenvalues with a zero real part. There are eleven such types in three-dimensional flows. Six of these have all real eigenvalues and are of the form $(0, -, -)$, $(+, 0, -)$, $(+, +, 0)$, $(0, 0, -)$, $(+, 0, 0)$, and $(0, 0, 0)$. Five have one real and a complex conjugate pair of eigenvalues, only two of which have nonzero real eigenvalues. The stability of the systems that do not have an eigenvalue with a positive real part cannot be determined from the eigenvalues, so it requires nonlinear analysis in general.

First, consider nine types of such systems, in which at least one eigenvalue is real and zero. With $\lambda = 0$, Eq. (3) shows that $a_3 = 0$, and thus according to Eq. (3), an equilibrium is present only if $a_9 = 0$. In this case, there is a line of equilibria along the z-axis, $(0, 0, z)$. Such cases have been studied [30], including those in the form of Eq. (1). But, they fall outside the scope of the present book, which involves chaotic systems with a single equilibrium. Thus, nine of the eleven possible non-hyperbolic isolated equilibria cannot occur in system (1), although this does not imply that they cannot exist in other systems.

The remaining two cases have a complex conjugate pair of eigenvalues of the form $\lambda = 0 \pm i\omega$. Substituting them into Eq. (1) gives $\omega_2 = 1 - a_9$ and $a_9 = 1 - a_3^2$ or $a_9 = 0$. The remaining real eigenvalue can be negative or positive.

Examples of these two types of chaotic systems are discussed below.

3.3.1 Equilibrium Type 9

This type of system has a single equilibrium with one real negative eigenvalue and a complex conjugate pair with zero real parts.

Chaotic systems of this type have been known, such as the Sprott E system (see Table 1).

A system of this type, in the form of Eq. (1), is

$$\begin{cases} \dot{x} = y \\ \dot{y} = -x + yz \\ \dot{z} = -z - 4xy + xz\,. \end{cases} \tag{22}$$

It has an equilibrium with eigenvalues $\lambda = (-1, 0 \pm i)$ and a chaotic attractor with Lyapunov exponents $(0.0394, 0, -1.4067)$. The equilibrium at the origin is nonlinearly unstable, and the chaotic attractor is self-excited. Initial conditions close to the attractor are $(0, 1, 0.4)$, and the basin of attraction is relatively small.

Table 3 Eight types of hyperbolic equilibria in three-dimensional flows

Type	Equations	Eigenvalues	LEs	(x_0, y_0, z_0)
Index-0 node	$\begin{cases} \dot{x} = y \\ \dot{y} = -x + yz \\ \dot{z} = -z - 8xy + 0.3xz - 3 . \end{cases}$	(−0.381966 − 1 − 2.618034)	(0.0505, 0, −17.2283)	(−2, 1, 0.7)
Index-1 saddle point	$\begin{cases} \dot{x} = y \\ \dot{y} = -x + yz \\ \dot{z} = 0.5z - y^2 + 5 . \end{cases}$	(0.5 − 0.101021 − 9898980)	(0.0141, 0, −0.3030) (0, −0.1239, −0.2336) (0, −0.0264, −0.0264)	(±0.9, 0, −2) (0.43, 2, 0.18) (±0.4, ±3, 1)
Index-2 saddle point	$\begin{cases} \dot{x} = y \\ \dot{y} = -x + yz \\ \dot{z} = -x - 0.1z - y^2 + 0.3 . \end{cases}$	(2.618034 0.381966 −0.1)	(0.02191, 0, −0.3181)	(0, 0.1, 0)
Index-3 repellor	$\begin{cases} \dot{x} = y \\ \dot{y} = -x + yz \\ \dot{z} = z + 8.894x^2 - y^2 - 4 . \end{cases}$	(3.732051 1 0.267949)	(0.1767, 0, −0.9158)	(0, 3.8, 0.7)
Index-0 spiral node	$\begin{cases} \dot{x} = y \\ \dot{y} = -x + yz \\ \dot{z} = 2x - 2z + y^2 - 0.3 . \end{cases}$	$(-2 - 0.075 \pm 0.997184i)$	(0.0203, 0, −2.4751)	(0.9, 0, 0.7)
Index-1 spiral saddle	$\begin{cases} \dot{x} = y \\ \dot{y} = -x + yz \\ \dot{z} = 0.28z - xy + 0.48 . \end{cases}$	$(0.28 - 0.857143 \pm 0.515079i)$	(0.0677, 0, −1.5020)	(0, ±4, 2)
Index-2 spiral saddle	$\begin{cases} \dot{x} = y \\ \dot{y} = -x + yz \\ \dot{z} = -z + xy + 0.39 . \end{cases}$	$(-10.195 \pm 0.980803i)$	(0.0820, 0, −0.6920)	(±1.4, ±1, 1)
Index-3 spiral repellor	$\begin{cases} \dot{x} = y \\ \dot{y} = -x + yz \\ \dot{z} = 0.2z + 0.1y^2 - xy - 0.08 . \end{cases}$	$(0.20.2 \pm 0.979796i)$	(0.1083, 0, −3.2555)	(1, −2, 0.4)

3.3.2 Equilibrium Type 10

This type of system has a single equilibrium point with one real positive eigenvalue and a complex conjugate pair with zero real parts.

A typical system of this type is

$$\begin{aligned}\dot{x} &= y\\ \dot{y} &= -x + yz\\ \dot{z} &= 0.23z + y2 - 10xy\,.\end{aligned} \tag{23}$$

It has an equilibrium with eigenvalues $\lambda = (0.23, 0 \pm i)$ and a chaotic attractor with Lyapunov exponents $(0.1241, 0, -2.4424)$. The system is invariant under the transformation $(x, y, z) \to (-x, -y, z)$, and the chaotic attractor is symmetric and self-excited. Initial conditions close to the attractor are $(0, 1, 1)$, and the basin of attraction is very large.

4 Conclusions

Autonomous systems in the form of Eq. (1) admit chaotic solutions, with one or more chaotic attractors, in the presence of a single hyperbolic equilibrium, for each of the eight types shown in Fig. 10. Two of these systems have a chaotic attractor coexisting with a stable equilibrium, three of them have a symmetric pair of chaotic attractors, and one has two chaotic attractors coexisting with three limit cycles. Five of the eight cases have hidden attractors.

There are eleven types of non-hyperbolic equilibria that can occur in three-dimensional systems. However, nine of the eleven types cannot occur in isolation of a system in the form of Eq.1. The other two cases admit chaotic solutions with a single self-excited attractor, as shown in Fig. 11. All of the nine systems have complicated and interesting basins of attraction [38], with a rich set of bifurcations as the parameters are varied.

Given a system in the form of Eq. (1), which can have chaos in the absence of equilibria as in system (13), it is not surprising to see that chaos can coexist with isolated equilibria of all types, although in most cases the equilibria are relatively close to the attractors and, thus, would be expected to influence the system dynamics.

The results presented in this chapter support the idea that any dynamics not explicitly forbidden by theory may occur in an appropriately designed dynamical system. One only needs to search carefully to find suitable examples.

An interesting question is whether chaotic attractors can occur in systems with the fourteen types of hyperbolic equilibria that occur in four dimensions. The answer is YES according to Ref. [40].

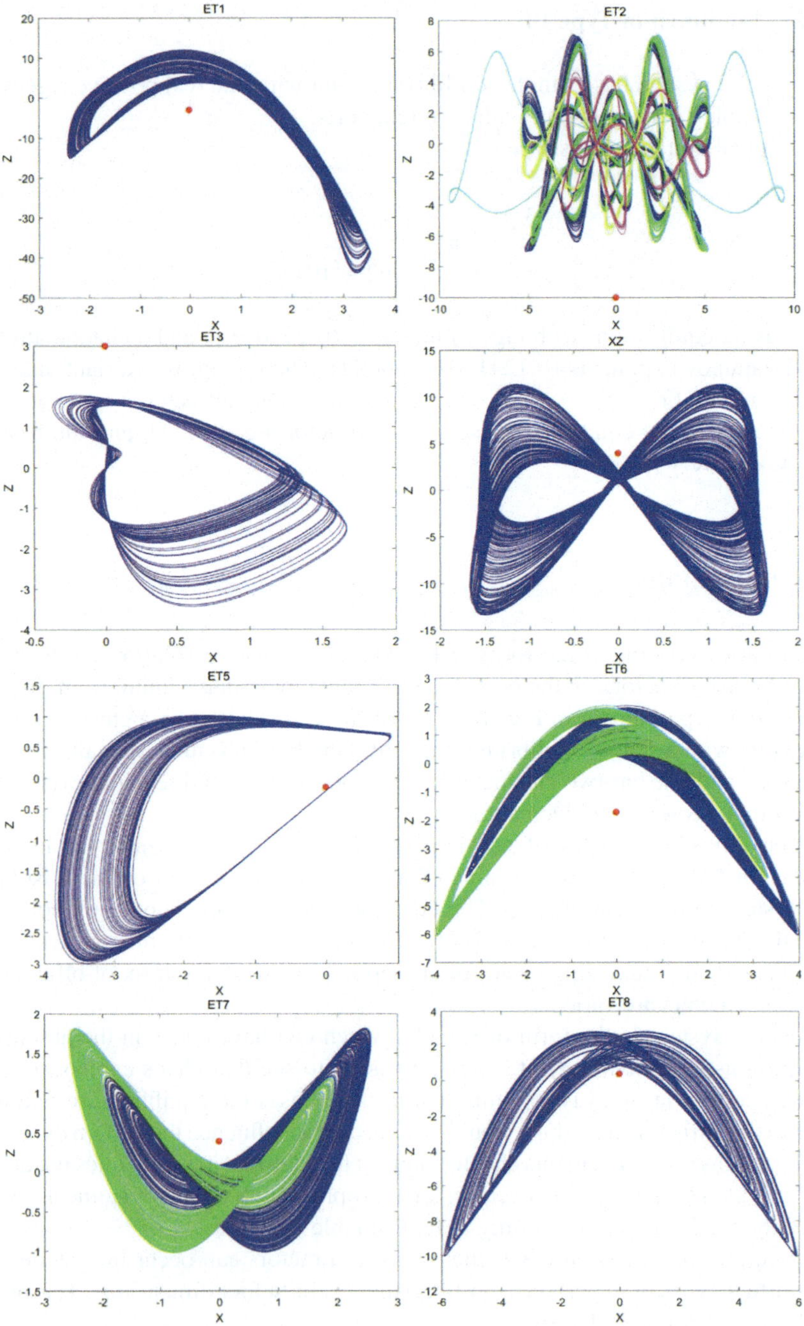

Fig. 10 Attractors for systems with a single hyperbolic equilibrium, for each of the eight types of systems in Table 3, projected onto the x-z plane. The equilibria are indicated by red dots, which lie in the $y = 0$ plane [39]

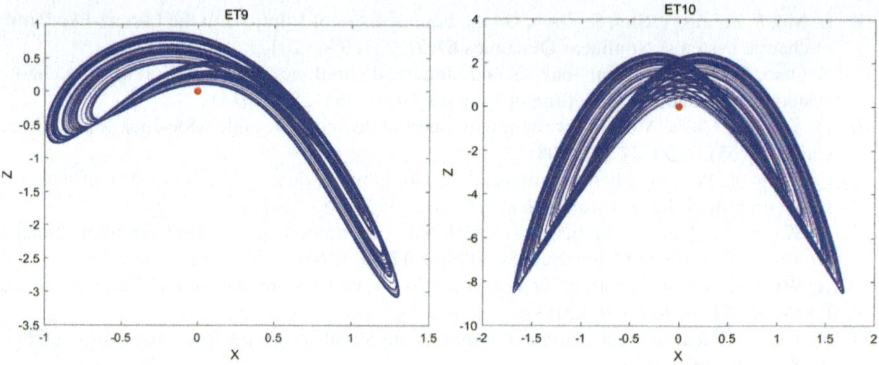

Fig. 11 Attractors for systems with a single non-hyperbolic equilibrium, for two of the eleven types projected onto the x-z-plane. The equilibria are indicated by red dots, which lie on the $y = 0$ plane [39]

References

1. L.P. Shilnikov, Acontribution to the problem of the structure of an extended neighborhood of Arough equilibrium state of saddle-focus type. Mathematics of the USSR-Sbornik **10**(1), 92–103 (1970)
2. I.M. Ovsyannikov, L.P. Shil'nikov, On systems with a saddle-focus homoclinic curve. Mathematics of the USSR-Sbornik **58**(4), 552–570 (1987)
3. A.L. Shil'Nikov, L.P. Shil'nikov, D.V. Turaev, Normal forms and Lorenz attractors. Int. J. Bifurc. Chaos **3**(5), 1123–1139 (1993)
4. L.P. Shilnikov, A.L. Shilnikov, D.V. Turaev, L.O. Chua, *Methods of Qualitative Theory in Nonlinear Dynamics* (World Scientific, Singapore, 1998)
5. B. Chen, T. Zhou, G. Chen, An extended Šil'nikov homoclinic theorem and its applications. Int. J. Bifurc. Chaos **19**(05), 1679–1693 (2009)
6. B. Chen, T. Zhou, Šil'nikov homoclinic orbits in two classes of 3d autonomous nonlinear systems. Int. J. Modern Phys. B **25**(20), 1110112 (2011)
7. J.C. Sprott, Automatic generation of strange attractors. Computers & Graphics **17**(3), 325–332 (1993)
8. J.C. Sprott, Some simple chaotic flows. Phys. Rev. E **50**(2), R647 (1994)
9. J.C. Sprott, Simplest dissipative chaotic flow. Physics Letters A **228**(4–5), 271–274 (1997)
10. J.C. Sprott, S.J. Linz, Algebraically simple chaotic flows. International Journal of Chaos Theory and Applications **5**(2), 1–20 (2000)
11. G. Chen, T. Ueta, Yet another chaotic attractor. Int. J. Bifurc. Chaos **9**(07), 1465–1466 (1999)
12. T. Ueta, G. Chen, Bifurcation analysis of Chen's equation. Int. J. Bifurc. Chaos **10**(8), 1917–1931 (2000)
13. E.N. Lorenz, Deterministic nonperiodic flow. Journal of Atmospheric Sciences **20**(2), 130–141 (1963)
14. I. Grigorenko, E. Grigorenko, Chaotic dynamics of the fractional Lorenz system. Physical Review Letters **91**(3), 034101 (2003)
15. T. Zhou, G. Chen, S. ?elikovsky, Ši'lnikov chaos in the generalized Lorenz canonical form of dynamical systems. Nonlinear Dyn. **39**(4), 319–334 (2005)
16. T. Li, G. Chen, Y. Tang, L. Yang, Hopf bifurcation of the generalized Lorenz canonical form. Nonlinear Dynamics **47**(4), 367–375 (2006)
17. X. Li, Q. Ou, Dynamical properties and simulation of a new Lorenz-like chaotic system. Nonlinear Dyn. **65**(3), 255–270(16) (2011)

18. C. Mu, F. Zhang, Y. Shu, S. Zhou, On the boundedness of solutions to the Lorenz-like family of chaotic systems. Nonlinear Dynamics **67**(2), 987–996 (2012)
19. Z. Qiao, X. Li, Dynamical analysis and numerical simulation of a new Lorenz-type chaotic system. Mathematical Modelling of Systems **20**(3), 264–283 (2014)
20. Q. Yang, G. Chen, A chaotic system with one saddle and two stable node-foci. Int. J. Bifurc. Chaos **18**(05), 1393–1414 (2008)
21. Q. Yang, Z. Wei, G. Chen, An unusual 3d autonomous quadratic chaotic system with two stable node-foci. Int. J. Bifurc. Chaos **20**(04), 1061–1083 (2010)
22. X. Wang, G. Chen, A chaotic system with only one stable equilibrium. Communications in Nonlinear Science and Numerical Simulation **17**(3), 1264–1272 (2012)
23. X. Wang, G. Chen, Constructing a chaotic system with any number of equilibria. Nonlinear Dynamics **71**(3), 429–436 (2013)
24. X. Zhang, Constructing a chaotic system with any number of attractors. Int. J. Bifurc. Chaos **27**(8), 1750118 (2017)
25. L.P. Shilnikov, A.L. Shilnikov, *Methods of Qualitative Theory in Nonlinear Synamics* (World Scientific, Singapore, 2001)
26. G.A. Leonov, N.V. Kuznetsov, V.I. Vagaitsev, Localization of hidden Chua's attractors. Physics Letters A **375**(23), 2230–2233 (2011)
27. G.A. Leonov, N.V. Kuznetsov, Hidden attractors in dynamical systems. from hidden oscillations in Hilbert–Kolmogorov, Aizerman, and Kalman problems to hidden chaotic attractor in Chua circuits. Int. J. Bifurc. Chaos **23**(1), 1330002 (2013)
28. N. Shuichi, Constant temperature molecular dynamics methods. Progress of Theoretical Physics Supplement **103**(103), 1–46 (1991)
29. W.G. Hoover, Remark on 'some simple chaotic flows. Physical Review E **51**(1), 759–760 (1995)
30. M. Molaie, S. Jafari, J.C. Sprott, S.M.R.H. Golpayegani, Simple chaotic flows with one stable equilibrium. Int. J. Bifurc. Chaos **23**(11), 1350188 (2013)
31. J. Sprott, S. Jafari, V.-T. Pham, Z.S. Hosseini, A chaotic system with a single unstable node. Physics Letters A **379**(36), 2030–2036 (2015)
32. Q. Wang, S. Yu, C. Li, J. Lü, X. Fang, C. Guyeux, J.M. Bahi, Theoretical design and FPGA-based implementation of higher-dimensional digital chaotic systems. IEEE Transactions on Circuits & Systems I: Regular Papers **63**(3), 401–412 (2016)
33. Z. Hua, S. Yi, Y. Zhou, C. Li, Y. Wu, Designing hyperchaotic cat maps with any desired number of positive Lyapunov exponents. IEEE Transactions on Cybernetics **48**(2), 463–473 (2018)
34. J.C. Sprott, W.G. Hoover, C.G. Hoover, Heat conduction, and the lack thereof, in time-reversible dynamical systems: Generalized nosé-hoover oscillators with a temperature gradient. Physical Review E **89**(4), 042914 (2014)
35. A. Politi, G. Oppo, R. Badii, Coexistence of conservative and dissipative behavior in reversible dynamical systems. Physical Review A **33**(6), 4055 (1986)
36. J.C. Sprott, A dynamical system with a strange attractor and invariant tori. Physics Letters A **378**(20), 1361–1363 (2014)
37. O.E. Rössler, An equation for continuous chaos. Physics Letters A **57**(5), 397–398 (1976)
38. U. Chaudhuri, A. Prasad, Complicated basins and the phenomenon of amplitude death in coupled hidden attractors. Physics Letters A **378**(9), 713–718 (2014)
39. J.C. Sprott, Strange attractors with various equilibrium types. Eur. Phys. J. Special Topics **224**(8), 1409–1419 (2015)
40. Z.C. Wei, J.C. Sprott, H. Chen, Elementary quadratic chaotic flows with a single non-hyperbolic equilibrium. Phys. Lett. A **379**(37), S0375960115005599 (2015)

Part III

Hyperchaotic Systems with Hidden Attractors

Yu-Ming Chen

1 Introduction

Recently, research focus has been shifted from classifying periodic and chaotic attractors to self-excited and hidden attractors [1–20].

Most well-known chaotic and hyperchaotic systems have one to three equilibria, such as the classical Lorenz, Rössler, Chen, and Lü systems [21–24], whose chaotic attractors with typical parameters are self-excited attractors. The basin of attraction of a self-excited attractor intersects with a small neighborhood of some unstable equilibrium. However, the basin of attraction of a hidden attractor does not intersect with any sufficiently small neighborhood of arbitrary equilibrium. The precise mathematical definition of hidden attractors was first defined by Leonov and Kuznetsov [1, 2].

It is obvious that the periodic and chaotic attractor of any chaotic system with only stable equilibrium or without equilibrium are hidden attractors [6–9]. In addition to the above two kinds of systems, in 2013 Jafari and Sprott studied a class of three-dimensional chaotic systems with a line of equilibria [25]. In this kind of systems, although the basin of attraction of a chaotic attractor may intersect with any small neighborhood of a length of the line of equilibria, there will still be uncountable equilibria falling outside the basin of attraction. Therefore, by selecting the initial conditions in the small domain of an unstable equilibrium, the periodic and chaotic attractors of this kind of systems may not be found easily. In other words, from the numerical calculation point of view, any attractor of this kind of systems is a hidden attractor.

Many complex dynamical systems [26–28], from climate and ecosystem to the stock market and applied engineering systems, have the complex characteristic of

Y.-M. Chen (✉)
Department of Mathematics and Computer Science, Gannan Normal University, Ganzhou 341000, P.R. China
e-mail: blkhpz@126.com

X. Wang et al. (eds.), *Chaotic Systems with Multistability and Hidden Attractors*, Emergence, Complexity and Computation 40,
https://doi.org/10.1007/978-3-030-75821-9_8

coexisting attractors. This kind of characteristic of a system is referred to as the multi-stability. A system with this kind of characteristic is neither completely stable nor completely unstable, but frequently switches back and forth between multiple states. Recent studies have shown that the multi-stability of a system is related to the existence of hidden attractors. A multi-stable system is extremely sensitive to the initial states, noise and system parameters, so that the state of the system is easily changed under sudden disturbances, switching from an ideal state to another perhaps undesirable state.

A self-excited attractor can be found smoothly by some standard computational procedure. However, there is no standard method to find hidden attractors and coexisting behaviors in a multi-stable system, which makes the study of coexisting behaviors of hidden attractors more complicated and challenging.

In this chapter, we summarize some four-dimensional (4D) hyperchaotic systems with hidden attractors, including hyperchaotic systems without equilibrium [6–8], hyperchaotic systems with limited equilibria [9–13], hyperchaotic systems with lines or curves of equilibria [14–16, 18], and hyperchaotic systems with planes or surfaces of equilibria [19, 20].

2 Hyperchaotic Systems with No Equilibria

2.1 Example 1

In [6], a new 4D chaotic system is described:

$$\begin{cases} \dot{x} = y, \\ \dot{y} = -x + yz + axzw, \\ \dot{z} = 1 - y^2, \\ \dot{w} = z + bxz + cxyz\,, \end{cases} \tag{1}$$

where a, b, and c are real parameters.

Let $\dot{x} = 0$, $\dot{y} = 0$, $\dot{z} = 0$ and $\dot{w} = 0$, namely

$$\begin{cases} y = 0, \\ -x + yz + axzw = 0, \\ 1 - y^2 = 0, \\ z + bxz + cxyz = 0\,. \end{cases} \tag{2}$$

Then, if there are equilibria in the above system, they can be solved. It is not hard to find that the third equation of (2), $y = \pm 1$, which is inconsistent with the first equation of (2). Hence, system (1) has no equilibrium, thus there are no pitchfork bifurcation, Hopf bifurcation, and so on. Moreover, there is no sink for this system as there is no equilibrium.

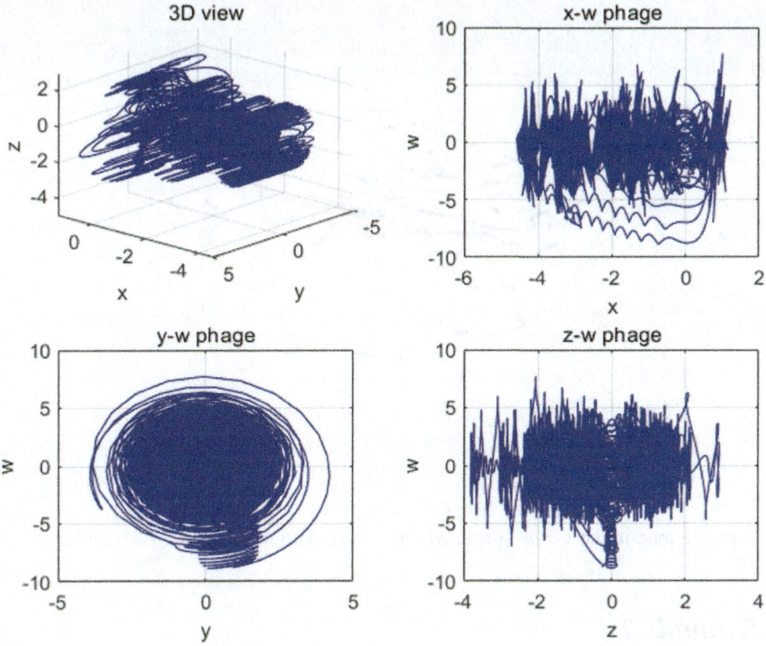

Fig. 1 Hyperchaotic attractor without equilibrium, with $a = 8$, $b = -2.5$, and $c = -30$

Let parameters $a = 8$, $b = -2.5$, and $c = -30$, and the initial condition be $(0.1, 0.1, 0.1, 0.1)$. The projections of its phase portrait are shown in Fig. 1, and the Poincaré map in the y-z-w space with $x = 0$ is shown in Fig. 2.

Under the above setting of parameters and initial conditions, the Lyapunov exponents of system (1) are $\lambda_1 = 0.87$, $\lambda_2 = 0.03$, $\lambda_3 = 0.00$, and $\lambda_4 = -1.01$, which show that the system is hyperchaotic. The Lyapunov exponents of system (1) are calculated by the efficient QR-based method, where the simulation time is 15000 s and the step length is 0.0005 s.

In order to verify the realizability of system (1), an electronic circuit was designed in [6]. To prevent the operational amplifiers and analog multipliers from saturating, under the transformation $x = 2x_m$, $y = 2y_m$, $z = 2z_m$ and $w = 2w_m$, the system (1) is transformed to

$$\begin{cases} \dot{x_m} = y_m, \\ \dot{y_m} = -x_m + 2y_m z_m + 4a x_m z_m w_m, \\ \dot{z_m} = 1 - 2y_m^2, \\ \dot{w_m} = z_m + 2b x_m z_m + 4c x_m y_m z_m, \end{cases} \tag{3}$$

which possesses the analogous properties with system (1). The designed circuit realizing (3) is presented in Fig. 3, and the phase diagrams on the y_m-w_m and z_m-w_m planes are shown in Fig. 4, which are similar to Fig. 1.

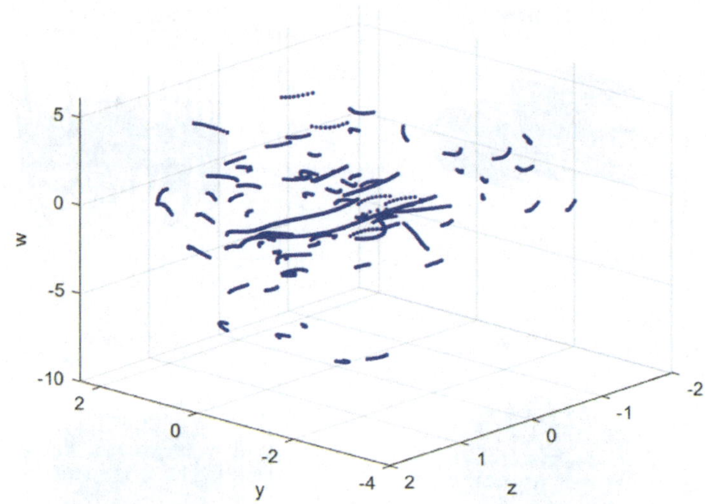

Fig. 2 Poincaré map in the y-z-w space, where $x = 0$, with $a = 8$, $b = -2.5$, and $c = -30$

2.2 *Example 2*

By adding one more state variable, w, to the following generalized diffusionless Lorenz equations [29]:

$$\begin{cases} \dot{x} = a(y - x), \\ \dot{y} = -xz - cy, \\ \dot{z} = -b + xy, \end{cases} \tag{4}$$

Wei [8] obtained a new 4D hyperchaotic system, expressed as

$$\begin{cases} \dot{x} = a(y - x), \\ \dot{y} = -xz - cy + kw, \\ \dot{z} = -b + xy, \\ \dot{w} = -my, \end{cases} \tag{5}$$

where a, b, c, k, m are real parameters, with $km \neq 0$.

When the parameters of system (4) satisfy $a = 10, b = 100, c = 11.2$, the system has a chaotic attractor, coexisting with two stable equilibria, $(\pm 10, \pm 10, -11.2)$.

When $b \neq 0$, system (5) has no equilibria, which means that the system has no such phenomena as pitchfork, Hopf or homoclinic bifurcations, which are common to systems with equilibria.

Under the transformation $(x, y, z, w) \rightarrow (-x, -y, z, -w)$, the new hyperchaotic system (5) is invariant, which means that system (5) is symmetrical about the z-axis. It is not hard to find that the z-axis is an orbit of this hyperchaotic system, and the trajectory on the z-axis will tend to the origin if $b > 0$, as $t \rightarrow \infty$.

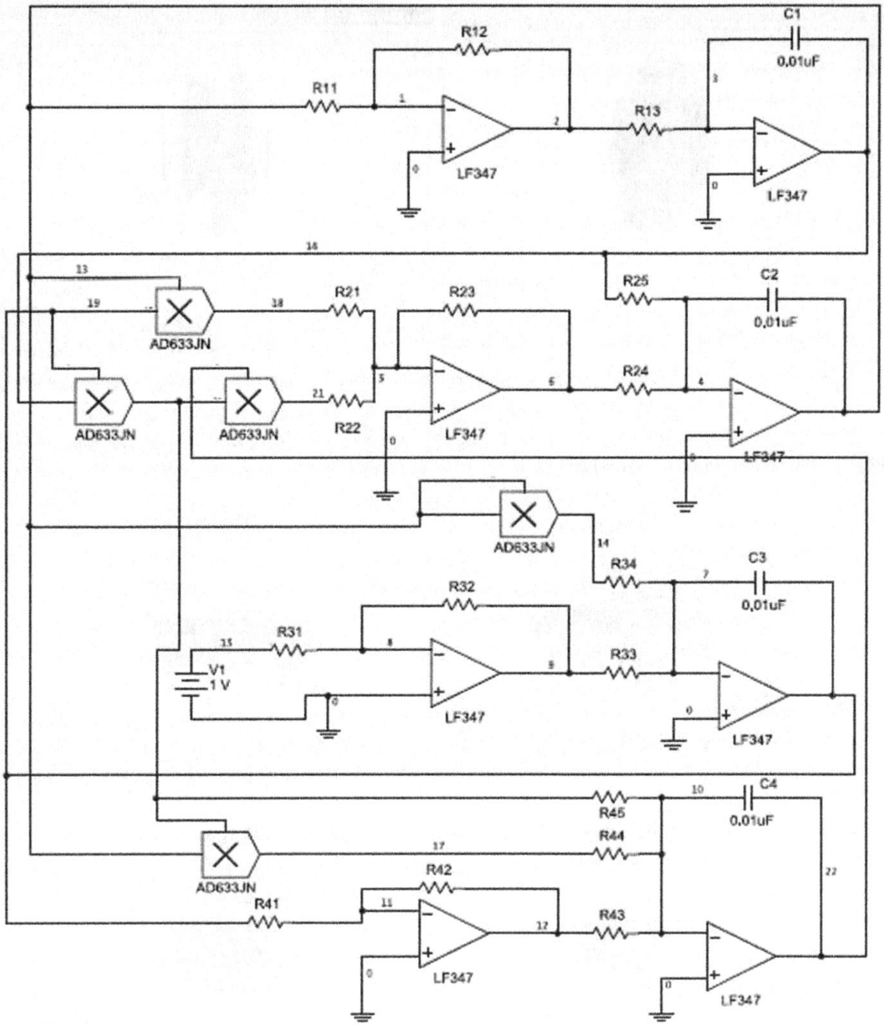

Fig. 3 Circuitry realization of the chaotic system (3)

Let the parameters be $a = 10$, $b = 25$, $c = -2.5$, $k = 1$, $m = 1$, and choose the initial condition $(0.2, 0.1, 0.75, -2)$. Then, a two-scroll hyperchaotic attractor of system (5) can be found, as shown in Fig. 5. The corresponding Lyapunov exponents are $L_1 = 0.9115$, $L_2 = 0.0224$, $L_3 = 0$, and $L_4 = -8.4330$, and the Kaplan–Yorke dimension is $D_L = 3.1107$.

Using the above parameter values and initial condition, it can be found that the Poincaré image of the system (5) on the cross section $x = y$ has no regular limbs, as shown in Fig. 6. This means that the new hyperchaotic system has extremely rich

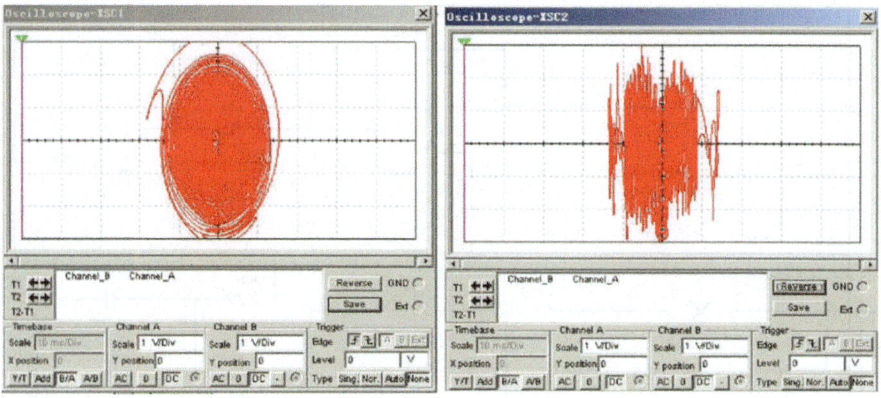

(a) Projection on the $y-w$ plane

(b) Projection on the $z-w$ plane

Fig. 4 Phase diagram of system (3)

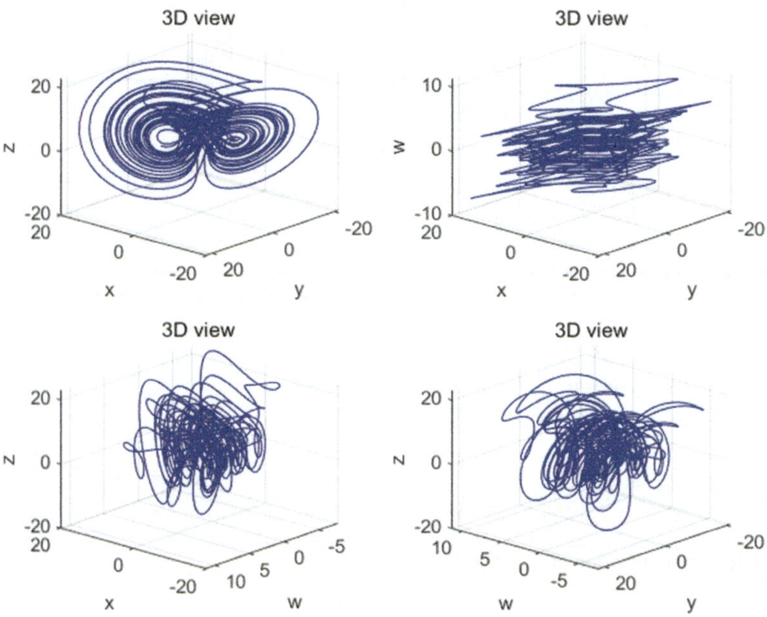

Fig. 5 Different perspectives on the two-scroll hyperchaotic attractor of the 4D system (5) without equilibria, with $a = 10$, $b = 25$, $k = 1$, $m = 1$ and $c = -2.5$

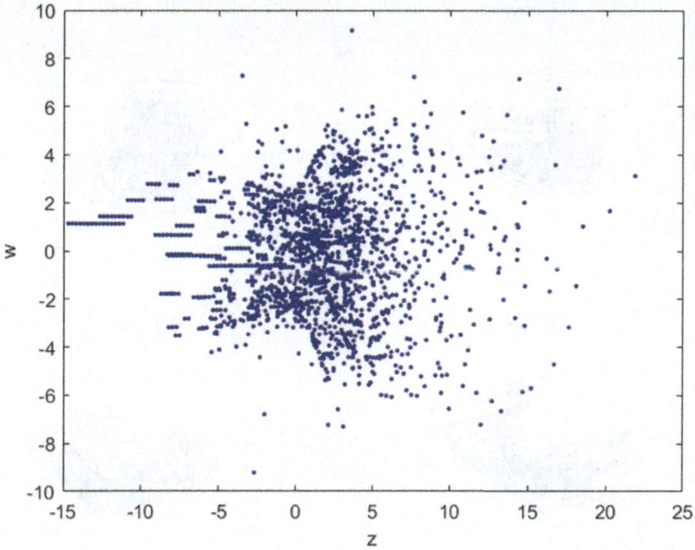

Fig. 6 Poincaré maps of the hyperchaotic generalized diffusionless Lorenz equations (5), with parameters $a = 10$, $b = 25$, $k = 1$, $m = 1$ and $c = -2.5$, on the cross section $x = y$

dynamics, which are different from the normal hyperchaotic systems with equilibria. This phenomenon could be due to the loss of limits from the equilibria.

3 Hyperchaotic Systems with a Limited Number of Equilibria

3.1 Hyperchaotic System with One Equilibrium

Based on system (4) and the 4D Lorenz–Stenflo system [30]

$$\begin{cases} \dot{x} = \sigma(y - x) + \beta w, \\ \dot{y} = rx - xz - y, \\ \dot{z} = -bz + xy, \\ \dot{w} = -x - \sigma w, \end{cases} \tag{6}$$

Wei [9] added an extra term yz to system (6), so as to obtain a new 4D hyperchaotic system in the following form:

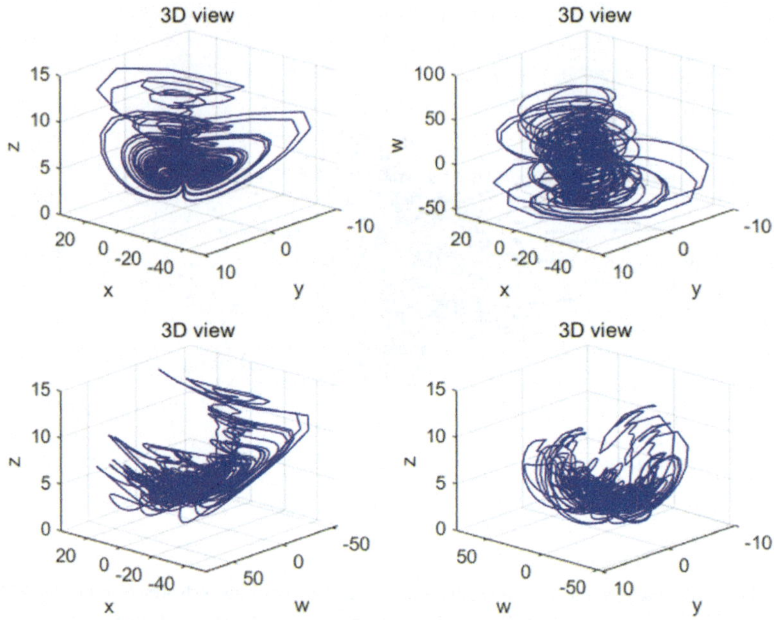

Fig. 7 Hidden hyperchaotic attractor for the 4D hyperchaotic system (), with parameter values $\sigma = 35, k = -17, b = 0.8, s = 35, n = 4, v = -20, s = 35$ and $q = 0$ (7)

$$\begin{cases} \dot{x} = \sigma(y - x) + syz + w, \\ \dot{y} = -xz - ky, \\ \dot{z} = -n - bz + xy, \\ \dot{w} = -vx - qw, \end{cases} \tag{7}$$

where $\sigma > 0, n > 0, sbvk \neq 0$ and $q \in \mathbb{R}$. Although the algebraic form of system (7) is similar to the Lorenz–Stenflo system (6), system (7) has some interesting dynamics that were not found from other 4D autonomous systems.

Let the parameters be $\sigma = 35$, $k = -17$, $b = 0.8$, $s = 35$, $n = 4$, $v = -20$ and $q = 0$, and choose the initial condition $(0, 1, -0.5, 0)$. Then, the system has a two-scroll-like hyperchaotic attractor coexisting with a stable equilibrium, as shown in Fig. 7. The corresponding Lyapunov exponents are $L_1 = 1.2908$, $L_2 = 0.2086$, $L_3 = 0$ and $L_4 = -20.2994$, and the Kaplan–Yorke dimension is $D_L = 3.0739$.

Under the above parametric conditions, system (7) has no homoclinic (heteroclinic) orbits. It is obvious that the Poincaré image of the system (7) on the cross section $y = w$ has no regular limbs, as shown in Fig. 8. This means that the new hyperchaotic system has extremely rich dynamics, which are different from the normal hyperchaotic systems with unstable equilibria.

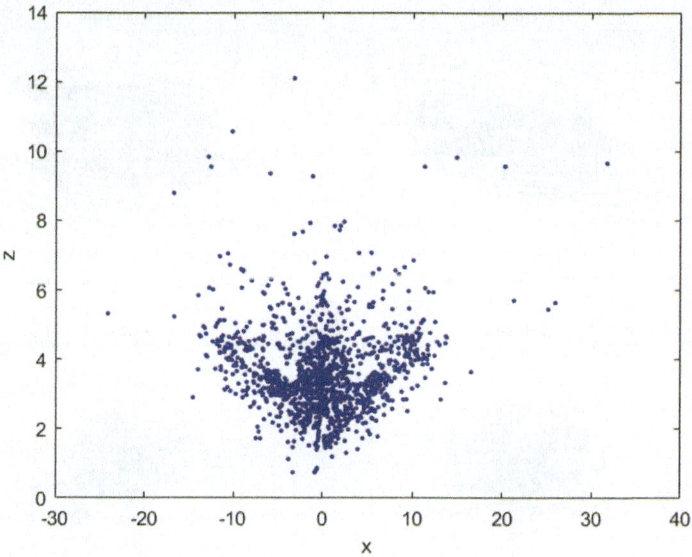

Fig. 8 Poincaré maps of the hyperchaotic attractor of system (7), with parameter values $\sigma = 35$, $k = -17$, $b = 0.8$, $s = 35$, $n = 4$, $v = -20$ and $q = 0$

3.2 Hyperchaotic System with Two Equilibria

Wei [13] added two state variables, u and v, into an existing dynamo model, obtaining a 5D autonomous system as

$$\begin{cases} \dot{x} = r(y - x) + u, \\ \dot{y} = -(1 + m)y + xz - v, \\ \dot{z} = g(1 + mx^2 - (1 + m)xy), \\ \dot{u} = 2(1 + m)u + xz - k_1 x, \\ \dot{v} = -mv + k_2 y, \end{cases} \tag{8}$$

where parameters r, m, and g are positive constants, and k_1, k_2 are variational control parameters.

Under the transformation $(x, y, z, u, v) \to (-x, -y, z, -u, -v)$, system (8) is invariant, which means that the system has a rotational symmetry around the z-axis. Since the divergence of system (8) is $1 - r$, system (8) is dissipative when $r > 1$.

The following theorem excludes some parameter sets that cannot make system (8) to produce bounded chaotic or hyperchaotic solutions.

Theorem 8.1 ([13]) *Consider the five-parameter family of systems (8), with two real parameters l_1, l_2. If parameters $(r, g) \in (0, +\infty) \times (0, +\infty)$ and m, k_1, k_2, l_1, l_2 satisfy*

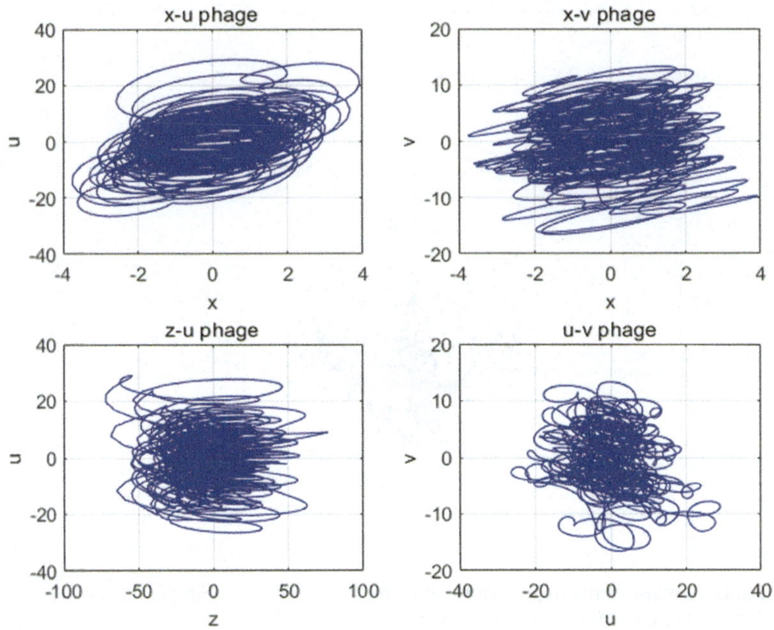

Fig. 9 Hidden hyperchaotic attractor of system (8), with parameters $m = 0.04,\ g = 140.6,\ k_2 = 12$, and $r = 7,\ k_1 = 34$ and initial condition $(0.05, -0.5, 0.1, -1, 2)$

$$
\begin{aligned}
m &= -2 + l_1 + \tfrac{1}{l_2} > 0, \\
k_1 &= \tfrac{l_1(-2+l_2(2-l_1+r))}{l_2}, \\
k_2 &= \tfrac{3-l_2(3+l_1(-2+r))}{l_2^2}, \\
l_1 &< 2 + 2m\,,
\end{aligned}
$$

then system (8) has no bounded chaotic or hyperchaotic solutions.

In order to find hidden attractors, parameters were chosen so that the system (8) has only stable equilibria. Considering that the equilibria $E_{1,2}$ of (8) are symmetric about the z-axis, the stability of equilibrium $E_1(x_0, y_0, z_0, u_0, v_0)$ is analyzed as follows.

When parameters satisfy $m = 0.04,\ g = 140.6,\ k_2 = 12$, and $r = 7,\ k_1 = 34$, system (8) has a hidden hyperchaotic attractor, with Lyapunov exponents $(0.9616, 0.5477, 0.1425, 0.0000, -7.6518)$, and Kaplan–Yorke dimension $D_{KY} = 4.2159$, as shown in Fig. 9.

It is worth noting that system (8) has an important feature, which is the existence of hidden hyperchaos for a range of parameters in the asymptotically stable regions of $E_{1,2}$ (see Fig. 10).

To show evidence of multi-stability and to examine the robustness of the existing hyperchaos, let $3 < r < 8$ and $k_1 = 34$. The dynamics of system (8) are shown in

Fig. 11, with three choices of initial conditions, where the equilibria $E_{1,2}$ are always stable.

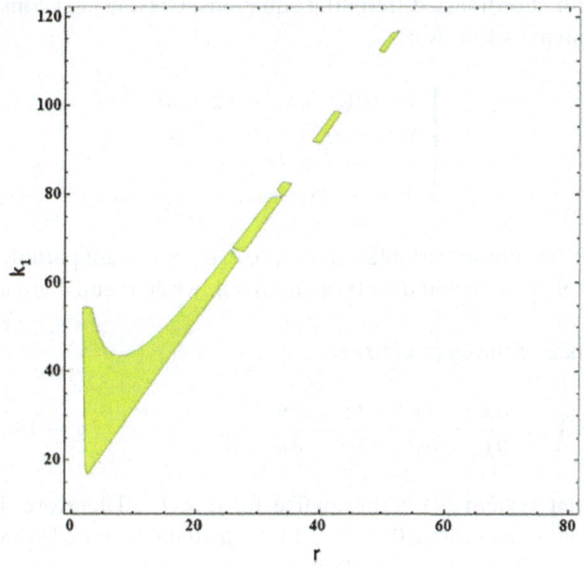

Fig. 10 The equilibria $E_{1,2}$ of system (8) are asymptotically stable in the yellow regions

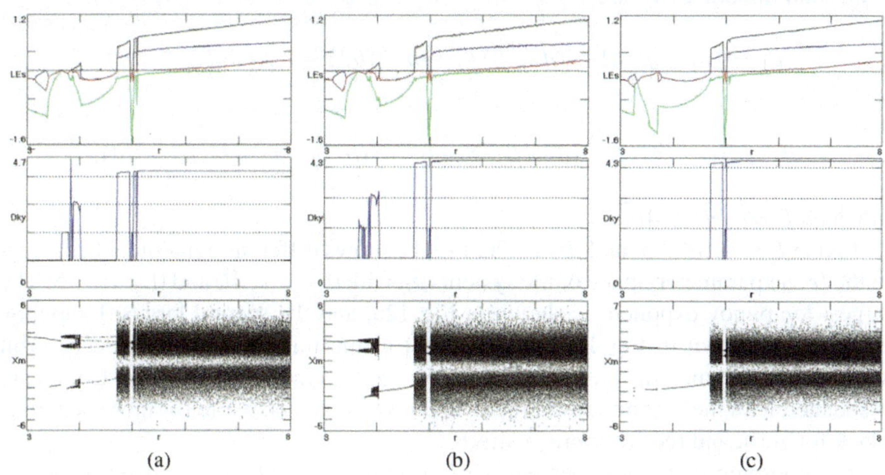

Fig. 11 The first four Lyapunov exponents (the fifth is large and negative), the Kaplan–Yorke dimension and bifurcation diagrams of system (8) versus parameters $m = 0.04$, $g = 140.6$, $k_1 = 34$, $k_2 = 12$: Left: fixed initial condition $(0.05, -0.5, 0.1, -1.2)$; Middle: varying initial condition for increasing r from 3 to 8; Right: varying initial condition for decreasing r from 8 to 3

3.3 Hyperchaotic System with Three Equilibria

By introducing a state feedback control to the first equation of a nonlinear autonomous first-order system of ordinary differential equations, Correia [11] constructed a four-dimensional system as follows:

$$\begin{cases} \dot{x} = a(y - x) - yz + w, \\ \dot{y} = -by + xz, \\ \dot{z} = -cz + dx + xy, \\ \dot{w} = -e(x + y), \end{cases} \tag{9}$$

where x, y, z, w are phase variables, and $a, b, c, d, e > 0$ are parameters. Here, the parameters $b = 9,\ c = 5$, and $d = 0.06$ are fixed, while a and e are simultaneously varied.

The divergence of the system (9) is

$$\nabla V = \frac{\partial \dot{x}}{\partial x} + \frac{\partial \dot{y}}{\partial y} + \frac{\partial \dot{z}}{\partial z} + \frac{\partial \dot{w}}{\partial w} = a - b - c = a - 14,$$

which means that system (9) is dissipative for $a < 14$. Therefore, if one chooses adequately the parameter range $0 < a < 14$, then all the bounded system trajectories finally settle onto an attractor in the four-dimensional phase space.

The system (9) has three equilibria, where obviously the origin $P_0 = (0, 0, 0, 0)$ is one and the other two are

$$P_1 = ((d + \beta)/2, -(d + \beta)/2, -b, \beta(b/2 - a) + bd/2 - ad)$$

and

$$P_2 = ((d - \beta)/2, -(d - \beta)/2, -b, -\beta(b/2 - a) + bd/2 - ad),$$

in which $\beta = \sqrt{d^2 + 4bc}$.

Fixing $b = 9,\ c = 5$, and $d = 0.06$, Fig. 12 shows different dynamical behaviors in the (e, a) parameter-space of the system (9), with $0 < e < 10$ and $0 < a < 5$. The largest Lyapunov exponent is shown in Fig. 12a, and the second largest Lyapunov exponent is shown in Fig. 12b. With respect to the initial conditions, every orbit starts from the same $(x_0, y_0, z_0, w_0) = (5, 5, -5, -5)$ in the phase space. Colors are associated with the magnitudes of the Lyapunov exponents, white for more negative, black for zero, and red for more positive.

The complex dynamical behaviors of system (9) are investigated by means of numerical simulations in [31]. Bifurcation diagrams, Lyapunov exponents, hyper-chaotic attractors, power spectra, and time charts are mapped out through theoretical analysis and numerical simulations.

The chaotic and hyper-chaotic attractors exist and alter over a wide range of parameter values according to the variety of Lyapunov exponents and bifurcation

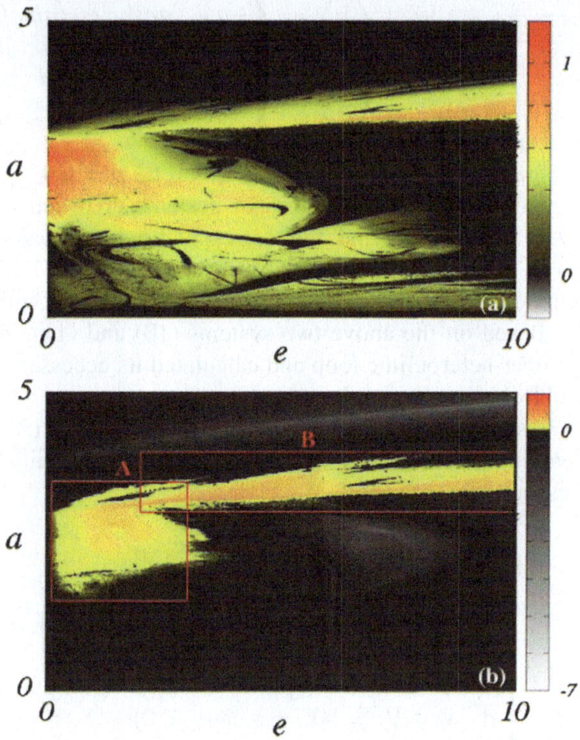

Fig. 12 Regions of different dynamical behaviors in the (e, a) parameter-space of the system (9), with $b = 9,\ c = 5$, and $d = 0.06$. **a** The largest Lyapunov exponent. **b** The second largest Lyapunov exponent. In each plot, colors are associated with the magnitudes of the Lyapunov exponents, as shown in the respective column at right. (For interpretation of the references to colors in this figure legend, the reader is referred to the web version of this article.)

diagrams. Furthermore, linear feedback controllers are designed for stabilizing the hyperchaos to the unstable equilibria; thus, the goal of a second control is achieved, which will be useful in potential applications.

3.4 Hyperchaotic Systems with Limited Number of Equilibria

Based on two basic four-dimensional linear systems, Zhang [12] developed a novel approach for constructing four-dimensional piecewise-linear multi-wing hyperchaotic dynamic systems.

Consider the following two four-dimensional linear systems:

$$\begin{pmatrix} \dot{x} \\ \dot{y} \\ \dot{z} \\ \dot{u} \end{pmatrix} = \begin{pmatrix} -a & ac & \sqrt{abc} & 0 \\ -1 & c & 0 & 1 \\ 0 & 2\sqrt{abc} & -b & 0 \\ e & p & 0 & q \end{pmatrix} \begin{pmatrix} x \\ y \\ z \\ u \end{pmatrix} \tag{10}$$

and

$$\begin{pmatrix} \dot{x} \\ \dot{y} \\ \dot{z} \\ \dot{u} \end{pmatrix} = \begin{pmatrix} -a & ac & -\sqrt{abc} & 0 \\ -1 & c & 0 & 1 \\ 0 & -2\sqrt{abc} & -b & 0 \\ e & p & 0 & q \end{pmatrix} \begin{pmatrix} x \\ y \\ z \\ u \end{pmatrix}, \tag{11}$$

where $a = 20,\ b = 5,\ c = 10,\ e = 0.6,\ p = 0.4,\ q = 1.2$.

Notice that the equilibria of systems (10) and (11) are $O_1 = (0, 0, 0, 0)$ and $O_2 = (0, 0, 0, 0)$, and the corresponding eigenvalues at O_1 and O_2 are same, as $\gamma_1 = -18.2194$, $\gamma_2 = 2.1701$, and $\sigma \pm j\omega = 1.1246 \pm j10.2943$, respectively. Obviously, O_1 and O_2 are saddle-focus equilibria with index-2.

Based on the above two systems (10) and (11), Zhang [12] furthermore found a super-heteroclinic loop and calculated its necessary parameters using a switching control strategy.

Assuming that the switch hyperplane is $S = \{(x, y, z, u)|y = 0\}$, and by using switching controller $F(x, y, z, u)$ to shift the equilibria of systems (10) and (11), respectively, one gets

$$\begin{cases} \text{if} \quad V \in V_1 = \{(x, y, z, u)|y > 0\} \\ \begin{pmatrix} \dot{x} \\ \dot{y} \\ \dot{z} \\ \dot{u} \end{pmatrix} = \begin{pmatrix} -a & ac & \sqrt{abc} & 0 \\ -1 & c & 0 & 1 \\ 0 & 2\sqrt{abc} & -b & 0 \\ e & p & 0 & q \end{pmatrix} \left(\begin{pmatrix} x \\ y \\ z \\ u \end{pmatrix} - F(x, y, z, u) \right) \\ \text{if} \quad V \in V_2 = \{(x, y, z, u)|y < 0\} \\ \begin{pmatrix} \dot{x} \\ \dot{y} \\ \dot{z} \\ \dot{u} \end{pmatrix} = \begin{pmatrix} -a & ac & -\sqrt{abc} & 0 \\ -1 & c & 0 & 1 \\ 0 & -2\sqrt{abc} & -b & 0 \\ e & p & 0 & q \end{pmatrix} \left(\begin{pmatrix} x \\ y \\ z \\ u \end{pmatrix} - F(x, y, z, u) \right), \end{cases} \tag{12}$$

where the switching controller is

$$F(x, y, z, u) = (f_1(x, y, z, u), f_2(x, y, z, u), f_3(x, y, z, u), f_4(x, y, z, u))^T,$$

and its expressions are determined by the conditions in forming the super-heteroclinic loop.

Under the control of $F(x, y, z, u)$, the equilibria of system (12) become $P_1(x_1, y_1, z_1, u_1) \in V_1$ and $P_2(x_2, y_2, z_2, u_2) \in V_2$, which are located at different sides of the hyperplane $S = \{(x, y, z, u)|y = 0\}$. The characteristic lines of 1D stable manifolds $E^S(P_1)$ and $E^S(P_2)$, and the characteristic planes of 2D unstable manifolds $E^U(P_1)$ and $E^U(P_2)$, are shown in Fig. 13.

Note that the switch hyperplane is $S = \{(x, y, z, u)|y = 0\}$, so the function "sgn(y)" can be introduced to system (12). Then, the following system is obtained:

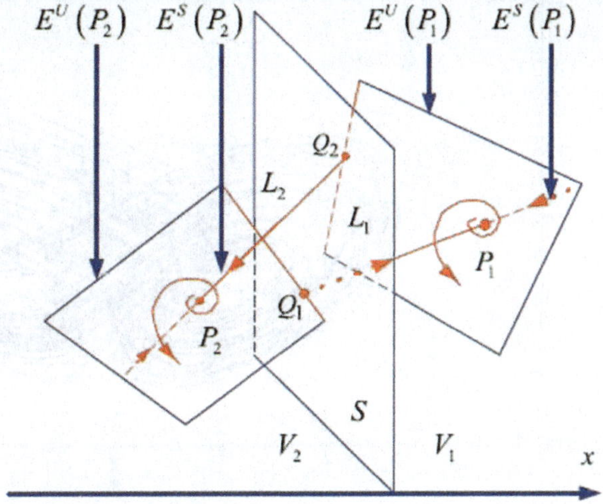

Fig. 13 Heteroclinic loop connecting two equilibria in the double-wing hyperchaotic system

$$\begin{pmatrix} \dot{x} \\ \dot{y} \\ \dot{z} \\ \dot{u} \end{pmatrix} = \begin{pmatrix} -a & ac & sgn(y)\sqrt{abc} & 0 \\ -1 & c & 0 & 1 \\ 0 & 2sgn(y)\sqrt{abc} & -b & 0 \\ e & p & 0 & q \end{pmatrix} \left(\begin{pmatrix} x \\ y \\ z \\ u \end{pmatrix} - F(x, y, z, u) \right). \quad (13)$$

System (13) is called a two-piecewise-linear hyperchaotic system, where $a = 20$, $b = 5$, $c = 10$, $e = 0.6$, $p = 0.4$, $q = 1.2$ are fixed parameters, and the controller is $F(x, y, z, u) = (x_0 sgn(y), y_0 sgn(y), 0, 0)$, with $x_0 = 10.1023$, $y_0 = 1$. From system (13), a double-wing hyperchaotic attractor can be obtained, as shown in Fig. 14.

Based on system (13), by shifting the transformation of two four-dimensional linear systems, a four-dimensional multi-piecewise-linear hyperchaotic system and a multi-wing hyperchaotic attractor can be generated. It should be noted that each shifting linear system has unique corresponding equilibria $P_{mn}(m, n = \pm1, \pm2, \cdot)$. Based on the system (13), Zhang [12] further obtained the characteristic lines of 1D stable manifolds $E^S(P_{mn})$ and the characteristic planes of 2D unstable manifolds $E^U(P_{mn})$ of the four-dimensional multi-piecewise-linear hyperchaotic system, as shown in Fig. 15.

Based on the above theory and analysis, in order to generate a multi-wing hyperchaotic attractor, the controller $F(x, y, z, u)$ in system (13) can be selected as

$$F(x, y, z, u) = \begin{pmatrix} x_0 sgn(y) + \Sigma_{m=1}^{M} x_0(sgn(y + 2my_0) + sgn(y - 2my_0)), \\ y_0 sgn(y) + \Sigma_{m=1}^{M} y_0(sgn(y + 2my_0) + sgn(y - 2my_0)), \\ 0 \\ 0 \end{pmatrix} \quad (14)$$

where $x_0 = 10.1023$, $y_0 = 1$, and M is a positive integer.

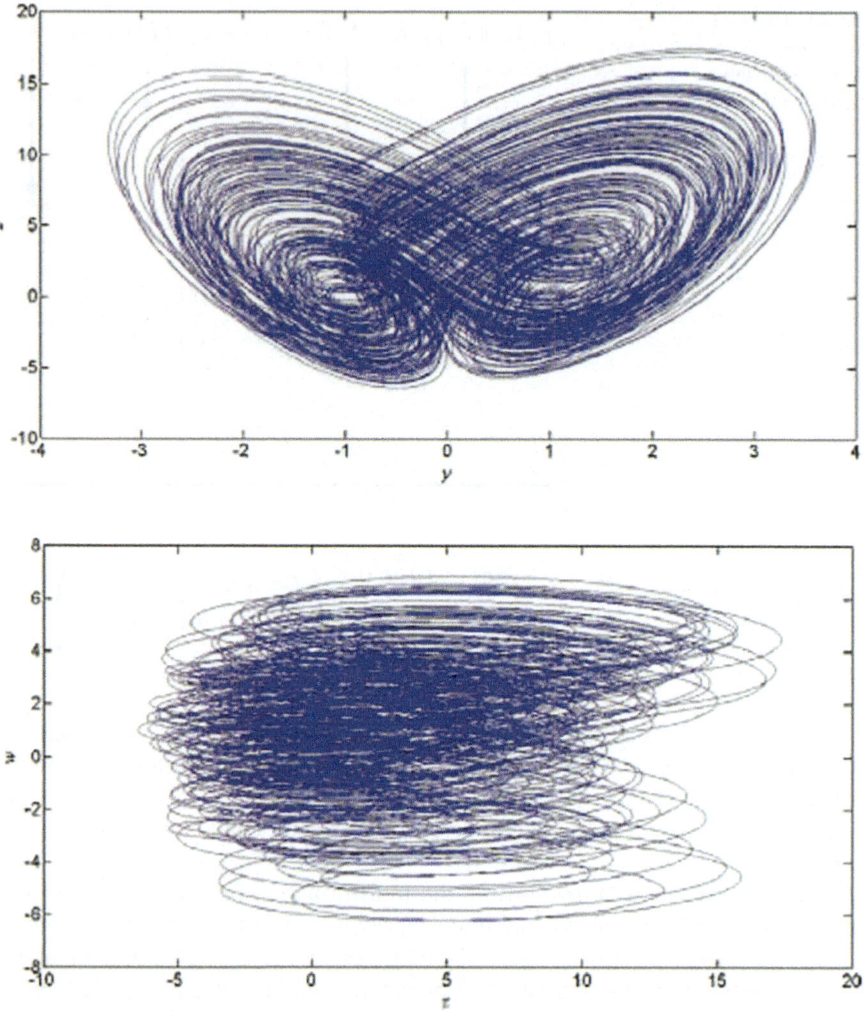

Fig. 14 Double-wing hyperchaotic attractor

According to (13) and (14), a $(2M + 2)$-wing hyperchaotic attractor can be generated. For example, with $x_0 = 10.1023$, $y_0 = 1$ and $M = 5$, a 12-wing hyperchaotic attractor is generated, as shown in Fig. 16.

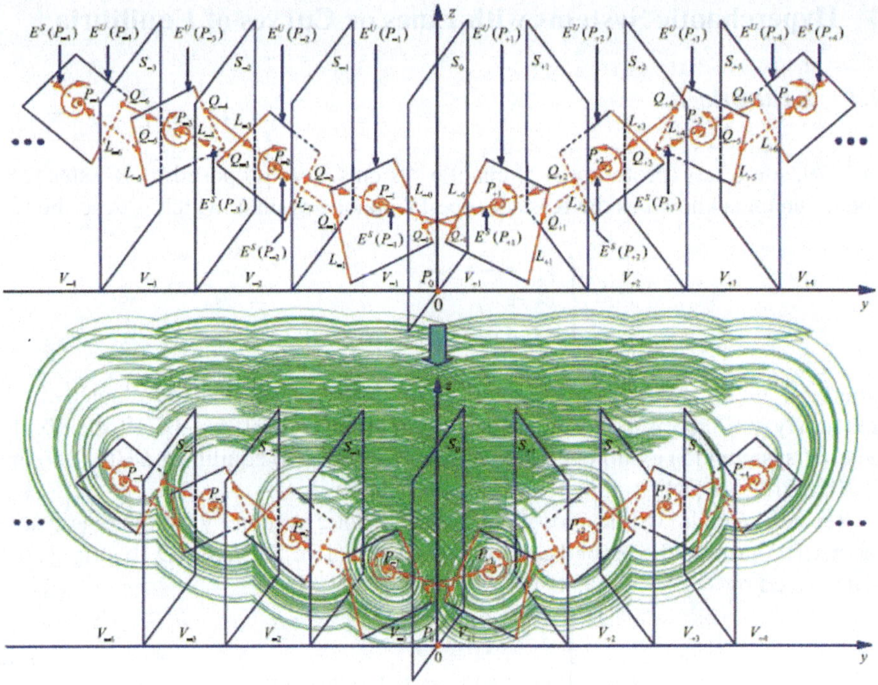

Fig. 15 Multiple heteroclinic loops in the multi-piecewise-linear hyperchaotic system

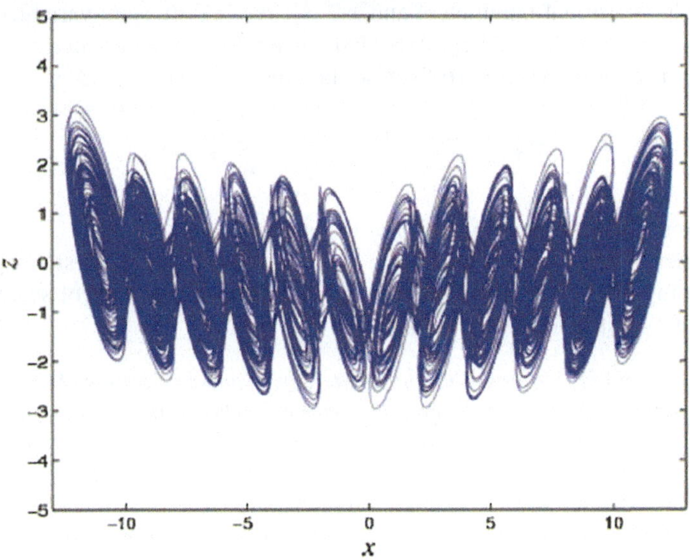

Fig. 16 A 12-wing hyperchaotic attractor

4 Hyperchaotic Systems with Lines or Curves of Equilibria

4.1 Example 1

In 1979, some sets (prototypes) of chaotic systems were proposed by Rössler [21], one of which is known as the Rössler type-IV chaotic system, which is described by

$$\begin{cases} \dot{x_1} = -x_2 - x_3, \\ \dot{x_2} = x_1, \\ \dot{x_3} = a(x_2 - x_2^2) - bx_3, \end{cases} \tag{15}$$

where $x_1,\ x_2,\ x_3$ are the state variables, and $a,\ b$ are real parameters. When parameters satisfy $a = 0.386,\ b = 0.2$, and initial condition $(0.7, -0.6, -0.7)^T$, the Rössler type-IV system (15) exhibits chaotic behaviors, with two equilibria $E_1(0, 0, 0)$ and $E_2(0, 1.518, -1.581)$.

Recently, Singh [15] developed a new 4D chaotic system by introducing a state feedback control to the second equation of system (15). The new 4D chaotic system is described by

$$\begin{cases} \dot{x_1} = -x_2 - x_3, \\ \dot{x_2} = x_1 - x_4, \\ \dot{x_3} = a(x_2 - x_2^2) - bx_3, \\ \dot{x_4} = cx_2, \end{cases} \tag{16}$$

where $a,\ b,\ c$ are real parameters and $x_1,\ x_2,\ x_3,\ x_4$ are state variables. In [15], parameters $a = b = 0.5$ are kept fixed, but c is a bifurcation parameter.

The divergence of system (16) is described by

$$\nabla V = \frac{\partial \dot{x}}{\partial x} + \frac{\partial \dot{y}}{\partial y} + \frac{\partial \dot{z}}{\partial z} + \frac{\partial \dot{w}}{\partial w} = -b\,.$$

Thus, system (16) is dissipative for $b > 0$.

Considering parameter $b = 0.5$, the system (16) has a rate of state-space contraction being $-1/2$. The boundedness of the chaotic trajectories of system (16) is guaranteed by in the following theorem.

Theorem 8.2 ([15]) *Suppose that parameters $a,\ b$ and c of system (16) are positive. Then, the orbits of system (16), including chaotic orbits, are confined in a bounded region.*

Under any coordinates, with any plane and space transformations, system (16) is not invariant, thus this system is asymmetrical about its coordinates, planes and spaces.

Let the state equation of (16) be zero. Then, it is not hard to find that $E = (x_1^*, 0, 0, x_1^*)$ is an equilibrium of system (16) for any x_1^*, namely, system (16) has a line of equilibria.

Sl. no.	Eigenvalues	Eigenvectors
1	$\lambda_1 = 0.1509 + 1.1105i$	$(0.7186, 0.0892 - 0.6420i, -0.1976 - 0.1560i, -0.0078 - 0.0022i)$
2	$\lambda_2 = 0.1509 - 1.1105i$	$(0.7186, 0.0892 + 0.6420i, -0.1976 + 0.1560i, -0.0078 + 0.0022i)$
3	$\lambda_3 = -0.8018$	$(-0.3898, -0.4758, 0.7884, 0.0083)$
4	$\lambda_4 = 0$	$(0.7071, 0, 0, 0.7071)$

Fig. 17 The parameters and their corresponding eigenvectors of the new system with $c = 0.014$ at equilibria E

Sl. no.	Property	System (1)	System (2)
1	Divergence	$\nabla v < 0$	$\nabla v < 0$
2	Symmetrical	Asymmetry	Asymmetry
3	Equilibrium point	Two	Infinite number (line of equilibria)
4	Dynamical behaviour	Chaotic, periodic, quasi-periodic	Chaotic, chaotic 2- torus, periodic, quasi-periodic
5	Types of attractors	Self-excited attractors	Hidden attractors
6	Lyapunov dimension (D_{KY})	$2 < D_{KY} < 2.1$	$3 < D_{KY} < 3.2$

Fig. 18 Comparison between the Rössler type-IV system and the new system

The Jacobian matrix of system (16) at equilibria E is

$$J = \begin{pmatrix} 0 & -1 & -1 & 0 \\ 1 & 0 & 0 & -1 \\ 0 & a & -b & 0 \\ 0 & c & 0 & 0 \end{pmatrix}. \tag{17}$$

It is noted that this Jacobian matrix is irrelevant to the constant x_1^*. The characteristic equation of (17) is

$$\lambda(\lambda^3 + k_1\lambda^2 + k_2\lambda + k_3) = 0,$$

where $k_1 = b,\ k_2 = c,\ k_3 = bc$. It is obvious that there is a zero eigenvalue for all equilibria of system (16). When parameters $a = b = 0.5,\ c = 0.014$, the eigenvalues and their corresponding eigenvectors of the system can be obtained, as shown in Fig. 17.

A comparison between system (16) and (15) is given in Fig. 18.

The self-excited chaotic or hyperchaotic system with countable equilibria can be found numerically based on the knowledge of the location of the equilibria. However, using the above method, it is difficult to locate the attractors of the system that has a line of equilibria.

Let the parameter c be varied, while keeping other parameters fixed. The bifurcation diagrams of system (16) are shown in Figs. 19 and 20, respectively, for $c \in [0.0, 0.0629]$ and $c \in [0.063, 0.26]$. The corresponding Lyapunov spectra are as shown in Figs. 21 and 22, respectively, for $c \in [0.0, 0.0629]$ and $c \in [0.063, 0.26]$.

According to the Lyapunov spectra, shown in Figs. 21 and 22, it is obvious that system (16) has chaotic, chaotic 2-torus and quasi-periodic (including 2-torus and 3-torus) behaviors.

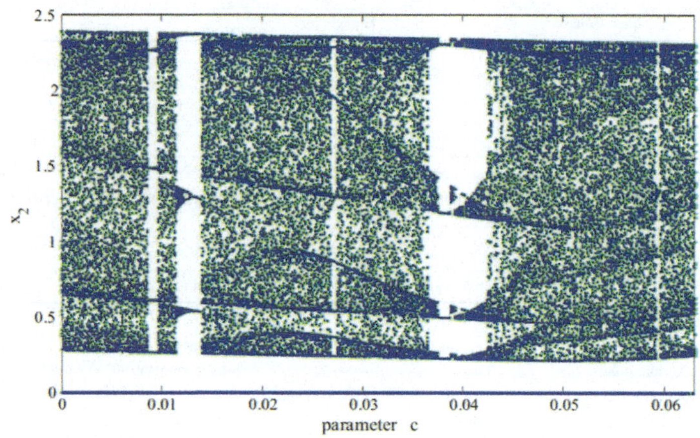

Fig. 19 Bifurcation diagram of system (16), with $c \in [0.0, 0.0629]$ and $a = b = 0.5$

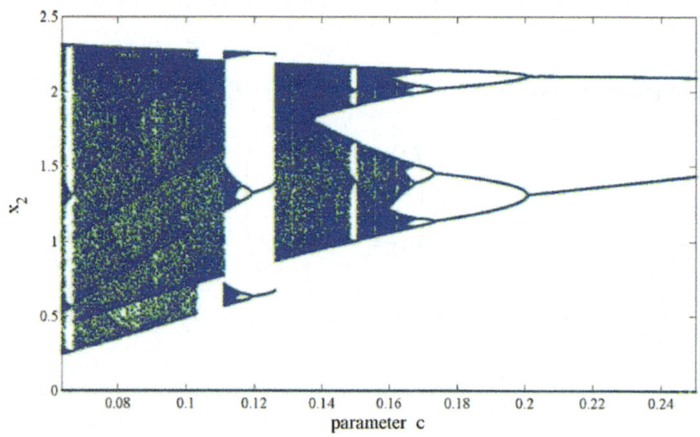

Fig. 20 Bifurcation diagram of system (16), with $c \in [0.063, 0.25]$ and $a = b = 0.5$

In [15], the $0 - 1$ test method is used to classify chaotic or periodic behaviors of system (16). The chaotic and periodic behaviors of system (16) are defined based on the values of the asymptotic growth rate (k_c) of the mean square displacement of the trajectories.

The translation variables are defined as

$$\begin{cases} p_c(n) = \Sigma_{j=1}^{n} x(j) cos(jc), \\ q_c(n) = \Sigma_{j=1}^{n} x(j) sin(jc), \end{cases}$$

where $c \in (0, 2\pi)$ is a variable constant, which is selected randomly, and $x(j)$ is the time series of any state variable. If the space of translation variables has a random Brownian-like motion, then it means that the selected trajectory belongs to a chaotic

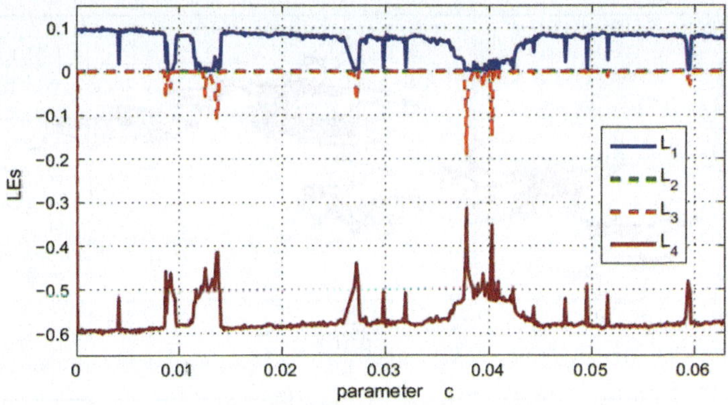

Fig. 21 Lyapunov spectrum of system (16), with $c \in [0.0, 0.0629]$ and $x(0) = (0.01, 0.001, 0.001, 0.1)^T$

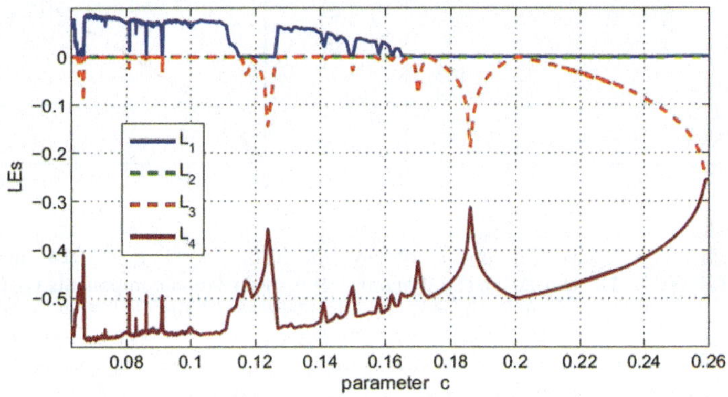

Fig. 22 Lyapunov spectrum of system (16), with $c \in [0.063, 0.25]$ and $x(0) = (0.01, 0.001, 0.001, 0.1)^T$

attractor. Whereas if the space of translation variables has a regular motion, then the trajectory belongs to a regular attractor.

Based on the translation variables $(p_c(n), q_c(n))$, the mean square displacement $M_c(n)$ can be defined as

$$M_c(n) = lim_{n\to\infty}\frac{1}{n}\Sigma_{j=1}^{n}\{[p_c(j+n) - p_c(j)]^2 + [q_c(j+n) - q_c(j)]^2\}.$$

Clearly, $M_c(n)$ grows exponentially for chaotic motions, whereas it varies periodically for regular and periodic behaviors.

Furthermore, based on the mean square displacement $M_c(n)$, the asymptotic growth rate (k_c) can be defined as

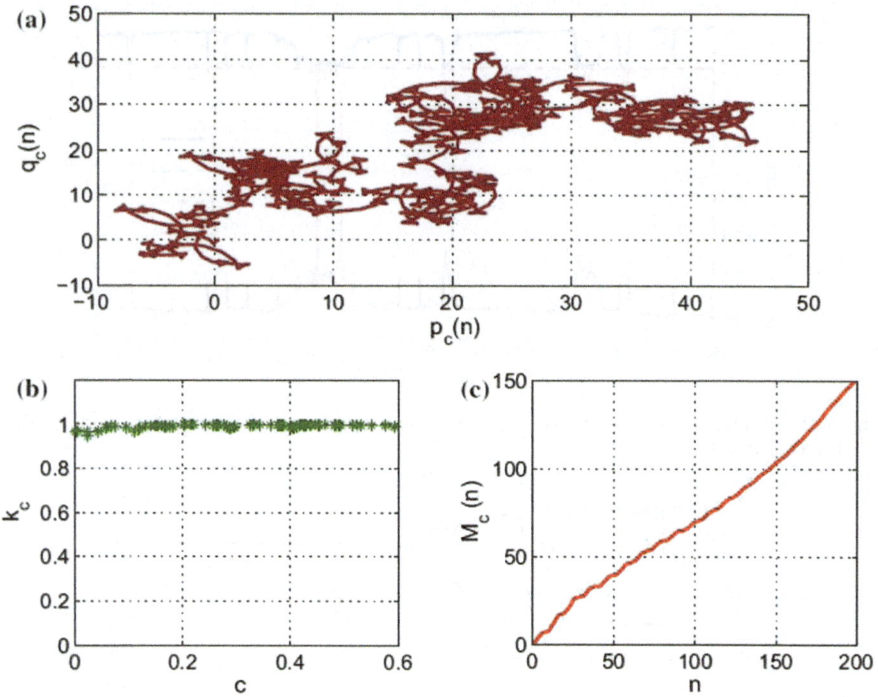

Fig. 23 The $0-1$ test on $x_2(t)$ of system (16), with $c = 0.014$ and $x(0) = (0.01, 0.001, 0.001, 0.1)^T$ and $\Delta t = 1$: **a** dynamics of the translation components $(p_c(n), q_c(n))$, **b** asymptotic growth rate (k_c), and **c** mean square displacement $M_c(n)$

$$k_c = lim_{n\to\infty}\frac{log M_c(n)}{logn}.$$

The value of $k_c \approx 1$ determines the chaotic behavior, and $k_c \approx 0$ indicates the periodic or regular behaviors.

Here, the $x_2(t)$ signal of system (16) is used for the $0-1$ test. The translation variables, asymptotic growth rate and mean square displacement are shown in Fig. 23. It was found that $k_c = 0.99691 \approx 1$ for $a = b = 0.5,\ c = 0.014$, which confirms the chaotic nature of the new system.

4.2 *Example 2*

By adding a controller to the classical Lorenz system, Chen [16] obtains a new 4-D Lorenz-type hyperchaotic system with a curve of equilibria. This new 4-D autonomous system can display periodic, quasi-periodic, chaotic and hyperchaotic dynamic behaviors, and the singular degenerate heteroclinic cycles.

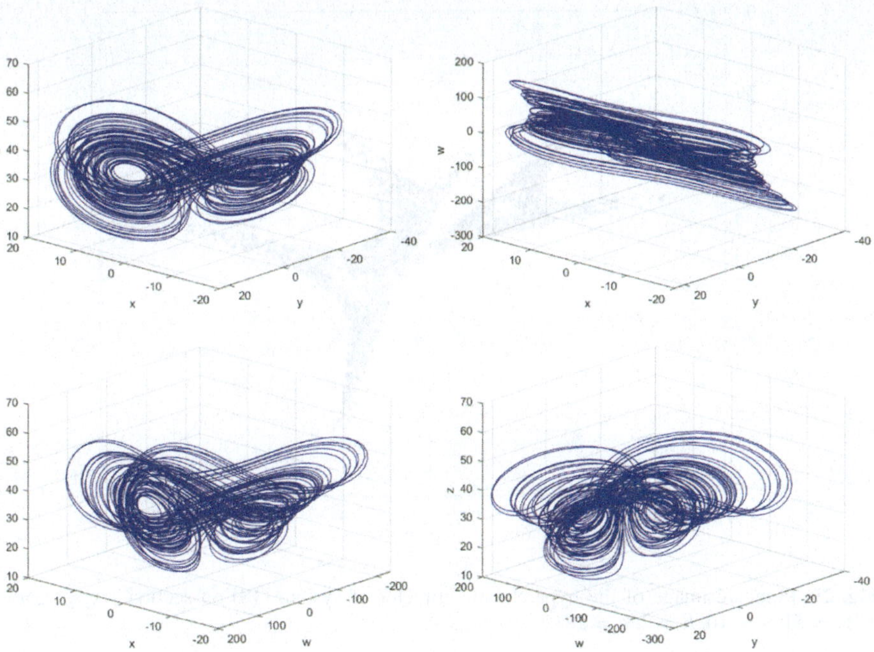

Fig. 24 Three-dimensional projections of the hyperchaotic attractor of system (18): $a = 10, b = 2, c = 28$

This system is described by

$$\begin{cases} \dot{x} = a(y - x), \\ \dot{y} = cx - y - xz + w, \\ \dot{z} = -bz + xy, \\ \dot{w} = (c - 1)y + w - \frac{x^3}{b}. \end{cases} \tag{18}$$

Let parameters be $a = 10$, $b = 2$ and $c = 28$, and choose initial condition $(1, 0, 0, 0)$, system (18) can display hyperchaotic dynamic behavior, as shown in Figs. 24 and 25. The Lyapunov exponents of its corresponding hyperchaotic attractor are

$$\lambda_{LE_1} = 0.5697, \quad \lambda_{LE_2} = 0.0453, \quad \lambda_{LE_3} = 0.0005, \quad \lambda_{LE_4} = -12.6078.$$

Thus, the Lyapunov dimension of this hyperchaotic attractor is $D_L = 3.0488$.

The divergence of this 4D system (18) is

$$\nabla V = \frac{\partial \dot{x}}{\partial x} + \frac{\partial \dot{y}}{\partial y} + \frac{\partial \dot{z}}{\partial z} + \frac{\partial \dot{w}}{\partial w} = -(a + b),$$

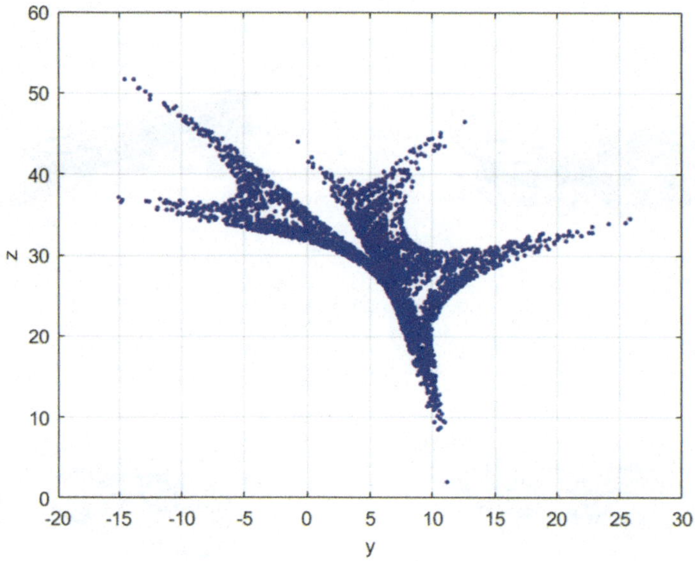

Fig. 25 Poincaré image of the hyperchaotic attractor of system (18) on section $\{(x, y, z, w) \in \mathbb{R}^4 | x = 5\}$: $a = 10, b = 2, c = 28$

thus system (18) is dissipative when $a + b > 0$.

Under the transformation $(x, y, z, w) \to (-x, -y, z, -w)$, system (18) is invariant. Thus, system (18) is symmetrical about the z-axis, and the z-axis is an orbit of the system.

It is worth noting that system (18) has a curve of equilibria:

$$\left\{(x, y, z, w) \in \mathbb{R}^4 \,\middle|\, y = x,\ z = \frac{x^2}{b},\ w = \frac{x^3}{b} - (c-1)x \right\}. \tag{19}$$

Let $E = (x^*, x^*, x^{*2}/b, x^{*3}/b - (c-1)x^*)$, $x^* \in \mathbb{R}$, which belongs to the curve of equilibria (19). The Jacobian matrix of system (18) at equilibrium E is

$$A = \begin{pmatrix} -a & a & 0 & 0 \\ c - \frac{x^{*2}}{b} & -1 & -x^* & 1 \\ x^* & x^* & -b & 0 \\ -\frac{3x^{*2}}{b} & c-1 & 0 & 1 \end{pmatrix}.$$

The corresponding characteristic equation is

$$\lambda \left(\lambda^3 + c_1\lambda^2 + c_2\lambda + c_3\right) = 0,$$

where

$$\begin{aligned} c_1 &= a + b, \\ c_2 &= ab - c - ac + x^{*2} + \tfrac{ax^{*2}}{b}, \\ c_3 &= -bc - abc - x^{*2} + 3ax^{*2} + \tfrac{2ax^{*2}}{b}. \end{aligned}$$

There is always one zero eigenvalue for any equilibrium E, and its corresponding eigenvector is

$$\left(1, 1, \frac{2x^*}{b}, \frac{3x^{*2}}{b} - (c-1)\right).$$

This vector always tangents to the curve of equilibria (19) at any equilibrium E.

The local stability of equilibrium E is guaranteed by the following theorem.

Theorem 8.3 ([16]) *If the following conditions are satisfied:*

$$\begin{cases} a + b > 0, \\ b\left(b(1+b)x^{*2} + a^2\left(b^2 - bc + x^{*2}\right) + a\left(b^3 - 2x^{*2} - b\left(c + x^{*2}\right)\right)\right) > 0, \\ -b\left((1+a)b^2c - 2ax^{*2} + (1-3a)bx^{*2}\right) > 0, \end{cases}$$

then the equilibrium E has a three-dimensional stable manifold. Otherwise, equilibrium E has other eigenvalues with zero real part, or has an unstable manifold.

According to the characteristic equation, when fixing parameters as $a = 14, b = 2$ and $c = 28$, expect the inherent one zero eigenvalue, the features of other eigenvalues of equilibrium E are as follows:

(1) Equilibrium E has one positive and two negative real eigenvalues for $|x^*| < 2\sqrt{\frac{42}{11}}$;

(2) Equilibrium E has one negative and two positive real eigenvalues for $2\sqrt{\frac{42}{11}} < |x^*| < 4.9047$;

(3) Equilibrium E has one negative real eigenvalue and a pair of conjugate complex eigenvalues with positive real part for $4.9047 < |x^*| < 2\sqrt{\frac{1358}{73}}$;

(4) Equilibrium E and has one negative real eigenvalue and a pair of conjugate complex eigenvalues with negative real part for $|x^*| > 2\sqrt{\frac{1358}{73}}$.

Based on detailed numerical studies of system (18) with parameters $a = 14$, $b = 2$ and $c = 28$, an infinite set of singular degenerate heteroclinic cycles can be observed, each one of which connects an equilibrium E_1 with $|x^*| < 4.9047$ to a focus equilibrium E_2 with $|x^*| > 2\sqrt{\frac{1358}{73}}$. Choosing initial conditions $(x^* - 0.00001, x^*, x^{*2}/b, x^{*3}/b - (c-1)x^*)$ with $x^* \in \{0, 1, 2, 3, 4\}$, the corresponding trajectories of system (18) are shown in Fig. 26a; choosing initial conditions $(x^* + 0.00001, x^*, x^{*2}/b, x^{*3}/b - (c-1)x^*)$ with $x^* \in \{0, -1, -2, -3, -4\}$, the corresponding trajectories of system (18) are shown in Fig. 26b.

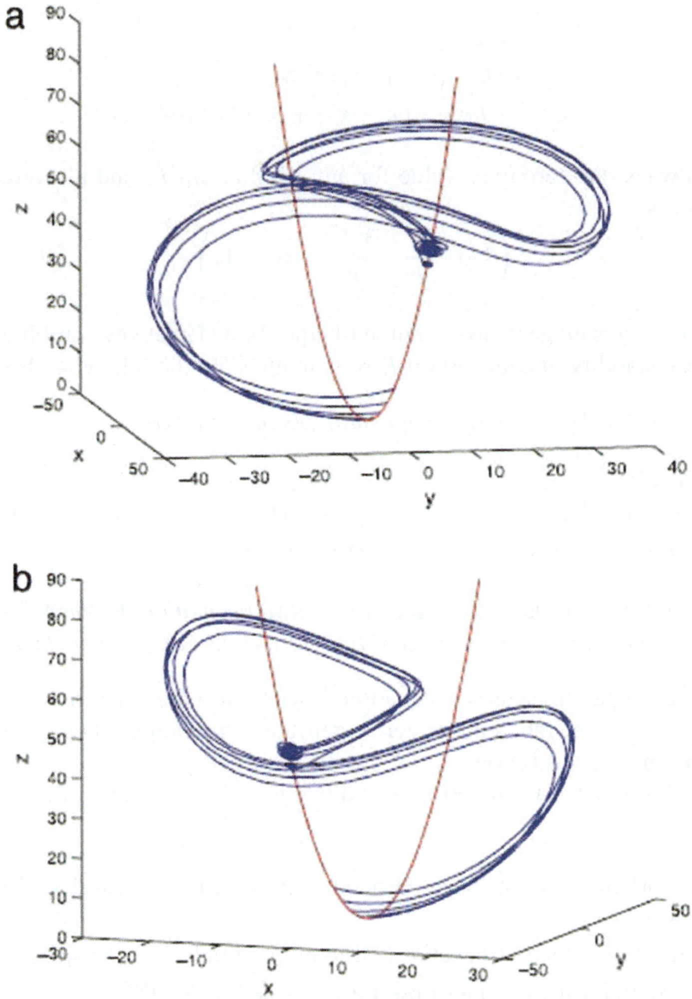

Fig. 26 The singular degenerate heteroclinic cycles of system (18): $a = 14$, $b = 2$, $c = 28$

Let parameters $b = 2, c = 28$ and initial condition $(1, 0, 0, 0)$ be fixed, and choose parameter $a \in [4, 15]$. Then, the Lyapunov exponent spectrum and bifurcation diagram of system (18) are shown in Figs. 27 and 28, where the Poincaré section is chosen at $x = 5$.

It should be noted that, under the parameter condition $b = 2$, $c = 28$ and initial condition $(1, 0, 0, 0)$, the trajectory of system (18) will tend to infinity for $a \in (13.76, 15]$.

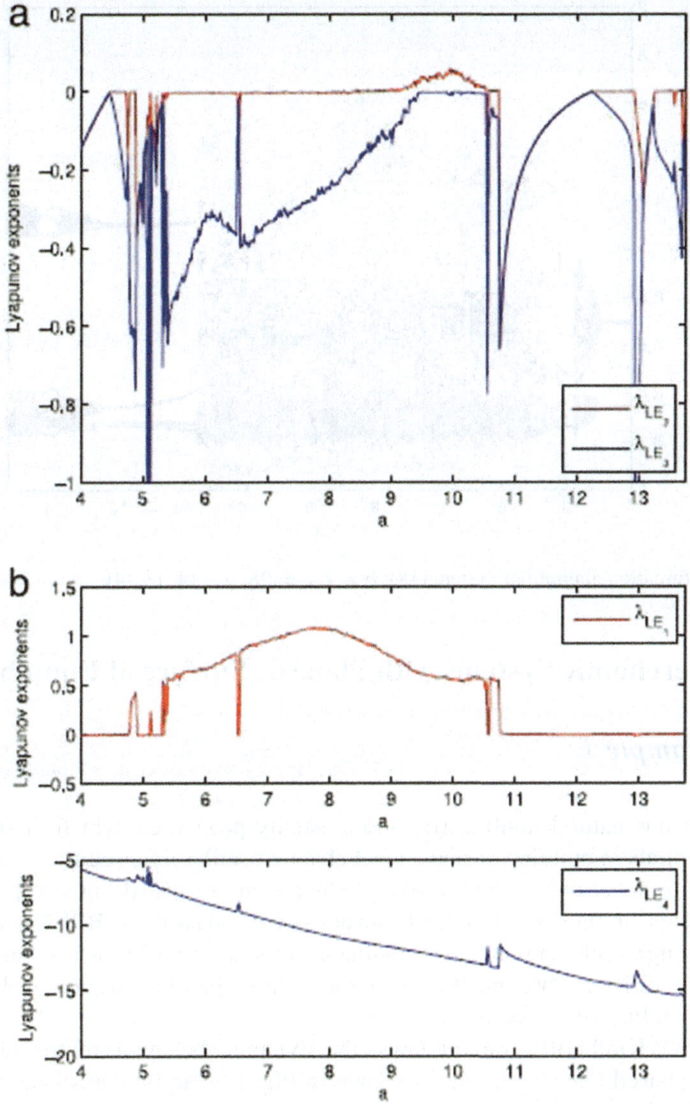

Fig. 27 Lyapunov exponent spectrum of system (18): $b = 2$, $c = 28$, $a \in [4, 13.74]$

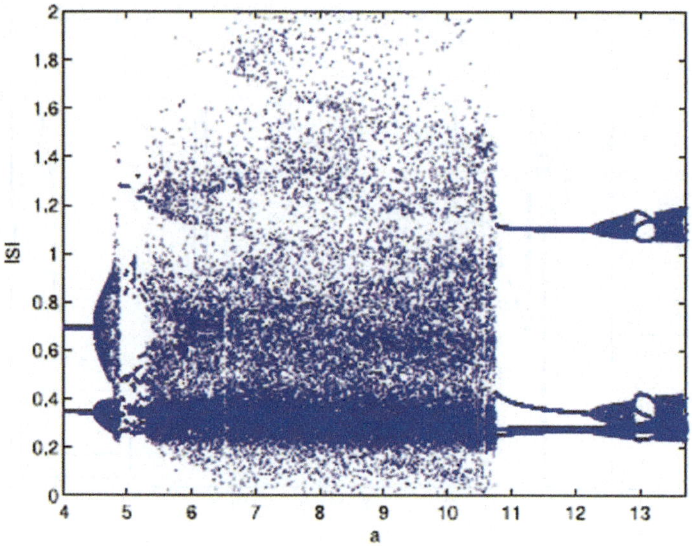

Fig. 28 Bifurcation diagram of system (18): $b = 2, c = 28, a \in [4, 13.74]$

5 Hyperchaotic Systems with Plane or Surface of Equilibria

5.1 *Example 1*

Memristor has natural nonlinearity and plasticity properties, which marks it have great potentials in building various novel chaotic oscillatory circuits.

In [19], a novel two-memristor-based Chua's circuit was designed, as shown in Fig. 29a. This circuit is synthesized by an active band-pass filter (BPF)-based Chua's circuit through replacing an active nonlinear resistor and a linear resistor with two different memristors, W_4 and W_5. The equivalent realization circuits of W_4 and W_5 are shown in Fig. 29b, c, respectively.

Applying Kirchhoff's circuit laws, the dynamic behavior of the novel two-memristor-based Chua's circuit, as shown in Fig. 29, can be described by the following equations:

$$\begin{cases} \frac{dV_1}{dt} = -\frac{1}{RC_1}(V_1 - V_2) + \frac{1}{R_bC_1}\left(1 - g_1V_4^2\right)V_1, \\ \frac{dV_2}{dt} = -\frac{1}{RC_2}(V_1 - V_2) + \frac{1}{R_dC_2}\left(1 - g_2V_5^2\right)V_2 - \frac{2k+1}{(k+1)R_1C_2}V_3, \\ \frac{dV_3}{dt} = -\frac{k+1}{RC_3}(V_1 - V_2) + \frac{k+1}{R_dC_3}\left(1 - g_2V_5^2\right)V_2 - \frac{2}{R_1C_3}V_3, \\ \frac{dV_4}{dt} = -\frac{1}{R_aC_4}V_1, \\ \frac{dV_5}{dt} = -\frac{1}{R_cC_5}V_2, \end{cases} \tag{20}$$

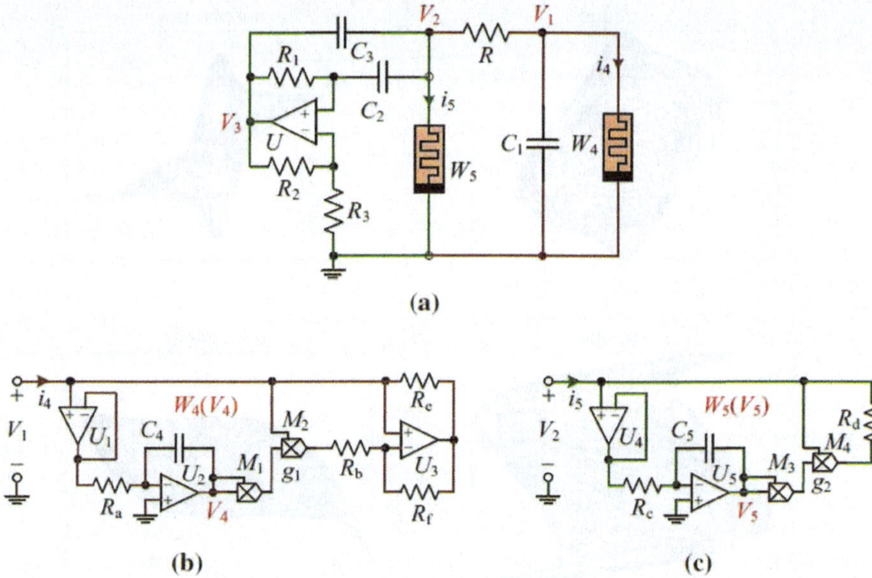

Fig. 29 Two-memristor-based Chua's circuit: **a** circuit schematic of active BPF-based memristive Chua's circuit, **b** equivalent circuit for the memristor W_4, **c** equivalent circuit for the memristor W_5

where V_1, V_2, V_3, V_4, and V_5 are five node voltages, C_1, C_2, and C_3 are three capacitors, W_4 and W_5 are two memristors, and $k = R3/R2$.

Under the following transformation:

$$x = V_1, y = V_2, z = V_3, w = V_4, u = V_5, C_2 = C_3 = C, R_a C_4 = R_c C_5,$$
$$\tau = \frac{t}{RC}, a = \frac{C}{C_1}, b = \frac{RC}{R_b C_1}, c = \frac{R}{R_d}, d = \frac{R}{R_1}, e = \frac{RC}{R_a C_4},$$

system (20) can be rewritten as

$$\begin{cases} \dot{x} = -a(x-y) + b(1 - g_1 w^2)x, \\ \dot{y} = -k(x-y) + kc(1 - g_2 u^2)y - \frac{2k+1}{k+1} dz, \\ \dot{z} = -(k+1)(x-y) + (k+1)c(1 - g_2 u^2)y - 2dz, \\ \dot{w} = -ex, \\ \dot{u} = -ey, \end{cases} \tag{21}$$

where $a, b, c, d, e, g_1, g_2, k$ are eight system parameters in a dimensionless form.

System (21) also describes the dynamic behaviors of the novel two-memristor-based Chua's hyperchaotic circuit, as shown in Fig. 29.

When the parameters of system (21) are $a = 20$, $b = 150/7$, $c = 15$, $d = 0.15$, $e = 3$, $k = 0.05$, $g_1 = g_2 = 0.1$, with initial condition (10.9, 0, 0, 0, 0), system (21) displays hyperchaotic dynamics. The corresponding Lyapunov exponents

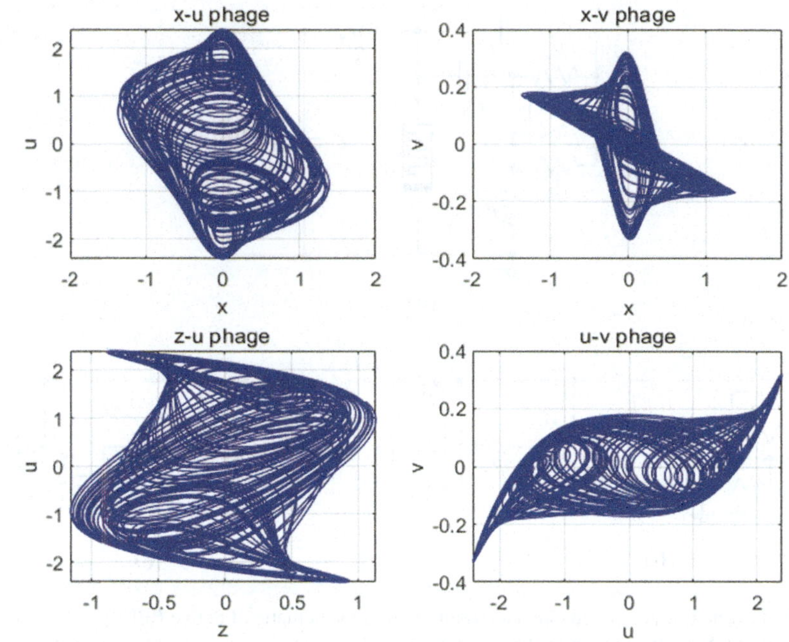

Fig. 30 Numerically simulated phase portraits of typical hyperchaotic attractors in two different planes: **a** phase portrait on the x-y plane and **b** phase portrait on the w-y plane

are $L_1 = 0.2341$, $L_2 = 0.0104$, $L_3 = 0$, $L_4 = -0.0024$, and $L_5 = -4.1459$. The phase portrait of this double-scroll hyperchaotic attractor is shown in Fig. 30.

It is obvious that system (21) has a plane of equilibria, represented by

$$P = \{(x, y, z, w, u)|x = y = z = 0, w = \mu, u = \eta\},$$

where μ, η are arbitrary constants. The Jacobian matrix of system (21) at the equilibrium P is

$$\begin{pmatrix} -a + h_1 & a & 0 & 0\,0 \\ -k & k + kh_2 & -\frac{2k+1}{k+1}d & 0\,0 \\ -k - 1 & (k+1)(1+h_2) & -2d & 0\,0 \\ -e & 0 & 0 & 0\,0 \\ 0 & -e & 0 & 0\,0 \end{pmatrix}, \tag{22}$$

where $h_1 = b(1 - g_1\mu^2)$, $h_2 = c(1 - g_2\eta^2)$. The corresponding characteristic equation is

$$F(\lambda) = \lambda^2(\lambda^3 + m_1\lambda^2 + m_2\lambda + m_3), \tag{23}$$

where

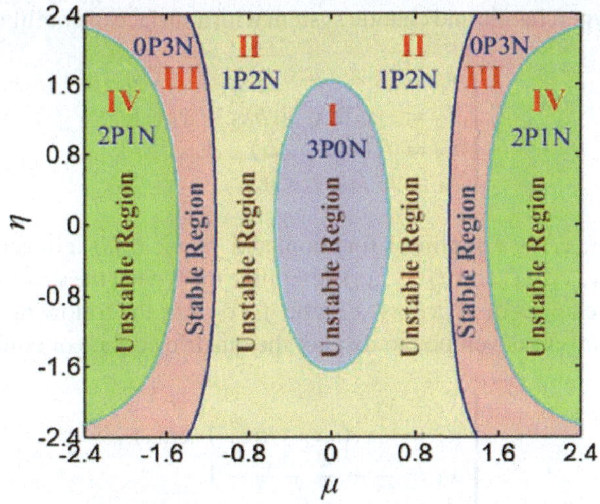

Fig. 31 Stability regions corresponding to the three nonzero eigenvalues distributed in the μ-η plane, with $a = 20,\ b = 150/7,\ c = 15,\ d = 0.15,\ e = 3,\ k = 0.05$, and $g_1 = g_2 = 0.1$

$$
\begin{aligned}
m_1 &= 2d - k + a - h_1 - kh_2, \\
m_2 &= 2ad + d + (k - 2d)h_1 + (d - ka + kh_1)h_2, \\
m_3 &= -dh_1 + dh_2(a - h_1).
\end{aligned}
$$

According to the characteristic equation (23), it is obvious that the Jacobian matrix (22) has two inherent zero eigenvalues. Based on the Routh–Hurwitz principle, the sign of other three eigenvalues of the Jacobian matrix (22) can be studied. If

$$
m_1 > 0, m_3 > 0, m_1 m_2 - m_3 > 0\,, \tag{24}
$$

then the equilibrium P is stable; whereas if any one of the conditions (24) is not satisfied, then the equilibrium P is unstable, which means that the trajectory starts from nearby P will tend to a periodic or chaotic attractor of system (21).

According to the condition (24), let parameters $a = 20,\ b = 150/7,\ c = 15,\ d = 0.15,\ e = 3,\ k = 0.05$, and $g_1 = g_2 = 0.1$ be fixed, with $\mu \in [-2.4, 2.4]$ and $\eta \in [-2.4, 2.4]$. The distribution diagram of different stability regions corresponding to the three nonzero eigenvalues in the μ-η parameter plane is shown in Fig. 31. It is not hard to notice that regions I, II and IV are unstable, while region III is stable.

5.2 Example 2

Singh [20] studied some 4D hyperchaotic and chaotic systems with quadric surfaces of equilibria.

A new 4D hyperchaotic and chaotic system with a surface of equilibria is described by

$$\begin{cases} \dot{x_1} = x_3, \\ \dot{x_2} = -x_3(ax_2 + bx_2^2 + x_1x_3), \\ \dot{x_3} = f_1(x_1, x_2, x_4) - 1, \\ \dot{x_4} = x_3 f_2(x_2, x_4), \end{cases} \tag{25}$$

where $f_1(x_1, x_2, x_4)$ is a nonlinear function, and $f_2(x_2, x_4)$ is a linear function. The choices of $f_1(x_1, x_2, x_4)$ and $f_2(x_2, x_4)$ determine the type of the surface of equilibria. Under suitable choices of $f_1(x_1, x_2, x_4)$ and $f_2(x_2, x_4)$, the following three different types of system were developed to display the quadric surface of equilibria.

$$\begin{cases} \dot{x_1} = x_3, \\ \dot{x_2} = -x_3(ax_2 + bx_2^2 + x_1x_3), \\ \dot{x_3} = \frac{x_1^2}{a^2} + \frac{x_2^2}{b^2} + \frac{x_4^2}{c^2} - 1, \\ \dot{x_4} = -gx_3x_4\,. \end{cases} \tag{26}$$

$$\begin{cases} \dot{x_1} = x_3, \\ \dot{x_2} = -x_3(ax_2 + bx_2^2 + x_1x_3), \\ \dot{x_3} = \frac{x_1^2}{a^2} + \frac{x_2^2}{a^2} - \frac{x_4^2}{b^2} - 1, \\ \dot{x_4} = -gx_2x_3\,. \end{cases} \tag{27}$$

$$\begin{cases} \dot{x_1} = x_3, \\ \dot{x_2} = -x_3(ax_2 + bx_2^2 + x_1x_3), \\ \dot{x_3} = \frac{x_1^2}{a^2} + \frac{x_2^2}{b^2} - 1, \\ \dot{x_4} = -gx_3x_4\,. \end{cases} \tag{28}$$

Based on systems (26)–(28), eight different cases (named as $QS1$ to $QS8$) were developed to display the quadric surfaces of equilibria with hyperchaotic and chaotic behaviors, as shown in Fig. 32, including the types of dynamics, shapes of the surfaces of equilibria, Lyapunov exponents, Kaplan–Yorke dimension, initial conditions for simulation, etc.

The shapes of surfaces of equilibria of these eight cases are shown in Fig. 33. The first six of them have ellipsoid, spheroid, or hyperboloid type of equilibria, and the rest two have cylinder type of equilibria. Cases $QS1$, $QS2$, $QS3$, $QS4$ and $QS6$ belong to system (26), Case $QS5$ belongs to system (27), and Cases $QS7$ and $QS8$ belong to system (28). Cases $QS1$, $QS4$ and $QS7z$ display hyperchaotic behaviors and the rest five cases have chaotic behaviors.

The stability of the equilibria of cases $QS1$, $QS5$ and $QS7$ are analyzed in [20], while the other five cases can be discussed in a similar manner.

Only the complex dynamic behaviors of case $QS1$, which is a hyperchaotic case, is presented in detail to avoid repetitions of similar figures. Bifurcation diagram and Lyapunov exponent spectrum of the case $QS1$ are shown in Fig. 34, where the

Case	System dynamics	Equilibria	Shape	LEs and nature	D_{KY}	Initial condition
Non-degenerate real quadric surfaces						
QS1	$\dot{x}_1 = x_3$ $\dot{x}_2 = -x_3(hx_2 + dx_2^2 + x_1x_3)$ $\dot{x}_3 = \frac{x_1^2}{a^2} + \frac{x_2^2}{b^2} + \frac{x_4^2}{c^2} - 1$ $\dot{x}_4 = -gx_3x_4$ $h = 5,\ d = 3,\ g = 1,\ \frac{1}{a^2} = 2.7,\ b = 1,\ 1/c^2 = 1.5$	$\left(\frac{x_1^2}{a^2} + \frac{x_2^2}{b^2} + \frac{x_4^2}{c^2} = 1, x_3 = 0\right)$	Ellipsoid	$LE = \begin{pmatrix} 0.0684 \\ 0.0047 \\ 0 \\ -0.2542 \end{pmatrix}$ and hyperchaotic	3.28	$x(0) = \begin{pmatrix} 0.1 \\ 0.01 \\ 0.01 \\ 0.1 \end{pmatrix}^T$
QS2	$\dot{x}_1 = x_3$ $\dot{x}_2 = -x_3(hx_2 + dx_2^2 + x_1x_3)\dot{x}_3 = \frac{x_1^2}{a^2} + \frac{x_2^2}{a^2} + \frac{x_4^2}{b^2} - 1$ $\dot{x}_4 = -gx_3x_4$ $h = 5,\ d = 3,\ g = 1,\ a = 1,\ 1/b^2 = 2$	$\left(\frac{x_1^2}{a^2} + \frac{x_2^2}{a^2} + \frac{x_4^2}{b^2} = 1, x_3 = 0\right)$	Spheroid	$LE = \begin{pmatrix} 0.0403 \\ 0 \\ -0.0024 \\ -0.0909 \end{pmatrix}$ and chaotic	3.41	$x(0) = \begin{pmatrix} 0.1 \\ 0.01 \\ 0.01 \\ 0.1 \end{pmatrix}^T$
QS3	$\dot{x}_1 = x_3$ $\dot{x}_2 = -x_3(hx_2 + dx_2^2 + x_1x_3)\dot{x}_3 = \frac{x_1^2}{a^2} + \frac{x_2^2}{a^2} + \frac{x_4^2}{a^2} - 1$ $\dot{x}_4 = -gx_3x_4$ $h = 5,\ d = 3,\ g = 1,\ a = 1$	$\left(\frac{x_1^2}{a^2} + \frac{x_2^2}{a^2} + \frac{x_4^2}{a^2} = 1, x_3 = 0\right)$	Sphere	$LE = \begin{pmatrix} 0.03547 \\ 0 \\ -0.0030 \\ -0.0776 \end{pmatrix}$ and chaotic	3.42	$x(0) = \begin{pmatrix} 0.1 \\ 0.01 \\ 0.01 \\ 0.1 \end{pmatrix}^T$
QS4	$\dot{x}_1 = x_3$ $\dot{x}_2 = -x_3(hx_2 + dx_2^2 + x_1x_3)\dot{x}_3 = \frac{x_1^2}{a^2} + \frac{x_2^2}{b^2} - \frac{x_4^2}{c^2} - 1$ $\dot{x}_4 = -gx_3x_4$ $h = 5,\ d = 3,\ g = 1,\ 1/a^2 = 2.5, b = 1,\ 1/c^2 = 2.5$	$\left(\frac{x_1^2}{a^2} + \frac{x_2^2}{b^2} - \frac{x_4^2}{c^2} = 1, x_3 = 0\right)$	Elliptic hyperboloid of one sheet	$LE = \begin{pmatrix} 0.08756 \\ 0.0036 \\ 0 \\ -0.264 \end{pmatrix}$ and hyperchaotic	3.34	$x(0) = \begin{pmatrix} 0.01 \\ 0.01 \\ 0.01 \\ 0.1 \end{pmatrix}^T$
QS5	$\dot{x}_1 = x_3$ $\dot{x}_2 = -x_3(hx_2 + dx_2^2 + x_1x_3)\dot{x}_3 = \frac{x_1^2}{a^2} + \frac{x_2^2}{a^2} - \frac{x_4^2}{b^2} - 1$ $\dot{x}_4 = -gx_2x_3 h = 5,\ d = 3,\ g = 0.01,\ a = 1, b = 1$	$\left(\frac{x_1^2}{a^2} + \frac{x_2^2}{a^2} - \frac{x_4^2}{b^2} = 1, x_3 = 0\right)$	Circular hyperboloid of one sheet	$LE = \begin{pmatrix} 0.0312 \\ 0 \\ -0.0014 \\ -0.0737 \end{pmatrix}$ and chaotic	3.40	$x(0) = \begin{pmatrix} 0.01 \\ 0.01 \\ 0.01 \\ 0.1 \end{pmatrix}^T$
QS6	$\dot{x}_1 = x_3$ $\dot{x}_2 = -x_3(hx_2 + dx_2^2 + x_1x_3)\dot{x}_3 = \frac{x_1^2}{a^2} + \frac{x_2^2}{a^2} - \frac{x_4^2}{b^2} - 1$ $\dot{x}_4 = -gx_3x_4$ $h = 5,\ d = 3,\ g = 1,\ a = 1,\ 1/b^2 = 0.5$	$\left(\frac{x_1^2}{a^2} + \frac{x_2^2}{a^2} - \frac{x_4^2}{b^2} = 1, x_3 = 0\right)$	Circular hyperboloid of one sheet	$LE = \begin{pmatrix} 0.03080 \\ 0 \\ -0.0028 \\ -0.0686 \end{pmatrix}$ and chaotic	3.40	$x(0) = \begin{pmatrix} 0.01 \\ 0.01 \\ 0.01 \\ 0.1 \end{pmatrix}^T$
Degenerate quadric surfaces						
QS7	$\dot{x}_1 = x_3$ $\dot{x}_2 = -x_3(hx_2 + dx_2^2 + x_1x_3)\dot{x}_3 = \frac{x_1^2}{a^2} + \frac{x_2^2}{b^2} - 1$ $\dot{x}_4 = -gx_3x_4$ $h = 5,\ d = 3,\ g = 1,\ \frac{1}{a^2} = 2.78,\ b = 1$	$\left(\frac{x_1^2}{a^2} + \frac{x_2^2}{b^2} = 1, x_3 = 0\right)$	Elliptic cylinder	$LE = \begin{pmatrix} 0.1005 \\ 0.0040 \\ 0 \\ -0.2881 \end{pmatrix}$ and hyperchaotic	3.36	$x(0) = \begin{pmatrix} 0.01 \\ 0.01 \\ 0.01 \\ 0.1 \end{pmatrix}^T$
QS8	$\dot{x}_1 = x_3$ $\dot{x}_2 = -x_3(hx_2 + dx_2^2 + x_1x_3)\dot{x}_3 = \frac{x_1^2}{a^2} + \frac{x_2^2}{a^2} - 1$ $\dot{x}_4 = -gx_3x_4$ $h = 6,\ d = 3,\ g = 1,\ a = 1$	$\left(\frac{x_1^2}{a^2} + \frac{x_2^2}{a^2} = 1, x_3 = 0\right)$	Circular cylinder	$LE = \begin{pmatrix} 0.0337 \\ 0 \\ -0.0030 \\ -0.0714 \end{pmatrix}$ and chaotic	3.42	$x(0) = \begin{pmatrix} 0.01 \\ 0.01 \\ 0.01 \\ 0.1 \end{pmatrix}^T$

Fig. 32 Three hyperchaotic and five chaotic Cases of systems (26)–(28), with quadric surfaces of equilibria

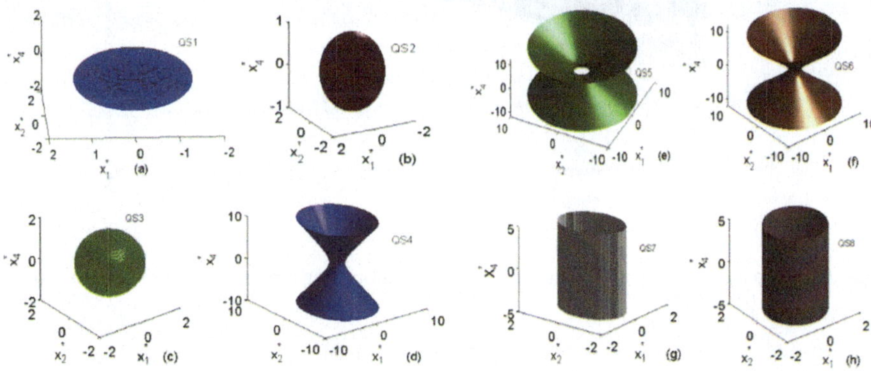

Fig. 33 Shapes of the surfaces of equilibria of the cases shown in Fig. 32

Lyapunov exponent spectrum of the case $QS1$ is shown for the parameter a varying in the range of $2.615 < \frac{1}{a^2} < 3.025$.

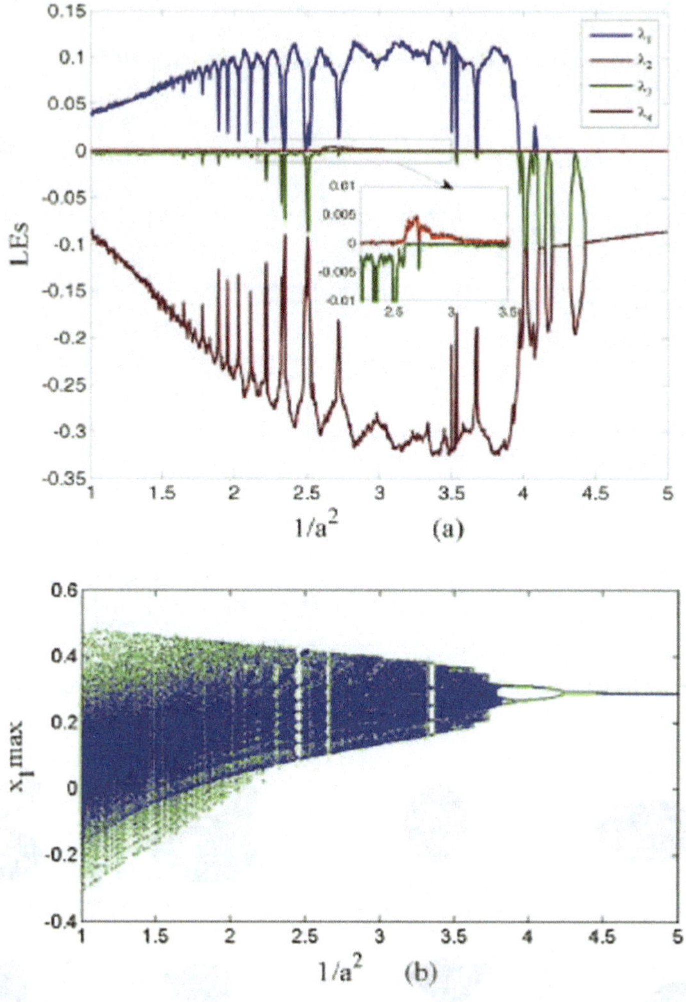

Fig. 34 Lyapunov exponent spectrums and bifurcation diagram of case $QS1$, with $x(0) = (0.01, 0.01, 0.01, 0.01)^T$

6 Coexistence of Different Attractors

6.1 Coexistence of Chaotic Attractors with No Equilibria

All the hyperchaotic attractors reported in [6–8] are hidden attractor, since there are no equilibria in these systems, as shown in Figs. 1 and 5.

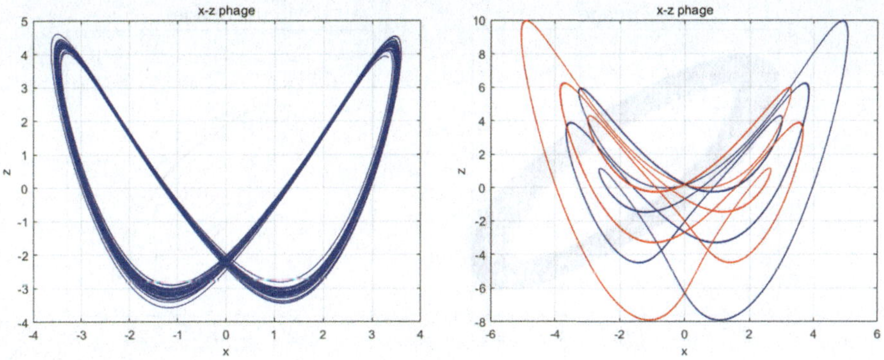

Fig. 35 Projection onto the x-z plane of the trajectory (red and blue attractors correspond to two symmetric initial conditions), with $a = 5$, $b = 0.5$. A symmetric pair of limit cycles coexists with another symmetric limit cycle. **a** Initial conditions $(0, -1, 1, -1)$ with Lyapunov exponents $(0, -0.0180, -0.0180, -0.9639)$ and **b** initial conditions $(\pm 0.2, \pm 5.3, 0, \pm 2)$ with Lyapunov exponents $(0, -0.0520, -0.0589, -0.8891)$

6.1.1 Example 1

Li [7] investigated a perhaps the simplest four-dimensional hyperchaotic system, proposed in [32], resulting in

$$\begin{cases} \dot{x} = y - x, \\ \dot{y} = -xz + w, \\ \dot{z} = xy - a, \\ \dot{w} = -by, \end{cases} \tag{29}$$

where a and b are real parameters.

Since the fine coefficients of the seven terms can be normalized to be ± 1 through a linear re-scaling of the four variables and time, without loss of generality, system (29) has only two independent parameters, namely a and b, which characterize the system dynamics completely.

In some regions of the parameter space, it was found that system (29) has three coexisting limit cycles and Arnold tongues (Fig. 35), while in other regions of the parameter space, system (29) has an attracting torus coexisting with either a symmetric pair of chaotic attractors or a symmetric pair of limit cycles (Fig. 36), whose basin boundaries have an intricate fractal structure (Fig. 37).

6.1.2 Example 2

Wei [8] investigated, for certain parameter values, the coexisting attractors of system (5), including the coexistence of hyperchaotic and chaotic attractors, and the coexistence of hyperchaotic and periodic attractors.

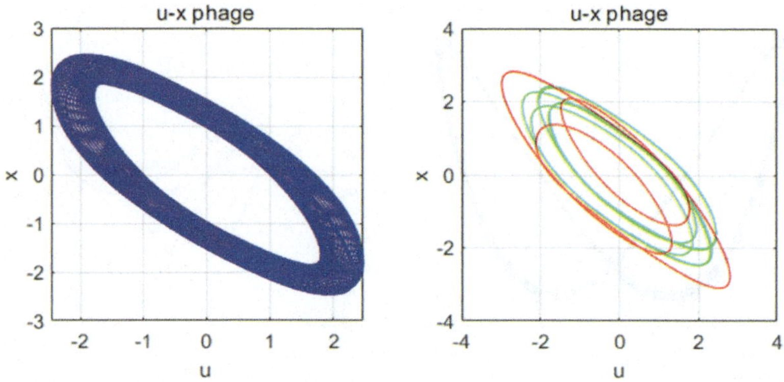

Fig. 36 Torus coexisting with two limit cycles, with $a = 2,\ b = 0.8$ (green and red attractors correspond to two symmetric initial conditions). **a** Initial conditions $(4, -1, 1, -1)$ with Lyapunov exponents $(0, 0, -0.0400, -0.9600)$ and **b** initial conditions $(-5, -1, 1, -1)$ with Lyapunov exponents $(0, -0.0381, -0.1615, -0.8004)$

Fig. 37 Cross-section for $y = 0$ and $w = 0$ of the basins of attraction for the torus (blue) and the symmetric pair of limit cycles (red and green) for system (29), with $a = 2,\ b = 0.8$

Fixing parameters $a = 10,\ b = 25,\ k = 1,\ m = 1,\ c = -4.66$ and choosing different initial conditions, the trajectories of the system (5) may tend to different attractors in the long run:

(I) For initial condition $(0.2, 0.1, 0.75, -2)$, the Lyapunov exponents of the system are $L_1 = 1.1819,\ L_2 = 0.0131,\ L_3 = -0.0007$, and $L_4 = -6.5343$, and the corresponding Kaplan–Yorke dimension is 3.1828, which means that the trajectories tend to a hyperchaotic attractor.

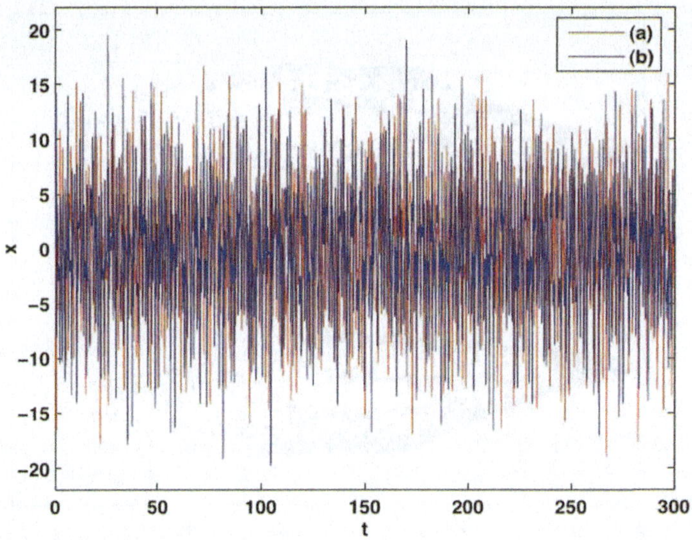

Fig. 38 The x-coordinates of the solutions of the hyperchaotic generalized diffusionless Lorenz equations (5), with same parameter values of $(a, b, m, k, c) = (10, 25, 1, 1, -4.66)$, starting from two initial values: **a** $(x(0), y(0), z(0), u(0)) = (0.2, 0.1, 0.75, -2)$ and **b** $(x(0), y(0), z(0), w(0)) = (0.2, 0.8, 0.75, -2)$

(II) For initial condition $(0.2, 0.8, 0.75, -2)$, the Lyapunov exponents of the system are $L_1 = 1.1955,\ L_2 = 0,\ L_3 = -0.0084$ and $L_4 = -6.5275$, thus the trajectories will tend to a chaotic attractor. These numerical results imply that the trajectories of system (5), with different initial conditions, may converge to different types of attractors, such as hyperchaotic and chaotic attractors, as shown in Fig. 38.

When parameters $a = 10,\ b = 25,\ k = 1,\ m = 1,\ c = 2$, choosing different initial conditions, another type of coexistence of attractors of system (5) can be discovered:

(a) For initial values $(0.2, 0.8, 0.75, -2)$, the Lyapunov exponents of system (5) are $L_1 = 0.0845,\ L_2 = 0.0151,\ L_3 = 0$ and $L_4 = -12.0976$, and the corresponding Kaplan–Yorke dimension is 3.0082. The trajectories will tend to a hyperchaotic attractor.

(b) For initial values $(0.2, 0.82, 0.75, -2)$, the Lyapunov exponents are $L_1 = 0,\ L_2 = -0.1148,\ L_3 = -0.1162$ and $L_4 = -11.7711$. With these initial values, the trajectories of the system (5) will converge to a stable period orbit. The Poincaré map with this group of parameter values are as shown in Fig. 39.

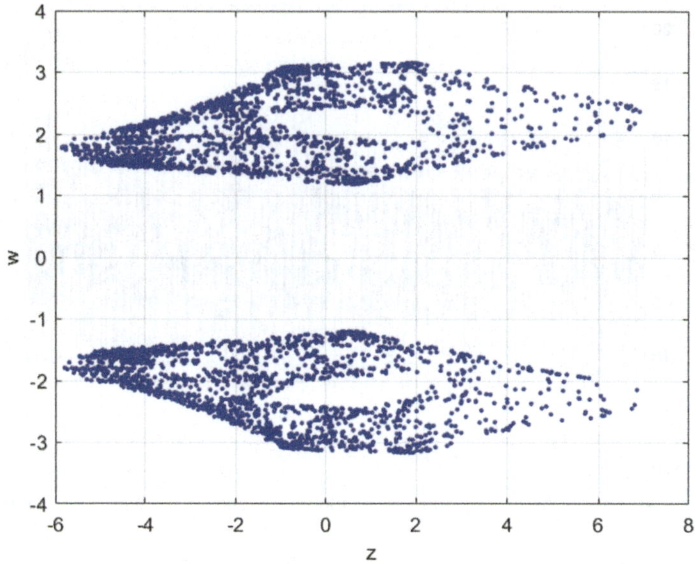

Fig. 39 Poincaré maps of the 4D system (5), with same parameter values $(a, b, m, k, c) = (10, 25, 1, 1, 2)$, starting from two initial values, **a** $(x(0), y(0), z(0), w(0)) = (0.2, 0.8, 0.75, -2)$ and crossing section $3x = y$, and **b** $(x(0), y(0), z(0), w(0)) = (0.2, 0.82, 0.75, -2)$ and crossing section $x = y$

6.2 *Coexistence of Attractors with a Limited Number of Equilibria*

6.2.1 Example 1

Wei [9] studied some extraordinary dynamical behaviors of the non-Šil'nikov system (7). Although the algebraical form of system (7) is similar to that of the Lorenz–Stenflo system, they are actually different. A typical difference is that system (7) has only one stable equilibrium under certain parameter conditions. The extraordinary dynamical behaviors of system (7) includes the coexistence of attracting sets under some parameter conditions with different initial values, including hyperchaotic attractor and stable equilibrium, hyperchaotic attractor and periodic attractor.

Case I. Coexistence of attracting sets, under parameter conditions $\sigma = 35$, $k = -17$, $b = 0.8$, $s = 35$, $n = 4$, $v = -20$ and $q = 0$.

(1) For initial value $(0, 1, -0.5, 0)$, the Lyapunov exponents of system (7) are $L_1 = 1.2908$, $L_2 = 0.2086$, $L_3 = 0$, $L_4 = -20.2994$. The trajectories tend to a hyperchaotic attractor.

(2) For initial values $(0, 1, -1, 0)$, the corresponding Lyapunov exponents are $L_1 = -0.7997$, $L_2 = -2.2035$, $L_3 = -2.2479$, $L_4 = -13.5489$, which means that the trajectories converge to a stable equilibrium. Thus, under these parame-

ter conditions, there is a hidden hyperchaotic attractor, coexisting with only one equilibrium, as shown in Fig. 7.

Case II. Coexistence of attracting sets, under parameter conditions $\sigma = 35$, $k = -17$, $b = 0.8$, $s = 32$, $n = 4$, $v = -20$ and $q = 0$.

(1) For initial values $(1, 1, 5, 0)$, the Lyapunov exponents of system (7) are $L_1 = 1.3115$, $L_2 = 0.1402$, $L_3 = 0$, $L_4 = -20.2517$. The trajectories tend to a hyperchaotic attractor, which is generated by an unstable equilibrium.

(2) For initial values $(0.1, 0, -6, 0)$, the Lyapunov exponents are $L_1 = 0$, $L_2 = -0.3562$, $L_3 = -0.4003$, $L_4 = -18.0435$, corresponding to a periodic attractor. Thus, under these parameter conditions, there is a hidden hyperchaotic attractor, coexisting with a stable periodic attractor.

6.2.2 Example 2

Wei [13] also investigated the 5D system (8), which has hidden hyperchaotic attractors, with certain parameters, system (8) has only two stable equilibria. Because the missing of unstable equilibria in system (8), in some parameter regions this system has multi-stability.

An extraordinary feature of system (8) is the coexistence of a hyperchaotic attractor with two stable equilibria $E_{1,2}$ in a certain range of parameter values, as shown in Fig. 10. For example, for parameters $m = 0.04$, $g = 140.6$, $k_2 = 12$, and $r = 7$, $k_1 = 34$, system (8) has a hyperchaotic attractor along with two stable equilibria, as shown in Fig. 9. The Lyapunov exponents of this hyperchaotic system are $(0.9616, 0.5477, 0.1425, 0.0000, -7.6518)$, and the corresponding Kaplan–Yorke dimension is $D_{KY} = 4.2159$.

To show evidence of multi-stability and to examine the robustness of the hyperchaos, let $3 < r < 8$ and $k_1 = 34$. Then, the dynamics of system (8) were as shown in Fig. 11, with three different initial conditions, where the equilibria $E_{1,2}$ are always stable. In the left panel, the initial conditions are kept constant at $(0.05, -0.5, 0.1, -1.2)$, while in the middle panel, r is increased without re-initializing, and in the right panel, it is decreased without re-initializing. Figure 11 clearly shows the periodic windows, quasiperiodic regions, chaotic and hyperchaotic regions, multi-stability and also hysteresis.

In the region of parameter space where all the equilibria of system (8) are stable, there are other kinds of multi-stability. With $r = 3.5$ and $k_1 = 34$, there are two periodic orbits coexisting in the space, with initial values $(2, 1, 2, 0, 0)$ and $(0, 2, 6, 11, 21)$, respectively, as shown in Fig. 40. With $r = 2.9$ and $k_1 = 29.12$, there is a periodic orbit coexisting with a chaotic attractor, as shown in Fig. 41. With $r = 3.75$ and $k_1 = 32$, there is a periodic orbit coexisting with a quasiperiodic attracting torus, as shown in Fig. 42. Additional evidence shows that this is a torus rather than a high-periodic limit cycle or a thin chaotic attractor as can be seen from Fig. 43. The figure shows that the cross-section of the attractor for $y = 0$ consists of five closed loops. In addition, the Kaplan–Yorke dimension in this region is precisely 2.0.

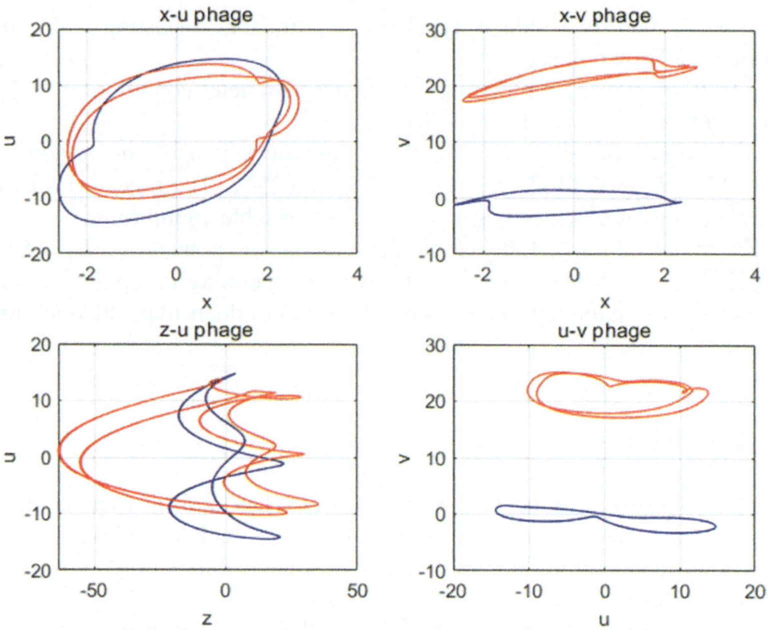

Fig. 40 Coexisting attractors of system (8), with $r = 3.5$ and $k_1 = 34$: period-1 attractor in blue from initial condition (2, 1, 2, 0, 0); period-2 attractor in red from initial condition (0, 2, 6, 11, 21)

6.3 Coexistence of Attractors with Lines or Curves of Equilibria

6.3.1 Example 1

The seemingly simple 4D dissipative autonomous chaotic system (16) reported by Singh [15] has a line of equilibria, with many extraordinary properties, such as chaotic 2-torus, quasiperiodic motions and multi-stability behaviors.

In order to observe the phenomena of multi-stability of system (16), the bifurcation diagram under the variation of parameter c is plotted in Fig. 44. In the evolutionary process, the final state of the system for each value of parameter c is used as the initial value for the next value of c. By comparing Figs. 20 and 44, one can observe that the dynamic behavior of system (16) has changed for different initial values within the bifurcation parameter range $c \in [0.12, 0.14]$. For example, when $c = 0.12$, Fig. 44 shows that system (16) has chaotic behavior, whereas Fig. 20 shows that the system has periodic behavior. It is also can be observed from Fig. 20 that system (16) has, in addition to a reverse period-doubling route to chaos, a cascaded reverse period-doubling route to chaos within the bifurcation parameter range $c \in [0.11, 0.125]$, but there is no cascaded reverse periodic period-doubling route to chaos observable from Fig. 44 within the same parameter range. Furthermore, there are four jump routes to

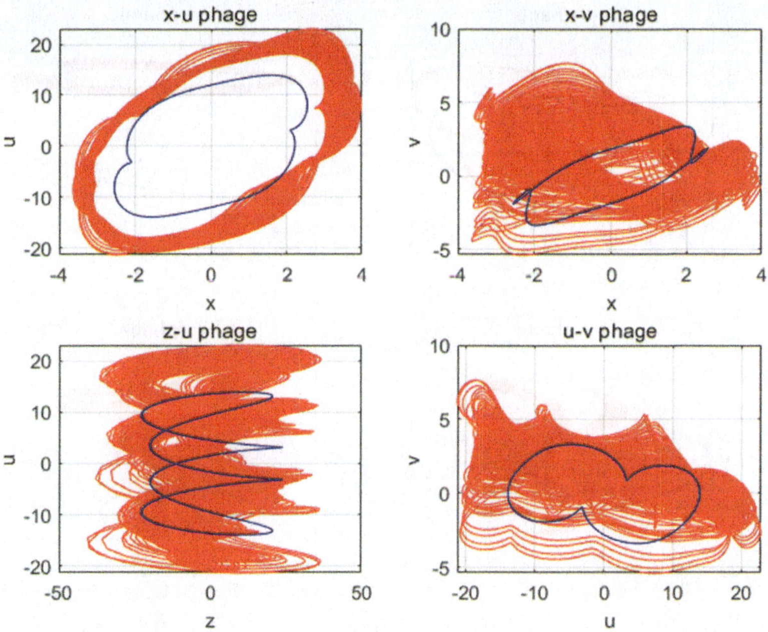

Fig. 41 Coexisting attractors of system (8), with $r = 2.9$ and $k_1 = 29.12$: periodic-1 attractor (blue) from initial condition (2, 2, 5, 2, 2); hidden chaos with Lyapunov exponents (0.0628, 0, −0.1497, −0.4999, −1.3133) (red) from initial condition (2, 0, 1, −4, −2)

chaos in Fig. 44. This means that system (16) has coexistence of different attractors according to different initial values.

The coexistence of (i) quasi-periodic (2-torus) orbit with chaotic attractor and (ii) two different periodic orbits of system (16) are shown in Fig. 45 for $c = 0.123$. The coexistence of (i) quasi-periodic (2-torus) with chaotic attractor and (ii) two different periodic orbits of system (16) are shown in Fig. 46 for $c = 0.119$.

6.3.2 Example 2

Based on the classical Lorenz system, a new 4D Lorenz-type of hyperchaotic system with a curve of equilibria was reported by Chen [16]. This new 4D hyperchaotic system possesses various multi-stabilities, such as the coexistence of (i) chaotic attractor and quasi-periodic attractor, (ii) chaotic attractor and singular degenerate heteroclinic cycle, (iii) periodic attractor and singular degenerate heteroclinic cycle, and (iv) different periodic attractors.

Considering the symmetry of system (18), for any orbit Γ of the system, there is an orbit Γ' symmetrical with Γ about the z-axis. In the following, only one of them is discussed.

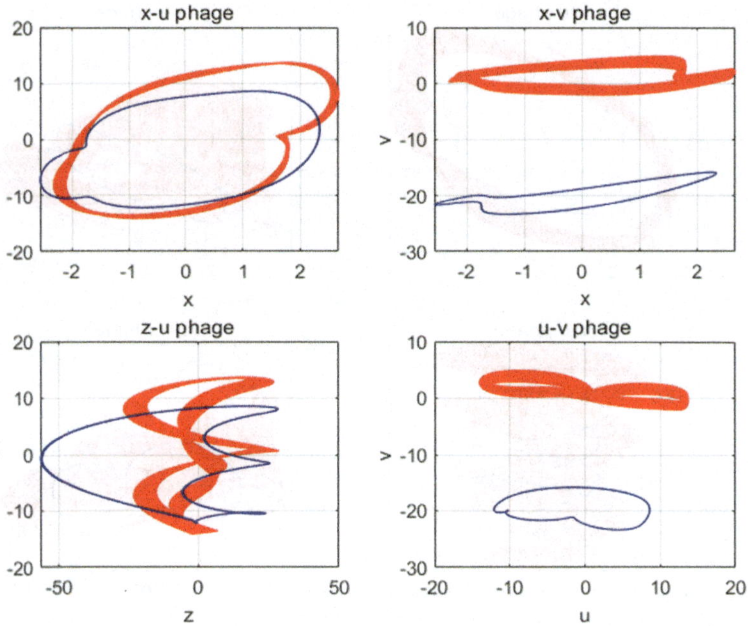

Fig. 42 Coexisting attractors of system (8), with $r = 3.75$ and $k_1 = 32$: periodic-1 attractor (blue) from initial condition $(1.5, 4.5, 7.2, 8.5, -19)$; hidden quasi-periodic orbit with two zero Lyapunov exponent (red) from initial condition $(0.05, -0.5, 0.1, -1, 2)$

Fig. 43 Quasi-periodic on an attracting torus for cross-section $y = 0$ with parameters $r = 3.75,\ k_1 = 32,\ m = 0.04,\ g = 140.6,\ k_2 = 12$ of system (8) from initial condition $(0.05, -0.5, 0.1, -1, 2)$

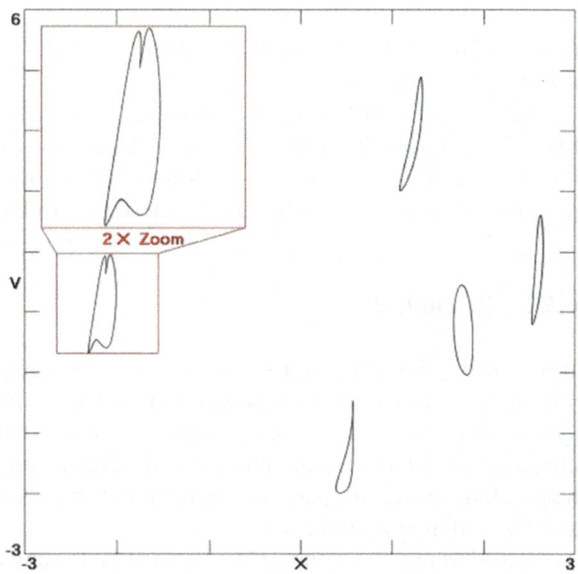

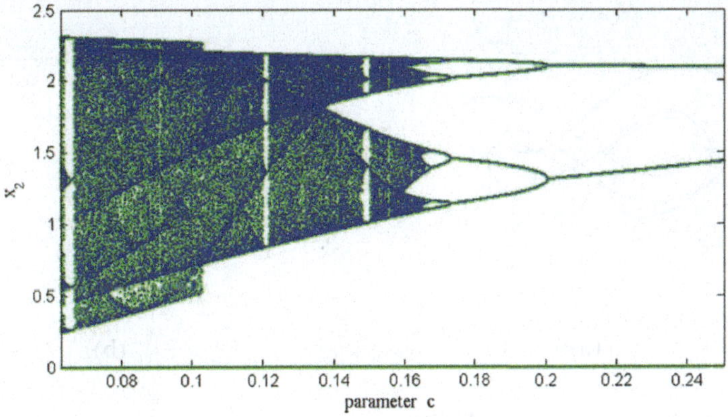

Fig. 44 Bifurcation diagram of system (16), with $a = b = 0.5$

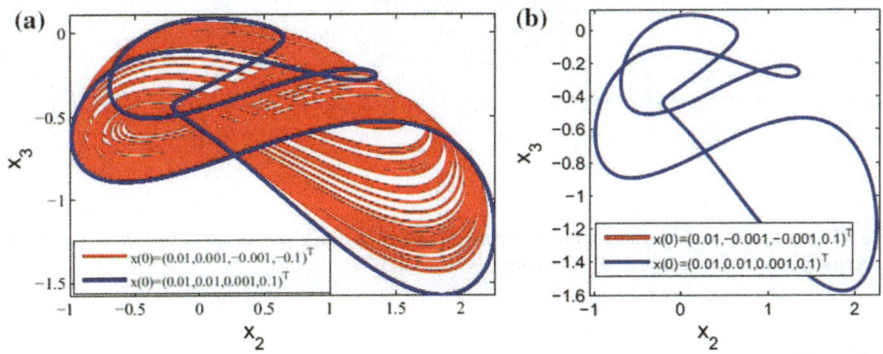

Fig. 45 **a** Coexistence of 2-torus with chaotic attractor and **b** coexistence of 2-torus with 2-torus of system (16), with $a = b = 0.5,\ c = 0.123$

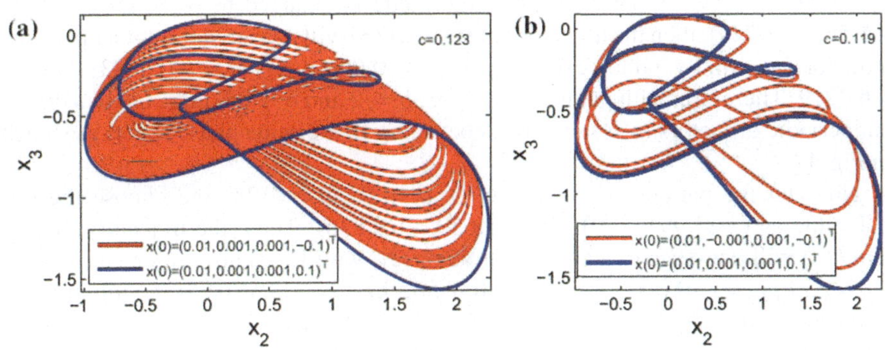

Fig. 46 **a** Coexistence of 2-torus with chaotic attractor for $c = 0.123$ and **b** coexistence of 2-torus with 3-torus for $c = 0.119$ of system (16), with $a = b = 0.5$

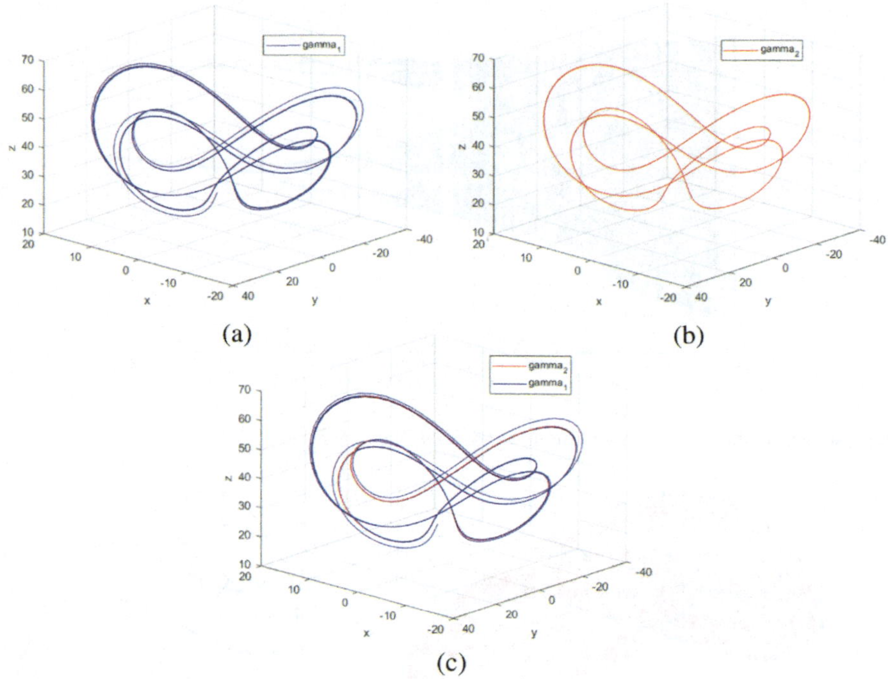

Fig. 47 Projections of coexisting attractors of system (18) in the x-y-z space: $a = 5, b = 2, c = 28$. **a** Periodic attractor Γ_1, **b** periodic attractor Γ_2, **c** Γ_1 and Γ_2

Let the parameters be $a = 5$, $b = 2$ and $c = 28$, and choose the initial value as $(-16.06,\ 12.978,\ 14.118,\ -8.899)$. Then, the trajectory of system (18) will converge to a periodic attractor Γ_1, with Lyapunov exponents $\lambda_{LE_1} = 0.0003$, $\lambda_{LE_{2,3}} = -0.1809$, $\lambda_{LE_4} = -6.6367$. If the initial value is changed to $(-16.281,\ 16.808,\ -5.562, -7.779)$, then the trajectory of system (18) will converge to another periodic attractor Γ_2, with Lyapunov exponents $\lambda_{LE_1} = 0.0003$, $\lambda_{LE_{2,3}} = -0.3383$, $\lambda_{LE_4} = -6.3219$. Therefore, with parameters $a = 5$, $b = 2$ and $c = 28$, there are two different periodic attractors in system (18), coexisting in the same phase space, as shown in Fig. 47.

Now, let the parameters be $a = 6.55$, $b = 2$ and $c = 28$, and choose the initial value as $(-23.997,\ -10.146,\ -1.650,\ -29.836)$. Then, the trajectory of system (18) will converge to a chaotic attractor Ω_1, with Lyapunov exponents $\lambda_{LE_1} = 0.0742$, $\lambda_{LE_2} = -0.0004$, $\lambda_{LE_3} = -0.0556$, $\lambda_{LE_4} = -8.5610$. If the initial value is changed to $(-5.400,\ 21.928,\ 2.985,\ -21.027)$, then the trajectory of system (18) will converge to a quasi-periodic attractor Ω_2, with Lyapunov exponents $\lambda_{LE_{1,2}} = -0.0002$, $\lambda_{LE_3} = -0.0179$, $\lambda_{LE_4} = -8.5188$. Therefore, with parameters $a = 6.55, b = 2$ and $c = 28$, there are chaotic and quasi-periodic attractors in system (18), coexisting in the same phase space, as shown in Figs. 48 and 49.

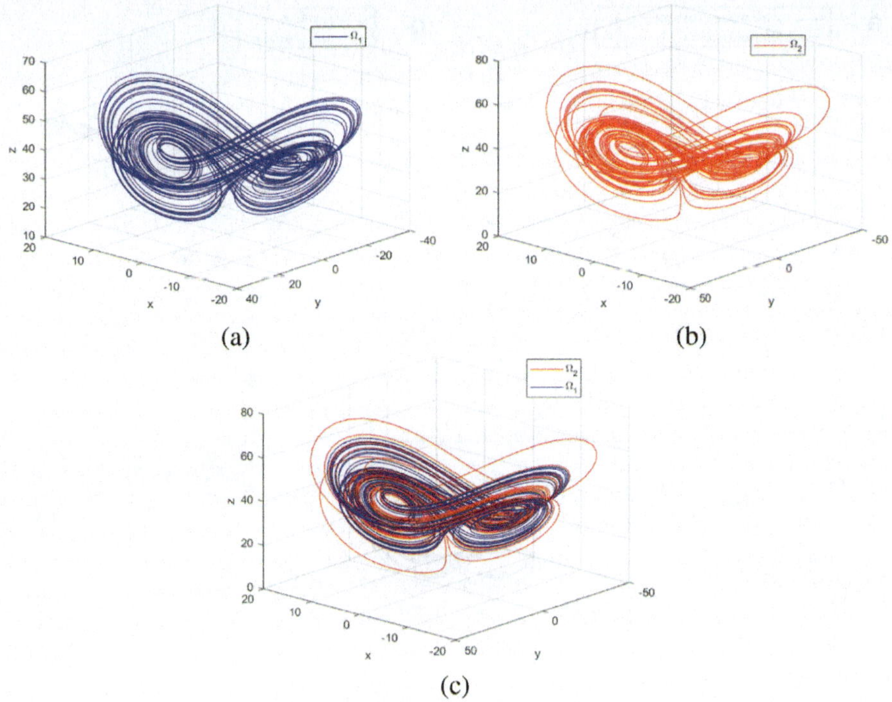

Fig. 48 Projections of coexisting attractors of system (18) in the x-y-z space: $a = 6.55$, $b = 2$, $c = 28$. **a** Chaotic attractor Ω_1, **b** quasi-periodic attractor Ω_2, **c** Ω_1 and Ω_2

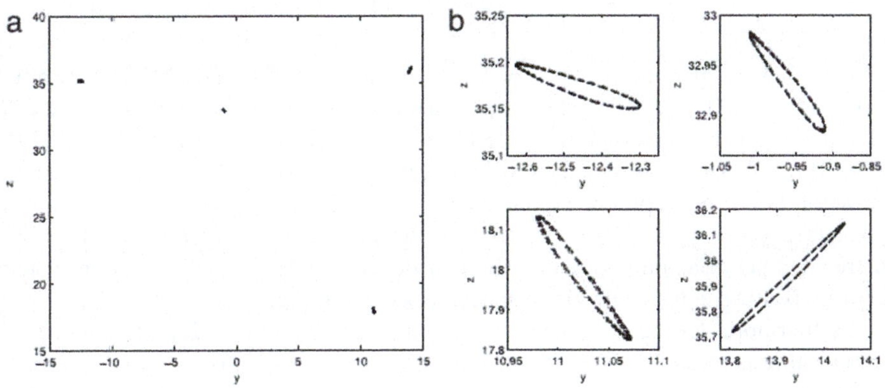

Fig. 49 **a** Poincaré image of the quasi-periodic attractor of system (18) on the y-z plane, **b** the local magnification of **a** $a = 6.55$, $b = 2$, $c = 28$

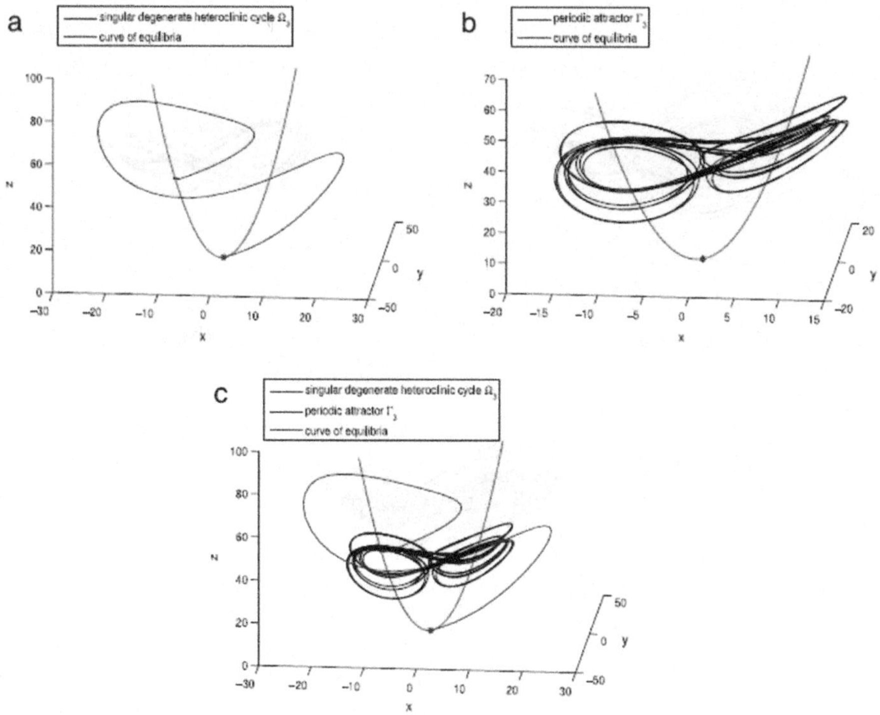

Fig. 50 Projections of coexisting solutions of system (18) in the x-y-z space: $a = 14$, $b = 2$, $c = 28$. **a** Singular degenerate heteroclinic cycle Ω_3, **b** periodic attractor Γ_3, **c** Ω_3 and Γ_3

Next, let the parameters be $a = 14$, $b = 2$ and $c = 28$, and choose the initial value as (0.00001, 0, 0, 0). Then, the trajectory of system (18) will converge to a singular degenerate heteroclinic cycle Ω_3. If the initial value is changed to $(-7.346, -16.911, 4.728, 6.853)$, then the trajectory of system (18) will converge to a periodic attractor Γ_3, with Lyapunov exponents $\lambda_{LE_1} = 0.0004$, $\lambda_{LE_2} = -0.0863$, $\lambda_{LE_3} = -0.3081$, $\lambda_{LE_4} = -15.6040$. Therefore, with parameters $a = 14, b = 2$ and $c = 28$, there are a periodic attractor and a singular degenerate heteroclinic cycle in system (18), coexisting in the same phase space, as shown in Fig. 50.

Furthermore, let the parameters be $a = 15$, $b = 2$ and $c = 28$, and choose the initial value as $(0.00001, 0, 0, 0)$. Then, the trajectory of system (18) will converge to a singular degenerate heteroclinic cycle Ω_4. If the initial value is changed to (27.982, 15.548, 29.480, 21.994), then the trajectory of system (18) will converge to a chaotic attractor Ω_5, with Lyapunov exponents $\lambda_{LE_1} = 0.8482$, $\lambda_{LE_2} = 0.0000$, $\lambda_{LE_3} = -0.8424$, $\lambda_{LE_4} = -16.9971$. Therefore, with parameters $a = 15, b = 2$ and $c = 28$, there are a chaotic attractor and a singular degenerate heteroclinic cycle in system (18), coexisting in the same phase space, as shown in Fig. 51.

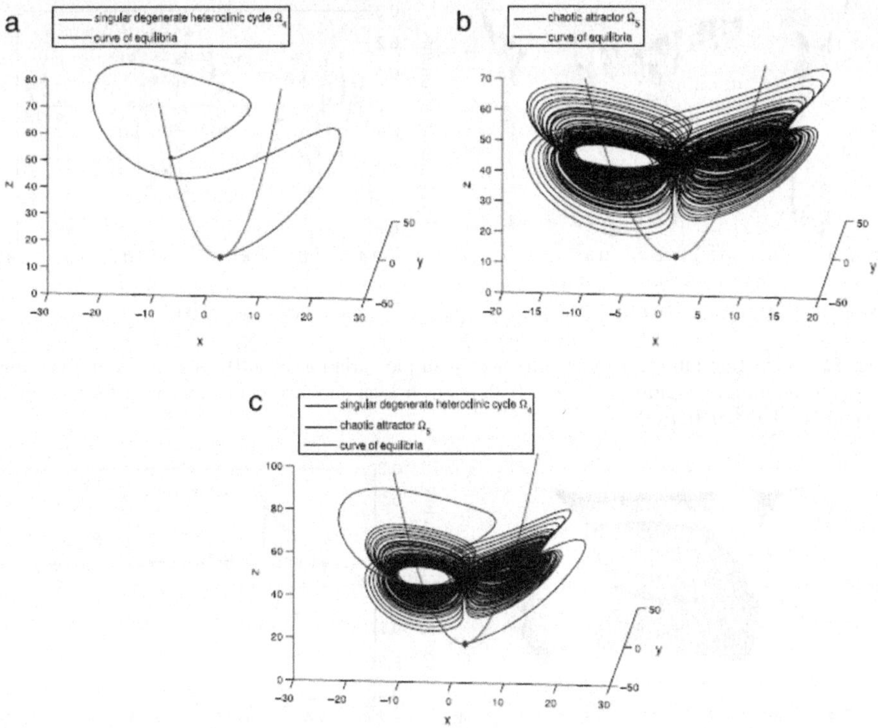

Fig. 51 Projections of coexisting solutions of system (18) in the x-y-z space: $a = 15$, $b = 2$, $c = 28$. **a** Singular degenerate heteroclinic cycle Ω_4, **b** chaotic attractor Ω_5, **c** Ω_4 and Ω_5

6.4 Coexistence of Attractors with a Plane of Equilibria

The 5D two-memristor-based Chua's hyperchaotic circuit (21), which has a plane of equilibria with complex multi-stability, was reported in [19]. Bifurcation diagrams and Lyapunov exponent spectra are used to measure the multi-stability of system (21). Let the parameters be $a = 20,\ b = 150/7,\ c = 15,\ d = 0.15,\ e = 3,\ k = 0.05,\ g_1 = g_2 = 0.1$, the initial values be $y(0) = 0$, $z(0) = 0$, and $x(0) = \pm 10^{-9}$, and the two-memristor initial conditions be $w(0) = \mu$ and $u(0) = \eta$, which are adjustable.

Firstly, choose initial values $x(0) = \pm 10^{-9}$, $w(0) = \mu$, and $u(0) = \eta = 0$. When μ is monotone increasing in the range $[-2.3, 2.3]$, the bifurcation diagrams of the system with $x(0) = \pm 10^{-9}$ and the first four Lyapunov exponents for the system with $x(0) = 10^{-9}$ are as shown in Fig. 52.

As can be seen from Fig. 52, as the value of μ is varied, system (21) displays equilibrium, periodic, chaotic, and hyperchaotic behaviors, along with period-doubling bifurcation routes to chaos and boundary crisis, and so on. In particular,

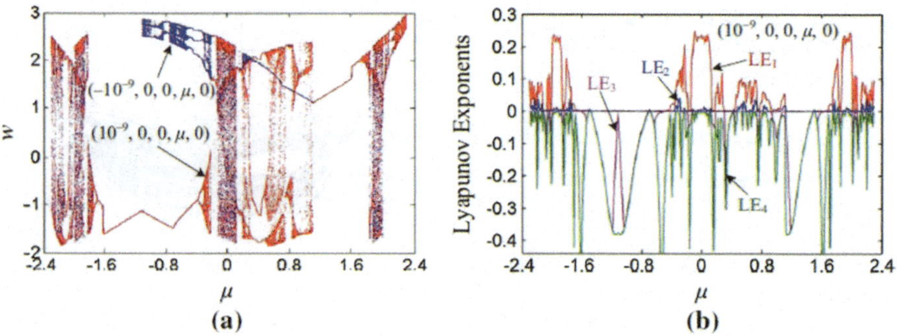

Fig. 52 Coexisting infinitely many attractors with the variation of $w(0)$: **a** bifurcation diagrams of w for the initial conditions $(\pm 10^{-9}, 0, 0, \mu, 0)$, **b** Lyapunov exponent spectrum for the initial conditions $(10^{-9}, 0, 0, \mu, 0)$

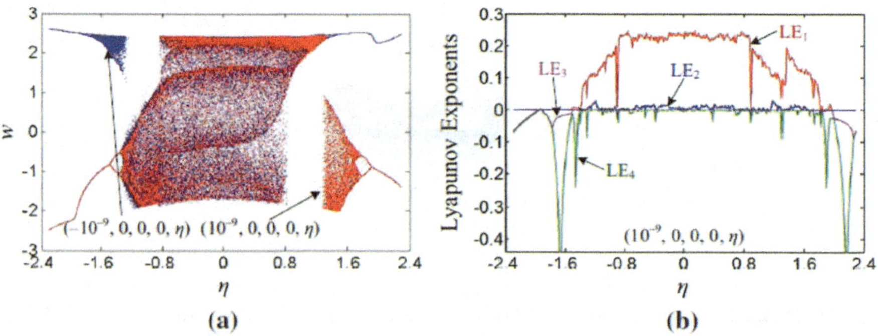

Fig. 53 Coexisting infinitely many attractors' behaviors of the system, with the variation of $u(0)$: **a** bifurcation diagrams of w for the initial conditions $(\pm 10^{-9}, 0, 0, 0, \eta)$, **b** Lyapunov exponent spectrum of the system for the initial conditions $(10^{-9}, 0, 0, 0, \eta)$

for $-1.11 \leq \mu \leq -0.21$ and $0.21 \leq \mu \leq 1.11$, the dynamical behaviors of system (21) are extremely sensitive to the initial value $x(0)$.

Secondly, let the initial values be fixed as $x(0) = \pm 10^{-9}$, $w(0) = 0$, whereas $u(0) = \eta$ be variable. When η is monotone increasing in the range $[-2.3, 2.3]$, the bifurcation diagrams of the system with $x(0) = \pm 10^{-9}$ and the first four Lyapunov exponents of the system with $x(0) = 10^{-9}$ are shown in Fig. 53. As can be seen from Fig. 53, as the value of η is varied, the forward and reversed period-doubling bifurcation routes to chaos and the extreme sensitivity to the disturbance coming from the tiny initial conditions of $x(0)$ appear clearly, which further verify the coexistence of infinitely many attractors in the two-memristor-based Chua's hyperchaotic circuit.

References

1. G.A. Leonov, N.V. Kuznetsov, Prediction of hidden oscillations existence in nonlinear dynamical systems: analysis and simulation, in *Advances in Intelligent Systems and Computing*, vol. 210, ed. by J. Kacprzyk (Springer, New York, 2013), pp. 5–13
2. G.A. Leonov, N.V. Kuznetsov, T.N. Mokaev, Homoclinic orbits, and self-excited and hidden attractors in a Lorenz-like system describing convective fluid motion. Eur. Phys. J. Spec. Top. **224**, 1421–1458 (2015)
3. N. Stankevich, N.V. Kuznetsov, G.A. Leonov, L.O. Chua, Scenario of the birth of hidden attractors in the Chua circuit. Int. J. Bifurc. Chaos **27**(12), 1730038 (2017)
4. G. Chen, N.V. Kuznetsov, G.A. Leonov, T.N. Mokaev, Hidden attractors on one path: Glukhovsky-Dolzhansky, Lorenz, and Rabinovich systems. Int. J. Bifurc. Chaos **27**(8), 1750115 (2017)
5. N.V. Kuznetsov, G.A. Leonov, T.N. Mokaev, A. Prasad, Shrimali, M.D.: finite-time Lyapunov dimension and hidden attractor of the Rabinovich system. Nonlinear Dyn. **92**(2), 267–285 (2018)
6. Z. Wang, S. Cang, E.O. Ochola, Y. Sun, A hyperchaotic system without equilibrium. Nonlinear Dyn. **69**, 531–537 (2011)
7. C. Li, J.C. Sprott, Coexisting hidden attractors in a 4-D simplified Lorenz system. Int. J. Bifurc. Chaos **24**(3), 1450034 (2014)
8. Z. Wei, R. Wang, A. Liu, A new finding of the existence of hidden hyperchaotic attractors with no equilibria. Math. Comput. Simul. **100**, 13–23 (2014)
9. Z. Wei, W. Zhang, Hidden hyperchaotic attractors in a modified Lorenz-Stenflo system with only one stable equilibrium. Int. J. Bifurc. Chaos **24**(10), 1450127 (2014)
10. J.P. Singh, B.K. Roy, A novel asymmetric hyperchaotic system and its circuit validation. Int. J. Control Theory Appl. **8**(3), 10005-1013 (2015)
11. M.J. Correia, P.C. Rech, Hyperchaotic states in the parameter-space. Appl. Math. Comput. **218**, 6711–6715 (2012)
12. C. Zhang, Theoretical design approach of four-dimensional piecewise-linear multi-wing hyperchaotic differential dynamic system. Optik **127**, 4575–4580 (2016)
13. Z. Wei, T. Moroz, J.C. Sprott, A. Akgul, W. Zhang, Hidden hyperchaos and electronic circuit application in a 5D self-exciting homopolar disc dynamo. Chaos **27**, 033101 (2017)
14. Q. Li, H. Zeng, J. Li, Hyperchaos in a 4D memristive circuit with infinitely many stable equilibria. Nonlinear Dyn. **79**, 2295–2308 (2015)
15. J.P. Singh, B.K. Roy, The simplest 4-D chaotic system with line of equilibria, chaotic 2-torus and 3-torus behaviour. Nonlinear Dyn. **89**, 1845–1862 (2017)
16. Y. Chen, Q. Yang, A new Lorenz-type hyperchaotic system with a curve of equilibria. Math. Comput. Simul. **112**, 40–55 (2015)
17. Y. Chen, Dynamics of a Lorenz-type multistable hyperchaotic system. Math. Methods Appl. Sci. (2018). https://doi.org/10.1002/mma.5171
18. J.P. Singh, B.K. Roy, Coexistence of asymmetric hidden chaotic attractors in a new simple 4-D chaotic system with curve of equilibria. Optik **145**, 209–217 (2017)
19. B. Bao, T. Jiang, G. Wang, P. Jin, H. Bao, M. Chen, Two-memristor-based Chua's hyperchaotic circuit with plane equilibrium and its extreme multistability. Nonlinear Dyn. **89**, 1157–1171 (2017)
20. J.P. Singh, B.K. Roy, S. Jafari, New family of 4-D hyperchaotic and chaotic systems with quadric surfaces of equilibria. Chaos, Solitons Fractals **106**, 243–257 (2018)
21. O.E. Rössler, Continuous chaos: four prototype equations. Ann. N. Y. Acad. Sci. **316**, 376–392 (1979)
22. E.N. Lorenz, Deterministic non-periodic flow. J. Atmos. Sci. **20**, 130–141 (1963)
23. G. Chen, T. Ueta, Yet another chaotic attractor. Int. J. Bifurc. Chaos **9**, 1465–1466 (1999)
24. J. Lü, G. Chen, A new chaotic attractor coined. Int. J. Bifurc. Chaos **12**, 659–661 (2002)
25. S. Jafari, J.C. Sprott, Simple chaotic flows with a line equilibrium. Chaos, Solitons Fractals **57**, 79–84 (2013)

26. M. Shrimali, A. Prasad, R. Ramaswamy, U. Feudel, The nature of attractor basins in multistable systems. Int. J. Bifurc. Chaos **18**(6), 1675–1688 (2008)
27. T. Kapitaniak, Stochastic response with bifurcations to non-linear Duffing's oscillator. J. Sound Vib. **102**(3), 440–441 (1985)
28. A. Chudzik, P. Perlikowski, A. Stefanski, T. Kapitaniak, Multistability and rare attractors in van der Pol-Duffing oscillator. Int. J. Bifurc. Chaos **21**(7), 1907–1912 (2011)
29. Q. Yang, Z. Wei, An unusual 3D autonomous quadratic chaotic system with two stable node-foci. Int. J. Bifurc. Chaos **20**, 1061–1083 (2010)
30. L. Stenflo, Generalized Lorenz equations for acoustic-gravity waves in the atmosphere. Phys. Scr. **53**, 83–84 (1996)
31. J. Ma, Y. Yang, Hyperchaos numerical simulation and control in a 4D hyperchaotic system. Discret. Dyn. Nat. Soc. **2013**, 980578 (2013)
32. Z. Gao, C. Zhang, A novel hyperchaotic system. J. Jishou Univ. (Nat. Sci. Ed.) **32**, 65–68 (2011)

Fractional-Order Chaotic Systems with Hidden Attractors

Xiong Wang and Guanrong Chen

1 Introduction

The notion of fractional-order calculus is as old as the classical calculus, which formed a branch in mathematics lately, since the time of Leibniz in the 17th century [1–3]. An outline of the simple history of fractional calculus was offered by Machado et al. [4]. Factional calculus plays an important role in physics [5], electrical and electronic engineering, especially circuit theory [6], control systems [7], signal processing [8], and even chemical mixing [1].

Today, a number of techniques are available for calculating or approximating the fractional-order derivatives and integrals [7, 9]. In the study of chaotic systems and complex dynamics, two most commonly-used fractional derivative formulas are the Riemann-Liouville derivative and the Caputo derivative.

The Riemann-Liouville derivative of order α, with the lower limit 0, is defined by

$$\frac{\partial^\alpha}{\partial t^\alpha} f(t) = \frac{1}{\Gamma(m-\alpha)} \frac{d^m}{dt^m} \int_0^t (t-\tau)^{m-\alpha-1} f(\tau) d\tau ,$$

where $m - 1 \le \alpha < m \in Z_+$ and Γ is the gamma function.

X. Wang (✉)
Institute for Advanced Study, Shenzhen University,
Shenzhen, Guangdong 518060, People's Republic of China
e-mail: wangxiong8686@szu.edu.cn

G. Chen
Department of Electrical Engineering, City University of Hong Kong, Hong Kong SAR 999077, People's Republic of China
e-mail: eegchen@cityu.edu.hk

X. Wang et al. (eds.), *Chaotic Systems with Multistability and Hidden Attractors*, Emergence, Complexity and Computation 40,
https://doi.org/10.1007/978-3-030-75821-9_9

The Caputo derivative of order α, with the lower limit 0, is defined by

$$\frac{\partial^\alpha}{\partial t^\alpha} f(t) = \frac{1}{\Gamma(m-\alpha)} \int_0^t (t-\tau)^{m-\alpha-1} f^{(m)}(\tau) d\tau ,$$

where $m - 1 < \alpha \leq m \in Z_+$.

In the last two decades, many chaotic systems were modified to be in the setting of fractional calculus. By varying the so-called system "effective order", it has been demonstrated that the chaotic or hyperchaotic systems of effective order less than three or four can indeed display chaos or hyperchaos, as well as other complex dynamical behaviors. There is a trend of growing interest in the study of chaos and hyperchaos in fractional-order dynamical systems.

In this chapter, both three-dimensional and four-dimensional fractional-order chaotic models are introduced, including the fractional-order systems modified from the classical integral-order systems, and some new systems with infinite equilibria or without equilibria, or with stable equilibria. It is noted that these systems have neither homoclinic nor heteroclinic orbits, and thus the Šil'nikov criteria [10] are not applicable to show the existence of chaos in such systems.

First, several fractional-order systems will be introduced, which come directly from the integral-order counterparts. To date, this category of fractional-order systems includes the fractional version of Chua's circuit, the fractional-order Lorenz system, the fractional-order Chen system, the fractional-order Lü system, the fractional-order chaotic Rössler system and its hyperchaotic version, the fractional-order systems with multi-scroll attractors, and so forth [11–15]. A collection of such systems were provided by Petráš and Bednárová [16].

On the other hand, by applying the predictor-corrector numerical algorithm [17, 18], fraction-order dynamical systems with an infinite number of equilibria or without any equilibria can be discussed. Moreover, by utilizing some control schemes for stabilization [19, 20], fractional-order systems with stable equilibria can be analyzed and discussed.

Numerical simulations and synchronization results of various fractional-order systems will be presented in this Chapter, for their potential applications in secure communications and control processing. For example, Li and Deng [21] investigated the synchronization of the chaotic fractional-order Chua circuit by applying the method of Ref. [22]. Other systems such as financial systems [23], cellular neural networks [24], and the van der Pol oscillator [25], etc., will not be considered in this chapter for brevity of presentation.

2 Classical Fractional-Order Chaotic Systems

2.1 *Fractional-order Chua's Circuit*

Fractional-order Chua's circuit is described by Ref. [27]

$$\begin{cases} \dfrac{d^q x}{dt^q} &= \alpha[y - x - f(x)] \\ \dfrac{d^q y}{dt^q} &= x - y + z \\ \dfrac{d^q z}{dt^q} &= -\beta y - \gamma z, \end{cases} \tag{1}$$

where

$$f(x) = m_1 x + \frac{1}{2}(m_0 - m_1) \times (|x+1| - |x-1|), \tag{2}$$

with real parameters $\alpha,\ \beta,\ \gamma,\ m_0,\ m_1$.

Hartley et al. [26] presented a fractional-order form of Chua's circuit. The fractional-order Chua-Hartley system replaces the piecewise-linear resistor in the classical Chua's circuit [28, 29] by an appropriate cubic nonlinearity that can produce similar chaotic behaviors, in the following form:

$$\begin{cases} \dfrac{d^q x}{dt^q} &= \alpha[y + \dfrac{x - 2x^3}{7}] \\ \dfrac{d^q y}{dt^q} &= x - y + z \\ \dfrac{d^q z}{dt^q} &= -\beta y, \end{cases} \tag{3}$$

in which α is allowed to be varied but $\beta = \frac{100}{7}$ is fixed.

Within some specific values of q, for example, $0.9,\ 1.0,\ 1.1$, system (3) can have chaotic dynamics in systems of effective order less than 3, verified by Lyapunov exponents. In Ref. [30], a useful method is provided for studying such state-space configurations. Phase projections for such a state-space system are illustrated in Fig. 1.

Table 1 shows the simulations of an appropriate qth ($q < 1$) integrals. For the cases of $q > 1$, the approximation equation for $\frac{1}{s^q}$ can be obtained by choosing $\frac{1}{s}$ times $\frac{1}{s^{q-1}}$ from the table.

In addition, Table 2 shows the largest Lyapunov exponents and their corresponding system effective order $q = 0.9,\ 1.0,\ 1.1$. Notice that the second largest Lyapunov exponent for each case is almost zero and the positive one indicates that the system can generate chaos. In Ref. [26], numerical simulations are given to show that the lower limit of the vector fractional derivative q for such systems to have chaos is

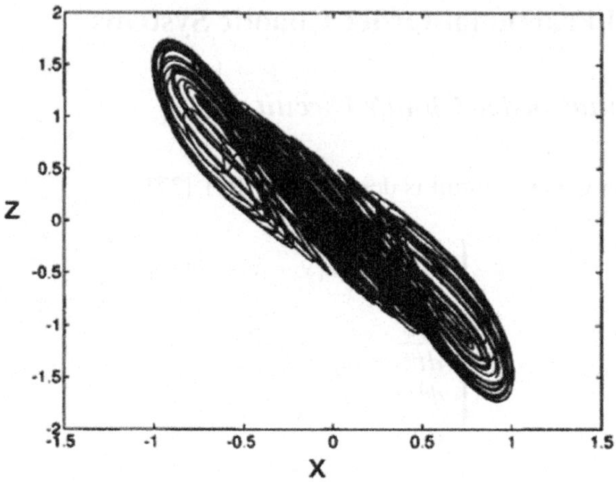

Fig. 1 Double-scroll attractor of Chua's circuit (3) in 2D, when the system effective order is 3.3 [26]

Table 1 List of integer-order approximations to fractional-order operators. Each has error of approximately 2 dB, from $w = 10^{-2}$ to 10^2 rad/s [26]

$$\frac{1}{s^{0.1}} \approx \frac{220.4s^4 + 5004s^3 + 5038s^2 + 234.5s + 0.4840}{s^5 + 359.8s^4 + 5742s^3 + 4247s^2 + 147.7s + 0.2099}$$

$$\frac{1}{s^{0.2}} \approx \frac{60.95s^4 + 816.9s^3 + 582.8s^2 + 23.24s + 0.04934}{s^5 + 134.0s^4 + 956.5s^3 + 383.5s^2 + 8.953s + 0.01821}$$

$$\frac{1}{s^{0.3}} \approx \frac{23.76s^4 + 224.9s^3 + 129.1s^2 + 4.733s + 0.01052}{s^5 + 64.51s^4 + 252.2s^3 + 63.61s^2 + 1.104s + 0.002267}$$

$$\frac{1}{s^{0.4}} \approx \frac{25.00s^4 + 558.5s^3 + 664.2s^2 + 44.15s + 0.1562}{s^5 + 125.6s^4 + 840.6s^3 + 317.2s^2 + 7.428s + 0.02343}$$

$$\frac{1}{s^{0.5}} \approx \frac{15.97s^4 + 593.2s^3 + 1080s^2 + 135.4s + 1}{s^5 + 134.3s^4 + 1072s^3 + 543.4s^2 + 20.10s + 0.1259}$$

$$\frac{1}{s^{0.6}} \approx \frac{8.579s^4 + 255.6s^3 + 405.3s^2 + 35.93s + 0.1696}{s^5 + 94.22s^4 + 472.9s^3 + 134.8s^2 + 2.639s + 0.009882}$$

$$\frac{1}{s^{0.7}} \approx \frac{5.406s^4 + 177.6s^3 + 209.6s^2 + 9.197s + 0.01450}{s^5 + 88.12s^4 + 279.2s^3 + 33.30s^2 + 1.927s + 0.0002276}$$

$$\frac{1}{s^{0.8}} \approx \frac{5.235s^3 + 1453s^2 + 5306s + 254.9}{s^4 + 658.1s^3 + 5700s^2 + 658.2s + 1}$$

$$\frac{1}{s^{0.9}} \approx \frac{1.766s^2 + 38.27s + 4.914}{s^3 + 36.15s^2 + 7.789s + 0.01000}$$

ranging from 0.8 to 0.9. In Table 2, the lowest system effective order to generate chaos is 2.7, using the fractional vector derivative $q = 0.9$. It is interesting to see that there is no upper bound for the effective order of the system.

Table 2 Largest Lyapunov exponents found in the state-space configuration, for $q = 0.9, 1.0, 1.1$, which gives a system effective order of 2.7, 3.0, 3.3, respectively

Effective order	Effective order approximation	α	LEs
2.7	9	12.75	(0.338, −0.000201, −0.132)
3.0	3	9.50	(0.248, −0.00412, −3.07)
3.5	18	7.00	(0.318, *, *)

2.2 *Fractional-Order Lorenz System*

Recall the classical Lorenz system described by Ref. [31]

$$\begin{cases} \dot{x} = -\sigma(y - x) \\ \dot{y} = x(r - z) - y \\ \dot{z} = xy - bz\,, \end{cases} \tag{4}$$

where parameters $\sigma,\ r,\ b > 0$. Physically, σ is the Prandtl number and r is the Rayleigh number.

Grigorenko [32] proposed a generalized form of the fractional-order Lorenz system, using the Caputo derivative, as follows:

$$\begin{cases} \dfrac{d^\alpha x}{dt^\alpha} = \sigma(y - x)\ , \\ \dfrac{d^\beta y}{dt^\beta} = \rho x - y - xz^r\ , \\ \dfrac{d^\gamma z}{dt^\gamma} = -xy - bz\ . \end{cases} \tag{5}$$

For $0 < \alpha,\ \beta,\ \gamma \le 1$, and $r \ge 1$, define the *effective order* or *effective dimension* of system (5) by $\Sigma = \alpha + \beta + \gamma$.

Set $\sigma = 10,\ \rho = 28,\ b = \frac{8}{3}$, so that in the case of $\alpha = \beta = \gamma = r = 1$, system (5) becomes the original Lorenz system (4).

By integrating the system for different values of the parameters $\alpha,\ \beta,\ \gamma,\ r$, and initial conditions, some interesting phenomena and properties of the system can be found.

The first finding is that such a fractional-order Lorenz system can generate chaos with the effective order Σ lower than 3. By setting the initial condition at $t = 0$ as $(x_0, y_0, z_0) = (10, 0, 10)$ and using the parameters $\alpha = \beta = \gamma = 0.99$, $r = 1$, the effective order of the system will be $\Sigma = 2.97 < 3$.

The phase portrait, in this case, is shown in Fig. 2, which is similar to the phase portrait of the original Lorenz system. In the figure, one could also characterize a

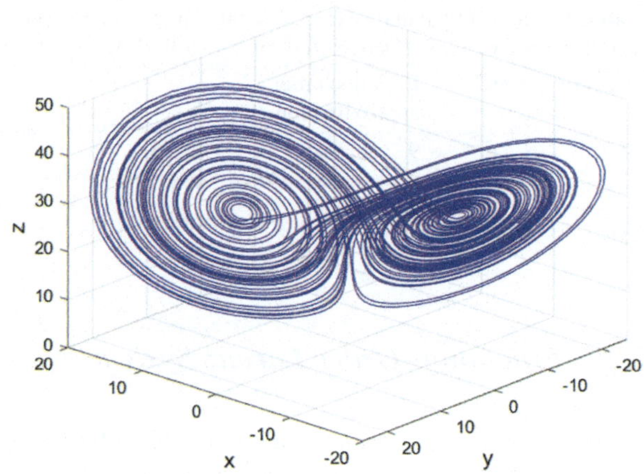

Fig. 2 Phase portrait of the generalized fractional-order Lorenz system, with the parameters $\alpha = \beta = \gamma = 0.99$, $r = 1$, and initial condition $(x_0, y_0, z_0) = (10, 0, 10)$ at $t = 0$. The effective order is $\Sigma = 2.97$ [32]

set of points as a strange attractor, but is slightly deformed compared to the original Lorenz chaotic attractor.

Note that it is rather time consuming to calculate Lyapunov exponents of system (5). Using an implicit procedure of the TISEAN package [33], the largest Lyapunov exponent of the above system was obtained, as $\lambda \approx 0.85$, signifying the existence of chaos.

Interestingly, some effective damping exists in system (5), when the effective order Σ decreases. In Ref. [32], it is concluded that the decrease of the parameters $\alpha,\ \beta,\ \gamma$ can result in a further decreasing of the largest Lyapunov exponent. At a specific critical dimension, for any initial condition, the dynamics of the system have some qualitative changes and finally become regular. The numerical solution and synchronization of the fractional-order Lorenz system are studied in Ref. [16, 34].

The lowest effective order that still is able to generate chaos is found to be 2.91. The corresponding parameter values are $\alpha \approx 0.91,\ \beta = \gamma = 1$, which indicates that the first linear differential equation of system (5) is less "sensitive" to the damping than the other two equations.

2.3 *Fractional-Order Chen System*

Recall the Chen system [35]

$$\begin{cases} \dot{x} = a(y - x) \\ \dot{y} = (c - a)x - xz + cy \\ \dot{z} = xy - bz\,, \end{cases} \tag{6}$$

where $a = 35$, $b = 3$, $c = 28$, with three equilibria and a chaotic attractor.

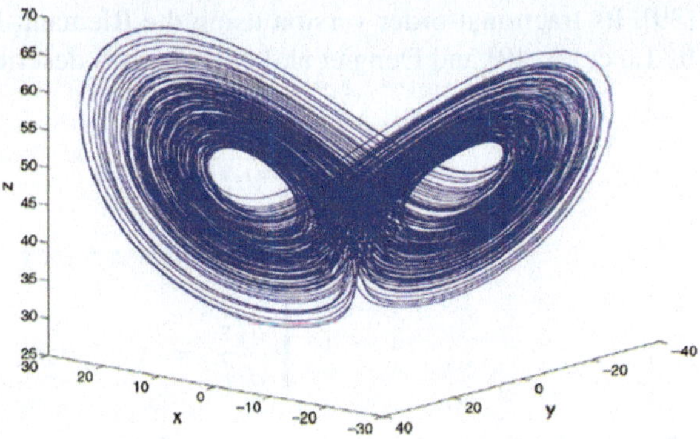

Fig. 3 Chaotic attractor of the fractional-order Chen system, with effective order $\Sigma = 2.1$ and $a = 35, b = 3, c = 40$ [38]

By replacing the ordinary derivative with the Riemann-Liouville derivative, the fractional-order Chen system is obtained as

$$\begin{cases} \dfrac{d^{\alpha_1} x}{dt^{\alpha_1}} = a(y - x) \\ \dfrac{d^{\alpha_2} y}{dt^{\alpha_2}} = (c - a)x - xz + cy \\ \dfrac{d^{\alpha_3} z}{dt^{\alpha_3}} = xy - bz , \end{cases} \tag{7}$$

where α_i is the fractional order with effective order $\Sigma = \alpha_1 + \alpha_2 + \alpha_3$. When $\alpha = 1$, the system becomes the original Chen system of integer order.

Numerical simulations on system (7) is studied in Ref. [16]. For $\alpha = 0.9,\ 0.8,\ 0.7,\ 0.6$, simulation results show that there exists a chaotic attractor with effective order less than 3. In particular, when $\alpha = 0.6$, no chaotic dynamics are found. However, chaotic attractors are clearly found for the cases of $\alpha = 0.9,\ 0.8,\ 0.7$ and their chaotic portraits are very similar to the original Chen system. Therefore, one can conclude that the lowest limit of the effective order for system (7) to be chaotic is $\Sigma = 1.8 \sim 2.1$.

The phase portrait of system (7), with effective order $\Sigma = 2.1$ and $a = 35,\ b = 3,\ c = 40$, is illustrated in Fig. 3.

Chaos and hybrid projective synchronization of the fractional-order Chen system are studied in Ref. [34, 36, 37], where a fractional-order hyperchaotic Chen system is also investigated.

2.4 *Fractional-order Lü System*

The integer-order Lü system is considered as the bridge between the Lorenz system and the Chen system, according to the study proposed by Vaněček and Čelikovský

[39]. Its fractional-order version using the Riemann-Liouville derivative is studied by Lu et al. [40] and Deng et al. [41], which is described by

$$\begin{cases} \dfrac{d_1^q x}{dt_1^q} = a(y-x) \\ \dfrac{d_2^q y}{dt_2^q} = -xz + cy \\ \dfrac{d_3^q z}{dt_3^q} = xy - bz\,, \end{cases} \tag{8}$$

where $a,\ b,\ c$ are system parameters and $0 < q_1,\ q_2,\ q_3 \leq 1$.

This fractional-order Lü system is discussed with $q = (q_1, q_2, q_3)$, including chaos synchronization, topological horseshoe analysis and circuit realization. In numerical simulations, bifurcation diagrams, numerical simulations and largest Lyapunov exponents were computed [16, 41, 42].

System (8) has three equilibria:

$$E_1 = (0,0,0), \quad E_2 = \left(\sqrt{bc}, \sqrt{bc}, c\right), \quad E_3 = \left(-\sqrt{bc}, -\sqrt{bc}, c\right).$$

Thus, the Jacobian matrix at the equilibria can be found:

$$J_{E^*} = \begin{bmatrix} -a & -a & 0 \\ -z^* & c & -x^* \\ y^* & x^* & -b \end{bmatrix}. \tag{9}$$

Through numerical simulations, with $a = 36,\ b = 3,\ c = 20$, one can obtain the eigenvalues $\lambda_1 = -3,\ \lambda_2 = 20,\ \lambda_3 = -36$ for saddle equilibrium E_1, and with $\lambda_1 \approx -22.6516,\ \lambda_{2,3} \approx 1.8258 \pm 13.6887j$ for the saddle-focus equilibrium $E_2(7.7460, 7.7460, 20)$, and with $\lambda_1 \approx -22.6516,\ \lambda_{2,3} \approx 1.8258 \pm 13.6887j$ for equilibrium $E_3\ (-7.7460, -7.7460, 20)$.

The lowest effective order for the fractional-order Lü system to generate chaos is q_i.0.9156, as can be calculated from the above eigenvalues.

Figure 4 shows the phase portraits for effective order $q_1 = q_2 = q_3 = 0.95$, with simulation time $90s$ and time step $h = 0.005$, when $a = 36$, $b = 3$, $c = 20$.

2.5 Fractional-Order Rössler System

The original Rössler system [43] has seven terms, with two equilibria. It is described by

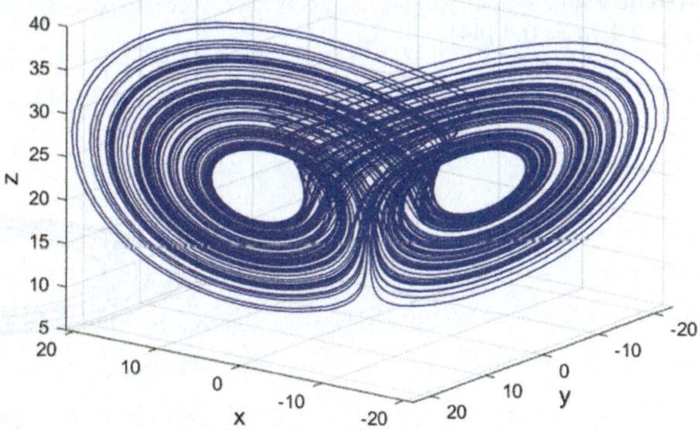

Fig. 4 Phase portrait of the fractional-order Lü system in state space, with parameters $a = 36,\ b = 3,\ c = 20$ and orders $q_1 = q_2 = q_3 = 0.95$ [16]

$$\begin{cases} \dot{x} = -(y + z) \\ \dot{y} = x + ay \\ \dot{z} = b + xz - cz\,, \end{cases} \tag{10}$$

where $a = b = 0.2,\ c = 5.7$, which can generate chaos.

Li and Chen [44] introduced the first factional-order Rössler system, utilizing the Riemann-Liouville derivative. This system has the following form:

$$\begin{cases} \dfrac{d^\alpha x}{dt^\alpha} = -(y + z) \\ \dfrac{d^\alpha y}{dt^\alpha} = x + ay \\ \dfrac{d^\alpha z}{dt^\alpha} = b + xz - cz\,, \end{cases} \tag{11}$$

where α is the fractional order. In Ref. [44], the parameters were chosen as $b = 0.2$ and $c = 10$, while a is allowed to be varied, giving

$$\begin{cases} \dfrac{d^\alpha x}{dt^\alpha} = -(y + z) \\ \dfrac{d^\alpha y}{dt^\alpha} = x + ay \\ \dfrac{d^\alpha z}{dt^\alpha} = 0.2 + xz - 10z\,. \end{cases} \tag{12}$$

In Ref. [44], all variables have the same fractional order α although different variables (i.e., $x,\ y,\ z$) can be calculated with different fractional orders. Clearly, system (12) is equivalent to the classical Rössler system if and only if the fractional order $\alpha = 1$.

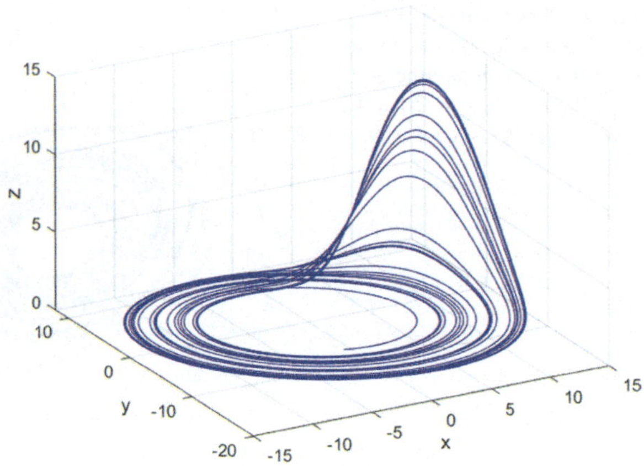

Fig. 5 Chaotic attractor of the fractional-order Rössler system, with $\alpha = 0.9,\ a = 0.4$ [44]

Numerical solution and simulation result from the fractional Rössler system were studied in Ref. [16, 44]. Considering the case with $a = 0.5,\ b = 0.2$ and $c = 10$, system (12) has two equilibria:

$$E_1 = (9.98998, -19.97997, 19.97997), \quad E_2 = (0.10010, -0.20020, 0.20020),$$

with the corresponding eigenvalues $\lambda_1 = 0.47595,\ \lambda_{2,3} = 0.007017 \pm 4.57910$ for the unstable focus-node equilibrium E_1 and $\lambda_1 = -9.98800,\ \lambda_{2,3} = 0.249007 \pm 0.96808$ for the unstable saddle-focus equilibrium E_2. With these eigenvalues, the system is chaotic with the minimal effective order of $\alpha > 0.839$.

Moreover, simulations were carried out using $\alpha = 0.7,\ 0.8$, and 0.9, respectively. The simulation results indicate that there exists chaotic behavior in the fractional-order Rössler system with effective order less than 3. In particular, Li and Chen [44] found that the lowest effective order for system (12) to generate chaos is 2.4. Since no chaotic behavior was found when $\alpha = 0.7$, the lowest limit of the fractional order should be between 0.7 and 0.8. Hence, 2.4 is the lowest effective dimension.

On the other hand, chaotic behaviors were found when $\alpha = 0.8,\ 0.9$. Figure 5 shows the phase portrait for the case of $\alpha = 0.9,\ a = 0.4$, which is very similar to the integer-order Rössler system [43].

The dynamical behavior of system (11) is further investigated in [45]. Synchronization and delayed feedback control of the fractional-order Rössler system were studied in [46, 47]. Studies found that there are spiral-type and screw-type chaotic attractors represented in the fractional-order Rössler system, while these types of attractors were only found in the integer-order Rössler system. The bifurcation diagram and largest Lyapunov exponent are analyzed, as well as the period-doubling route to chaos in the fractional-order Rössler system.

Moreover, Li and Chen [44] introduced a four-dimensional model of the fractional-order Rössler system after the 3D model (11). This hyperchaotic system is given by

$$\begin{cases} \dfrac{d^\alpha x}{dt^\alpha} &= -(y+z) \\ \dfrac{d^\alpha y}{dt^\alpha} &= x + ay + w \\ \dfrac{d^\alpha z}{dt^\alpha} &= 3 + xz \\ \dfrac{d^\alpha w}{dt^\alpha} &= -0.5z + 0.05w\,. \end{cases} \tag{13}$$

Notice that such a fractional-order system (13) is equivalent to the integer-order Rössler system [48] if and only if $\alpha = 1$. Interesting properties of the fractional-order Rössler hyperchaotic system were also discussed in Ref. [16, 49].

2.6 Fractional-Order Liu System

Liu et al. [50] introduced the following system:

$$\begin{cases} \dot{x} = -ax - ey^2 \\ \dot{y} = by - kxz \\ \dot{z} = -cz + mxy\,, \end{cases} \tag{14}$$

which has chaotic behavior with parameters $a = e = 1,\ b = 2.5,\ c = 5,\ k = m = 4$, and initial condition $(0.2, 0, 0.5)$. This system can be obtained from the classical Shimizu–Morioka chaotic system via an appropriate transformation of state variables and time rescaling [102].

Gejji and Bhalekar [51] introduced a fractional-order Liu system, as follows:

$$\begin{cases} \dfrac{d^{\alpha_1} x}{dt^{\alpha_1}} &= -ax - ey^2 \\ \dfrac{d^{\alpha_2} y}{dt^{\alpha_2}} &= by - kxz \\ \dfrac{d^{\alpha_3} z}{dt^{\alpha_3}} &= -cz + mxy\,, \end{cases} \tag{15}$$

where $0 < \alpha_1,\ \alpha_2,\ \alpha_3 \le 1$.

If $\alpha_1 = \alpha_2 = \alpha_3 = \alpha$, the system (15) has a non-integer effective order.

System (14) has three real equilibria with corresponding eigenvalues $\lambda_1 = -5,\ \lambda_2 = 2.5,\ \lambda_3 = -1$ at equilibrium $E_1(0, 0, 0)$, eigenvalues $\lambda_1 = -4.3878$, $\lambda_{2,3} = 0.4439 \pm 3.3464i$ at $E_2(-0.8839, 0.9402, -0.6648)$, and $\lambda_1 = -4.3878$, $\lambda_{2,3} = 0.4439 \pm 3.3464i$ at $E_2(-0.8839, -0.9402, 0.6648)$. The equilibria E_1 and E_2 are both saddle nodes of index 2. Therefore, there is a two-scroll chaotic attractor in the fractional-order Liu system (15).

Fig. 6 Chaotic attractor of the fractional-order Liu system, with parameters $a = e = 1,\ b = 2.5,\ c = 5,\ k = m = 4$, orders $\alpha_1 = \alpha_2 = \alpha_3 = \alpha = 0.95$, and initial condition $(0.2, 0, 0.5)$ [16]

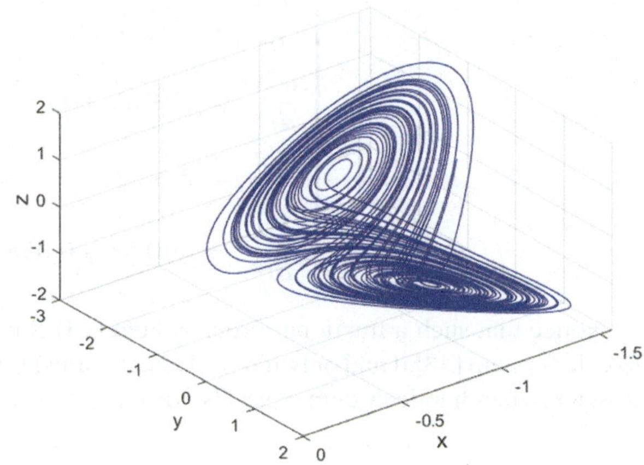

Fig. 7 Chaotic attractor of the fractional-order Liu system, with parameters $a = e = 1,\ b = 2.5,\ c = 5,\ k = m = 4$, orders $\alpha_1 = 1.0,\ \alpha_2 = 0.9,\ \alpha_3 = \alpha = 0.8$ and initial condition $(0.2, 0, 0.5)$ [16]

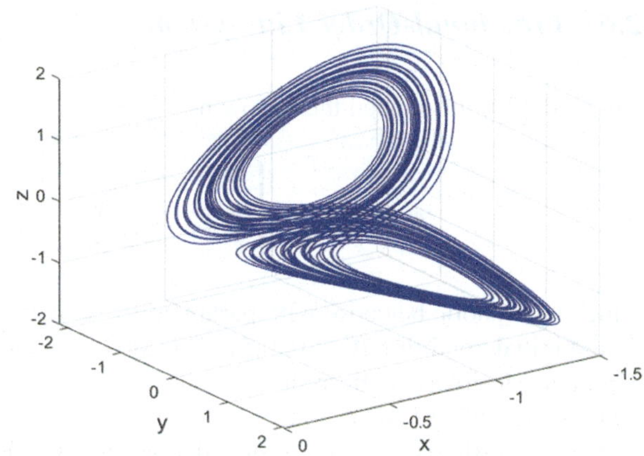

Now, consider an effective order of system (15), with $\alpha_1 = \alpha_2 = \alpha_3 = \alpha = 0.95$ and with parameters $a = e = 1,\ b = 2.5,\ c = 5,\ k = m = 4$. By linearizing the system, its characteristic equation can be obtained, as

$$\lambda^{285} + 3.5\lambda^{190} + 7.5\lambda^{95} + 50 = 0\,,$$

giving $\lambda_{1,2} = 1.0128 \pm 0.0153j$. Hence, the minimal order α is determined as $\alpha > 0.916$. Figure 6 shows the phase trajectory of the fractional-order system.

On the other hand, by setting $\alpha_1 = 1.0,\ \alpha_2 = 0.9$ and $\alpha_3 = 0.8$, with parameters $a = e = 1,\ b = 2.5,\ c = 5,\ k = m = 4$, the corresponding characteristic equation of the fractional-order system is obtained as

$$\lambda^{27} + 5\lambda^{19} - 2.5\lambda^{18} + \lambda^{17} + 5\lambda_9 + 2.5\lambda^8 + 50 = 0\,,$$

giving $\lambda_{1,2} = 1.1224 \pm 0.1770j$. Figure 7 illustrates the phase trajectory of the fractional-order system.

Numerical results show the existence of chaos in the fractional-order Liu system and give the lowest effective order of the system (15) [16]. Synchronization and time-delay control of this system are also discussed in Refs. [51–53].

2.7 *Fractional-Order System with Multi-Scroll Attractors*

Two examples that demonstrate multi-scroll chaotic attractors in fractional-order systems are discussed.

2.7.1 Example 1

Fractional-order systems with multi-directional multi-scroll chaotic attractors include one-directional (1D) n-scroll, two-directional (2D) $n \times m$-grid scroll and three-directional (3D) $n \times m \times l$-grid scroll chaotic attractors, which were studied by Deng and Lü in Refs. [54, 55].

Using the Caputo derivative, the fractional-order chaotic system is described by

$$\begin{bmatrix} \frac{d^{\alpha_1} x}{dt^{\alpha_1}} \\ \frac{d^{\alpha_2} y}{dt^{\alpha_2}} \\ \frac{d^{\alpha_3} z}{dt^{\alpha_3}} \end{bmatrix} . = \begin{bmatrix} 0 & 1 & 0 \\ 0 & 0 & 1 \\ -a & -b & -c \end{bmatrix} . = \begin{bmatrix} x \\ y \\ z \end{bmatrix} . = A \begin{bmatrix} x \\ y \\ z \end{bmatrix}, \tag{16}$$

where $a,\ b,\ c$ are positive constants, A is the coefficient matrix, and α_i is the fractional order that satisfies $0 < \alpha_i < 1, i = 1, 2, 3$.

The characteristic equation of system (16) is

$$\lambda^3 + c\lambda^2 + b\lambda + a = 0\,,$$

and the stability of the equilibria can be easily determined [56].

A generalization of the predictor-corrector scheme [18] is used for the numerical computation of system (16) and a systematic design method is used to construct the fractional-order multi-directional multi-scroll chaotic system in Ref. [54].

Applying a saturated function series as the controller, the controlled system is given by

$$D_*^{\alpha} X \quad = \quad AX + BU(X)\,, \tag{17}$$

where

$$B = \begin{bmatrix} 0 & -\frac{d_2}{b} & 0 \\ 0 & 0 & -\frac{d_3}{c} \\ d_1 & d_2 & d_3 \end{bmatrix}.$$

and $X = (x, y, z)^T$, $D_*^{\alpha} X = (D_*^{\alpha_1} x, D_*^{\alpha_2} y, D_*^{\alpha_3} z)$, with the multi-directional controllers given by the following.

(i) 1D n-scroll controller:

$$U(X) = \begin{bmatrix} f(x; k_1, h_1, p_1, q_1) \\ 0 \\ 0 \end{bmatrix}. \tag{18}$$

(ii) 2D $n \times m$-grid scroll controller:

$$U(X) = \begin{bmatrix} f(x; k_1, h_1, p_1, q_1) \\ f(y; k_2, h_2, p_2, q_2) \\ 0 \end{bmatrix}. \tag{19}$$

(iii) 3D $n \times m \times l$-grid scroll controller:

$$U(X) = \begin{bmatrix} f(x; k_1, h_1, p_1, q_1) \\ f(y; k_2, h_2, p_2, q_2) \\ f(z; k_3, h_3, p_3, q_3) \end{bmatrix}. \tag{20}$$

In the above, a, b, c, d_1, d_2, d_3 are positive constants and

$$f(x; k_1, h_1, p_1, q_1),\ f(y; k_2, h_2, p_2, q_2),\ f(z; k_3, h_3, p_3, q_3)$$

are defined by the following piecewise linear function [57]:

$$f(x; k, h, p, q) = \sum_{i=-p}^{q} f_i(x; k, h). \tag{21}$$

Simulation results show that a 1D 6-scroll chaotic system, with fractional order $\alpha_1 = 0.85$, $\alpha_2 = 0.9$, $\alpha_3 = 0.9$ and parameters $a = 2$, $b = 1$, $c = 0.6$, $d_1 = 2$, $k_1 = 10$, has Lyapunov exponents $(LE_1, LE_2, LE_3) = (0.2251, 0, -1.2143)$. A 2D 6×6-grid scroll chaotic system with fractional order $\alpha_1 = 0.8$, $\alpha_2 = 0.9$, $\alpha_3 = 1.0$ and parameters $a = 2$, $b = 1$, $c = 0.5$, $d_1 = 2$, $d_2 = 1$, $k_1 = k_2 = 50$ has Lyapunov exponents $(LE_1, LE_2, LE_3) = (0.2131, 0, -1.5718)$. Moreover, a 3D $6 \times 6 \times 6$-grid scroll chaotic attractor are illustrated in Fig. 8. In this case, the fractional order is $\alpha_1 = \alpha_2 = \alpha_3 = 0.9$ with parameters $a = 2.2$, $b = 1.3$, $c = 0.6$, $d_1 = 2.2$, $d_2 = 1.3$, $d_3 = 0.6$, $,_1 = 100$, $k_2 = k_3 = 40$. Lyapunov exponents for the three-directional multi-scroll chaotic system are (LE_1, LE_2, LE_3)=$(0.3629, 0, -1.2742)$.

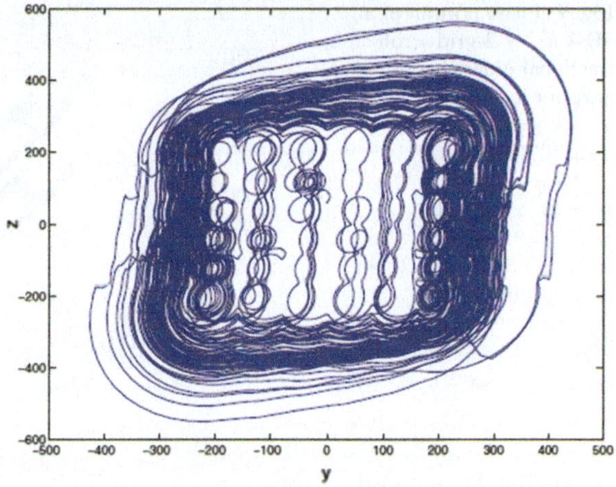

Fig. 8 A 3D 6 × 6 × 6-grid scroll attractor, with factional order $\alpha_1 = \alpha_2 = \alpha_3 = 0.9$ on the y-z plane [54]

2.7.2 Example 2

Based on the chaotic system of Lü et al. [58], Fei et al. [59] introduced a series of trigonometric functions that can generate multi-directional multi-scroll fractional-order chaos. Three models were proposed, for generating 1D, 2D, and 3D multi-scroll chaotic attractors. Here, only the 3D fractional-order $m \times n \times l$-grid scroll chaotic model is discussed, which is given by

$$\begin{cases} \dfrac{d^{\alpha_1}x}{dt^{\alpha_1}} = y - \dfrac{\tau_2}{2}[(r_2 - s_2) + \displaystyle\sum_{j=-r_2}^{s_2} \dfrac{2}{\pi} tan^{-1}(y - j\tau_2)] \ , \\ \dfrac{d^{\alpha_2}y}{dt^{\alpha_2}} = z - \dfrac{\tau_3}{2}[(r_3 - s_3) + \displaystyle\sum_{j=-r_3}^{s_3} \dfrac{2}{\pi} tan^{-1}(z - j\tau_3)] \ , \\ \dfrac{d^{\alpha_3}z}{dt^{\alpha_3}} = -ax - by - cz \\ \quad +a\frac{\tau_1}{2}[(r_1 - s_1) + \sum_{j=-r_1}^{s_1} \frac{2}{\pi} tan^{-1}(z - j\tau_1)] \\ \quad +b\frac{\tau_2}{2}[(r_2 - s_2) + \sum_{j=-r_2}^{s_2} \frac{2}{\pi} tan^{-1}(y - j\tau_2)] \\ \quad +c\frac{\tau_3}{2}[(r_3 - s_3) + \sum_{j=-r_3}^{s_3} \frac{2}{\pi} tan^{-1}(z - j\tau_3)], \end{cases} \tag{22}$$

where $a, b, c, \tau_1, \tau_2, \tau_3$ are positive constants, and $r_1, r_2, r_3, s_1, s_2, s_3$ are non-negative integers, with the fractional order α_i satisfying $0 < \alpha_i < 1, i = 1, 2, 3$.

Note that system (22) has 3D $(r_1 + s_1 + 2) \times (r_2 + s_2 + 2) \times (r_3 + s_3 + 2)$ chaotic attractors, where $m = r_1 + s_1 = 2, n = r_2 + s_2 + 2$, and $l = r_3 + s_3 + 2$.

In generating such a fractional-order 3D multi-scroll chaotic system (22), inverse trigonometric function $\tan^{-1}(x)$ was used in [60], instead of the piecewise-linear function, which is different form the previous example (16). The $tan^{-1}(x)$ function

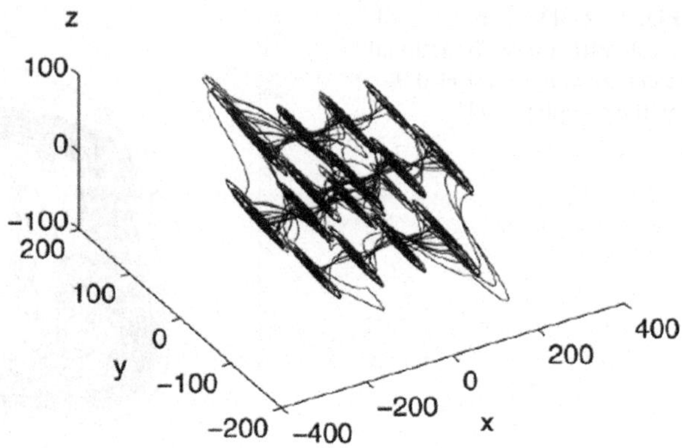

Fig. 9 Phase portrait of a 3D $4 \times 2 \times 2$-grid scroll fractional-order chaotic attractor in system (22), with $\alpha_1 = \alpha_2 = 0.99$, $\alpha_3 = 0.998$, $a = b = c = 0.5$, $\tau_1 = \tau_2 = \tau_3 = 100$, $r_1 = s_1 = 1$, and $r_2 = r_3 = s_2 = s_3 = 0$, on the x-y-z plane [59]

series can also be used:

$$f(x) = \frac{2}{\pi} \sum_{j=-r}^{s} tan^{-1}(x + j\tau), \tag{23}$$

in which τ is a real parameter, and r, s are non-negative integers.

By selecting the fractional orders as $\alpha_1 = \alpha_2 = 0.99, \alpha_3 = 0.998$, and the parameters as $a = b = c = 0.5, \tau_1 = \tau_2 = \tau_3 = 100, r_1 = s_1 = 1$, and $r_2 = r_3 = s_2 = s_3 = 0$, the chaotic trajectory of the system (22) is a 3D $4 \times 2 \times 2$-grid scroll chaotic attractor, as shown in Fig. 9.

3 Fractional-Order Chaotic System with a Limited Number of Equilibria

3.1 3D Examples

3.1.1 Example 1

Based on the Rikitake system [14], a fractional-order 3D chaotic system was proposed in Ref. [11] by using the Caputo derivative, the new fractional-order system is described by

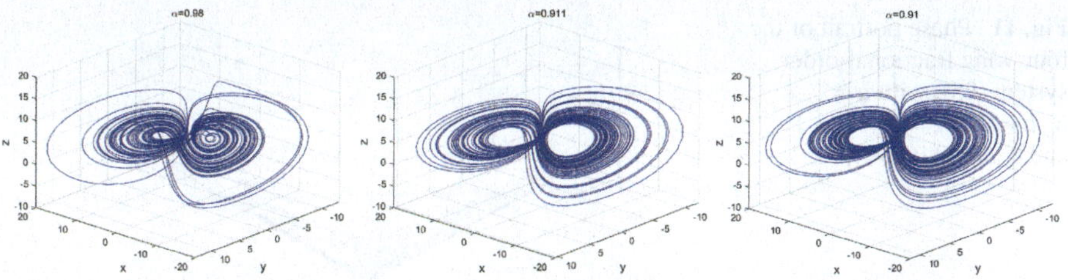

Fig. 10 Phase portraits of the fractional-order system (24), with initial condition (0.2, 0, 0.5) in the 3D space: (a) $\alpha = 0.98$; (b) $\alpha = 0.911$; (c) $\alpha = 0.91$ [11]

$$\begin{cases} \dfrac{d^\alpha x}{dt^\alpha} = -\mu x + yz \\ \dfrac{d^\alpha y}{dt^\alpha} = -\mu y + (z-a)x \\ \dfrac{d^\alpha z}{dt^\alpha} = z - xy\,, \end{cases} \tag{24}$$

which has five equilibria. By analyzing the Jacobian matrix and the corresponding eigenvalues, Wu and Wang [11] found that the equilibrium $E_0(0, 0, 0)$ is an unstable saddle point, while all the other four equilibria are unstable saddle-foci.

In numerical simulations, choose $\mu = 2$ and $a = 5$ for system (24). The result shows that the system can generate chaos with fractional order less than 3, namely, $0.911 \le \alpha \le 1$. Especially, when $\alpha = 0.98$, the phase portrait with the largest Lyapunov exponent $LE_{max} = 0.3104$ is shown in Fig. 10a; when $\alpha = 0.911$, the phase portrait with the largest Lyapunov exponent $LE_{max} = 0.1653$ is shown in Fig. 10b. Moreover, system (24) has no chaotic behavior $\alpha < 0.911$, as shown in Fig. 10c. Therefore, the minimal fractional order for the system to exhibit chaos is $\alpha = 0.91 - 0.911$. The lowest effective order of system (24) is 2.733.

3.1.2 Example 2

A fractional-order four-wing chaotic system was introduced by Jia et al. [61], based on the fractional transfer function approximation in frequency domain.

Using the Riemann-Liouville derivative, the new chaotic system is described by

$$\begin{cases} \dfrac{d^\alpha x}{dt^\alpha} = ax + dy - yz \\ \dfrac{d^\alpha y}{dt^\alpha} = -by - z + xz \\ \dfrac{d^\alpha z}{dt^\alpha} = -x - cz + xy\,, \end{cases} \tag{25}$$

where $0 < \alpha < 1$ and $a,\ b,\ c,\ d$ are real parameters.

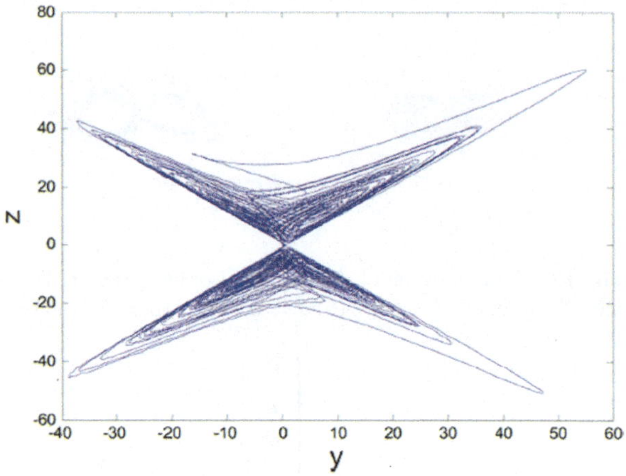

Fig. 11 Phase portrait of the four-wing fractional-order system (25), with $a = 6,\ b = 12,\ c = 5,\ d = 1$ and $\alpha = 0.9$ [61]

By selecting parameters $a = 6,\ b = 12,\ c = 5,\ d = 1$ and $\alpha = 0.9$, system (25) can display a rare four-wing chaotic attractor. A chaotic attractor is also found when the effective order of the system is varied between 1.5 and 2.7. Hence, different chaotic behaviors occur when the system parameters $a,\ b,\ c,\ d$ are varied. In Ref. [61], some bifurcation diagrams and phase portraits are illustrated, in which the four-wing chaotic attractor and two-wing attractor are included. Figure 11 shows a typical case, which confirms the chaotic dynamics of the system.

Moreover, the topological entropy of the system was investigated, to prove that the topological entropy of system (25) is non-zero, and for an m-shift map g, one has $ent(f) \geq ent(g) = log\, m$, where $m > 1$.

3.1.3 Example 3

Using the Caputo derivative, Ma et al. [62] introduced a fractional-order chaotic system, described by

$$\begin{cases} \dfrac{d^{\alpha_1} x}{dt^{\alpha_1}} = a(y - x) \\ \dfrac{d^{\alpha_2} y}{dt^{\alpha_2}} = bx - xz \\ \dfrac{d^{\alpha_3} z}{dt^{\alpha_3}} = -dz + xy + fx^2, \end{cases} \tag{26}$$

in which the fractional order satisfies $0 < \alpha_i \leq 1, i = 1, 2, 3$.

In Ref. [62], system (26) was found to exhibit chaotic dynamics if and only if the system fractional order satisfies $0.93 < \alpha_i \leq 1, i = 1, 2, 3$, according to the stability theory in fractional calculus [63].

By selecting parameters $a = 6$, $b = 16$, $d = 2$, $f = 1$ and fractional orders $\alpha_1 = \alpha_2 = \alpha_3 = 0.92$, the phase portrait of the system is shown in Fig. 12, which exhibits no chaotic dynamics. Here, the numerical simulation results show the minimal fractional order $0.92\alpha \leq 0.93$. Therefore, the lowest effective order of the system is 2.79.

Moreover, linear feedback control and synchronization of the factional-order system (26) were investigated. A controller for stabilizing system (26) to its unstable equilibrium $(4, 4, 16)$ was formulated by applying the linear feedback control method. On the other hand, a nonlinear control method and the Lyapunov stability theory were used to realize chaos synchronization.

3.1.4 Example 4

Motivated by Ref. [64], Kingni et al. [65] proposed a 3D chaotic system with circular equilibria and its fractional-order version. The fractional-order chaotic system is

$$\begin{cases} \dfrac{d^\alpha x}{dt^\alpha} = z \\ \dfrac{d^\alpha y}{dt^\alpha} = z^3 + z^2 + 3xz \\ \dfrac{d^\alpha z}{dt^\alpha} = x^2 + y^2 - r^2 - 4yz^2 , \end{cases} \tag{27}$$

where r is a real parameter, and the Caputo derivative is used.

This system has circular equilibria as

$$E = (x^*, \pm\sqrt{r^2 - (x^*)2}, 0) . \tag{28}$$

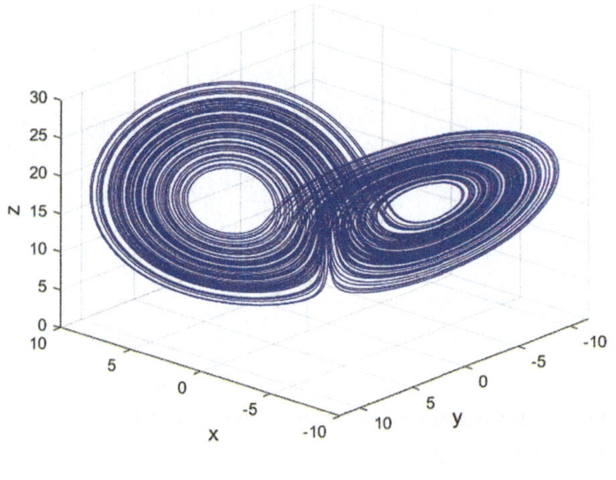

Fig. 12 Phase portrait of the fractional-order system (26), with $a = 6$, $b = 16$, $d = 2$, $f = 1$, $\alpha = 0.92$, and initial condition $(1, 0, -1)$ [62]

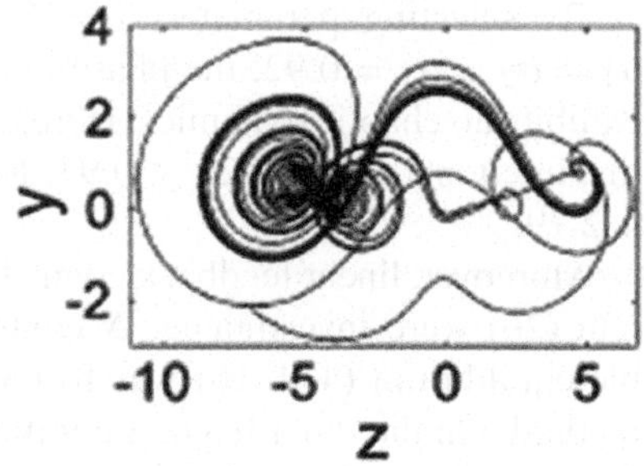

Fig. 13 Phase portrait of the fractional-order system (27), with $\alpha = 0.92$ [67]

By evaluating the eigenvalues of the corresponding Jacobian matrix at each equilibrium, it was found that the equilibria are not saddle points of index 2 [66]. Hence, there is no analytical method to determine the minimal effective order of the fractional-order system (27) for generating chaos. Also, the effect of fractional derivation for the case of $r = 2$ can only be numerically investigated.

In Fig. 13, the phase portrait of the system with $\alpha = 0.92$ is illustrated. By analyzing the dynamic behavior and the autocorrelation, it was found that the system (27) converges to no equilibria when $\alpha = 0.92$ or $\alpha = 0.87$.

3.2 4D Examples

Next, several 4D examples of fractional-order hyperchaotic systems are discussed.

3.2.1 Example 5

Based on the investigation in Ref. [68], Wu et al. [69] studied a fractional-order hyperchaotic system defined by the Caputo derivative. The hyperchaotic system is given by

$$\begin{cases} \dfrac{d^\alpha x}{dt^\alpha} = a(y - x) \\ \dfrac{d^\alpha y}{dt^\alpha} = dx - xz + cy - w \\ \dfrac{d^\alpha z}{dt^\alpha} = dxy - bz \\ \dfrac{d^\alpha w}{dt^\alpha} = x + k\,, \end{cases} \tag{29}$$

where the fractional order α satisfies $0 < \alpha \leq 1$.

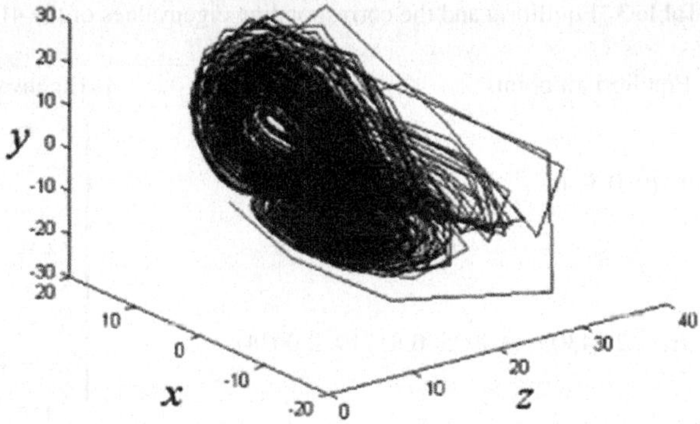

Fig. 14 Hyperchaotic attractors of the fractional-order hyperchaotic system (29), with the fractional order $\alpha = 0.9$ [69]

Numerical simulation and synchronization were investigated in Ref. [69]. By selecting $a = 36$, $b = 3$, $c = 28$, $d = -16$, $k = 0.5$, both system (29) and its integer-order version are hyperchaotic, displaying the hyperchaotic behaviors. In particular, this fractional-order hyperchaotic system can generate chaos with effective order less than 4. The Lyapunov exponents can be calculated by the familiar Wolf algorithm.

When $\alpha = 0.9$, there exhibits chaos with two largest Lyapunov exponents $\lambda_1 = 12.3014$ and $\lambda_2 = 0.2318$, as shown in Fig. 14. Moreover, hyperchaotic attractors were found with the two largest Lyapunov exponents $\lambda_1 = 8.2130$ and $\lambda = 0.1015$, with the fractional order $\alpha = 0.72$.

When $\alpha < 0.72$, however, no hyperchaotic behaviors were found in system (29). Therefore, the lowest limit of the fractional order for the hyperchaotic system (29) to generate hyperchaotic dynamics is $\alpha = 0.71 - 0.72$, which indicates that the lowest effective order for the system to yield hyperchaos is 2.88.

3.2.2 Example 6

Chen et al. [70] proposed another fractional-order hyperchaotic system, described by

$$\begin{cases} \dfrac{d^\alpha x}{dt^\alpha} = a(y - x) + byz + w \\ \dfrac{d^\alpha y}{dt^\alpha} = -cx - dxz^2 + gy \\ \dfrac{d^\alpha z}{dt^\alpha} = y^2 - kz \\ \dfrac{d^\alpha w}{dt^\alpha} = by - w\,, \end{cases} \tag{30}$$

Table 3 Equilibria and the corresponding eigenvalues of the 4D fractional-order system (30) [70]

Equilibrium points	Eigenvalues
$E_0(0, 0, 0, 0)$	$\begin{cases} \lambda_1 = -30.847 \\ \lambda_2 = 23.823 \\ \lambda_3 = -0.97638 \\ \lambda_4 = -0.33333 \end{cases}$
$E_1(-2.0430, -0.9925, 0.81719, 2.0034)$	$\begin{cases} \lambda_1 = -29.546 \\ \lambda_2 = 21.738 \\ \lambda_3 = 0.4585 \\ \lambda_4 = -0.98297 \end{cases}$
$E_2(2.0430, 0.9925, 0.81719, 2.0034)$	$\begin{cases} \lambda_1 = -28.786 \\ \lambda_2 = 20.924 \\ \lambda_3 = 0.51142 \\ \lambda_4 = -0.98249 \end{cases}$
$E_3(-2.2653i, -0.81142i, -0.90611i, -2.4631)$	$\begin{cases} \lambda_1 = -32.849 - 0.14436i \\ \lambda_2 = 26.323 + 0.13977i \\ \lambda_3 = -0.85676 + 0.0077466i \\ \lambda_4 = -0.95049 - 0.0031571i \end{cases}$
$E_4(2.2653i, 0.81142i, 0.90611i, -2.4631)$	$\begin{cases} \lambda_1 = -32.849 + 0.14436i \\ \lambda_2 = 26.323 - 0.13977i \\ \lambda_3 = -0.85676 - 0.0077466i \\ \lambda_4 = -0.95049 + 0.0031571i \end{cases}$
$E_5(-5.5.38i, -0.72293e - i, -2.2015i, -14.540)$	$\begin{cases} \lambda_1 = -41.265 + 0.90343i \\ \lambda_2 = 34.349 - 0.89941i \\ \lambda_3 = -1.0238 - 0.0012513i \\ \lambda_4 = -0.39305 - 0.0027599i \end{cases}$
$E_6(5.5.38i, 0.72293e - i, 2.2015i, -14.540)$	$\begin{cases} \lambda_1 = -41.265 - 0.90343i \\ \lambda_2 = 34.349 + 0.89941i \\ \lambda_3 = -1.0238 + 0.0012513i \\ \lambda_4 = -0.39305 + 0.0027599i \end{cases}$

where a, b, c, d, g, k are non-negative constants and α is the fractional order satisfying $0 < \alpha < 1$.

This 4D fractional-order system has seven equilibria. Through numerical simulation, the corresponding eigenvalues can calculated as given in Table 3. Moreover, the results of numerical simulation and circuit simulation are in good agreement. With $\alpha = 0.8$ and $a = 35,\ b = 2.5,\ c = 7,\ d = 4,\ g = 28,\ k = \frac{1}{3}$ with initial condition $(0.1, 0.1, 2.1, 0.1)^T$, the state trajectory of system (30) is shown in Fig. 15.

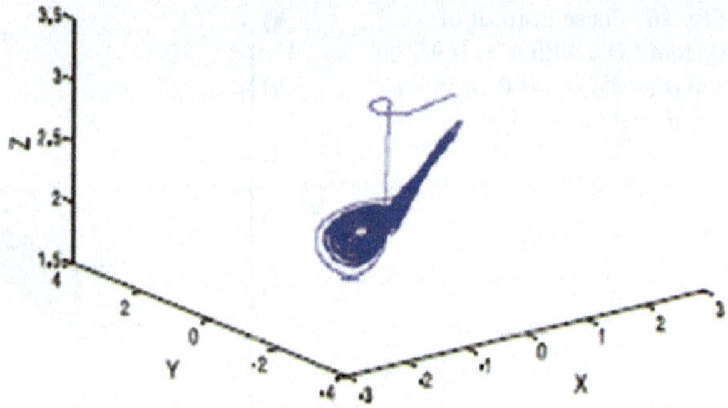

Fig. 15 SDate trajectory of system (30), with $\alpha = 0.8$ [70]

3.2.3 Example 7

Inspired by Refs. [44, 71–73], Gao et al. [74] constructed a fractional-order hyperchaotic system based on the Liu system [50].

The new fractional-order system is given by

$$\begin{cases} \dfrac{d^\alpha x}{dt^\alpha} = -ax + by \\ \dfrac{d^\alpha y}{dt^\alpha} = cx - xz - y + w \\ \dfrac{d^\alpha z}{dt^\alpha} = x^2 - d(x + z) \\ \dfrac{d^\alpha w}{dt^\alpha} = -gx\,, \end{cases} \tag{31}$$

where a, b, c, d, g are constant parameters and α is the fractional order satisfying $0 < \alpha \leq 1$.

Since one cannot estimate the lowest order for the system (31) to exhibit hyperchaotic behaviors, according to the necessary condition for the existence of chaotic attractor in a fractional-order system [75, 76], numerical method was used in Ref. [44] with the predictor-corrector algorithm under condition $a = 25,\ b = 60,\ c = 40,\ d = 4,\ g = 5$. When $\alpha = 0.95$, the four Lyapunov exponents are obtained as $\lambda_1 = 2.3057,\ \lambda_2 = 0.1424,\ \lambda_3 = -0.4631,\ \lambda_4 = -47.0718$.

Therefore, the lowest limit of the fractional order for the system to have hyperchaos is $\alpha = 0.83 - 0.84$ based on simulation results. Thus, the hyperchaotic behavior does occur in the fractional-order hyperchaotic system with order less than 4. Figure 16 illustrates the phase portrait of the system, with $\alpha = 0.95$ and $a = 25,\ b = 60,\ c = 40,\ d = 4,\ g = 5$.

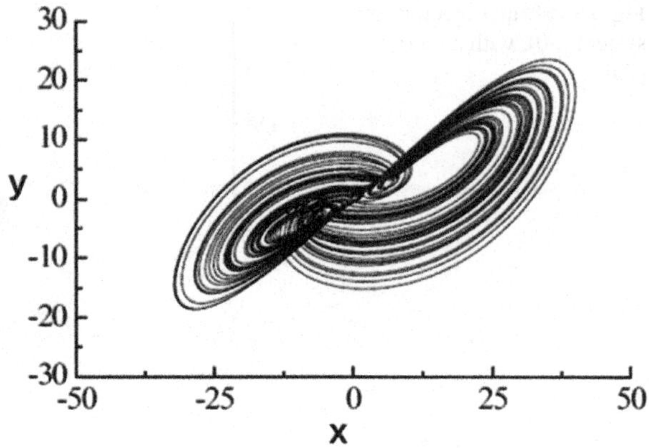

Fig. 16 Phase portrait of system (31), with $\alpha = 0.95$ and $a = 25,\ b = 60,\ c = 40,\ d = 4,\ g = 5$ [74]

4 Fractional-Order Systems with an Infinite Number of Equilibria

Motivated by the research on chaotic systems with an infinite number of equilibria [67, 77–79], Kingni et al. [80] presented a 3D fractional-order autonomous system with an infinite number of equilibria. Based on the predictor-corrector scheme [18] and using the Caputo derivative, this system is described by

$$\begin{cases} \dfrac{d^\alpha x}{dt^\alpha} &= -z \\ \dfrac{d^\alpha y}{dt^\alpha} &= -xz^2 + a\,sign(z) \\ \dfrac{d^\alpha z}{dt^\alpha} &= x - x^2 y + z(by^2 - z^2)\,, \end{cases} \tag{32}$$

where α is the fractional order satisfying $0 < \alpha \leq 1$.

Through numerical simulation, it is demonstrated that the fractional-order system (32) has an infinite number of equilibria, located on a line $E_1(0, y^*, 0)$ and a hyperbolic curve $E_2(\frac{1}{y^*}, y^*, 0)$.

The corresponding characteristic equations for both equilibrium curves E_1 and E_2 are:

$$\lambda[\lambda^2 - b(y^*)^2\lambda + 1] = 0 \tag{33}$$

and

$$\lambda[\lambda^2 - b(y^*)^2\lambda - 1 - 4a\delta(z^*)(\frac{1}{y^*})^2] = 0\,. \tag{34}$$

Thus, eigenvalues of system (32) can be calculated and it was found that two equilibria E_1 and E_2 are both not saddle points of index 2 [75]. Therefore, one cannot determined the minimal fractional order of the system (32) to generate hyperchaos.

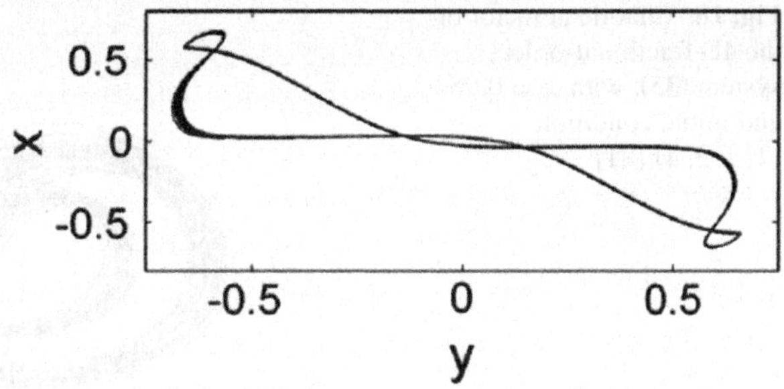

Fig. 17 Phase portrait of system (32), with $\alpha = 0.6$ and $a = 0.25,\ b = 5.95$ and initial condition $(x_0.y_0.z_0) = (0.1, 0.1, 0.1)$ [80]

Figure 17 shows the phase portrait of the fractional-order system with an infinite number of equilibria, when $\alpha = 0.6$ and $a = 0.25,\ b = 5.95$, with initial condition $(x_0.y_0.z_0) = (0.1, 0.1, 0.1)$. The time series of the results and the projective synchronization of the system were also considered in [80].

On the other hand, another 4D fractional-order chaotic system was also introduced, as

$$\begin{cases} \dfrac{d^\alpha x}{dt^\alpha} &= 10(y - x) + w \\ \dfrac{d^\alpha y}{dt^\alpha} &= 15x - xz \\ \dfrac{d^\alpha z}{dt^\alpha} &= 4x^2 - 2.5z \\ \dfrac{d^\alpha w}{dt^\alpha} &= -10y - w\,, \end{cases} \tag{35}$$

where α is the fractional order. Let

$$\begin{cases} \dfrac{d^\alpha x}{dt^\alpha} &= 0 \\ \dfrac{d^\alpha y}{dt^\alpha} &= 0 \\ \dfrac{d^\alpha z}{dt^\alpha} &= 0 \\ \dfrac{d^\alpha w}{dt^\alpha} &= 0\,. \end{cases} \tag{36}$$

Then, the equilibria of system (35) are obtained, as $E = (0, y, 0, -10y)$, which shows an infinite number of equilibria in system (35).

The Jacobian matrix at the equilibria is

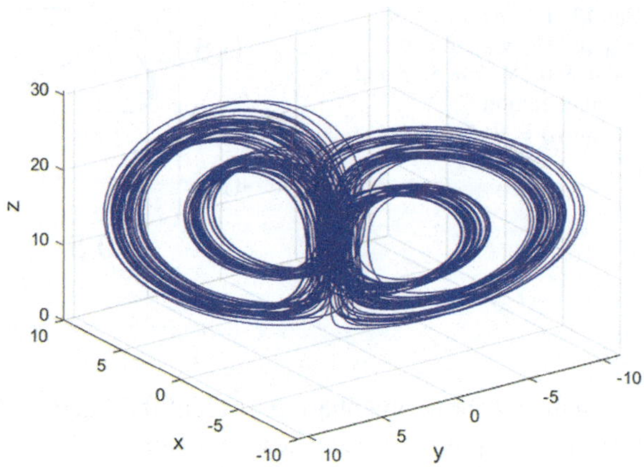

Fig. 18 Chaotic attractor of the 4D fractional-order system (35), with $\alpha = 0.95$ and initial condition $(1, 2, 3, 4)$ [81]

$$J = \begin{bmatrix} -10 & 10 & 0 & 1 \\ 15 & 0 & 0 & 0 \\ 0 & 0 & -2.5 & 0 \\ 0 & -10 & 0 & -1 \end{bmatrix},$$

and its corresponding eigenvalues are

$$\lambda_1 = -18.548, \quad \lambda_2 = -2.5, \quad \lambda_3 = 0, \quad \lambda_4 = 7.548\,,$$

which are all unstable.

By simulation, the largest Lyapunov exponent of the system (35) is found to be 0.8939, which implies that the system has chaotic behaviors. Figure 18 shows the chaotic attractor of the system. However, when $\alpha = 1$, the system becomes integer-order, so no chaotic behaviors can be found from the system.

5 Fractional-Order Systems with Stable Equilibria

5.1 Lorenz-like system with Two Stable Node-foci

Using the Caputo derivative, a fractional-order system with two stable node-foci was introduced in [82]. This system has a Lorenz-like form, as

$$\begin{cases} \frac{d^{\alpha_1}x}{dt^{\alpha_1}} &= a(y-x) \\ \frac{d^{\alpha_2}y}{dt^{\alpha_2}} &= -cy - xz \\ \frac{d^{\alpha_3}z}{dt^{\alpha_3}} &= xy - b\,, \end{cases} \tag{37}$$

where $a,\ b,\ c$ are non-negative constants and α_i are the fractional orders satisfying $0 < \alpha_i < 1, i = 1, 2,)$.

The solution of system (37) is invariant under the transformation $T(x, y, z) \to (-x, -y, z)$, thus any non-invariant orbit has its "twin" orbit in the system [82].

The equilibria of this fractional-order system are given by

$$E^{\pm} = (\pm\sqrt{b}, \pm\sqrt{b}, -c)\,.$$

By linearizing the system at the equilibria, the Jacobian matrix is obtained, as

$$J = \begin{bmatrix} -a & a & 0 \\ c & -c & \mp\sqrt{b} \\ \pm\sqrt{b} & \pm\sqrt{b} & 0 \end{bmatrix},$$

and its corresponding characteristic equation is

$$\lambda^3 + (a+c)\lambda^2 + b\lambda + 2ab = 0\,,$$

which has three roots with negative real parts. According to the Routh-Hurwitz criterion, the equilibria $E^{\pm}$ are asymptotically stable.

Numerical simulations were performed using the predictor-corrector scheme [18]. When parameters $a = 10,\ b = 100,\ c = 11.2$, with initial condition $(x_0, y_0, z_0) = (0.98, -1.82, -0.49)$, and the fractional orders $\alpha_1 = \alpha_2 = \alpha_3 = 0.9999$, the system has the largest Lyapunov exponents as shown in Fig. 19. The effective order of the system in this case is $\sum = 2.9997 < 3$.

To obtain the lowest effective order of the system, the fractional orders α_i, $i = 1, 2, 3$, are re-set. Choosing $\alpha_1 = \alpha_2 = 1$, the minimal fractional order α_3 for the system to generate chaos is 0.9943. Also, taking $\alpha_1 = \alpha_3 = 1$, one has $0.99 \le \alpha_2 \le 1$; while taking $\alpha_2 = \alpha_3 = 1$, one has $0.8485 \le \alpha_1 \le 1$. Therefore, the lowest effective order for the system to generate chaos with stable equilibria is 2.8454.

5.2 A Chaotic System with One Stable Equilibrium

Moreover, a globally exponentially stable controller and the non-existence of finite-time stable equilibria of the fractional-order chaotic systems were discussed in

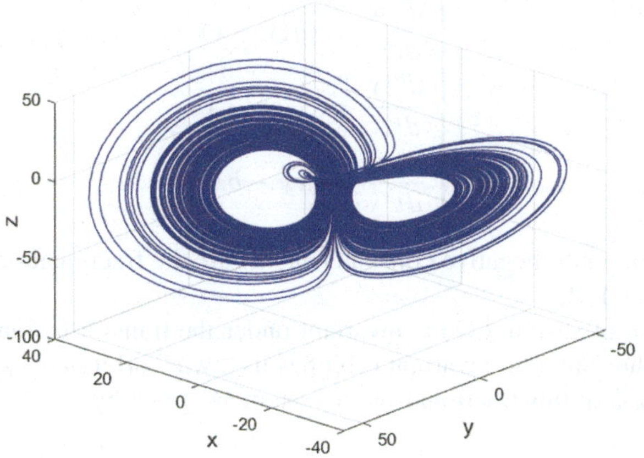

Fig. 19 Chaotic attractor of the fractional-order Lorenz-like chaotic system (37), with parameters $a = 10,\ b = 100,\ c = 11.2$ and initial condition $(x_0, y_0, z_0) = (0.98, -1.82, -0.49)$, when $\alpha_1 = \alpha_2 = \alpha_3 = 0.9999$. In this case, the simulation step length is 0.005, iteration time is 10, 000, where the first 2,000 data points are omitted [82]

Ref. [20, 83]. In Ref. [84], another 3D fractional-order chaotic system with only one stable equilibrium was introduced.

Based on the Caputo derivative, the new fractional-order system is described by

$$\begin{cases} \dfrac{d^\alpha x}{dt^\alpha} &= (y + \dfrac{1}{16})z \\ \dfrac{d^\alpha y}{dt^\alpha} &= x^2 + \dfrac{1}{2}x - y \\ \dfrac{d^\alpha z}{dt^\alpha} &= -2x\,, \end{cases} \tag{38}$$

where α is the fractional order satisfying $0 < \alpha < 1$.

The system (38) has only one equilibrium $E(0, 0, 0)$. By calculating its Jacobian matrix at the equilibrium, the eigenvalues are obtained, as

$$\lambda_1 = -1 \quad , \quad \lambda_{2,3} = \pm 0.25\sqrt{2}i\ ,$$

and this equilibrium $E(0, 0, 0)$ is proved locally asymptotically stable, according to a lemma in Refs. [63, 85].

Figure 20 shows the phase portrait of the system (38), with initial condition $(x_0, y_0, z_0) = (-1.2, 1, 1)$ and fractional order $\alpha = 0.958$. Simulation results give the largest Lyapunov exponents as $LEs = 0.0022$.

The system (38) shows chaos with only one stable equilibrium, for the fractional order varying between 0.958 and 1.

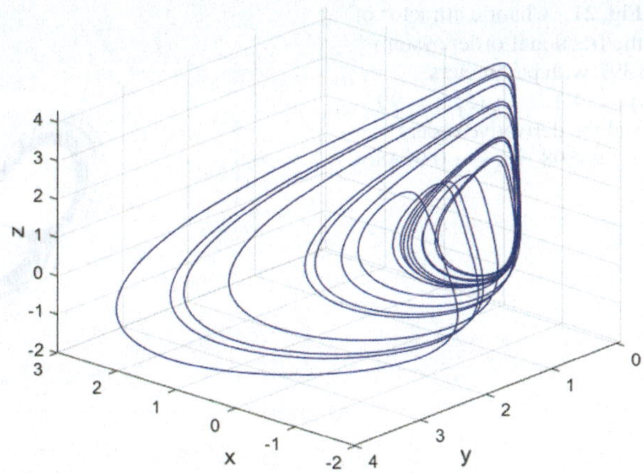

Fig. 20 Chaotic attractor of the fractional-order system (38), with only one stable equilibrium for $\alpha = 0.958$ [84]

Utilizing the Caputo derivative, Wang et al. [86] introduced a 3D fractional-order chaotic system with two stable equilibria. The model is described by

$$\begin{cases} \dfrac{d^{\alpha_1} x}{dt^{\alpha_1}} &= a(z - x) \\ \dfrac{d^{\alpha_2} y}{dt^{\alpha_2}} &= -xz + e \\ \dfrac{d^{\alpha_3} z}{dt^{\alpha_3}} &= xy - bz \,, \end{cases} \tag{39}$$

where a, b, e are positive parameters and α_i are the fractional orders satisfying $0 < \alpha_i < 1, i = 1, 2, 3$.

The system (39) has two equilibria:

$$E_1 = (\sqrt{e}, b, \sqrt{e}) \,, \quad E_2 = (-\sqrt{e}, b, -\sqrt{e}) \,,$$

which are stable for $b > a$ according to the Routh-Hurwitz stability criterion.

Applying the Adam-Bashforth-Moulton predictor-corrector method, proposed in Refs. [18, 63, 87], numerical solutions were performed. Figure 21 shows the chaotic attractor in system (39), with $a = 14.1$, $b = 15$, $e = 90$ and $\alpha_1 = 0.98$, $\alpha_{2,3} = 0.99$. Two equilibria were observable in Fig. 21.

In a further study, synchronization schemes including the full-state hybrid projective synchronization and the inverse full-state hybrid projective synchronization were investigated in Refs. [86, 88].

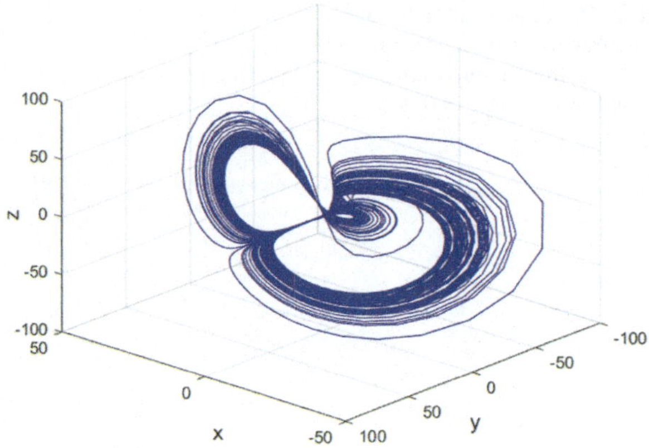

Fig. 21 Chaotic attractor of the fractional-order system (39), with parameters $a = 14.1,\ b = 15,\ e = 90$ and the derivative orders $\alpha_1 = 0.98,\ \alpha_{2,3} = 0.99$ [86]

6 Fractional-Order Systems without Equilibria

Chaos in fractional-order systems without equilibria will be studied in both 3D and 4D models, and several examples will be discussed. Since there is no equilibria in these systems, the Melnikov and Šil'nikov methods cannot be used to mathematically verify the existence of chaos. Hence, the numerical predictor-corrector schemes will be used.

6.1 3D Examples

6.1.1 Example 1

Utilizing the Caputo derivative, Cafagna and Grassi [89] first introduced an 3D "elegant" fractional-order chaotic system without equilibria, based on Refs. [90, 91].

The chaotic system is described by

$$\begin{cases} \dfrac{d^\alpha x}{dt^\alpha} = y \\ \dfrac{d^\alpha y}{dt^\alpha} = z \\ \dfrac{d^\alpha z}{dt^\alpha} = -y - xz - yz + a\,, \end{cases} \tag{40}$$

where α_i is the fractional order satisfying $0 < \alpha < 1$ and parameter $a \neq 0$.

This system is considered "elegant" since its corresponding differential equations contain very few terms, only the necessary ones. It can be easily verified that the system (40) has chaos without equilibria [89, 92].

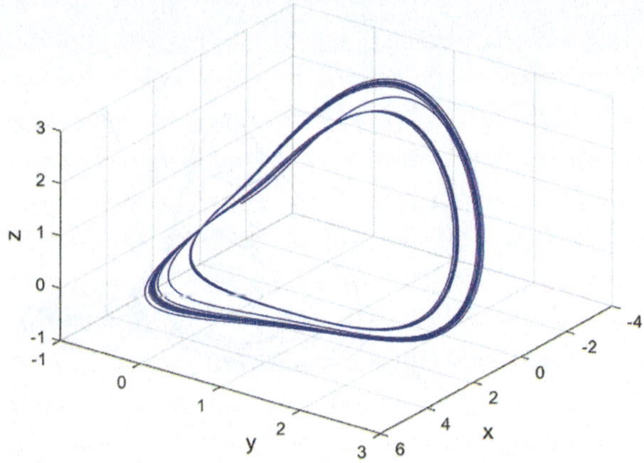

Fig. 22 Chaotic attractor of the "elegant" fractional-order system (40), with parameter $a = -0.757$, initial condition (0, 3, 0.1) and derivative order $\alpha = 0.98$ [89]

Based on the method in Ref. [93], a transformation is applied to calculate the Lyapunov exponents of system (40) yielding

$$\lambda_1 = 0.013, \quad \lambda_2 = 0, \quad \lambda_3 = -3.018\,.$$

By setting parameter $a = -0.757$ with initial condition (0,3,0.1) and derivative order $\alpha = 0.98$, the phase portrait of the chaotic system is shown in Fig. 22. The minimal effective order of system (40) to generate chaos, in the absence of equilibria, is 2.94.

InRef. [94], the observation-based synchronization of system (40) is discussed.

6.1.2 Example 2

Pham et al. [95] expanded the list of 3D fractional-order systems without equilibria. By using the Caputo derivative, the additional system is described by

$$\begin{cases} \dfrac{d^\alpha x}{dt^\alpha} = y \\ \dfrac{d^\alpha y}{dt^\alpha} = -x - yz \\ \dfrac{d^\alpha z}{dt^\alpha} = xy + ax^2 - b\,, \end{cases} \tag{41}$$

where a, b are positive parameters and α is the fractional order satisfying $0 < \alpha < 1$.

Note that the system (41) is proposed based on the integer-order NE_8 model listed in Ref. [91]. There are no equilibria in this fractional-order system. Moreover, a rotational symmetry with respect to the z-axis is founded in system (41), as it is invariant under the coordinate transformation $(x, y, z) \to (-x, -y, z)$.

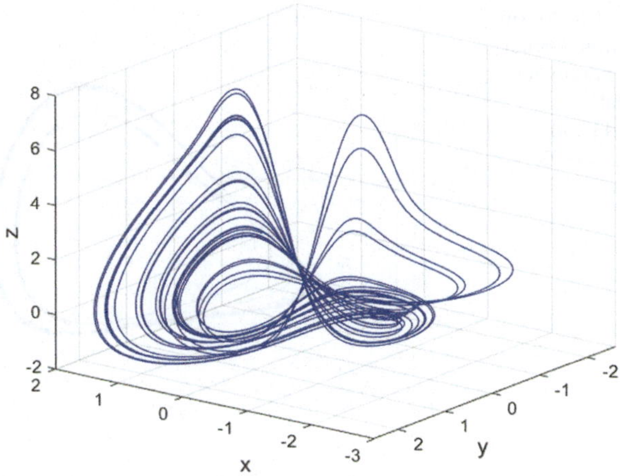

Fig. 23 Chaotic attractor of the fractional-order system (41), with $a = 1.5,\ b = 1.3$, initial condition $(x_0, y_0, z_0) = (0, 0.1, 0)$, and $\alpha = 0.9$ [95]

By selecting $a = 1.5,\ b = 1.3$ with initial condition $(x_0, y_0, z_0) = (0, 0.1, 0)$ and $\alpha = 0.9$, chaotic behavior can be observed, as illustrated in Fig. 23. Applying the Adams-Bashforth-Moulton method [18], it can be found that there is a route from non-chaotic dynamics to chaotic ones as the parameter b decreases from 1.2 to 1.4, in the case of $a = 1.5$ and $\alpha = 0.9$.

By applying the numerical method reported in Ref. [96], the largest Lyapunov exponent is obtained as 0.1374, for the case shown in Fig. 23. Also, circuit implementation, chaos control and synchronization were studied in Ref. [95].

6.1.3 Example 3

Using the Caputo derivative, Borah and Roy [97] introduced a fractional-order chaotic system without equilibria, described by

$$\begin{cases} \dfrac{d^\alpha x}{dt^\alpha} = ay - x + yz \\ \dfrac{d^\alpha y}{dt^\alpha} = -bxz + cx + yz + d \\ \dfrac{d^\alpha z}{dt^\alpha} = e + gxy - x^2 , \end{cases} \tag{42}$$

where $a,\ b,\ c,\ d,\ e,\ g$ are positive parameters.

Note that system (42) does not have any symmetry under any transformation with respect to any principal coordinate axis. Thus, it is asymmetrical, different from example (41).

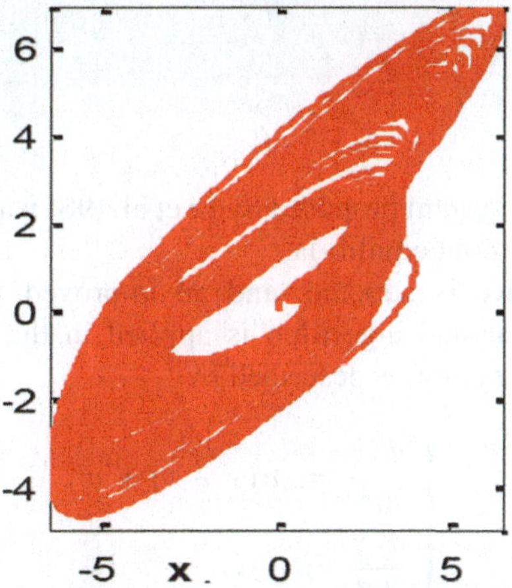

Fig. 24 Chaotic attractor of the fractional-order system (42), with $a = 2.8,\ b = 0.2,\ c = 1.4,\ d = 1,\ e = 10,\ g = 2$, initial condition $(x_0, y_0, z_0) = (0, 0, 0)$ and $\alpha = 0.92$ on the x-y plane [97]

Table 4 Equilibria and the corresponding eigenvalues of the 3D fractional-order system (42) [97]

Equilibrium points	Eigenvalues
$E_1(2.707 + 0.573j, 0.413 - 0.661j, -1.584 + 3.332j)$	$\lambda_1 = -1.0678 - 2.1487j$ $\lambda_2 = -2.5371 + 2.5417j$ $\lambda_3 = 1.0209 + 2.9390j$
$E_2(2.707 = 0.573j, 0.413 + 0.661j, -1.584 - 3.332j)$	$\lambda_1 = -1.0678 + 2.1487j$ $\lambda_2 = -2.5371 - 2.5417j$ $\lambda_3 = 1.0209 - 2.9390j$
$E_3(-2.962 - 0.739j, -0.107 + 0.766j, -3.215 + 3.924j)$	$\lambda_1 = 1.2894 + 3.0774j$ $\lambda_2 = -1.1770 - 1.5003j$ $\lambda_3 = -4.3273 + 2.3469j$
$E_4(-2.962 + 0.739j, -0.107 - 0.766j, -3.215 - 3.924j)$	$\lambda_1 = 1.2894 - 3.0774j$ $\lambda_2 = -1.1770 + 1.5003j$ $\lambda_4 = -4.3273 - 2.3469j$

By selecting $a = 2.8,\ b = 0.2,\ c = 1.4,\ d = 1,\ e = 10,\ g = 2$ with initial condition $(x_0, y_0, z_0) = (0, 0, 0)$ for any value of α in $(0.92, 1]$, its Lyapunov exponents have the signs of $(+, 0, -)$. Therefore, this 3D fractional-order system is chaotic. Phase portrait of the chaotic attractor of the system, with $\alpha = 0.92$, is shown in Fig. 24. Numerical simulation results indicate that the lowest effective order of the system (42) to generate chaos is 2.73.

Moreover, the equilibria and the corresponding eigenvalues of the system are shown in Table 4. The system is verified as a chaotic system with hidden attractors since all their equilibria, as listed in Table 4, are purely imaginary.

6.2 4D Examples

6.2.1 Example 4

The 4D fractional-order system proposed by Li et al. [98] was the first 4D fractional-order chaotic system without equilibria.

The Caputo derivative is adopted, and an improved version of the Adams-Bashforth-Moulton numerical algorithm is applied in the investigation. This 4D fractional-order chaotic system is described by

$$\begin{cases} \frac{d^\alpha x}{dt^\alpha} &= a(y-x)+w \\ \frac{d^\alpha y}{dt^\alpha} &= -xz \\ \frac{d^\alpha z}{dt^\alpha} &= -b+xy \\ \frac{d^\alpha w}{dt^\alpha} &= -kx\,, \end{cases} \tag{43}$$

where $a,\ b,\ k$ are positive constants and α_i is the fractional order satisfying $0 < \alpha < 1$.

Through calculating the Lyapunov exponent spectrum for varying α, the lowest effective order of this 4D fractional-order system to have chaos, in the absence of equilibria, is 3.28. Figure 25 shows the chaotic attractor for the case of $\alpha = 0.9$, with step size $h = 0.005$ and initial condition $(x_0, y_0.z_0, w_0) = (0.12, 0.2, 0.11, 0.3)$ in simulations.

Moreover, a one-way coupling configuration was used, with a drive system and a response system, to investigate the synchronization of system (43) in Ref. [98].

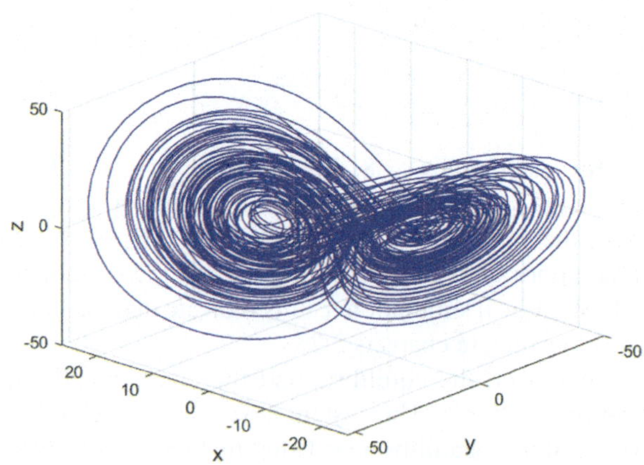

Fig. 25 Phase portrait of the chaotic attractor of the 4D fractional-order system (43), with $\alpha = 0.9$, $a = 5$, $b = 90$, $k = 10$, step size $h = 0.005$, and initial condition $(x_0, y_0.z_0, w_0) = (0.12, 0.2, 0.11, 0.3)$ on the x-y plane [98]

6.2.2 Example 5

Inspired by Li et al. [98], Zhou [99] suggested another 4D fractional-order chaotic system withour equilibria, which is described by

$$\begin{cases} \dfrac{d^\alpha x}{dt^\alpha} &= 5(z-x)+5w \\ \dfrac{d^\alpha y}{dt^\alpha} &= 10-xz \\ \dfrac{d^\alpha z}{dt^\alpha} &= -60+xy \\ \dfrac{d^\alpha w}{dt^\alpha} &= -10y\,, \end{cases} \tag{44}$$

with the fractional order satisfying $0 < \alpha < 1$.

Note that the integer-order version of system (44), with $\alpha = 1$, has no chaotic behavior, since numerical simulations show that the largest Lyapunov exponents of the integer-order system, with $\alpha = 1$, are $(0, 0, -0, 011, -4.99)$.

When $\alpha = 0.9$, the largest Lyapunov exponent of system (44) is 1.6267; when $\alpha = 0.8$, the largest Lyapunov exponent is 0.0218. The phase portrait of the system is shown in Fig. 26. The lowest effective order for this system to generate chaos is 3.2.

6.2.3 Example 6

By modifying the integer-order hyperchaotic system without equilibria proposed by Wang et al. [100], Cafagna and Grassi [100] introduced a fractional-order system, described by

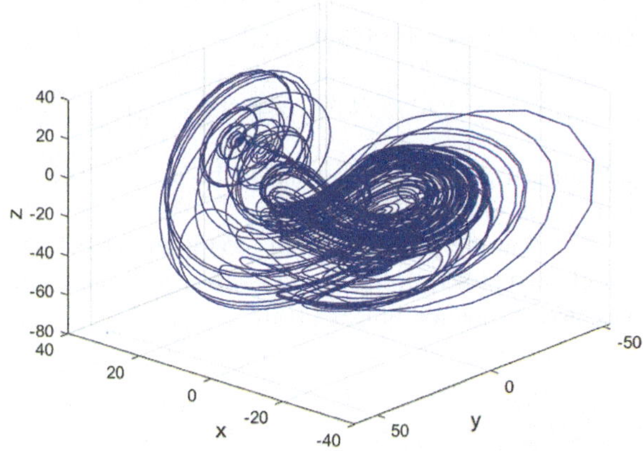

Fig. 26 Phase portrait of the hyperchaotic attractor of the 4D fractional-order system (44), with $\alpha = 0.8$ and initial condition (0.12, 0.2, 0.11, 0.3) in the x-y-z space [99]

Fig. 27 Phase portrait of the hyperchaotic attractor of the 4D fractional-order system (45), with $\alpha = 0.96$ and initial condition $(x_0, y_0, z_0, w_0) = (0.1, 0.1, 0.1, 0.1)$ on the y-w plane [100]

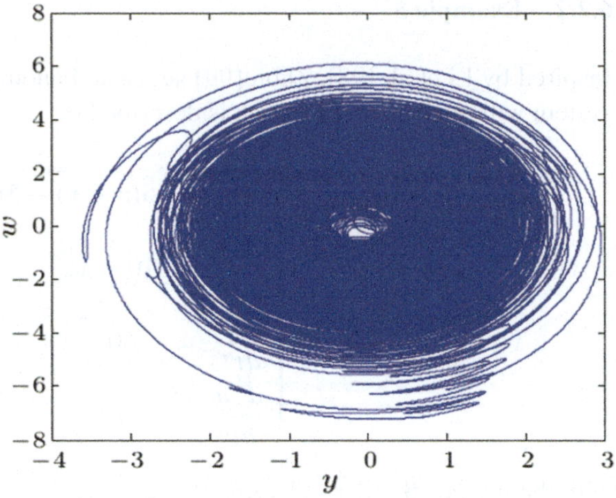

$$\begin{cases} \dfrac{d^\alpha x}{dt^\alpha} &= y \\ \dfrac{d^\alpha y}{dt^\alpha} &= -x + yz + axzw \\ \dfrac{d^\alpha z}{dt^\alpha} &= 1 - y^2 \\ \dfrac{d^\alpha w}{dt^\alpha} &= z + bxz + cxyz\,, \end{cases} \tag{45}$$

where the Caputo derivative is used, with real parameters $a,\ b,\ c$, in which the fractional order satisfies $0 < \alpha < 1$. This system has no equilibria for any values of parameters $a,\ b,\ c$.

By the predictor-corrector algorithm [18], the system (45) is simulated. By selecting the constant parameters $a = 8,\ b = -2.5,\ c = -30$, with initial condition $(x_0, y_0, z_0, w_0) = (0.1, 0.1, 0.1, 0.1)$ and $\alpha = 0.96$, the Lyapunov exponent spectrum of the system is obtained:

$$\lambda_1 = 0.89,\ \ \lambda_2 = 0.21,\ \ \lambda_3 = 0,\ \ \lambda_4 = -1.54\,.$$

indicating that hyperchaos exists in this setting. The phase portrait of the hyperchaotic attractor is shown in Fig. 27.

Note that the plot of system (45) is in perfect agreement with the results given in Ref. [101] for its integer-order version. Further dynamical analysis of system (45) was also performed in Ref. [100].

References

1. K.B. Oldham, J. Spanier, The fractional calculus. Math. Gazette **56**(247), 396–400 (1974)
2. A.C. Mcbride, *Fractional Calculus and Integral Transforms of Generalized Functions* (Pitman Advanced Publishing, Boston, 1979)
3. K.S. Nisar, Fractional-order calculus and its application. ICM Conference Proceedings. Springer, Berlin (2017)
4. J.T. Machado, V. Kiryakova, F. Mainardi, Recent history of fractional calculus. Commu. Nonlinear Sci. Numer. Simul. **16**(3), 1140–1153 (2011)
5. R. Hilfer, *Applications of Fractional Calculus in Physics* (World Scientific, Singapore, 2000)
6. P. Arena, R. Caponetto, L. Fortuna, D. Porto, *Nonlinear Noninteger Order Circuits and Systems: An Introduction* (World Scientific, Singapore, 2000)
7. I. Podlubny, *Fractional Differential Equations* (Elsevier, Amsterdam, 1998)
8. C.C. Tseng, *Design of FIR and IIR Fractional Order Simpson Digital Integrators* (Elsevier North-Holland, Amsterdam, 2014)
9. H. Sun, A. Abdelwahab, B. Onaral, Linear approximation of transfer function with a pole of fractional power. IEEE Trans. Autom. Control **29**(5), 441–444 (1984)
10. J. Gluckheimer, P. Holmes, K.K. Lee, Nonlinear oscillations, dynamical systems, and bifurcations of vector fields. Phys. Today **38**(11), 102–105 (1993)
11. X. Wu, H. Wang, A new chaotic system with fractional order and its projective synchronization. Nonlinear Dyn. **61**(3), 407–417 (2010)
12. I. Petráš, A note on the fractional-order Volta's system. Commu. Nonlinear Sci. Numer. Simul. **15**(2), 384–393 (2010)
13. S.K. Agrawal, M. Srivastava, S. Das, Synchronization of fractional order chaotic systems using active control method. Chaos, Solitons & Fractals **45**(6), 737–752 (2012)
14. N. Nyamoradi, M. Javidi, Dynamic analysis of a fractional-order Rikitake system. Dyn. Contin. Discrete Impulsive Syst. **2**(2), 189–204 (2014)
15. J.G. Lu, Chaotic dynamics and synchronization of fractional-order Arneodo's systems. Chaos, Solitons & Fractals **26**(4), 1125–1133 (2005)
16. I. Petráš, D. Bednárová, *Fractional-Order Chaotic Systems* (Springer, Berlin, 2011)
17. R. Caponetto, G. Dongola, L. Fortuna, I. Petráš, *Fractional Order Systems: Modeling and Control Applications* (World Scientific, Singapore, 2010)
18. D. Kai, N.J. Ford, A.D. Freed, A predictor-corrector approach for the numerical solution of fractional differential equations. Nonlinear Dyn. **29**(1–4), 3–22 (2002)
19. Q. Huang, C. Dong, Q. Chen, Control of the fractional-order Chen chaotic system via fractional-order scalar controller and its circuit implementation. Math. Probl. Eng. **2014**(3), 1–9 (2014)
20. S. Kamal, B. Bandyopadhyay, S. Spurgeon, Stabilization of a fractional-order chain of integrators: a contraction-based approach. IMA J. Math. Control Inf. **32**(2), 291 (2015)
21. C.P. Li, W.H. Deng, D. Xu, Chaos synchronization of the Chua system with a fractional order. Phys. A **360**(2), 171–185 (2012)
22. S. Boccaletti, J. Kurths, G. Osipov, D.L. Valladares, C.S. Zhou, The synchronization of chaotic systems. Phys. Rep. **366**(1), 1–101 (2002)
23. Z. Wang, X. Huang, G. Shi, Analysis of nonlinear dynamics and chaos in a fractional order financial system with time delay. Comput. Math. Appl. **62**(3), 1531–1539 (2011)
24. P.D. Thanh, C.P. Thuong, Chaos in fractional order CNN and its synchronization. International Conference on Control, Automation and Systems. IEEE, New York (2015)
25. T. Kanamaru, *Van Der Pol Oscillator* (Springer, New York, 2007)
26. T.T. Hartley, C.F. Lorenzo, H.K. Qammer, Chaos in a fractional order Chua's system. IEEE Trans. Circuits Syst. I: Fundamental Theory Appl. **42**(8), 485–490 (1995)
27. I. Petráš, A note on the fractional-order Chua's system. Chaos, Solitons & Fractals **38**(1), 140–147 (2008)
28. T.T. Hartley, The duffing double scroll. American Control Conference. IEEE, New York (1989), pp. 419–424

29. T.T. Hartley, F. Mossayebi, Control of Chua's circuit. J. Circuits Syst. Comput. **3**(01), 173–194 (1993)
30. G. Benettin, L. Galgani, A. Giorgilli, J.M. Strelcyn, Lyapunov characteristic exponents for smooth dynamical systems and for Hamiltonian systems; a method for computing all of them. Part 1: Theory. Meccanica **15**(1), 9–20 (1980)
31. E.N. Lorenz, Deterministic nonperiodic flow. J. Atmos. Sci. **20**(2), 130–141 (1963)
32. I. Grigorenko, E. Grigorenko, Chaotic dynamics of the fractional Lorenz system. Phys. Rev. Lett. **91**(3), 034101 (2003)
33. R. Hegger, H. Kantz, T. Schreiber, Practical implementation of nonlinear time series methods: The Tisean package. Chaos **9**(2), 413–435 (1999)
34. C. Li, J. Yan, The synchronization of three fractional differential systems. Chaos, Solitons & Fractals **32**(2), 751–757 (2007)
35. J. Lü, T. Zhou, G. Chen, S. Zhang, Local bifurcations of the Chen system. Int. J. Bifurc. Chaos **12**(10), 2257–2270 (2002)
36. H. Zhu, S. Zhou, Z. He, Chaos synchronization of the fractional-order Chen's system. Chaos, Solitons & Fractals **41**(5), 2733–2740 (2009)
37. C.-M. Chang, H.-K. Chen, Chaos and hybrid projective synchronization of commensurate and incommensurate fractional-order Chen-Lee systems. Nonlinear Dyn. **62**(4), 851–858 (2010)
38. C. Li, G. Chen, Chaos in the fractional order Chen system and its control. Chaos, Solitons & Fractals **22**(3), 549–554 (2004)
39. A. Vaněček, S. Čelikovský, *Control Systems: From Linear Analysis to Synthesis of Chaos* (Prentice-Hall, London, 1996)
40. J.G. Lu, Chaotic dynamics of the fractional-order Lü system and its synchronization. Phys. Lett. A **354**(4), 305–311 (2006)
41. W.H. Deng, C.P. Li, Chaos synchronization of the fractional Lü system. Phys. A **353**(1), 61–72 (2005)
42. H.Y. Jia, Z.Q. Chen, G.Y. Qi, Topological horseshoe analysis and circuit realization for a fractional-order lü system. Nonlinear Dyn. **74**(1–2), 203–212 (2013)
43. O.E. Rössler, An equation for continuous chaos. Phys. Lett. A **57**(5), 397–398 (1976)
44. C. Li, G. Chen, Chaos and hyperchaos in the fractional-order Rössler equations. Phys. A **341**(1–4), 55–61 (2004)
45. W. Zhang, S. Zhou, H. Li, H. Zhu, Chaos in a fractional-order Rössler system. Chaos, Solitons & Fractals **42**(3), 1684–1691 (2009)
46. T. Zhou, C. Li, Synchronization in fractional-order differential systems. Physica D **212**(1), 111–125 (2005)
47. A. Gjurchinovski, T. Sandev, V. Urumov, Delayed feedback control of fractional-order chaotic systems. J. Phys. A: Math. Theor. **43**(44), 3045–3067 (2011)
48. O.E. Rössler, An equation for hyperchaos. Phys. Lett. A **71**(2), 155–157 (1979)
49. D. Cafagna, G. Grassi, Hyperchaos in the fractional-order Rössler system with lowest-order. Int. J. Bifurc. Chaos **19**(01), 339–347 (2009)
50. C. Liu, L. Ling, L. Tao, A novel three-dimensional autonomous chaos system. Chaos, Solitons & Fractals **39**(4), 1950–1958 (2011)
51. S. Bhalekar, V. Daftardar-Gejji, Fractional ordered Liu system with time-delay. Commu. Nonlinear Sci. Numer. Simul. **15**(8), 2178–2191 (2010)
52. X.Y. Wang, M.J. Wang, Dynamic analysis of the fractional-order Liu system and its synchronization. Chaos **17**(3), 304–311 (2007)
53. S. Bhalekar, V. Daftardar-Gejji, Synchronization of different fractional order chaotic systems using active control. Commu. Nonlinear Sci. Numer. Simul. **15**(11), 3536–3546 (2010)
54. W. Deng, J. Lü, Design of multi-directional multi-scroll chaotic attractors based on fractional differential systems. Chaos **16**(4), 1 (2006)
55. W. Deng, J. Lü, Generating multi-directional multi-scroll chaotic attractors via a fractional differential hysteresis system. Phys. Lett. A **369**(5–6), 438–443 (2007)
56. W. Deng, Y. Wu, C. Li, Stability analysis of differential equations with time-dependent delay. Int. J. Bifurc. Chaos **16**(02), 465–472 (2006)

57. J. Lü, G. Chen, Generating multiscroll chaotic attractors: Theories, methods and applications. Int. J. Bifurc. Chaos **16**(04), 775–858 (2006)
58. J. Lu, G. Chen, X. Yu, H. Leung, Design and analysis of multiscroll chaotic attractors from saturated function series. IEEE Trans. Circuits Syst. I: Regular Papers **51**(12), 2476–2490 (2004)
59. F. Xu, P. Yu, X. Liao, Synchronization and stabilization of multi-scroll integer and fractional order chaotic attractors generated using trigonometric functions. Int. J. Bifurc. Chaos **23**(8), 1350145 (2013)
60. M. Forti, Some extensions of a new method to analyze complete stability of neural networks. IEEE Trans. Neural Networks **13**(5), 1230–8 (2002)
61. H.Y. Jia, Z.Q. Chen, G.Y. Qi, Chaotic characteristics analysis and circuit implementation for a fractional-order system. IEEE Trans. Circuits Syst. I: Regular Papers **61**(3), 845–853 (2017)
62. S. Ma, Chaos control and synchronization of a new fractional order chaotic system. J. Inf. Comput. Sci. **11**(10), 3469–3479 (2014)
63. D. Kai, N.J. Ford, Analysis of fractional differential equations. J. Math. Anal. Appl. **265**(2), 229–248 (2002)
64. T. Gotthans, J. Petržela, New class of chaotic systems with circular equilibrium. Nonlinear Dyn. **81**(3), 1–7 (2015)
65. S. Kingni, S. Jafari, H. Simo, P. Woafo, Three-dimensional chaotic autonomous system with only one stable equilibrium: Analysis, circuit design, parameter estimation, control, synchronization and its fractional-order form. Eur. Phys. J. Plus **129**(5), 76 (2014)
66. S. Westerlund, L. Ekstam, Capacitor theory. IEEE Trans. Dielectrics Electrical Insulation **1**(5), 826–839 (1994)
67. S.T. Kingni, V.T. Pham, S. Jafari, G.R. Kol, P. Woafo, Three-dimensional chaotic autonomous system with a circular equilibrium: Analysis, circuit implementation and its fractional-order form. Circuits Syst. Sig. Process. **35**(6), 1933–1948 (2016)
68. T. Gao, Z. Chen, Z. Yuan, D. Yu, Adaptive synchronization of a new hyperchaotic system with uncertain parameters. Chaos, Solitons & Fractals **33**(3), 922–928 (2007)
69. X. Wu, H. Lu, S. Shen, Synchronization of a new fractional-order hyperchaotic system. Phys. Lett. A **373**(27), 2329–2337 (2009)
70. D. Chen, C. Liu, C. Wu, Y. Liu, X. Ma, Y. You, A new fractional-order chaotic system and its synchronization with circuit simulation. Circuits Syst. Sig. Process. **31**(5), 1599–1613 (2012)
71. X. Wu, Y. Lu, Generalized projective synchronization of the fractional-order Chen hyperchaotic system. Nonlinear Dyn. **57**(1–2), 25 (2009)
72. S. Dadras, H.R. Momeni, G. Qi, Z.L. Wang, Four-wing hyperchaotic attractor generated from a new 4d system with one equilibrium and its fractional-order form. Nonlinear Dyn. **67**(2), 1161–1173 (2012)
73. T. Wang, X. Wang, Generalized synchronization of fractional order hyperchaotic Lorenz system. Mod. Phys. Lett. B **23**(17), 2167–2178 (2011)
74. Y. Gao, C. Liang, Q. Wu, H. Yuan, A new fractional-order hyperchaotic system and its modified projective synchronization. Chaos Solitons & Fractals **76**(9), 190–204 (2015)
75. M.S. Tavazoei, M. Haeri, A necessary condition for double scroll attractor existence in fractional-order systems. Phys. Lett. A **367**(1), 102–113 (2007)
76. M.S. Tavazoei, M. Haeri, Chaotic attractors in incommensurate fractional order systems. Physica D **237**(20), 2628–2637 (2008)
77. S. Jafari, J. Sprott, Simple chaotic flows with a line equilibrium. Chaos, Solitons & Fractals **57**, 79–84 (2013)
78. V.-T. Pham, S. Jafari, C. Volos, S. Vaidyanathan, T. Kapitaniak, A chaotic system with infinite equilibria located on a piecewise linear curve. Optik: Int. J. Light Electron Opt. **127**(20), 9111–9117 (2016)
79. V. Pham, S. Jafari, C. Volos, T. Kapitaniak, A gallery of chaotic systems with an infinite number of equilibrium points. Chaos, Solitons & Fractals **93**, 58–63 (2016)
80. S.T. Kingni, V.T. Pham, S. Jafari, P. Woafo, A chaotic system with an infinite number of equilibrium points located on a line and on a hyperbola and its fractional-order form. Chaos Solitons Fractals **99**, 209–218 (2017)

81. P. Zhou, K. Huang, C.D. Yang, A fractional-order chaotic system with an infinite number of equilibrium points. Discret. Dyn. Nat. Soc. **2013**(3), 331–372 (2013)
82. Z. Alam, L. Yuan, Q. Yang, Chaos and combination synchronization of a new fractional-order system with two stable node-foci. IEEE/CAA J. Autom. Sinica **3**(2), 157–164 (2016)
83. J. Shen, J. Lam, Non-existence of finite-time stable equilibria in fractional order nonlinear systems. Automatica **50**(2), 547–551 (2014)
84. S. Shen, M. Ke, P. Zhou, A 3D fractional-order chaotic system with only one stable equilibrium and controlling chaos. Discret. Dyn. Nat. Soc. **2017**, 1–5 (2017)
85. E. Ahmed, A.M.A. El-Sayed, H.A.A. El-Saka, Equilibrium points, stability and numerical solutions of fractional-order predator-prey and rabies models. J. Math. Anal. Appl. **325**(1), 542–553 (2007)
86. X. Wang, A. Ouannas, V.-T. Pham, H.R. Abdolmohammadi, A fractional-order form of a system with stable equilibria and its synchronization. Adv. Differ. Equ. **2018**(1), 20 (2018)
87. D. Kai, N.J. Ford, A.D. Freed, Detailed error analysis for a fractional Adams method. Numerical Algorithms **36**(1), 31–52 (2004)
88. Y. Zhao, R. Wu, Chaos and synchronisation of a new fractional order system with only two stable equilibria. Int. J. Dyn. Syst. Differ. Equ. **6**(3), 187 (2016)
89. D. Cafagna, G. Grassi, Elegant chaos in fractional-order system without equilibria. Math. Probl. Eng. **2013**(5), 1–7 (2013)
90. J.C. Sprott, *Elegant Chaos: Algebraically Simple Chaotic Flows* (World Scientific, Singapore, 2010)
91. S. Jafari, J. Sprott, S.M.R.H. Golpayegani, Elementary quadratic chaotic flows with no equilibria. Phys. Lett. A **377**(9), 699–702 (2013)
92. D. Cafagna, G. Grassi, Chaos in a new fractional-order system without equilibrium points. Commu. Nonlinear Sci. Numer. Simul. **19**(9), 2919–2927 (2014)
93. R. Caponetto, S. Fazzino, A semi-analytical method for the computation of the Lyapunov exponents of fractional-order systems. Commu. Nonlinear Sci. Numer. Simul. **18**(1), 22–27 (2013)
94. D. Cafagna, G. Grassi, Observer-based synchronization for a class of fractional chaotic systems via a scalar signal: Results involving the exact solution of the error dynamics. Int. J. Bifurc. Chaos **21**(3), 1102874 (2011)
95. V.T. Pham, S.T. Kingni, C. Volos, S. Jafari, T. Kapitaniak, A simple three-dimensional fractional-order chaotic system without equilibrium: Dynamics, circuitry implementation, chaos control and synchronization. AEU: Int. J. Electron. Commun. **78**, 220–227 (2017)
96. M.T. Rosenstein, J.J. Collins, C.J.D. Luca, A practical method for calculating largest Lyapunov exponents from small data sets. Physica D-nonlinear Phenomena **65**(1–2), 117–134 (1993)
97. M. Borah, B.K. Roy, Hidden attractor dynamics of a novel non-equilibrium fractional-order chaotic system and its synchronisation control. Indian Control Conference. IEEE, New York (2017), pp. 450–455
98. H. Li, X. Liao, M. Luo, A novel non-equilibrium fractional-order chaotic system and its complete synchronization by circuit implementation. Nonlinear Dyn. **68**(1–2), 137–149 (2012)
99. P. Zhou, K. Huang, A new 4D non-equilibrium fractional-order chaotic system and its circuit implementation. Commu. Nonlinear Sci. Numer. Simul. **19**(6), 2005–2011 (2014)
100. D. Cafagna, G. Grassi, Fractional-order systems without equilibria: the first example of hyperchaos and its application to synchronization. Chin. Phys. B **24**(8), 224–232 (2015)
101. Z. Wang, S. Cang, E.O. Ochola, Y. Sun, A hyperchaotic system without equilibrium. Nonlinear Dyn. **69**(1–2), 531–537 (2012)
102. S. Celikovsky, G. Chen, Secure synchronization of a class of chaotic systems from a nonlinear observer approach. IEEE Trans. Autom. Control **50**(1), 76–82 (2005)

Memristive Chaotic Systems with Hidden Attractors

Yicheng Zeng

1 Introduction

The concept of memristor was first introduced by Chua in 1971 [1]. Subsequently, solid-state implementation of memristor was fabricated by the research group of Hewlett–Packard [2]. This momentous development is successfully attracting tremendous attention to an intensive and extensive study of the memristor.

As a fundamental nonlinear circuit element with adjustable resistance or conductance in memory, the memristor is a passive two-terminal component, which has added a new dimension to the new development of analogous circuit designs and engineering applications. Many memristor emulators with different nonlinearities and many new properties have been proposed [3–7]. In most cases, memristor-based circuits or systems were constructed by introducing one or more of memristor into some existing linear or nonlinear circuits or systems, to replace its original self-variable resistor or linear coupling resistor, resulting in such as the new memristive Chua-like circuit [8], memristive Wien-bridge circuit [9], memristive jerk system [5], memristive Lü system [10], a memristive circuit based on active band-pass filter [11], and so forth.

Naturally, constructing a memristive system that can generate hidden attractors is interesting and important but also challenging. This chapter reviews this subject of hidden attractors in memristive chaotic systems, which were developed in recent years.

Y. Zeng (✉)
Department of Microelectronic Science and Engineering,
Xiangtan University, Xiangtan 411105, China
e-mail: yichengz@xtu.edu.cn

X. Wang et al. (eds.), *Chaotic Systems with Multistability and Hidden Attractors*,
Emergence, Complexity and Computation 40,
https://doi.org/10.1007/978-3-030-75821-9_10

2 Memristive Chua-Like Circuits

Many memristive chaotic circuits are designed based on Chua's circuit. Some examples are collected in this section.

2.1 Memristive Chua's Circuit

In [8], a novel memristive chaotic circuit is presented. The circuit is implemented with a first-order memristive diode bridge, used to replace Chua's diode in the classical Chua's circuit. The circuit can generate self-excited attractors and hidden attractors in different parameter settings.

The memristor-based Chua's circuit and its corresponding generalized memristor are shown in Figs. 1 and 2, respectively.

The mathematical model of the memristor shown in Fig. 2 is described by

$$i = g(v_C,\ v) = 2I_S e^{\rho v_C sinh(\rho v)}$$

$$\frac{dv_C}{dt} = f(v_C, v) = \frac{2I_S e^{\rho v_C sinh(\rho v)}}{C} - \frac{v_C}{RC} - \frac{2I_S}{C},$$

where $\rho = 1/(2nV_T)$, V_T and I_S represent the emission coefficient, thermal voltage of the diode, and reverse saturation current, respectively.

The state equation of the system is expressed as

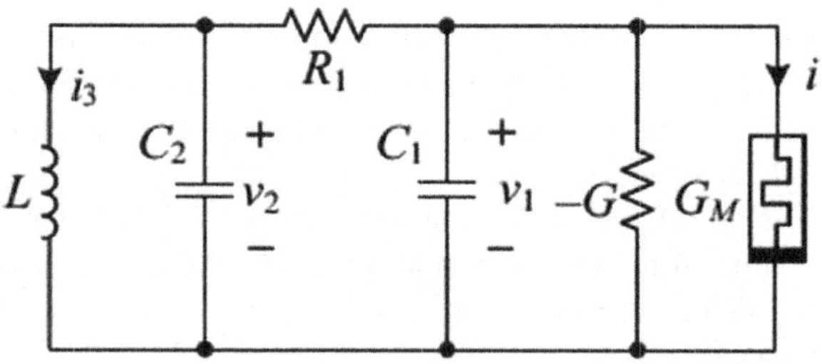

Fig. 1 Generalized memristor-based Chua's circuit

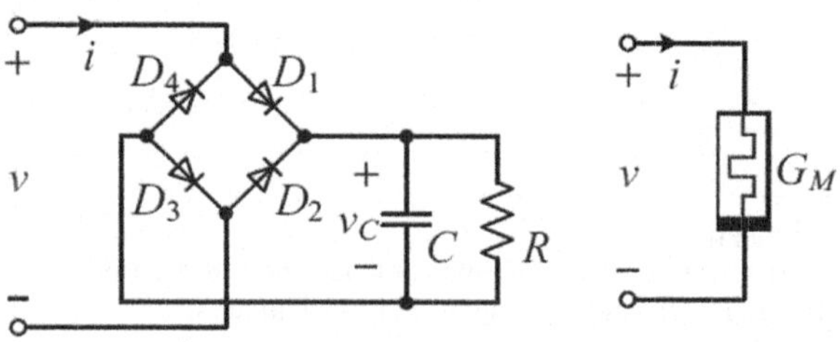

Fig. 2 Generalized memristor realized by memristive diode bridge with a parallel RC filter

Table 1 Circuit parameters for generating hidden attractors

Parameters	Significations	Values
R_1	Resistance	1.96 kΩ
G	Conductance	0.56 mS
L	Inductance	80 mH
C_1	Capacitance	22 nF
C_2	Capacitance	113 nF
R	Resistance	0.8 kΩ
C	Capacitance	1 uF

$$\begin{cases} \frac{dv_1}{t} = \frac{(R_1G-1)v_1}{R_1C_1} + \frac{v_2}{R_1C_1} - \frac{2I_S e^{-\rho v_C} sinh(\rho v_1)}{C_1} \\ \frac{dv_2}{t} = \frac{v_1-v_2}{R_2C_2} + \frac{i_3}{C_2} \\ \frac{dv_3}{t} = \frac{v_2}{L} \\ \frac{dv_4}{t} = \frac{2I_S e^{-\rho v_C} sinh(\rho v_1)}{C} - \frac{v_C}{RC} - \frac{2I_S}{C} . \end{cases} \tag{1}$$

When the new circuit parameters are chosen as listed in Table 1, three equilibria can be easily obtained as

$$S_0 = (0, 0, 0, 0), \quad S_1 = (0.8324, 0, 0.00042, 0.0343)$$
$$S_2 = (-0.8324, 0, -0.00042, 0.0343).$$

The four eigenvalues are obtained as

$$S_0 : \lambda_1 = 5939.07.\lambda_2 = -1250.15, \lambda_{3,4} = -1250.15 \pm j5024.42$$

$$S_{1,2} : \lambda_{1,2} = -109.55 \pm j9394.59, \lambda_3 = -1191.43, \lambda_4 = -23830.48$$

It can be easily verified that S_0 is an unstable saddle point and $S_{1,2}$ are two stable saddle-foci.

When the initial conditions are set as $v_1(0) = 0.01V$, $v_2(0) = 0.01V$, $i_3(0) = 0A$, and $v_c(0) = 0V$, the memristive Chua's circuit is chaotic with a double-scroll chaotic attractor, as shown in Fig. 3. The four Lyapunov exponents are $LE1 = 402.33$, $LE2 = -39.29$, $LE3 = -1126.54$ and $LE4 = -15796.97$, respectively. Clearly, the chaotic attractors are not excited from unstable equilibria, thus the circuit has hidden attractors.

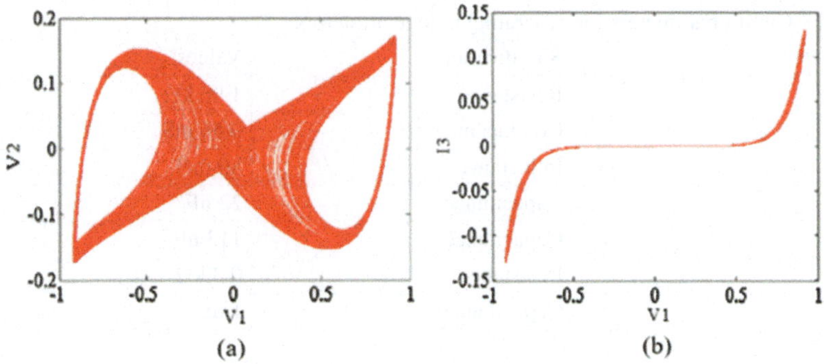

Fig. 3 Hidden chaotic attractors in a phase portrait on **a** the $v_1(0)$-$v_2(0)$ plane and **b** the $v_1(0)$-$i_3(0)$ plane

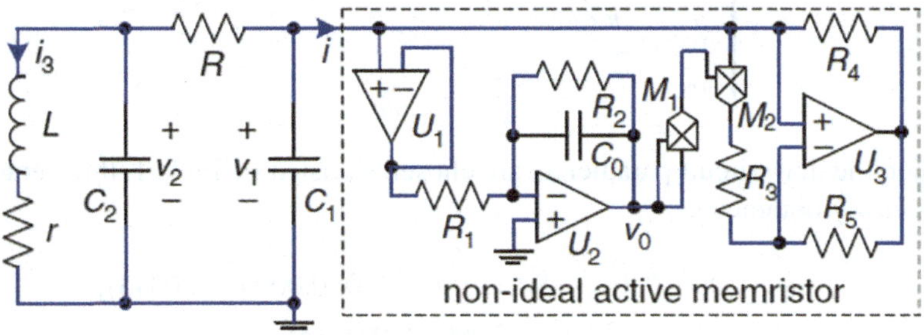

Fig. 4 Modified memristive Chua's circuit [12]

2.2 *Modified Memristive Chua's Circuit*

Chen [12] proposed a modified memristive Chua's circuit with hidden dynamics. This memristive circuit is constructed by substituting Chua's diode with an improved voltage-controlled memristor.

The modified memristive Chua's circuit has three equilibria, including one zero saddle point and two non-zero saddle-foci. When appropriate parameters are selected, the two non-zero saddle-foci are stable. In this case, hidden attractors exist.

The circuit is shown in Fig. 4, where the equivalent realization circuit of voltage-controlled memristor is marked by a dotted-line box.

The mathematical model of the memristor is given by

$$i = W(v_0)v_1 = \left(-\frac{1}{R_3} + \frac{g_1 g_2}{R_3} v_0^2\right) v_1$$

$$\frac{dv_0}{dt} = -\frac{v_1}{R_1 C_0} - \frac{v_0}{R_2 C_0},$$

where g_1 and g_2 are the variable scale factors in multipliers M_1 and M_2, respectively.

The state equations of the circuit is given by

$$\begin{cases} \dfrac{dv_1}{t} = \dfrac{v_2 - v_1}{RC_1} + \dfrac{(1 - g_1 g_2 v_0^2)v_1}{R_3 C_1} \\ \dfrac{dv_2}{t} = \dfrac{v_1 - v_2}{R_2 C_2} + \dfrac{i_3}{C_2} \\ \dfrac{dv_3}{t} = \dfrac{v_2}{L} - \dfrac{ri_3}{L} \\ \dfrac{dv_4}{t} = -\dfrac{v_1}{R_1 C_0} - \dfrac{v_0}{R_2 C_0} \, . \end{cases} \tag{2}$$

Let the right-hand side of system (2) be zero. Then, three equilibria can be obtained as

$$S_0 = (0, 0, 0, 0)$$

$$\S_{1,2} = \left(\pm\frac{R_1 \xi}{R_2}, \pm\frac{r R_1 \xi}{r R_2 + R R_2}, \pm\frac{R_1 \xi}{r R_2 + R R_2}, \mp\xi \right),$$

where

$$\xi = \sqrt{(g_1 g_2)^{-1} - [g_1 g_2 (R + r)/R_3]^{-1}} \, .$$

Obviously, for any given parameters of the non-ideal active memristor, the distributions of the two non-zero equilibria in system (2) are determined by the circuit parameter R.

Fix the circuit parameters to be $L = 12.6$ mH, $r = 2\ \Omega$, $C_1 = 6.8$ nF, $C_2 = 68$ nF, $R_1 = 4\ \text{k}\Omega$, $C_0 = 1$ nF, $R_2 = 10\ \text{k}\Omega$, $g_1 = 1$, $g_2 = 0.1$, $R_3 = 1.43\ \text{k}\Omega$, $R_4 = R_5 = 2$ kΩ, $R = 2.177$ kΩ, and $R = 2.2$ kΩ, respectively. Then the two non-zero equilibria ${}^1S_{1,2}$, ${}^2S_{1,2}$ and their corresponding four eigenvalues can be obtained as

$$\begin{aligned} {}^1S_{1,2} &= (\pm 0.7514 \quad \pm 6.8 \times 10^{-4} \quad \pm 3.4 \times 10^{-4} \quad \mp 1.854)\,, \\ {}^1S_0 &: \lambda_1 = 41275,\ \lambda_{2,3} = -6451 \pm j30965,\ \lambda_4 = -100000\,, \\ {}^1S_{1,2} &: \lambda_{1,2} = -1.34 \pm j33505,\ \lambda_{3,4} = -53486 \pm j67052\,; \end{aligned} \tag{3}$$

$$\begin{aligned} {}^2S_{1,2} &= (\pm 0.749 \quad \pm 6.8 \times 10^{-4} \quad \pm 3.4 \times 10^{-4} \quad \mp 1.8724)\,, \\ {}^2S_0 &: \lambda_1 = 41848,\ \lambda_{2,3} = -6349 \pm j31083,\ \lambda_4 = -100000\,, \\ {}^2S_{1,2} &: \lambda_{1,2} = -86.73 \pm j33563,\ \lambda_{3,4} = -53365 \pm j68046\,. \end{aligned} \tag{4}$$

It can be seen that, for the two specified parameters of $R = 2.177$ kΩ and $R = 2.2$ kΩ, the zero equilibrium is always an unstable saddle point, whereas the two non-zero equilibria are stable saddle-foci.

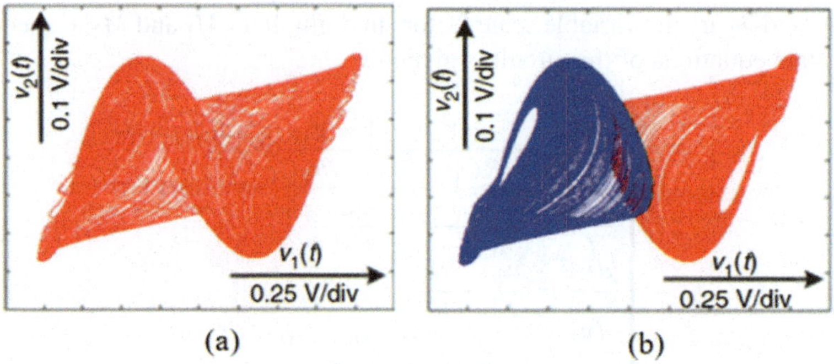

Fig. 5 Phase portraits of hidden attractors on the $v_1(t)$-$v_2(t)$ plane, with **a** one double-scroll hidden attractor at $R = 2.177\ \mathrm{k}\Omega$, **b** two coexisting hidden attractors at $R = 2.2\ \mathrm{k}\Omega$, where the blue hidden attractor starting from $v_1(0) = 0V$, $v_2(0) = -0.01V$, $i_0(0) = 0\ A$ and $v_0(0) = 0V$, and the red hidden attractor starting from $v_1(0) = 0V$, $v_2(0) = 0.01V$, $i_3(0) = 0A$ and $v_0(0) = 0V$ [12]

Hidden attractors generated from this memristive circuit at $R = 2.177\ \mathrm{k}\Omega$ and $R = 2.2\ \mathrm{k}\Omega$, which were numerically found, are depicted in Fig. 5.

2.3 *Memristive Self-oscillating Circuit*

Based on the three-dimensional self-excited oscillations, Bao et al. [13, 14], constructed a three-dimensional memristive self-oscillating circuit by introducing a non-ideal voltage-controlled memristor to replace a linear coupling resistor of the original circuit.

The simplified mathematical model of the new circuit is described by

$$\begin{cases} \ddot{x} = (x + x^2 - \alpha x^4)\dot{x} - \omega_0^2 x \\ \dot{z} = \beta - x^2 , \end{cases} \tag{5}$$

where α, β and ω are real parameters.

Let $y = \dot{x}$. Then, system (5) can be converted to a system of first-order ordinary differential equations as follows:

$$\begin{cases} \dot{x} = y \\ \dot{y} = (x + x^2 - \alpha x^4)y - \omega_0^2 x \\ \dot{z} = \beta - x^2 , \end{cases} \tag{6}$$

Where ω_0 represents the oscillation frequency, and α and β are two control parameters.

The circuit realization of system (6) is shown in Fig. 6.

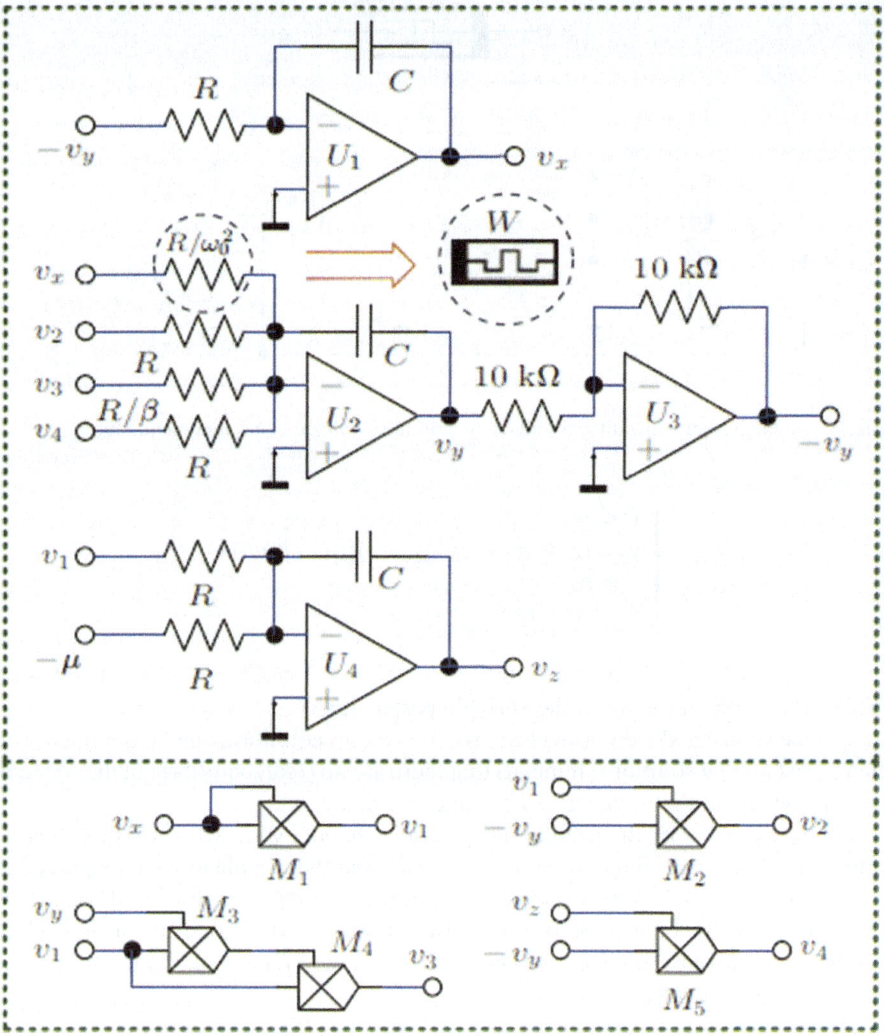

Fig. 6 Realization circuit of system (6) and the memristive self-oscillating circuit

The memristive self-oscillating circuit can be implemented by using a non-ideal voltage-controlled memristor to replace the linear coupling resistor, which is connected with the inverting input of the operational amplifier U2, as shown in Fig. 7.

An equivalent realization circuit of the non-ideal voltage-controlled memristor is displayed in Fig. 7, and the mathematical model of the voltage-controlled memristor is given by

$$\begin{cases} i = W(v_w)v_x = (a + b|v_w|)v_x \\ \dot{v}_w = v_x - v_w \,. \end{cases} \tag{7}$$

The dimensionless equations of the memristive self-oscillating circuit can be easily derived as

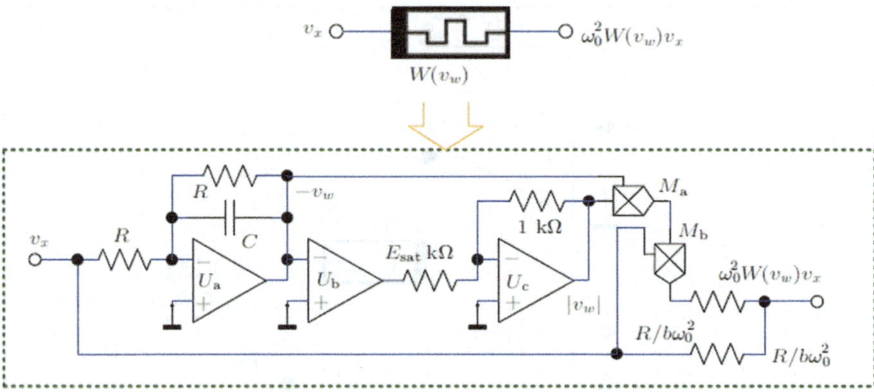

Fig. 7 An equivalent realization circuit of the non-ideal voltage-controlled memristor

$$\begin{cases} \dot{x} = y \\ \dot{y} = (z + x^2 - \alpha x^4)y - \omega_0^2(a + b|w|)x \\ \dot{z} = \beta - x^2 \\ \dot{w} = x - w\,. \end{cases} \tag{8}$$

Fix the system parameters $\alpha = 0.5$, $\beta = 0.9$, $a = 1$ and $b = 0.1$. The oscillation frequency parameter is set as the variable parameter.

Let the right-hand side of (8) be zero, the system equilibria can be obtained. But there exist no real solutions. It means that there are no (real) equilibria in this system. So, the chaotic attractor generated by this system is hidden.

When $w_0 = 2.035$, the oscillating system is chaotic, with two coexisting hidden attractors. The phase diagram on the x-y plane and the x-z plane were displayed in Fig. 8a, b, respectively, where the red attractor was starting from the initial conditions (2, 0, 0, 0) and the blue attractor was stating from (−2, 0, 0, 0). The corresponding Lyapunov exponents are $LE_1 = 0.0419$, $LE_2 = 0$, $LE_3 = -0.3705$, and $LE_4 = -1.2118$, respectively.

3 Memristive Hyperjerk Circuit

A memristive hyperjerk circuit with hidden chaotic attractors is considered.

Prousalis [15] introduced a new hyperjerk memristive system with hidden attractors, which has an infinite number of equilibria.

Recall the memristive jerk system:

$$\begin{cases} f_m = (1 + x_m)u_m \\ \dot{x}_m = u_m\,, \end{cases} \tag{9}$$

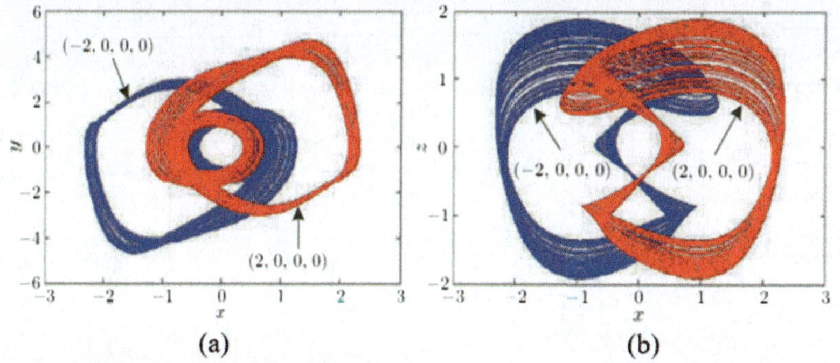

Fig. 8 Phase diagram of the hidden attractor, with $w_0 = 2.035$, **a** on the x-y plane, **b** on the x-z plane

Fig. 9 Attractor of the hyperjerk memristive system (10) on the x-y plane

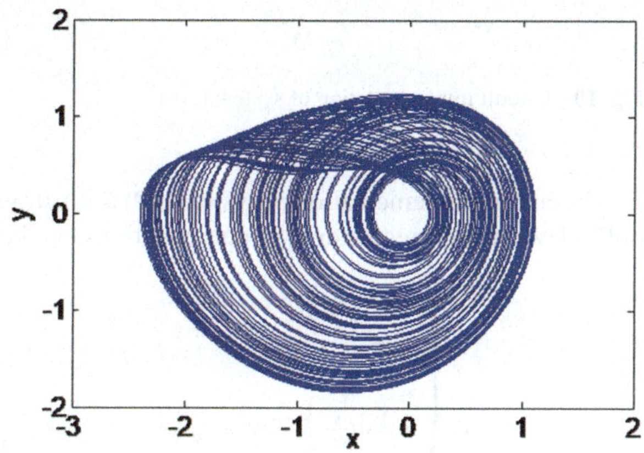

where f_m, x_m and u_m denote the state, output and input of system, respectively.

Based on the memristive system (9), the following hyperjerk dynamical system is introduced [15]:

$$\begin{cases} \dot{x} = y \\ \dot{y} = z \\ \dot{z} = w \\ \dot{w} = -z - aw - bz^2wf \,, \end{cases} \tag{10}$$

where a, b are real parameters, and $x = x_m$, $y = u_m$, $f = f_m = (1 + x)y$.

By setting the right-hand side of the hyperjerk memristive system to be zero, one can find infinitely many equilibria, $E(x, 0, 0, 0)$. According to [15], this chaotic hyperjerk memristive system has hidden attractors, because it is impossible to verify the chaotic attractors by choosing arbitrary initial conditions in any vicinities of the unstable equilibria. With $a = 0.5$, $b = 0.1$, the system (10) has four Lyapunov exponents $L_1 = 0.08945$, $L_2 = 0$, $L_3 = 0$, $L_4 = -0.89978$. The phase portrait of the hidden attractor is depicted in Fig. 9.

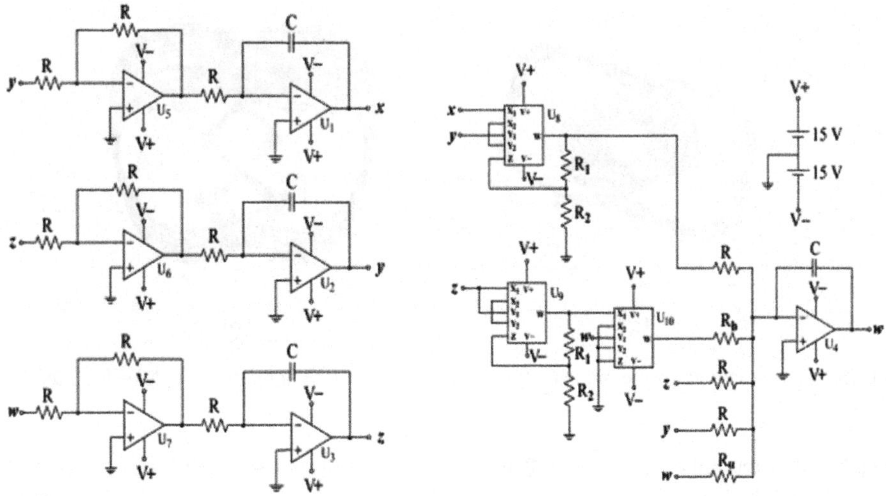

Fig. 10 Circuit implementation of system (10)

The circuit implementation of system (10) is illustrated in Fig. 10. Based on Kirchhoff's laws, the corresponding circuit equations can be obtained as

$$\begin{cases} \dot{x} = \frac{1}{R_1 C} y \\ \dot{y} = \frac{1}{R_1 C} z \\ \dot{z} = \frac{1}{RC} w \\ \dot{w} = \frac{1}{RC}\left[-xy - \frac{R}{R_b 10V} z^2 w - z - y - \frac{R}{R_a} W\right], \end{cases} \tag{11}$$

where $W = \frac{xy}{10V} + (\frac{R_2}{R_1+R_2})w$, $a = \frac{R}{R_a}$, $b = \frac{R}{10R_b}$.

4 Hidden Attractors in Memristive Hyperchaotic Systems

4.1 4D Memristive Hyperchaotic System

Based on the memristive system [16], Pham et al. [17] introduced a memristive element in the following form:

$$\begin{cases} f = (1 + 0.24w^2 - 0.0016w^4)y \\ \dot{w} = y. \end{cases} \tag{12}$$

Using the memristive element (12), a novel four-dimensional system is proposed as follows:

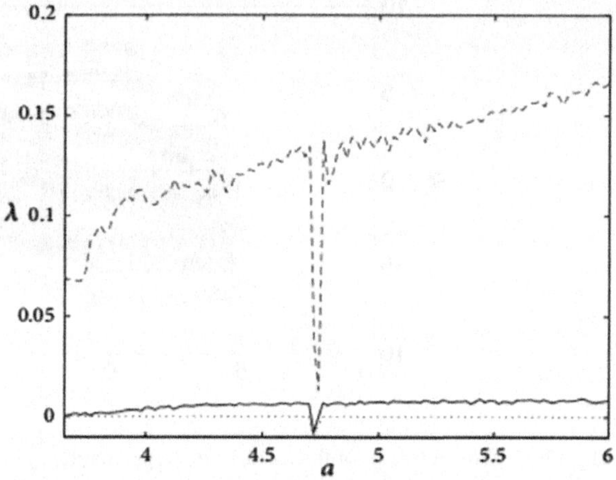

Fig. 11 Lyapunov exponents of system (13)

$$\begin{cases} \dot{x} = -10x - ay - yz \\ \dot{y} = -6x + 1.2xz + 0.1f - b \\ \dot{z} = -z - 1.2xy \\ \dot{w} = y\,, \end{cases} \tag{13}$$

where a and b are positive real parameters and f is the output of the memristive element (12).

When $b = 0$, system (13) has a line of equilibria, $E(0, 0, 0, w_m)$.

When $a = 5,\ b = 0$ and the initial conditions are chosen as $(0, 0.01, 0.01, 0)$, the memristive system (13) exhibits hyperchaos. The corresponding Lyapunov exponents are $L_1 = 0.1364,\ L_2 = 0.0071,\ L_3 = 0,\ L_4 = -10.8564$. This hyperchaotic system can be considered as a dynamical system with hidden attractor because it is impossible to verify the chaotic attractor by choosing arbitrary initial conditions in any vicinities of the unstable equilibria.

When $b \neq 0$, the system (13) has no equilibria, so all the attractors generated from system (13) are hidden.

When $b = 0.001$ and the initial conditions are chosen as $(0, 0.01, 0.01, 0)$, the system exhibits periodic, chaotic, and hyperchaotic dynamics in the region of $1 \leq a \leq 6$. The corresponding Lyapunov exponents are plotted in Fig. 11.

When $a = 5$, the system generates a hyperchaotic attractor, as shown in Fig. 12.

An equivalent circuit diagram of the above memristive element and the circuit implementation of system (13) are shown in Figs. 13 and 14, respectively.

Based on Kirchhoff's laws, the state equations of the circuit can be written as

$$\begin{cases} \dot{x} = -\frac{1}{R_1C_1}x - \frac{1}{R_2C_1}y - \frac{1}{10R_3C_1}yz \\ \dot{y} = -\frac{1}{R_4C_2}x + \frac{1}{10R_5C_2}xz + \frac{1}{R_6C_2}f - \frac{1}{R_7C_2}V_b \\ \dot{z} = -\frac{1}{R_8C_3}z - \frac{1}{10R_9C_3}xy \\ \dot{w} = \frac{1}{R_{10}C_4}y\,, \end{cases} \tag{14}$$

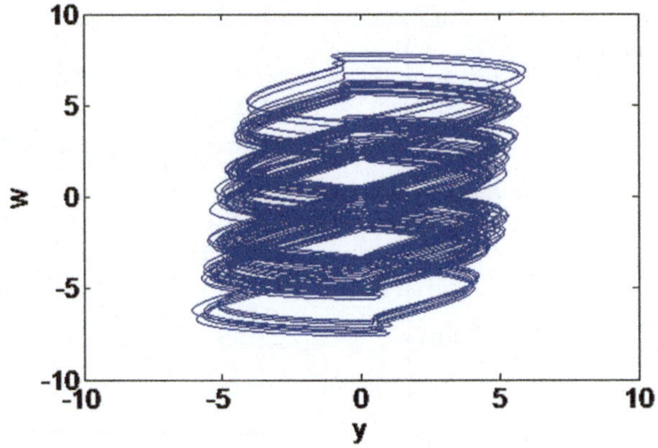

Fig. 12 Hidden attractor of the memristive hyperchaotic system (13) on the y-w plane

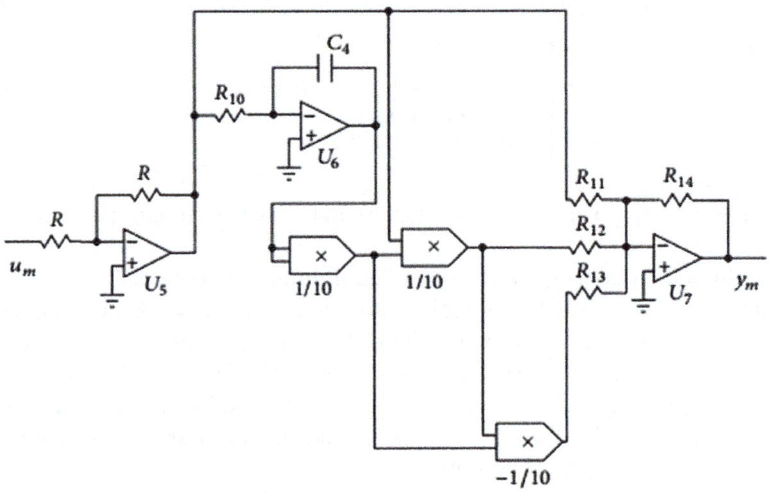

Fig. 13 The equivalent circuit diagram of the memristive element

where $f = (\frac{R_{14}}{R_{11}} + \frac{R_{14}}{100R_{12}}w^2 - \frac{R_{14}}{10^4 R_{13}}w^4)y$.

Using a flux-controlled memristor to substitute a coupling resistor in the realization circuit of a three-dimensional chaotic system [18], a new memristive hyperchaotic system with coexisting infinitely many hidden attractors is proposed by Bao et al. [19], which is illustrated in Fig. 15a.

For the memristor W, the relationship between the terminal voltage v and the terminal current i is described by

$$i = W(\varphi)v, \quad \varphi = v\,, \tag{15}$$

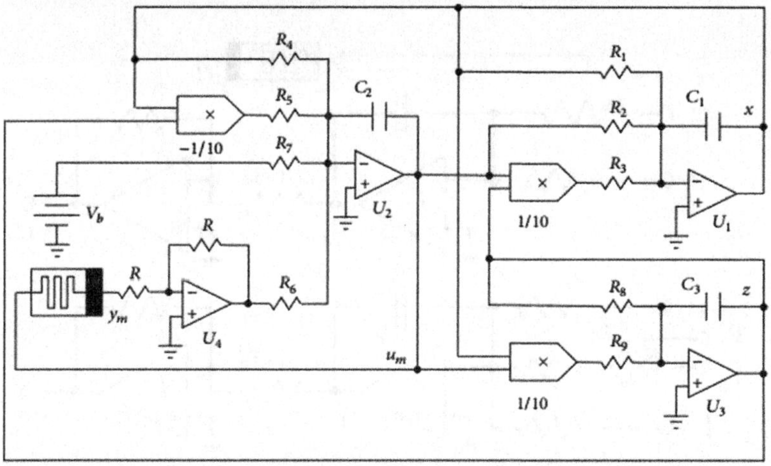

Fig. 14 The circuit implementation of system (13)

where the memductance function $W(\varphi) = \alpha + \beta\varphi^2$, and α and β are two positive constant parameters.

The equivalent circuit diagram of the memristor is shown in Fig. 15b.

From Fig. 15, one can see that the state equations of the memristive circuit is described by

$$\begin{cases} C\dot{v_x} = (v_y/R_c + v_w^2 v_y/R_b) - v_x/R_2 \\ C\dot{v_y} = v_x/R_3 - v_x v_z/R_4 + V_1//R_5 \\ C\dot{v_z} = v_x v_y/R_6 - v_z/R_7 \\ C\dot{v_w} = v_y/R_a\,. \end{cases} \tag{16}$$

When $R_a = R$, $d\alpha = R/R_C$ and $d\beta = 0.0025R/R_b$, the system (16) can be transformed to a dimensionless form, as follows:

$$\begin{cases} \dot{x} = dW(w)y - ax \\ \dot{y} = cx - xz + u \\ \dot{z} = xy - bz \\ \dot{w} = y\,, \end{cases} \tag{17}$$

where $W(w) = \alpha + \beta w^2$, a, b, c, d are positive parameters, and u is a non-zero control constant.

By setting the right-hand side of system (17) to zero, the resulting equations have no real solutions, so the system does not have any equilibrium. Thus, the attractors generated from system (17) are all hidden.

With $a = 35$, $b = 35$, $d = 40$, $u = \alpha = 1$, $\beta = 0.02$, and the initial condition $(0.1, 0, 0, 0)$, the memristive system (17) generates a double-scroll attractor, as depicted in Fig. 16. The corresponding Lyapunov exponents are $L_1 = 0.5881$, $L_2 =$

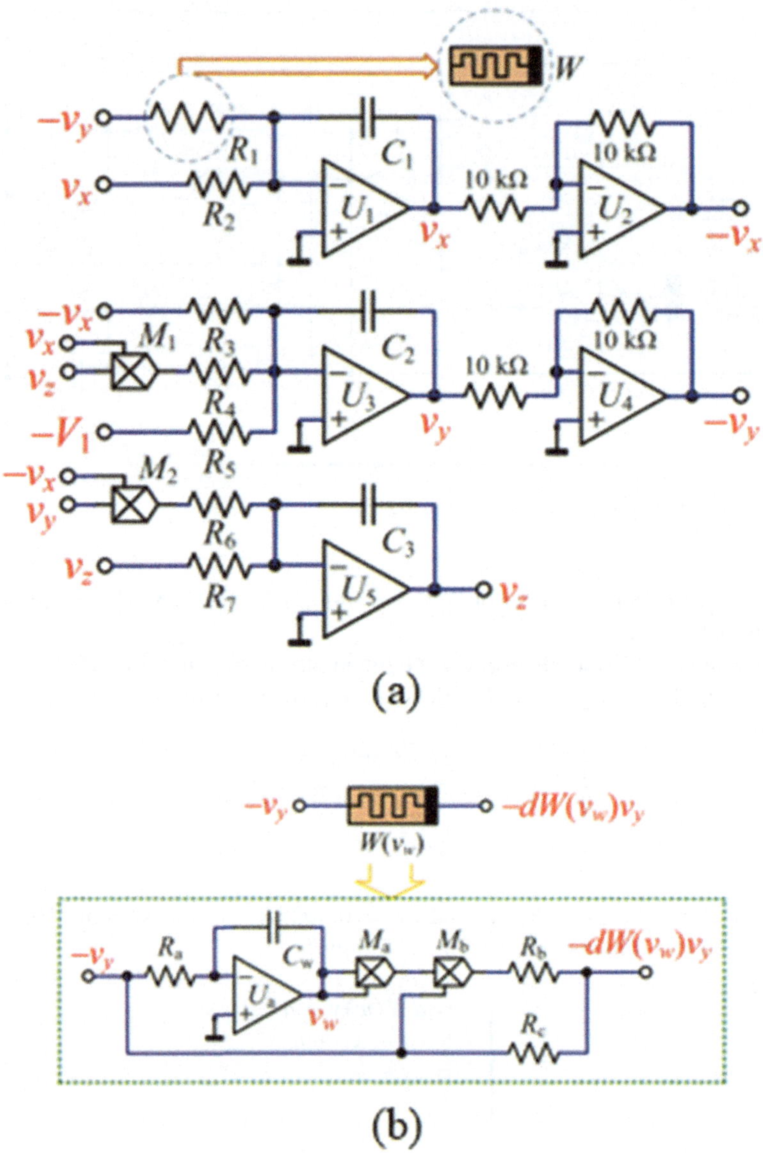

Fig. 15 Circuit implementation of the memristive hyperchaotic system: **a** Realization circuit of the memristive hyperchaotic system, **b** The equivalent realization circuit of the non-ideal flux-controlled memristor

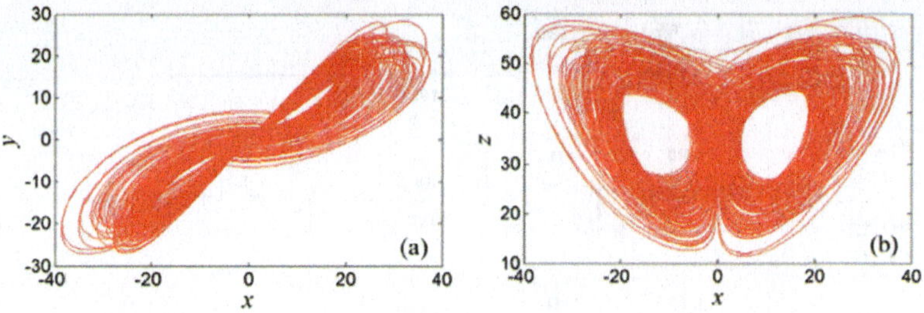

Fig. 16 Hidden attractor of the memristive hyperchaotic system (17)

0.1306, $L_3 = 0$, $L_4 = -37.7922$. Since there are two positive Lyapunov exponents, the memristive system is hyperchaotic.

4.2 5D Memristive Hyperchaotic Systems

Wang [20] proposed two kinds of flux-controlled smooth multi-piecewise quadratic nonlinearity memristor models, and used them to design a memristive multi-scroll Chua's circuit, which can generate $2N$-scroll and $(2N + 1)$-scroll chaotic attractors. Inspired by Wang [20], two kinds of non-ideal voltage-controlled multi-piecewise cubic nonlinear memristors and their mathematical models are introduced here.

By adding the memristor into a 4-D chaotic system, a new 5-D memristive hyperchaotic system is obtained. The relationship between the terminal voltage v and the terminal current i of the meristor is defined as

$$\begin{cases} i = W(w)v = (p + qf^2(w))v \\ \dot{u} = v - f(w)\,, \end{cases} \tag{18}$$

where $f(w)$ is the memristive internal state variable function.

The first kind of memristive memductance is described by

$$W_1(w) = p + qf_1^2(w),\ f_1(w) = w - sgn\ w - \sum_{\dot{n}=1}^{N} sgn(w - 2n) - \sum_{n=1}^{N} sgn(w + 2n)\,; \tag{19}$$

and the second memristive memductance is defined by

$$W_2(w) = p + qf_2^2(w),\ f_2(w) = w - sgn\ w - \sum_{\dot{n}=1}^{N} sgn(w - (2m - 1)) - \sum_{n=1}^{N} sgn(w + (2m - 1))\,, \tag{20}$$

where $p > 0$, $q > 0$, n, N and m, M are control parameters, which are positive integers.

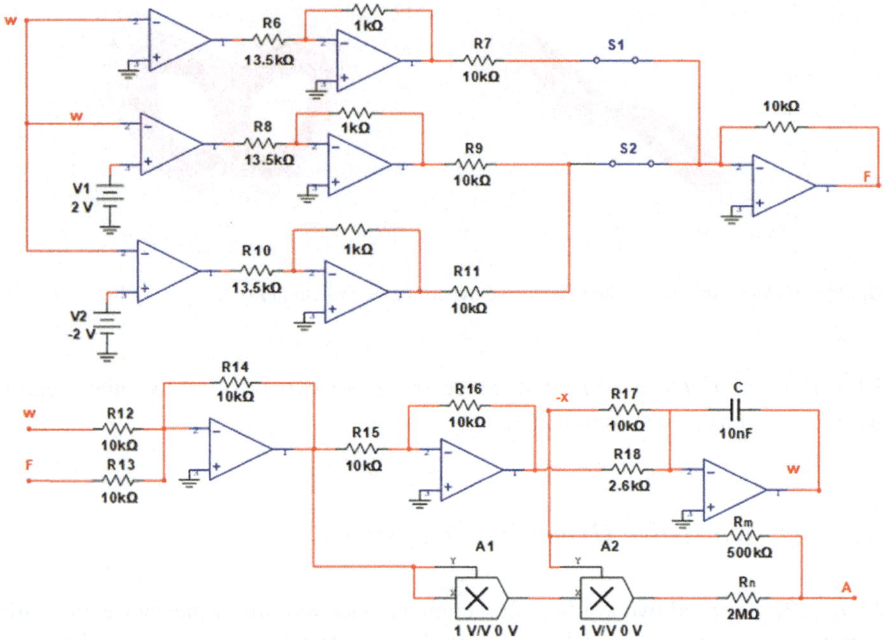

Fig. 17 Equivalent realization circuit of the non-ideal voltage-controlled memristor, with $N = 1$

Now, consider the first kind of memristor, with $N = 1$.

The equivalent circuit diagram of the memristor is shown in Fig. 17.

By adding the memristor into the circuit of a 4D chaotic system, a novel 5D memristive hyperchaotic system can be attained. The circuit realization of this new system is shown in Fig. 18, and its state equation of the system is

$$\begin{cases} \dot{v_x} = \frac{1}{R_1 C} v_y + W(-v_x) \\ \dot{v_y} = \frac{1}{R_2 C} v_2 + \frac{1}{R_3 C} \\ \dot{v_z} = -\frac{1}{R_a C}(v_x + v_y + v_z) + \frac{1}{R_4 C} sgn(v_x) \\ \dot{v_w} = v_y - \frac{1}{R_c C} f(v_w) \\ \dot{v_u} = \frac{1}{R_5 C} v_x , \end{cases} \tag{21}$$

where $W = \frac{1}{R_p C} + \frac{1}{R_q C} f^2(v_x)$, $f(v_w)$ is the memristive internal state variable function.

By introducing the following variables: $x = v_x,\ y = v_y,\ z = v_z,\ w = v_w,\ u = v_u$, a dimensionless of state equations of the system is obtained as

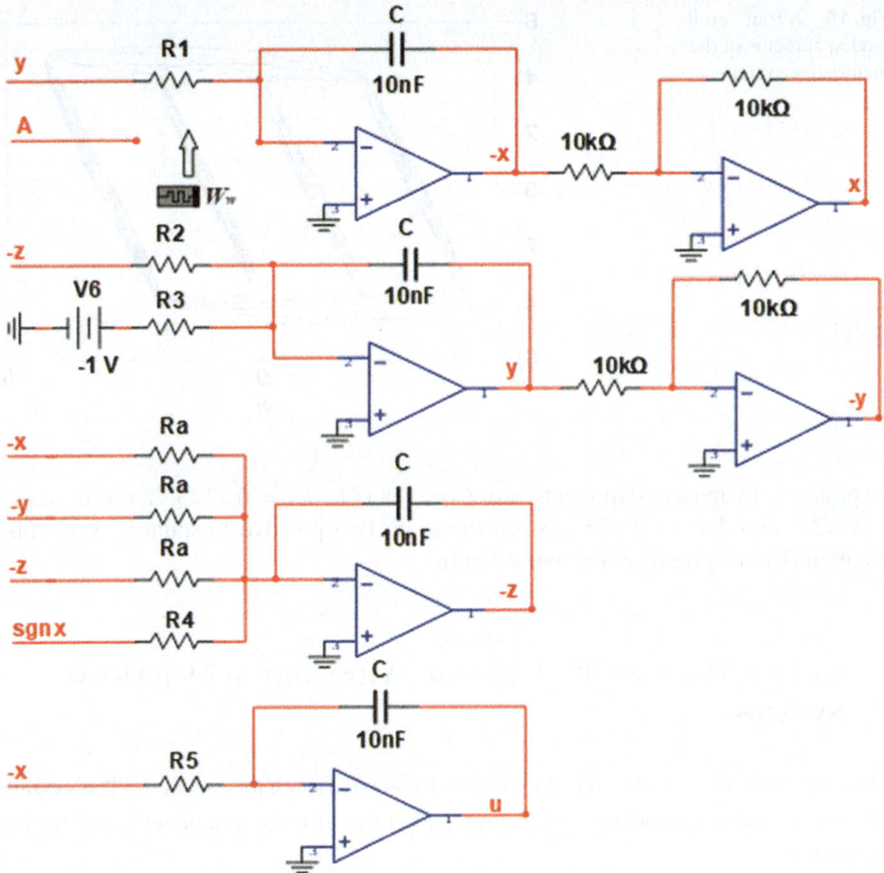

Fig. 18 Realization circuit of the memristive hyperchaotic system

$$\begin{cases} \dot{x} = y - kxW(w) \\ \dot{y} = z + b \\ \dot{z} = -a(x + y + z) + sgn(x) \\ \dot{w} = x - cf(w) \\ \dot{u} = x\,, \end{cases} \tag{22}$$

where $W(w) = p + qf^2(w)$, $a,\ c,\ q,\ m,\ n$ are positive parameters and b is a non-zero control constant.

By setting the right-hand side of the above system to zero, the resulting equations have no real solutions, so system (22) has no equilibria. Thus, all attractors generated from system (22) are hidden.

When the parameters are fixed as $a = 0.5,\ b = 0.1,\ c = 3.8,\ p = 0.1,\ n = 0.03,\ k = 0.2$, and the initial condition is chosen as (0.1, 0.1, 0.1, 0.1), the memristive system (22) generates a four-scroll hidden attractor, as shown in Fig. 19. The cor-

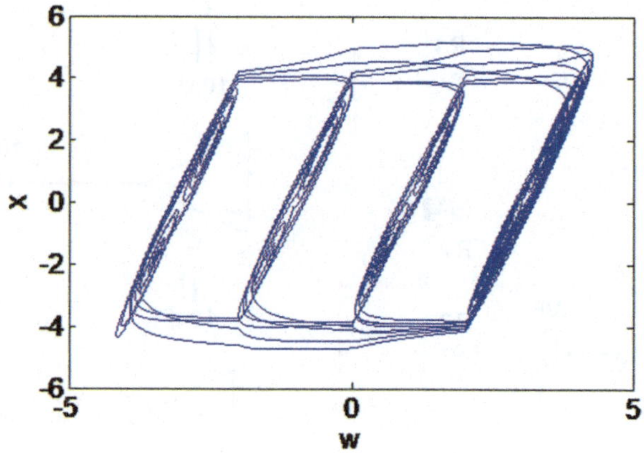

Fig. 19 A four-scroll hidden attractor of the memristive system

responding Lyapunov exponents are $L_1 = 0.1121$, $L_2 = 0.1123$, $L_3 = 0$, $L_4 = -0.7731$, and $L_5 = -2.9751$. Since there are two positive Lyapunov exponents, the memristive system (22) is hyperchaotic.

5 Hidden Multi-scroll/Multi-wing Attractors in Memristive Systems

Hu et al. [21] presented a five-dimensional memristive chaotic system. The system can generate multi-scroll and multi-wing hidden attractors simultaneously in different phase space.

A flux-controlled memristor was considered, where the relation between terminal current i and terminal voltage v is described by

$$i = W(\varphi)v$$

and

$$\dot{\varphi} = v\,,$$

where $W(\varphi) = g + h\varphi^2$. Based on that, the new memristive system is written as

$$\begin{cases} \dot{x} = ay \\ \dot{y} = by - z + c\ \sin(2\pi dx) \\ \dot{z} = y - ez \\ \dot{u} = -xy - (g + h\varphi^2)v + k \\ \dot{\varphi} = v\,. \end{cases} \tag{23}$$

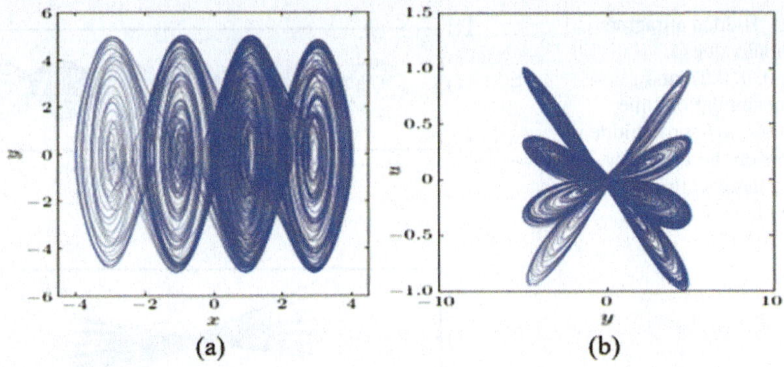

Fig. 20 Hidden attractors with initial values (0.2, 0, 0, 0, 0.2), and transient simulation time $t =$ 3000 s; **a** 4-scroll hidden attractors on the x-y plane, **b** 8-wing hidden attractors on the y-u plane [21]

The equilibria of the system can be obtained by setting the right-hand side of the above system to zero, namely,

$$\begin{cases} ay = 0 \\ by - z + c\ \sin(2\pi dx) = 0 \\ y - ez = 0 \\ -xy - (g + h\varphi^2)v + k = 0 \\ v = 0\,, \end{cases} \tag{24}$$

where the parameter k is a key factor to determine the type of the solutions. There are two scenarios.

Case A: when $k \neq 0$, clearly the system has no solutions.

Case B: when $k = 0$, the equilibria of the system are $(n/2d, 0, 0, 0, \varphi^*)$, where the parameter n is an integer and φ^* is an arbitrary real constant. In this case, the system has multiple lines of equilibria.

From the above analysis, one can see that the attractors in the system are all hidden.

Now, consider Case A. It is worth noting that, because of the existence of the sine function in Eqs. (23), the system's nonlinearity is unbound. Thus, the number of multi-scroll hidden attractors and multi-wing hidden attractors are affected by the transient time.

The system parameters are chosen as $a = 0.25,\ b = 0.4,\ c = 2,\ d = 0.5,\ e = 0.5,\ g = 15,\ h = 0.01$, and $k = 0.05$. Set the initial condition be (0.2, 0, 0, 0, 0.2) and the transient simulation time be $t = 3000$. The system can generate 4-scroll hidden attractors on the x-y plane and generate 8-wing hidden attractors on the y-u phase plane, simultaneously. The corresponding phase portraits are shown in Fig. 20a, b, respectively. When the transient time is changed to $t = 5000$, 7-scroll

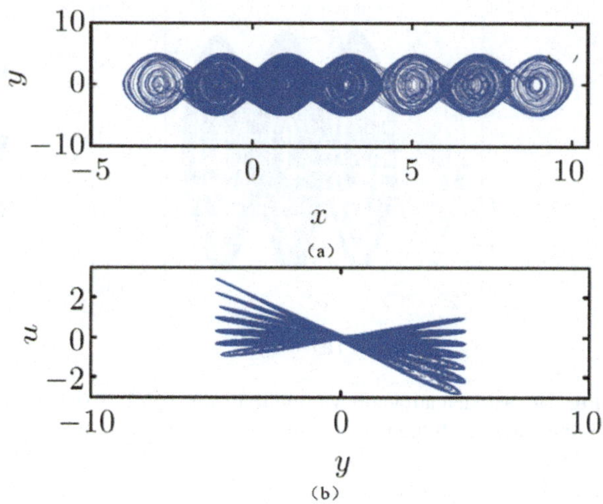

Fig. 21 Hidden attractors with initial values (0.2, 0, 0, 0, 0.2), and transient simulation time $t = 5000$ s; **a** 7-scroll hidden attractors on the x-y plane, **b** 14-wing hidden attractors on the y-u plane [21]

hidden attractors and 14-wing hidden attractors are generated by the system, as shown in Fig. 21a, b, respectively.

6 Hidden Attractors in Fractional-Order Memristive Chaotic Systems

6.1 4D Example for Hidden Chaos

Based on the Sprott N system (see Table 1) [22], a new memristive chaotic system without equilibria was proposed by introducing a flux-controlled memristor, described by

$$\begin{cases} \dot{x} = -2y \\ \dot{y} = z^2 + x - b(1 + 0.5w^2)z \\ \dot{z} = 1 - 2z + y \\ \dot{w} = az\,, \end{cases} \tag{25}$$

where a and b are real parameters.

From system (25), using the Caputo derivative [23], a fractional-order memristive system is constructed as follows:

$$\begin{cases} \frac{d^q x}{dt^q} = -2y \\ \frac{d^q y}{dt^q} = z^2 + x - b(1 + 0.5w^2)z \\ \frac{d^q z}{dt^q} = 1 - 2z + y \\ \frac{d^q w}{dt^q} = az\,, \end{cases} \tag{26}$$

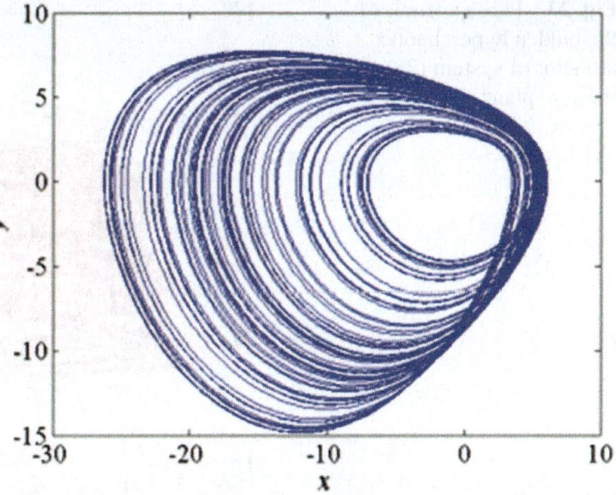

Fig. 22 Phase portrait of the one-scroll hidden attractor of system (26) on the x-y plane

where a and b are real parameters, and q is the fractional order.

Select the parameters as $a = 0.001,\ b = 0.02,\ q = 0.9$ and the initial condition as $(0.1, 0.1, 0.1, 0.1)$, system (26) can generate a one-scroll hidden attractor as shown in Fig. 22.

6.2 4D Example for Hidden Hyperchaos

Pham et al. [17] presented a memristive hyperchaotic system with no equilibria, described by

$$\begin{cases} \dot{x} = -10 - ay - yz \\ \dot{y} = -6x + 1.2xz + 0.1(1 + 0.24w^2 - 0.0016w^4)y - b \\ \dot{z} = -z - 1.2xy \\ \dot{w} = y\,, \end{cases} \tag{27}$$

where a and b are real parameters.

Applying the Caputo derivative, system (27) becomes a fractional-order system without equilibria, namely,

$$\begin{cases} \dfrac{d^q x}{dt^q} = -10 - ay - yz \\ \dfrac{d^q y}{dt^q} = -6x + 1.2xz + 0.1(1 + 0.24w^2 - 0.0016w^4)y - b \\ \dfrac{d^q z}{dt^q} = -z - 1.2xy \\ \dfrac{d^q w}{dt^q} = y\,, \end{cases} \tag{28}$$

where a and b are real parameters, and q is the fractional order.

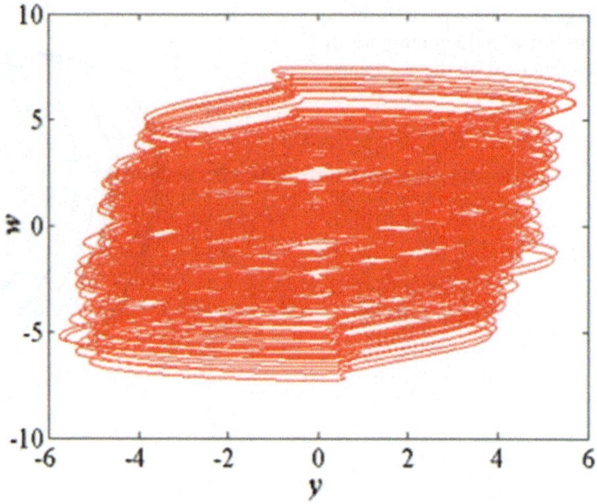

Fig. 23 Phase portrait of the hidden hyperchaotic attractor of system (28) on the y-w plane

With parameters $a = 5,\ b = 0.001,\ q = 0.9$, and the initial condition (0, 0.01, 0.01, 0), system (28) generates a hidden hyperchaotic attractor, as shown in Fig. 23.

7 Applications of Memristive Chaotic Systems

Due to the unusual rich dynamics of chaotic systems, they are widely applied in various fields, such as neural networks, image encryption, and secure communications, to name a few. This section introduces some applications of memristive chaotic systems.

Pham et al. [24] presented a memristive system with hidden attractors, and used it for image encryption. Using a memristor, a new memristive system was introduced in [17], as follows:

$$\begin{cases} \dot{x} = -10 - ay - yz \\ \dot{y} = -6x + 1.2xz + 0.1h(w, y) + b \\ \dot{z} = -z - 1.2xy \\ \dot{w} = y\,, \end{cases} \tag{29}$$

where a and b are two positive parameters, and $h(w, y)$ is the output of the memristive element described by

$$\begin{cases} h(w, y) = (1 + 0.24w^2 - 0.0016w^4)y \\ \dot{w} = y\,. \end{cases} \tag{30}$$

When $b \neq 0$, the system (30), which has no equilibria, has hidden attractors.

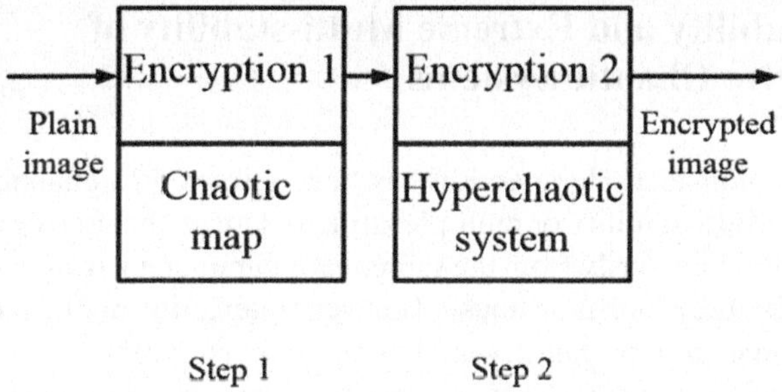

Fig. 24 Block diagram of the encryption scheme in two steps

Fig. 25 Simulation results: **a** Presentation of the plain image; **b** Presentation of the encrypted image; **c** Presentation of the decrypted image

When $b = 0$, the system (30) has an infinite number of equilibria, $E(0, 0, 0, w)$. The system belongs to a new class of systems with hidden attractors [25].

When $a = 5$, $b = 0$, and the initial condition is $(0, 0.01, 0.01, 0)$, the memristive system (30) is hyperchaotic, with Lyapunov exponents $L_1 = 0.1364$, $L_2 = 0.0071$, $L_3 = 0$, $L_4 = -10.8564$.

Then, the above system is applied to the encryption scheme of Gao and Chen [26]. The encryption procedure has two steps, as illustrated in Fig. 24.

Step 1 is to shuffle the pixel positions of the plain image, using a chaotic map, so as to derive an image total shuffling matrix, $P_{i,j}^{hl}$.

Step 2 is to encrypt the shuffled image, based on a hyperchaotic system. Here, the hyperchaotic memristive system (30) is used to group the states (B_1, B_2, B_3) to perform encryption, and then the XOR operation is applied between three bytes of the image total shuffling matrix $P_{i,j}^{hl}$ and three bytes of the selected group of three states (B_1, B_2, B_3).

Finally, continue performing the above two encryption steps until the whole image is encrypted [26].

By applying the above image encryption scheme to the plain-image as shown in Fig. 25a, and setting a secret key, the encrypted image is obtained as illustrated in Fig. 25b, while the decrypted image is shown in Fig. 25c. Cryptanalysis was also performed in [26].

8 Multi-stability and Extreme Multi-stability of Memristive Chaotic Systems

In recent years, considerable attention has been attracted to chaotic systems with multi-stability. Multi-stability or multiple attractors mean that two or more attractors are generated simultaneously from the same set of parameters in a system with different initial conditions, which is an unusual but very interesting nonlinear phenomenon, which has become an important research topic in recent years.

Multi-stability commonly exists in many nonlinear systems, such as power systems [27], biological systems [28], electrochemical models [29], food chains [30], and neural networks [31–34]. Notably, the appearance of multiple coexisting attractors (multi-stability) indicates an extra source of randomness in chaotic systems, with great potential for chaos-based applications such as secure communication and random signal generation. Such systems have a rich diversity of states, offering a great flexibility for different purposes of applications [35–41]. For example, if a chaotic system with coexisting attractors is influenced by noise or other disturbances, it can switch between different models to keep its normal operation. However, multi-stability may not be desirable in many situations, therefore needs to be controlled. Especially, when the number of coexisting attractors generated from a chaotic system rises to infinity, the coexistence of infinitely many attractors, depending on initial conditions and is named extreme multi-stability, are usually undesirable [42, 43]. Very recently, even mega-stability is being studied, which means that a system generates an infinite number of coexisting nested attractors [44, 45].

In this section, a rough classification on systems with multi-stability is presented; that is, systems with self-excited multi-stability [46–57], systems with self-excited extreme multi-stability [11, 58–73], systems with hidden multi-stability [12, 74–81], systems with hidden extreme multi-stability [19, 82–86], and systems with mega-stability [45, 87]. But the focus is on the multi-stability and extreme multi-stability in memristive chaotic systems.

8.1 *Memristive Chaotic Systems with Self-excited Multi-stability*

Liu and Chen [86] reported a three-dimensional chaotic system with a pseudo-four-wing attractor, described by

$$\begin{cases} \dot{x} = ax - byz \\ \dot{y} = -cy + xz \\ \dot{z} = xy - dz\,, \end{cases} \tag{31}$$

where $a,\ b,\ c$ and d are real numbers.

By introducing a flux-controlled memristor into system (31), Zhou et al. [56] presented a new 4D memristive chaotic system, described by

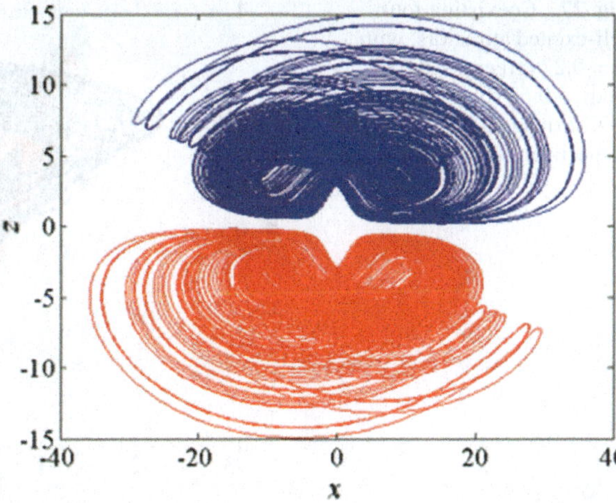

Fig. 26 Coexisting two-scroll self-excited attractors, with initial conditions (1, 10, 0, 0) (blue) and (−1, 10, 0, 0) (red), respectively

$$\begin{cases} \dot{x} = ax - byz \\ \dot{y} = -cy + xz \\ \dot{z} = xy - dz - kW(u)x \\ \dot{u} = ex\,, \end{cases} \tag{32}$$

where a, b, c, d, e and k are real parameters, and $W(u)$ is a memductance function with $W(u) = m + 3nu^2$.

It can be easily verified that this system has a line of equilibria and all the equilibria are unstable. Therefore, the resulting attractors are all self-excited attractors. For example, when the parameters are set as $a = 4$, $b = 6$, $c = 10$, $d = 5$, $e = 0.48$, $k = 0.01$, $m = 1$ and $3n = 0.03$, system (32) generates coexisting two-wing chaotic attractors from initial conditions $(1, 10, 0, 0)$ and $(-1, 10, 0, 0)$, respectively, as plotted in Fig. 26.

8.2 *Memristive Chaotic Systems with Self-excited Extreme Multi-stability*

Li et al. [59] introduced a charge-controlled memristive chaotic system, described by

$$\begin{cases} \dot{x} = a(z - xW(w)) \\ \dot{y} = z - u \\ \dot{z} = -\beta(y + x + z) \\ \dot{u} = \gamma y \\ \dot{w} = x\,, \end{cases} \tag{33}$$

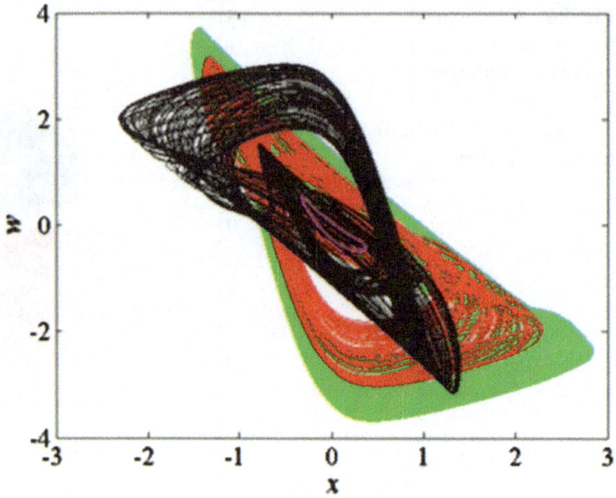

Fig. 27 Coexisting four self-excited attractors, with $k = 0.27$ (green), -0.63 (red), -0.63 (black) and -0.27 (magenta), respectively

where $\alpha,\ \beta,\ \gamma,\ a$ and b are real parameters, and $W(w)$ is a memductance function with $W(u) = -a + bw^2$.

Fix the parameters $\alpha = 9,\ \beta = 30,\ \gamma = 17,\ a = 1.2$ and $b = 1.2$, and the initial condition $(10^{-6}, 0, 0, k)$ and $v(0) = k$. Varying k in the region of $[-1, 1]$, system (33) can exhibit coexisting infinitely many attractors, namely possessing the self-excited extreme multi-stability. Typical phase portrait of the multiple coexisting attractors is shown in Fig. 27, with different values of k [59].

8.3 *Memristive Chaotic Systems with Hidden Multi-stability*

Chen et al. [12] reported an improved memristive Chua's circuit, exhibiting coexisting hidden attractors. The circuit diagram is plotted in Fig. 28, from which one can write out the circuit equations, as follows:

$$\begin{cases} \dfrac{dv_1}{dt} = \dfrac{v_2 - v_1}{RC_1} + \dfrac{(1 - g_1 g_2 v_0^2)v_1}{R_3 C_1} \\ \dfrac{dv_2}{dt} = \dfrac{v_1 - v_2}{RC_1} - \dfrac{i_3}{C_2} \\ \dfrac{di_3}{dt} = \dfrac{v_2}{L} - \dfrac{ri_3}{L} \\ \dfrac{dv_0}{dt} = -\dfrac{v_1}{R_1 C_1} - \dfrac{v_0}{R_2 C_0} \end{cases} . \tag{34}$$

Fix the circuit parameters $L = 12.6$ mH, $r = 2\ \Omega$, $C_1 = 6.8$ nF, $C_2 = 68$ nF, $R_1 = 4\ \text{k}\Omega$, $C_0 = 1$ nF, $R_2 = 10\ \text{k}\Omega$, $g_1 = 1,\ g_2 = 0.1,\ R_3 = 1.43\ \text{k}\Omega$ and $R_4 = R - 5 = 2\ \text{k}\Omega$.

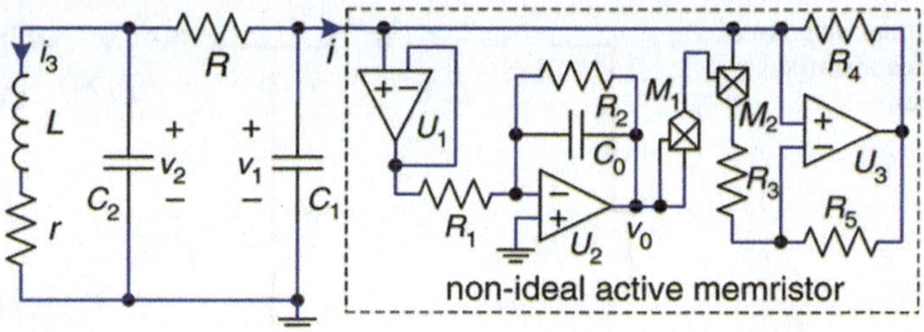

Fig. 28 Improved memristive Chua's circuit

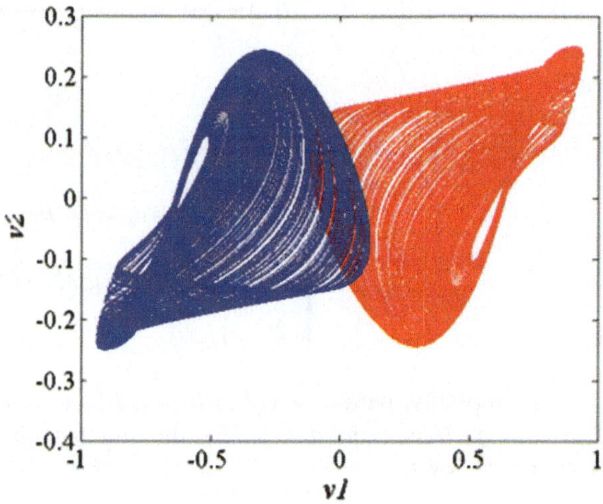

Fig. 29 Coexisting one-scroll hidden attractors, with initial conditions (0, 0.01, 0, 0) (blue) and (0, −0.01, 0, 0) (red), respectively

When $R = 2.2\,\mathrm{kW}$, with the initial conditions (0, 0.01, 0, 0) and (0, −0.01, 0, 0), respectively, system (34) generates two coexisting hidden attractors as illustrated in Fig. 29.

8.4 Memristive Chaotic Systems with Hidden Extreme Multi-stability

By introducing a tiny perturbation to a memristive chaotic oscillator, Wang et al. [83] derived a new memristive chaotic system without equilibria, which has hidden extreme multi-stability.

The circuit diagram is plotted in Fig. 30, in which M denotes the charge-controlled memristor. From Fig. 30, and according to Kirchhoff's laws, one can write out the circuit equation as follows:

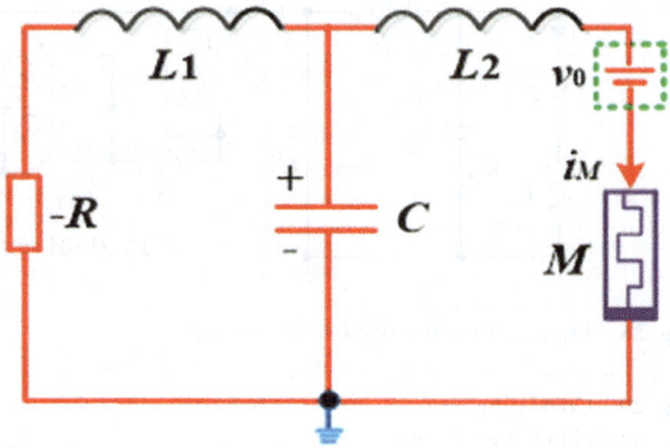

Fig. 30 Circuit diagram of tiny-perturbed memristive chaotic circuit

$$\begin{cases} C\dfrac{dv_c}{dt} = i_{L_1} - i_{L_2} \\ L_1\dfrac{di_{L_1}}{dt} = -v_c + i_{L_1}R \\ L_2\dfrac{di_{L_2}}{dt} = v_c - i_{L_2}M(q) - v_0 \\ \frac{dq}{dt} = i_{L_2}. \end{cases} \tag{35}$$

For simplicity, with $a = 1/L_1,\ b = 1/L_2,\ c = R/L_1,\ x = v_C,\ y = iL1,\ z = i_{L_2},\ w = q,\ C = 1$ and $d = v_0/L_2$, the dimensionless form of system (35) is presented as follows:

$$\begin{cases} \dot{x} = y - z \\ \dot{y} = -ax + cy \\ \dot{z} = b(x - z(\alpha + 3\beta w^2)) - d \\ \dot{w} = z, \end{cases} \tag{36}$$

where $a,\ b,\ c$ and d are positive parameters. When the system parameters are set to $a = 1,\ b = 5,\ c = 0.65,\ d = 0.001,\ \alpha = -0.4,\ \beta = 0.8$, and the initial conditions are set as $(0.01, 0, 0, w(0))$, system (36) can exhibit extreme multi-stability for different values of $w(0)$, as displayed in Fig. 31.

8.5 *Chaotic Systems with Mega-stability*

Sprott et al. [44] presented a chaotic system with an infinite number of nested coexisting attractors, referred to as the mega-stability.

The system is expressed as

$$\begin{cases} \dot{x} = y \\ \dot{y} = -(0.33)^2 y + y\ \cos(x) + \sin(0.73t) \end{cases} \tag{37}$$

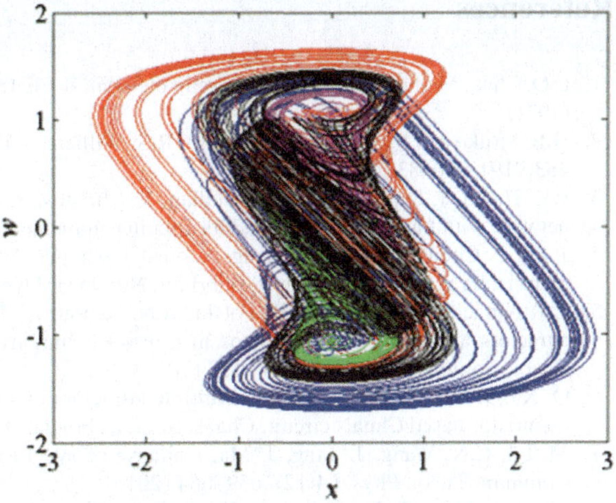

Fig. 31 Coexisting five hidden attractors, with $w(0) = 0$ (black), 0.4 (blue), -0.3 (red), 1.06 (green) and -1.06 (magenta), respectively

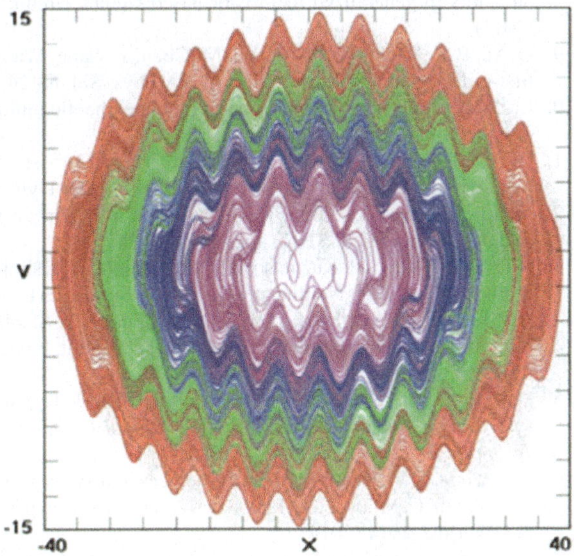

Fig. 32 Coexisting four attractors, with initial conditions $(5p, 0)$ (magenta), $(7p, 0)$ (blue), $(9p, 0)$ (green) and $(11p, 0)$ (red), respectively

The initial conditions are chosen as $(n\pi, 0)$.

When selecting different values of n, system (37) displays mega-stability, as exhibited in Fig. 32.

References

1. L.O. Chua, Memristor-the missing circuit element. IEEE Trans. Circuit Theory **18**(5), 507–519 (1971)
2. D.B. Strukov, G.S. Snider, D.R. Stewart, R.S. Williams, The missing memristor found. Nature **453**(7191), 80–83 (2008)
3. P.V. Thanh, J. Sajad, V. Sundarapandian, V. Christos, X. Wang, A novel memristive neural network with hidden attractors and its circuitry implementation. Sci. China **59**(3), 358 (2016)
4. J. Ma, Z. Chen, Z. Wang, Q. Zhang, A four-wing hyper-chaotic attractor generated from a 4-D memristive system with a line equilibrium. Nonlinear Dyn. **81**(3), 1275–1288 (2015)
5. Z.T. Njitacke, J. Kengne, H.B. Fotsin, A.N. Negou, D. Tchiotsop, Coexistence of multiple attractors and crisis route to chaos in a novel memristive diode bridge-based Jerk circuit. Chaos, Solitons Fractals **91**, 180–197 (2016)
6. Q. Xu, Y. Lin, B. Bao, M. Chen, Multiple attractors in a non-ideal active voltage-controlled memristor based Chua's circuit. Chaos, Solitons Fractals **83**, 186–200 (2016)
7. M. Lü, C.N. Wang, J. Tang, J. Ma, Collapse of synchronization in a memristive network. Commun. Theor. Phys. **64**(12), 659–664 (2015)
8. M. Chen, M. Li, Q. Yu, B. Bao, Q. Xu, J. Wang, Dynamics of self-excited attractors and hidden attractors in generalized memristor-based Chua's circuit. Nonlinear Dyn. **81**(1–2), 215–226 (2015)
9. Q. Yu, B.C. Bao, F.W. Hu, Q. Xu, M. Chen, J. Wang, Wien-bridge chaotic oscillator based on first-order generalized memristor. ACTA Phys. Sin. **63**(24), 240 505–240 505 (2014)
10. L. Zhou, C. Wang, L. Zhou, Generating hyperchaotic multi-wing attractor in a 4D memristive circuit. Nonlinear Dyn. **85**(4), 1–11 (2016)
11. B. Bao, T. Jiang, Q. Xu, M. Chen, H. Wu, Y. Hu, Coexisting infinitely many attractors in active band-pass filter-based memristive circuit. Nonlinear Dyn. **86**(3), 1711–1723 (2016)
12. M. Chen, J. Yu, B.C. Bao, Finding hidden attractors in improved memristor-based Chua's circuit. Electron. Lett. **51**(6), 462–464 (2015)
13. A.P. Kuznetsov, S.P. Kuznetsov, N.V. Stankevich, A simple autonomous quasiperiodic self-oscillator. Commun. Nonlinear Sci. Numer. Simul. **15**(6), 1676–1681 (2010)
14. A.P. Kuznetsov, S.P. Kuznetsov, E. Mosekilde, N.V. Stankevich, Generators of quasiperiodic oscillations with three-dimensional phase space. Eur. Phys. J.: Spec. Top. **222**(10), 2391–2398 (2013)
15. D.A. Prousalis, C.K. Volos, I.N. Stouboulos, I.M. Kyprianidis, A 4D hyperjerk memristive system with hidden attractors, in *International Conference on Modern Circuits and Systems Technologies* (2017), pp. 1–4
16. L.O. Chua, S.M. Kang, Memristive devices and systems. Proc. IEEE **64**(2), 209–223 (1976)
17. V.-T. Pham, C. Volos, L.V. Gambuzza, A memristive hyperchaotic system without equilibrium. Sci. World J. **2014** (2014)
18. Q. Yang, G. Chen, A chaotic system with one saddle and two stable node-foci. Int. J. Bifurc. Chaos **18**(05), 1393–1414 (2008)
19. B.C. Bao, H. Bao, N. Wang, M. Chen, Q. Xu, Hidden extreme multistability in memristive hyperchaotic system. Chaos, Solitons Fractals **94**, 102–111 (2017)
20. C. Wang, X. Liu, H. Xia, Multi-piecewise quadratic nonlinearity memristor and its 2n-scroll and (2n+1)-scroll chaotic attractors system. Chaos **27**(3), 033114 (2017)
21. X. Hu, C. Liu, L. Liu, Y. Yao, G. Zheng, Multi-scroll hidden attractors and multi-wing hidden attractors in a 5-dimensional memristive system. Chin. Phys. B **26**(11), 120–126 (2017)
22. J.C. Sprott, Some simple chaotic flows. Phys. Rev. E **50**(2), R647 (1994)
23. M. Caputo, Linear models of dissipation whose q is almost frequency independent-ii. Geophys. J. R. Astron. Soc. **13**(5), 529–539 (1967)
24. V.T. Pham, S. Jafari, C. Volos, A. Giakoumis, S. Vaidyanathan, T. Kapitaniak, A chaotic system with equilibria located on the rounded square loop and its circuit implementation. IEEE Trans. Circuits Syst. II: Express Briefs **63**(9), 878–882 (2017)

25. S. Jafari, J. Sprott, Simple chaotic flows with a line equilibrium. Chaos, Solitons Fractals **57**, 79–84 (2013)
26. T. Gao, Z. Chen, A new image encryption algorithm based on hyper-chaos. Phys. Lett. A **372**(4), 394–400 (2008)
27. V. Venkatasubramanian, W. Ji, Coexistence of four different attractors in a fundamental power system model. IEEE Trans. Circuits Syst. I: Fundam. Theory Appl. **46**(3), 405–409 (1999)
28. J.M. Cushing, S.M. Henson, C.C. Blackburn, Multiple mixed-type attractors in a competition model. J. Biol. Dyn. **1**(4), 347–362 (2007)
29. A. Massoudi, M.G. Mahjani, M. Jafarian, Multiple attractors in Koper–Gaspard model of electrochemical periodic and chaotic oscillations. J. Electroanal. Chem. **647**(1), 74–86 (2010)
30. R.K. Upadhyay, Multiple attractors and crisis route to chaos in a model food-chain. Chaos, Solitons Fractals **16**(5), 737–747 (2003)
31. Z. Zeng, W.X. Zheng, Multistability of neural networks with time-varying delays and concave-convex characteristics. IEEE Trans. Neural Netw. Learn. Syst. **23**(2), 293–305 (2012)
32. Z. Zeng, T. Huang, W.X. Zheng, *Multistability of Recurrent Neural Networks with Time-Varying Delays and the Piecewise Linear Activation Function* (IEEE Press, 2010)
33. P. Liu, Z. Zeng, J. Wang, Multistability analysis of a general class of recurrent neural networks with non-monotonic activation functions and time-varying delays. Neural Netw. **17**(3), 1447–1459 (2016)
34. F. Zhang, Z. Zeng, Multistability and instability analysis of recurrent neural networks with time-varying delays. Neural Netw. **97**, 116–126 (2018)
35. F.R. Ruiz-Oliveras, A.N. Pisarchik, Synchronization of semiconductor lasers with coexisting attractors. Phys. Rev. E **79**(1 Pt 2), 016202 (2009)
36. V.N. Chizhevsky, Coexisting attractors in a CO_2 laser with modulated losses. J. Opt. B: Quantum Semiclassical Opt. **2**(6), 711 (2000)
37. F.T. Arecchi, R. Meucci, G. Puccioni, J. Tredicce, Experimental evidence of subharmonic bifurcations, multistability, and turbulence in a q-switched gas laser. Phys. Rev. Lett. **49**(17), 1217 (1982)
38. J.A. Kelso, Multistability and metastability: understanding dynamic coordination in the brain. Philos. Trans. R. Soc. Lond. **367**(1591), 906–918 (2012)
39. F.C. Moon, P.D. Stiefel, Coexisting chaotic and periodic dynamics in clock escapements. Philos. Trans. A: Math. Phys. Eng. Sci. **364**(1846), 2539–2563 (2006)
40. S. Morfu, B. Nofiele, P. Marquié, On the use of multistability for image processing. Phys. Lett. A **367**(3), 192–198 (2007)
41. Y.A. Chen, T.R. Huang, Multistability of the brain network for self-other processing. Sci. Rep. **7**, 43313 (2017)
42. C. Hens, S.K. Dana, U. Feudel, Extreme multistability: attractor manipulation and robustness. Chaos **25**(5), 1607–218 (2015)
43. M.S. Patel, U. Patel, A. Sen, G.C. Sethia, C. Hens, S.K. Dana, U. Feudel, K. Showalter, C.N. Ngonghala, R.E. Amritkar, Experimental observation of extreme multistability in an electronic system of two coupled Rössler oscillators. Phys. Rev. E **89**(2), 022918 (2014)
44. J.C. Sprott, S. Jafari, A.J.M. Khalaf, T. Kapitaniak, Megastability: coexistence of a countable infinity of nested attractors in a periodically-forced oscillator with spatially-periodic damping. Eur. Phys. J.: Spec. Top. **226**(9), 1979–1985 (2017)
45. P. Prakash, K. Rajagopal, J.P. Singh, B.K. Roy, Megastability in a quasi-periodically forced system exhibiting multistability quasi-periodic behaviour, and its analogue circuit simulation. AEU: Int. J. Electron. Commun. **92**, 111–115 (2018)
46. Z.H. Guan, Q. Lai, M. Chi, X.M. Cheng, F. Liu, Analysis of a new three-dimensional system with multiple chaotic attractors. Nonlinear Dyn. **75**(1–2), 331–343 (2014)
47. M. Borah, B.K. Roy, Can fractional-order coexisting attractors undergo a rotational phenomenon? ISA Trans. (2017)
48. B.C. Lai, J.J. He, Dynamic analysis, circuit implementation and passive control of a novel four-dimensional chaotic system with multiscroll attractor and multiple coexisting attractors. Pramana **90**(3), 33 (2018)

49. Q. Lai, A. Akgul, X.W. Zhao, H. Pei, Various types of coexisting attractors in a new 4D autonomous chaotic system. Int. J. Bifurc. Chaos **27**(9), 1750142 (2017)
50. Q. Lai, T. Nestor, J. Kengne, X.W. Zhao, Coexisting attractors and circuit implementation of a new 4D chaotic system with two equilibria. Chaos, Solitons Fractals **107**, 92–102 (2018)
51. Q. Lai, S. Chen, Coexisting attractors generated from a new 4D smooth chaotic system. Int. J. Control Autom. Syst. **14**(4), 1124–1131 (2016)
52. L. Wang, 3-scroll and 4-scroll chaotic attractors generated from a new 3-D quadratic autonomous system. Nonlinear Dyn. **56**(4), 453–462 (2009)
53. Q. Lai, S. Chen, Research on a new 3D autonomous chaotic system with coexisting attractors. Opt.: Int. J. Light Electron Opt. **127**(5), 3000–3004 (2016)
54. J. Kengne, T.Z. Njitacke, T.V. Kamdoum, N.A. Nguomkam, Periodicity, chaos, and multiple attractors in a memristor-based Shinriki's circuit. Chaos **25**(10), 103126 (2015)
55. J. Kengne, A.N. Negou, D. Tchiotsop, Antimonotonicity, chaos and multiple attractors in a novel autonomous memristor-based Jerk circuit. Int. J. Bifurc. Chaos **27**(7), 1–20 (2017)
56. L. Zhou, C. Wang, L. Zhou, Generating four-wing hyperchaotic attractor and two-wing, three-wing, and four-wing chaotic attractors in 4D memristive system. Int. J. Bifurc. Chaos **27**(2), 1750027 (2017)
57. S. Zhang, Y. Zeng, Z. Li, One to four-wing chaotic attractors coined from a novel 3D fractional-order chaotic system with complex dynamics. Chin. J. Phys. **56** (2018)
58. C.N. Ngonghala, U. Feudel, K. Showalter, Extreme multistability in a chemical model system. Phys. Rev. E **83**(5 Pt 2), 056206 (2011)
59. C. Li, F. Min, Q. Jin, H. Ma, Extreme multistability analysis of memristor-based chaotic system and its application in image decryption. AIP Adv. **7**(12), 125204 (2017)
60. B. Bao, T. Jiang, G. Wang, P. Jin, H. Bao, M. Chen, Two-memristor-based Chua's hyperchaotic circuit with plane equilibrium and its extreme multistability. Nonlinear Dyn. **60**, 1–15 (2017)
61. H. Bao, N. Wang, B. Bao, M. Chen, P. Jin, G. Wang, Initial condition-dependent dynamics and transient period in memristor-based hypogenetic jerk system with four line equilibria. Commun. Nonlinear Sci. Numer. Simul. **57**, 264–275 (2018)
62. H. Bao, T. Jiang, K. Chu, M. Chen, Q. Xu, B. Bao, Memristor-based canonical Chua's circuit: extreme multistability in voltage-current domain and its controllability in flux-charge domain. Complexity **2018**, ID 5935637 (2018)
63. B.C. Bao, Q. Xu, H. Bao, M. Chen, Extreme multistability in a memristive circuit. Electron. Lett. **52**(12), 1008–1010 (2016)
64. M. Chen, M. Sun, B. Bao, H. Wu, Q. Xu, J. Wang, Controlling extreme multistability of memristor emulator-based dynamical circuit in flux-charge domain. Nonlinear Dyn. **91**(2), 1395–1412 (2018)
65. G. Peng, F. Min, Multistability analysis, circuit implementations and application in image encryption of a novel memristive chaotic circuit. Nonlinear Dyn. **90**(3), 1607–1625 (2017)
66. D.J.W. Simpson, Sequences of periodic solutions and infinitely many coexisting attractors in the border-collision normal form. Int. J. Bifurc. Chaos **24**(06), 201–279 (2014)
67. C. Li, J.C. Thio, J.C. Sprott, H.H.C. Iu, Y. Xu, Constructing infinitely many attractors in a programmable chaotic circuit. IEEE Access **99**, 1–1 (2018)
68. G. Wang, C. Shi, X. Wang, F. Yuan, Coexisting oscillation and extreme multistability for a memcapacitor-based circuit. Math. Probl. Eng. **2017**(5), 1–13 (2017)
69. J.C. Sprott, C. Li, Comment on 'how to obtain extreme multistability in coupled dynamical systems'. Phys. Rev. E **89**(6), 066902 (2014)
70. C.R. Hens, R. Banerjee, U. Feudel, S.K. Dana, How to obtain extreme multistability in coupled dynamical systems. Phys. Rev. E Stat. Nonlinear Soft Matter Phys. **85**(2), 035202 (2012)
71. S. Jafari, A. Ahmadi, S. Panahi, K. Rajagopal, Extreme multi-stability: when imperfection changes quality. Chaos, Solitons Fractals **108**, 182–186 (2018)
72. V.R.F. Signing, J. Kengne, Reversal of period-doubling and extreme multistability in a novel 4D chaotic system with hyperbolic cosine nonlinearity. Int. J. Dyn. Control **23**, 1–13 (2018)
73. J. Hizanidis, N. Lazarides, G. Tsironis, Flux bias-controlled chaos and extreme multistability in squid oscillators. Chaos **28**(6), 063117 (2018)

74. P.R. Sharma, M.D. Shrimali, A. Prasad, N.V. Kuznetsov, G.A. Leonov, Control of multistability in hidden attractors. Eur. Phys. J.: Spec. Top. **224**(8), 1485–1491 (2015)
75. Z.T. Zhusubaliyev, E. Mosekilde, Multistability and hidden attractors in a multilevel Dc/Dc converter. Math. Comput. Simul. **109**, 32–45 (2015)
76. N. Kuznetsov, G. Leonov, Hidden attractors in dynamical systems: systems with no equilibria, multistability and coexisting attractors. IFAC Proc. **47**(3), 5445–5454 (2014)
77. J.P. Singh, B.K. Roy, Multistability and hidden chaotic attractors in a new simple 4D chaotic system with chaotic 2-torus behaviour. Int. J. Dyn. Control **6**(2), 529–538 (2018)
78. S. Zhang, Y. Zeng, Z. Li, M. Wang, X. Zhang, D. Chang, A novel simple no-equilibrium chaotic system with complex hidden dynamics. Int. J. Dyn. Control **23**, 1–12 (2018)
79. T. Kapitaniak, G.A. Leonov, Multistability: uncovering hidden attractors. Eur. Phys. J.: Spec. Top. **224**(8), 1405–1408 (2015)
80. C. Li, J.C. Sprott, Coexisting hidden attractors in a 4D simplified Lorenz system. Int. J. Bifurc. Chaos **24**(03), 1450034 (2014)
81. V.T. Pham, C. Volos, S. Jafari, T. Kapitaniak, Coexistence of hidden chaotic attractors in a novel no-equilibrium system. Nonlinear Dyn. **87**(3), 2001–2010 (2017)
82. F. Yuan, G. Wang, X. Wang, Extreme multistability in a memristor-based multi-scroll hyper-chaotic system. Chaos **26**(7), 507–519 (2016)
83. L. Wang, S. Zhang, Y.C. Zeng, Z.J. Li, Generating hidden extreme multistability in memristive chaotic oscillator via micro-perturbation. Electron. Lett. **54**(13), 808–810 (2018)
84. S. Zhang, Y. Zeng, Z. Li, M. Wang, L. Xiong, Generating one to four-wing hidden attractors in a novel 4D no-equilibrium chaotic system with extreme multistability. Chaos **28**(1), 013113 (2018)
85. S. Jafari, A. Ahmadi, A.J.M. Khalaf, H.R. Abdolmohammadi, V.T. Pham, F.E. Alsaadi, A new hidden chaotic attractor with extreme multi-stability. AEU: Int. J. Electron. Commun. **89**, 131–135 (2018)
86. W. Liu, G. Chen, A new chaotic system and its generation. Int. J. Bifurc. Chaos **13**(01), 261–267 (2003)
87. M. Borah, B.K. Roy, An enhanced multi-wing fractional-order chaotic system with coexisting attractors and switching hybrid synchronisation with its nonautonomous counterpart. Chaos, Solitons Fractals **102**, 372–386 (2017)

Chaotic Jerk Systems with Hidden Attractors

Xiong Wang and Guanrong Chen

1 Introduction

During the last two decades, there has been an immense effort with success towards a new class of minimal chaotic flows. Since Poincaré deduced some of the qualitative properties of chaotic systems, including sensitive dependence on initial conditions [1], Lorenz [2], Rössler [3] and Chua [4, 5] had identified some very simple examples with quadratic or piecewise linear nonlinearities.

Lately, Sprott [6] found fourteen chaotic systems with six terms and only one quadratic nonlinearity, and five cases with five terms with two quadratic nonlinearities, which were simpler than the other three-dimensional autonomous quadratic chaotic systems. These Sprott models can be written in the following form:

$$\dddot{x} = J(\ddot{x}, \dot{x}, x)\,, \tag{1}$$

where J is a *jerk function*, with the third parametric derivative $\dddot{x}$ of the displacement x, namely the time derivative of acceleration, known as a "jerk" [7].

To define the *jerk function*, set an system of n coupled first-order ODEs that is not explicitly dependent on time t:

$$\dot{x} = V(x), \tag{2}$$

X. Wang (✉)
Institute for Advanced Study, Shenzhen University,
Shenzhen 518060, Guangdong, People's Republic of China
e-mail: wangxiong8686@szu.edu.cn

G. Chen
Department of Electrical Engineering, City University of Hong Kong, Hong Kong, SAR 999077, People's Republic of China
e-mail: eegchen@cityu.edu.hk

X. Wang et al. (eds.), *Chaotic Systems with Multistability and Hidden Attractors*, Emergence, Complexity and Computation 40,
https://doi.org/10.1007/978-3-030-75821-9_11

where $x = (x, y, z)^T$ and $x(t)$ represents its orbit in the phase space.

Referring to [8], any nth-order autonomous ODE that is given in an explicit form can be recast to an n-dimensional system of dynamical equations. In fact, one can transform the function (1) to the dynamical system (2) by selecting

$$\dot{x} = v \, , \quad \dot{v} = a \, , \quad \text{and} \quad \dot{a} = J(x, v = \dot{x}, a = \ddot{x}) \, .$$

Apparently, the jerk systems are a restricted class of third-order ODEs, and the jerk function must depend explicitly on x. Particularly, for some constraints, the function (1) is interpreted as the derivative of a one-dimensional Newtonian equation with a memory term depending on the history of motion. Meanwhile, it must satisfy the condition that the acceleration $\ddot{x}$ enters only linearly into the jerky dynamics [9].

Gottlieb [10] pointed out that the Sprott A system (see Table 1 of Chap. 'Introduction') can be recast to the jerk form, and posed the question of finding the simplest jerk function that still exhibits chaos.

Another fundamental question was also raised: what is the characteristic of a three-dimensional dissipative system with less than five quadratic terms that can generate chaos?

In response to Gottlieb's challenge, Sprott [6] proposed two jerk systems with three terms and two quadratic nonlinearities [11], and one simple system with three terms and a single quadratic nonlinearity [12]. At the same time, minimal polynomial dissipative and conservative jerky dynamical systems that can generate chaos had also been constructed by Linz [13], on the basis of the Sprott R system (see Table 1 of Chap. 'Introduction') [6], the Lorenz system [2] and the Rössler [3] model, but in somewhat more complicated formulations.

These jerk dynamical systems are the simplest ones with a single quadratic nonlinearity since Fu and Heidel [14], with a technical correction by Gascon [15]. It had been proved that there are no other dissipative systems that can generate chaos with fewer than five terms.

Since the jerky dynamics show all major features of the vector field and are conceptually simple, they might serve an important role in further studying chaotic and non-chaotic dynamical behaviors, including various routes to chaos.

This chapter reviews the development of jerk functions that generate chaos. Starting from the simple Sprott system, symbolic jerky dynamical systems found in the last two decades are introduced. Then, a systematic method for constructing such a jerk system is illustrated. Finally, several related topics such as the hyperjerk chaotic system, the coexistence of multiple attractors, and the jerk system with hidden attractors are discussed.

2 Simple Jerk Function that Generates Chaos

2.1 Simplest Jerk Function for Generating Chaos

Sprott [6] proposed fourteen algebraically simple three-dimensional ODEs with chaotic solutions (see Table 1 of Chap. 'Introduction'). These systems have six terms with one nonlinearity. Particularly, one of these cases is volume-conservative, while the others are dissipative.

Concurrently, Hoover [16] pointed out that the conservative Sprott A system (see Table 1 of Chap. 'Introduction'), namely

$$\begin{cases} \dot{x} = y \\ \dot{y} = -x + yz \\ \dot{z} = 1 - y^2. \end{cases} \tag{3}$$

is a special case of the Nose-Hoover thermostated dynamical system, that exhibits time-reversible Hamiltonian chaos.

System (3) can be transformed to an explicit jerk version [10]:

$$\dddot{x} + \dot{x}^3 - \ddot{x}(x + \ddot{x})/\dot{x} = 0\,. \tag{4}$$

Sprott I system (see Table 1 of Chap. 'Introduction') can also be also recast into the jerk form:

$$\dddot{x} + \ddot{x} + \frac{1}{5}\dot{x} + 5\dot{x}^2 + \frac{2}{5}x = 0\,. \tag{5}$$

Though the explicit ODE can be cast into a system of n coupled first-order ODEs, it is not common for the conservative case to do so. In fact, the final would ODE look quite different after if it was recast into the jerk form. Therefore, Gottlieb [10] asked the question "What is the simplest jerk function that gives chaos?"

Answer 1

Based on the Lorenz system [2] and the Rössler system [3], two simpler jerk functions were introduced by Linz [13]. Subsequently, the Lorenz model can be reduced to the jerk form, as

$$\dddot{x} + [1 + \sigma + b - \dot{x}/x]\ddot{x} + [b(1 + \sigma + x^2) - (1 + \sigma)\dot{x}/x]\dot{x} - b\sigma(r - 1 - x^2)x = 0\,, \tag{6}$$

which has eight terms with multiple nonlinearities.

The jerk Rössler model reads

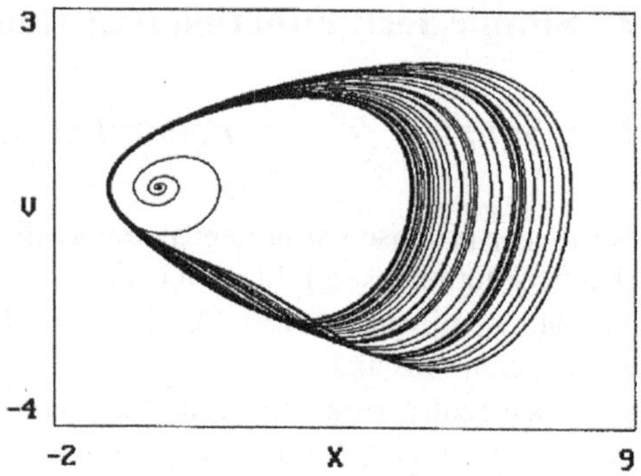

Fig. 1 Chaotic dynamics of jerk function (10), with $A = 2.017$, on the x-v plane [11]

$$\dddot{x} + [c - \varepsilon + \varepsilon x - \dot{x}]\ddot{x} + [1 - \varepsilon c - (1 + \varepsilon^2)x + \varepsilon \dot{x}]\dot{x} + (\varepsilon x + c)x + \varepsilon = 0\,, \tag{7}$$

which is a rather complicated quadratic polynomial with ten terms. System (7) generates chaos when $\varepsilon = 0.2$ and $c = 5.7$.

In addition, Linz also modified the Sprott R system (see Table 1 of Chap. 'Introduction'):

$$\dddot{x} + \ddot{x} - x\dot{x} + ax + b = 0\,, \tag{8}$$

which is also a polynomial but with only five terms and one quadratic nonlinearity. Interestingly, its chaotic attractor is a folded band, which is similar to the Rössler attractor, for $a = 0.9$ and $b = 0.4$ in system (8).

Answer 2

Obviously, new cases should be at least as simple as the 19 Sprott system shown in Table 1 of Chap. 'Introduction'.

To find chaotic solutions with the fewest non-zero coefficients and with the fewest nonlinearities, inspired by system (4) and (5) Sprott [11] proposed the most general second-degree polynomial jerk function:

$$J = (a_1 + a_2 x + a_3 \dot{x} + a_4 \ddot{x})\ddot{x} + (a_5 + a_6 x + a_7 \dot{x})\dot{x} + (a_8 + a_9 x)x + a_{10}\,. \tag{9}$$

Through numerically detecting chaos by randomly choosing three or four of the coefficients from a_1 to a_{10}, and by using its characteristic sensitive dependence on initial conditions, or calculating the Lyapunov exponents by Wolf's scheme [17], Sprott concluded that chaos indeed exists in some algebraically very simple jerk systems.

Subsequently, the simplest dissipative chaotic system based on system (9) was found with all coefficients equal to zero except for a_1, a_7 and a_8. Additionally, one coefficient can be arbitrary while the other two can be set to unity without loss of generality. Thus, one may select $a_1 = -A$, $a_7 = 1$ and $a_8 = -1$, so that the jerk function is specifically

$$\dddot{x} + A\ddot{x} - \dot{x}^2 + x = 0\,, \tag{10}$$

which contains only one single dissipative quadratic nonlinearity.

Function (10) is a special case of function (5), with the $\dot{x}$ term absent, and so can be written as an equivalent system of three equations having a total of five terms:

$$\dot{x} = v\,,\ \dot{v} = a\,,\ \dot{a} = -Aa + v^2 - x\,.$$

Numerical simulation shows that the bounded solutions to function (10) is given by 2.017... $< A <$ 2.082..., the largest Lyapunov exponent is 2.017 and the Kaplan-Yorke dimension is $D_{KY} = 2.0265$. The corresponding characteristic equation $\lambda^3 + A\lambda^2 + 1 = 0$ gives eigenvalues $\lambda = -2.24, 0.1 \pm 0.66i$ over the range of A for which bounded solution exist, which gives a single fixed unstable saddle-focus with an instability index of 2 at the origin.

The projection of the attractor of (10) is illustrated in Fig. 1, which shows the time history of x and its first three time-derivatives, after the trajectory has settled onto the attractor, for $A = 2.017$ [12].

Similarly, the $\dot{x}^2$ term in (10) can be replaced by $x\dot{x}$, which yields another function:

$$\dddot{x} + A\ddot{x} - x\dot{x} + x = 0\,, \tag{11}$$

with which the system is chaotic over the same range of A as in Eq. (10).

The projection of the attractor is illustrated in Fig. 2. Both trajectories of Eqs. (10) and (11) contain spirals outward to the attractor from an initial condition near but not at the unstable saddle-focus origin.

By adjusting the above two cases, other two cases with three terms and two quadratic nonlinearities in the jerk function can be found, which are described by

$$\dddot{x} + Ax\ddot{x} - \dot{x}^2 + x = 0\,, \tag{12}$$

with $A = 0.645$, and

$$\dddot{x} + Ax\ddot{x} - x\dot{x} + x = 0\,, \tag{13}$$

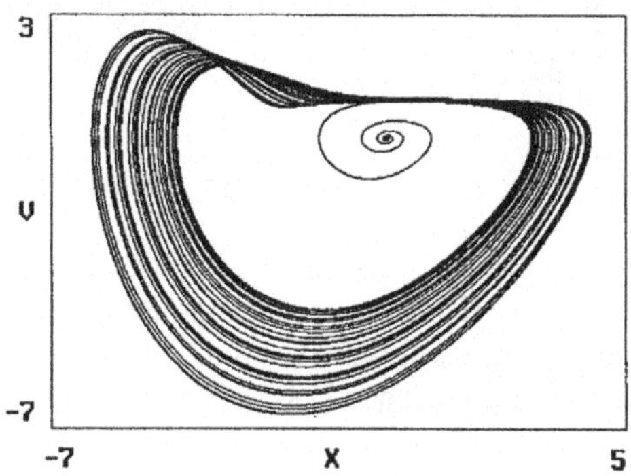

Fig. 2 Chaotic dynamics of jerk function (11), with $A = 2.017$, on the x-v plane [11]

with $A = -0.113$.

Concurrently, a more general form of the jerk function can be proposed, based on Eq. (10), as follows:

$$\dddot{x} + A\ddot{x} \pm \dot{x}^2 + x = 0 , \tag{14}$$

which has three terms and a single quadratic nonlinearity.

Equation 14 can also be written as

$$\dddot{x} + A\ddot{x} \pm x\dot{x} + x = 0 . \tag{15}$$

Referring to Fu and Heidel [14, 15], it was already proved that no three-dimensional, dissipative, dynamical systems with less than five terms and quadratic nonlinearities are able to generate chaos. Their result lends credence to claim that Eq. (14) is the simplest possible jerky system with one quadratic term that can generate chaos.

Furthermore, other types of general jerk functions were possible, as introduced and discussed by Sprott in [11].

2.2 Newtonian Jerky Dynamics

Considering that the jerk function is related to the time derivative of the acceleration of a particle11 with mass m, by Newton's second law, one has

$$\frac{dF}{dt} = mJ . \tag{16}$$

Assuming that the force has an explicit dependence on x, $\dot{x}$ and time t, which is $F = F(\dot{x}, x, t)$, and J has dependence on $\ddot{x}$, one can write

$$M = \int^t G(x(\tau))d\tau . \tag{17}$$

Thus, by differentiating Newton's second law with a force depending explicitly on the instantaneous position, velocity and time, Sprott [11] introduced a Newtonian jerk function as follows:

$$J = (a_1 + a_2 x + a_3 \dot{x})\ddot{x} + (a_5 + a_6 x + a_2 \dot{x})\dot{x} + a_{10} , \tag{18}$$

which produces no example, however, even if all six coefficients are non-zero.

Rather than concerning the motion of physical bodies, the aforementioned Linz and Sprott's works have actually transformed the conventional study of jerky dynamics to a new science [18].

2.3 Jerk Function with Cubic Nonlinearities

More interestingly, a general jerk function with only cubic nonlinearities was also introduced by Sprott [11] as follows:

$$J = (b_1 + b_2 x^2 + b_3 \dot{x}^2 + b_4 \ddot{x}^2 + b_5 x\dot{x} + b_6 x\ddot{x} + b_7 \dot{x}\ddot{x})\ddot{x} \\ + (b_8 + b_9 x^2 + b_{10}\dot{x}^2 + b_{11} x\dot{x})\dot{x} + (b_{12} + b13x^2)x\,, \quad (19)$$

which contains terms higher than linear in $\ddot{x}$.

Chaotic solutions of Eq. (19) were identified. Also, eight functionally distinct forms were found with three terms and two cubic nonlinearities, and four examples with four terms and one cubic nonlinearity were coined as well. But all of these examples are more complicated than Eqs. (10) and (11).

Three of these jerk functions with cubic nonlinearities are

$$\dddot{x} + \ddot{x}^3 + x^2\dot{x} + Ax = 0\,, \quad (20)$$

with $A = 0.25$,

$$\dddot{x} + A\ddot{x} - x\dot{x}^2 + x^3 = 0\,, \quad (21)$$

with $A = 3.6$, and

$$\dddot{x} + x^2\dot{x} - A(1 - x^2)x = 0\,, \quad (22)$$

with $A = 0.01$.

Moreover, a special form of the system, related to the old Moore-Spiegel oscillator [19], which models the inviscid convection of a rotating fluid, has four terms and one nonlinearity, described by

$$\dddot{x} + \ddot{x} + (T - R + Rx^2)\dot{x} + Tx = 0\,, \quad (23)$$

where T is analogous to the Prandtl number [20] times the Taylor number, while R is analogous to the Prandtl number times the Rayleigh number [21, 22].

For $T = 6$ and $R = 20$, the aperiodic behavior of this system is illustrated in Fig. 3.

Malasoma [25] searched for simple chaotic jerk functions with cubic nonlinearities, and introduced a new example system

$$\dddot{x} + A\ddot{x} - x\dot{x}^2 + x = 0\,, \quad (24)$$

which is invariant under the parity transformation $x \to -x$. Equation (24) is the simplest jerk function with cubic nonlinearities, which is chaotic when $A = 2.05$.

The chaotic attractor on the x-$\dot{x}$ plane is illustrated in Fig. 4.

When A decreases, the usual period-doubling route to chaos occurs, with a boundary crisis and unbounded solutions. Therefore, the range of A to allow chaos to exist

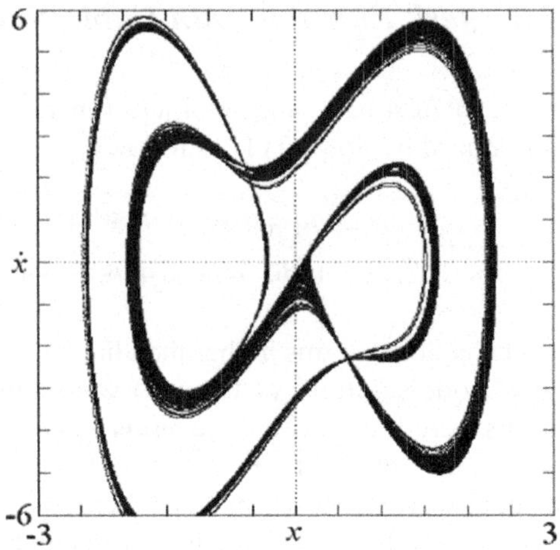

Fig. 3 Attractor for the Moore-Spiegel oscillator in Eq. (23) with $T = 6$ and $R = 20$ [23]

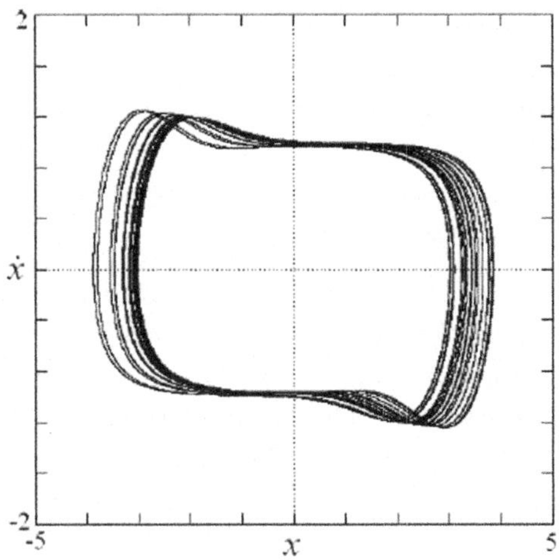

Fig. 4 Attractor for the simple cubic flow in Eq. (24), with $A = 2.05$ [24]

is narrow: $2.0278\cdots < A < 2.0840\cdots$. For $A = 2.05$, the Lyapunov exponents are $(0.0541, 0, -2.1041)$ and the Kaplan-Yorke dimension is $D_{KY} = 2.0257$.

2.4 Piecewise-Linear Jerk Functions

After confirming the simplest chaotic jerk function with a quadratic nonlinearity or a cubic nonlinearity, it is natural to ask if there are jerk functions in piecewise linear form.

One accessible method was introduced by Sprott [26]. Let $\dot{x}^2$ be replaced with $|\dot{x}|$ in Eq. (14). Then, consider systems of the following form:

$$\dddot{x} = a_1\ddot{x} + a_2\varphi(\ddot{x}) + a_3\dot{x} + a_4\varphi(\dot{x}) + a_5x + a_6\varphi(x) + a_7\,, \tag{25}$$

where $\varphi(x)$ is a simple nonlinear function and $a_1 - a_7$ are non-zero coefficients with random values. Note that the coefficients can range from minus infinity to plus infinity. When they are of order unity, the chaotic solutions occur in groups. Consequently, unbounded solutions are generated when $|\ddot{x}|$, $|\dot{x}|$ and $|x|$ never exceed 10.

One simple example is to use $\varphi(x) = |\dot{x}|$ as a piecewise-linear approximation to $\varphi(x) = x^2$, with the nonlinearity confined to the point $x = 0$.

Based on the verification method introduced and discussed in [6, 11, 12, 27], a general class of piecewise-linear jerk functions is proposed by Linz and Sprott [28], as

$$\dddot{x} + A_1\ddot{x} + A_2\dot{x} + A_3|x| + A_4 = 0\,, \tag{26}$$

where A_i $i = 1, 2, 3, 4$, are control parameters.

Note that Eq. (26) is different from the linear jerky function with the replacement of x(t) by its modulus, but not by the addition of any nonlinearity, and it is not reversible.

By choosing coefficients $A_1 = A$ and $A_2 = -A_3 = A_4 = 1$, a non-polynomial chaotic jerk function, which is probably the most elementary piecewise-linear, is introduced as follows:

$$\dddot{x} = A\ddot{x} + \dot{x} - |x| + 1 = 0\,, \tag{27}$$

which can also be written as

$$\dddot{x} = A\ddot{x} + \dot{x} \pm x + 1 = 0\,. \tag{28}$$

Thus, its exact solution can be obtained, segment by segment, for $x(t) < 0$ and $x(t) > 0$ along with the appropriate matching conditions for $x(t) = 0$. Through numerical calculation, Eq. (27) can be solved, which has two equilibria $x^\star = \pm 1$. The corresponding eigenvalues obtained from the characteristic equation are $\lambda_1 = -1$ and $\lambda_2 = \lambda_3 = i$. According to the Routh-Hurwitz criterion [29], $x^\star = +1$ is unstable for any value of A while $x^\star = -1$ is stable for $A > 1$. At $A = 1$, $x^\star = -1$ becomes unstable via a Hopf bifurcation.

As a typical case, Fig. 5 shows a stereoscopic view of the chaotic attractor generated by Eq. (27) with $A = 0.6$. It resembles the attractor of the minimal polynomial jerky dynamics, as well as the structure known from the Rössler system.

This system shows a periodic-doubling route to chaos as A is decreasing from 0.8 to 0.64085. On the other hand, there is a representative chaotic band structure disrupted by the occurrence of periodic windows for a small value of A. Particularly, at $A \simeq 0.547$, the periodic-three dynamics becomes unstable and the long-time evolution of Eq. (27) diverges to infinity.

Linz [31] proved that no chaotic dynamics exhibits in Eq. (27) if any of its terms are set to zero. Also, the term $|x|$ can be replaced with $|x^n|$, $|x|^n$, or x^{2n} where n is a positive integer, or more generally with any inversion of symmetric function

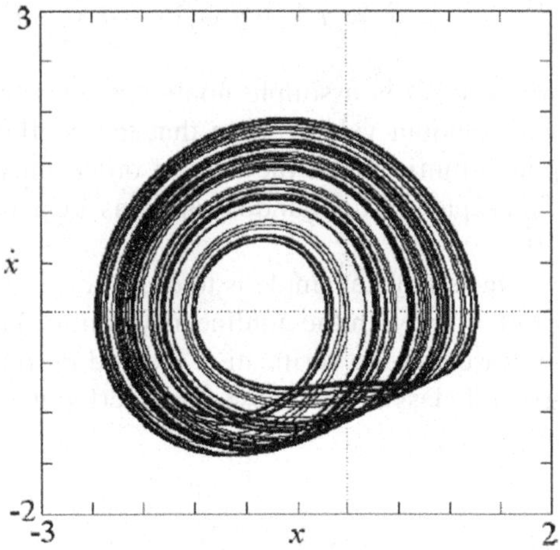

Fig. 5 Attractor for the simplest piecewise-linear flow in Eq. (26), with $A = 0.6$ [30]

$f(x) = -f(x)$, for all x, while this system can still have chaos. Numerical experiments show that solutions $f(x) = |x|^n$ exist for any integer $n > 0$.

More recently, another piecewise-linear jerk function was introduced by Fischer et al. [32], as follows:

$$\dddot{x} + A\ddot{x} + B\dot{x} + x - |x| + C = 0\,, \tag{29}$$

which is intended to model an experimental example discovered in an electronic circuit. Chaos exhibits in Eq. (29) with $B = 0.3$ and $A = C = 0.1$. Obviously it contains one more term and one more coefficient than Eq. (27), thus not leading to a "maximally simple" function, although the term $x - |x|$ can be written as $2\min(x, 0)$.

2.5 *Jerky Dynamics Accompanied with Many Driving Functions*

Actually, both Eqs. (27) and (29) can be obtained from a more general function. Sprott [26] found an equation in the general jerk form for most of the dissipative chaotic systems, which is

$$\dddot{x} + A\ddot{x} + \dot{x} = G(x)\,, \tag{30}$$

where the driving function $G(x)$ is a second-degree (or higher degree) polynomial, such as $x^2 - b$ or $x(x - b)$, in which A and b are constants. By integrating each term of Eq. (30), it can be found that this system is a damped harmonic oscillator driven by a nonlinear memory term, including the integral of $G(x)$.

Table 1 Some simple forms of function $G(x)$ that produce chaos in Eq. (30) with $A = 0.6$. The constant C is arbitrary, which only affects the size of the attractor [36]

G(x)	B	LEs
$\pm(B\|x\| - C)$	1.0	0.036
$-B\, max(x, 0) + C$	6.0	0.093
$Bx - C\, sgn(x)$	1.2	0.657
$-Bx + C\, sgn(x)$	1.2	0.162
$\pm B(x^2/C - C)$	0.58	0.073
$Bx(x^2/C - 1)$	1.6	0.103
$-Bx(x^2/C - 1)$	0.9	0.126
$-B[x - 2tanh(CX)/C]$	2.2	0.221
$\pm B\, sin(Cx)/c$	2.7	0.069
$\pm B\, cos(Cx)/c$	2.7	0.069

In other studies, chaos in numerical simulations were performed on the above equation with a cubic form [33] or special piecewise-linear form [65, 66] of $G(x)$, and an RLC circuit with an unspecified nonlinear amplifier [34, 35].

Coullet et al. [33] observed chaotic dynamics for the above equation with a cubic driving term $G(x) = Bx(1 - x^2)$, with $A = 0.1$ and $B = 0.44$, as well as a special piecewise-linear form of the driving function

$$G(x) = \begin{cases} -bx - b - c & if\ x \leq -1 \\ cx & if\ |x| \leq 1 \\ -bx + b + c & if\ x \geq 1\,, \end{cases} \tag{31}$$

which generates chaos with $a = 0.1$, $b = 0.2061612$, and $c = 0.2171604$. Equation (31) generates a homoclinic orbit that satisfies the Šil'nikov criterion.

Note that Eq. (30) with a nonlinear $G(x)$ has not been well studied. Sprott [36] listed some elementary examples in the sense of having the algebraically simplest structure of ODE with few terms of nonlinearity. By selecting

$$G(x) = B|x| - C\,,$$

these systems are specified as in Table 1, where typical values of B for arbitrary values of C, accompanied with the largest Lyapunov exponent, are given. Besides these examples, other accessible $G(x)$ are possible, for example, a delta function or a hysteretic function.

In all the cases studied here, the case of $G(x) = Bx - C\ \mathrm{sgn}(x)$ has the largest Lyapunov exponent. By applying a variant of simulated annealing, the value of coefficients $A = 0.55$ and $B = 2.84$ are adjusted for maximizing the Lyapunov exponents, yielding $(1.055, 0, -1.655)$. The Kaplan-Yorke dimension is $D_{KY} = 2.637$. From the simulation, it is easy to see that the attractor contains a thin torus that nearly

Table 2 Some additional simple functions $G(x)$ with appropriate values of a and b [30]

Functions	
$\dddot{x} = a\ddot{x} - bsgn(\ddot{x}) - \dot{x} - x$	$\dddot{x} = -a\ddot{x} - exp(\dot{x} - x)$
$\dddot{x} = -a\dot{x} - bx + cosh(x) - 1$	$\dddot{x} = -a\ddot{x} - b\dot{x} + cos(\dot{x}) - x$
$\dddot{x} = -a\dot{x} \pm (\|x\| - 1)$	$\dddot{x} = -a\ddot{x} - b\dot{x} + x^2 - 1$
$\dddot{x} = -a\dot{x} \pm (x - x^3)$	$\dddot{x} = -a\ddot{x} - b\dot{x} + x - cosh(x)$
$\dddot{x} = -a\dot{x} \pm b(x - max(x, 0) + 1)$	$\dddot{x} = -a\ddot{x} - b\dot{x} + x \pm exp(x)$
$\dddot{x} = -a\dot{x} \pm b(x - min(x, 0) - 1)$	$\dddot{x} = -a\ddot{x} - b\dot{x} - min(x, 0) - 1$
$\dddot{x} = -a\dot{x} \pm (x - sinh(x))$	$\dddot{x} = -a\ddot{x} - b\dot{x} - x \pm cosh(x)$
$\dddot{x} = -a\dot{x} \pm x + x^2$	$\dddot{x} = -a\ddot{x} - b\dot{x} \pm (cosh(x) - 1)$
$\dddot{x} = -a\dot{x} \pm bx - cosh(x) + 1$	$\dddot{x} = -a\ddot{x} - b\dot{x} \pm cosh(\dot{x}) - x$
$\dddot{x} = -a\ddot{x} + b\dot{x} - \dot{x}^3 - x$	$\dddot{x} = -a\ddot{x} - b\dot{x} \pm x - x^2$
$\dddot{x} = -a\ddot{x} + b\dot{x} - sinh(\dot{x}) - x$	$\dddot{x} = -a\ddot{x} \pm b(cosh(\dot{x}) - 1) - x$

reaches the boundary of its small basin of attraction; therefore, initial conditions must be chosen carefully to produce bounded solutions. One appropriate initial condition is $(x_0, y_0, z_0) = (0.03, -0.33, -0.3)$.

To determine the least nonlinear form of $G(x)$, consider the two-part piecewise-linear function with the smallest bend at the knee "θ" so as to obtain $G(x) = \pm(B|x| - C)$ with $A = 0.025$ and $B = 0.468$. In this case, $\theta = 2tan^{-1}B$, which is equivalent to 50.2°. In this case, the basin of attraction is very small, and the chaotic attractor coexists with a nearby limit cycle.

Since Table 1 is not exhaustive of simple chaotic jerk functions, Sprott and Linz [30] proposed an equation in a different form, as follows:

$$\dddot{x} = a\ddot{x} + b\varphi(\ddot{x}) + c\dot{x} + d\varphi(\dot{x}) + ex + f\varphi(x) + g\,, \tag{32}$$

where $\varphi(x)$ is a nonlinear function.

In this form, a dozen of algebraically distinct cases with three terms, and several hundred cases with four terms on the right-hand side. Some of these additional examples are listed in Table 2.

Table 2 includes examples that are a superset of a simpler equation, as well as those with multiple nonlinearities. However, these cases have not been carefully verified, thus providing no values for the coefficients a and b. Note that most cases in the table are conservative with very small Lyapunov exponents, and they require a careful choice of initial conditions to generate chaos.

More recently, Patidar and Sud [67] investigated a special jerky system by selecting

$$G(x) = B(x^2 - 1)\,,$$

yielding

$$\dddot{x} + A\ddot{x} + \dot{x} = B(x^2 - 1)\,, \tag{33}$$

which, with some appropriate values of coefficients A and B, can generate chaos.

To investigate the effect of weakening the nonlinearity on the jerk dynamics of Eq. (33), the term x^2 is replaced by $|x|$, leading to a new jerk system:

$$\dddot{x} + A\ddot{x} + \dot{x} = B(|x| - 1)\,. \tag{34}$$

It was found that both Eqs. (33) and (34) possess equilibria $(\pm 1, 0, 0)$, and they share some common properties, such as: for $A > 0$, $B < 0$ and $A > -2B$, the equilibrium $(1, 0, 0)$ is stable, but when $A = -2B$ this equilibrium becomes unstable via a Hopf bifurcation, while a state limit cycle emerges. On the other hand, for $A > 0$, $B > 0$, and $A > 2B$, the equilibrium $(-1, 0, 0)$ is stable, but with $A = 2B$ this equilibrium becomes unstable via a Hopf bifurcation, while a stable limit cycle emerges. Moreover, it can be verified that these jerk functions are dissipative, possessing stable chaotic attractors, for all positive values of A, but it is diverging with no stable attractors for all negative values of A.

2.6 Multi-scroll Chaotic Jerk System

Multi-scroll attractors were well developed and documented in the literature [37–40, 68], yet the aforementioned general jerk function that can generate multi-scroll attractors had not been investigated until Xie et al. [41] who attempted to modify a jerk function to generate multi-directional multi-scroll attractors.

The modified jerk system reads

$$\begin{cases} \dot{x} = y - F_2(y) \\ \dot{y} = z = F_3(z) \\ \dot{z} = a(-x - y - z + F_1(x) + F_2(y) + F_3(z))\,, \end{cases} \tag{35}$$

where $a \in [0.47,\ 0.96]$ and $F_1(x)$, $F_2(y)$, $F_3(z)$ are the staircase function series, described by

$$F_1(x) = \begin{cases} \xi[sgn(x) + \sum_{m=1}^{M} sgn(x - 2\xi m) + \sum_{m=1}^{M} sgn(x + 2\xi m)] \\ \xi[\sum_{m=1}^{M} sgn(x - \xi(2m-1)) + \sum_{m=1}^{M} sgn(x + \xi(2m-1))]\,, \end{cases}$$

$$F_2(y) = \begin{cases} \xi[sgn(y) + \sum_{n=1}^{N} sgn(y - 2\xi n) + \sum_{n=1}^{N} sgn(y + 2\xi n)] \\ \xi[\sum_{n=1}^{N} sgn(y - \xi(2n-1)) + \sum_{n=1}^{N} sgn(y + \xi(2n-1))]\,, \end{cases}$$

$$F_3(z) = \begin{cases} \xi[sgn(z) + \sum_{l=1}^{L} sgn(z - 2\xi n) + \sum_{l=1}^{L} sgn(z + 2\xi n)] \\ \xi[\sum_{l=1}^{L} sgn(z - \xi(2n-1)) + \sum_{n=l}^{L} sgn(z + \xi(2n-1))]\,. \end{cases}$$

By solving $\dot{x} = \dot{y} = \dot{z} = 0$, the equilibrium (x_{eq}, y_{eq}, z_{eq}) can be found and the associated Jacobian matrix is given by

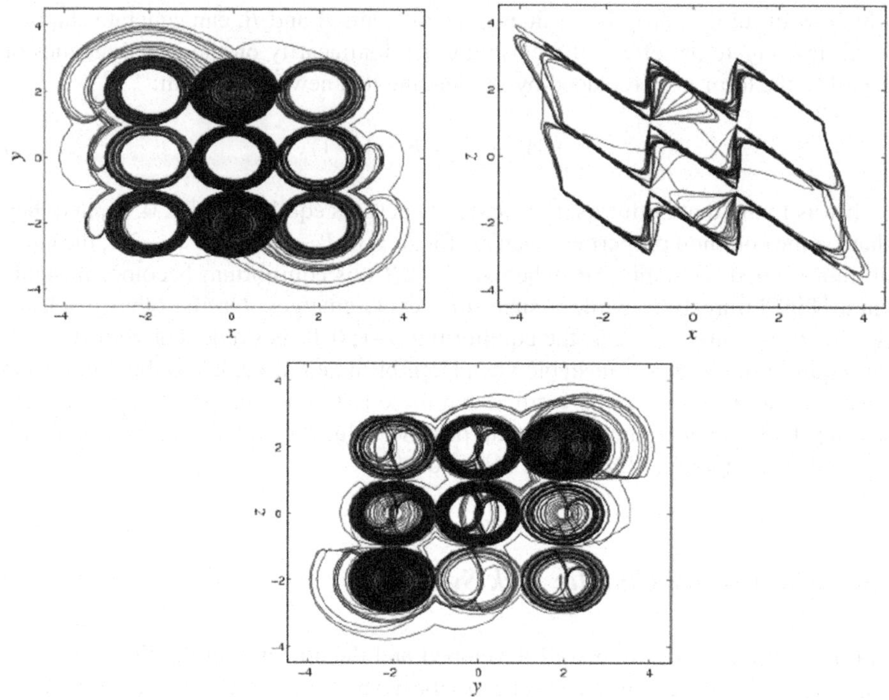

Fig. 6 $3 \times 3 \times 3$ grid-scroll attractors of system (35) with $\xi > 0$ [41]

$$J = \begin{bmatrix} 0 & 1 & 0 \\ 0 & 0 & 1 \\ -a & -a & -a \end{bmatrix}.$$

The corresponding characteristic is

$$\lambda^3 + \lambda^2 + \lambda a - a = 0,$$

which is independent of M, N and L. For $a = 0.65$, the eigenvalues are given by $\lambda_1 = -0.8217$, $\lambda_{2,3} = 0.0858 \pm 0.8853i$.

An example of $3 \times 3 \times 3$ grid-scroll attractors generated from system (35) with $\xi > 0$ is shown in Fig. 6.

Subsequently, Bao et al. [42] introduced a chaotic jerk system that can generate multi-scroll attractors. Driven by a step function series, the new system is given by

$$\dddot{x} = -\mu[\ddot{x} + \dot{x} + x - f(x)], \tag{36}$$

or, in a state-space form,

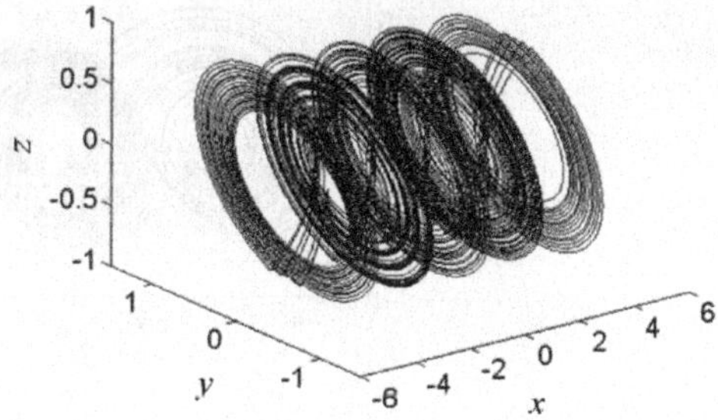

Fig. 7 Projection of 5-scroll chaotic attractor in the x-y-z space [42]

$$\begin{cases} \dot{x} = y \\ \dot{y} = z \\ \dot{z} = -\mu[x + y + z - f(x)], \end{cases} \tag{37}$$

where μ is a positive constant and $f(x)$ is a step function series in the following form:

$$f(x) = \sum_{n=1}^{N} sgn[x + (2n - 1)] + \sum_{n=1}^{N} sgn[x - (2n - 1)].$$

By numerical computation, the Lyapunov exponents of Eq. (36) with $\mu = 0.9$ are found as $(0.1117, 0, -1.0129)$, and the Lyapunov dimension is 2.1103, for initial condition $(x_0, y_0, z_0) = (0.8, 0.8, 0)$.

The projection of the 5-scroll chaotic attractor is shown in Fig. 7. It is clear from the equation that there are $(2N + 1)$ scroll equilibria located on the x-axis, where a one-scroll equilibrium is corresponding to a one-scroll attractor.

More recently, several chaotic systems based on the jerk model are reported by Liu et al. [43]. Consider the jerk system, proposed by Sprott [26, 36], in the following form :

$$\begin{cases} \dot{x} = y \\ \dot{y} = z \\ \dot{z} = -x - y - az + f(x)]. \end{cases} \tag{38}$$

Set

$$f(x) = \text{sgn}(x) + \text{sgn}(x + 2) + \text{sgn}(x - 2).$$

For $a = 0.6$, a four-scroll chaotic attractor is found and the equilibria are obtained as $(1, 0, 0)$, $(-1, 0, 0)$, $(3, 0, 0)$, $(-3, 0, 0)$, with $f(x) = x$. The phase projection is shown in Fig. 8.

Based on the mechanism generating the four-scroll attractor, a n-scroll jerky functional form can be deduced. By adjusting the nonlinear function in Eq. (38), the characteristic value of the equilibria will satisfy Šil'nikov theorem with the slope kept to one. Six more examples with either odd or even properties are listed in Table 3.

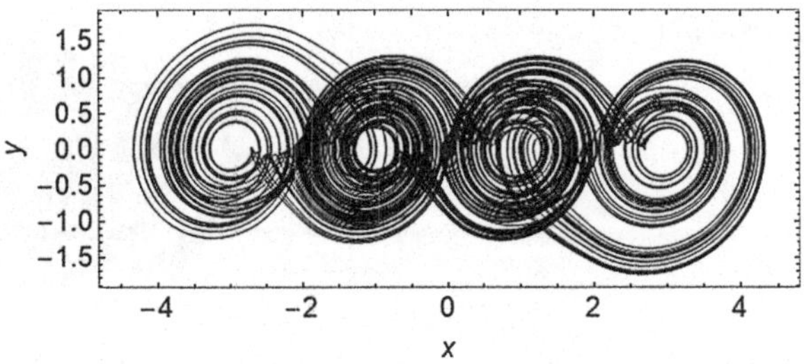

Fig. 8 Projection of 4-scroll chaotic attractor onto the x-y plane [43]

Table 3 Three even-scroll and three odd-scroll examples with different $f(x)$ [43]

Odd scrolls	$f(x)$
3-scroll	$f(x) = sgn(x+1) + sgn(x-1)$
5-scroll	$f(x) =$ $sgn(x+1) + sgn(x-1) + sgn(x+3) + sgn(x-3)$
7-scroll	$f(x) = sgn(x+1) + sgn(x-1) + sgn(x+3) +$ $sgn(x-3)$ $+sgn(x+5) + sgn(x-5)$
Even scrolls	
6-scroll	$f(x) = sgn(x) + sgn(x+2) + sgn(x-2) + sgn(x+$ $4) + sgn(x-4)$
8-scroll	$f(x) = sgn(x) + sgn(x+2) + sgn(x-2) + sgn(x+4)$ $+sgn(x-4) + sgn(x+6) + sgn(x-6)$
10-scroll	$f(x) = sgn(x) + sgn(x+2) + sgn(x-2) + sgn(x+$ $4) + sgn(x-4)$ $+sgn(x+6) + sgn(x-6) + sgn(x+8) + sgn(x-8)$
$\dddot{x} = -a\ddot{x} + b\dot{x} - sinh(\dot{x}) - x$	$\dddot{x} = -a\ddot{x} \pm b(cosh(\dot{x}) - 1) - x$

2.7 *Other Examples*

2.7.1 Example 1

Thomas [44] introduced a set of symmetric equations in the following form:

$$\begin{cases} \dot{x} = -ax + f(y) \\ \dot{y} = -ay + f(z) \\ \dot{z} = -az + f(x)\,, \end{cases}$$

which exhibits chaos with cubic polynomial f or $f(x) = sin(x)$, where $a = 0.18$.

It was shown that this system is simple and particularly elegant, since its trajectory percolates chaotically within the infinite three-dimensional lattice of unstable steady states. Therefore, it is called "Labyrinth Chaos". However, the jerk form of this system is so complicated due to the three nonlinearities, which will not be detailed here.

On the other hand, considered the terms in the Jacobian matrix as feedback loops and deduced candidate chaotic systems and the signs of the coefficients, other two simpler jerk systems can be formulated. One is

$$\dddot{x} + (c - a)\ddot{x} + (1 - ac)\dot{x} + 2x\dot{x} - ax^2 + cx = 0\,, \tag{39}$$

where $a = 0.385$, $c = 2$, which generates chaos with several equilibria. The other is

$$\dddot{x} + (c - a)\ddot{x} + (1 - ac)\dot{x} + 2x\dot{x} + cx = 0\,, \tag{40}$$

which exhibits chaos with only one equilibrium at the origin. Note that this example is actually a generalization of Eq. (15).

2.7.2 Example 2

The van der Pol oscillator [46] is well known, since it was the first example to be analyzed for chaotification. The van der Pol oscillator is described by

$$\ddot{x} - \varepsilon\dot{x}(1 - x^2) + x = 0\,, \tag{41}$$

where ε is a real parameter.

Zuppa [47] proposed a van der Pol jerk function by adding a new state variable to it:

$$\dddot{x} + \ddot{x} - \varepsilon\dot{x}(1 - x^2) + x = 0\,. \tag{42}$$

By setting $\varepsilon = 1$, the jerk function can be written as

$$\begin{cases} \dot{x} = y \\ \dot{y} = z \\ \dot{z} = -x + y - z - x^2 y\,, \end{cases} \tag{43}$$

where $y(t) = \dot{x}(t)$ and $z(t) = \ddot{x}$.

Ding and Zhang [45] further added four innovative terms to system (43) and change the coefficients of these two terms, to obtain

$$\begin{cases} \dot{x} = ax + by + cz \\ \dot{y} = dx + ey + fz \\ \dot{z} = -x + y - z - x^2 y\,. \end{cases} \tag{44}$$

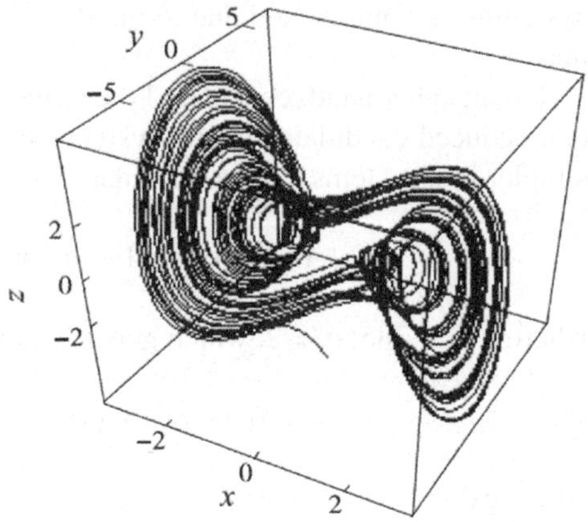

Fig. 9 Phase space portrait of system (44) [45]

By solving the equation $\dot{x} = \dot{y} = \dot{z} = 0$, the equilibria of system (44) are obtained. The Jacobian matrix for this system at the equilibria is given by

$$J = \begin{bmatrix} a & b & c \\ d & e & f \\ -1 & 1 & -1 \end{bmatrix}.$$

For the case of $a = 0.1,\ b = 1.5,\ c = -0.4,\ d = e = -0.5,\ f = 20$, with initial condition $(x_0, y_0, z_0) = (1.6, -8, 0.2)$, the corresponding eigenvalues are obtained:

$$\lambda_1 = -5.74915, \quad \lambda_{2,3} = 2.12458 \pm 0.874763i\ .$$

The phase portrait of this case is shown in Fig. 9. The result exhibits a homoclinic orbit that satisfies the Šil'nikov theorem.

2.7.3 Example 3

Motivated by Sprott's work, Malasoma [25] found many distinct classes of dissipative chaotic systems with five terms, including one quadratic nonlinearity and their jerk form. By listing all possible combinations of variables in a systematic way, 75 types of nonlinear systems that are neither asymptotic to a two-dimensional surface nor deducible to a two-dimensional autonomous system were proposed. Based on the spectrum of the Lyapunov exponents, six types of chaotic jerky dynamics are classified, as listed in Table 4.

The example of $\dddot{x} = -ax - a\ddot{x} + x\dot{x} + \dot{x}\ddot{x}/x$ can also be reformulated as

Table 4 Jerk equations for the minimal chaotic systems in [25]

$\dddot{x} = -x - a\ddot{x} + x\dot{x}$
$\dddot{x} = x + x^2 - a\dot{x} + (\dot{x}\ddot{x} + a\dot{x}^2)/x$
$\dddot{x} = -x - a\ddot{x} + \dot{x}^2$
$\dddot{x} = -ax - a\ddot{x} + x\dot{x} + \dot{x}\ddot{x}/x$
$\dddot{x} = -x - a\ddot{x} + x\dot{x} + \dot{x}\ddot{x}/x$

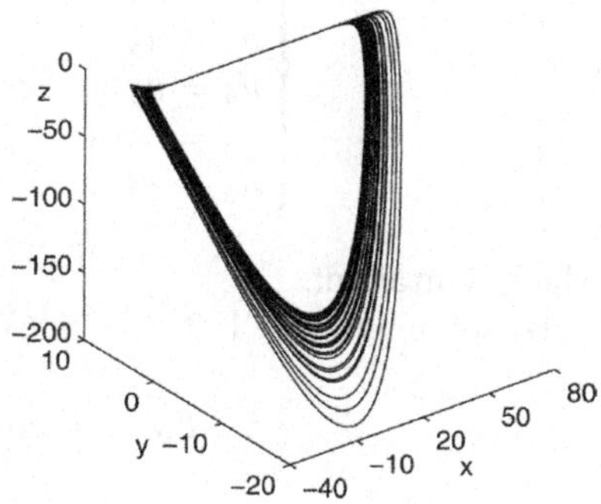

Fig. 10 Phase portrait of system (44) [45]

$$\begin{cases} \dot{x} = z \\ \dot{y} = -ay + z \\ \dot{z} = -x + xy\,, \end{cases} \tag{45}$$

where a is a real parameter.

For $a = 10.285$ with initial condition $(x_0, y_0, z_0) = (-22, -12, -109)$, the phase portrait of the attractor is illustrated in Fig. 11. This system is not sensitive to the iteration step size based on numerical analysis (Fig. 10).

2.7.4 Example 4

Mahnoud and Ahmed [48] introduced a complex jerk function of the following form:

$$\dddot{x} = J(x, \bar{x}, \dot{x}, \dot{\bar{x}}, \ddot{x}, \ddot{\bar{x}})\,,$$

where x is a complex variable while $\bar{x}$ is its complex conjugate variable.

An example of this form is

$$\dddot{x} + \alpha\ddot{x} + \beta\dot{x} + \upsilon x + \eta x^2\bar{x} = 0\,, \tag{46}$$

where $\alpha,\ \beta,\ \eta$ are positive parameters and υ is a negative constant. Equation (46) can be written as

$$\begin{cases} \dot{x} = y \\ \dot{y} = z \\ \dot{z} = -\alpha z - \beta y - \upsilon x - \eta x^2 \bar{x} , \end{cases} \tag{47}$$

where $x = u_1 + iu_2$, $y = u_3 + iu_4$ and $z = u_5 + iu_6$. Therefore, Eq. (47) has a real version in the form of

$$\begin{cases} \dot{u_1} = u3 \\ \dot{u_2} = u_4 \\ \dot{u_3} = u_5 \\ \dot{u_4} = u_6 \\ \dot{u_5} = -\alpha u_5 - \beta u_3 - \upsilon u_1 - \eta u_1(u_1^2 + u_2^2) \\ \dot{u_6} = -\alpha u_6 - \beta u_4 - \upsilon u_2 - \eta u_2(u_1^2 + u_2^2) , \end{cases} \tag{48}$$

which is invariant.

By solving $\dot{u_i}(i = 1, 2, 3, 4, 5, 6) = 0$, the equilibrium of Eq. (47) is obtained as

$$E_0 = (r \ \cos\theta, r \ \sin\theta, 0, 0, 0, 0).$$

By linearizing the system, the Jacobian matrix at the equilibrium is obtained, as

$$J = \begin{bmatrix} 0 & 0 & 1 & 0 & 0 & 0 \\ 0 & 0 & 0 & 1 & 0 & 0 \\ 0 & 0 & 0 & 0 & 1 & 0 \\ 0 & 0 & 0 & 0 & 0 & 1 \\ -\upsilon & 0 & -\beta & 0 & -\alpha & 0 \\ 0 & -\upsilon & 0 & -\beta & 0 & -\alpha \end{bmatrix}.$$

The corresponding characteristic equation is

$$(\lambda^3 + \alpha\lambda^2 + \beta\lambda + \upsilon)^2 = 0 .$$

Thus, the real parts of the roots λ are negative. The system is stable, by the Routh-Hurwitz criterion, if and only if

$$\alpha > 0, \ \beta > 0, \ \upsilon > 0, \ \alpha\beta - \upsilon > 0 .$$

Otherwise, the equilibrium E_0 is unstable with $\upsilon < 0$. The chaotic attractor generated by $\alpha = \eta = 1, \ \beta = 4, \ \upsilon = -5$ and initial conditions $(t_0, u_1, u_2, u_3, u_4, u_5, u_6) = (0, 4, 1, -2, 2, -1, 1)$ is shown in Fig. 11.

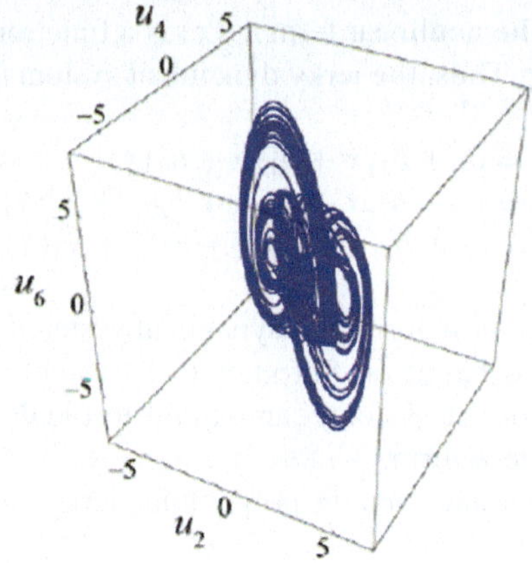

Fig. 11 Chaotic attractor of Eq. (48), with $\alpha = \eta = 1,\ \beta = 4,\ \upsilon = -5$ and initial condition $(t_0, u_1, u_2, u_3, u_4, u_5, u_6) = (0, 4, 1, -2, 2, -1, 1)$ [48]

3 Systematic Method for Constructing a Simple 3D Jerk System

Since all the previously mentioned chaotic systems have either one quadratic nonlinearity or multiple nonlinearities that can be recast to the form of jerk functions, Eichorn et al. [9] summarized the situations with three-dimensional jerk functions and proposed a systematic method of finding such systems, based on the method introduced by Becker and Thomas [49].

Assume a dynamical system of the following form:

$$\begin{cases} \dot{x} = c_1 + b_{11}x + b_{12}y + b_{13}z + n_1(x) \\ \dot{y} = c_2 + b_{21}x + b_{22}y + b_{23}z + n_2(x) \\ \dot{z} = c_3 + b_{31}x + b_{32}y + b_{33}z + n_3(x)\,, \end{cases} \tag{49}$$

with n_i, $i = 1, 2, 3$, being nonlinear functions that can be recast to a jerky equation $\dddot{x} = J(x, \dot{x}, \ddot{x})$. Assume that the following condition is satisfied:

$$b_{12}n_2(x) + n_{13}n_3(x) = f(x, b_{12}y + b_{13}z)\,, \tag{50}$$

with f being any well-defined function and

$$b_{12}^2 b_{23} - b_{12}^2 b_{32} + b_{12}b_{13}(b_{33} - b_{22}) \neq 0\,. \tag{51}$$

Note that the variables y and z are allowed to enter only linearly into the $\dot{x}$ equation, and $b_{13} = 0$ gives an important special case of Eq. (49). Thus, the condition (51) reduces to

$$b_{12} \neq 0\ \ ,\quad b_{23} \neq 0\,.$$

Clearly, in Eq. (50), the nonlinear term $n_2(x)$ is a function of x and y, and $n_3(x)$ can be any function of x. Thus, the jerky dynamical system (49) can be rewritten as

$$\begin{cases} \dot{x} = c_1 + b_{11}x + b_{12}y + n_1(x) \\ \dot{y} = c_2 + b_{21}x + b_{22}y + b_{23}z + n_2(x, y) \\ \dot{z} = c_3 + b_{31}x + b_{32}y + b_{33}z + n_3(x) . \end{cases} \tag{52}$$

By applying the above transformable dynamical systems, all the examples introduced in [6], constructed based on Sprott A to S systems (see Table 1 of Chap. 'Introduction') or the Rössler system, can exhibit toroidal chaos (donated by TR) and can be recast to the jerk form.

Seven examples of simple chaotic polynomial jerky dynamics were listed in Table 5.

More recently, in finding jerk functions that can generate chaos in the form of

$$\dddot{x} + \ddot{x} + x = f(\dot{x}) ,$$

Munmuangsaen et al. [50] applied another systematic numerical method from the space of control parameters embedded in the nonlinear function $f(\dot{x})$ with initial conditions suitable to find positive Lyapunov exponents. By this method, many more jerk functions were found that can generate chaos, following the systems proposed earlier by Sprott [6, 11, 12, 26]. The expression of such jerk functions are listed in Table 6, and the corresponding projections of the attractors of these systems are illustrated in Fig. 12.

Table 5 Classification of simple chaotic polynomial jerk systems [9]

Models	Equations
JD_1	$\dddot{x} = a\ddot{x} + \dot{x}^2 - x$
JD_2	$\dddot{x} = a\ddot{x} + bx + x\dot{x} - 1$
JD_3	$\dddot{x} = a\ddot{x} + b\dot{x} + cx^2 + x\dot{x} - 1$
JD_4	$\dddot{x} = a\ddot{x} + b\dot{x} + cx^2 + x\ddot{x} - 1$
JD_5	$\dddot{x} = a\dot{x} + bx^2 + \dot{x}^2 - x\dot{x}$
JD_6	$\dddot{x} = a\ddot{x} + b\dot{x} + cx^2 + d\dot{x}^2 + x\ddot{x} - 1$
JD_7	$\dddot{x} = a\ddot{x} + b\dot{x} + cx^2 + d\dot{x}^2 + ex\dot{x} + x\ddot{x} - 1$

Table 6 Some functions $f(\dot{x})$ that can produce chaos in $\dddot{x} + \ddot{x} + x = f(\dot{x})$ [9]

Cases	$f(\dot{x})$
GS_1	$\pm 0.1\ exp(\mp\dot{x})$
GS_2	$\pm exp(\mp\dot{x} - 2)$
GS_3	$\pm 5.1\ cos(\pm\dot{x} + 0.1)$
GS_4	$\pm 0.2\ tan(\mp\dot{x})$
GS_5	$\pm\,\mathrm{sgn}(1 \mp 4\dot{x})$
GS_6	$\pm\dot{x}^2 - 0.2\dot{x}^3$
GS_7	$\pm 1/(\dot{x} \pm 2)^2$
GS_8	$-5\dot{x} \pm \|1 \pm 5\dot{x}\|$
GS_9	$\pm 0.4/\| \pm\dot{x} + 1\|$
GS_10	$\pm 1/\| \pm\dot{x} + 1\|^{0.5}$
GS_11	$\pm 4\ sin(\pm\dot{x} + 1) - 2.2\dot{x}$
GS_12	$\pm cosh(\dot{x}) - 0.6\dot{x}$

4 Chaotic Hyperjerk Systems

By integrating each term of Eq. (30), one obtains

$$\ddot{x} + A\dot{x} + x = \int g(x)dt\,. \tag{53}$$

It can be generalized to a higher-dimensional form, as

$$\frac{d^{(n)}x}{dt^{(n)}} + a\frac{d^{(n-1)}x}{dt^{(n-1)}} + \frac{d^{(n-2)}x}{dt^{(n-2)}} = g(x)\,. \tag{54}$$

When $n > 4$, the system is called a "hyperjerk system", which is in the form of

$$\frac{d^{(n)}x}{dt^{(n)}} = f\left(\frac{d^{(n-1)}x}{dt^{(n-1)}}, \frac{d^{(n-2)}x}{dt^{(n-2)}}, \ldots, x\right),$$

since it involves time derivatives of a jerk function.

Next, some examples are presented.

4.1 Example 1

By modifying Eq. (30), Chlouverakis et al. [51] presented a generalized jerk function in the following form:

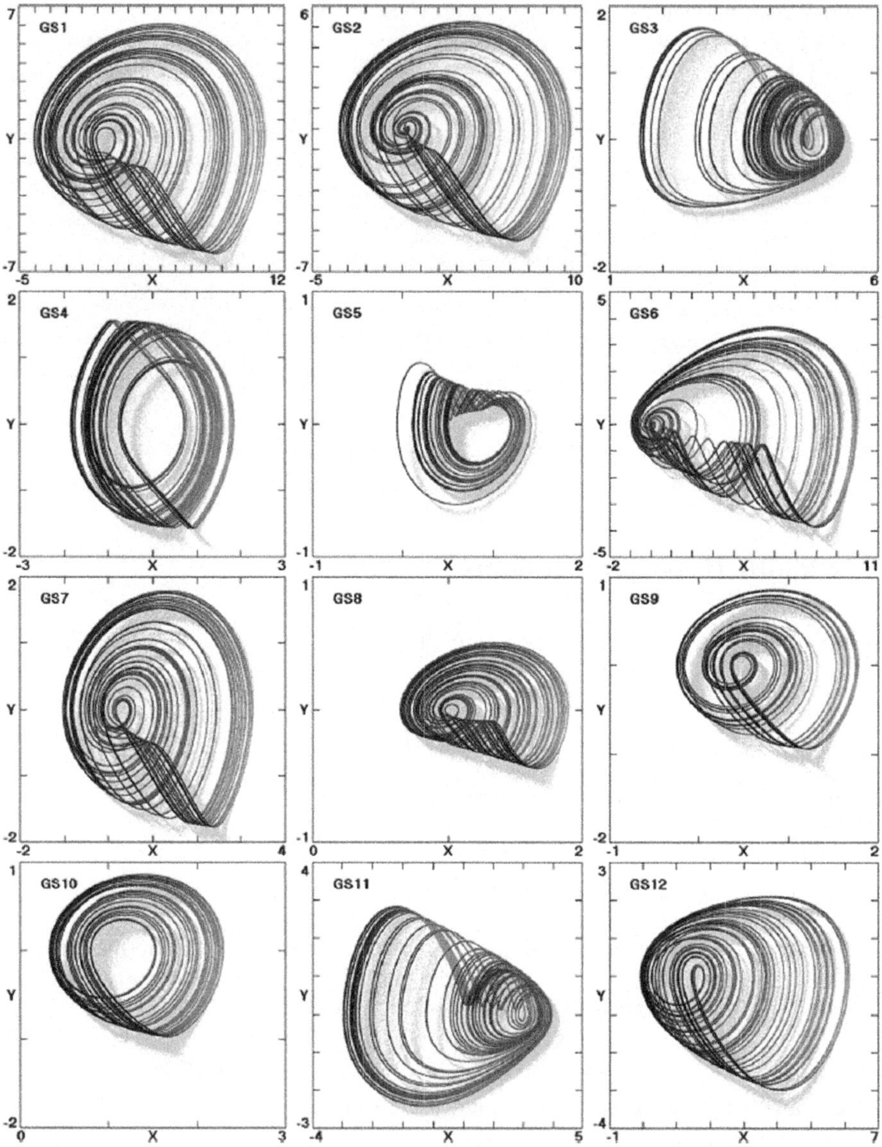

Fig. 12 Attractors for all nonlinear functions in Table 6 [9]

$$\frac{d^4x}{dt^4} + a_0\frac{d^3x}{dt^3} + a_1\frac{d^2x}{dt^2} + a_2\frac{dx}{dt} = g\left(x, \frac{dx}{dt}, \frac{d^2x}{dt^2}\right), \tag{55}$$

which leads to six hyperjerk systems that can exhibit chaos, as listed in Table 7.

The simplest form of hyperjerk functions reported in [51] is

Table 7 Some cases of Eq. (55) with $n = 4$ [51]

(a_0, a_1, a_2)	$g(x, \dot{x}, \ddot{x})$	ICs	$(\lambda_1, \lambda_2, \lambda_3, \lambda_4)$
$(1, 5.2, 2.7)$	$4.5(x^2 - 1)$	$(0.01, 0.01, 0.01, 4)$	$(0.185, 0, -0.483, -0.7)$
$(1, 10.4, 8.4)$	$9.3x^3 + 2.4x^2 + 13.6x - 1$	$(0, 0, 0, 5)$	$(0.4, 0, -0.23, -1.18)$
$(1, 2.6, 2.4)$	$1.9x - tam^{-1}(200x)$	$(0.02, -0.33, -0.27, 0.25)$	$(0.2, 0, -0.165, -1.03)$
$(1, 1.87, 2.34)$	$5.45x + x^2\ddot{x}$	$(-0.32, 0.15, -0.39, -0.36)$	$(0.37, 0, -0.63, -0.84)$
$(1, 3, 2.95)$	$-3.93sin(x)\ddot{x} - 1.47$	$(-0.37, -0.11, 0, 0.9)$	$(1.94, 0, -0.4, -2.54)$
$(0.25, 2.2, 1.2)$	$-1.9x - 3.2x^2\dddot{x}$	$(-0.32, 0.15, 0.39, -0.36)$	$(0.284, 0, -0.108, -0.425)$

$$\ddddot{x} + \dddot{x} + A\ddot{x} + \dot{x} + x = 0\,. \tag{56}$$

4.2 Example 2

Chlouverakis and Sprott [51] introduced another simple hyperchaotic hyperjerk system, described by

$$\ddddot{x} + \dddot{x}x^4 + A\ddot{x} + \dot{x} + x = 0\,, \tag{57}$$

which can also be expressed as

$$\begin{cases} \dot{x_1} = x_2 \\ \dot{x_2} = x_3 \\ \dot{x_3} = x_4 \\ \dot{x_4} = -x_1 - x_2 - Ax_3 - x_1^4 x_4\,, \end{cases} \tag{58}$$

where A is a real parameter.

By modifying Eq. (58), four 4D hyperjerk systems were found, which were analyzed in different papers.

(i) Vaidyanathan's model [52]:

$$\begin{cases} \dot{x_1} = x_2 \\ \dot{x_2} = x_3 \\ \dot{x_3} = x_4 \\ \dot{x_4} = -x_1 - x_2 - ax_3 - bx_1^3 - cx_1^4 x_4\,, \end{cases} \tag{59}$$

where $a,\ b$ and c are positive parameters. Select $(a, b, c) = (3.7, 0.2, 1.5)$ with initial condition $(0.1, 0.1, 0.1, 0.1)$, the Lyapunov exponents are obtained as $(0.1448, 0.0328, 0, -1.1294)$ and the Kaplan-Yorke dimension is $D_{KY} = 3.1573$. The 3D projection of Eq. (59) is shown in Fig. 13.

(ii) Vaidyanathan's model 2 [52]:

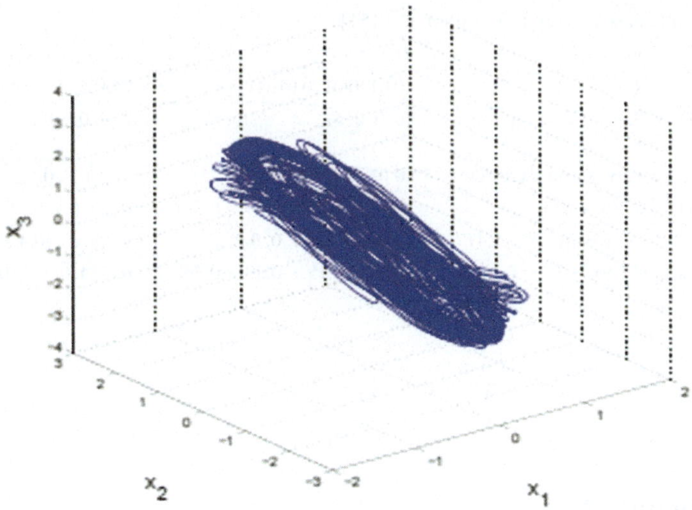

Fig. 13 3D projection of Eq. (59) on the (x_1, x_2, x_3) space [53]

$$\begin{cases} \dot{x_1} = x_2 \\ \dot{x_2} = x_3 \\ \dot{x_3} = x_4 \\ \dot{x_4} = -x_1 - x_2 - ax_3 - bx_2^3 - cx_1^4 x_4 , \end{cases} \tag{60}$$

where a, b and c are positive parameters. Select $(a, b, c) = (3.7, 0.05, 1.3)$ with initial condition $(0.5, 0.5, 0.5, 0.5)$, the Lyapunov exponents are obtained as $(0.13403, 0.03849, 0, -1.20579)$. The 3D projection of Eq. (60) is shown in Fig. 14.

(iii) Daltzis's model [54]:

$$\begin{cases} \dot{x_1} = x_2 \\ \dot{x_2} = x_3 \\ \dot{x_3} = x_4 \\ \dot{x_4} = -x_1 - x_2 - ax_3 - b|x_1| - cx_1^4 x_4 , \end{cases} \tag{61}$$

where a, b and c are positive parameters. Select $(a, b, c) = (3.7, 0.1, 1.5)$ with initial condition $(0.1, 0.1, 0.1, 0.1)$, the Lyapunov exponents are obtained as $(0.1555, 0.0330, 0, -1.6100)$ and the Kaplan-Yorke dimension is $D_{KY} = 3.1171$. The 2D projection of Eq. (61) is shown in Fig. 15.

(iv) Daltzis's model 2 [54]:

$$\begin{cases} \dot{x_1} = x_2 \\ \dot{x_2} = x_3 \\ \dot{x_3} = x_4 \\ \dot{x_4} = -x_1 - x_2 - ax_3 - b|x_2| - cx_1^4 x_4 , \end{cases} \tag{62}$$

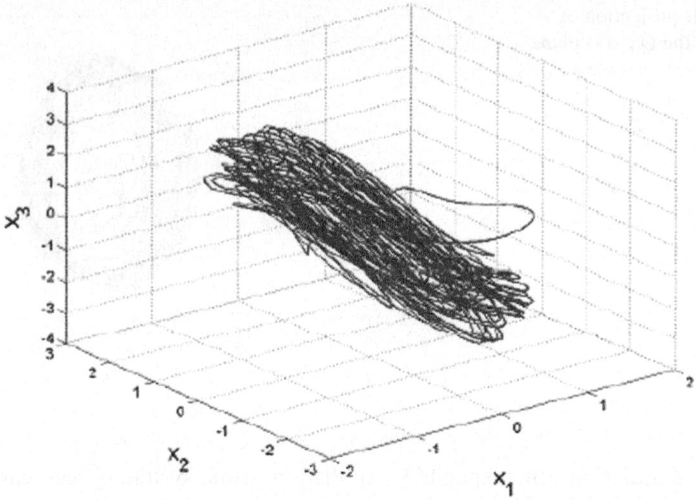

Fig. 14 3D projection of Eq. (60) on the (x_1, x_2, x_3) space [52]

Fig. 15 Projection of Eq. (59) on the (x_1, x_2) plane [54]

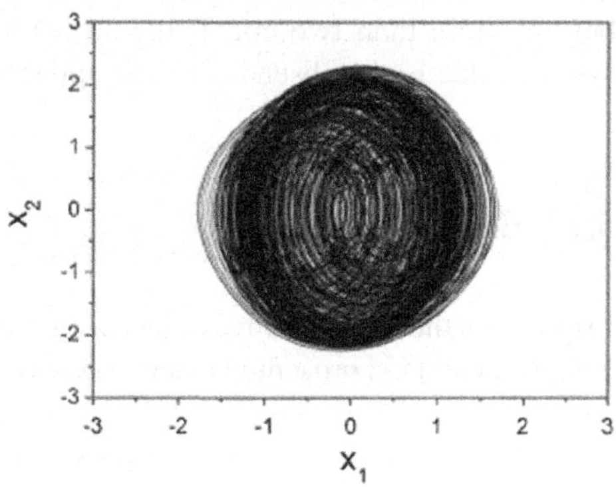

where a, b and c are positive parameters. Select $(a, b, c) = (3.8, 0.1, 1.5)$ with initial condition $(0.1, 0.1, 0.1, 0.1)$, the Lyapunov exponents are obtained as (0.1909, 0.06462, 0, -1.81846) and the Kaplan-Yorke dimension is $D_{KY} = 3.1405$. The 2D projection of Eq. (62) is shown in Fig. 16.

5 Coexisting Attractors in Jerk Systems

The dynamics of simple jerk functions that generate coexistence of chaotic attractors have not been well studied. The first studied examples were the JD_1 system: $\dddot{x} = a\ddot{x} + \dot{x}^2 - x$, and JD_2 system: $\dddot{x} = a\ddot{x} + bx + x\dot{x} - 1$, by Eichhorn et al. [56]. Both systems have chaotic behaviors in some specific parameter ranges. Based on

Fig. 16 2D projection of Eq. (62) on the (x_1, x_2) plane [55]

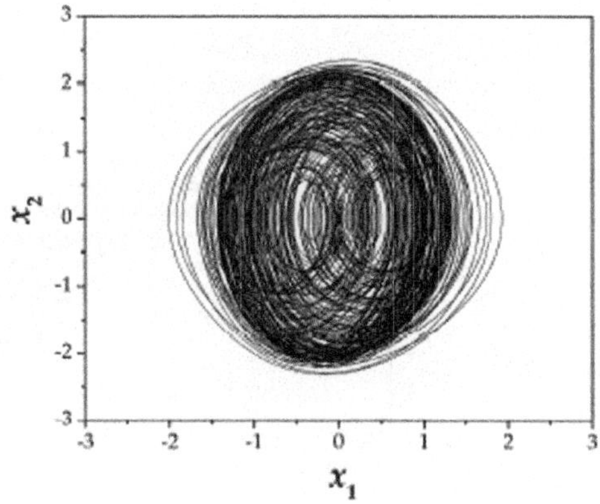

numerical simulations, the dependence of the long-time system dynamical behavior on the system parameters were determined by evaluating their Lyapunov spectra. For example, the dependence of initial conditions is addressed using forward and backward bifurcation diagrams. The coexistence of two stable attractors was found, but not more than two coexisting attractors were found from both JD_1 and JD_2 systems, due to the absence of their apparent symmetry.

5.1 Example 1

Inspired by the previous investigation on the coexistence of attractors in jerk systems, Kengne et al. [57] presented a new jerk system based on the Sprott MO5 system [30]:

$$\dddot{x} = -a\gamma\dot{x} - \ddot{x} + ax - ax^3\,, \tag{63}$$

which is invariant under the transformation

$$(x_1, x_2, x_3) = (-x_1, -x_2, -x_3)\,.$$

Based on numerical simulations, there are three equilibria of Eq. (63): $E_0 = (0, 0, 0)$ and $E_{1,2} = (\pm 1, 0, 0)$. Since these equilibria share the same stability, the Jacobian matrix

$$J = \begin{bmatrix} 0 & 1 & 0 \\ 0 & 0 & a \\ 1 - 3x_1^2 & -\gamma & -1 \end{bmatrix}. \tag{64}$$

gives the corresponding characteristic equation:

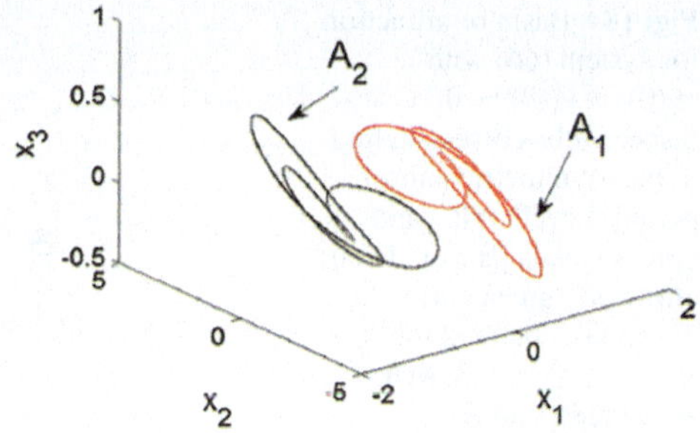

Fig. 17 Coexistence of four different attractors (a pair of period-3 limit cycles and a pair of strange attractors) for $\gamma = 0.7250$, $a = 18$, with initial conditions $(0, \pm 5, \pm 5)$ [57]

$$\lambda^3 + \lambda^2 + a\gamma\lambda - a(1 - 3x_1^2) = 0\,, \tag{65}$$

thus, the eigenvalues can be found.

Notice that $E_0(0, 0, 0)$ is always unstable, while the stability of $E_{1,2}$ depends on the two control parameters a and γ. Accordingly, $E_{1,2}$ are stable only for $\gamma > 2$.

For $\gamma = 0.7250$, $a = 18$, with initial conditions $(0, \pm 5, \pm 5)$, up to four different attractors can be obtained, as shown in Fig. 17.

5.2 *Example 2*

Njitacke et al. [58] also presented another jerk system in the form of a fourth-order smooth differential equation:

$$\begin{cases} \dot{x_1} = x_2 \\ \dot{x_2} = ax_3 \\ \dot{x_3} = x_1 - x_3 - \gamma x_2 - \beta exp(-\rho x_4) sinh(\rho_1 x_1) \\ \dot{x_4} = \varepsilon(\beta(exp(-\rho x_4) cosh(\rho_1 x_1) - 1) - \eta x_4)\,, \end{cases} \tag{66}$$

where the over dot represents differentiation with respect to the dimensionless time τ. Note that all the state variables of Eq. (66) are real and are determined in real experiments with a standard oscilloscope. The system is invariant.

By setting all the terms on the right-hand side be zero, the equilibria of Eq. (66) can be obtained: $E_0(0, 0, 0)$ is trivial, while the non-trivial symmetric equilibria $E_{1,2}$ are defined by the three control parameters β, η and ρ.

By linearizing system (66) at any equilibrium, the Jacobian matrix is obtained, as

$$J = \begin{bmatrix} 0 & 1 & 0 & 0 \\ 0 & 0 & a & 0 \\ 1 - \beta\rho_1 x & -\gamma & -1 & \beta\rho\phi \\ \varepsilon\beta\rho_1\phi & 0 & 0 & -\varepsilon(\beta\rho\chi + \eta) \end{bmatrix}. \tag{67}$$

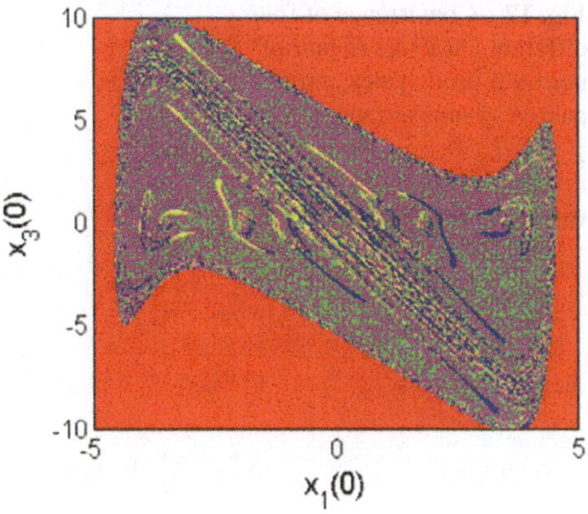

Fig. 18 Basin of attraction for system (66) with $x_2(0) = x_4(0) = 0$, respectively, corresponding to the asymmetric pair of period-4 cycle (blue and yellow) and a pair of chaotic attractors (green and magenta), obtained for $\gamma = 2.2,\ \rho = 9.3$. Red zones represent the unbounded solutions [58]

For $a = 5,\ \gamma = 2.2,\ \varepsilon = 0.002,\ \eta = 100,\ \beta = 5.36 \times 10^{-5},\ \rho = 10.121$, eigenvalues at $E_0(0, 0, 0, 0)$ are $\lambda_1 = -0.2,\ \lambda_{2,3} = -0.7152 \pm 3.3323i,\ \lambda_4 = 0.4304$, while eigenvalues at $E_{1,2}$ are $\lambda_{1,2} = 1.0734 \pm 4.0747i,\ \lambda_3 = -3.1469$, $\lambda_4 = -0.273$. Numerically, for values of a within the range $8.952 \le a \le 9.4$, the long-term dynamical behavior of Eq. (66) depends on the initial states. Therefore, there occurs an interesting phenomenon of coexisting multiple attractors. Figure 18 shows the basin of attraction of the four coexisting attractors.

5.3 *Example 3*

Tamba et al. [59] added a parameter a to the van der Pol–Duffing jerk oscillator and presented a new system that can exhibit coexisting chaotic attractors. This jerk system is described by

$$\begin{cases} \dot{x} = y \\ \dot{y} = az \\ \dot{z} = -z + \varepsilon(1 - x^2)y - x + \beta x^3 , \end{cases} \tag{68}$$

which is dissipative and invariant.

Based on numerical analysis, the three equilibria are found to be $E_0(0, 0, 0)$ and $E_{1,2}(\pm 1/\sqrt{\beta}, 0, 0)$. The corresponding characteristic equation is

$$\lambda^3 + \lambda^2 - a\varepsilon[1 - (x^*)^2]\lambda + a[1 + 2\varepsilon(x^*)(y^*) - 3\beta(x^*)^2] = 0 .$$

According to the Routh-Hurwith criteria, E_0 is stable if and only if $a > 0$ and $\varepsilon < -1$, and $E_{1,2}$ are stable if and only if $a > 0$ and $\varepsilon(1 - 1/\beta) < 2$.

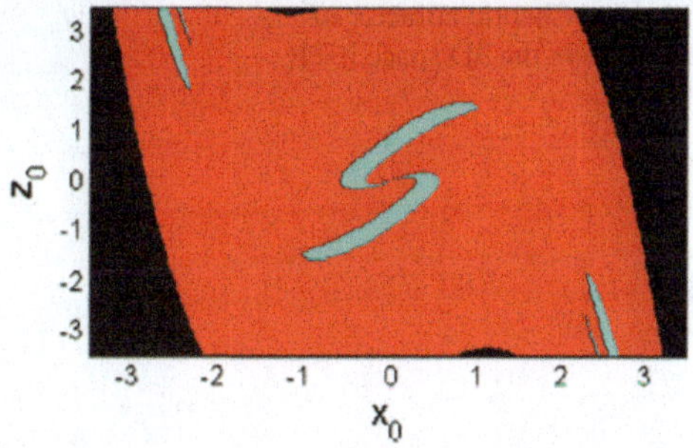

Fig. 19 Cross section of the basin of attraction of Eq. (68) on the x-z-plane, with $\beta = 0.125$, $\varepsilon = 1.0$ and $a = 8.1$. Initial conditions in the black regions generate unbounded orbits; those in the light blue region lead to periodic attractors; those in the red region lead to chaotic attractors [59]

When $\beta = 0.125$, Eq. (68) has a pair of period-1 attractors and chaotic attractors. As can be seen from Fig. 19, the possibility of having a chaotic attractor is much greater than the ones to have periodic attractors.

6 Chaotic Jerk Systems with Hidden Attractors

6.1 Example 1

Motivated by [60, 61], Wang et al. [62] proposed a new quadratic jerk system with hidden attractors, as follows:

$$\begin{cases} \dot{x} = y \\ \dot{y} = z \\ \dot{z} = -ax - by - cz + y^2 + bxy\,, \end{cases} \tag{69}$$

where $a,\ b,\ c$ are real parameters with $a \neq 0$.

Set $\dot{x} = \dot{y} = \dot{z} = 0$. Them=n, it can be found that the origin $E_0(0, 0, 0)$ is the only equilibrium of Eq. (69). The Jacobian matrix at the origin is

$$J = \begin{bmatrix} 0 & 1 & 0 \\ 0 & 0 & 1 \\ -a & -b & -c \end{bmatrix}, \tag{70}$$

and the characteristic equation is

$$\lambda^3 + c\lambda^2 + b\lambda + a = 0\,. \tag{71}$$

Based on the Routh-Hurwitz criterion, there are three roots with negative real parts for the case of $a > 0,\ c > 0,\ bc - a > 0$. Thus, system (73) has a stable node or two stable node-foci for $a > 0,\ c > 0,\ bc - a > 0$. Besides, the system has two

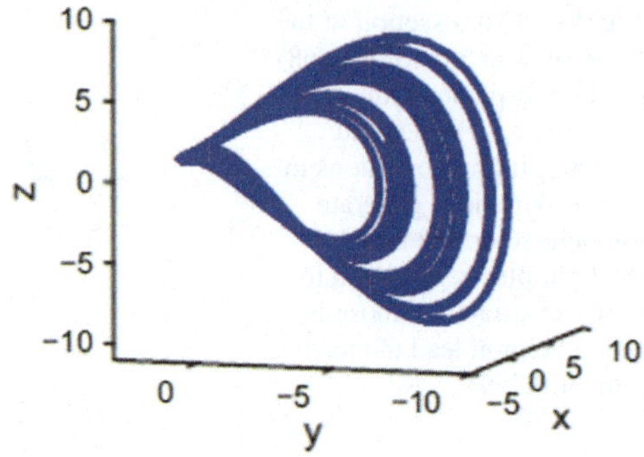

Fig. 20 Chaotic attractor of Eq. (69) in the 3D space [62]

saddle-foci for $a > 0$, $c > 0$, $bc - a < 0$. For $a = 3.4$, the Lyapunov exponents are (0.062, 0, −4.059) and the chaotic attractor is shown in Fig. 20.

6.2 Example 2

Modifying the Sprott R system, the first piecewise-linear jerk system with hidden attractor was introduced by Li et al. [63].

Interestingly, the circuit implementation of this system shows smooth orbits subject to the inherent electronic noise, even without using any equipment to set the initial conditions.

With the discontinuous signum function, this new jerk system is given by

$$\begin{cases} \dot{x} = y \\ \dot{y} = z \\ \dot{z} = -a \ \mathrm{sgn}(x) - y \ \mathrm{sgn}(x) - z + b \, . \end{cases} \tag{72}$$

Note that for $a = 2.65$, $b = 0.4$, Eq. (72) restores chaotic solutions while for $a = 1.325$, $b = 0.2$, the system generates chaotic oscillations. The corresponding Lyapunov exponents are $(0.016, 0, -1.016)$ and the Kaplan-Yorke dimension is 2.016.

Figure 21 shows the basin of attraction for the hidden attractor. Analytically, the coexistence of a constant and a signum function makes the equilibria disappear. Therefore, the corresponding chaotic attractor should be hidden, which cannot be found by choosing initial conditions according to any equilibria.

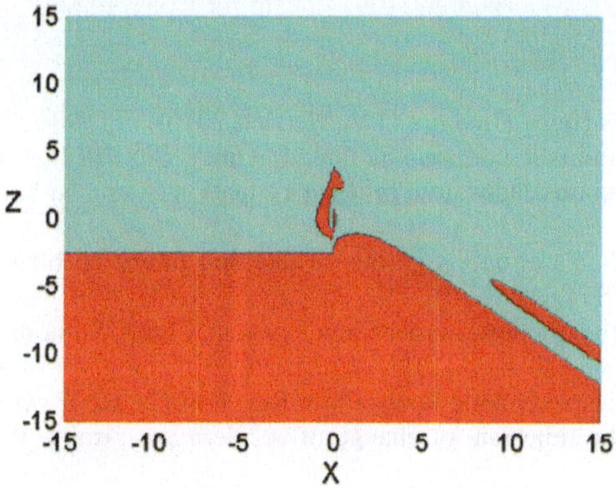

Fig. 21 Basin of attraction of Eq. (72) for the hidden attractor (light blue), with $a = 1.325,\ b = 0.2$. Red region lead to unbounded solutions [63]

Fig. 22 Chaotic attractor of Eq. (73) in the 3D space, with $a = c = 1,\ b = 1.08,\ e = -1,\ f = -0.4,\ g = h = 0$ [64]

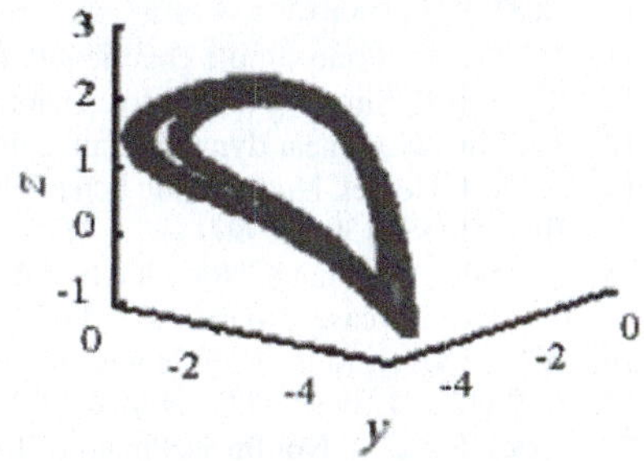

6.3 Example 3

Wang and Lin [64] found another chaotic jerk system with hidden attractors, which reads

$$\begin{cases} \dot{x} = y \\ \dot{y} = z \\ \dot{z} = -ax - by - cz + ez^2 - fxy - gxz - hyz\,. \end{cases} \tag{73}$$

Numerical analysis gives the only equilibrium of Eq. (73): $E_0(0, 0, 0)$. The characteristic equation is $\lambda^3 + c\lambda^2 + b\lambda + a = 0$.

There are three roots with negative real parts for $a > 0,\ c > 0,\ bc - a > 0$. Thus, according to the Routh-Hurwitz criterion, system (73) has a stable node or two stable node-foci for $a > 0,\ c > 0,\ bc - a > 0$, and the system has two saddle-foci for $a > 0,\ c > 0,\ bc - a < 0$.

For $a = c = 1,\ b = 1.08,\ e = -1,\ f = -0.4,\ g = h = 0$, the Lyapunov exponents are $(0.03, 0, -1.03)$ and the chaotic attractor is shown in Fig. 22.

References

1. A. Schubbach, *Jules Henri Poincaré* (J.B. Metzler, 2016)
2. E.N. Lorenz, Deterministic nonperiodic flow. J. Atmos. Sci. **20**(2), 130–141 (1963)
3. O.E. Rössler, Continuons chaos: four prototype equation. Annu. N. Y. Acad. Sci. **316**, 376–392 (1979)
4. L.P. Shil'Nikov, Chua's circuit: rigorous results and future problems. Int. J. Bifurc. Chaos **04**(03), 9400037 (2014)
5. I. Petráš, A note on the fractional-order Chua's system. Chaos, Solitons Fractals **38**(1), 140–147 (2008)
6. J.C. Sprott, Some simple chaotic flows. Phys. Rev. E **50**(2), R647 (1994)
7. S.H. Schot, Jerk: the time rate of change of acceleration. Am. J. Phys. **46**(46), 1090–1094 (1978)
8. L. Perko, *Differential Equations and Dynamical Systems* (Springer, Berlin, 1996)
9. R. Eichhorn, S.J. Linz, P. Hänggi, Transformations of nonlinear dynamical systems to jerky motion and its application to minimal chaotic flows. Phys. Rev. E **58**(6), 7151–7164 (1998)
10. H.P.W. Gottlieb, Question 38. What is the simplest jerk function that gives chaos? Am. J. Phys. **64**(5), 525 (1996)
11. J.C. Sprott, Some simple chaotic jerk functions. Am. J. Phys. **65**(6), 537–543 (1997)
12. J.C. Sprott, Simplest dissipative chaotic flow. Phys. Lett. A **228**(4–5), 271–274 (1997)
13. S.J. Linz, Nonlinear dynamical models and jerky motion. Am. J. Phys. **65**(6), 523–526 (1997)
14. Z. Fu, J. Heidel, Non-chaotic behaviour in three-dimensional quadratic systems. Nonlinearity **10**(3), 1289–1303 (1997)
15. J. Heidel, F. Zhang, Non-chaotic behaviour in three-dimensional quadratic systems ii. the conservative case. Nonlinearity **12**(3), 617–633 (1999)
16. W.G. Hoover, Remark on 'some simple chaotic flows'. Phys. Rev. E **51**(1), 759 (1995)
17. A. Wolf, J.B. Swift, H.L. Swinney, J.A. Vastano, Determining lyapunov exponents from a time series. Phys. D: Nonlinear Phenom. **16**(3), 285–317 (1985)
18. H.C.V. Baeyer, All shook up. Sciences **38**(1), 12–14 (1998)
19. D.W. Moore, E.A. Spiegel, A thermally excited non-linear oscillator. Astrophys. J. **143**(3), 871 (1966)
20. J.W. Gooch, *Prandtl Number* (Springer, New York, 2011)
21. S. Grossmann, D. Lohse, Prandtl and rayleigh number dependence of the reynolds number in turbulent thermal convection. Phys. Rev. E **66**(2), 016305 (2002)
22. E.K. Waldorm, F. Dynamics, D. Quantity, *Taylor Number* (Psycho Publishing, 2011)
23. T.Y. Li, J.A. Yorke, Period three implies chaos. Am. Math. Mon. **82**(10), 985–992 (2004)
24. J.M. Malasoma, What is the simplest dissipative chaotic jerk equation which is parity invariant? Phys. Lett. A **264**(5), 383–389 (2000)
25. J.M. Malasoma, A new class of minimal chaotic flows. Phys. Lett. A **305**(1), 52–58 (2002)
26. J.C. Sprott, Simple chaotic systems and circuits. Am. J. Phys. **68**(8), 758–763 (2000)
27. J.C. Sprott, Automatic generation of strange attractors. Comput. Graph. **17**(3), 325–332 (1993)
28. S.J. Linz, J.C. Sprott, Elementary chaotic flow. Phys. Lett. A **259**(3–4), 240–245 (1999)
29. A. Mishra, Routh-Hurwitz criterion. Control Design
30. J.C. Sprott, S.J. Linz, Algebraically simple chaotic flows. Int. J. Chaos Theory Appl. **5**(2), 1–20 (2000)
31. S.J. Linz, No-chaos criteria for certain jerky dynamics. Phys. Lett. A **275**(3), 204–210 (2000)
32. S. Fischer, A. Weiler, D. Froehlich, O.E. Roessler, Kleiner-attractor in a piecewise-linear c1-system. Zeitschrift Fur Naturforschung A **54**(3), 268–269 (1999)
33. P. Coullet, C. Tresser, A. Arnéodo, Transition to stochasticity for a class of forced oscillators. Phys. Lett. A **72**(4), 268–270 (1979)
34. N.F. Rul'kov, A.R. Volkoskii, M.G. Velarde, Mutual synchronization of chaotic self-oscillators with dissipative coupling. Int. J. Bifurc. Chaos **2**(03), 669–676 (1992)

35. E.D. Rössler, M.G. Velarde, A.R. Volkovskii, Experimental evidence for synchronous behavior of chaotic nonlinear oscillators with unidirectional or mutual driving. Int. J. Bifurc. Chaos **4**(04), 1003–1009 (1994)
36. J.C. Sprott, A new class of chaotic circuit. Phys. Lett. A **266**(1), 19–23 (2000)
37. M.E. Yalcin, J.A.K. Suykens, J. Vandewalle, Experimental confirmation of 3- and 5-scroll attractors from a generalized chua's circuit. IEEE Trans. Circuits Syst. I: Fundam. Theory Appl. **47**(3), 425–429 (2002)
38. S.M. Yu, Z.G. Ma, S.S. Qiu, S.G. Peng, Q.H. Lin, Generation and synchronization of n-scroll chaotic and hyperchaotic attractors in fourth-order systems. Chin. Phys. B **13**(3), 317–328 (2004)
39. M.E. Yalcin, J.A.K. Suykens, J.V.S. Özoguz, Families of scroll grid attractors. Int. J. Bifurc. Chaos **12**(01), 23–41 (2002)
40. A.S. Elwakil, S. Ozoguz, Multiscroll chaotic oscillators: the nonautonomous approach. IEEE Trans. Circuits Syst. II: Express Briefs **53**(9), 862–866 (2006)
41. G. Xie, P. Chen, M. Liu, Generation of multidirectional multiscroll attractors under the third-order jerk system, in *International Symposium on Information Science and Engieering* (2008), pp. 145–149
42. B. Bao, Z. Liu, X. Wang, J. Xu, Generalized projective synchronization of n-scroll chaotic jerk system, in *International Conference on Communications, Circuits and Systems* (2009), pp. 833–836
43. C. Liu, J. Yi, X. Xi, L. An, Y. Qian, Y. Fu, Research on the multi-scroll chaos generation based on jerk mode. Procedia Eng. **29**(4), 957–961 (2012)
44. R. Thomas, Deterministic chaos seen in terms of feedback circuits: Analysis, synthesis, "labyrinth chaos". Int. J. Bifurc. Chaos **9**(10), 1889–1905 (1999)
45. Y. Ding, Q.C. Zhang, Impulsive homoclinic chaos in van der Pol jerk system. Trans. Tianjin Univ. **16**(6), 457–460 (2010)
46. J. Guckenheimer, K. Hoffman, W. Weckesser, The forced van der pol equation I. SIAM J. Appl. Dyn. Syst. **2**(1), 1–35 (2003)
47. L.A. Zuppa, J.C.R. Garrido, S.B. Escobar, A chaotic oscillator using the van der pol dynamic immersed into a jerk system, in *Proceedings of the WSEAS International Conference*, Athens (2003)
48. G.M. Mahmoud, M.E. Ahmed, Chaotic and hyperchaotic complex jerk equations. Int. J. Mod. Nonlinear Theory Appl. **1**(1), 6–13 (2012)
49. T. Becker, V. Weispfenning, H. Kredel, *Grobner Bases: A Computational Approach to Commutative Algebra* (Springer, Berlin, 1993)
50. B. Munmuangsaen, B. Srisuchinwong, J.C. Sprott, Generalization of the simplest autonomous chaotic system. Phys. Lett. A **375**(12), 1445–1450 (2011)
51. K.E. Chlouverakis, J.C. Sprott, Chaotic hyperjerk systems. Chaos Solitons Fractals **28**(3), 739–746 (2006)
52. S. Vaidyanathan, A novel hyperchaotic hyperjerk system with two nonlinearities, its analysis, adaptive control and synchronization via backstepping control method. Adv. Chaos Theory Intell. Control **9**(1), 257–278 (2016)
53. V. Sundarapandian, V. Christos, P. Vietthanh, M. Kavitha, Analysis, adaptive control and synchronization of a novel 4-d hyperchaotic hyperjerk system and its spice implementation. Arch. Control Sci. **25**(1), 135–158 (2015)
54. P. Daltzis, S. Vaidyanathan, V. Pham, C. Volos, E. Nistazakis, G. Tombras, Hyperchaotic attractor in a novel hyperjerk system with two nonlinearities. Circuits Syst. Signal Proces. **37**(2), 1–23 (2017)
55. P.A. Daltzis, C.K. Volos, H.E. Nistazakis, A.D. Tsigopoulos, G.S. Tombras, Analysis, synchronization and circuit design of a 4D hyperchaotic hyperjerk system. Computation **6**(1), 14 (2018)
56. R. Eichhorn, S.J. Linz, P. Hanggi, Simple polynomial classes of chaotic jerky dynamics. Chaos, Solitons Fractals **13**(1), 1–15 (2002)

57. J. Kengne, Z.T. Njitacke, H.B. Fotsin, Dynamical analysis of a simple autonomous jerk system with multiple attractors. Nonlinear Dyn. **83**(1–2), 751–765 (2016)
58. Z.T. Njitacke, J. Kengne, H.B. Fotsin, A.N. Negou, D. Tchiotsop, Coexistence of multiple attractors and crisis route to chaos in a novel memristive diode bidge-based jerk circuit. Chaos, Solitons Fractals **91**, 180–197 (2016)
59. V.K. Tamba, S.T. Kingni, G.F. Kuiate, H.B. Fotsin, P.K. Talla, Coexistence of attractors in autonomous van der Pol-Duffing jerk oscillator: analysis, chaos control and synchronisation in its fractional-order form. Pramana **91**(1), 12 (2018)
60. Y. Song, J. Wei, Bifurcation analysis for Chen's system with delayed feedback and its application to control of chaos. Chaos, Solitons Fractals **22**(1), 75–91 (2004)
61. W. Yu, J. Cao, Stability and hopf bifurcation analysis on a four-neuron bam neural network with time delays. Phys. Lett. A **351**(1), 64–78 (2006)
62. Z. Wang, W. Sun, Z. Wei, S. Zhang, Dynamics and delayed feedback control for a 3d jerk system with hidden attractor. Nonlinear Dyn. **82**(1–2), 577–588 (2015)
63. P. Li, T. Zheng, C. Li, X. Wang, W. Hu, A unique jerk system with hidden chaotic oscillation. Nonlinear Dyn. **86**(1), 1–7 (2016)
64. X. Wang, V.T. Pham, C. Volos, Dynamics, circuit design, and synchronization of a new chaotic system with closed curve equilibrium. Complexity **2017**, 1–9 (2017)
65. A. Aliamiri, J. Stalnaker, E.L. Miller, Oscillators with chaotic behavior: an illustration of a theorem by Shil'nikov. J. Stat. Phys. **27**(1), 171–182 (1982)
66. A. Arneodo, P. Coullet, C. Tresser, Possible new strange attractors with spiral structure. Commun. Math. Phys. **79**(4), 573–579 (1981)
67. V. Patidar, K.K. Sud, Bifurcation and chaos in simple jerk dynamical systems. Pramana **64**(1), 75–93 (2005)
68. J. Lü, G. Chen, Generating multi-scroll chaotic attractors: theories, methods and applications. Int. J. Bifurc. Chaos **16**(4), 775–858 (2006)

Part IV

Multi-Stability in Symmetric Systems

Chunbiao Li and Julien Clinton Sprott

1 Introduction

The structure of a dynamical system is the key factor for investigating its multi-stability. Generally, a system can be divided into two categories; that is, a symmetric one and an asymmetric one. A dynamical system $\dot{X} = F(X) = (f_1(X), f_2(X), ..., f_N(X))$ $(X = (x_1, x_2, ..., x_N)^T)$ is symmetric if there exists a variable substitution: $u_1 = -x_1, u_2 = -x_2, ..., u_k = -x_k, u_i = x_i$, (here $k, i \in Z^+, i \in k+1, k+2, ..., N$) satisfies $\dot{U} = F(U)(U = (u_1, u_2, ..., u_N))$. Specifically, for a three-dimensional dynamical system, $\dot{X} = F(X)$ $(X = (x_1, x_2, x_3)^T)$: If $x_i = -u_i$ $(i \in 1, 2, 3)$ is subject to the same governing equation, the system is reflection symmetric; if $x_i = -u_i,\ x_j = -u_j,\ x_k = u_k$ $(i, j \in \{1, 2, 3\}, i \neq j, k \in \{1, 2, 3\} \backslash \{i, j\})$ is subject to the same governing equation, the system is rotationally symmetric; if $x_1 = -u_1,\ x_2 = -u_2,\ x_3 = -u_3$ is subject to the same governing equation, the system is inversion symmetric.

For symmetric systems, it is common to find multiple coexisting attractors [1–10], because a pair of attractors will appear when the symmetry is broken [9, 10]. Even more specifically, a symmetric pair of attractors may coexist with another regime of symmetric attractors. When the dimension of a dynamical system increases, the coexistence becomes more complicated, and other coexisting attractors may show up. Symmetric bifurcations describe the dynamical evolvement in the symmetric systems and also imply a new phenomenon of attractor merging. In Fig. 1, a variety of coexistence of symmetric attractors are illustrated.

C. Li (✉)
Nanjing University of Information Science & Technology, Nanjing, China
e-mail: goontry@126.com; chunbiaolee@nuist.edu.cn

J. C. Sprott
University of Wisconsin–Madison, Madison, Wisconsin, USA
e-mail: sprott@physics.wisc.edu

X. Wang et al. (eds.), *Chaotic Systems with Multistability and Hidden Attractors*,
Emergence, Complexity and Computation 40,
https://doi.org/10.1007/978-3-030-75821-9_12

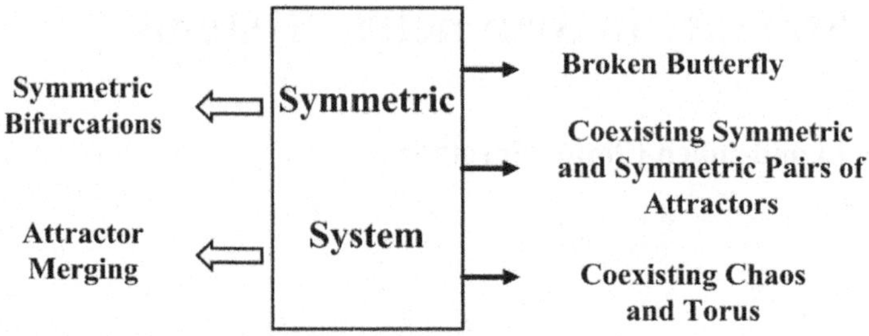

Fig. 1 Coexistence of attractors in a symmetric system

2 Broken Butterfly

One way to show the phenomenon of broken butterfly is to find the multi-stability in the classic Lorenz system [11], which is described by

$$\begin{cases} \dot{x} = \sigma(y - x), \\ \dot{y} = rx - xz - y, \\ \dot{z} = xy - bz. \end{cases} \tag{1}$$

The Lorenz system includes seven terms, five of which are linear, and two are quadratic nonlinearities. Because the variables x, y, z, and time t can be linearly rescaled, four of the seven coefficients can be set to ± 1, leaving exactly three parameters to completely determine the dynamics of the system. The selected parameters (σ, r, b) were originally chosen to characterize the atmospheric convection. When $\sigma = 10$, $r = 28$, $b = 8/3$, there is a single symmetric double-wing chaotic attractor that resembles a butterfly. Interestingly, this chaotic solution is globally attracting, and the attractor is robust to relatively large variations of the parameter values.

Almost all studies assumed that these three parameters are positive, due to the fact that the system has attractors only if $\sigma + b + 1 > 0$, and it is globally attracting only if $\sigma > 0$ and $b > 0$, so that all three dynamic variables are damped [12].

Normally, a positive r is required to provide the energy for maintaining oscillations from damping by corresponding positive feedback. However, negative values of the parameters can still lead to periodic and chaotic solutions. Although, in most cases, there are unbounded regions that refute globally attracting. For negative b, the anti-damping in the $\dot{Z}$ equation can provide the necessary energy. Since chaos can survive for negative b regardless of positive or negative r, we focus on the case with $r = 0$ because it has features common to the two regimes and it reduces the parameter space to two dimensions, allowing it to be completely explored.

When r is zero, the Lorenz system remains the rotational symmetry about the z-axis, correspondingly three real equilibrium points, i.e., $P_1 = (0, 0, 0)$, $P_2 = (\sqrt{-b}, \sqrt{-b}, -1)$, and $P_3 = (-\sqrt{-b}, -\sqrt{-b}, -1)$ are also symmetric with respect to the z-axis, if the parameter b is negative. The dynamical regions in σb space, when r is

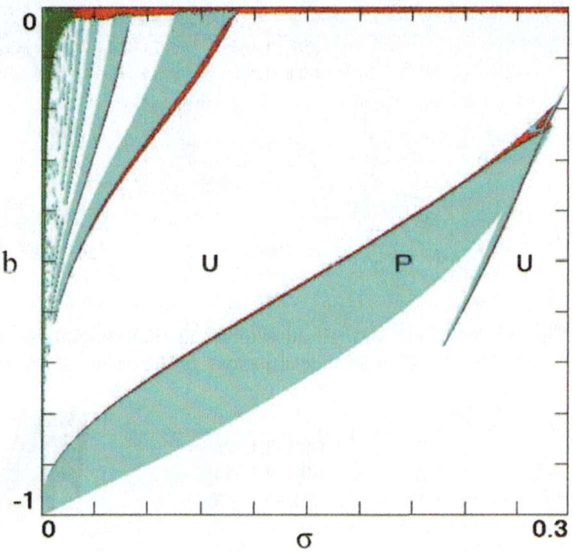

Fig. 2 Dynamical regions when $r = 0$. The chaotic regions are indicated in red, the periodic regions are shown in cyan, and the unbounded regions are shown in white

zero, show us an overall perspective of the properties, where each pixel is calculated for a random initial condition taken from a Gaussian distribution with mean zero and unit variance, and therefore the dotted regions are candidates for coexisting attractors. As shown in Fig. 2, the majority of solutions are unbounded, except some separate areas of periodic and chaotic solutions. The narrow area in the vicinity of the line $b = 2.17\sigma - 0.866$ has bounded solutions. While the green part at small σ and the red area at small b are numerical artifacts that come about because the orbit remains for long times near the equilibrium point P_1 at the origin. Although the parameter region giving a chaotic solution is very tiny, we found many regimes of multi-stability.

For example, when $r = 0, \sigma = 0.192, b = 0.45$, a symmetric pair of strange attractors coexist, as shown in Fig. 2a, with basins of attraction shown in Fig. 3b [13]. The basins of attraction for the two strange attractors are indicated by pink and light blue, respectively. The basins have the expected symmetry about the z-axis and an intricate fractal structure. Most initial conditions, as expected, lead to unbounded orbits as indicated in white. Moreover, let $r = 0$, $b = 0.30$, some other patterns of coexistence can be observed, including a typical symmetric pair of strange attractors, a pair of nearly touching strange attractors or limit cycles, a symmetric limit cycle with a symmetric pair of limit cycle, and even strikingly a symmetric strange attractor with a symmetric pair of limit cycles, as shown in Fig. 4a. The basins of attraction of the three coexisting attractors are of the fractal structure shown in Fig. 4b [13].

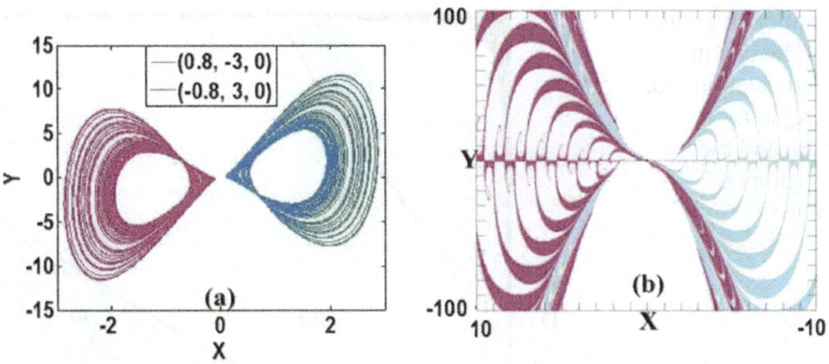

Fig. 3 Coexisting strange attractors and their fractal basins of attraction when $r = 0$, $\sigma = 0.192$, $b = -0.45$. **a** Strange attractors. **b** The basins of attraction at $z = 0$

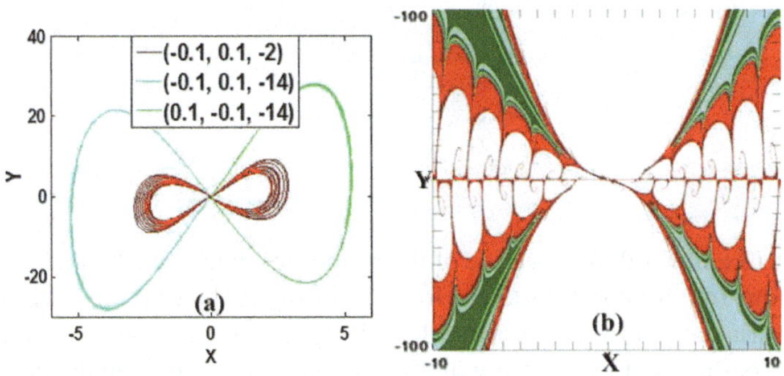

Fig. 4 Coexisting attractors of system (1) and their fractal basins of attraction when $r = 0$, $b = -0.3$, $\sigma = 0.279$. **a** Coexisting three attractors. **b** Basin of attraction at $z = 0$

3 Symmetric Bifurcations

The Lorenz system can obtain its linearization and be further reduced to Ref. [14]

$$\begin{cases} \dot{x} = y - x, \\ \dot{y} = -z\,\mathrm{sgn}(x), \\ \dot{z} = x\,\mathrm{sgn}(y) - b\,, \end{cases} \tag{2}$$

which is exactly a piecewise linearized version of the quadratic diffusionless Lorenz system [15, 16]

$$\begin{cases} \dot{x} = y - x, \\ \dot{y} = -zx, \\ \dot{z} = xy - R. \end{cases} \tag{3}$$

System (2) has two saddle-focus equilibria at $(b, b, 0)$ and $(-b, -b, 0)$, with eigenvalues $(-1.4656, 0.2328 \pm 0.7926i)$, and thus it satisfies the Šil'nikov condition for the existence of chaos [17, 18].

Note that the parameter b in system (2) is the only term not of first order, and hence it is an amplitude parameter and can be set to unity without loss of generality, unlike the case of system (3), where R is a bifurcation parameter.

Since system (2) has five terms, four of which can be scaled to ± 1 by a linear rescaling of the variables x, y, z, and time variable t, it should have one bifurcation parameter, just like system (3). In fact, a linear transformation of system (3) with $x \to \sqrt{R}x$, $y \to \sqrt{R}y$, $z \to Rz$, $t \to t$ moves the parameter R from the $\dot{z}$ equation to the $\dot{y}$ equation whose linearized form is then

$$\begin{cases} \dot{x} = y - x, \\ \dot{y} = -Rz\,\mathrm{sgn}(x), \\ \dot{z} = x\,\mathrm{sgn}(y) - 1\,. \end{cases} \tag{4}$$

The amplitude parameter b can be inserted back into system (4) if desired, but there is no loss of generality to take $b = 1$. Now, the bifurcation parameter R can be used to illustrate bifurcation and find multi-stability as shown in Fig. 5. Note that the bifurcation diagram shows the same evolvement of dynamical behavior under a unified Lyapunov exponent spectrum, indicating the symmetric bifurcation in system (4). In this figure, the plot labeled Xm shows the local maxima of x in red and the negative of the local minima of x in green. Since the equations are symmetric under the transformation $(x, y) \to (-x, -y)$, any attractor must either share that symmetry or have a symmetric pair of them. In the former case, the two colors are intermingled, and in the latter case, they are separated. Four coexisting attractors show up as predicted in the bifurcation diagram, as shown in Fig. 6. Since the ordinary differential equation solver ode23s is applicable for the stiff problem, here we take this solver replacing regular ode45.

The two equilibrium points have eigenvalues that satisfy $\lambda^3 + \lambda^2 + R = 0$ and hence are unstable for all values of R. For $R = 7$, there is a symmetric pair of limit cycles that undergo period-doubling as R decreases, forming a pair of strange attractors that merge into a single symmetric attractor at approximately $R = 5.81$ that persists with periodic windows down to approximately $R = 0.63$. The behavior at small R is similar, except that there is a presumably fractal succession of ever smaller pairs of limit cycles that period-double into a pair of strange attractors that merge into ever smaller symmetric strange attractors. Typical coexisting attractors are shown in Fig. 6. Similar behavior has been observed in system (3) [10]. There is no evidence for hysteresis or the coexistence of attractors other than the symmetric pair shown in red and green in Fig. 6.

4 Coexisting Symmetric and Symmetric Pairs of Attractors

Three coexisting attractors are observed in the above classic Lorenz system. Considering that more different nonlinearities may bring out more unexpected coexistences, here we introduce a new three-dimensional system with four quadratic terms [9]:

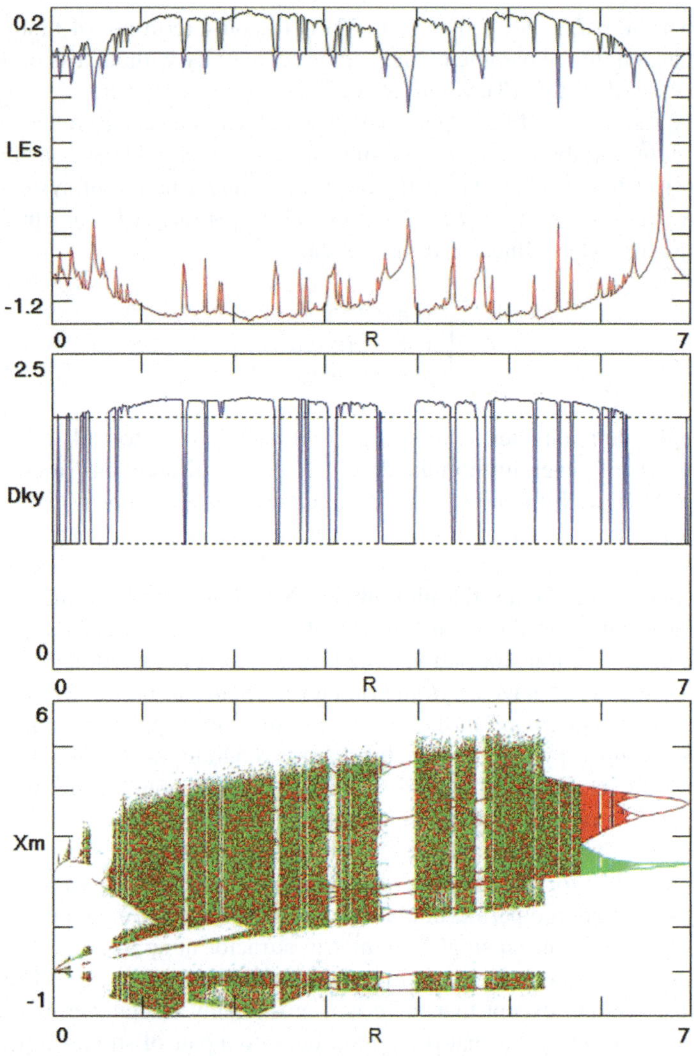

Fig. 5 Lyapunov exponents (LEs), Kaplan-Yorke dimension (Dky), and maximum values of x (Xm) as a function of the parameter R for system (4)

$$\begin{cases} \dot{x} = y + yz, \\ \dot{y} = -xz + yz, \\ \dot{z} = -az - xy + b\,. \end{cases} \tag{5}$$

From the coordinate transformation $(x, y, z) \to (-x, -y, z)$, we know system (5) is also symmetric with respect to a 180° rotation about the z-axis. When $a + b > 0$ and $a \neq 0$, system (5) has three real equilibrium points: $P_1 = (0, 0, b/a)$, $P_{2,3} = (\pm\sqrt{a+b}, \pm\sqrt{a+b}, -1)$. According to the Routh-Hurwitz criterion, the equilib-

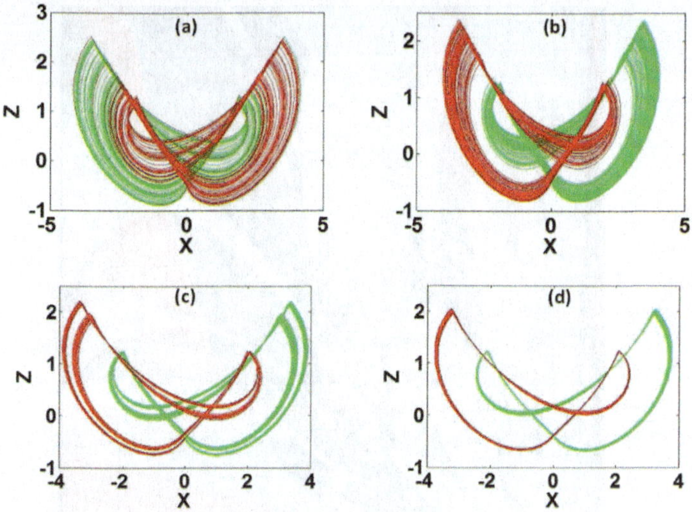

Fig. 6 Coexisting attractors in system (4) on the x-z plane, $(-1, -1, 1)$ is green, $(1, 1, 1)$ is red. **a** $R = 5.7$. **b** $R = 5.7$. **c** $R = 5.7$. **d** $R = 5.7$

rium point P_1 is stable provided $f(a, b) = (a^2 - b)^2 + ab + a^2b < 0$, which can be seen in the characteristic equation $\lambda^3 + ((a^2 - b)/a)\lambda^2 + ((b^2 + ab - a^2b)/a^2)\lambda + ((b^2 + ab)/a) = 0$. Virtually, the equilibrium P_1 is always unstable when the parameters a and b are positive. Equilibrium points P_2 and P_3 are symmetric with respect to the z-axis, and their stability depends on the location of parameters a and b.

The dynamical regions are shown in Fig. 7, and the typical chaotic oscillation is shown in Fig. 8, where $a = 0.6, b = 3$. The attractor exhibits four small butterfly wings embedded in an outer double-wing. The Lyapunov exponents are $(0.1528, 0, -0.7828)$, and the Kaplan-Yorke dimension is $D_{KY} = 2.1952$. The dotted regions of red and black suggest a strange attractor coexisting with a symmetric pair of stable equilibrium points, but further inspection finds that the chaos is transient and the flow eventually attracts to one of the stable equilibria. The dotted regions of red, blue, and black represent coexisting point attractors, limit cycles and strange attractors, another example of which has been reported with three such attractors [8], whereas the present case has five coexisting attractors of three types. As reported in Ref. [9], system (5) has eight regimes of multi-stability, including a symmetric pair of point attractors, two-point attractors with a symmetric limit cycle, a symmetric pair of limit cycles, two-point attractors with a symmetric pair of limit cycles, two-point attractors with a symmetric limit cycle, two-point attractors with a symmetric limit cycle, two-point attractors with a limit cycle and a symmetric pair of strange attractors, and two-point attractors with a limit cycle and a symmetric pair of limit cycles.

As we mentioned above, the basins of attraction show the initial conditions that lead to the respective coexisting attractors. As indicated in Fig. 7, for $a = 0.9$ and $b = 4$, the system (5) has a symmetric pair of point attractors coexisting with a

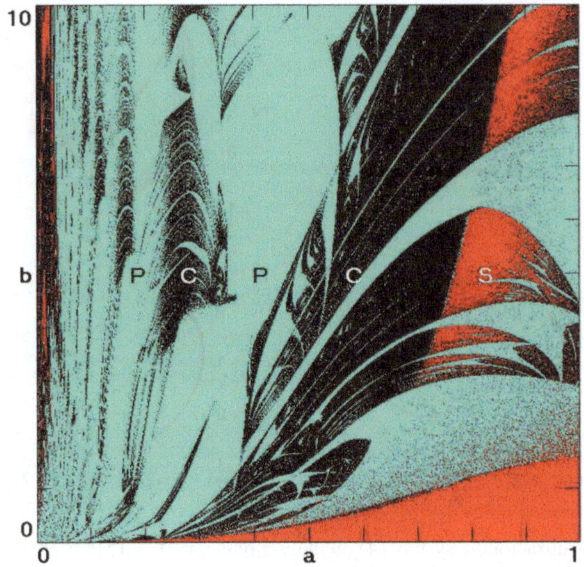

Fig. 7 Dynamical regions in the parameter space of a and b. The chaotic and transiently chaotic regions are shown in black, the periodic regions are shown in blue, and the stable equilibrium regions are shown in red

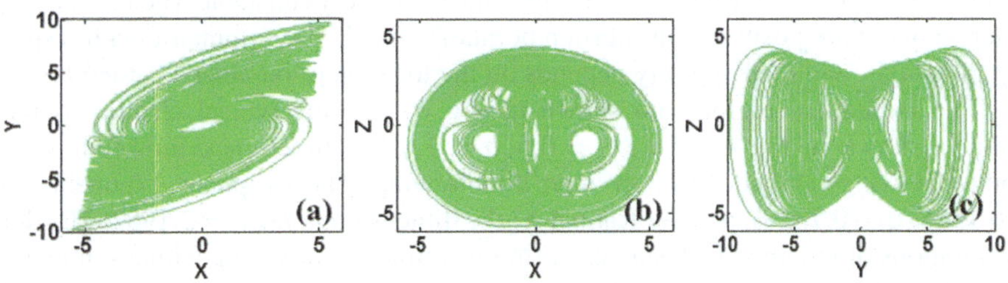

Fig. 8 Butterfly chaotic attractor from Eq. (5) with $a = 0.6, b = 3$ for $(x_0, y_0, z_0) = (1, -1, 1)$: (a) x-y plane, (b) x-z plane, (c) y-z plane

symmetric pair of limit cycles shown in Fig. 9a, the basins in the $z = -1$ plane are shown in Fig. 10a. The basins have the expected symmetry about the z-axis and a fractal boundary. When $a = 0.55$ and $b = 0.8$, the system (5) has five coexisting attractors, as shown in Fig. 9b, a symmetric limit cycle coexists with a symmetric pair of point attractors and a symmetric pair of strange attractors, and whose basins of attraction are shown in Fig. 10b. Here, two coexisting strange attractors are linked together, correspondingly the basins for these two attractors involve in a complex nested structure. The basins for point attractors are separated clearly, and the other initial conditions lead to a symmetric limit cycle like water around an island filling all the rest space of the basins. It is also very interesting that there is not any part of the basins leading to an unbounded solution.

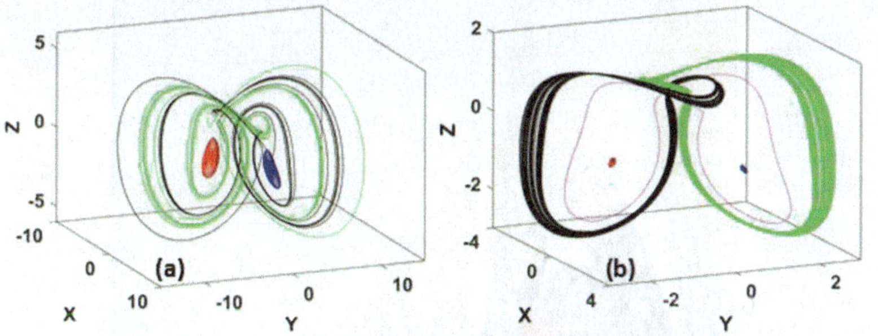

Fig. 9 Coexisting attractors for system (5) on the x-z plane: (a) four attractors when $a = 0.9$, $b = 4$, $(1, -1, 1)$ is black, $(-1, 1, 1)$ is green, $(6.5, -1, 1)$ is blue, $(-6.5, 1, 1)$ is red, (b) five attractors when $a = 0.55$, $b = 0.8$, $(1, 1, -1)$ is blue, $(-1, -1, -1)$ is red, $(0.4, 0, 1)$ is green, $(-0.4, 0, 1)$ is black, $(0.8, 0.3, 0.5)$ is pink

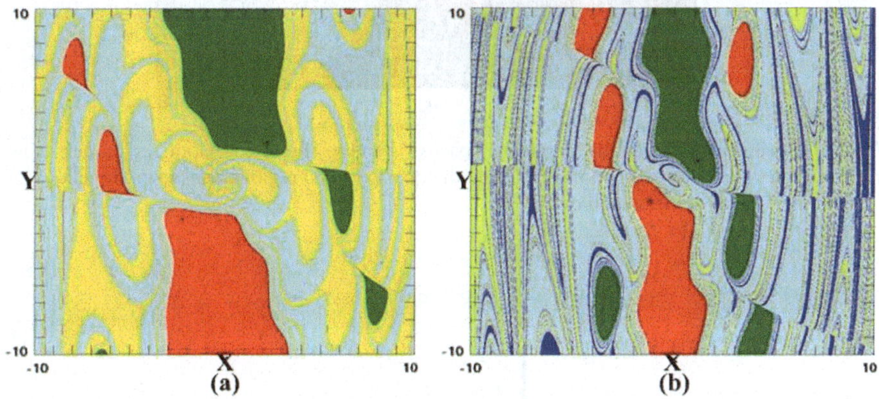

Fig. 10 The basins of attraction on $z = -1$ for system (5), red and green for the symmetric pair of point attractors. (a) $a = 0.9$, $b = 4$, light blue and yellow for the limit cycles, (b) $a = 0.55$, $b = 0.8$, light blue for the symmetric limit cycle and yellow and blue for a symmetric pair of strange attractors

5 Coexisting Chaos and Torus

We can also find multi-stability in four-dimensional systems, where the structure of symmetry can also turn to be broken and bursts out a symmetric pair of coexisting attractors. Basically, the dynamical system of higher-dimension generates more diversifications of multi-stability since it may degenerate to becoming lower dimensional under some certain circumstances. It is reasonable to expect more interesting multi-stability in higher-dimensional systems; here, we just look into some four-dimensional systems to give an observation of the existence of multi-stability.

We examine a simple 4D system derived from the diffusionless Lorenz system [15, 16]

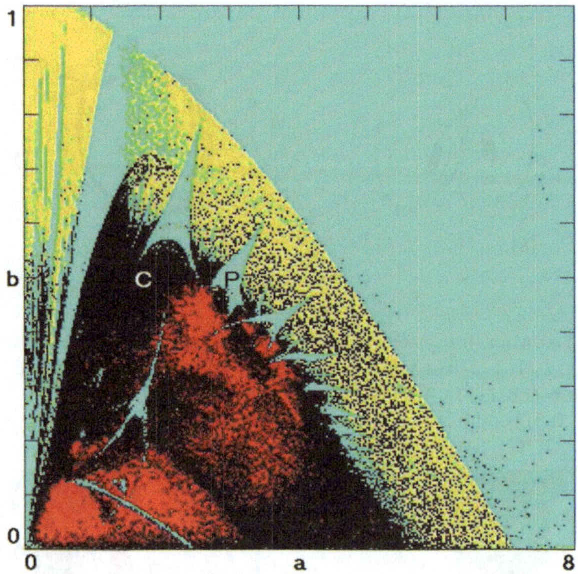

Fig. 11 Dynamical regions of the parameter space of a and b. The hyperchaotic regions are shown in red, the chaotic regions are shown in black, the quasiperiodic regions are shown in yellow, and the periodic regions are shown in cyan

$$\begin{cases} \dot{x} = y - x, \\ \dot{y} = -xz + u, \\ \dot{z} = xy - a, \\ \dot{u} = -by. \end{cases} \tag{6}$$

The system (6) is dissipative with solutions as time goes to infinity that contract onto an attractor of zero measure in the 4D state space, since the rate of hypervolume contraction is -1. System (6) has no equilibrium points, and also has rotational symmetry with respect to the z-axis as evidenced by its invariance under the coordinate transformation$(x, y, z, u) \to (-x, -y, z, -u)$. Linearly rescaling system (6) according to $x \to \sqrt{a}x,\ y \to \sqrt{a}y,\ z \to az$, and $u \to \sqrt{a}u$, two dimensions revise to be $\dot{y} = -axz + u,\ \dot{z} = xy - 1$, and we see that the system (6) separates into a 2D linear oscillator $(y - u)$ and a 2D parametrically forced linear response system $(x - z)$ in the limit $a \to 0$. This time, there is no attractor, since the amplitude of the oscillation is determined by the initial conditions. The parameter b is the square of the oscillation frequency, and a is a measure of the coupling of the response system back to the oscillator, as well as the nonlinearity that gives rise to a second often chaotic oscillation. These two oscillations give rise to the quasi-periodicity that occurs over much of the parameter space as well as Arnold tongues [19], where the two systems mode-lock. The abundant dynamical behavior identified by the Lyapunov exponents is shown in Fig. 11 [20].

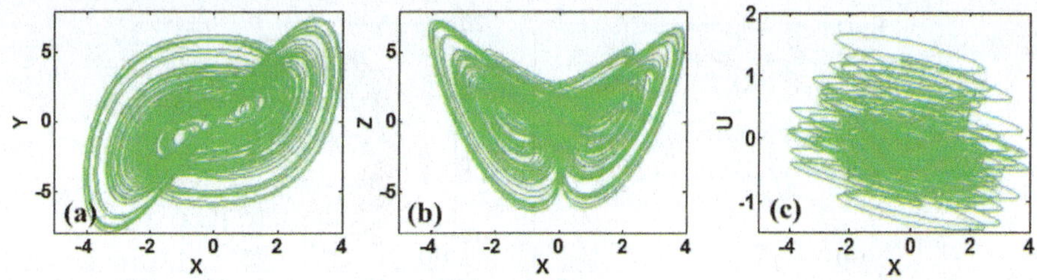

Fig. 12 Hyperchaotic attractor for system (6) with $a = 2$, $b = 0.1$ and initial condition $(1, -1, 1, -1)$. **a** x-y plane. **b** x-z plane. **c** x-u plane

Four dynamical behaviors are shown with red, black, cyan, and yellow, depending on the parameters a and b. When $a = 2$ and $b = 0.1$, system (6) is hyperchaotic with Lyapunov exponents $(0.2077, 0.0071, 0, -1.2148)$, and the Kaplan-Yorke dimension $D_{KY} = 3.1768$, whose phase trajectory is shown in Fig. 12. We can also see that the limit cycle is the dominant solution. Different limit cycles exist both at large a and in a window between the torus and the chaotic region dominated by Arnold tongues, as well as in windows amidst the chaos and in the region with coexisting attractors. Besides limit cycles, system (6) has quasi-periodicity, chaos, and hyperchaos; its behavior can be controlled by the parameters and initial conditions to make sure it stays in a specific preferred oscillation. Obviously, the dotted dynamical regions give us some information on multi-stability of different types. There are some regions for multi-stability with three attractors, where a limit cycle coexists with two strange attractors, and specifically a torus coexists with two limit cycles or with two strange attractors, as shown in Fig. 13.

However, despite the dots within the red/black region, it does not appear that there are regions in which hyperchaos coexist with chaos or other solutions. Rather, the boundary between chaos and hyperchaos is a fuzzy one where it is difficult to determine numerically with confidence, which of the two small Lyapunov exponents is actually zero. The basins of attraction of the different attracting sets are shown in Fig. 11. Note that the basins, in this case, are four-dimensional regions and have to be plotted in a two-dimensional cross-section. The basins in the $y = 0,\ u = 0$ plane for $a = 2, b = 0.8$, where a torus coexists with a symmetric pair of limit cycles, are shown in Fig. 11a. The basins of the two limit cycles are indicated by red and green, respectively, and the basin of the torus is shown in cyan. The basins have the expected symmetry about the z-axis and an intricate fractal boundary [20]. The basins in the $y = 0,\ u = 0$ plane for $a = 6,\ b = 0.1$, where a torus coexists with a symmetric pair of chaotic attractors are shown in Fig. 14b. The basins resemble those in Fig. 14a with symmetry and intricate fractal boundaries.

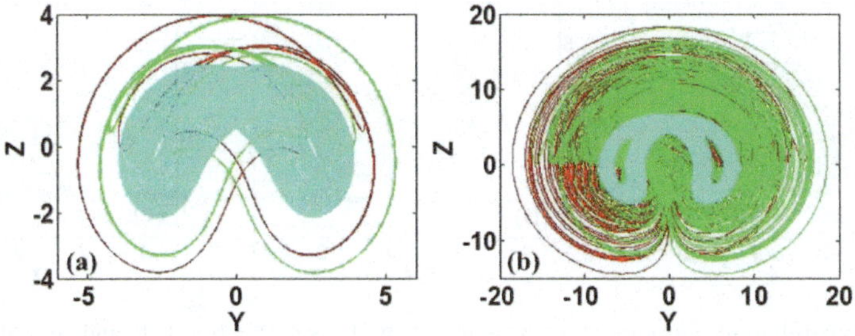

Fig. 13 Three coexisting attractors: **a** Quasi-periodic torus coexisting with two limit cycles at $a = 2,\ b = 0.8$, initial condition $(4, -1, 1, -1)$ with LEs $(0, 0, -0.0400, -0.9600)$, and initial condition $(\mp 5, \mp 1, 1, \mp 1)$ with LEs $(0, -0.0381, -0.1615, -0.8004)$. **b** Quasi-periodic torus coexisting with a symmetric pair of chaotic attractors at $a = 6,\ b = 0.1$, initial condition $(1, -1, 1, -1)$ with LEs $(0, 0, -0.1696, -0.8304)$ and initial condition $(0, \pm 4, 0, \mp 5)$ with LEs $(0.2520, 0, -0.0052, -1.2467)$

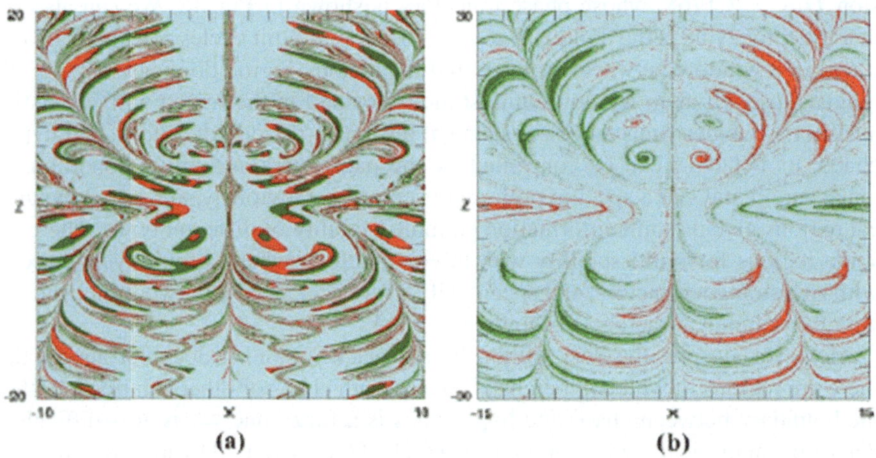

Fig. 14 The basins of attraction at $y = 0$ and $u = 0$, light blue for the quasi-periodic torus. **a** $a = 2,\ b = 0.8$, red and green for the symmetric pair of limit cycles. **b** $a = 6,\ b = 0.1$, red and green for the symmetric pair of strange attractors

6 Attractor Merging

Based on a new 3D flow of six-term with single linearity [21], a 4D hyperchaotic system with an infinite line of equilibrium points is designed [22]:

$$\begin{cases} \dot{x} = y - xz - yz + u, \\ \dot{y} = axz, \\ \dot{z} = y^2 - bz^2, \\ \dot{u} = -cy\,. \end{cases} \tag{7}$$

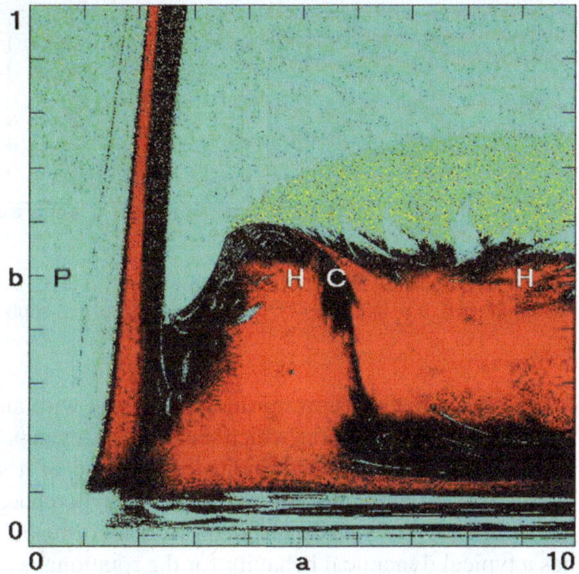

Fig. 15 Regions of various dynamical behaviors as a function of the bifurcation parameters a and b

The system has the only real line of equilibria $(x, 0, 0, 0)$ if the parameters a, b, c are not zero, and the corresponding eigenvalues are $(0, 0, 0, 0)$, which indicates that the system structure happens to stay on the bifurcation parameter point whose stability depends on the parameter and variable space of nonlinear terms. In the parameter space, where $a, b, c \neq 0$ and $z \neq 0$, the system 7 always has a full rank because the determinant of Jacobian matrix is $2abcz^2$, representing that the system (7) is a real 4D system. The rate of hypervolume contraction is $-(2b+1)z$, which shows that system (7) is dissipative with solutions as time goes to infinity that contract onto an attractor of zero measure in 4D state space when the time average of z is positive and $b > -0.5$. System (7) also has rotational symmetry with respect to the z-axis as evidenced by its invariance under the same coordinate transformation $(x, y, z, u) \to (-x, -y, z, -u)$. System (7) has eight terms, and thus three parameters. Selecting $c = 0.05$, various dynamical bifurcations can be observed in the figure of the dynamical region, as shown in Fig. 15, where the hyperchaotic regions are shown in red, the chaotic regions are shown in black, the quasiperiodic regions are shown in yellow, and the periodic regions are shown in cyan.

When $a = 5$, $b = 0.28$, $c = 0.05$, the system (7) is significantly hyperchaotic because it has four distinct Lyapunov exponents, two of which are positive. As a matter of fact, the degree of hyperchaos can be quantified by the value of the second Lyapunov exponent normalized to the most negative exponent. By this criterion, the maximum hyperchaos is obtained exactly with these parameters where the Lyapunov exponents are $(0.0750, 0.0366, 0, -1.6617)$, and correspondingly the Kaplan-Yorke dimension is $D_{KY} = 3.0672$. The hyperchaotic attractor in different projection is shown in Fig. 16.

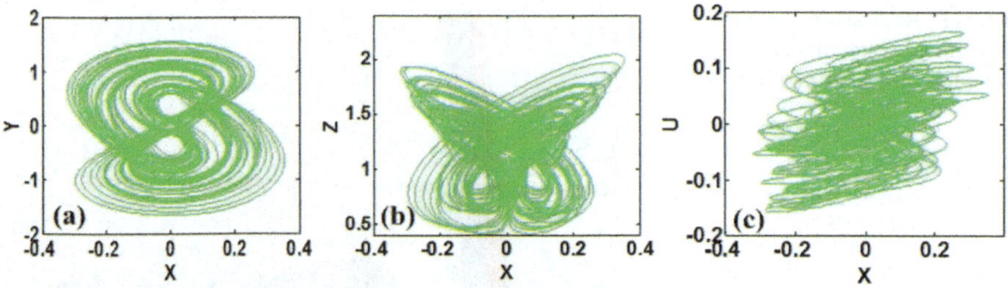

Fig. 16 Hyperchaotic attractor of system (4) with $a = 5,\ b = 0.28,\ c = 0.05$ under the initial condition (0, 0, 0.8, 0.02). **a** x–y plane. **b** x–z plane. **c** x–u plane

There is a distinct upward thick black line with an embedded red column in the figure of dynamical region, which separates the cyan periodic region into two different parts. The left area in cyan indicates the state of a symmetrical limit cycle, from which the symmetrical solution evolves to hyperchaos through chaos. Interestingly, the right part in cyan corresponds to a mixture of limit cycles. This system (7) has a typical dynamical behavior for the rotational symmetrical flow, where chaotic and hyperchaotic regions are surrounded by periodic region and hence the chaotic attractors usually can come from asymmetrical limit cycles. Then two asymmetrical coexisting strange attractors linked together, and merge somewhere, until eventually the system becomes hyperchaos. For example, set $b = 0.28$, $c = 0.05$, as a increases, a symmetric limit cycle splits at $a \approx 1.08$ into a symmetric pair of limit cycles that evolve into a symmetric pair of strange attractors that merge into a single symmetric strange attractor, which then unmerges before remerging and becoming hyperchaotic at about $a = 3.4$, as shown in Fig. 17.

7 Other Regimes of Coexisting Symmetric Attractors

All the above-mentioned chaotic systems are of rotational symmetry; moreover, many other regimes of symmetric structures still give coexisting attractors with their specific geometric characteristics. Inversion invariant systems are those which are invariant with respect to changes in all three-dimensional space. The system differential equation in this case only contains terms of odd powers of the variables. Since linear terms cannot exhibit chaos, those relatively simple examples of such systems are those that contain linear and cubic terms. One of the oldest such systems is the Moore-Spiegel [23] system but the system appears not to have coexisting attractors for any choice of the parameters. A variant system studied by Coullet et al. [24] admits coexisting attractors, limit cycles or chaotic solutions, as shown in Fig. 18. The equation is

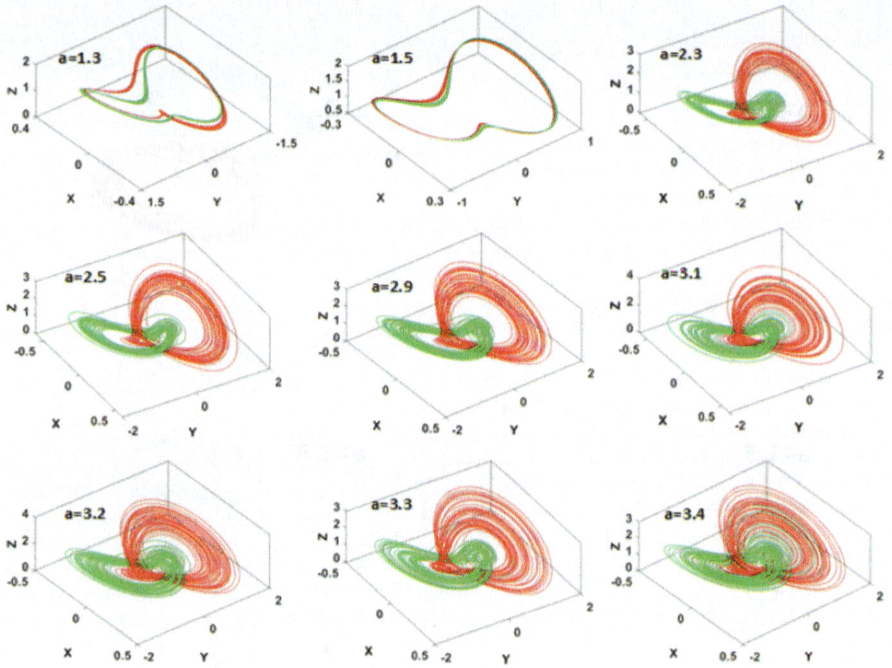

Fig. 17 Attractor merging when $b = 0.28$, $c = 0.05$

$$\begin{cases} \dot{x} = y, \\ \dot{y} = z, \\ \dot{z} = -z - ay + bx - x^3 . \end{cases} \tag{8}$$

Chaotic system of reflection symmetry admits coexisting attractors with symmetry to a vertical coordinate plane. The simplest such system proposed by Sprott [10] with the fewest number of terms is

$$\begin{cases} \dot{x} = x - xy, \\ \dot{y} = z, \\ \dot{z} = x^2 - y - az . \end{cases} \tag{9}$$

The above system exhibits coexisting chaos when $a = 0.3$. We can obtain the three equilibria, one at the origin $(0, 0, 0)$ and the others at $(\pm 1, 1, 0)$. The equilibrium point at the origin is unstable while the other two are stable if $a > 2$. As the parameter a decreases, a supercritical Hopf bifurcation occurs at $a = 2$ [10], at which point a symmetric pair of stable limit cycles are born. With the decrease of a, the symmetric pair of limit cycles undergo separate successive period-doublings, resulting in a symmetric pair of strange attractors at about $a = 0.7263$ that nearly touch one, another named as attractor kissing, on either side of the $x = 0$ plane as shown in Fig. 19.

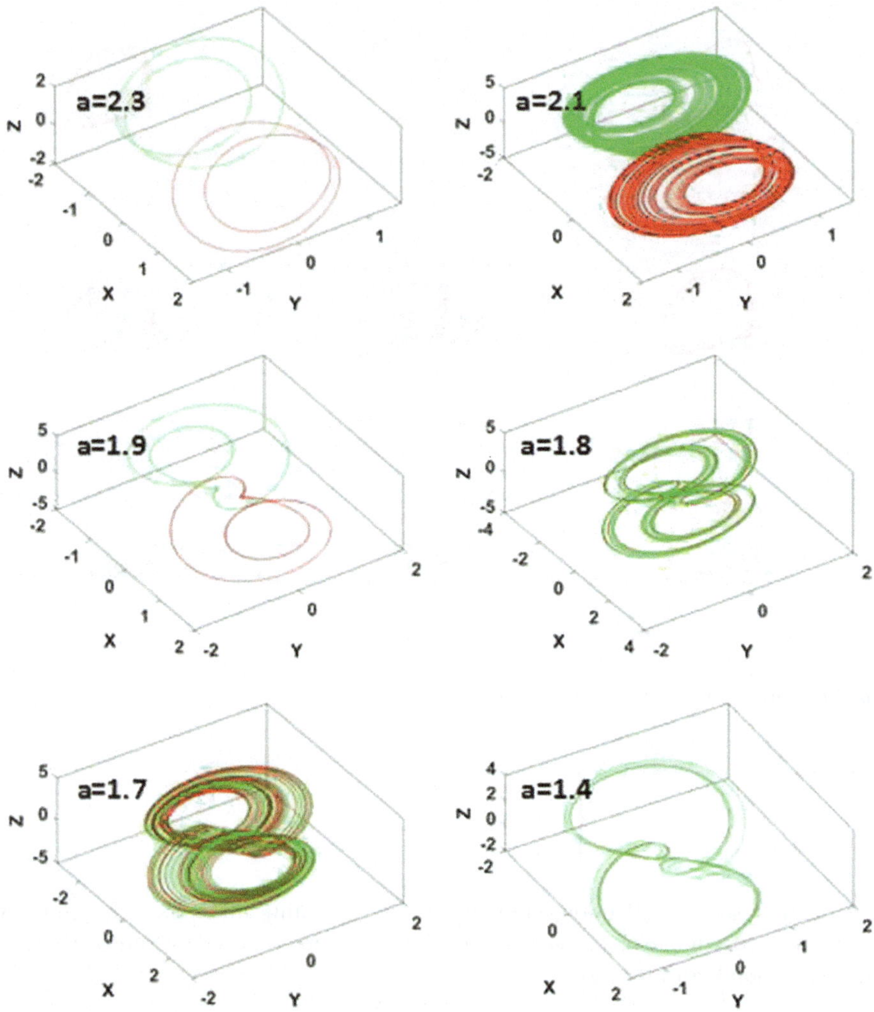

Fig. 18 Coexisting limit cycles and strange attractors before attractor merging in system (8) with $b = 2$, $(1.520.13 - 2.31)$ is red, $(-1.520.132.31)$ is green

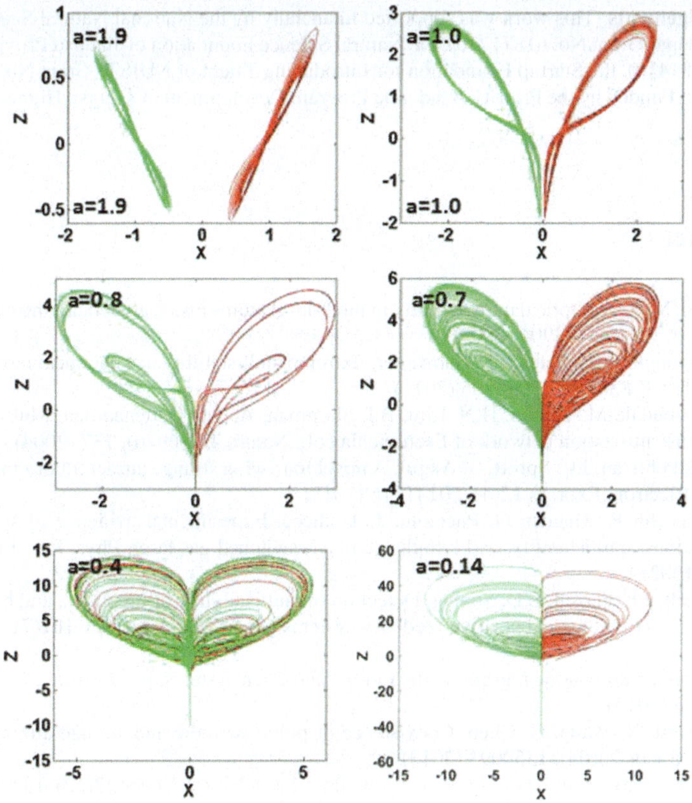

Fig. 19 Coexisting limit cycles and strange attractors before attractor merging in system (8) with $b = 2$

8 Conclusions

The symmetric system has been considered as the firstly-selected one to provide coexisting solutions. For the reason that the symmetry is prone to be broken and after that, the system exhibits a symmetric pair of attractors. More distinctively, under some circumstances, symmetric pairs of coexisting attractors may merge together to be a single symmetric one, or even further stand with other symmetric attractors. As results, symmetric strange attractor, limit cycle, and quasi-periodic torus coexisting with another symmetric pair of point attractors or limit cycles were demonstrated. Coexisting symmetric bifurcations associated with those coexisting attractors show us a clear description of the evolvement of dynamical behavior. Multi-stability is checked in the 3D chaotic Lorenz system or Lorenz-like system and the 4D hyperchaotic Lorenz-like system for demonstrating the coexisting dynamical evolvement.

Acknowledgements This work was supported financially by the National Natural Science Foundation of China (Grant No. 61871230), the Natural Science Foundation of Jiangsu Province (Grant No. BK20181410), the Startup Foundation for Introducing Talent of NUIST (Grant No. 2016205), and a Project Funded by the Priority Academic Program Development of Jiangsu Higher Education Institutions.

References

1. A. Joshi, M. Xiao, Optical multistability in three-level atoms inside an optical ring cavity. Phys. Rev. Lett. **91**(14), 143904 (2003)
2. C.N. Ngonghala, U. Feudel, K. Showalter, Extreme multistability in a chemical model system. Phys. Rev. E **83**(5 Pt 2), 056206 (2011)
3. E.M. Ozbudak, M. Thattai, H.N. Lim, B.I. Shraiman, A. Van Oudenaarden, Multistability in the lactose utilization network of Escherichia coli. Nature **427**(6976), 737 (2004)
4. C. Li, I. Pehlivan, J.C. Sprott, A. Akgul, A novel four-wing strange attractor born in bistability. IEICE Electron. Express **12**(4), 20141116 (2015)
5. F.T. Arecchi, R. Meucci, G. Puccioni, J. Tredicce, Experimental evidence of subharmonic bifurcations, multistability, and turbulence in a q-switched gas laser. Phys. Rev. Lett. **49**(17), 1217 (1982)
6. D. Angeli, J.E. Ferrell, E.D. Sontag, Detection of multistability, bifurcations, and hysteresis in a large class of biological positive-feedback systems. Proc. Natl. Acad. Sci. **101**(7), 1822–1827 (2004)
7. J.C. Sprott, New chaotic regimes in the Lorenz and Chen systems. Int. J. Bifurc. Chaos **25**(02), 1550033 (2015)
8. J.C. Sprott, X. Wang, G. Chen, Coexistence of point, periodic and strange attractors. Int. J. Bifurc. Chaos **23**(05), 1350093 (2013)
9. C. Li, J.C. Sprott, Multistability in a butterfly flow. Int. J. Bifurc. Chaos **23**(12), 1350199 (2013)
10. J.C. Sprott, Simplest chaotic flows with involutional symmetries. Int. J. Bifurc. Chaos **24**(01), 1450009 (2014)
11. E.N. Lorenz, Deterministic nonperiodic flow. J. Atmos. **20**(2), 130–141 (1963)
12. J.C. Sprott, Elegant Chaos: Algebraically Simple Chaotic Flows. Dedicated to the memory of Edward Norton Lorenz (World Scientific, Singapore, 2010)
13. C. Li, J.C. Sprott, Multistability in the Lorenz system: A broken butterfly. Int. J. Bifurc. Chaos **24**(10), 1450131 (2014)
14. C. Li, J.C. Sprott, W. Thio, Linearization of the Lorenz system. Phys. Lett. A **379**(10–11), 888–893 (2015)
15. B. Munmuangsaen, B. Srisuchinwong, A new five-term simple chaotic attractor. Phys. Lett. A **373**(44), 4038–4043 (2009)
16. V.D.S. Gerard, L.R.M. Maas, The diffusionless Lorenz equations; shil'nikov bifurcations and reduction to an explicit map. Physica D **141**(1–2), 19–36 (2000)
17. B. Deng, M. Han, S.-B. Hsu, Numerical proof for chemostat chaos of Shilnikov's type. Chaos **27**(3), 033106 (2017)
18. J.C. Sprott, Strange attractors with various equilibrium types. Eur. Phys. J.: Special Topics **224**(8), 1409–1419 (2015)
19. V. Paar, N. Pavin, Intermingled fractal Arnold tongues. Phys. Rev. E **57**(2), 1544–1549 (1998)
20. C. Li, J.C. Sprott, Coexisting hidden attractors in a 4-d simplified Lorenz system. Int. J. Bifurc. Chaos **24**(3), 1450034 (2014)
21. C. Li, J.C. Sprott, Chaotic flows with a single nonquadratic term. Phys. Lett. A **378**(3), 178–183 (2014)
22. C. Li, J.C. Sprott, W. Thio, Bistability in a hyperchaotic system with a line equilibrium. J. Exp. Theor. Phys. **118**(3), 494–500 (2014)

23. D.W. Moore, E.A. Spiegel, A thermally excited non-linear oscillator. Astrophys. J. **143**, 871 (1966)
24. P. Coullet, C. Tresser, A. Arnéodo, Transition to stochasticity for a class of forced oscillators. Phys. Lett. A **72**(4), 268–270 (1979)

Multi-Stability in Asymmetric Systems

Chunbiao Li and Julien Clinton Sprott

1 Introduction

Symmetric pairs of attractors are usually found in symmetric systems. However, it does not mean that all the coexisting behavior comes from symmetric structures. Compared with symmetric structure, asymmetric topology seems more common. Typically, for a dynamical system $\dot{X} = F(X) = (f_1(X), f_2(X), \cdots, f_N(X))$ $(X = (x_1, x_2, \cdots, x_N)^T)$, it is not easy to find a variable substitution: $u_1 = -x_1, u_2 = -x_2, ..., u_k = -x_k, u_i = x_i$ (here, $k, i \in Z^+, i \in \{k+1, k+2, ..., N\}$) satisfying $\dot{U} = F(U)$ $(U = (u_1, u_2, ..., u_N))$. More generally, any polarity reverse of the variables will destroy the polarity balance and the system solution appears in an asymmetric face.

It has been shown that the stability of an asymmetric system presents a greater variety of dynamics, and the hidden mechanism needs greater effort to disclose. The multi-stability in asymmetric systems has not been explored as frequently as in symmetric systems. Nevertheless, much effort has been made in finding that both systems seem to have coexisting attractors. Here, in this chapter, we show some examples of multi-stability in asymmetric systems. As shown in Fig. 1, an example with coexisting attractors is given by the Rössler system, where its mechanism was analyzed. Furthermore, dimension expanding and polarity balance breaking from symmetric structure to exhibit multi-stability is demonstrated for showing the coexisting attractors. Specific bifurcations are given simultaneously. Asymmetric system is transformed back to the symmetric structure, returning those symmetric pairs of attractors.

C. Li (✉)
Nanjing University of Information Science & Technology, Nanjing, China
e-mail: goontry@126.com; chunbiaolee@nuist.edu.cn

J. C. Sprott
University of Wisconsin–Madison, Madison, WI, USA
e-mail: sprott@physics.wisc.edu

X. Wang et al. (eds.), *Chaotic Systems with Multistability and Hidden Attractors*, Emergence, Complexity and Computation 40,
https://doi.org/10.1007/978-3-030-75821-9_13

2 Coexisting Attractors in Rössler System

The familiar Rössler system [1, 2] is given by

$$\begin{cases} \dot{x} = -y - z, \\ \dot{y} = x + ay, \\ \dot{z} = b + z(x - c). \end{cases} \tag{1}$$

When $a = b = 0.2,\ c = 5.7$, the system is chaotic, as shown in Fig. 1, with LEs (0.0714, 0, −5.3943) and Kaplan-Yorke dimension $D_{KY} = 2.0132$. There two equilibrium points, $E_{1,2} = (\frac{c\mp\sqrt{\Delta}}{2}, \frac{-c\pm\sqrt{\Delta}}{2a}, \frac{c\mp\sqrt{\Delta}}{2a})$, which exist when $a \neq 0$, $\Delta = c^2 - 4ab \geqslant 0$, one is close to the origin. Specifically, when $a = b = 0.2$, c=5.7, these two equilibrium points are spiral saddles at (0.0070, −0.0351, 0.0351) and (5.6930, −28.4649, 28.4649), with eigenvalues ($0.0970 \pm 0.9952i$, −5.6870), (0.1930, $-0.0000 \pm 5.4280i$), indicating that they are of index-2 and index-1, respectively.

The stability property of the system is changing, associated with the equilibria. When $a = 0.29,\ b = 0.14,\ c = 4.52$, those two equilibrium points are $E_1 =$ (0.0090, −0.0310, 0.0310) and $E_2 \approx$ (4.5110, −15.5552, 15.5552), which are a saddle focus with index-2 and a spiral repellor with index-3, indicated by their respective eigenvalues ($0.1417 \pm 0.9896i$, −4.5044) and (0.2720, $0.0045 \pm 4.0682i$). In this case, system (1) has two coexisting strange attractors [3], as shown in Fig. 2. It appears that the equilibrium point E_1 lies on the boundary between two strange attractors, presumably with its unstable manifold tangent to the boundary. Initial con-

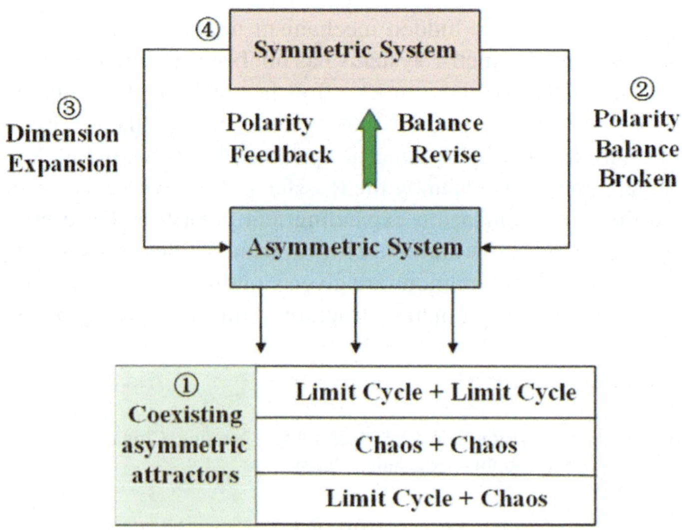

Fig. 1 Multistability in asymmetric structure

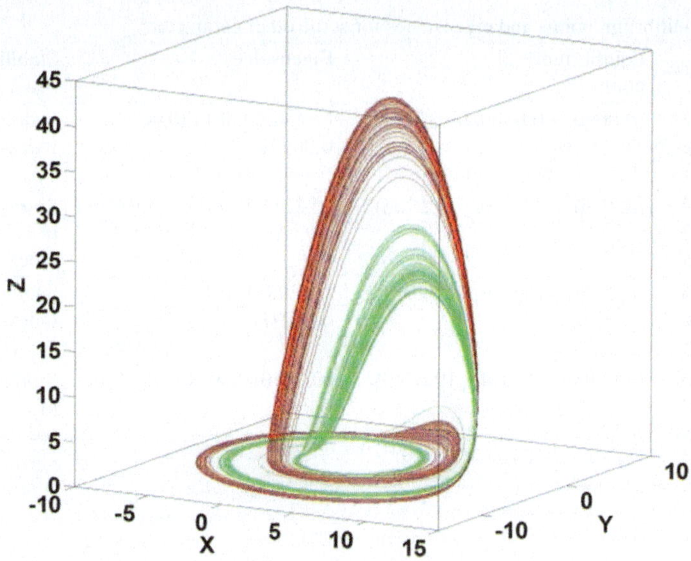

Fig. 2 Asymmetric strange attractors in the Rössler system (1) for $a = 0.29,\ b = 0.14,\ c = 4.52$. Red and green attractors correspond to initial conditions IC1 $= (-1.25, -0.72, -0.10)$ and IC2 $= (0.72, 1.28, 0.21)$, respectively. The two equilibrium points are shown as blue dots

ditions in the immediate vicinity of the equilibrium are about equally likely to end up on either attractor. Thus, the attractors are self-excited but could easily be missed if no other measurements are taken into consideration, since these two attractors look similar in almost the same shape. The other equilibrium point E_2 lies on the basin boundary of one of the strange attractors, with some initial conditions in its vicinity going to the attractor, with others going to infinity.

When $a = 0.2927,\ b = 0.1382,\ c = 4.4846$, the two equilibrium points changes, but with the similar stability listed in Table 1, and system (1) exhibits coexisting chaos and limit cycle, as shown in Fig. 3. Furthermore, when $a = 0.2536,\ b = 0.1643,\ c = 4.9979$, both coexisting attractors turn to be limit cycles, as shown in Fig. 4.

3 Introducing Additional Feedback for Breaking the Symmetry

It seems that symmetric system is more convenient to find coexisting symmetric pair of coexisting attractors. We guess that some minor feedback may break the symmetry but not disperse the coexisting attractors promptly. Further checking proves that additional feedback does deprive the symmetric structure of the system but leave coexisting asymmetric attractors.

Table 1 Equilibrium points and eigenvalues for some other parameters

Parameter	Equilibrium points	Eigenvalues	Stability
$a = 0.2927$ $b = 0.1382$ $c = 4.4846$	(0.0090, −0.0309, 0.0309)	$(-4.4690, 0.1430 + 0.9894i)$	Saddle focus of index-2
$a = 0.2927$ $b = 0.1382$ $c = 4.4846$	(4.4756, −15.2906, 15.2906)	$(0.2742, 0.0047 + 4.0356i)$	Spiral repellor of index-3
$a = 0.2536$ $b = 0.1382$ $c = 4.4846$	(0.0084, −0.0329, 0.0329)	$(-4.9831, 0.1236 + 0.9921i)$	Saddle focus of index-2
$a = 0.2536$ $b = 0.1382$ $c = 4.4846$	(4.9895, −19.6749, 19.6749)	$(0.2410, 0.0021 + 4.5466i)$	Spiral repellor of index-3

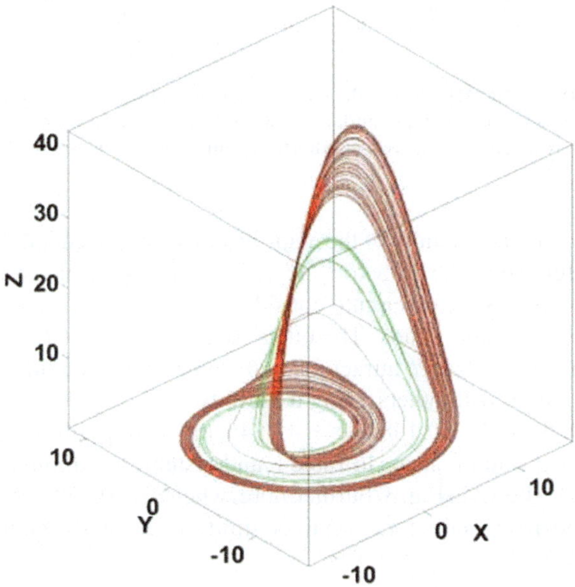

Fig. 3 Coexisting strange attractor and limit cycle in the Rössler system for $a = 0.2927$, $b = 0.1382$, $c = 4.4846$. Red and green attractors correspond to two different initial conditions $(-2, 1.28, 0.21)$ and $(1.5, 1.28, 0.21)$

We know the diffusionless Lorenz system [4, 5],

$$\begin{cases} \dot{x} = y - x, \\ \dot{y} = -xz, \\ \dot{z} = xy - a. \end{cases} \tag{2}$$

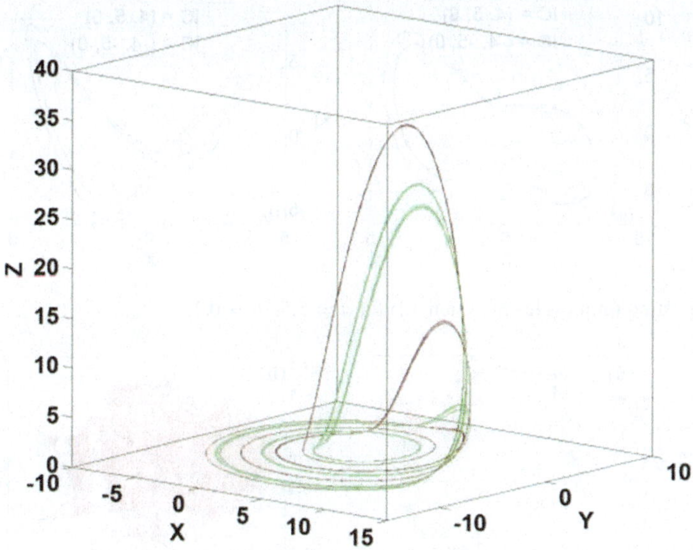

Fig. 4 Coexisting limit cycles in the Rössler system for $a = 0.2536,\ b = 0.1643,\ c = 4.9979$. Red and green attractors correspond to two different initial conditions (0, 7.2, 0.3) and (3, 2.31, 0.1)

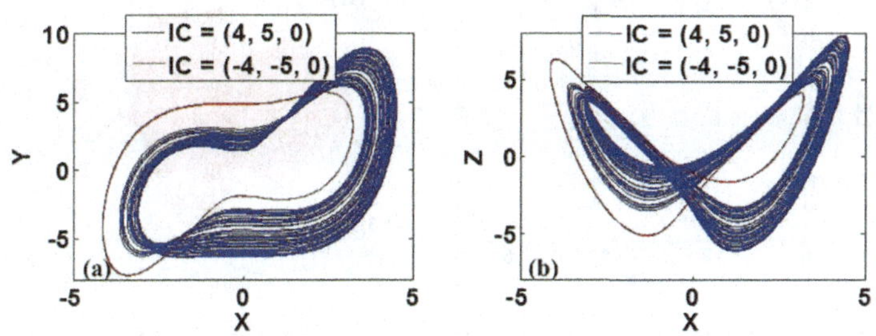

Fig. 5 Coexisting limit cycle and strange attractor in system (3) for $a = 5,\ b = 0.1$

When $a = 5$, system (2) has a symmetric pair of strange attractors [6]. An additional feedback can break the symmetry, and give coexisting asymmetric attractors, as in the following equation:

$$\begin{cases} \dot{x} = y - x, \\ \dot{y} = -xz + b|z|, \\ \dot{z} = xy - a. \end{cases} \tag{3}$$

When $a = 5, b = 0.1$, system (3) has a chaotic solution coexisting with a limit cycle, as shown in Fig. 5. While when $a = 5.5,\ b = 0.1$, two asymmetric limit cycles appear, as shown in Fig. 6. Further exploration shows that there are two parallel bifurcations in system (3). When $b = 0.1$, a varies in [0, 7] or $a = 5$, b varies in [0, 2.5], system (3) changes its evolvement undergoing different chaotic or periodic

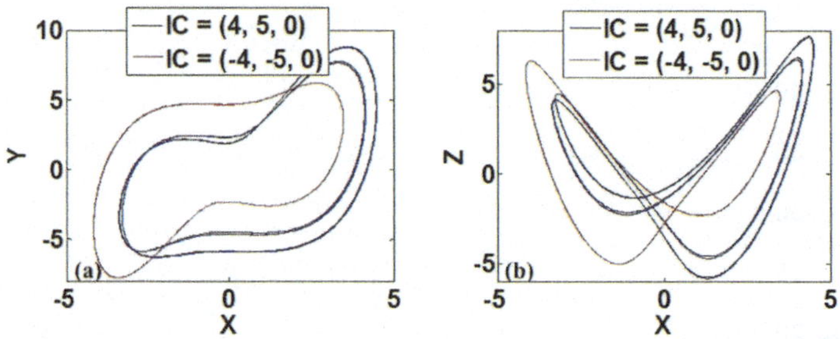

Fig. 6 Coexisting limit cycles in system (3) for $a = 5.5,\ b = 0.1$

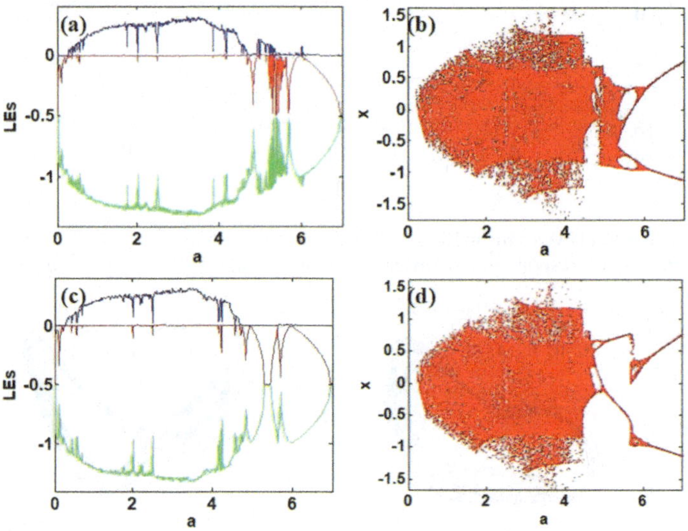

Fig. 7 Lyapunov exponent spectrum and bifurcation diagram (Cross section $z = -1$) in system (3) with $b = 0.1$ when a varies in [0, 7]: **a** and **b** are for initial condition IC1 $= (4, 5, 0)$. **c** and **d** are for IC2 $= (1, 1, 1)$

oscillations, as shown in Figs. 7 and 8. Readers can check those coexisting attractors if they are interested in it.

4 Dimension Expansion for Breaking the Symmetry

A new approach for introducing feedback can resort to additional dimension [7, 8]. The feedback is introduced from a new dimension, as

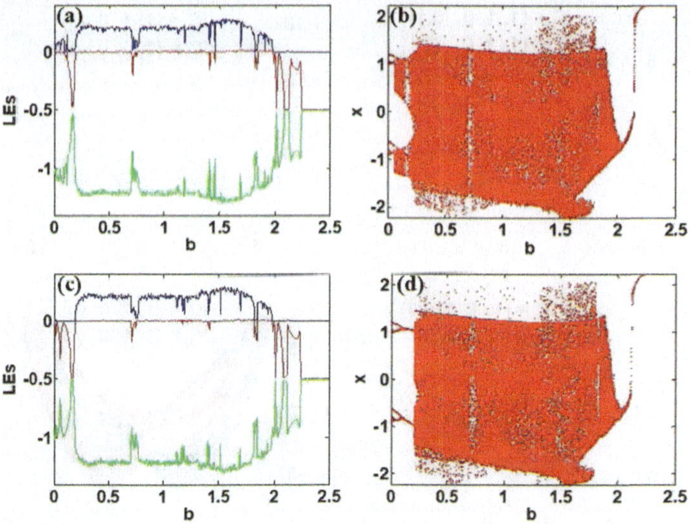

Fig. 8 Lyapunov exponent spectrum and bifurcation diagram (Cross section $z = -1$) in system (3) with $a = 5$ when b varies in [0, 2.5]: **a** and **b** are for initial condition IC1 = (4, 5, 0). **c** and **d** are for IC2 = (1, 1, 1)

$$\begin{cases} \dot{x} = y - x, \\ \dot{y} = -xz, \\ \dot{z} = xy - a - u. \\ \dot{u} = bz. \end{cases} \tag{4}$$

Almost infinitely many attractors show up as shown in Fig. 9, which is called as extreme multi-stability [9–11]. The reason being hidden is that the new dimension modifies the bifurcation parameter, and thus the initial condition changes the system solution. Further proof for this phenomenon is identified in the diagrams of Lyapunov exponent spectrum and bifurcation, as shown in Fig. 10. Different initial conditions give different Lyapunov exponents and bifurcation diagrams, showing multi-stability.

5 A Bridge Between Symmetry and Asymmetry

Since the symmetric structure of a dynamical system is associated with the multi-stability when the symmetry is broken, it is natural to design multi-stability from symmetry construction. Any asymmetric system can be transformed to a symmetric one when the polarity of some of the variables is modified to accept the polarity balance. Our exploration is based on the fact that many dynamical systems can survive chaos even after some new terms are introduced, which will revise the structure leading to different symmetry. Here, we try to revise the asymmetric Rössler system

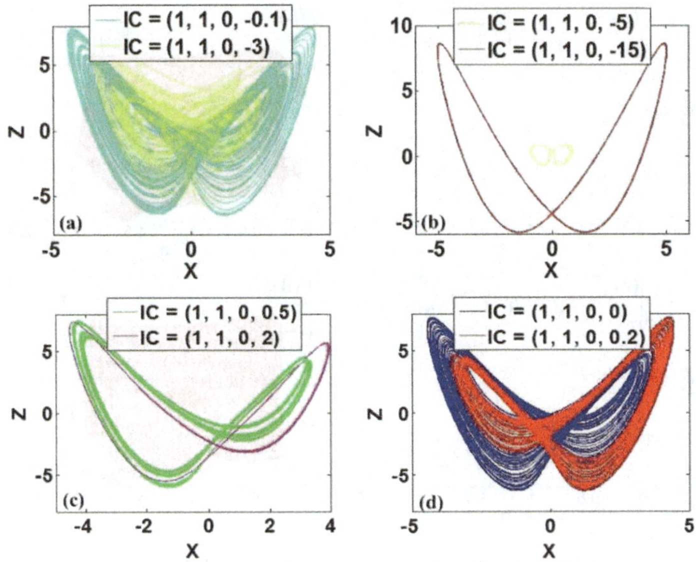

Fig. 9 Coexisting attractors in system (4) for $a = 5,\ b = 0.1$

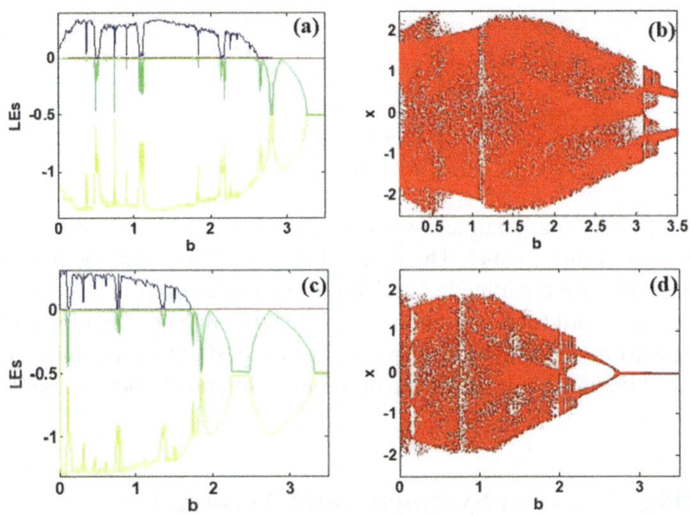

Fig. 10 Lyapunov exponent spectrum and bifurcation diagram (Cross section $z = 1$) in system (4) with $a = 5$ when b varies in [0, 3.5]: **a** and **b** are for initial condition IC1 = $(1, 1, 0, -0.1)$. **c** and **d** are for IC2 = $(1, 1, 0, -2)$

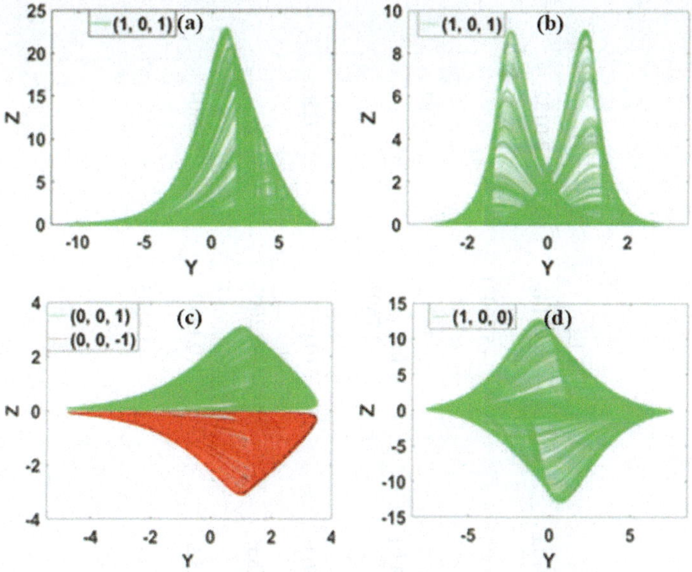

Fig. 11 Strange attractors for symmetric versions of the Rössler system in the y-z plane. **a** Original system. **b** Rotational symmetric system. **c** Reflectional symmetric system. **d** Central symmetric system

[12] to be symmetric for multi-stability, and we observe the evolvement of the basins of attraction leading to multiple states.

The classic Rössler system (1) has seven terms, including one constant, one quadratic term, and correspondingly three parameters. Some other versions are presented in Table 1. The classic Rössler system is of asymmetric structure, and it can be altered by revising either the linear or the nonlinear terms to be symmetric [12]. To construct a rotational invariant system corresponding to a 180° rotation about the z-axis, consider the system that is invariant under the transformation $(x, y, z) \rightarrow (-x, -y, z)$. Similarly, for constructing a reflection invariant system corresponding to symmetry about the $z = 0$ plane, the invariance under the transformation $(x, y, z) \rightarrow (x, y, -z)$ should be preserved. While for constructing an inversion invariant system corresponding to symmetry about the origin, the invariance under the transformation $(x, y, z) \rightarrow (-x, -y, -z)$ should be obtained. Thus, we multiply some of the terms in system (1) by appropriate choices of the variables and the symmetric candidates are obtained, as shown in Table 1, and the strange attractors are shown in Fig. 11. The symmetric version of dynamical systems gives birth to bistability as expected, as shown in Fig. 11c. Symmetric pairs of coexisting attractors exist as predicted at some specific parameter and are also found in those cases of rotational symmetry and central symmetry, as shown in Fig. 12, whose basins of attraction are given in Fig. 13.

Table 2 Rössler system and its symmetric versions

Cases	Equations	Parameters	Equilibria	Eigenvalues	LEs	D_{KY}
R0	$\dot{x} = -y - z$ $\dot{y} = x + ay$ $\dot{z} = b + z(x - c)$	$a = 0.2$ $b = 0.2$ $c = 5.7$	(0.0070,−0.0351, 0.0351) (5.6930,−28.4649, 28.4649)	(−5.6870,0.0970± 0.9952i) (0.1930,-0.000005 ±5.4280i)	(0.0714, 0, −5.3943)	2.0132
R1	$\dot{x} = -y - yz$ $\dot{y} = x + ay$ $\dot{z} = b + z(x^2 - c)$	$a = 0.2$ $b = 0.2$ $c = 6.5$	(0,0,0.0308) (±2.5884, ∓12.9422,−1)	(−6.5,0.1±1.0103i) (0.2,0.1±8.1847i)	(0.1639, 0, −2.7607)	2.0594
R2	$\dot{x} = -y - z^2$ $\dot{y} = x + ay$ $\dot{z} = b\,sgn(z) + z(x - c)$	$a = 0.2$ $b = 0.2$ $c = 2.5$	(2.4428,−12.2139, ±3.4948) (0.0013,-0.0064, ±0.0800)	(0.1899,-0.0236± 5.0423i) (-2.4942,0.0978± 0.9956i)	(0.0538, 0, −2.2601)	2.0258
R3	$\dot{x} = -y - z$ $\dot{y} = x + ay$ $\dot{z} = bx + x\|z\| - cz$	$a = 0.5$ $b = 0.2$ $c = 5.7$	(0,0,0) (±5.6,∓11.2, ±11.2)	(-5.6657,0.2328± 0.9665i) (0.4526,-0.0263± 3.5175i)	(0.1597, 0, −1.9865)	2.0804

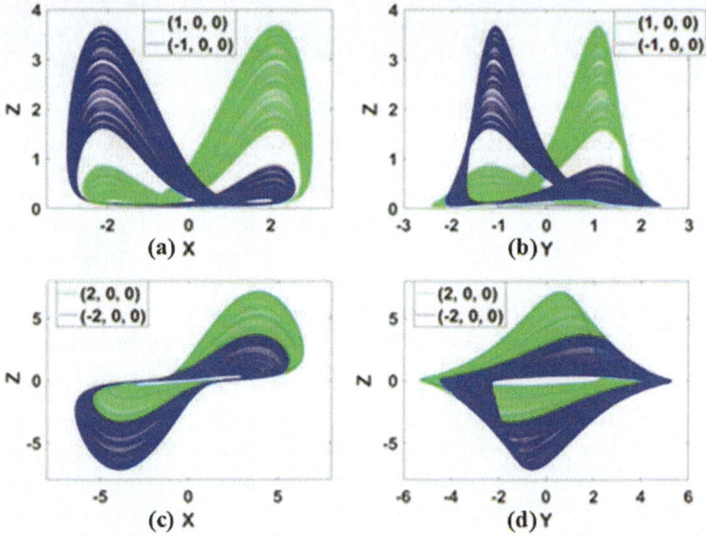

Fig. 12 Symmetric pair of coexisting attractors in system R1 and R3. **a** and **b** R1 with $a = b = 0.2$, $c = 4.6$. **c** and **d** R3 with $a = 0.5,\ b = 0.2,\ c = 4$

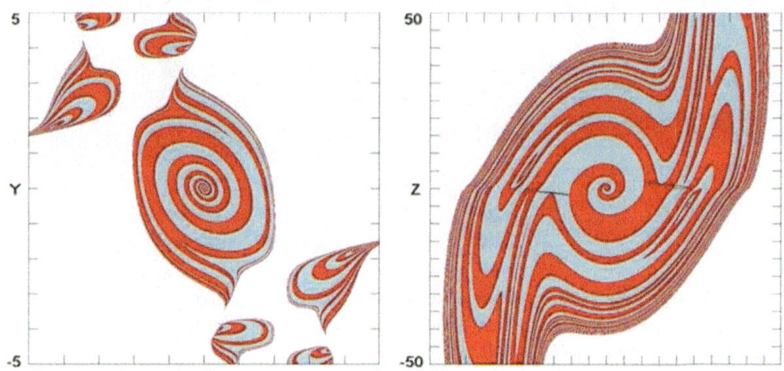

Fig. 13 Basins of attraction for the symmetric pair of strange attractors: **a** at $z = -1$ for R1 with $a = b = 0.2,\ c = 4.6$. **b** at $x = 0$ for R3 at $a = 0.5,\ b = 0.2,\ c = 4$

Following the routine mentioned above, the following four-dimensional Rössler equation [13]

$$\begin{cases} \dot{x} = -y - z, \\ \dot{y} = x + ay + u, \\ \dot{z} = b + xz, \\ \dot{u} = cu - dz, \end{cases} \tag{5}$$

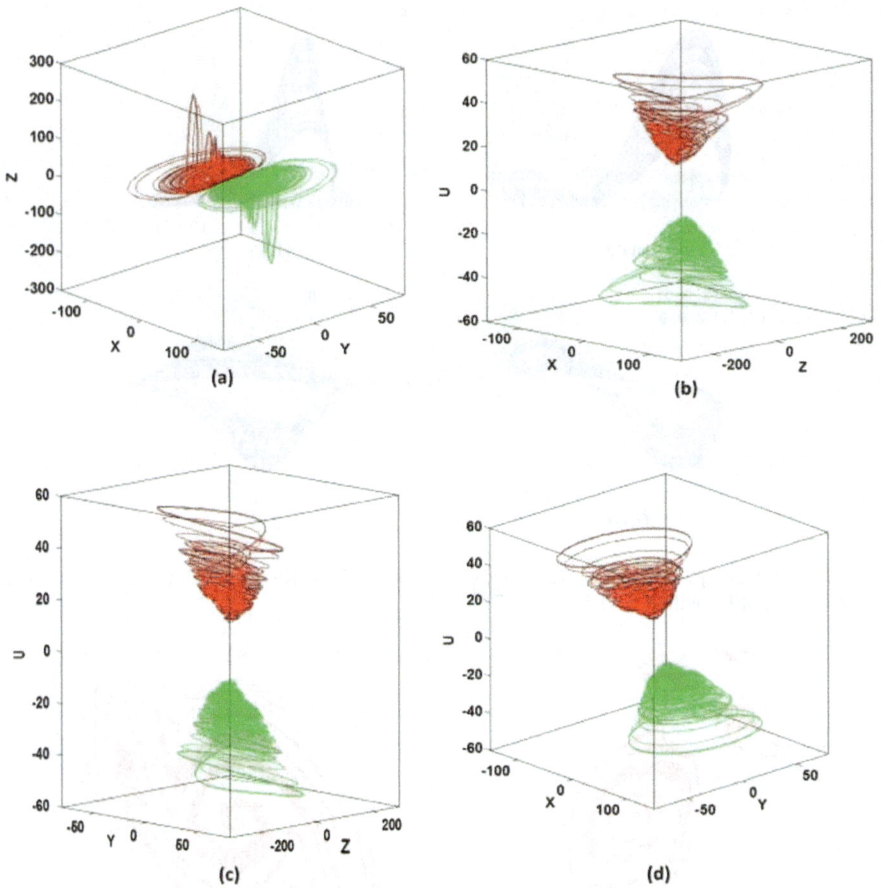

Fig. 14 Coexisting hyperchaotic attractors of system (6) for $a = 0.25,\ b = 3$ and $c = 0.05,\ d = 0.5$. **a** x-y-z phase space. **b** x-z-u phase space. **c** y-z-u phase space. **d**x-y-u phase space, red for $(-6, 0, 0.5, 14)$ and green for $(6, 0, -0.5, -14)$

can also be revised to be of symmetric structure, when the third dimension is modified to be $\dot{z} = b\,\mathrm{sgn}(z) + x|z|$, as follows:

$$\begin{cases} \dot{x} = -y - z, \\ \dot{y} = x + ay + u, \\ \dot{z} = b\,sgn(z) + x|z|. \\ \dot{u} = cu - dz. \end{cases} \tag{6}$$

System (6) has inversion symmetry with respect to the original point based on the invariance under the variable transformation $(x, y, z, u) \to (-x, -y, -z, -u)$, and hence it may give a symmetric pair of coexisting hyperchaotic attractors. When $a = 0.25,\ b = 3,\ c = 0.05,\ d = 0.5$, system (6) provides coexisting hyperchaotic attractors with LEs $(0.1121, 0.0213, 0, -24.9268)$ and Kaplan-Yorke dimension $D_{KY} = 3 - (\lambda_1 + \lambda_2)/\lambda_4 \approx 3.0054$, as shown in Fig. 14.

Compared with the original system (5), the symmetric system (6) has symmetric equilibrium points. When $a = 0.25,\ b = 3,\ c = 0.05,\ d = 0.5$, the system has three equilibrium points: one at the origin (0, 0, 0, 0) with eigenvalues $(0, 0.05, 0.125 \pm 0.9922i)$, which is an unstable node; and two other equilibrium points at $(\pm 5.4083, \pm 0.5547, \mp 0.5547, \mp 5.5470)$ with eigenvalues $(-5.3090, 0.1019, 0.0494 \pm 0.9987i)$, which are four-dimensional saddle-foci. This indicates that the coexisting attractors are self-excited rather than hidden [14–20].

6 Conclusion

Coexisting attractors may be found easily in symmetric or asymmetric systems. In this chapter, by giving such examples, different regimes of attractor coexistence are demonstrated. Typically two attractors coexist in similar shapes. Additional dimension sometimes indicates extreme multi-stability for the entanglement of parameter and system variable. Different feedback or polarity balance may transform an asymmetric system to be a symmetric one and thus produce symmetric pairs of attractors. All those coexisting attractors governed by a differential equation seem to have certain similarity in the phase structure.

Acknowledgements This work was supported financially by the National Natural Science Foundation of China (Grant No. 61871230), the Natural Science Foundation of Jiangsu Province (Grant No. BK20181410), the Startup Foundation for Introducing Talent of NUIST (Grant No. 2016205), and a Project Funded by the Priority Academic Program Development of Jiangsu Higher Education Institutions.

References

1. O.E. Rössler, An equation for continuous chaos. Phys. Lett. A **57**(5), 397–398 (1976)
2. O.E. Rössler, Continuous chaos four prototype equations. Ann. N. Y. Acad. Sci. **316**(1), 376–392 (1979)
3. J.C. Sprott, Asymmetric bistability in the Rössler system. Acta Phys. Polonica B **48**(1), 97 (2017)
4. B. Munmuangsaen, B. Srisuchinwong, A new five-term simple chaotic attractor. Phys. Lett. A **373**(44), 4038–4043 (2009)
5. V.D.S. Gerard, L.R.M. Maas, The diffusionless lorenz equations; shil'nikov bifurcations and reduction to an explicit map. Physica D **141**(1–2), 19–36 (2000)
6. J.C. Sprott, Simplest chaotic flows with involutional symmetries. Int. J. Bifurc. Chaos **24**(01), 1450009 (2014)
7. J.C. Sprott, C. Li, Comment on "how to obtain extreme multistability in coupled dynamical systems". Phys. Rev. E **89**(6), 066901 (2014)
8. C. Hens, R. Banerjee, U. Feudel, S. Dana, How to obtain extreme multistability in coupled dynamical systems. Phys. Rev. E **85**(3), 035202 (2012)

9. B. Bao, T. Jiang, G. Wang, P. Jin, H. Bao, M. Chen, Two-memristor-based chua's hyperchaotic circuit with plane equilibrium and its extreme multistability. Nonlinear Dyn. **89**(60), 1157–1171 (2017)
10. B.C. Bao, Q. Xu, H. Bao, M. Chen, Extreme multistability in a memristive circuit. Electron. Lett. **52**(12), 1008–1010 (2016)
11. B.C. Bao, H. Bao, N. Wang, M. Chen, Q. Xu, Hidden extreme multistability in memristive hyperchaotic system. Chaos Solitons Fractals **94**, 102–111 (2017)
12. C. Li, W. Hu, J.C. Sprott, X. Wang, Multistability in symmetric chaotic systems. Eur. Phys. J.: Special Topics **224**(8), 1493–1506 (2015)
13. O.E. Rössler, An equation for hyperchaos. Phys. Lett. A **71**, 155–157 (1979)
14. D.A. Prousalis, C.K. Volos, I.N. Stouboulos, I.M. Kyprianidis, Hyperchaotic memristive system with hidden attractors and its adaptive control scheme. Nonlinear Dyn. **90**(3), 1681–1694 (2017)
15. C. Li, J.C. Sprott, W. Thio, Bistability in a hyperchaotic system with a line equilibrium. J. Exp. Theor. Phys. **118**(3), 494–500 (2014)
16. C. Li, J.C. Sprott, Coexisting hidden attractors in a 4-d simplified lorenz system. Int. J. Bifurc. Chaos **24**(3), 1450034 (2014)
17. Z. Wang, S. Cang, A hyperchaotic system without equilibrium. Nonlinear Dyn. **69**(1–2), 531–537 (2012)
18. Q. Li, S. Hu, S. Tang, G. Zeng, Hyperchaos and horseshoe in a 4d memristive system with a line of equilibria and its implementation. Int. J. Circuit Theory Appl. **42**(11), 1172–1188 (2014)
19. C. Shen, S. Yu, J. Lü, G. Chen, Constructing hyperchaotic systems at will. Int. J. Circuit Theory Appl. **43**(12), 2039–2056 (2015)
20. Q. Li, X. S. Yang, Hyperchaos from two coupled wien-bridge oscillators. Int. J. Circuit Theory Appl. **36**(1), 19–29 (2008)

Multi-Stability in Conditional Symmetric Systems

Chunbiao Li and Julien Clinton Sprott

1 Introduction

Some of the coexisting attractors in symmetric or asymmetric systems share the geometric structure even with unified Lyapunov exponents [1–10]. In symmetric systems, many of the coexisting attractors stand in separate subspace respecting to some of the coordinate-axis [1–7]. Coexisting attractors may be embedded in each other in asymmetric systems [8–10]. Dynamical systems may exhibit coexisting attractors with other relative positions, specifically some of which in phase space may have some property of symmetry according to a super-plane but need extra offset boosting. The symmetry mentioned above is obtained from a necessary transformation and therefore, is defined as conditional symmetry. As shown in Fig.1, an asymmetric system can be revised to be symmetric one by polarity revise in the feedback term while conditional symmetric system needs a function-based polarity balance.

Any transformation in a dynamical system should obey the basic law of polarity balance to maintain its basic dynamical properties. A symmetric system keeps its polarity balance even when some of the variables get polarity reversed. More generally, the polarity imbalance in a dynamical system can be induced by the polarity reversal of any of the variables, or from the function in the feedback, or even from the time dimension. Polarity reverse from some of the variables can retain its polarity balance in a symmetric system while an asymmetric one loses its balance of polarity. It shows that the polarity balance can be restored in some specific asymmetric systems when some of the variables are offset boosted in the feedback function where the offset boosting does not change the polarity of the left-hand side of the differ-

C. Li (✉)
Nanjing University of Information Science & Technology, Nanjing, China
e-mail: goontry@126.com; chunbiaolee@nuist.edu.cn

J. C. Sprott
University of Wisconsin–Madison, Madison, WI, USA
e-mail: sprott@physics.wisc.edu

X. Wang et al. (eds.), *Chaotic Systems with Multistability and Hidden Attractors*,
Emergence, Complexity and Computation 40,
https://doi.org/10.1007/978-3-030-75821-9_14

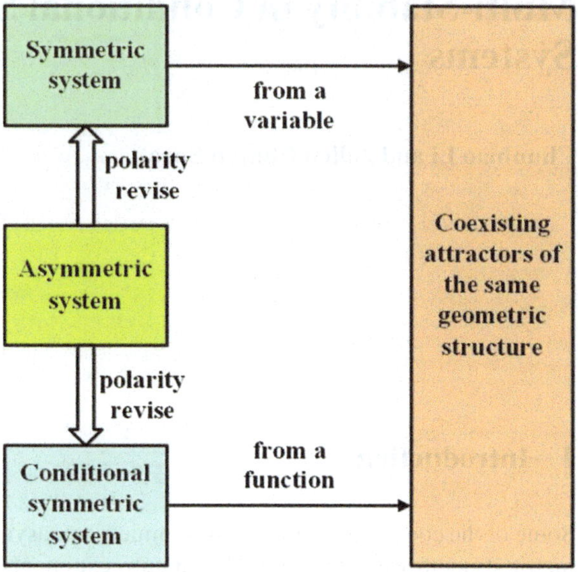

Fig. 1 System structure connected with symmetry

ential equation but give a negative sign on the right-hand side. In the following, we discuss the conception of conditional symmetry and the approach for constructing a conditional symmetric system with newly found examples. Here, the fundamental factor is polarity control. This control sometimes is combined with other control like frequency control, amplitude control making the attractor(s) with different geometric and distribution characteristics.

2 Conception of Conditional Symmetry

Offset boosting is the key factor for understanding conditional symmetry since it does not bring any change with derivative operation but can bring a negative sign in a nonmonotonic function. For a dynamical system $\dot{X} = F(X)$ $(X = (x_1, x_2, ..., x_N))$, a variable substitution of $x_j \to x_j + d_j$ (here, $1 \leqslant j \leqslant N, i \in \{1, 2, ..., N\} \setminus \{j\}$ makes the variable x_j in system $\dot{X} = F(X)$ exhibit **offset boosting**, which means that the average of the variable x_j is boosted by the new introduced constant d_j. Specifically, if the above substitution only introduces a separate constant d_j in one dimension on the right-hand side of the equations, then the system can be regarded as a **offset-boostable system** [11]. In electrical circuit, the variable x_j represents a circuit signal, and the newly introduced constant d_j is a direct current source. Similarly, a dynamical system $\dot{X} = F(X) = (f_1(X), f_2(X), ..., f_N(X))$ $(X = (x_1, x_2, ..., x_N))$ can be defined as an **N-D offset-boostable system** [11–14] if there exist n variable substitutions recover its governing equation, except for n additional constants allowing offsets with those variables. As reported in Ref. [14], the nonmonotonic function like an absolute value function or a trigonometric function may result in a polarity reversal, which can be applied to reconstruct polarity balance for conditional symmetry.

For a differential equation $\dot{X} = F(X) = (f_1(X), f_2(X), ..., f_N(X))$ $(X = (x_1, x_2, ... , x_N))$, if there exists a variable substitution including polarity reversal and offset boosting like $u_{i_1} = -x_{i_1}$, $u_{i_2} = -x_{i_2}$, ..., $u_{i_k} = -x_{i_k}$, $u_{j_1} = -x_{j_1} + d_{j_1}$, $u_{j_2} = -x_{j_2} + d_{j_2}$, ..., $u_{j_l} = -x_{j_l} + d_{j_l}$, $u_i = x_i$ (here, $1 \leqslant i_1, ..., i_k \leqslant N$, $1 \leqslant j_1, ..., j_l \leqslant N$, $i_1, ..., i_k$and $j_1, ..., j_l$ arc not identical, $i \in \{1, 2, ..., N\} \setminus \{i_1, ..., i_k, j_1, ..., j_l\}$), the derived equation retains its balance of polarity on the two sides of the equation and satisfies $\dot{U} = F(U)$ $(U = (u_1, u_2, ..., u_N))$. The corresponding system $\dot{X} = F(X)$ $(X = (x_1, x_2, ..., x_N))$ is defined as one of l-dimensionally conditional symmetry, since the polarity balance needs l-dimensional offset boosting [12, 13]. Specifically, for a three-dimensional dynamical system, $\dot{X} = F(X)$ $(X = (x_1, x_2, x_3))$, there exist only conditional rotational symmetry in 1-dimension and conditional reflection symmetry in 1-dimension or 2-dimension. Note that offset boosting does not produce a minus sign on the left-hand side of the equation, since $d(x_{jm} + d_{jm}) = d(x_{jm})$.

3 Constructing Conditional Symmetry from Single Offset Boosting

A jerk system is a simple structure to pass the polarity, which consequently provides an easy way to consider polarity balance. Such a case can be found even in a hypogenetic chaotic jerk flow JH5 with absolute value nonlinearities [15], as follows:

$$\begin{cases} \dot{x} = |y| - b, \\ \dot{y} = z, \\ \dot{z} = f(x) - y - az. \end{cases} \tag{1}$$

When $a = 0.6$, $b = 1$, $f(x) = |x| - c$, $c = 2$, the system (1) is chaotic and has two coexisting symmetric attractors, with Lyapunov exponents (0.0534, 0, −0.6534) and Kaplan-Yorke dimension $D_{KY} = 2.0817$, as shown in Fig. 2. System (1) has four equilibrium points, (±3, 1, 0), (±1, −1, 0), which may consist of the centers of different attractors. The symmetric coexistence of chaotic solutions can be explained by a transformation. Even the structure of system (1) has no symmetry, a variable boosting in the variable x can return a symmetry-like transformation, which is now defined as conditional symmetry and correspondingly give birth to symmetric bistability. Specifically, let $x = u + d$, $y = -v$, $z = -w$(here, d is a new introduced constant). If the condition $c = (|x| + |u + d|)/2$ is satisfied, the new deduced system has a conditional rotational symmetry because the symmetry-like transformation $\dot{(u)} = |v| - b$, $\dot{(v)} = w$, $\dot{(w)} = -|u + d| - v - aw + c$ can be obtained. In the new

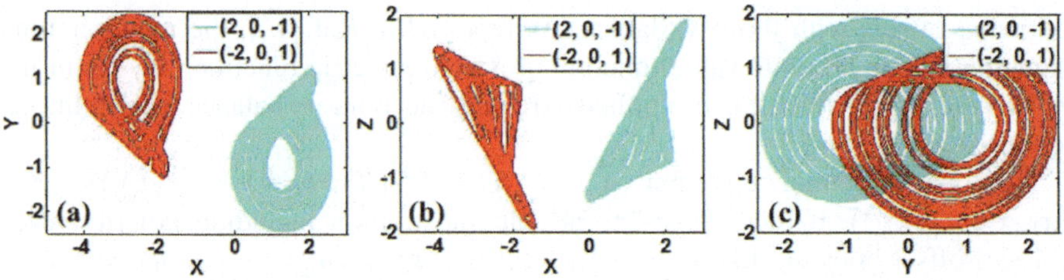

Fig. 2 Symmetric strange attractors for system (1) with $a = 0.6,\ b = 1,\ c = 2$. **a** x-y plane. **b** x-z plane. **c** y-z plane

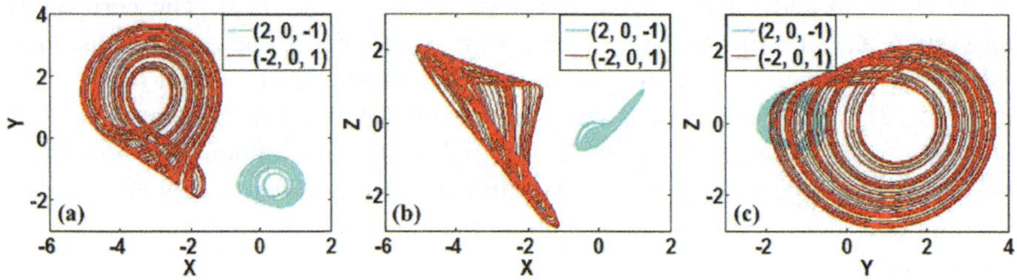

Fig. 3 Asymmetric strange attractors for system (1) with $a = 0.6,\ b = 1.5,\ c = 2$. **b** x-y plane. **b** x-z plane. **c** y-z plane

space of variables u, v, w, the newly derived equations are conditionally identical to the original one, when $c - |u + d| = |x| - c$. Here, the symmetry is obtained on the condition equation in the z-dimension. The basins of attraction are shown in Fig. 4a, as predicted, although the regions in light blue and red representing two different attractor basins are asymmetric, the strange attractors, represented in cross section by black lines, are symmetric and nearly touch their basin boundaries. Two of the equilibrium points are within the attracting basins, while the other two are on the basin boundary. As shown in Fig. 3, when $b = 1.5$, two asymmetric strange attractors coexist, whose attracting basins are shown in Fig. 4b. Here, the black lines of cross section are asymmetric.

A similar chaotic system of conditional rotational symmetry can be obtained if the absolute value function of y is replaced by a quadratic function rewritten as

$$\begin{cases} \dot{x} = y^2 - a, \\ \dot{y} = bz, \\ \dot{z} = -y - z + f(x). \end{cases} \tag{2}$$

The revised system keeps on the polarity balance required for the symmetry transformation from the polarity reversal of the function given by offset boosting. The system (2) yields coexisting symmetric attractors according to the y-axis and z-axis, as shown in Fig. 5. System (2) has two groups of equilibria with the same stability when $a = 1.22,\ b = 8.48,\ f(x) = |x| - 3$, which are $P_{11} = (4.1045, 1.1045, 0),\ P_{12} =$

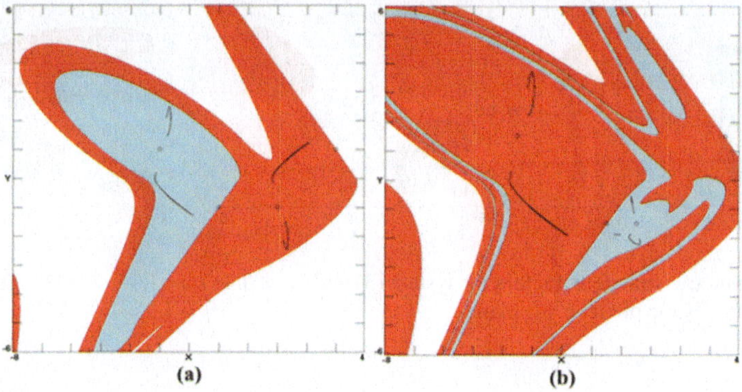

Fig. 4 Cross section $z = 0$ of the basins of attraction for the symmetric strange attractors of system (5). **a** $a = 0.6,\ b = 1,\ c = 2$. **b** $a = 0.6,\ b = 1.5,\ c = 2$.

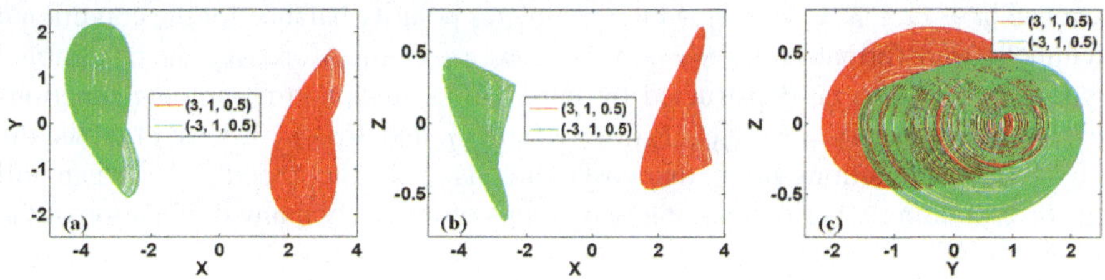

Fig. 5 Symmetric strange attractors for system (2) with $a = 1.22,\ b = 8.48,\ f(x) = |x| - 3$. **a** x-y plane. **b** x-z plane. **c** y-z plane

(1.8955, 1.1045, 0), $P_{21} = (1.8955, 1.1045, 0)$, $P_{22} = (4.1045, 1.1045, 0)$. When $|u + c| - 3 = 3 - |x|$, the transformation $x \to u + c$, $y \to -v$, $z \to -w$ in Eq. (2) is subject to the same governing equation and therefore generates coexisting rotational symmetric attractors. The basin of attraction for the coexisting attractors shows that the symmetric attractors lie in corresponding asymmetric basins [14].

As another case, we can construct a conditional reflection symmetric system by introducing the absolute value function for returning a polarity balance based on offset boosting [14], as follows:

$$\begin{cases} \dot{x} = y^2 - az^2, \\ \dot{y} = -z^2 - by + c, \\ \dot{z} = yz + f(x). \end{cases} \tag{3}$$

System (3) also has two groups of equilibria, which are P_{11}=(6.5576, 1.5, 2.3717), $P_{12} = (1.9881, 0.8, 1.2649)$, $P_{13} = (4.0119, 0.8, 1.2649)$, with eigenvalues $\lambda_{11} = (0.6029, 1.9265 \pm 3.7926i)$, $\lambda_{12} = (1.5526, 0.3013 \pm 1.9123i)$, and λ_{13}=(1.4116, $1.1808 \pm 1.6515i$), respectively, $P_{21} = (6.5576, 1.5, 2.3717)$, $P_{22} = (4.0119, 0.8, 1.2649)$, $P_{23} = (1.9881, 0.8, 1.2649)$, with the same eigenvalues $\lambda_{21} = (1.1397, 1.0551 \pm 2.9085i)$, $\lambda_{22} = (1.5526, 0.3013 \pm 1.9123i)$, and λ_{23}=(1.4116, $1.1808\pm$

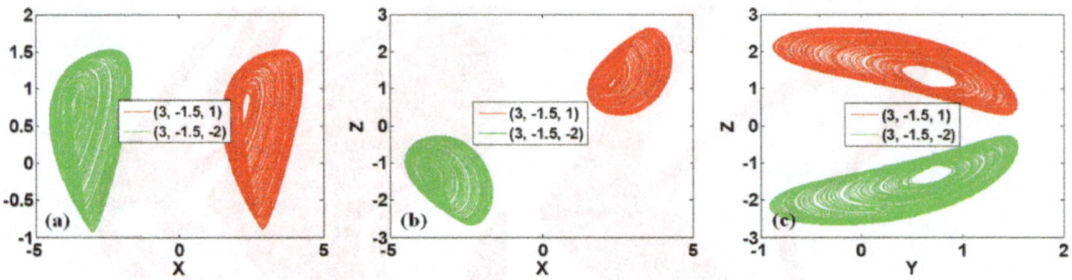

Fig. 6 Symmetric strange attractors for system (3) with $a = 0.4,\ b = 1.75,\ c = 3,\ f(x) = |x| - 3$. **a** x-y plane. **b** x-z plane. **c** y-z plane

$1.6515i$), respectively. The equilibrium point P_{21} is also a stable focus, and therefore it coexists with the other strange attractors [14]. The conditional symmetry of Eq. (3) is obtained by the offset boosting of the variable x resulting in a polarity reversal $|u + c| - 3 = 3 - |x|$, which completes polarity balance for the conditional symmetry transformation $x \to u + c,\ y \to v,\ z \to -w$. Moreover, the conditional symmetry can also be constructed by introducing other non-monotonic functions $F(x)$, such as $F_2(x) = 3 - |x|$, $F_3(x)$= 1.5sin(x), and $F_4(x)$= 1.5cos(x). Since the trigonometric functions sin(x) and cos(x) are also periodic, the attractor basin will be correspondingly periodic giving two groups of infinitely many duplication of the attractors [13].

4 Constructing Conditional Symmetry from Multiple Offset Boosting

More generally, conditional symmetry can be constructed from a structure with multiple offset boosting in addition to rigid variable-boostable systems [11]. Let us think, if the goal is to construct a conditional reflection symmetric system given the polarity reversal in the variable x_i, the polarity revise of the dimension of x_i on the left-hand side of x_i in turn requires the adjustment in the polarity with the terms on the right-hand side of x_i to get $-f_i(X)$. At the same time, all the other dimensions should keep the polarity balance regardless of that $-x_i$ may introduce a minus sign on the right-hand side, which further requires a new polarity reverse from offset boosting to cancel it. Here, the polarity balance of a polynomial equation $f_j(X)$ is saved by introducing new functions in $f_j(x_1, x_2, ..., -x_i, ..., F_{j_1}(x_{j_1}), F_{j_2}(x_{j_2}), ..., F_{j_l}(x_{j_l}), ..., x_N)$ $(1 \leqslant j_1, ..., j_l \leqslant N,\ j_1, ..., j_l$ are not identical, and l is an odd) for variables $(x_{j_1}, x_{j_2}, ..., x_{j_l})$, which admits offset boosting returning a new minus sign to cancel the one from $-x_i$, by $F_{jm}(x_{jm} + d_{jm} = -F_{jm})(x_j m)$ $(j_1 \leqslant j_m \leqslant j_l)$.

Four general equations for hosting conditional symmetry are designed for exhaustive computer searching [12]. We can conclude that Eqs. (4) and (5) can be transformed to exhibit conditional symmetry from 1D offset boosting, while Eqs. (6) and (7) can be modified to show conditional symmetry from 2D offset boosting. Equa-

Table 1 Conditional symmetric systems (CSS)

Systems	Equations	Parameters	LEs	D_{KY}	(x0,y0,z0)
CSS1	$\dot{x} = F(y),$ $\dot{y} = z,$ $\dot{z} = -x^2 - az +$ $b(F(y))^2 + 1,$ $F(y) = \|y\| - 4$	a= 2.6, b= 2	0.0463, 0 −2.6463	2.0175	0.5, 4, −1
CSS2	$\dot{x} = y,$ $\dot{y} = F(z),$ $\dot{z} = x^2 - ay^2 + bxy +$ $xF(z)$ $F(z) = \|z\| - 8$	a= 1.24, b= 1	0.0463, 0, −2.6463	2.0513	4, 0.8, −2
CSS3 (from VB6)	$\dot{x} = 1 - G(y)z,$ $\dot{y} = az^2 - G(y)z,$ $\dot{z} = F(x),$ $F(x) = \|x\| - 3$ $G(y) = \|y\| - 5$	a= 0.22	0.0729 0 −1.6732	2.0436	−1, 1, −1
CSS4	$\dot{x} = F(y),$ $\dot{y} = xG(z),$ $\dot{z} = -axF(y) -$ $bxG(z) - x^2 + (F(y)^2)$ $F(y) = \|y\| - 5$ $G(z) = \|z\| - 5$	a= 3, b= 1.2	0.0506 0 −0.2904	2.1735	0, −6, −6

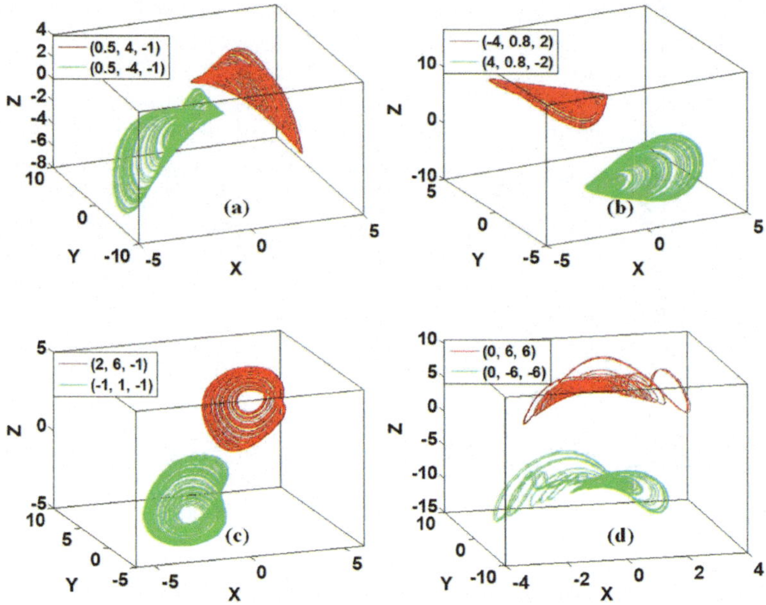

Fig. 7 Coexisting attractors. **a** CSS1 induced by 1D offset boosting in the y-dimension. **b** CSS2 induced by 1D offset boosting in the z-dimension. **c** CSS3 induced by 2D offset boosting in the x and y dimensions. **d** CSS4 induced by 2D offset boosting in the y and z dimensions

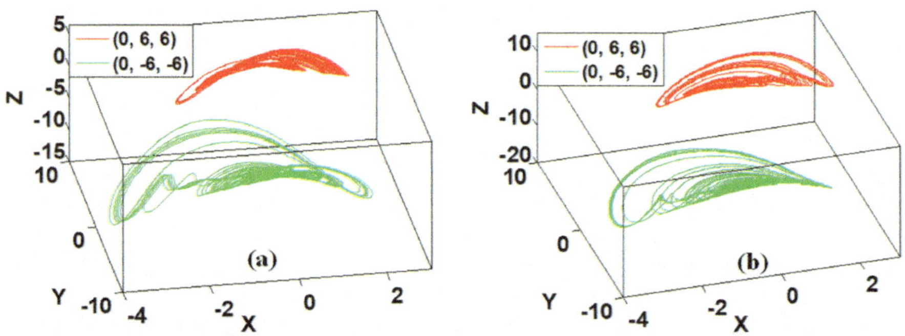

Fig. 8 Coexisting attractors in CSS4 induced by 2D offset boosting in the y and z dimensions. **a** $F(y) = |y| - 5$, and $G(z) = |z| - 3.3$. **b** $F(y) = |y| - 4.5$, and $G(z) = |z| - 5$

tions (4) and (5) are different from the above mentioned cases reported in Ref. [14], where a single 1D offset boosting is in the third dimension. Here, the new conditional symmetry comes from a jerk structure, where the reflection or rotational symmetry is broken by a neighbor variable, but the polarity balance can be restored by a general offset boosting. Equations (6) and (7) are the structures for hosting conditional reflection symmetry, where 2D offset boosting is necessary for restoring polarity bal-

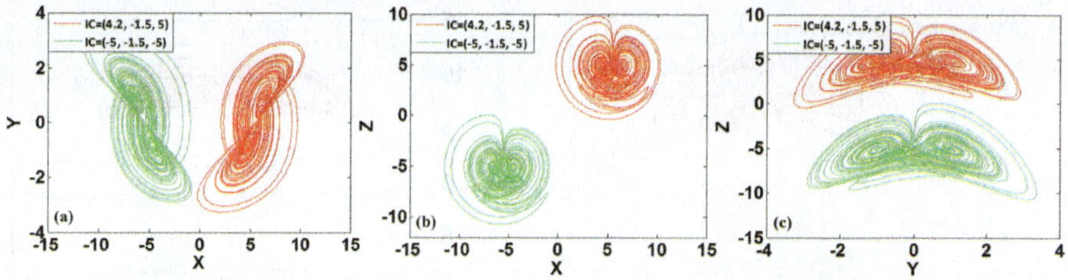

Fig. 9 Coexisting attractors for system (9) with $F(x) = |x| - 5.5,\ G(z) = |z| - 5$. **a** x-y plane. **b** x-z plane. **c** y-z plane. Red is for $IC = (4.2, -1.5, 5)$ and green is for $IC = (-5, -1.5, -5)$

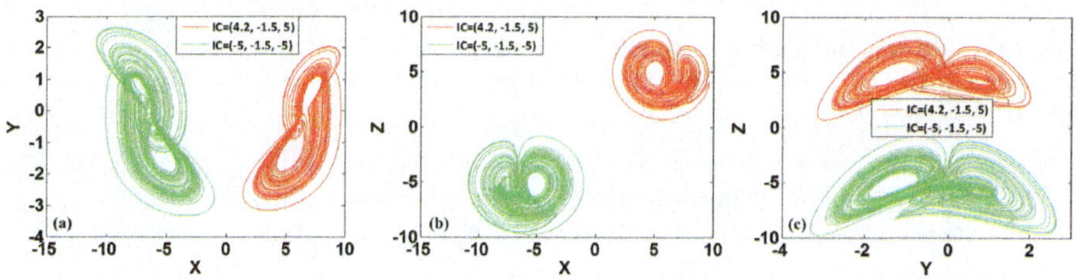

Fig. 10 Coexisting attractors for system (9) with $F_1 = |x| - 6.6,\ F_2 = |x| - 6$. **a** x-y plane. **b** x-z plane. **c** y-z plane. Red is for $IC = (4.2, -1.5, 5)$ and green is for $IC = (-5, -1.5, -5)$

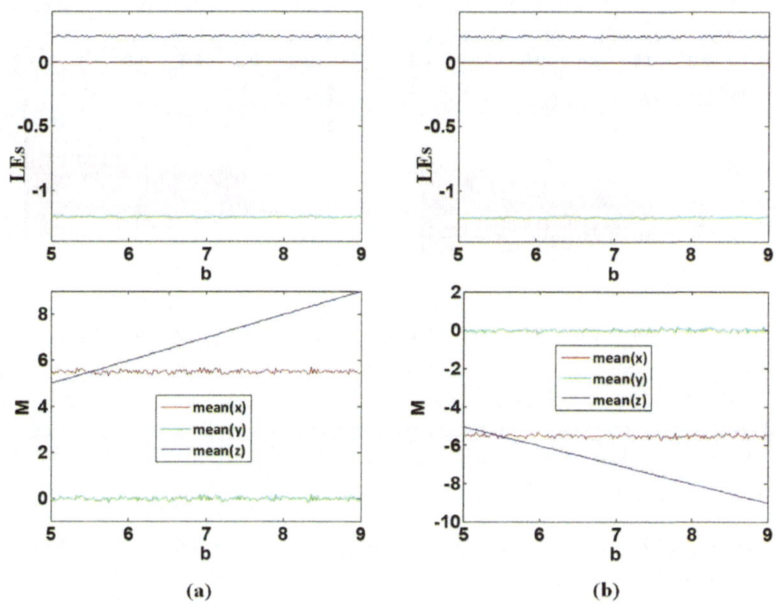

Fig. 11 Lyapunov exponents and average value for system (9) with $F(x) = |x| - 5.5,\ G(z) = |z| - b$. **a** $IC = (4.2, -1.5, 5 + b)$. **b** $IC = (-5, -1.5, -5 - b)$

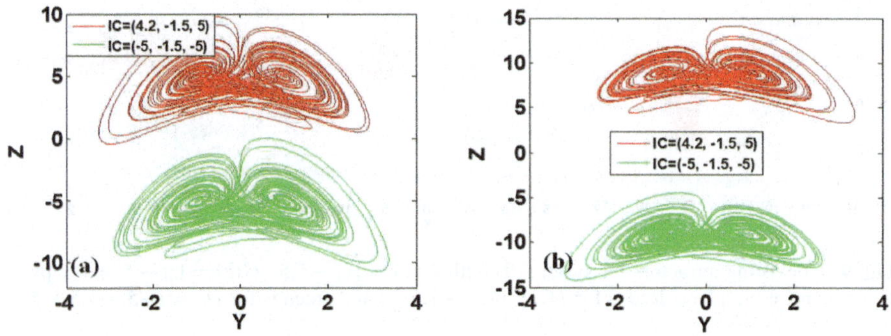

Fig. 12 Coexisting attractors for system (9) in the y–z plane. Red is for $IC = (4.2, -1.5, 5)$ and green is for $IC = (-5, -1.5, -5)$. **a** $F(x) = |x| - 5.5,\ \ G(z) = |z| - 5$. **b** $F(x) = |x| - 5.5,\ \ G(z) = |z| - 9$

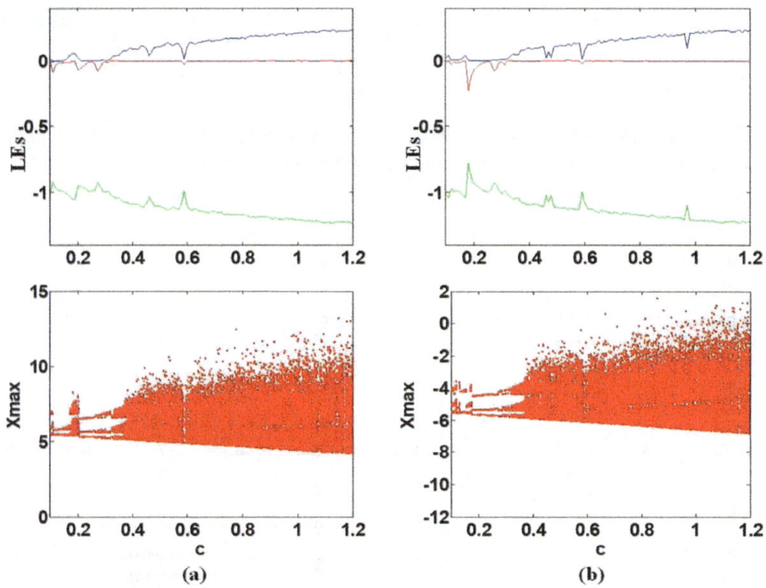

Fig. 13 Bifurcation diagrams and Lyapunov exponents of system (9) with $F(x) = |x| - 5.5,\ \ G(z) = |z| - 5$, when c is varied in [0.1, 1.2]. **a** $IC = (4.2, -1.5, 5)$. **b** $IC = (-5, -1.5, -5)$

ance. Newly constructed conditional symmetric systems (CSS) are listed in Table 1, exhibiting chaotic coexisting attractors.

$$\begin{cases} \dot{x} = y, \\ \dot{y} = z, \\ \dot{z} = a_1xy + a_2x^2 + a_3y^2 + a_4z + a_5z^2 + a_6. \end{cases} \tag{4}$$

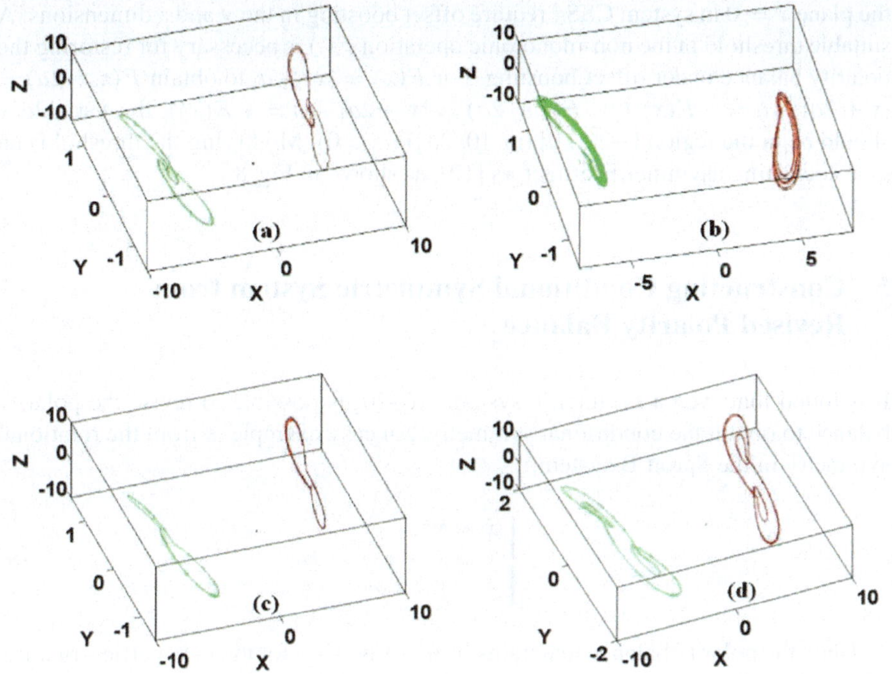

Fig. 14 Some other coexisting attractors of system (9) with $F(x) = |x| - 5.5,\ G(z) = |z| - 5$, where $IC = (4.2, -1.5, 5)$ is red, $IC = (-5, -1.5, -5)$ is green. **a** $c = 0.1$. **b** $c = 0.17$. **c** $c = 0.3$. **d** $c = 0.59$

$$\begin{cases} \dot{x} = y, \\ \dot{y} = z, \\ \dot{z} = a_1x^2 + a_2y^2 + a_3xy + a_4xz + a_5yz + a_6z^2 + a_7. \end{cases} \tag{5}$$

$$\begin{cases} \dot{x} = a_1z^2 + a_2xz + a_3yz + a_4, \\ \dot{y} = a_5z^2 + a_6xz + a_7yz + a_8, \\ \dot{z} = a_9z + a_{10}x + a_{11}y. \end{cases} \tag{6}$$

$$\begin{cases} \dot{x} = y, \\ \dot{y} = xz, \\ \dot{z} = a_1xy + a_2xz + a_3x^2 + a_4y^2 + a_5z^2 + a_6yz + a_7. \end{cases} \tag{7}$$

Symmetric pairs of coexisting attractors in the conditional symmetric versions are shown in Fig. 7. Figure 7a shows the coexisting attractors in CSS1 induced by 1D offset boosting in the y-dimension, while Fig. 7b plots the coexisting symmetric attractors in the plane $x = 0$, requiring an offset boosting in the z-dimension. Two symmetric attractors in the plane $z = 0$ in system CSS3 demand offset boosting in the x and y dimensions, as indicated in Fig. 7c. In Fig. 7d, two symmetric attractors in the plane $x = 0$ in system CSS4 require offset boosting in the y and z dimensions. A suitable threshold in the non-monotonic operation $F(.)$ is necessary for restoring the polarity balance under offset boosting. For $F(x) = |x| - a$, to obtain $F(x + 2a) = |x + 2a| - a = -F(x)$ (or, $F(x - 2a) = |x - 2a| - a = -F(x)$), the variable x should be in the region $[-2a\ , 0]$ (or, $[0, 2a]$) ($a \geqslant 0$). Modifying the threshold can give coexisting asymmetric attractors [12], as shown in Fig. 8.

5 Constructing Conditional Symmetric System from Revised Polarity Balance

It is found that even a symmetric system [16–20] is possible to revise the polarity balance to obtain the conditional symmetry. An easy example is from the rotational symmetry in the Sprott B system:

$$\begin{cases} \dot{x} = yz, \\ \dot{y} = x - y, \\ \dot{z} = 1 - xy. \end{cases} \tag{8}$$

Here, the polarity balance maintains from $-x$ and $-y$ for the symmetric structure. Meanwhile, the unique structure can still maintain its polarity balance when the polarity is changed from an extra function without revising its basic dynamics, as

$$\begin{cases} \dot{x} = G(z)y, \\ \dot{y} = F_1(x) - y, \\ \dot{z} = c - F_2(x)y. \end{cases} \tag{9}$$

When $F_1(x) = F_2(x) = F(x) = |x| - 5.5,\ G(z) = |z| - 5,\ c = 1$, system (9) exhibits coexisting conditional reflection symmetric attractors with Lyapunov exponents $(0.2101, 0, -1.2101)$ and Kaplan-Yorke dimension $D_{KY} = 2.173$, as shown in Fig. 9.

Mismatched offset boosting may result in asymmetric coexisting attractors. In system (8), the variable x appears twice, and so two functions are necessary for conditional symmetry. Moreover, these two functions should keep phase synchronization for returning the same polarity reverse. As shown in Fig. 10, when $F_1 = |x| - 6.6$, $F_2 = |x| - 6$ and $G(z) = |z| - 5$, two asymmetric strange attractors appear, as predicted. Offset controllers in different variables may keep independent from each other, as shown in Fig. 11. When $F(x) = |x| - 5.5$, $G(z) = |z| - b$ and b is varied, as in Ref. [5, 9], the average of the variable z changes smoothly without revising the averages of other variables x and y with the same Lyapunov exponents. As shown in Fig. 12, larger offset controller b makes the symmetric pair of coexisting attractors farther in the z-dimension.

Normal bifurcation parameter c in system (9) control the dynamical behavior dramatically. But in the special structure of conditional symmetry, system (9) displays coexisting bifurcations from different initial conditions, as shown in Fig. 13. The common bifurcation controller c may lead to different evolutions, giving asymmetric coexisting attractors, as shown in Fig. 14.

6 Discussions and Conclusions

Multi-stability shows that various profiles, specifically symmetric strange attractors and asymmetric ones, can coexist in asymmetric basins of attraction within a simple nonlinear structure due to conditional symmetry. Offset boosting is the key routine for restoring polarity balance and thereafter produces symmetric pairs of coexisting attractors. Offset-boostable system provides a compact structure for coining conditional symmetric system, while multiple dimensional offset boosting can find more cases of chaotic systems of conditional symmetry. Mismatched offset boosters may draw the system to exhibit asymmetric coexisting attractors. The symmetric system can also be transformed into a conditional symmetric one if the polarity balance is restored based on offset boosting.

Acknowledgements This work was supported financially by the National Natural Science Foundation of China (Grant No. 61871230), the Natural Science Foundation of Jiangsu Province (Grant No. BK20181410), the Startup Foundation for Introducing Talent of NUIST (Grant No. 2016205), and a Project Funded by the Priority Academic Program Development of Jiangsu Higher Education Institutions.

References

1. C. Li, J.C. Sprott, W. Thio, Linearization of the Lorenz system. Phys. Lett. A **379**(10–11), 888–893 (2015)
2. L. Zhou, C. Wang, L. Zhou, Cluster synchronization on multiple sub-networks of complex networks with nonidentical nodes via pinning control. Nonlinear Dyn. **83**(1–2), 1079–1100 (2016)
3. C. Li, I. Pehlivan, J.C. Sprott, A. Akgul, A novel four-wing strange attractor born in bistablity. IEICE Electron. Express **12**(4), 20141116 (2015)
4. J.C. Sprott, Simplest chaotic flows with involutional symmetries. Int. J. Bifurc. Chaos **24**(1), 1450009 (2013)
5. J.C. Sprott, New chaotic regimes in the Lorenz and Chen systems. Int. J. Bifurc. Chaos **25**(5), 1550033 (2015)
6. Q. Lai, S. Chen, Generating multiple chaotic attractors from Sprott B system. Int. J. Bifurc. Chaos **26**(11), 1650177 (2016)
7. Q. Lai, T. Nestor, J. Kengne, X.W. Zhao, Coexisting attractors and circuit implementation of a new 4d chaotic system with two equilibria. Chaos, Solitons Fractals **107**, 92–102 (2018)
8. R. Barrio, F. Blesa, S. Serrano, Qualitative analysis of the Rössler equations: Bifurcations of limit cycles and chaotic attractors. Physica D **238**(13), 1087–1100 (2009)
9. E. Freire, E. Gamero, A.J. Rodríguez-Luis, A. Algaba, A note on the triple-zero linear degeneracy: normal forms, dynamical and bifurcation behaviors of an unfolding. Int. J. Bifurc. Chaos **12**(12), 2799–2820 (2002)

10. J.C. Sprott, C. Li, Asymmetric bistability in the Rössler system. Acta Phys. Pol., B **48**(1), 699–792 (2017)
11. C. Li, J.C. Sprott, Variable-boostable chaotic flows. Optik: Int. J. Light Electron Optics **127**(22), 10389–10398 (2016)
12. C. Li, J.C. Sprott, Y. Liu, Z. Gu, J. Zhang, Offset boosting for breeding conditional symmetry. Int. J. Bifurc. Chaos **28**(14), 1850163 (2018)
13. C. Li, J.C. Sprott, W. Hu, Y. Xu, Infinite multistability in a self-reproducing chaotic system. Int. J. Bifurc. Chaos **27**(10), 1750160 (2017)
14. C. Li, J.C. Sprott, H. Xing, Constructing chaotic systems with conditional symmetry. Nonlinear Dyn. **87**(2), 1351–1358 (2017)
15. C. Li, J.C. Sprott, H. Xing, Hypogenetic chaotic jerk flows. Phys. Lett. A **380**(11–12), 1172–1177 (2016)
16. C. Li, W. Hu, J.C. Sprott, X. Wang, Multistability in symmetric chaotic systems. Eur. Phys. J.: Special Topics **224**(8), 1493–1506 (2015)
17. K. Rajagopal, S. Panahi, A. Karthikeyan, A. Alsaedi, V.-T. Pham, T. Hayat, Some new dissipative chaotic systems with cyclic symmetry. Int. J. Bifurc. Chaos **28**(13), 1850164 (2018)
18. X. Di, J. Li, H. Qi, L. Cong, H. Yang, A semi-symmetric image encryption scheme based on the function projective synchronization of two hyperchaotic systems. PLoS ONE **12**(9), e0184586 (2017)
19. X. Wang, A. Akgul, S. Cicek, V.-T. Pham, D.V. Hoang, A chaotic system with two stable equilibrium points: Dynamics, circuit realization and communication application. Int. J. Bifurc. Chaos **27**(08), 1750130 (2017)
20. J. Liu, J.C. Sprott, S. Wang, Y. Ma, Simplest chaotic system with a hyperbolic sine and its applications in DCSK scheme. IET Commun. **12**(7), 809–815 (2018)

Multi-Stability in Self-Reproducing Systems

Chunbiao Li and Julien Clinton Sprott

1 Introduction

As we discussed in the above chapters, many dynamical systems can produce similar attractors, specifically some of which [1–10] share the same Lyapunov exponents. As shown in Fig. 1, a symmetric pair of coexisting attractors may be found in symmetric systems [1–5] or conditional symmetric systems [6–10]. Those coexisting attractors nested with each other, who are in a similar structure, can be found commonly in asymmetric systems. As shown in Fig. 1, in a dynamical system like an attractor box, we can take out two or more coexisting attractors by selecting corresponding initial conditions. So, a bold question comes out, that is, can we obtain any number of attractors of the same Lyapunov exponents by giving different initial conditions? The direct idea is that we can clone the original attractor infinitely by a periodic function and then choose the initial data according to the period of the clone function [11–15]. A dynamical system with infinitely many cloned attractors is named as a self-reproducing system. In the following, we discuss how the attractor gets cloned based on offset boosting and after that, different self-reproducing systems are obtained in various dimensions.

C. Li (✉)
Nanjing University of Information Science & Technology, Nanjing, China
e-mail: goontry@126.com; chunbiaolee@nuist.edu.cn

J. C. Sprott
University of Wisconsin–Madison, Madison, WI, USA
e-mail: sprott@physics.wisc.edu

X. Wang et al. (eds.), *Chaotic Systems with Multistability and Hidden Attractors*,
Emergence, Complexity and Computation 40,
https://doi.org/10.1007/978-3-030-75821-9_15

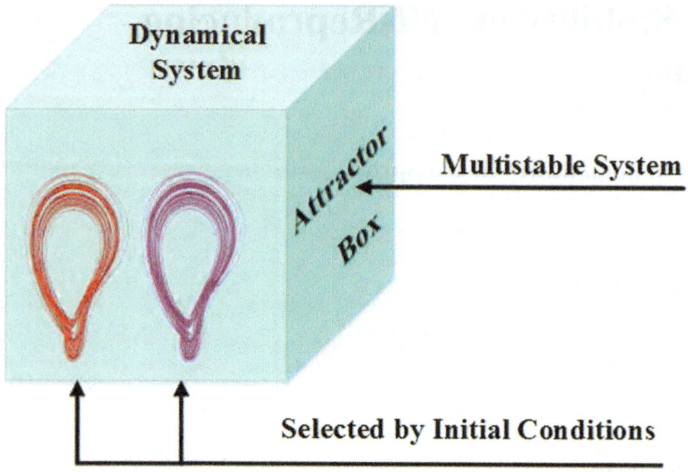

Fig. 1 Multistable system is like an attractor box

2 Concept of Self-Reproducing System

Here, offset boosting is the key technology to clone the original attractor. As defined in the last chapter, offset boosting refers to a method that can shift any attractor as well as its basin of attraction in the solution space of a dynamical system without altering the system solutions. Any attractor in a smooth dynamical system can obtain offset boosting by introducing an appropriate constant into any dimension of the system. Generally, an attractor is limited in the variable space. For a smooth dynamical system, $\dot{\boldsymbol{X}} = \boldsymbol{F}(\boldsymbol{X})$, $\boldsymbol{X} = (x_1, x_2, ..., x_n)$, introducing a constant vector $\boldsymbol{a} = (a_1, a_2, ..., a_n)$ into each variable will give the system offset boosting within the interval of $\boldsymbol{a}$.

If there is a power to make the attractor walk in the variable space, then that attractor can be cloned in any position and correspondingly the system outputs many maybe infinite coexisting attractors. As summarized in Fig. 2, the attractor reproducing process based on the core operation of offset boosting. If an attractor can climb along a specific route or axis under the action of a single constant, some of the footprints of the climbing attractor can be solidified and consist of coexisting attractors, which can be selected by initial conditions if the constant hides in a periodic or any other non-monotonic functions. This operation depends on the fact that the derivative of a constant is zero, a differential equation will not change its form if a constant is added to a variable.

Definition 1 Suppose that there is a dynamical system, $\dot{X} = F(X)$ $(X = (x_1, x_2, x_3, ..., x_i, ...)$ $(i \in N)$. If the offset boosting $x_i = u_i + k\mathrm{P}(k \in N, \mathrm{P} \in Z)$ makes the system recover its original governing equations, the system can reproduce its own attractor by the periodic offset boosting, consequently it is defined as a **self-reproducing system**. Self-reproducing systems keep their form when any of the

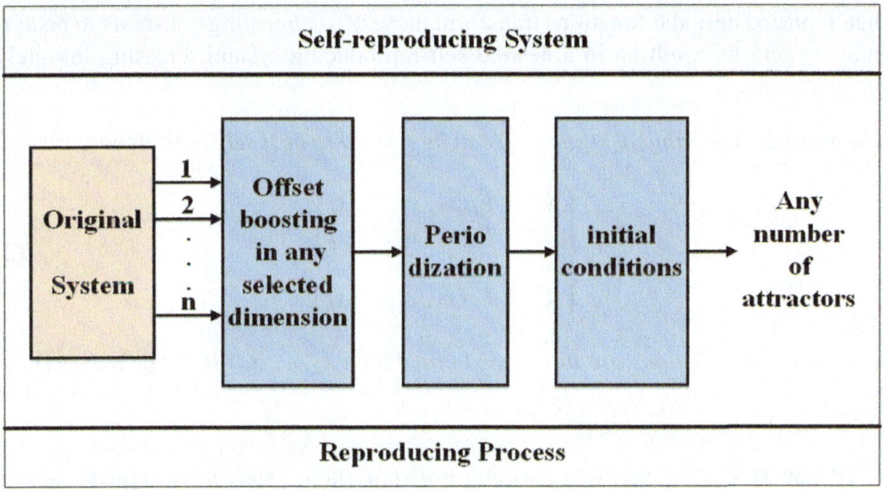

Fig. 2 Process of attractor self-reproducing

variables is boosted, which means that the variable only varies in its initial space while not altering the structure of the solution and not influencing its parameter space. Many functions can be introduced to transform the dynamical system to be a self-reproducing one, a typical example of which is the periodic function.

Theorem 1 *For a three-dimensional self-reproducing system,*

$$\begin{cases} \dot{x} = f(y, z), \\ \dot{y} = g(y, z), \\ \dot{z} = h(y, z) + F(x). \end{cases} \tag{1}$$

if the function $F(x)$ in the z-dimension is periodic and system (1) converges to an attractor in one period, system (1) has infinitely many attractors equally spaced along the x-axis with a period P.

[Proof 1] Since $F(x)$ is periodic, suppose $F(x) = F(x + \mathrm{P})$, $\mathrm{P} \neq 0 \in \mathrm{R}$ is the period of the function $F(x)$. Make a substitution $x = x + k\mathrm{P}$ $(k \in N)$ in system (1), leading to the same form of Eq. (1), therefore, system (1) produces infinitely many identical attractors when the variable x varies by an integer multiple of P.

The approach for coining a self-reproducing system can be applied to other ordinary differential equations. In this case, it becomes more complicated for introducing bloated functions to transform a normal dynamical system for attractor hatching. To boost the offset of any variable, more terms are necessary to change to adapt to that and consequently more introduced functions for periodic offset-boosting get coupled posing great threat on the dynamics [9]. Theoretically, any dimension in a dynamical system can get offset boosted from the introduction of separating constants. After that, replaced periodic functions transform those offset boosting constants to be normalized periods resulting in a desired self-reproducing system releasing infinitely many attractors.

Theorem 2 *A dynamical system (2) can be revised to be a self-reproducing system,*

$$\begin{cases} \dot{x}_1 = F_1(x_1, x_2, ..., x_n), \\ \dot{x}_2 = F_2(x_1, x_2, ..., x_n), \\ \cdots \\ \dot{x}_n = F_n(x_1, x_2, ..., x_n), \end{cases} \tag{2}$$

when periodic functions are introduced into $F_i(x_1, x_2, ..., x_n)$ $(i \in \{1, 2, ..., n\})$ *as:* $x_k = g_k(x_k)$, $(k \in \{j_1, j_2, ..., j_m\}$, $1 \leqslant j_1 < j_2 < \cdots < j_m \leqslant n)$ *and* $x_k = x_k$, $(k \in \{1, 2, ..., n\} \setminus \{j_1, j_2, ..., j_m\})$.

[Proof 2] Since $g_l(x_l)$ is a periodic function, there exists a constant $\mathrm{P_l}$ and an integer S_l, such that $g_l(x_l + \mathrm{S}_l\mathrm{P}_l) = g_l(x_l)$, $l \in \{j_1, j_2, ..., j_m\}$. With a variable substitution $x'_l = x_l + \mathrm{S}_l\mathrm{P}_l$, $x'_w = x_w$, $w \in \{1, 2, ..., n\} \setminus \{j_1, j_2, ..., j_m\}$, the following equation is obtained:

$$\begin{cases} \dot{x}'_1 = F_1(x'_1, x'_2, ..., x'_n), \\ \dot{x}'_2 = F_2(x'_1, x'_2, ..., x'_n), \\ \cdots \\ \dot{x}'_n = F_n(x'_1, x'_2, ..., x'_n). \end{cases} \tag{3}$$

Equation (3) recovers the same structure as Eq. (2), indicating that periodic-function-equipped system (2) is a self-reproducing system, releasing infinitely many attractors in phase space according to the period of the newly introduced function.

For better demonstration, here we give an example with easy offset boosting. As reported in Refs. [15–17], many chaotic systems have the freedom for offset boosting. The Sprott S system is such a simple one [15] (see Table 1):

$$\begin{cases} \dot{x} = -x - 4y, \\ \dot{y} = x + z^2, \\ \dot{z} = 1 + x. \end{cases} \tag{4}$$

This system has six terms, three of which are x-terms, one is y-term, one is constant and one is quadratic term. The system is chaotic with Lyapunov exponents $(0.2006, 0, -1.2006)$ and Kaplan-Yorke dimension $D_{\mathrm{KY}} = 2.1670$. Obviously, the offset boosting of the variable y can be obtained by a constant introduction in the first dimension. To make it simple, by variable substitution (variable exchange between x and y, and then a second exchange between y and z followed with additional amplitude rescaling in the variable x (to be 1/4 x)), and we transform Eq. (4) to the following:

$$\begin{cases} \dot{x} = 4y^2 + 4 * z, \\ \dot{y} = 1 + z, \\ \dot{z} = -x - z. \end{cases} \tag{5}$$

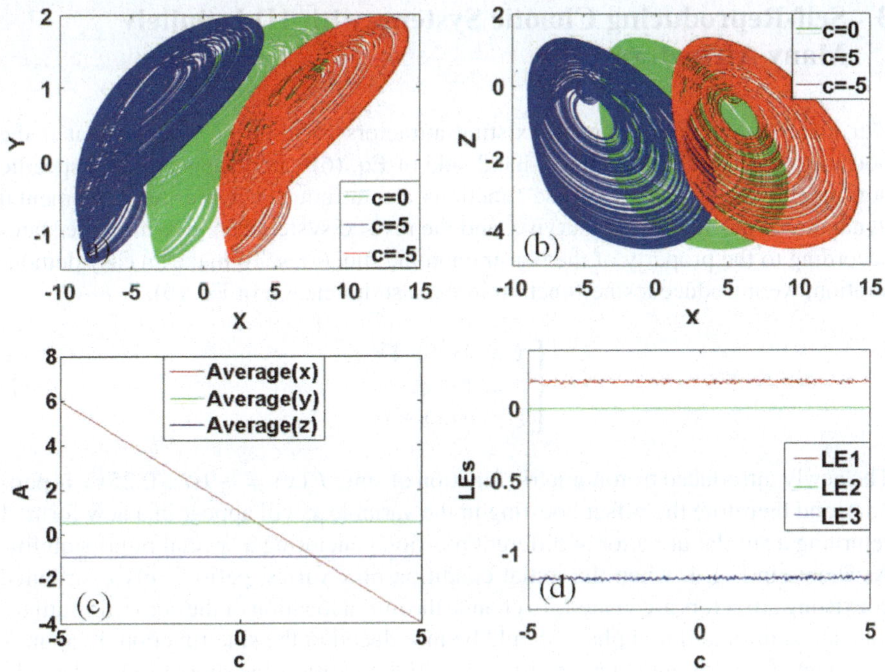

Fig. 3 Attractor boosting of system (6) without changing the Lyapunov exponents: **a** Boosted attractor in x-y plane. **b** Boosted attractor in x–z plane. **c** Average of the variables. **d** Lyapunov exponents

When x revise to be $x + c$, there will be a constant c in the last dimension, giving

$$\begin{cases} \dot{x} = 4y^2 + 4 * z, \\ \dot{y} = 1 + z, \\ \dot{z} = -x - z - c. \end{cases} \tag{6}$$

The constant c lifts the attractor along the x-axis without changing the fundamental dynamics, as shown in Fig. 3. Note that the x-coordinate of the initial condition $(4, 0.01, 0)$ should vary with the constant c for avoiding the unbounded solution. We can imagine that the last step to obtain a self-reproducing system from Eq. (6) could be resort to a period function substitution. In the following, we will give further demonstration on the process of attractor hatching.

3 Self-Reproducing Chaotic Systems with 1D Infinitely Many Attractors

Our idea for obtaining many coexisting attractors is based on the fact that if the boosting controller c in the right-hand side of Eq. (6) can disappear when specific non-monotonic or other periodic functions are introduced while the fundamental dynamics of the system can survive, and the revised system can give multiple states according to the property of the non-monotonic functions. To make an easy demonstration, we introduce a sine function in the last dimension of Eq. (6),

$$\begin{cases} \dot{x} = 4y^2 + 4 * z, \\ \dot{y} = 1 + z, \\ \dot{z} = F(x) - c. \end{cases} \tag{7}$$

The newly introduced trigonometric function of sine, $F(x) = -9sin(0.25x)$, is periodic, and therefore the offset boosting in the variable x will appear in a new form of returning a similar attractor at different position generating a special multi-stability. As shown in Fig. 4, when the initial condition of x varies, periodically-distributed coexisting attractors are visited. To change the initial location of the coexisting attractors, an additional initial phase should be introduced in the sine function. Lyapunov exponents for system (7) under different initial conditions as shown in Fig. 5a indicate that all the attractors have the same Lyapunov exponents (0.1992, 0, −1.1992) and the corresponding Kaplan-Yorke dimension D_{KY} is 2.1662 [(slightly revised compared with Eq. (6], which are different from those for Eq. 1. When the initial condition varies as $(0 + 8k\pi, 0.01, 0(-50 \leqslant k \in Z \leqslant 50))$, the average variable of the variable x is revised accordingly as shown in Fig. 5, without any change in the Lyapunov exponents.

Furthermore, other trigonometric functions such as the tangent function can be applied for reproducing attractors. When $F(x) = -9tan(0.1x)$, system (7) generates

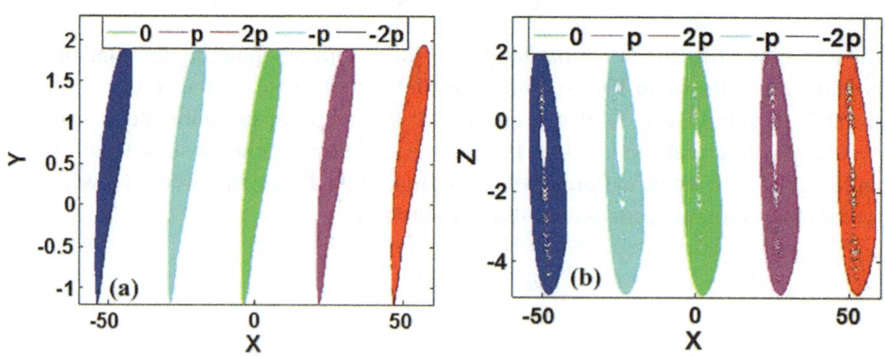

Fig. 4 Coexisting attractors of system (7) selected by the initial condition of x_0 with $y_0 = 0.01$, $z_0 = 0$, here $P = 8\pi$. **a** x–y plane. **b** x–z plane

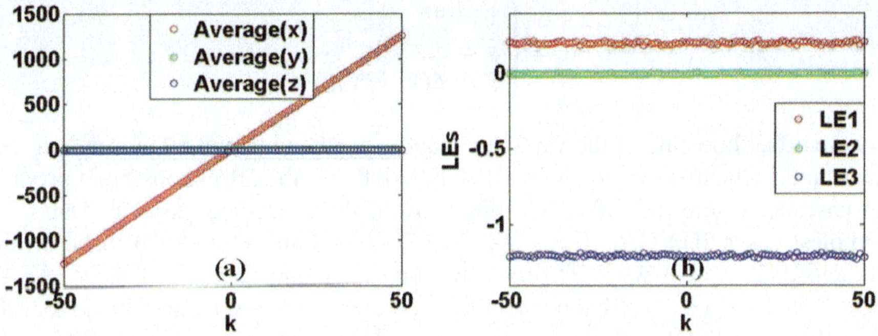

Fig. 5 Regulated offset of system (7) with $F(x) = -9\sin(0.25x)$ for initial conditions $(0{+}8k\pi, 0.01, 0)$ $(-50 \leqslant k \in Z \leqslant 50)$. **a** Regulated offset. **b** Invariant Lyapunov exponents

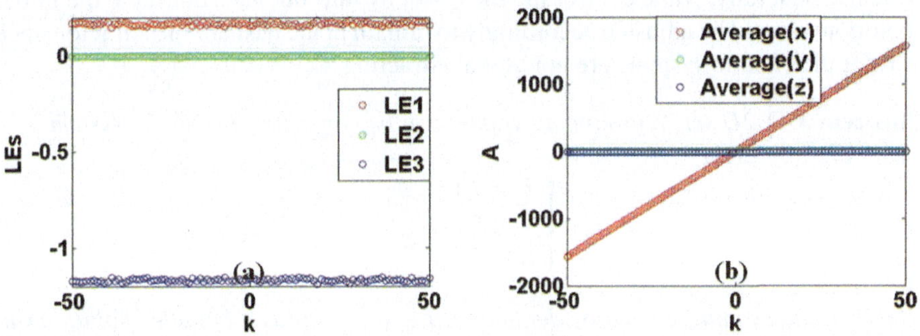

Fig. 6 Regulated offset of system (7) with $F(x) = -9\tan(0.1x)$ under initial conditions $(0{+}10k\pi, 0.01, 0)$ $(-50 \leqslant k \in Z \leqslant 50)$. **a** Regulated offset. **b** Invariant Lyapunov exponents

strange attractors with Lyapunov exponents (0.1679, 0, −1.1679) and a Kaplan-Yorke dimension of 2.1437. When the initial conditions vary in $(0 + 10k\pi, 0.01, 0)$ $(-50 \leqslant k \in Z \leqslant 50)$, the time-average of x change accordingly, while system (7) has the same strange attractor with invariant Lyapunov exponents as shown in Fig. 6. The distribution of the basins of the attraction in system (7) is determined by the chosen trigonometric function.

4 Self-Reproducing Chaotic Systems with 2D Lattices of Coexisting Attractors

If offset boosting extends to two or more dimensions, correspondingly coexisting attractors can be hatched when the offset boosting is realized by initial conditions. Interestingly, a jerk flow $\dddot{x} = (x, \dot{x}, \ddot{x})$ can be transformed to a 2D offset-boostable system [18, 19] by introducing two other variables y and z, like

$$\begin{cases} \dot{x} = y, \\ \dot{y} = z, \\ \dot{z} = f(\dot{x}, \dot{y}, x), \end{cases} \tag{8}$$

where offset boosting of the variables y and z can be obtained by introducing extra constants in the first two equations [19]. In jerk flows, the offset boosting is produced in cascade, where the offset boosting is from the preceding variable. One of the simplest cases JD0 [16], $\dddot{x} = -2.02\ddot{x} + \dot{x}^2 - x$), can be modified to be a three-dimensional system with 2D offset boosting of y and z: $\dot{x} = y + m$, $\dot{y} = z + n$, $\dot{z} = -2.02\dot{y} + \dot{x}^2 - x$. Chaotic memory oscillators [20] can be transformed for offset boosting as well. Conservative system does not have attractors but can exhibit chaos. Nose-Hoover oscillator [21, 22], $\dot{x} = y$, $\dot{y} = yz - x$, $\dot{z} = 1 - y^2$, can be written in jerk form [23] and thus admits 2D offset boosting in this type of conservative systems. Generally, while controlling the offset by introducing a constant, the initial conditions should be adjusted accordingly to remain in the basin of the attractor since most of the chaotic systems are not global attracting.

Theorem 3 *A 2D self-reproducing system can be constructed from the system*

$$\begin{cases} \dot{x} = F(y), \\ \dot{y} = G(z), \\ \dot{z} = f(\dot{x}, \dot{y}, x), \end{cases} \tag{9}$$

which produces infinitely many identical attractors if it has a bounded solution (an attractor) for one period and if the functions $F(y)$ and $G(z)$ are periodic.

[Proof 3] For the periodic functions $F(y)$ and $G(z)$, suppose that P_1 and P_2 are their respective periods, i.e., $F(y) = F(y + P_1)$ and $G(z) = G(z + P_2)$. For $x = u$, $y = v + kP_1$), $z = w + lP_2$) $(k, l \in Z)$, the system (9) becomes

$$\begin{cases} \dot{u} = F(v), \\ \dot{v} = G(w), \\ \dot{w} = f(\dot{u}, \dot{v}, u). \end{cases} \tag{10}$$

System (10) is identical to system (9), indicating that introducing the constants kP_1 and lP_2 does not change the property of system (9) but gives corresponding offset boosting in the dimensions of y and z, which consequently reproduces infinitely many attractors on a lattice in the y-z plane. The time derivatives on the left-hand side of Eq. (10) associated with offset can be easily removed.

The chaotic memory oscillator MO4 [20], $\dddot{x} + 0.5\ddot{x} + \dot{x} = x(x - 1)$, can be transformed to a self-reproducing system,

$$\begin{cases} \dot{x} = sin(y), \\ \dot{y} = 1.05sin(z), \\ \dot{z} = -\dot{x} - 0.5\dot{y} - x + x^2, \end{cases} \tag{11}$$

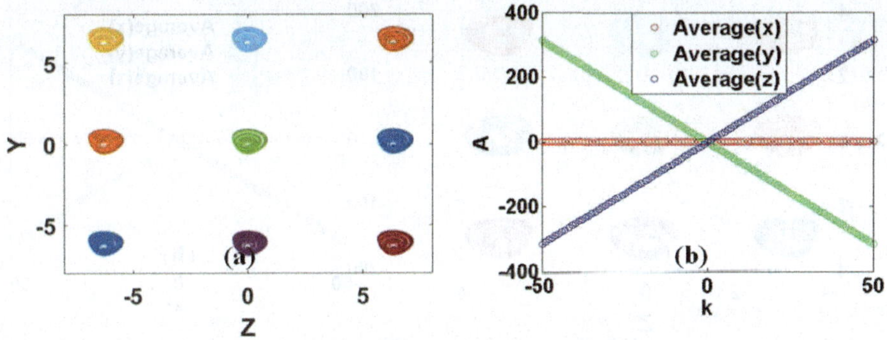

Fig. 7 Lattice of strange attractors from system (11): (a) coexisting strange attractors when initial conditions are $(0, 0.1 + 2k\pi, 0 + 2l\pi(-1 \leqslant k, l \in Z \leqslant 1))$, (b) regulated offset when initial conditions are $(0, 0.1 - 2k\pi, 0 + 2k\pi(-50 \leqslant k \in Z \leqslant 50))$

which has an infinite 2D lattice of strange attractors with Lyapunov exponents $(0.0890 \pm 0.0001, 0, -0.5808)$ and a Kaplan-Yorke dimension of 2.1534. Figure 8a shows nine of the coexisting attractors when the initial conditions are selected according to $(0, 0.1 - 2k\pi, 0 + 2l\pi(-1 \leqslant k,\ l \in Z \leqslant 1))$. Figure 8b shows that the time-averaged values of y and z on the various attractors are proportional to k for $-50 \leqslant k = l \in Z \leqslant 50$, while the average of x remains unchanged as expected.

Other trigonometric functions such as the tangent function can also be applied for offset periodization:

$$\begin{cases} \dot{x} = 1.1tan(y), \\ \dot{y} = 0.9tan(z), \\ \dot{z} = -\dot{x} - 0.5\dot{y} - x + x^2. \end{cases} \tag{12}$$

In this case, the attractors are arranged with a period of π. Comparing Fig. 8 with Fig. 7 shows that the spaces between adjacent attractors shrink, and the time-average of the variables y and z are also changed [19].

For a symmetric system, a symmetric pair of attractors can also be reproduced with the same distribution [19]. System MO5, $\dddot{x} + 0.7\ddot{x} + \dot{x} = x(1 - x^2)$, can be revised to be 2D self-reproducing system by introducing two sine functions, as

$$\begin{cases} \dot{x} = asin(y), \\ \dot{y} = bsin(z), \\ \dot{z} = -\dot{x} - 0.7\dot{y} + x - x^3. \end{cases} \tag{13}$$

System (13) preserves the dynamics of the original system. For $a = 1.55$, $b = 1.1$, system (13) has an infinite 2D lattice of symmetric strange attractor with Lyapunov exponents $(0.1122, 0, -0.8120)$ and a Kaplan-Yorke dimension of 2.1382. While when $a = 5$, $b = 1.1$, the solution turns to be an infinite 2-D lattice of symmetric pair of strange attractors with Lyapunov exponents $(0.1587, 0, -0.8795)$ and a Kaplan-

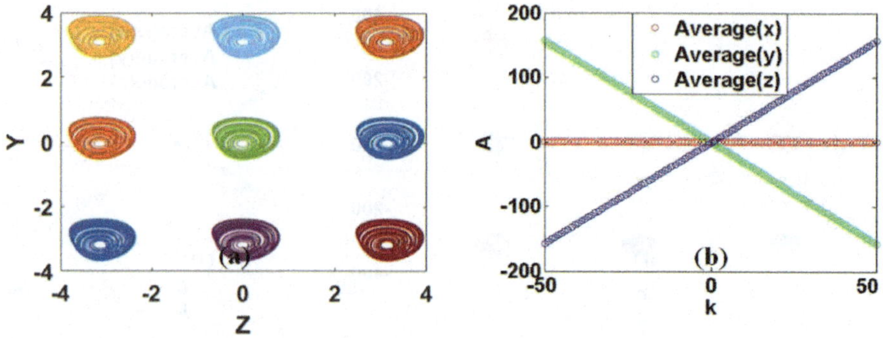

Fig. 8 Lattice of strange attractors from system (12). **a** Coexisting attractors when initial conditions are $(0, 0.1 + k\pi, 0 + l\pi(-1 \leqslant k, l \in Z \leqslant 1))$. **b** Regulated offset when initial conditions are $(0, 0.1 - k\pi, 0 + k\pi(-50 \leqslant k \in Z \leqslant 50))$

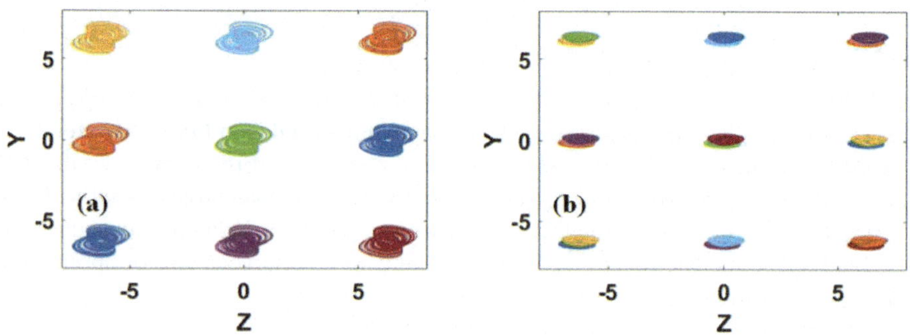

Fig. 9 Coexisting attractors of system (13) with initial conditions $(0, 0 + 2k\pi, \pm 0.1 + 2l\pi(-1 \leqslant k, l \leqslant Z \in 1)$. **a** Nine attractors when $a = 1.55$, $b = 1.1$. **b** Eighteen attractors when $a = 5$, $b = 1.1$

Yorke dimension of 2.1804. All the attractors can be picked out in a lattice using altered initial conditions as shown in Fig. 9.

5 Self-Reproducing Chaotic Systems with 3D Lattices of Coexisting Attractors

Similarly, a hyperjerk flow $\frac{d^4x}{dt^4} = (x, \dot{x}, \ddot{x}, \dddot{x})$ can produce its 3D offset-boosting [9] by introducing three other variables y, z and u, according to

$$\begin{cases} \dot{x} = y, \\ \dot{y} = z, \\ \dot{z} = u, \\ \dot{u} = f(\dot{x}, \dot{y}, \dot{z}, x). \end{cases} \tag{14}$$

Periodic functions can be introduced for coining a self-reproducing system, like

$$\begin{cases} \dot{x} = F(y), \\ \dot{y} = G(z), \\ \dot{z} = H(u), \\ \dot{u} = f(\dot{x}, \dot{y}, \dot{z}, x). \end{cases} \tag{15}$$

Therefore, infinitely many identical attractors can be hatched in the corresponding 3D space if system (15) has a bounded solution (an attractor) for one period. Here, the functions $F(y)$, $G(z)$, and $H(u)$ are periodic.

Chlouverakis and Sprott [20] found what may be the algebraically simplest hyperchaotic snap system given by $\ddddot{x} + x^4\dddot{x} + a\ddot{x} + \dot{x} + x = 0$, with a single parameter $a = 3.6$ and Lyapunov exponents (0.1310, 0.0358, 0, −1.2550) and a Kaplan-Yorke dimension of 3.1329. After two-step operation, one is offset boosting and the other is periodization. The above system obtains its new version of self-reproducing providing a 3D lattice of hyperchaotic strange attractors in a form with six-sinusoidal functions [9], as

$$\begin{cases} \dot{x} = F(y), \\ \dot{y} = G(z), \\ \dot{z} = H(u), \\ \dot{u} = -x^4\dot{z} - a\dot{y} - b\dot{x} - x. \end{cases} \tag{16}$$

When $a = 3.6$, $b = 1$, $F(y) = 2.5\sin(0.4y)$, $G(z) = 4\sin(0.25z)$, $H(u) = 8\sin(0.125u)$, System (14) exhibits an infinite 3D lattice of hyperchaotic strange attractors with Lyapunov exponents (0.1013, 0.0306, 0, −1.1340) and a Kaplan-Yorke dimension of 3.1163. Periodic initial condition takes out infinitely many hyperchaotic attractors, as shown in Fig. 10, the periods of $\sin(0.4y)$, $\sin(0.25z)$, and $\sin(0.125u)$, increase as $5k\pi$, $8k\pi$, $16k\pi$, determining the distances between two attractors of the lattice. As shown in Fig. 11, the offset of y, z and u is modulated according to initial conditions without influencing the Lyapunov exponents.

Other hyperchaotic systems can be transformed for releasing an infinite lattice of hyperchaotic attractors, such as Ref. [24]

$$\begin{cases} \dot{x} = y - x, \\ \dot{y} = -xz + u, \\ \dot{z} = xy - a, \\ \dot{u} = -by. \end{cases} \tag{17}$$

When $a = 2.6$, $b = 0.44$, system (17) has a hyperchaotic solution with Lyapunov exponents (0.0704, 0.0128, 0, −1.0832) and Kaplan-Yorke dimension $D_{KY} = 3.0768$ [24]. 1D lattice of hyperchaotic attractors can be easily obtained, since system (17) is a variable-boostable one in the u-dimension. Additional periodic functions can still reproduce the attractor in other dimensions in a bloated equation:

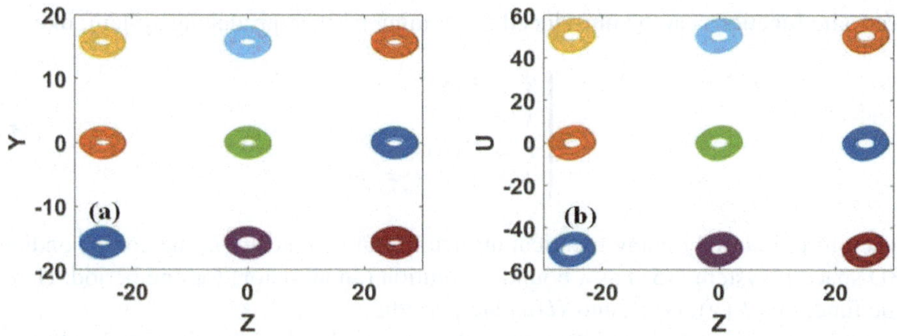

Fig. 10 3D lattice of coexisting hyperchaotic attractors in system (14). **a** z-y plane under the initial condition $(0, 5k\pi, -2 + 8k\pi, 0)$. **b** z-u plane under the initial condition $(0, 0, -2 + 8k\pi, 16k\pi)$ $(k \in \{-1, 0, 1\})$

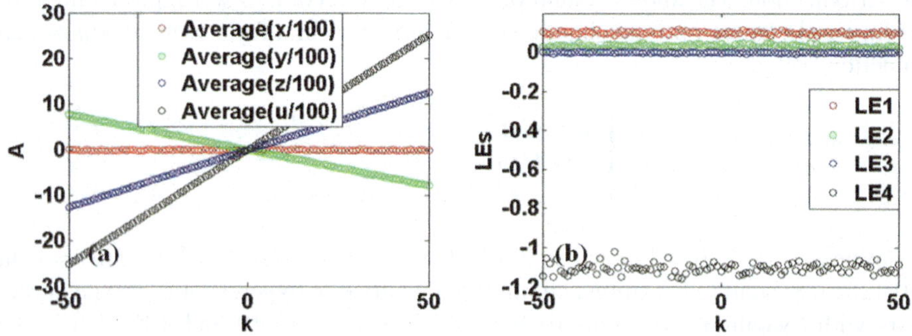

Fig. 11 Offset regulating of System (14) with invariable Lyapunov exponents. **a** Regulated offset with initial condition $(0, -5k\pi, -2 + 8k\pi, 16k\pi(-50 \leqslant k \leqslant 50))$. **b** Invariable Lyapunov exponents

$$\begin{cases} \dot{x} = g_2(y) - g_1(x), \\ \dot{y} = -g_1(x)g_3(z) + g_4(u), \\ \dot{z} = g_1(x)g_2(y) - a, \\ \dot{u} = -bg_2(y). \end{cases} \tag{18}$$

When $a = 2.6$, $b = 0.44$, $g_1(x) = x$, $g_2(y) = y$, $g_3(z) = z$, $g_4(u) = 2.5\sin(0.4u)$, system (18) gives an infinite 1D lattice of hyperchaotic strange attractors, with Lyapunov exponents $(0.0995, 0.0118, 0, -1.1113)$ and Kaplan-Yorke dimension of 3.1001. An infinite 2D lattice of hyperchaotic strange attractors can be coined by a further function substitution like $g_1(x) = x$, $g_2(y) = y$, $g_3(z) = 8\sin(0.125z)$, $g_4(u) = 2.5\sin(0.4u)$, releasing a lattice of hyperchaotic attractors, with Lyapunov exponents $(0.1097, 0.0081, 0, -1.1175)$ and Kaplan-Yorke dimension of 3.1052, as shown in Fig. 12.

More functions introduced may destroy the basic property for giving hyperchaos. When $a = 2.6$, $b = 0.44$, let $g_1(x) = 8\sin(\frac{x}{8})$, $g_2(y) = 8\sin(\frac{y}{8})$, $g_3(z) =$

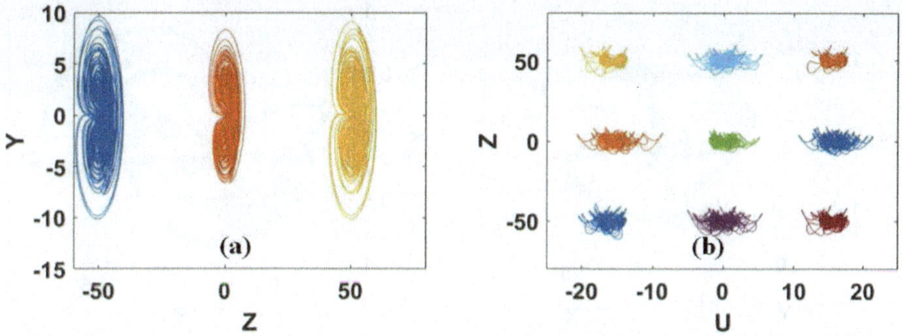

Fig. 12 2D lattice of hyperchaotic attractors from system (18). **a** z–y plane when the initial conditions are $(2, 4, 16k\pi, 0)$. **b** u–z plane when the initial conditions are $(2, 4, 16k\pi, 5k\pi)$ $(k \in \{-1, 0, 1\})$

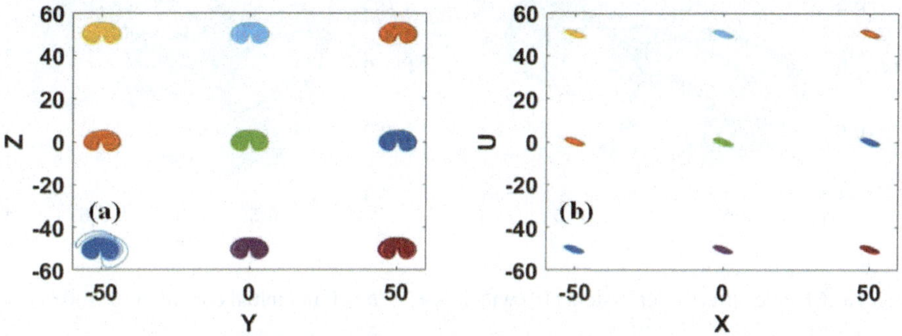

Fig. 13 4D Lattice of coexisting chaotic attractors from System (18). **a** y–z plane when the initial conditions are $(2 + 16k_1\pi, 4 + 16k_2\pi, 0, 0)$. **b** x–u plane when the initial conditions are $(2, 4, 16k_1\pi, 16k_2\pi)$ $(k_1, k_2 \in \{-1, 0, 1\})$

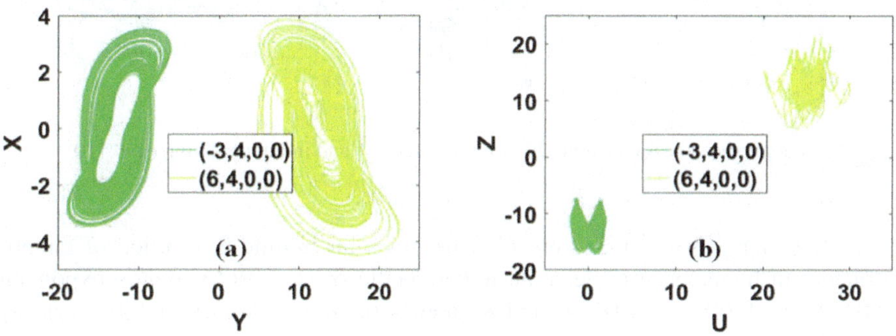

Fig. 14 Conditional symmetric coexisting attractors from system (18) when $g_2(y)$ and $g_3(z)$ are cosine functions. **a** y–x plane. **b** u–z plane

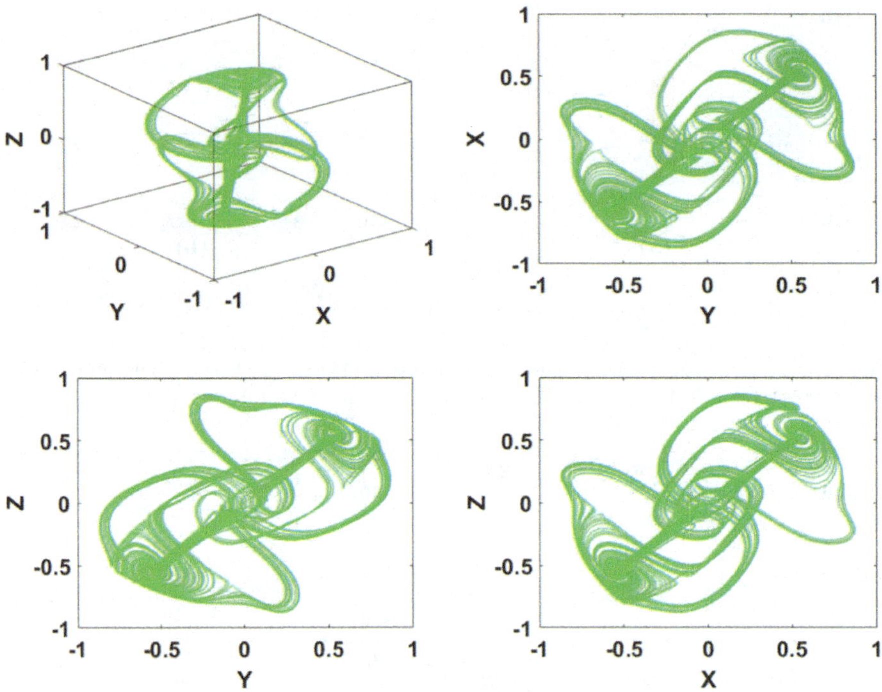

Fig. 15 Chaotic attractor for system (19) with $a = 4.75$, $b = 1$ and initial condition (0, 1, 0)

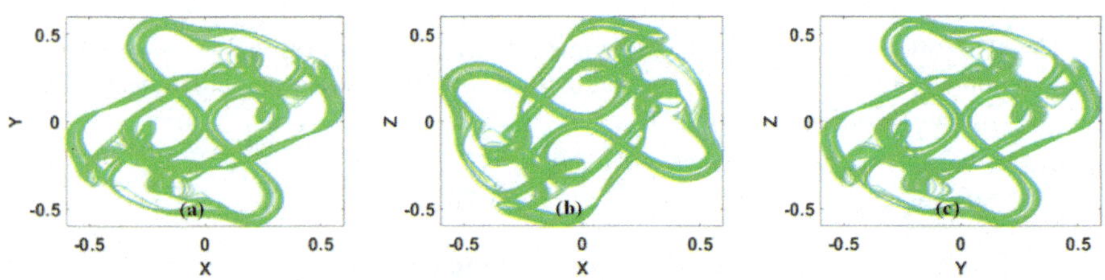

Fig. 16 Chaotic attractor for system (20) with $a = 8$, $b = 1$ and initial condition (0, 0.1, 0)

$8\sin(\frac{z}{8})$, $g_4(u) = 8\sin(\frac{u}{8})$, system (18) provides an infinite 4D lattice of chaotic attractors in hyperspace rather than hyperchaotic ones, with Lyapunov exponents (0.0462, 0, −0.0042, −1.0214) and Kaplan-Yorke dimension of 2.0450, as shown in Fig. 13.

System (18) also has no equilibria, and therefore all the coexisting attractors are hidden [25–28]. System (18) retains its rotational symmetry about the z-plane and therefore all the attractors are symmetric. Other trigonometric functions such as cosine, tangent, or cotangent functions can also be applied for attractor self-reproducing. When $g_1(x) = 8\sin(\frac{x}{8})$, $g_2(y) = 8\cos(\frac{y}{8})$, $g_3(z) = 8\cos(\frac{z}{8})$, $g_4(u) = 8\sin(\frac{u}{8})$, system (18) has a chaotic solution with Lyapunov exponents

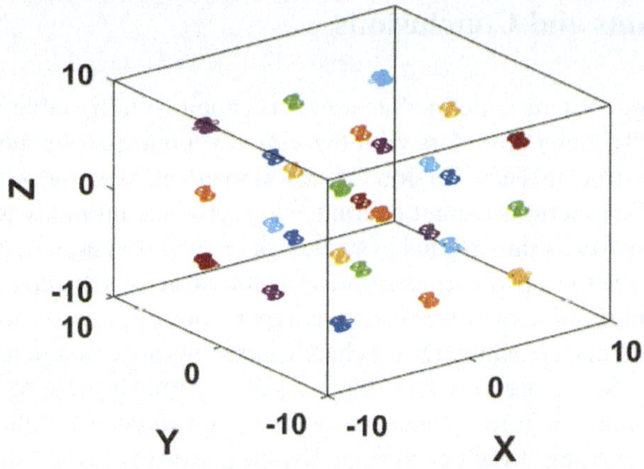

Fig. 17 Coexisting chaotic attractors for system (20) with $a = 8$, $b = 1$, under the initial condition $(0 + 2k_1\pi, 0.1 + 2k_2\pi, 0 + 2k_3\pi)$ $(k_1, k_2, k_3 \in \{-1, 0, 1\})$

(0.0462, 0, −0.0042, −1.0214) and Kaplan-Yorke dimension of 2.0450. A phase difference between the sine and cosine functions leads to a conditional symmetry, as shown in Fig. 14.

We can construct a perhaps simplest 3D self-reproducing system by introducing sine functions in the Thomas system with a cyclically symmetry, as

$$\begin{cases} \dot{x} = \sin(ay) - b\sin(x), \\ \dot{y} = \sin(az) - b\sin(y), \\ \dot{z} = \sin(ax) - b\sin(z). \end{cases} \tag{19}$$

Here, all the terms are spatially periodic. The resulting system admits multiple coexisting solutions depending on the initial conditions. When $a = 4.75$, $b = 1$, system (19) has a self-excited chaotic attractor near the origin, with Lyapunov exponents (0.2822, 0, −2.9562) and Kaplan-Yorke dimension of 2.0955 [9]. The strange attractor is shown in Fig. 15. The projections onto different orthogonal planes have the same shape, since system (19) has cyclical symmetry.

Other trigonometric functions such as tangent can also be introduced for attractor reproducing, e.g.,

$$\begin{cases} \dot{x} = \sin(ay) - b\tan(x), \\ \dot{y} = \sin(az) - b\tan(y), \\ \dot{z} = \sin(ax) - b\tan(z). \end{cases} \tag{20}$$

When $a = 8$, $b = 1$, system (20) has a chaotic attractor with Lyapunov exponents (0.2007, 0, −3.4812) and Kaplan-Yorke dimension of 2.0577, as shown in Fig. 16. When initial conditions vary, like $(0 + 2k_1\pi,\ 0.1 + 2k_2\pi,\ 0 + 2k_3\pi)$ $(k_1, k_2, k_3 \in \{-1, 0, 1\})$, a total of 27 coexisting attractors can be seen in Fig. 17.

6 Discussions and Conclusions

Self-reproducing system is defined as a new regime of multi-stable system, which releases infinitely many attractors with the same Lyapunov exponents. An effective approach to construct this new version is to resort to offset boosting. Although bloated equip of periodic functions cannot guarantee to reproduce infinitely many attractors of the same property as the original system, jerk or snap (hyperjerk) system brings a convenient chance to imbed periodic functions for attractor self-reproducing. In this chapter, an infinite lattice of 1D–3D chaotic/hyperchaotic attractors is obtained. Self-reproducing systems releasing 1D–4D chaotic attractors are constructed. Conditional symmetry can also be preserved in this kind of dynamical systems if the polarity balance is maintained from a function rather than a direct variable. The distance between two coexisting attractors is defined by the period of the reproducing function. When the distance is small enough, the boundary between two coexisting attractors becomes more fragile, which leads to the attractor number growth.

Acknowledgements This work was supported financially by the National Natural Science Foundation of China (Grant No. 61871230), the Natural Science Foundation of Jiangsu Province (Grant No. BK20181410), the Startup Foundation for Introducing Talent of NUIST (Grant No. 2016205), and a Project Funded by the Priority Academic Program Development of Jiangsu Higher Education Institutions.

References

1. J.C. Sprott, Simplest chaotic flows with involutional symmetries. Int. J. Bifurc. Chaos **24**(01), 1450009 (2014)
2. J.C. Sprott, New chaotic regimes in the Lorenz and Chen systems. Int. J. Bifurc. Chaos **25**(02), 1550033 (2015)
3. Q. Lai, S. Chen, Generating multiple chaotic attractors from Sprott B system. Int. J. Bifurc. Chaos **26**(11), 1650177 (2016)
4. Q. Lai, T. Nestor, J. Kengne, X.W. Zhao, Coexisting attractors and circuit implementation of a new 4d chaotic system with two equilibria. Chaos Solitons Fractals **107**, 92–102 (2018)
5. J. Sprott, A dynamical system with a strange attractor and invariant tori. Phys. Lett. A **378**(20), 1361–1363 (2014)
6. C. Li, Y. Xu, G. Chen, Y. Liu, J. Zheng, Conditional symmetry: Bond for attractor growing. Nonlinear Dyn. **95**(2), 1245–1256 (2019)
7. C. Li, J.C. Sprott, H. Xing, Constructing chaotic systems with conditional symmetry. Nonlinear Dyn. **87**(2), 1351–1358 (2017)
8. C. Li, J.C. Sprott, H. Xing, Hypogenetic chaotic jerk flows. Phys. Lett. A **380**(11–12), 1172–1177 (2016)
9. C. Li, J.C. Sprott, An infinite 3-d quasiperiodic lattice of chaotic attractors. Phys. Lett. A **382**(8), 581–587 (2018)
10. C. Li, J.C. Sprott, How to bridge attractors and repellors. Int. J. Bifurc. Chaos **27**(10), 1750149 (2017)
11. R. Barrio, F. Blesa, S. Serrano, Qualitative analysis of the Rössler equations: Bifurcations of limit cycles and chaotic attractors. Physica D: Nonlinear Phenom. **238**(13), 1087–1100 (2009)
12. J.C. Sprott, C. Li, Asymmetric bistability in the Rössler system. ACTA Physica Polonica B **48**(1), 97–107 (2017)

13. C. Li, J.C. Sprott, Y. Liu, Z. Gu, J. Zhang, Offset boosting for breeding conditional symmetry. Int. J. Bifurc. Chaos **28**(14), 1850163 (2018)
14. C. Li, J.C. Thio, J.C. Sprott, H.H.C. Iu, Y. Xu, Constructing infinitely many attractors in a programmable chaotic circuit. IEEE Access **99**, 1–1 (2018)
15. J.C. Sprott, Some simple chaotic flows. Phys. Rev. E **50**(2), R647 (1994)
16. J.C. Sprott, Some simple chaotic jerk functions. Am. J. Phys. **65**(6), 537 (1997)
17. J.C. Sprott, Strange attractors with various equilibrium types. Eur. Phys. J.: Special Topics **224**(8), 1409–1419 (2015)
18. C. Li, J.C. Sprott, Variable-boostable chaotic flows. Optik: Int. J. Light Electron Opt. **127**(22), 10389–10398 (2016)
19. C. Li, J.C. Sprott, W. Hu, Y. Xu, Infinite multistability in a self-reproducing chaotic system. Int. J. Bifurc. Chaos **27**(10), 1750160 (2017)
20. J.C. Sprott, *Elegant Chaos : Algebraically Simple Chaotic Flows* (World Scientific, Singapore, 2010)
21. N. Shuichi, Constant temperature molecular dynamics methods. Prog. Theor. Phys. Suppl. **103**(103), 1–46 (1991)
22. W.G. Hoover, Remark on 'some simple chaotic flows'. Phys. Rev. E **51**(1), 759 (1995)
23. H.P.W. Gottlieb, Question 38. What is the simplest jerk function that gives chaos? Am. J. Phys. **64**(5), 525–525 (1996)
24. C. Li, J.C. Sprott, Coexisting hidden attractors in a 4-d simplified Lorenz system. Int. J. Bifurc. Chaos **24**(3), 1450034 (2014)
25. G. Leonov, N. Kuznetsov, V. Vagaitsev, Localization of hidden Chua's attractors. Phys. Lett. A **375**, 2230–2233 (2011)
26. G. Leonov, N. Kuznetsov, V. Vagaitsev, Hidden attractor in smooth Chua systems. Physica D: Nonlinear Phenom. **241**(18), 1482–1486 (2012)
27. Z. Wei, W. Zhang, Hidden hyperchaotic attractors in a modified lorenz-stenflo system with only one stable equilibrium. Int. J. Bifurc. Chaos **24**(10), 1450127 (2014)
28. Z. Wei, I. Moroz, J. Sprott, A. Akgul, W. Zhang, Hidden hyperchaos and electronic circuit application in a 5d self-exciting homopolar disc dynamo. Chaos **27**(3), 033101 (2017)

Multi-Stability Detection in Chaotic Systems

Chunbiao Li and Julien Clinton Sprott

1 Introduction

Multi-stability is ubiquitous in dynamical systems, thereby attracting a great deal of interest for its potential threats or benefits to engineering systems and applications. A multi-stable system has much more complicated dynamics for its many-to-many mapping between system parameters and initial conditions. To date, identifying the basins of attraction depending on the ergodic initial conditions [1–10] has been applied widely for identifying multi-stability. For us, it is an urgent task to determine which kind of dynamical systems have coexisting attractors. However, there is a long way to go to find multi-stability theoretically, or practically but more efficiently.

From the existence analysis of multi-stability, we can get some clues for predicting the multi-stability. In fact, the complexity of multi-stability analysis can be reduced since an n-dimensional system has $(m\text{-}(n+1))$ independent parameters (m refers to the number of the terms in the equation) and any of which can provide an observation window for multi-stability. On the other hand, we have already seen that multi-stability can lead to the failure of amplitude control [11] and offset boosting, when the initial conditions are fixed. Therefore, in turn, the parameter for amplitude control or offset boosting provides a new method for detecting multi-stability, where the amplitude control or offset boosting with a fixed initial condition may involve in different solutions betraying the multi-stability with hops of Lyapunov exponents [12, 13].

As is well known, different kinds of attractors, including stable equilibria, periodicity, quasi-periodicity, chaos, and hyperchaos, have different largest-two Lyapunov

C. Li (✉)
Nanjing University of Information Science & Technology, Nanjing, China
e-mail: goontry@126.com; chunbiaolee@nuist.edu.cn

J. C. Sprott
University of Wisconsin–Madison, Madison, WI, USA
e-mail: sprott@physics.wisc.edu

X. Wang et al. (eds.), *Chaotic Systems with Multistability and Hidden Attractors*,
Emergence, Complexity and Computation 40,
https://doi.org/10.1007/978-3-030-75821-9_16

Fig. 1 Dynamic properties indicated by the second largest Lyapunov exponent. H represents hyperchaos, C represents chaos, T represents torus, LC represents limit cycle, SE represents stable equilibrium point

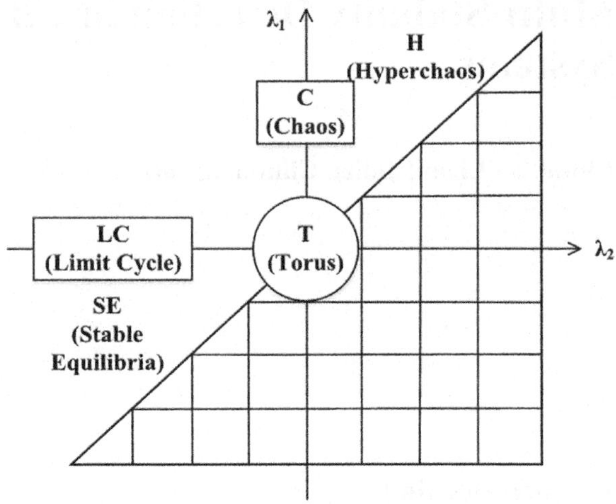

exponents. Therefore, a powerful method for identifying and quantifying the dynamic is to use the spectrum of Lyapunov exponents, whose number is equal to the number of dynamical variables and that are usually ordered from the largest (most positive) to the smallest (most negative) [14, 15].

Considering a two-dimensional space of the second largest Lyapunov exponent, λ_1 and λ_2 ($\lambda_1 \geqslant \lambda_2$), as shown in Fig. 1. From definition, the stable equilibria (SE) lie on or below the 45-degree line in the lower left quadrant, limit cycles (LC) locate along the negative λ_2-axis, toruses (T) locate at the origin, chaotic attractors (C) locate along the positive λ_1-axis, and hyperchaotic attractors (H) are in the upper right quadrant. For a dynamical system with different kinds of coexisting attractors, a scatter plot in the plane for different initial conditions will show clusters of points corresponding to the different coexisting attractors and will identify their types. Similarly, for a given initial condition, different settings of the amplitude control and offset boosting will generally cause the system to visit the basins of most if not all of the attractors, especially if the basin boundaries are fractal, which is common in nonlinear dynamical systems that are multi-stable. However, for some dynamical systems, the attracting basin is simple or symmetrical about some axis or original point, and so an appropriate initial condition must be chosen to increase the likelihood of visiting all the basins as the amplitude or offset is adjusted.

The above-mentioned alternative methods resort to non-bifurcation operations, where the basins of attractions are linearly transformed so as to generate a dynamical dispersion against the fixed initial condition. As shown in Fig. 2, amplitude control and offset boosting provide two basic linear transformations. Amplitude control corresponds to the rescaling of the basins of attraction, while the offset boosting shifts the basins of attraction in any direction. Therefore, the multiple coexisting attractors can be fished out by a fixed initial condition since the initial data floats into different basins of attraction under a linear transformation [12, 13].

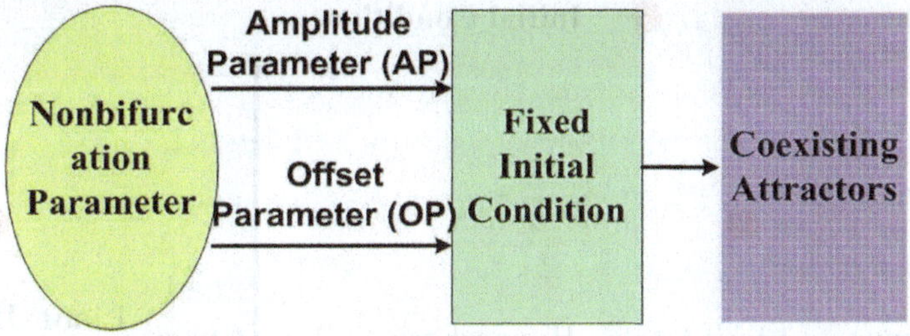

Fig. 2 Outputting coexisting attractors by non-bifurcation parameters

2 Multistability Identification by Amplitude Control

Amplitude control [16–20] makes the attractor larger or smaller by changing the scale of some or all of the variables without changing the dynamical and topological properties of the attractor.

Suppose that there is a p-dimensional multi-stable system with an amplitude parameter, $\dot{X} = F(a, X)$, where a is an amplitude controller. When $a = a_1$, the basins for coexisting strange attractor, torus, limit cycle and point attractor are Ω_{11}, Ω_{12}, Ω_{13}, Ω_{14}, while when $a = a_2$ these basins become Ω_{21}, Ω_{22}, Ω_{23}, Ω_{24}. Let $D_{ij} = \Omega_{1i} \cap \Omega_{2j}$ $(1 \le i \le 4, 1 \le j \le 4, i \ne j)$. For example, if the initial condition is set as $X_0 = (x_{10}, x_{20}, ..., x_{p0}) \in D_{ij}$, when the amplitude parameter a increases from a_1 to a_2, the second largest Lyapunov exponent can switch from positive to negative if a basin boundary is crossed between a hyperchaotic strange attractor and a limit cycle. In fact, if the amplitude controller rescales the basins of attraction without revising the corresponding initial conditions, different dynamical behaviors may occur for different settings of the amplitude control, which consists of a new method for identifying coexisting attractors. The above mechanism hidden in the process is shown in Fig. 3.

The method has been proved valid for checking the coexistence of the point attractor, limit cycle and strange attractor [21]. Here, we check the effectiveness of multi-stability diagnosis based on amplitude control in the revised Rössler system [22], which has a partial amplitude controller h, as

$$\begin{cases} \dot{x} = -y - yz, \\ \dot{y} = x + ay, \\ \dot{z} = b + z(hx^2 - c). \end{cases} \tag{1}$$

Here, the coefficient h controls the amplitudes of the variables x and y according to $1/\sqrt{h}$, while the amplitude of z is unchanged. When $a = 0.39$, $b = 2.23$, $c = 4.71$, the Lyapunov exponent spectrum and its distribution for initial condition (1, 0, 2) is shown in Fig. 4, showing the coexistence of a strange attractor and a limit cycle. However, a similar distribution of Lyapunov exponent spectrum is obtained when $a = 0.34$, $b = 1.7$, $c = 5.03$. Since the coexisting attractors are asymmetric while

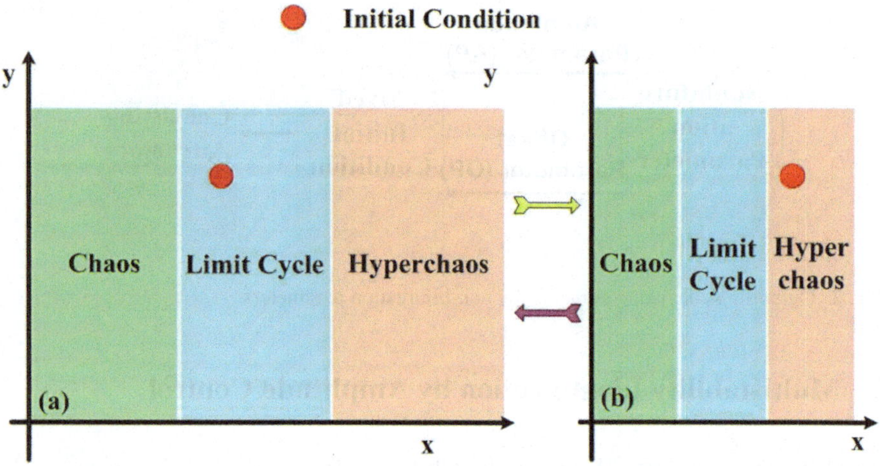

Fig. 3 Shrinkage or expansion of the basin of attraction under amplitude control

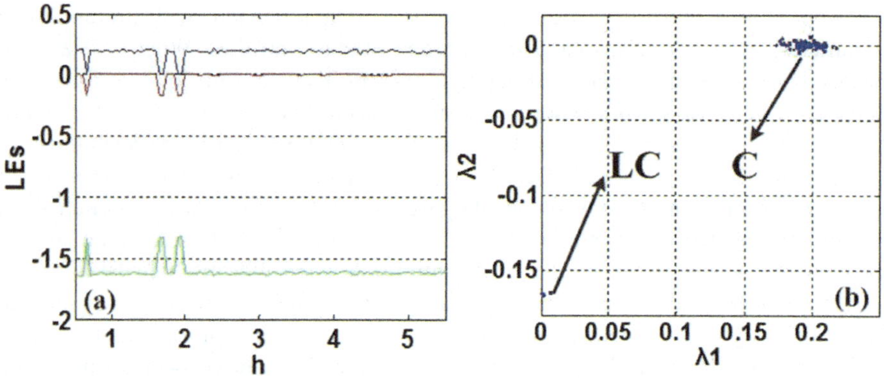

Fig. 4 Lyapunov exponent spectrum for system (1) with $a = 0.39$, $b = 2.23$, $c = 4.71$, and its distribution for initial condition (1, 0, 2)

the system is symmetric, we can easily conclude that the system has four coexisting attractors, including a symmetric pair of limit cycles and symmetric pairs of strange attractors, as shown in Fig. 5. Amplitude control with fixed initial conditions provides a tool for identifying coexisting attractors in a dynamical system that may be more convenient in practical applications than exploring all possible initial conditions. Note that the Lyapunov exponent calculation typically converges more slowly than other measures such as the location and size of an attractor, especially the second largest Lyapunov exponent, which may be one of the disadvantages of identifying multi-stability with amplitude control.

Amplitude rescaling for multi-stability still works for higher-dimensional dynamical systems if one of the Lyapunov exponents is used for stability identification, for

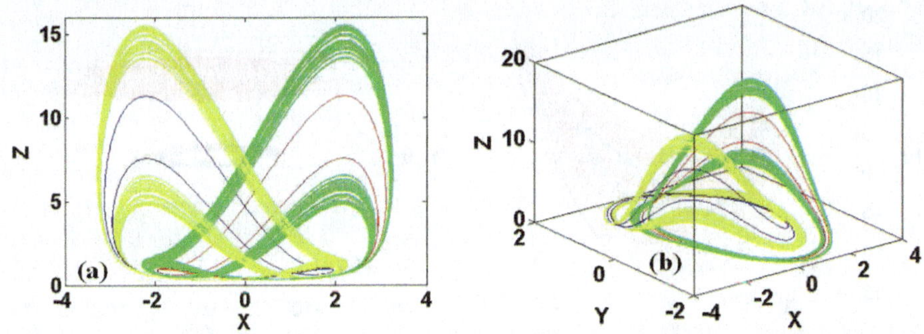

Fig. 5 Coexisting attractors of system (1) with $a = 0.34$, $b = 1.7$, $c = 5.03$, $h = 1$, where initial conditions $(\pm 0.4, \pm 1, 6.6)$ are shown in green and red, and $(\pm 2.8, 0, 2.8)$ shown in green and yellow, respectively. **a** x–z plane. **b** x–y–z space

example, in the following hyperchaotic system [23]:

$$\begin{cases} \dot{x} = y - x, \\ \dot{y} = -xz + u, \\ \dot{z} = cxy - a, \\ \dot{u} = -by. \end{cases} \tag{2}$$

When $a = 6$, $b = 0.1$, system (2) has solutions of a torus and a symmetric pair of coexisting chaotic attractors, as shown in Fig. 6, which can be identified by Lyapunov exponent spectrum and its distribution under a fixed initial condition like the case above. Here, we recognized that the symmetric coexistence is not captured until the special structure is considered. What is worse is that the method of rescaling the basins of attraction by amplitude control does not guarantee to fish up all the coexisting solutions. An obvious reason is that if the basin of attraction for a specific attractor extends in one direction or large enough, then the basin in that direction cannot be rescaled since $\pm\infty$ over a real parameter remains to be so. As a result, the relative motion of initial condition along that direction will not disclose any change of the basin, therefore cannot catch up the corresponding coexisting attractor. As an effective alternative method, the following approach based on offset boosting can be applied effectively to detect the coexisting cases, for which the amplitude control cannot especially find those coexisting symmetric attractors.

3 Multi-Stability Identification by Offset Boosting

Offset boosting [24–27] can shift the basin along any direction, which provides an effective way for fishing multiple attractors. Variable-boostable systems studied in Ref. [24] provide a special regime where some of the variables can get offset boosted

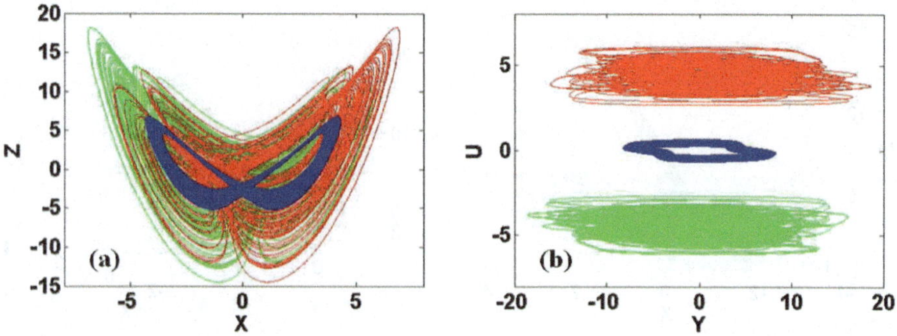

Fig. 6 Coexisting torus and a symmetric pair of chaotic attractors of system (2) with $a = 6$, $b = 0.1$: **a** x–z plane. **b** y–u plane

with a single independent constant. More generally, the offset boosting is from a variable substitution, which needs newly introduced constants, depending on how many times the variable shows up in the differential equation. But this concern is superfluous; the following practical tests show that offset boosting can catch up the coexisting attractors from even a single direction. Offset boosting does give a dynamical dispersion useful to see multiple solutions since a fixed initial condition visits different basins of attraction while the offset gets boosted.

Without loss of generality, consider the case where constants a, b are scalars. Suppose that the system has five coexisting attractors, i.e., a point attractor, two limit cycles A, B, a chaotic attractor, and a hyperchaotic attractor. If the shift constant a, b are positive, the basins of attraction will move along the negative direction, and vice versa. Here, the offset boosting of dimension x or y, combined with the whole basin of attraction, will be controlled according to the constant a or b. As shown in Fig. 7, the fixed initial condition A (red point) for limit cycle A will visit the basin of chaos when the offset booster a is negative through relative movement, while the initial condition B (purple point) still stay in the basin of point until the offset of y is shifted by the positive booster b and thereafter triggers another oscillation of limit cycle B. Here, the offset boosting in any direction reaches such an extent that the fixed initial condition can cross the boundary and visit different basins of attraction. By this relative movement, the fixed initial condition A or B can fish out different possible attractors and diagnose them thereafter. Different kinds of attractors correspond to their specific Lyapunov exponents, and the fixed-initial-condition offset boosting modifies the Lyapunov exponents accordingly, which shows various combinations of multi-stability, as shown in Fig. 1. The regime of coexistence with multiple attractors can be indicated from the distribution of the two largest Lyapunov exponents [13]. Moreover, since the offset boosting in any dimension revises the corresponding average of the variable, any newly introduced attractor can destroy the linearly scaled offset, betraying new coexisting attractors.

We can check the coexisting attractors following this routine based on offset boosting. In system (1), there is a quadratic term and another cubic term of z, a cubic

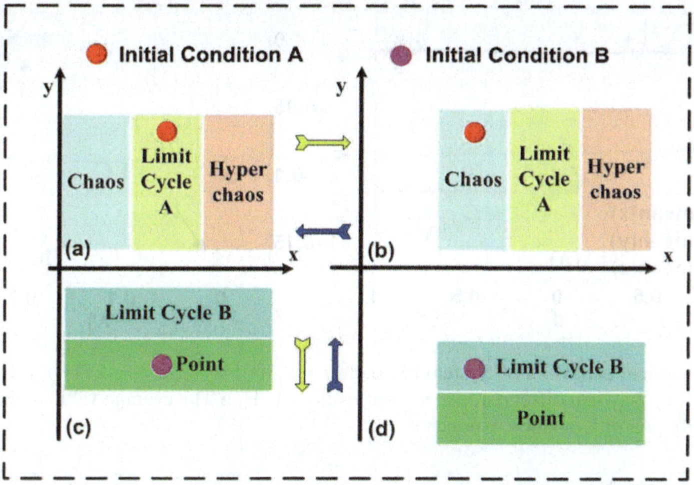

Fig. 7 Fixed initial condition visits different basins of attraction when different dimensions of offset boosting are executed. (**a** and **b**) for x-dimension. (**c** and **d**) for y-dimension

term of x; therefore, for simple offset boosting we select the dimension of y for identifying coexisting attractors. The newly introduced constant is in the following:

$$\begin{cases} \dot{x} = -(y+d) - (y+d)z, \\ \dot{y} = x + a(y+d), \\ \dot{z} = b + z(hx^2 - c). \end{cases} \tag{3}$$

As shown in Fig. 8, when the offset parameter d varies in $[-1, 1]$, the average value of variable y decreases accordingly, while the average values of variables x and z remain unchanged simultaneously, except some unexpected changes indicating multiple coexisting attractors. The offset boosting identifies the coexisting attractors effectively, giving two groups of the corresponding Lyapunov exponents like the distribution shown in Fig. 4b. Two-dimensional offset may provide a more effective diagnose despite that the equation has a heavy non-bifurcation parameter setting since we do not know how the basins of attraction is extending in any dimension. The scissors of offset boosting show the coexisting attractors with the same Lyapunov exponent distribution, as shown in Fig. 9.

In this case, the system is

$$\begin{cases} \dot{x} = -(y+d) - (y+d)(z-d), \\ \dot{y} = x + a(y+d), \\ \dot{z} = b + (z-d)(hx^2 - c). \end{cases} \tag{4}$$

Offset boosting can also be applied to the higher-dimensional systems. When $a = 6,\ b = 0.1$, system (2) has a symmetrical pair of strange attractors and a torus;

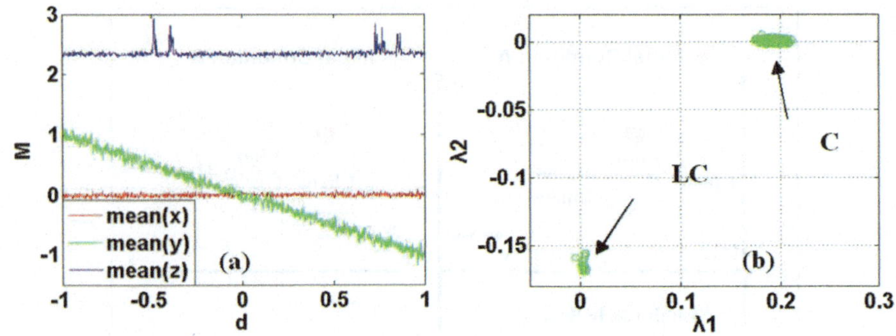

Fig. 8 Dynamical behaviors of System (3) with $a = 0.39$, $b = 2.23$, $c = 4.71$, $h = 1$, and initial condition (1, 0, 2), when offset parameter d varies in $[-1, 1]$. **a** The average value of the variables. **b** The distribution of Lyapunov exponents

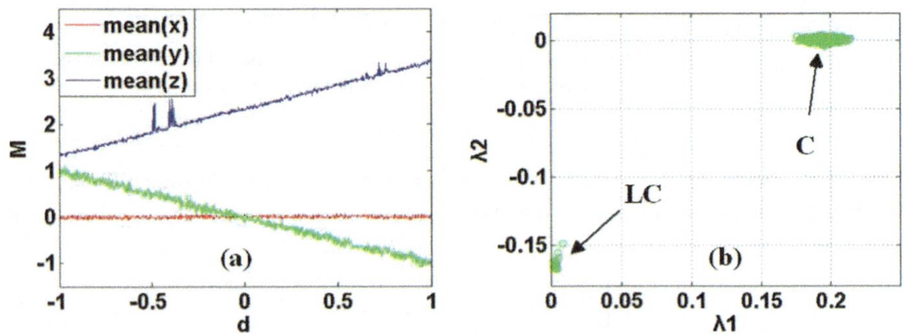

Fig. 9 Dynamical behaviors of System (3) with $a = 0.39$, $b = 2.23$, $c = 4.71$, $h = 1$, under two-dimensional offset boosting with initial condition (1, 0, 2), and d varies in $[-1, 1]$. **a** The average value of the variables. **b** The distribution of Lyapunov exponents

when $a = 7$, $b = 0.1$, system (2) has a symmetrical pair of strange attractors and a limit cycle. Without calculating the Lyapunov exponents, three coexisting attractors are caught up from the abnormal average value collapse. Here, the offset boosting process can be applied to three dimensions, y, z, u, as follows:

$$\begin{cases} \dot{x} = (y + d) - x, \\ \dot{y} = -x(z - d) + (u + d), \\ \dot{z} = x(y + d) - a, \\ \dot{u} = -b(y + d). \end{cases} \tag{5}$$

As shown in Fig. 10, for a system with two symmetric attractors, even their corresponding Lyapunov exponents are identical, the proposed method works very well.

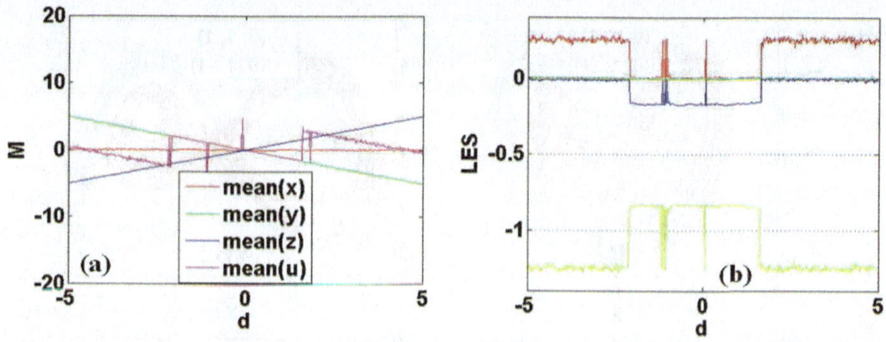

Fig. 10 Dynamical behaviors of System (5) with $a = 6$, $b = 0.1$, under the initial condition $(1, -1, 1, 0)$. **a** The average value of the variables. **b** Lyapunov exponents

4 Independent Amplitude Controller and Offset Booster

4.1 Constructing Independent Amplitude Controller

In order to realize the prediction of multi-stability with a direct amplitude controller, we need to find or even introduce a single knob into the dynamic system without influencing the fundamental dynamics of the original system. It happens that symmetric systems usually have multi-stability and simultaneously have one amplitude parameter for partial control or total control [16]. Besides, we can even obtain a total amplitude controller in a system by degree modification [28]. Our degree modification depends on the application of the signum function and the absolute function. This is based on the fact that many dynamical systems can preserve their fundamental solutions when the polarity of some of the variables is kept while the amplitude is removed, or some of the terms get an amplitude amplification while keeping its original polarity. A cute example is shown in Ref. [28] with the Sprott B system (Table 1), where six cases are obtained to unify the original system to be of the desired degree except one term. We know now that this transformation will preserve the original equilibria if ignoring the coordinate shifting. Under certain circumstances, one infinite line of equilibria or even two perpendicular lines of equilibria may be introduced, which will inevitably revise the basin of attraction when the multi-stability exists.

Take the revised Sprott B systems for examples:

$$\begin{cases} \dot{x} = z\ \mathrm{sgn}(y), \\ \dot{y} = x - y, \\ \dot{z} = a|x| - m\,xy\,, \end{cases} \tag{6}$$

and

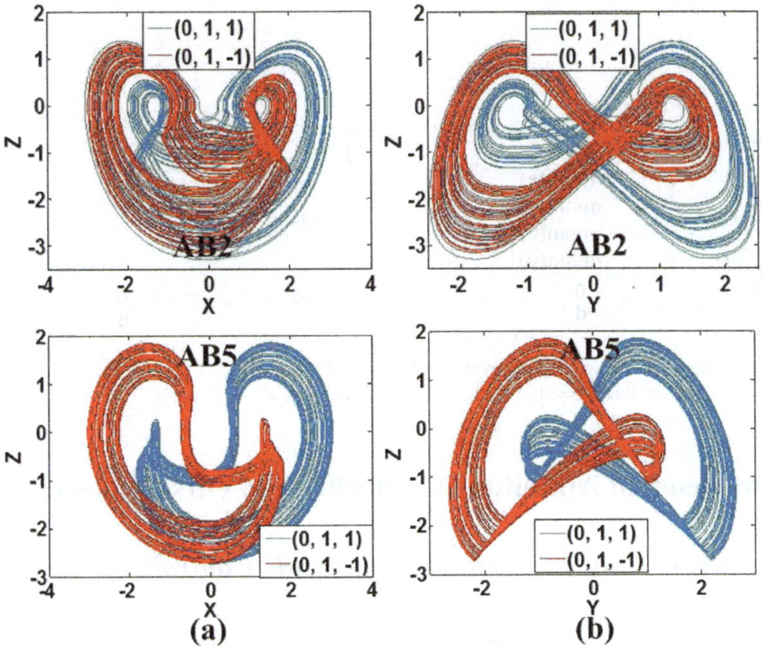

Fig. 11 Coexisting attractors in systems (6) and (7), with $a = 1.2$, $m = 1$. **a** x–z plane. **b** y–z plane

$$\begin{cases} \dot{x} = yz, \\ \dot{y} = x|x| - y|x|, \\ \dot{z} = m|x| - axy. \end{cases} \tag{7}$$

The parameter m is an amplitude parameter, while the only left parameter a is for bifurcation control. The system (6) (AB2) has a newly introduced line of equilibria, $(0, 0, z)$, with eigenvalue $(0, 0, -1)$, while the system (7) (AB5) has two perpendicular lines of equilibria, $(0, y, 0)$ and $(0, 0, z)$, with eigenvalues $(\sqrt{a} * |y|, 0, \sqrt{a} * |y|)$ and $(0, 0, 0)$, respectively. When $m = 1$ and $a = 1.2$, both systems have a symmetric pair of coexisting strange attractors, as shown in Fig. 11. However, the basins of attraction are rearranged, as shown in Fig. 12. Note that although the degree modification can preserve the basic properties of the original system and strip out an amplitude parameter, this does not mean that the original system gives the multi-stability under the same parameter. In fact, the original Sprott B system has the same coexisting solutions at the parameter near $a = 0.32$ [28]. From the above analysis, we may conclude that generally a direct variable substitution with the dynamical system, such as $x = ku,\ y = v,\ z = w$, the deduced parameter k will be a good detector for multi-stability, even it may exist in many terms rather than a single term.

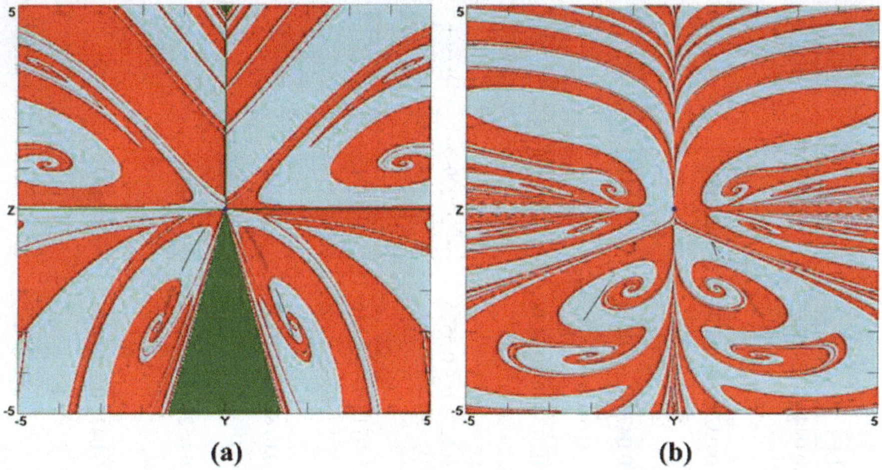

Fig. 12 Basins of attraction for the symmetric pair of strange attractors at $x = 0$ (light blue and red) with $a = 1.2$, $m = 1$. **a** System (6). **b** System (7)

4.2 *Finding Independent Offset Booster*

Typically, it is not easy to find a single independent offset booster in a chaotic system since different variables get tangled for coining chaos. But, we can still find some simple chaotic systems with a single independent offset booster by exhausting computer-based searching [24]. For such a dynamic system, if the variable x needs to be boosted by a single constant, the differential equation can be designed in the following form, where the extra constant is introduced in the z-dimension:

$$\begin{cases} \dot{x} = a_1 y + a_2 z + a_3 y^2 + a_4 z^2 + a_5 yz + a_{16}, \\ \dot{y} = a_6 y + a_7 z + a_8 y^2 + a_9 z^2 + a_{10} yz + a_{17}, \\ \dot{z} = a_{11} y + a_{12} z + a_{13} y^2 + a_{14} z^2 + a_{15} yz + a_{18} x\,. \end{cases} \tag{8}$$

Similarly, the variables y and z can be boosted by other newly introduced constants in other dimensions. However, all such cases can be written in the form of Eq. (8), without loss of generality, through a simple transformation of variables [24]. Twenty-four new cases were found [24], as shown in Table 1 any of which has a single independent offset booster c in the last dimension. The strange attractors are shown in Fig. 13. Offset booster c modifies the average value of the variable x linearly, except an extra attractor showing up, indicating multi-stability. As shown in Fig. 14, when c varies in $[-10, 10]$, the average value of the variable x is linearly scaled with occasionally hopping up and down in some other dimensions, indicating the coexistence of multiple attractors. In this case, the boosting controller c can fish out different solutions when it moves into the basin of attraction in the direction of the x-dimension. As shown in Fig. 14a, when $a = 3.6$, $c = 0$, VB5 has a symmetric

Table 1 Chaotic systems with an independent offset booster [24]

Mode1	Equations	Parameters	x_0 y_0 z_0	LEs	D_{KY}
VB1 (Sprott JD0)	$\dot{x} = y$ $\dot{y} = z$ $\dot{z} = -x + y^2 - az + c$	$a = 2.02$ $c = 0$	5 2 0	0.0448 0 -2.0648	2.0217
VB2	$\dot{x} = y + yz$ $\dot{y} = -z$ $\dot{z} = x - az + c$	$a = 2.04$ $c = 0$	1 1 −5	0.0327 0 −2.0727	2.0158
VB3	$\dot{x} = z^2 - y$ $\dot{y} = -z$ $\dot{z} = x - az + c$	$a = 2.02$ $c = 0$	1 5 1	0.0478 0 −2.0678	2.0231
VB4 (revised Sprott A)	$\dot{x} = -z$ $\dot{y} = z^2 - a$ $\dot{z} = x - yz + c$	$a = 1$ $c = 0$	0 0 5	0.0138 0 −0.0138	3
VB5	$\dot{x} = ayz$ $\dot{y} = 1 - z^2$ $\dot{z} = x - yz + c$	$a = 1$ $c = 0$	−1 1 −1	0.1271 0 −0.5526	2.2299
VB6	$\dot{x} = 1 - yz$ $\dot{y} = az^2 - yz$ $\dot{z} = x + c$	$a = 0.22$ $c = 0$	−1 1 −1	0.0717 0 −1.6646	2.0431
VB7 (Sprott J)	$\dot{x} = az$ $\dot{y} = -by + z$ $\dot{z} = -x + y + y^2 + c$	$a = 2$ $b = 2$ $c = 0$	−1 0.3 0	0.0787 0 −2.0787	2.0379
VB8 (revised Sprott L)	$\dot{x} = 1 - az$ $\dot{y} = bz^2 - y$ $\dot{z} = y + x + c$	$a = 3.9$ $b = 3.5$ $c = 0$	−1 0.01 0	0.0521 0 −1.0521	2.0495
VB9 (revised Sprott M)	$\dot{x} = a + y + bz$ $\dot{y} = -y - z^2$ $\dot{z} = -x + c$	$a = 1.7$ $b = 1.7$ $c = 0$	−1 0.3 0	0.0435 0 −1.0435	2.0417

Table 1 (continued)

Mode1	Equations	Parameters	$x_0\ y_0\ z_0$	LEs	D_{KY}
VB10 (revised Sprott N)	$\dot{x} = -az$ $\dot{y} = 1 - by + z$ $\dot{z} = x + y^2 + c$	$a = 2\ b = 2\ c = 0$	2 0 0.3	0.0746 0 −2.0746	2.0359
VB11 (revised Sprott P)	$\dot{x} = y + z$ $\dot{y} = y^2 - az$ $\dot{z} = x + by + c$	$a = 1\ b = 2.7\ c = 0$	0 0.3 0	0.0861 0 −0.4807	2.1790
VB12 (revised Sprott S)	$\dot{x} = ay^2 + bz$ $\dot{y} = 1 + z$ $\dot{z} = -x - z + c$	$a = 4\ b = 4\ c = 0$	4 0.01 0	0.1930 0 −1.1930	2.1617
VB13 (revised NE5)	$\dot{x} = az$ $\dot{y} = y^2 - z^2 + b$ $\dot{z} = -x - y + c$	$a = 4\ b = 2\ c = 0$	8 −1 −0.01	0.0130 0 −0.3615	2.0359
VB14	$\dot{x} = 1 - ayz$ $\dot{y} = z^2 - z$ $\dot{z} = x - bz + c$	$a = 3.55\ b = 0.5\ c = 0$	1 0 1	0.1510 0 −0.6510	2.2319
VB15	$\dot{x} = az - yz$ $\dot{y} = z^2 - by$ $\dot{z} = x - z + c$	$a = 4\ b = 0.4\ c = 0$	1 1 0	0.2229 0 −1.6229	2.1373
VB16	$\dot{x} = y$ $\dot{y} = -az - yz$ $\dot{z} = x + y^2 - bz + c$	$a = 4.2\ b = 3\ c = 0$	1 −1 1	0.0454 0 −3.0448	2.0149
VB17	$\dot{x} = yz$ $\dot{y} = az - y - z^2$ $\dot{z} = x - bz + c$	$a = 4\ b = 0.5\ c = 0$	1 0 −1	0.0947 0 −1.5948	2.0594
VB18	$\dot{x} = az + y^2 - 1$ $\dot{y} = byz$ $\dot{z} = -x - z + c$	$a = 2.8\ b = 4\ c = 0$	1 1 0	0.1155 0 −1.1155	2.1035

(continued)

Table 1 (continued)

Mode1	Equations	Parameters	$x_0\ y_0\ z_0$	LEs	D_{KY}
VB19	$\dot{x} = z^2 - ay - yz$ $\dot{y} = z$ $\dot{z} = x - bz + c$	$a = 1\ b = 0.9\ c = 0$	1 −1 1	0.1183 0 −1.0183	2.1162
VB20	$\dot{x} = z - 1$ $\dot{y} = az^2 - byz$ $\dot{z} = y^2 - x + c$	$a = 4\ b = 9\ c = 0$	1 1 1	0.2033 0 −9.2170	2.0221
VB21	$\dot{x} = ayz - y^2$ $\dot{y} = z - by - z^2$ $\dot{z} = x + c$	$a = 6\ b = 1\ c = 0$	1 1 0	0.1107 0 −1.1110	2.0996
VB22	$\dot{x} = y - ay^2 - yz$ $\dot{y} = bz^2 - y$ $\dot{z} = x + c$	$a = 1\ b = 0.25\ c = 0$	1 1 1	0.0689 0 −1.0689	2.0645
VB23	$\dot{x} = z^2 - 1$ $\dot{y} = az^2 + byz$ $\dot{z} = x - y^2 + c$	a=1 b=2 c=0	1 1 1	0.1223 0 −1.6965	2.0721
VB24	$\dot{x} = az^2 - yz$ $\dot{y} = z^2 - b$ $\dot{z} = x - z^2 + c$	$a = 1.5\ b = 0.75\ c = 0$	1 1 1	0.1144 0 −0.6994	2.1635

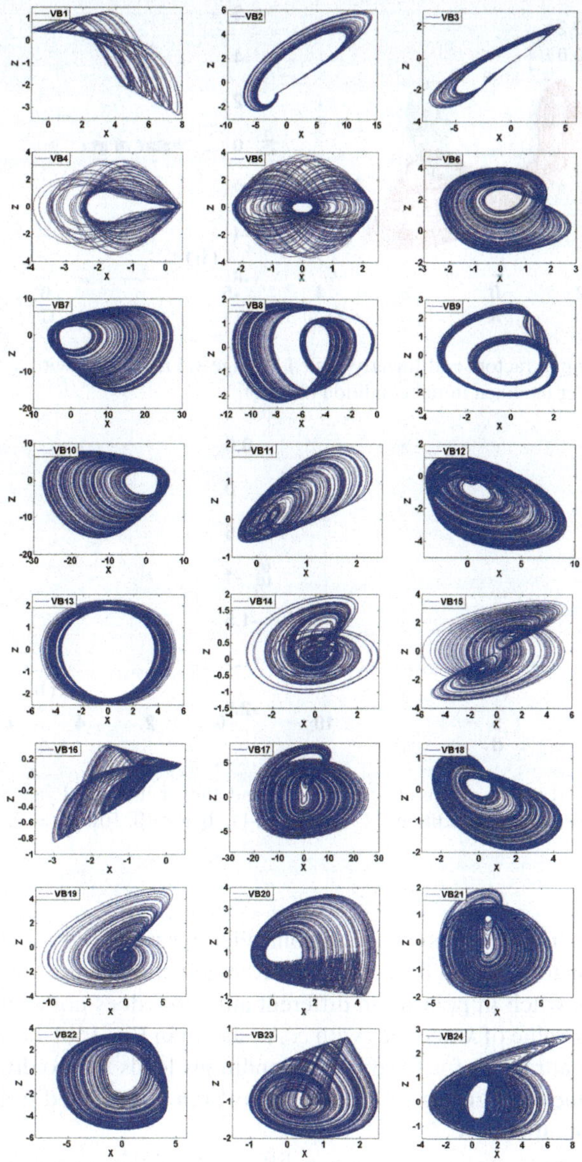

Fig. 13 Strange attractors in chaotic systems with an independent offset booster

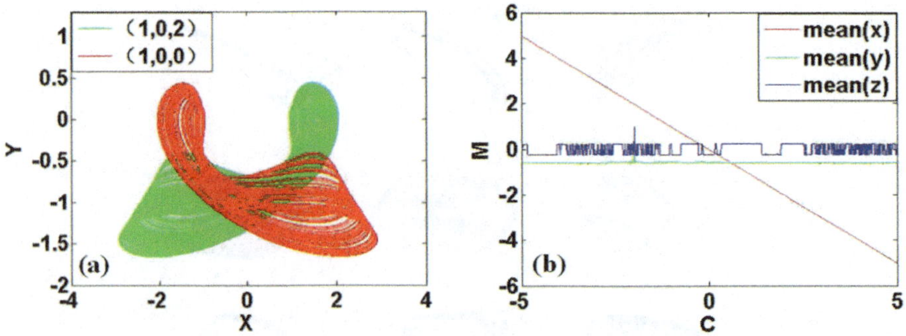

Fig. 14 Coexisting attractors and linearly scaled average value of VB5 with $a = 3.6$, c is varied from -5 to 5, under the fixed initial condition (2, 0, 1)

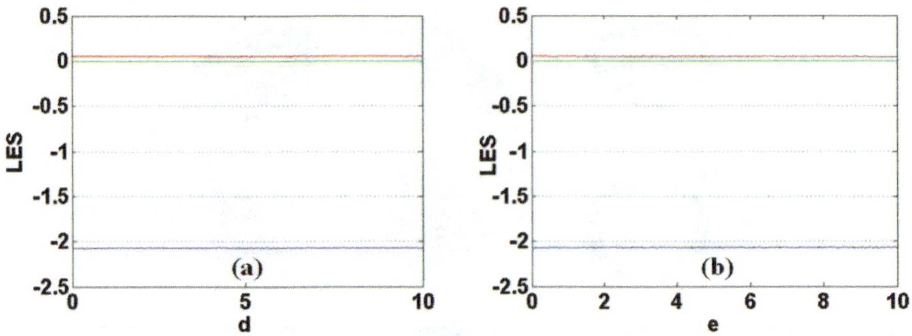

Fig. 15 Linearly rescaled offset of system (10) and the unchanged Lyapunov exponents. **a** $d \in [0, 10]$, $e = 0$, initial condition $IC = (1, -d, 1)$. **b** $e \in [0, 10]$, $d = 0$, initial condition $IC = (1, 0, 1 - e)$

pair of strange attractors. The boosting controller c varies from -5 to 5, giving two symmetric attractors, which have different average values of the variable z. For the variable x, the switch in polarity in different attractors does not change their sizes, and the average value of x declines with c. As shown in Fig. 15b, the switch between two symmetric attractors for fixed initial conditions leads to two different levels of the average value of z, giving an irregular waveform as expected from the complex fractal basins of attraction.

For a 3D chaotic system, it is more difficult to obtain two independent offset boosters. Fortunately, the jerk equation provides us a green channel for inserting two independent offset boosters. It is easy to verify that a jerk flow $\dddot{x} = (x, \dot{x}, \ddot{x})$ can obtain its two offset controllers [27], as

$$\begin{cases} \dot{x} = y + d, \\ \dot{y} = z + e, \\ \dot{z} = f(\dot{x}, \dot{y}, x). \end{cases} \tag{9}$$

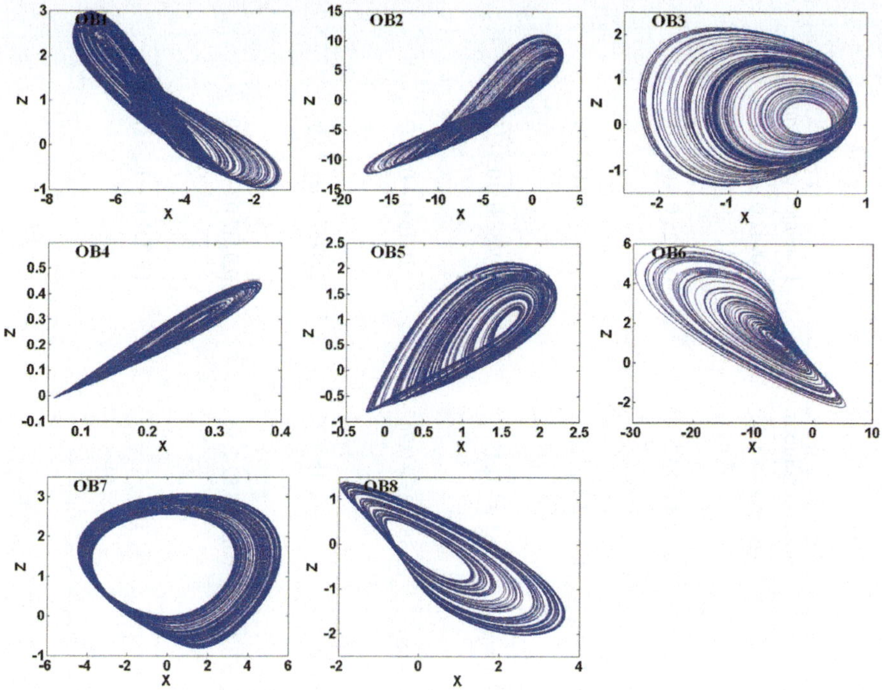

Fig. 16 2D offset-boostable strange attractors

Since $f(\dot{x},\ \dot{y},\ x)$ depends on the time derivatives of x and y, which is not altered by the constants e and d, Eq. (9) provides offset boosting of the variables y and z. In jerk flows, the offset boosting is produced in cascade, where the offset boosting is based on the preceding variable. The simplest case of this type is JD0 [29], $\dddot{x} = -2.02\ddot{x} + \dot{x}^2 - x$, which can be modified to be a 3D system with 2D offset boosting of y and z, as follows:

$$\begin{cases} \dot{x} = y + d, \\ \dot{y} = z + e, \\ \dot{z} = -2.02\dot{y} + \dot{x}^2 - x\,. \end{cases} \tag{10}$$

As shown in Fig. 16, when d and e are varied in $[-10, 10]$, the average values of y and z were rescaled linearly but the shifted attractor of system (10) has the same Lyapunov exponents.

Another regime of chaotic systems has two independent offset booster in interactive mode, where the offset boosting of two dimensions comes from a constant in the other dimension. Specifically, if the offset boosting of x is obtained in the y-dimension, while the offset boosting of y needs to be realized in the x-dimension, the general structure can be designed as

Table 2 Chaotic systems with two independent offset boosters [27]

Cases	Equations	Parameters	Initial conditions	Lyapunov exponents
OB1	$\dot{x} = y + n$ $\dot{y} = x + az + m$ $\dot{z} = -\dot{x} - b\dot{y} + z^2 - c$	$a = 4.98$ $b = 0.55$ $c = 2$ $m = 0$ $n = 0$	−0.86 −6.4 1.46	0.0942 0 −0.8376
OB2	$\dot{x} = y + n$ $\dot{y} = x - az + bz^2 + m$ $\dot{z} = c\dot{x} + \dot{y} + z$	$a = 1.65$ $b = 0.1$ $c = 1.3$ $m = 0$ $n = 0$	−2 1.4 −2.12	0.0626 0 −0.7125
OB3	$\dot{x} = y + az + n$ $\dot{y} = x + m$ $\dot{z} = -\dot{x} - b\dot{y} + cz - z^2$	$a = 2$ $b = 1.8$ $c = 2$ $m = 0$ $n = 0$	0.3 −2 1	0.0371 0 −0.8769
OB4	$\dot{x} = y - az + n$ $\dot{y} = x - z + m$ $\dot{z} = b\dot{x} - cz + z^2$	$a = 0.26$ $b = 1.64$ $c = 0.3$ $m = 0$ $n = 0$	−0.1 0 −0.25	0.0488 0 −0.3115
OB5	$\dot{x} = y - z + n$ $\dot{y} = x - az + m$ $\dot{z} = b\dot{x} + cz^2 - 1$	$a = 1.9$ $b = 2$ $c = 1$ $m = 0$ $n = 0$	0.35 1.25 −0.74	0.0876 0 −0.7573
OB6	$\dot{x} = y + az^2 + n$ $\dot{y} = x + bz + m$ $\dot{z} = -c\dot{y} + z - 1$	$a = 1.64$ $b = 5.83$ $c = 0.4$ $m = 0$ $n = 0$	−0.3 −6.18 0.97	0.0937 0 −1.4236
OB7	$\dot{x} = y + z + az^2 + n$ $\dot{y} = x + m$ $\dot{z} = -b\dot{y} - z + cz^2$	$a = 2.1$ $b = 0.7$ $c = 0.4$ $m = 0$ $n = 0$	0.05 −6.82 2.56	0.0737 0 −0.3489
OB8	$\dot{x} = y + az - z^2 + n$ $\dot{y} = x + bz + m$ $\dot{z} = -\dot{y} + cz$	$a = 1.38$ $b = 2.32$ $c = 1$ $m = 0$ $n = 0$	0 1 1	0.0331 0 −1.3531

$$\begin{cases} \dot{x} = a_1 y + a_2 z + a_3 z^2, \\ \dot{y} = a_4 x + a_5 z + a_6 z^2, \\ \dot{z} = a_7 \dot{x} + a_8 \dot{y} + a_9 z + a_{10} z^2 + a_{11} . \end{cases} \tag{11}$$

From an exhaustive computer search, eight chaotic cases restricted to no more than seven terms were captured with attractors, as shown in Fig. 16. The basic information including parameters, initial conditions and Lyapunov exponents, are shown in Table 2. As shown in Fig. 15, in this case, when m and n are varied, the average values of x and y are linearly rescaled, but without changing the Lyapunov exponents of the chaotic attractors, showing that the system OB1 has no coexisting attractors. A similar mathematic transformation can be explored according to the above two cases. For higher-dimensional chaotic systems, it is relatively easier to obtain two or more independent offset boosters.

5 Conclusions

Multiple coexisting attractors usually have different Lyapunov exponents and geometric properties. Basins of attraction show the coexistence from exhaustive computer-based searching, which takes more effort and time consumption. For an nD dynamical system, the initial condition in any dimension needs to vary in a suitable region for observing coexisting attractors. Basically, the basin of attraction of a dynamical system has a fractural structure, which leaves an opportunity for multi-stability identification even by a single fixed initial condition. When the amplitude or offset of the attractor is rescaled, the corresponding basins will get contraction, expansion, and translation. Therefore, the fixed initial data drop into different basins and trigger the corresponding attractor. In this chapter, two basic approaches of multi-stability identification are presented and explained, based on a fixed initial condition. Furthermore, an independent amplitude controller was obtained by degree modification, and chaotic systems with independent offset booster were classified.

References

1. I. Djellit, J.C. Sprott, M.R. Ferchichi, Fractal basins in the Lorenz model. Chin. Phys. Lett. **28**(6), 43–51 (2011)
2. C. Li, J.C. Sprott, Multistability in the Lorenz system: A broken butterfly. Int. J. Bifurc. Chaos **24**(10), 1450131 (2014)
3. C. Li, I. Pehlivan, J.C. Sprott, A. Akgul, A novel four-wing strange attractor born in bistablity. IEICE Electron. Express **12**(4), 2014116 (2015)
4. J. Sprott, A. Xiong, Classifying and quantifying basins of attraction. Chaos **25**(8), 083101 (2015)
5. J.C. Sprott, New chaotic regimes in the Lorenz and Chen systems. Int. J. Bifurc. Chaos **25**(02), 1550033 (2015)

6. D. Angeli, J.E. Ferrell, E.D. Sontag, Detection of multistability, bifurcations, and hysteresis in a large class of biological positive-feedback systems. Proc. Natl. Acad. Sci. **101**(7), 1822–1827 (2004)
7. R. Barrio, F. Blesa, S. Serrano, Qualitative analysis of the Rössler equations: Bifurcations of limit cycles and chaotic attractors. Physica D: Nonlinear Phenom. **238**(13), 1087–1100 (2009)
8. C. Li, J.C. Sprott, Multistability in a butterfly flow. Int. J. Bifurc. Chaos **23**(12), 1350199 (2013)
9. J.C. Sprott, Simplest chaotic flows with involutional symmetries. Int. J. Bifurc. Chaos **24**(1), 1450009 (2013)
10. Z. Elhadj, J.C. Sprott, About universal basins of attraction in high-dimensional systems. Int. J. Bifurc. Chaos **23**(12), 1350197 (2013)
11. C. Li, J.C. Sprott, H. Xing, Crisis in amplitude control hides in multistability. Int. J. Bifurc. Chaos **26**(14), 1650233 (2016)
12. C. Li, J.C. Sprott, Finding coexisting attractors using amplitude control. Nonlinear Dyn. **78**(3), 2059–2064 (2014)
13. C. Li, X. Wang, G. Chen, Diagnosing multistability by offset boosting. Nonlinear Dyn. **90**(2), 1335–1341 (2017)
14. N. Kuznetsov, The Lyapunov dimension and its estimation via the Leonov method. Phys. Lett. A **380**(25–26), 2142–2149 (2016)
15. N. Kuznetsov, T. Alexeeva, G. Leonov, Invariance of Lyapunov exponents and Lyapunov dimension for regular and irregular linearizations. Nonlinear Dyn. **85**(1), 195–201 (2016)
16. C. Li, J.C. Sprott, Amplitude control approach for chaotic signals. Nonlinear Dyn. **73**(3), 1335–1341 (2013)
17. W. Hu, A. Akgul, C. Li, T. Zheng, P. Li, A switchable chaotic oscillator with two amplitude-frequency controllers. J Circuit Syst. Comp. **26**(10), 1750158 (2017)
18. C. Li, J.C. Sprott, Chaotic flows with a single nonquadratic term. Phys. Lett. A **378**(3), 178–183 (2014)
19. A. Akgul, C. Li, I. Pehlivan, Amplitude control analysis of a four-wing chaotic attractor, its electronic circuit designs and microcontroller-based random number generator. J. Circuits Syst. Comput. **26**(12), 1750190 (2017)
20. C. Li, J.C. Sprott, W. Thio, H. Zhu, A new piecewise linear hyperchaotic circuit. IEEE Trans. Circuits Syst. II: Express Briefs **61**(12), 977–981 (2014)
21. J.C. Sprott, X. Wang, G. Chen, Coexistence of point, periodic and strange attractors. Int. J. Bifurc. Chaos **23**(05), 1350093 (2013)
22. C. Li, W. Hu, J.C. Sprott, X. Wang, Multistability in symmetric chaotic systems. Eur. Phys. J.: Special Topics **224**(8), 1493–1506 (2015)
23. C. Li, J.C. Sprott, Coexisting hidden attractors in a 4-d simplified Lorenz system. Int. J. Bifurc. Chaos **24**(3), 1450034 (2014)
24. C. Li, J.C. Sprott, Variable-boostable chaotic flows. Optik: Int. J. Light Electron Opt. **127**(22), 1038–10398 (2016)
25. C. Li, J.C. Sprott, H. Xing, Constructing chaotic systems with conditional symmetry. Nonlinear Dyn. **87**(2), 1351–1358 (2017)
26. C. Li, J.C. Sprott, Y. Liu, Z. Gu, J. Zhang, Offset boosting for breeding conditional symmetry. Int. J. Bifurc. Chaos **28**(14), 1850163 (2018)
27. C. Li, J.C. Sprott, W. Hu, Y. Xu, Infinite multistability in a self-reproducing chaotic system. Int. J. Bifurc. Chaos **27**(10), 1750160 (2017)
28. C. Li, J.C. Sprott, Z. Yuan, H. Li, Constructing chaotic systems with total amplitude control. Int. J. Bifurc. Chaos **25**(10), 1–14 (2015)
29. J.C. Sprott, Some simple chaotic jerk functions. Am. J. Phys. **65**(6), 537 (1997)

Part V

Complex Dynamics and Hidden Attractors in Delayed Impulsive Systems

Alexander N. Churilov, Alexander Medvedev, and Zhanybai T. Zhusubaliyev

1 Introduction

Differential equations with impulses (jumps) constitute a wide class of functional differential equations and play a significant role in various fields of science and technology (see, e.g., Refs. [1–3]). First papers on impulsive equations with delay date back to the 1980s [4, 5]. Later, the theory of delay impulsive equations was developed in depth in a number of publications, see e.g. Refs. [6–14].

Impulsive differential equations with delay have found numerous applications in mathematical modeling of biological systems, including cellular neural networks, models of population dynamics, dynamics of red blood cells and of hematopoiesis [15, 16].

In Ref. [17], the authors proposed to describe non-basal testosterone regulation in the male by means of impulsive differential equations. The impulsive model was based on a continuous-time one put forward by W.R. Smith [18, 19]. Notice that the

A. N. Churilov (✉)
Faculty of Mathematics and Mechanics, St. Petersburg State University, Stary Peterhof, Universitetsky av. 28, 198504 St. Petersburg, Russia

Laboratory of Multi-Agent, Distributed and Networked Control Systems, ITMO University, Kronverkskiy Pr. 49, 197101 St. Petersburg, Russia
e-mail: a_churilov@mail.ru

A. Medvedev
Department of Information Technology, Uppsala University, Lägerhyddsvägen 2, SE-751 05 Uppsala, Sweden
e-mail: alexander.medvedev@it.uu.se

Z. T. Zhusubaliyev
Department of Computer Science, International Scientific Laboratory for Dynamics of Non-Smooth Systems, Southwest State University, 50 Years of October Str. 94, 305040 Kursk, Russia
e-mail: zhanybai@hotmail.com

X. Wang et al. (eds.), *Chaotic Systems with Multistability and Hidden Attractors*, Emergence, Complexity and Computation 40,
https://doi.org/10.1007/978-3-030-75821-9_17

Smith model represents, in its turn, a version of a construct known in mathematical biology as Goodwin's oscillator [20]. The model in Ref. [17] incorporated the discrete dynamics due to the episodic release of hormones that are suitably captured by an impulsive feedback [21]. More recently, the impulsive modification of the Smith model was successfully validated on biological data [22]. It was shown to exhibit rich dynamics, including high multiplicity cycles and deterministic chaos [23].

In order to explicate the role of the transport phenomena in endocrine regulation and extend the parametric domain of self-sustained oscillations, the original (continuous) Smith model was augmented with a time delay, which construct was intensively studied over the years, see e.g. Refs. [19, 24–30]. Similarly, time delays were introduced in the impulsive model [31–33]. The main dynamical analysis tool employed there was a discrete Poincaré map [34] capturing the impulse-to-impulse evolution of the system states in the phase space. Poincaré maps are instrumental in bifurcation analysis of complex nonlinear dynamics due to their superior numerical properties compared to direct integration of differential equations [35].

The results on the impulsive model in Refs. [31–33] were obtained under limiting assumptions on the delay. It was found that the ratio between the time-delay value and the time interval between two consecutive impulses has a profound impact on the dynamics. For small time delays with the ratio values under one, the impulsive oscillator does not acquire new types of dynamical behavior compared to the delay-free case. For larger but limited to a certain interval time delays (see Refs. [36–39]), such complex nonlinear dynamics phenomena as multi-stability and quasi-periodic solutions have been discovered.

Multi-stability, i.e. coexistence of a set of attractors in the phase space of a dynamical system, is a typical nonlinear phenomenon. During the past six years, there has been a significant interest in multi-stability associated with the appearance of unpredictable attractors referred to as "hidden attractors" [40–42]. The historical background and a survey of the dynamical systems that display hidden attractors are outlined in the recent papers [42, 43]. By virtue of the particular numerical, experimental, and conceptual challenges they present, systems with hidden attractors are presently in the focus of intensive research, as evidenced by a rapidly growing number of publications in the last few years (see e.g. the references in Refs. [42, 43]).

Following the definition provided in Refs. [42] with respect to systems of smooth differential equations, an attractor is called hidden if its basin of attraction does not intersect with small neighborhoods of equilibria. By this definition, any attractor (periodic, quasi-periodic or chaotic) in a system without equilibria is deemed hidden [44]. An extensive list of references to practical examples of smooth systems without equilibria can be found in Ref. [44]. Another relevant scenario is a system excited by a periodic external signal [43].

Systems without equilibria are especially frequent among impulsive or switched systems (see e.g. Refs. [39, 45–47]). A characteristic feature of such systems is that they usually allow a discrete time (Poincaré) map that describes the evolution of the system's continuous state variables from impulse to impulse or from switch to switch [34]. Thus a study of the orbits in such a system is reduced to iterating of the discrete-time system constituted by the Poincaré map. In particular, any m-periodic solution of the discrete time system corresponding to a periodic orbit of the impulsive (or switched) system with m impulses (or switches) in the period (such orbit is called m-cycle).

Under amplitude and frequency pulse modulation in the feedback path, the model introduced in Ref. [17] lacks equilibrium points. However, as can be established by calculating the fixed points of the Poincaré map, the model always displays at least one, and in most cases three, (stable or unstable) 1-cycles. Under these conditions, the following definition of hidden attractor seems to be meaningful.

Definition 17.1 Consider an impulsive (or switched) system and its corresponding Poincaré map. An attractor of the system is called *hidden* if its basin of attraction does not overlap with small neighborhoods of the fixed points of the Poincaré map.

This definition (in equivalent terms) has appeared in Refs. [39, 45, 46] for different types of hybrid and switched systems of practical interest. A similar definition was proposed in Ref. [43] for a non-autonomous system with a periodic external action, where a stroboscopic discrete-time map was considered.

In this paper, multi-stability and hidden attractors in hybrid models of testosterone regulation are studied with special emphasis on situations where the delay time associated with the response of the testosterone secretion to the hypothalamic bursts exceeds the typical inter-burst interval in the hypothalamic activity. In the present model formulation, all the limitations on the time delay value are abolished. A Poincaré map is derived for a system of impulsive equations with a constant time delay of any finite duration.

Numerical analysis of the Poincaré map illustrates how, besides the fixed points of the map, a significant number of other modes arise and undergo a variety of bifurcations, including saddle-node and period-doubling bifurcations, subcritical and supercritical Neimark–Sacker bifurcations, as well as a quasi-periodic period-doubling bifurcation.

2 Preliminaries

Let $\mathbb{N}$ be the set of integers, $\mathbb{N}^+$ the set of nonnegative integers, $\mathbb{R}$ the set of real numbers, $\mathbb{R}^p$ the linear space of p-dimensional column vectors with elements from $\mathbb{R}$, and $\mathbb{R}^{p\times q}$ be the linear space of real $p \times q$ matrices.

Denote a finite interval as $[a, b] \subset \mathbb{R}$. The notation $PC\,([a, b], \mathbb{R}^p)$ corresponds to a set of functions $\sigma : [a, b] \to \mathbb{R}^p$ that are piecewise continuous and continuous from the left. In other words, $\sigma(t)$ is continuous at all points of (a, b) except a finite number of discontinuity points $t = \tilde{t}$, where both left hand and right hand limits $\sigma(\tilde{t}^-), \sigma(\tilde{t}^+)$ exist as finite vectors, and $\sigma(\tilde{t}) = \sigma(\tilde{t}^-)$. Moreover, suppose that finite $\sigma(a^+), \sigma(b^-)$ exist and $\sigma(b) = \sigma(b^-)$.

The set $PC\,([a, +\infty), \mathbb{R}^p)$ is defined as the set of functions σ such that $\sigma \in PC([a, b], \mathbb{R}^p)$ for every finite b, $b > a$.

For a function $\sigma \in PC$, define the operator $\Delta\sigma(t) = \sigma(t^+) - \sigma(t)$.

3 FD-Reducible Time Delay Systems

Consider the linear time-invariant autonomous system with a delayed state:

$$X' = \mathbf{A}_0 X(t) + \mathbf{A}_1 X(t-\tau), \tag{1}$$

where $' = \frac{d}{dt}$, $X(t) \in \mathbb{R}^p$, $\mathbf{A}_0, \mathbf{A}_1 \in \mathbb{R}^{p\times p}$, and τ is a constant time delay. A solution $X(t)$ to (1) is considered for $t \geqslant 0$.

Suppose that $X(t) = \varphi(t)$, $-\tau \leqslant t \leqslant 0^+$, where $\varphi \in PC([-\tau, 0], \mathbb{R}^p)$ and φ is extended to some finite value $\varphi(0^+)$.

Definition 17.2 *(*References [31, 32]*)* Time-delay linear system (1) is called finite-dimension reducible (*FD-reducible*) if there exists a constant matrix $\mathbf{D} \in \mathbb{R}^{p\times p}$ such that any solution $X(t)$ of (1) defined for $t \geqslant 0$ satisfies the ordinary differential equation

$$X' = \mathbf{D}X(t) \tag{2}$$

for $t \geqslant \tau$.

FD-reducibility means that the solutions of time-delay system (1) are indistinguishable from those of a finite-dimensional system of order p on the time interval $[\tau, +\infty)$. The theorem below summarizes the essential properties of FD-reducible systems, see [32] for the proof.

Theorem 17.1 *Any of the following statements (i), (ii) is necessary and sufficient for FD-reducibility of system* (1).

(i) The matrix coefficients of (1) *satisfy*

$$\mathbf{A}_1 \mathbf{A}_0^k \mathbf{A}_1 = 0 \quad \textit{for all} \quad k = 0, 1, \ldots, p-1. \tag{3}$$

(ii) There exists an invertible $p \times p$ matrix $\mathbf{S}$ *such that*

$$\mathbf{S}^{-1}\mathbf{A}_0\mathbf{S} = \begin{bmatrix} U & 0 \\ W & V \end{bmatrix}, \quad \mathbf{S}^{-1}\mathbf{A}_1\mathbf{S} = \begin{bmatrix} 0 & 0 \\ \bar{W} & 0 \end{bmatrix}, \tag{4}$$

where the blocks U, V are square and the sizes of the blocks W and $\bar{W}$ are equal. Moreover, the matrix $\mathbf{D}$ *in the FD-reduced system* (2) *for the FD-reducible system* (1) *is uniquely given by*

$$\mathbf{D} = \mathbf{A}_0 + \mathbf{A}_1 \mathrm{e}^{-\mathbf{A}_0 \tau}. \tag{5}$$

For $k = 0$, it follows from (3) that the matrix $\mathbf{A}_1$ is nilpotent of index 2. While condition (3) is easy to check, the alternative formulation in (4) provides an important insight into the structure of FD-reducible systems. This distinctive linear cascade (chain) structure often appears in biological, chemical, mechanical systems, as well as process engineering, and is sometimes parametrized in linear systems as input or output delay in virtue of linearity of the involved blocks.

4 A Time-Delay Impulsive System: Preliminary Results

Consider an impulsive functional differential system

$$\begin{aligned} X' &= \mathbf{A}_0 X(t) + \mathbf{A}_1 X(t-\tau) \quad \text{for} \quad t \neq t_n, \quad n \in \mathbb{N}, \ \Delta X(t_n) = \lambda_n \mathbf{B}, \\ t_{n+1} &= t_n + T_n, \quad T_n = \Phi(\mathbf{C}X(t_n)), \quad \lambda_n = \mathscr{F}(\mathbf{C}X(t_n)). \end{aligned} \tag{6}$$

Here, $\mathbf{B}$ is a column and $\mathbf{C}$ is a row such that $\mathbf{CB} = 0$, to guarantee that $\mathbf{C}X(t)$ is continuous for all $t > 0$, and $\tau > 0$ is a constant time delay. The continuous functions $\Phi(\cdot)$, $\mathscr{F}(\cdot)$ satisfy

$$0 < \Phi_1 \leqslant \Phi(\cdot) \leqslant \Phi_2, \qquad 0 < \mathscr{F}_1 \leqslant \mathscr{F}(\cdot) \leqslant \mathscr{F}_2, \tag{7}$$

for some constants $\Phi_i, \mathscr{F}_i, i = 1, 2$. The latter condition implies that system (6) has no equilibria.

Further, consider (6) for $t \geqslant 0$. Without loss of generality, let $t_0 = 0$, so that the initial function $\varphi \in PC([-\tau, 0], \mathbb{R}^p)$ is extended to $t = 0^+$ as

$$\varphi(0^+) = \varphi(0) + \mathscr{F}(\mathbf{C}\varphi(0))\mathbf{B}.$$

Thus, $X(t) = \varphi(t), -\tau \leqslant t \leqslant 0^+$, and $X \in PC([-\tau, +\infty), \mathbb{R}^p)$.

The right-hand side of (6) is linear except for the nonlinear functions $\Phi(\cdot)$, $\mathscr{F}(\cdot)$ describing impulse effect, and (6) can be termed quasilinear (see also [6]).

Previously, in Refs. [31–33], the case of small time delays, i.e. $\inf_y \Phi(y) > \tau$ was treated. Indeed, the inequality implies that the least distance in between two consecutive firings of the pulse-modulated feedback is greater than the time delay value. This corresponds to the intuitive notion of small-time delay in a sampled-data system that has to be under the sampling time. Furthermore, in Refs. [36, 38, 39], the analysis was extended to a less restrictive case of $2 \inf_\sigma \Phi(\sigma) > \tau$. The case of $3 \inf_\sigma \Phi(\sigma) > \tau$ was analyzed in Ref. [48]. In what follows, a general case of an arbitrary τ demanding further development of the involved mathematical tools is treated.

Pick a solution $(X(t), t_n)$ of hybrid system (6). Consider the ordered sets

$$\mathscr{T}_1 = \{t_0, t_1, \dots\}, \quad \mathscr{T}_2 = \{t_0 + \tau, t_1 + \tau, \dots\}.$$

It follows from (7) that $t_{k+1} - t_k \geqslant \Phi_1 > 0$ for all $k \geqslant 0$, so both sets $\mathscr{T}_i$, $i = 1, 2$ contain only isolated points.

As mentioned previously, the solution $x(t)$ has jumps at $t \in \mathscr{T}_1$, but is continuous for all $t \in \mathscr{T}_2 \backslash \mathscr{T}_1$. The delayed solution $x(t - \tau)$ has jumps at $t \in \mathscr{T}_2$, but is continuous for all $t \in \mathscr{T}_1 \backslash \mathscr{T}_2$.

For given sequences $\{t_k, \lambda_k\}|_{k=0}^{\infty}$ and $t \geqslant t_0$, introduce a function $G \in PC([0, +\infty), \mathbb{R}^p)$:

$$G(t) = \sum_{k=0}^{n} \lambda_k e^{\mathbf{A}_0(t-t_k)} \mathbf{B} \quad \text{if} \quad t_n < t < t_{n+1} \quad \text{for some} \quad n,$$

and $G(0) = 0$. Obviously, G obeys $\Delta G(t_n) = \lambda_n \mathbf{B}$ for $n \in \mathbb{N}$. Now, define another function $R \in PC([\tau, +\infty), \mathbb{R}^p)$:

$$R(t) = G(t) - e^{\mathbf{A}_0 \tau} G(t - \tau).$$

The function $R(t)$ has jumps at the points $t \in \mathcal{T}_1 \cup \mathcal{T}_2$ and is smooth otherwise. Namely:

- if $t \in \mathcal{T}_1 \backslash \mathcal{T}_2$, then $t = t_n$ for some n and $\Delta R(t) = \lambda_n \mathbf{B}$;
- if $t \in \mathcal{T}_2 \backslash \mathcal{T}_1$, then $t = t_r + \tau$ for some r and $\Delta R(t) = -\lambda_r e^{\mathbf{A}_0 \tau} \mathbf{B}$;
- if $t \in \mathcal{T}_1 \cap \mathcal{T}_2$, then $t = t_n = t_r + \tau$ for some n, r and $\Delta R(t) = \lambda_n \mathbf{B} - \lambda_r e^{\mathbf{A}_0 \tau} \mathbf{B}$.

For any $n \geqslant 0$, $r \leqslant n$, introduce

$$G_{n-r}^n(t) = \begin{cases} \sum_{k=n-r}^{n} \lambda_k e^{\mathbf{A}_0(t-t_k)} \mathbf{B} & \text{if } 0 \leqslant r \leqslant n, \\ 0 & \text{if } r < 0. \end{cases}$$

Lemma 17.1 *For the values of t that simultaneously satisfy the inequalities $t_n < t < t_{n+1}$ and $t_{n-r} + \tau < t < t_{n-r+1} + \tau$ with some integers n, r, it holds that*

$$R(t) = G_{n-r+1}^n(t). \tag{8}$$

Proof From the definition of $G_r^n(\cdot)$

$$\begin{aligned} e^{\mathbf{A}_0 \tau} G(t - \tau) &= G_0^{n-r}(t), \quad t_{n-r} + \tau < t < t_{n-r+1} + \tau, \\ G(t) &= G_0^n(t), \quad t_n < t < t_{n+1}. \end{aligned}$$

Then, since $G_{n-r+1}^n(t) = G_0^n(t) - G_0^{n-r}(t)$, equality (8) is true for $t_n < t < t_{n+1}$ and $t_{n-r} + \tau < t < t_{n-r+1} + \tau$.

Lemma 17.2 *Consider a solution $(X(t), t_n)$ of (6). Then the difference $z(t) = X(t) - R(t)$ satisfies the linear equation*

$$z' = \mathbf{D}z$$

for all points $t > \tau$, where $R(t)$ is continuous.

Proof In the special coordinate basis given by (4), system (1) can be rewritten as

$$u' = Uu, \qquad v' = Wu + Vv + \bar{W}u(t - \tau) \tag{9}$$

with $x^{\mathrm{T}} = [u^{\mathrm{T}}, v^{\mathrm{T}}]$, where $\cdot^{\mathrm{T}}$ denotes transpose. Thus $\mathbf{D}$ defined by (5) takes the form

$$\mathbf{D} = \begin{bmatrix} U & 0 \\ W + \bar{W}e^{-U\tau} & V \end{bmatrix}. \tag{10}$$

It is convenient now to assume, without loss of generality, that system (1) is readily represented in the form of (9). Introduce also the partitioning

$$\mathbf{B}^{\mathrm{T}} = \left[B_1^{\mathrm{T}}\ B_2^{\mathrm{T}}\right], \tag{11}$$

where the dimensions of the vectors B_1, B_2 correspond to those of u, v, respectively. If (4) is readily valid with $\mathbf{S}$ equal to the identity matrix, then the function $G_{n-r}^{n}(t)$ can represented as

$$G_{n-r}^{n}(t) = \begin{bmatrix} S_{n-r}^{n}(t) & 0 \\ * & * \end{bmatrix}, \tag{12}$$

where

$$S_{n-r}^{n}(t) = \begin{cases} \sum_{k=n-r}^{n} \lambda_k e^{U(t-t_k)} B_1, & \text{if } 0 \leqslant r \leqslant n, \\ 0, & \text{if } r < 0, \end{cases}$$

and the asterisks stand for arbitrary matrix blocks.

Define $u_n = u(t_n)$, $u_n^{+} = u(t_n^{+})$. Then $u_n^{+} = u_n + \lambda_n B_1$. It is seen from (9) that

$$u(t) = e^{U(t-t_0)} u_0 + \sum_{k=0}^{n} \lambda_k e^{U(t-t_k)} B_1 = e^{U(t-t_0)} u_0 + S_0^{n}(t) \tag{13}$$

when $t_n < t < t_{n+1}$. Hence,

$$u(t-\tau) = e^{U(t-\tau-t_0)} u_0 + S_0^{n-r}(t-\tau) = e^{-U\tau}\left[e^{U(t-t_0)} u_0 + S_0^{n-r}(t)\right] \tag{14}$$

when $t_{n-r} < t-\tau < t_{n-r+1}$.

Suppose that t simultaneously satisfies $t_n < t < t_{n+1}$ and $t_{n-r} < t-\tau < t_{n-r+1}$ with some n, r, $0 \leqslant r \leqslant n$. Then, from (13) and (14), it is concluded that

$$u(t) - e^{U\tau} u(t-\tau) = S_{n-r+1}^{n}(t). \tag{15}$$

As it was shown in [36],

$$X' = \mathbf{D}X(t) - (\mathbf{D} - \mathbf{A}_0)\eta(t), \tag{16}$$

with

$$\eta(t) = \begin{bmatrix} u(t) - e^{U\tau} u(t-\tau) \\ * \end{bmatrix},$$

where $*$ stands for any vector of a suitable size. Taking into account (12), (15), by a specific choice of the vector denoted as $*$, (16) can be rewritten as

$$X' = \mathbf{D}X(t) - (\mathbf{D} - \mathbf{A}_0)G^n_{r+1}(t). \tag{17}$$

From Lemma 17.1 and (17) the statement of Lemma 17.2 becomes evident.

The result obtained above plays a crucial role in the subsequent considerations. Introduce the notation: $\bar{X}_n = X(t_n)$. In particular, Lemma 17.2 implies that given the discrete time sequence $(\bar{X}_k, t_k)\big|_{k=0}^{\infty}$, the state vector $X(t)$ can be uniquely reconstructed at any point $t > \tau$. However, in the time interval $[0,\ \tau]$, the knowledge of the discrete time dynamics is insufficient: to calculate the inter-impulse behavior, the initial function $\varphi(t)$, $-\tau \leqslant t \leqslant 0$ is also needed.

5 Poincaré Map of a Time-Delay Impulsive System

For any pair of integers (r, s) such that $0 \leqslant r \leqslant s$, define a set

$$\Omega_{r,s} = \{\, (X, t) \in \mathbb{R}^p \times \mathbb{R} \,:\, t_r \leqslant t - \tau < t_{r+1},\ t_s < t - \tau + \Phi(\mathbf{C}X) \leqslant t_{s+1} \,\}.$$

Lemma 17.3 *Consider a pair $(\bar{X}_n, t_n)$ with $t_n \geqslant \tau$. Then, there exist integers r, s such that $1 \leqslant r \leqslant n$, $0 \leqslant s \leqslant r$, and $(\bar{X}_n, t_n) \in \Omega_{n-r,n-s}$.*

Proof With $t_0 = 0$, it can be concluded that there exist integers r, s such that

$$t_{n-r} \leqslant t_n - \tau < t_{n-r+1}, \tag{18}$$

$$t_{n-s} < t_{n+1} - \tau \leqslant t_{n-s+1}. \tag{19}$$

Due to $t_{n+1} = t_n + \Phi(\mathbf{C}\bar{X}_n)$, it is true that $(\bar{X}_n, t_n) \in \Omega_{n-r,n-s}$. From (18), (19), it follows that $t_{n-r} < t_n$, $t_{n-s} < t_{n+1}$ and $t_{n-r} < t_{n-s+1}$, so the statement of the lemma becomes evident.

Inequalities (18), (19) in Lemma 17.3 imply that the interval $(t_n,\ t_{n+1})$ contains exactly $r - s$ points of the set $\mathscr{T}_2$.

For any integer r, s, $0 \leqslant r \leqslant s$, introduce a vector-valued function

$$E_r^s(t) = \begin{cases} \sum_{k=r}^{s} \lambda_k \left(e^{\mathbf{D}(t-t_k-\tau)} - e^{\mathbf{A}_0(t-t_k-\tau)}\right) e^{\mathbf{A}_0\tau}\mathbf{B}, & \text{if } \ 0 \leqslant r \leqslant s, \\ 0, & \text{if } \ r > s. \end{cases} \tag{20}$$

Notice that if matrices $\mathbf{A}_0$, $\mathbf{D}$, $\mathbf{B}$ are represented in the block form of (4), (10), (11), then (20) can be rewritten as

$$E_r^s(t) = \sum_{k=r}^{s} \lambda_k \begin{bmatrix} 0 \\ K(t - t_k - \tau)B_1 \end{bmatrix}, \tag{21}$$

where

$$K(t) = \int_0^t e^{V(t-\theta)} \bar{W} e^{U\theta}\, d\theta. \tag{22}$$

Theorem 17.2 *Consider an integer n such that* $t_n \geqslant \tau$ *and define the integers* r, s, *such that* $1 \leqslant r \leqslant n$, $0 \leqslant s \leqslant r$, *and* $(\bar{X}_n, t_n) \in \Omega_{n-r,n-s}$. *Then*

$$\bar{X}_{n+1} = \alpha_n + e^{\mathbf{D}T_n} \beta^n_{n-r} - \gamma^n_{n-s}, \tag{23}$$

where

$$\alpha_n = e^{\mathbf{D}T_n} \left(\bar{X}_n + \lambda_n \mathbf{B}\right), \qquad \beta^n_{n-r} = E^n_{n-r+1}(t_n),$$
$$\gamma^n_{n-s} = E^n_{n-s+1}(t_{n+1}).$$

Proof Apply Lemma 17.2. In the interval (t_n, t_{n+1}), the function $z(t) = X(t) - R(t)$ has exactly $r - s$ jumps:

$$\Delta z(t_k + \tau) = \lambda_k e^{\mathbf{A}_0 \tau} \mathbf{B}, \quad k = n - r + 1, \ldots, n - s.$$

Then

$$z(t_{n+1}) = e^{\mathbf{D}T_n} z(t_n^+) + \delta_{n-r,n-s}, \tag{24}$$

where $\delta_{n-r,n-s}$ is defined by

$$\delta_{n-r,n-s} = \sum_{k=n-r+1}^{n-s} \lambda_k e^{\mathbf{D}(t_{n+1}-t_k-\tau)} e^{\mathbf{A}_0\tau} \mathbf{B}.$$

Equality (24) can be rewritten as

$$X(t_{n+1}) = R(t_{n+1}) + e^{\mathbf{D}T_n} \left[X(t_n^+) - R(t_n^+)\right] + \delta_{n-r,n-s}.$$

Since $X(t_{n+1}) = \bar{X}_{n+1}$, $X(t_n^+) = \bar{X}_n + \lambda_n \mathbf{B}$, $R(t_n^+) = G^n_{n-r+1}(t_n)$ and $R(t_{n+1}) = G^n_{n-s+1}(t_{n+1})$, one concludes that

$$\bar{X}_{n+1} = e^{\mathbf{D}T_n} \left[\bar{X}_n + \lambda_n \mathbf{B} - G^n_{n-r+1}(t_n)\right] + G^n_{n-s+1}(t_{n+1}) + \delta_{n-r,n-s}. \tag{25}$$

Since $\delta_{n-r,n-s} = \delta_{n-r,n} - \delta_{n-s,n}$, it can be easily shown that

$$\beta^n_{n-r} = e^{-\mathbf{D}T_n} \delta_{n-r,n} - G^n_{n-r+1}(t_n), \qquad \gamma^n_{n-s} = \delta_{n-s,n} - G^n_{n-s+1}(t_{n+1}),$$

so (23) follows from (25).

The result below facilitates the computation of the right-hand side of (23).

Remark 17.1 The defined in the proof of Theorem 17.2 quantities β^n_{n-r}, γ^n_{n-s} satisfy, for $r \geqslant 0$, $s \geqslant 0$, the following recurrent relationships

$$\beta_{n-r-1}^n = \beta_{n-r}^n + E_{n-r}^{n-r}(t_n), \qquad \gamma_{n-s-1}^n = \gamma_{n-s}^n + E_{n-s}^{n-s}(t_{n+1}) \tag{26}$$

with boundary conditions $\beta_n^n = 0$, $\gamma_n^n = 0$.

Poincaré maps with memory constitute an established tool for studying time-delay systems given by differential-difference equations. Periodic solutions of such a system correspond to fixed points of the map. Discrete maps with memory are even more useful in analysis of hybrid time-delay systems as iterating a map is easier and numerically more sound compared to direct solution of the system equations [35]. It worth recalling here that techniques based on analysis of equilibria are not applicable to system (6) as it lacks such.

Corollary 17.1 *The Poincaré map with memory*

$$Q : \left(\bar{X}_k\big|_{k=n-r_n}^n,\ t_n\right) \mapsto \bar{X}_{n+1}, \tag{27}$$

is given by (23)*, where the integer* $r = r_n$, $1 \leqslant r \leqslant n$*, satisfies the inequalities*

$$\sum_{i=n-r}^{n-1} \Phi(\mathbf{C}\bar{X}_i) \geqslant \tau, \qquad \sum_{i=n-r+1}^{n-1} \Phi(\mathbf{C}\bar{X}_i) < \tau. \tag{28}$$

Here, for the case $r = 1$*, it is assumed*

$$\sum_{i=n}^{n-1} \Phi(\mathbf{C}\bar{X}_i) = 0. \tag{29}$$

Moreover, the numerical value of the vector $\bar{X}_{n-r}$ *is not required in* Q*, as only existence of a solution to* (28) *has to be guaranteed.*

Proof The right-hand side of (23) depends on the values

$$\lambda_i,\ i = n-r+1, \ldots, n, \quad t_j,\ j = n-r+1, \ldots, n+1 \tag{30}$$

produced by the feedback modulation laws. Given $\bar{X}_{n-r+1}, \ldots, \bar{X}_n$ and t_n, the weights and times in (30) can be calculated from (6) as

$$\lambda_i = \mathscr{F}(\mathbf{C}\bar{X}_i), \quad i = n-r+1, \ldots, n,$$
$$t_j = t_{j+1} - \Phi(\mathbf{C}\bar{X}_j), \quad j = n-r+1, \ldots, n-1.$$

Since

$$t_n - t_{n-k} = \sum_{i=n-k}^{n-1} T_i = \sum_{i=n-k}^{n-1} \Phi(\mathbf{C}\bar{X}_i), \qquad (1 \leqslant k \leqslant n),$$

inequalities (18) can be rewritten as (28).

Similarly, (19) can be put as

$$\sum_{i=n-s}^{n} \Phi(\mathbf{C}\bar{X}_i) > \tau, \quad \sum_{i=n-s+1}^{n} \Phi(\mathbf{C}\bar{X}_i) \leqslant \tau, \tag{31}$$

where $s \leqslant r$.

Closed-form expressions for map (27) specialized to certain values of r can be found in Refs. [33, 36, 37].

Summing up, to calculate $\bar{X}_{n+1}$, r previous points $\bar{X}_{n-r+1}, \dots, \bar{X}_n$ are needed, where the value of r depends on n ($r = r_n$) and is determined from (28). The number r can be interpreted as a (discrete) memory span required to determine the value of $\bar{X}_{n+1}$ at the next step. Obviously, the actual memory span can vary with n, see (27).

Corollary 17.2 *Assume that*

$$m \inf_{\sigma} \Phi(\sigma) > \tau \tag{32}$$

for some integer m, $m \geqslant 1$. (Alternatively, $m\Phi_1 > \tau$, where Φ_1 is defined from (7).*) Then, $1 \leqslant r \leqslant m$ in* (27)*, so that the Poincaré map in* (27) *acts as*

$$Q : \left(\bar{X}_k \big|_{k=n-m+1}^{n}, \ t_n \right) \mapsto \bar{X}_{n+1}, \tag{33}$$

i.e. to find $\bar{X}_{n+1}$ it suffices to know at most m previous points $\bar{X}_{n-m+1}, \dots, \bar{X}_n$, $n \geqslant m - 1$.

Proof From (32), one has

$$\sum_{i=n-m}^{n-1} \Phi(\mathbf{C}\bar{X}_i) > \tau$$

for any $n \geqslant m$. Then, it follows from (29) that inequalities (28) are satisfied for some $1 \leqslant r \leqslant m$.

Thus, in case (32) holds, the memory span at every step does not exceed m [i.e., $r \leqslant m$ in (28)]. The integer s can be estimated in a similar manner. Finally, if (28), (31) are satisfied then

$$1 \leqslant r \leqslant m, \quad 0 \leqslant s \leqslant m - 1, \quad s \leqslant r. \tag{34}$$

It can be easily established that there exist $q = \frac{1}{2}m(m + 3) - 1$ different values for integer pairs (r, s) satisfying (34). Notice that the integer $m(m + 3)$ is always even.

Consider the following practically important special cases.

(i) Let $m = 1$ (see Refs. [31, 32]). Then, to find $\bar{X}_{n+1}$, it is sufficient to know only the point $\bar{X}_n$. Thus $q = 1$ and the only possible (r, s) satisfying (34) are $r = 1$, $s = 0$.

(ii) Let $m = 2$ (see Refs. [36–38]). Then, to find $\bar{X}_{n+1}$, at most two previous points $\bar{X}_{n-1}, \bar{X}_n$ are needed. Thus $q = 4$ and there are four possible cases: $r = 1, s = 0$; $r = 1, s = 1$; $r = 2$, $s = 0$; $r = 2$, $s = 1$.

With the increase of the maximal memory length m, the number of the domains $\Omega_{n-r,n-s}$ may grow significantly. For instance, for $m = 10, q = 64$ different domains exist. Yet, for computational purposes, the corresponding formulae can be obtained rather simply by means of the recurrent relationships in (26).

Making use of the bounds in (7) for the function $\Phi(y)$, closer estimates for r, s can be obtained. For example, from (18), (19), it follows that $t_n < t_{n-r+1} + \tau < t_{n-s} + \tau < t_{n+1}$, and so $t_{n-s} - t_{n-r+1} < t_{n+1} - t_n$. Hence, $(r - s - 1)\Phi_1 < \Phi_2$ and $r < s + \Phi_2/\Phi_1 + 1$. In particular, if $\Phi_1 = \Phi_2$ (no frequency modulation), then either $s = r$, or $s = r - 1$.

The following important property of the map Q is implied by Remark 17.1.

Theorem 17.3 *The map defined by* (23) *is continuous in its domain of definition.*

Proof Obviously, the map (23) is piecewise continuous and may have gaps only on the boundaries of the sets $\Omega_{n-r,n-s}$, i.e., when either $t = t_{n-r} + \tau$ holds for some r, or $t + \Phi(\mathbf{C}X) = t_{n-s} + \tau$ for some s. Considering $t = t_n$ and $X = \bar{X}_n$, this results in either $t_n = t_{n-r} + \tau$ or $t_{n+1} = t_{n-s} + \tau$.

From Remark 17.1, it follows that

$$\beta^n_{n-r} - \beta^n_{n-r-1}\Big|_{t_n=t_{n-r}+\tau} = 0, \quad \gamma^n_{n-s} - \gamma^n_{n-s-1}\Big|_{t_{n+1}=t_{n-s}+\tau} = 0.$$

This implies continuity.

Consider for illustration some simple special cases. From Remark 17.1, one has

$$\beta^n_{n-1} = \lambda_n \left(\mathrm{e}^{-\mathbf{D}\tau} - \mathrm{e}^{-\mathbf{A}_0\tau}\right) \mathrm{e}^{\mathbf{A}_0\tau}\mathbf{B},$$
$$\gamma^n_{n-1} = \lambda_n \left(\mathrm{e}^{\mathbf{D}(T_n-\tau)} - \mathrm{e}^{\mathbf{A}_0(T_n-\tau)}\right) \mathrm{e}^{\mathbf{A}_0\tau}\mathbf{B}.$$

Thus, as straightforwardly calculated in Ref. [36], for $r = s = 1$, it holds that

$$\bar{x}_{n+1} = \alpha_n + \mathrm{e}^{\mathbf{D}T_n}\beta^n_{n-1} - \gamma^n_{n-1} = \mathrm{e}^{\mathbf{D}T_n}\bar{x}_n + \lambda_n \mathrm{e}^{\mathbf{A}_0 T_n}\mathbf{B};$$

and, for $r = 1, s = 0$, the state vector of the continuous part satisfies

$$\bar{x}_{n+1} = \alpha_n + \mathrm{e}^{\mathbf{D}T_n}\beta^n_{n-1} - \gamma^n_n = \mathrm{e}^{\mathbf{D}T_n}\bar{x}_n + \lambda_n \mathrm{e}^{\mathbf{D}(T_n-\tau)}\mathrm{e}^{\mathbf{A}_0\tau}\mathbf{B}.$$

As it was mentioned previously, the discrete-time representation given by (23) entirely defines the dynamics of the impulsive continuous time system in (6) for $t > \tau$. Therefore, to investigate asymptotical behaviors of (6), it suffices to study discrete map (23) and its iterations. Some examples of such an analysis can be found in the next sections and in Refs. [37, 38].

6 Time-Delay Impulsive Model of Testosterone Regulation

In this section, the theoretical tools developed above are applied to a delayed system with impulsive feedback that originates from mathematical modeling of the non-basal testosterone regulation in the human male [19, 24–28].

In what follows, only the dominating delay due to the transport from pituitary, where LH (luteinizing hormone) is secreted, to testes, where Te (testosterone) is produced, is considered. The impulsive feedback is implemented via episodic secretion of GnRH (gonadotropin-releasing hormone) in response to the concentration of Te. The delay value τ takes into account not only the transport of LH from pituitary to testes, but also an additional time required for steroid synthesis in the testes [24]. Since this delay substantially exceeds the other delays in this endocrine axis, the latter ones can be neglected.

Consider a specialization of system (6) with

$$\begin{aligned} &x' = -b_1 x, \quad y' = -b_2 y + g_1 x, \quad z' = -b_3 z + g_2 y(t-\tau), \\ &t_0 = 0, \quad t_{n+1} = t_n + T_n, \quad \Delta x(t_n) = \lambda_n, \\ &T_n = \Phi(z(t_n)), \quad \lambda_n = \mathscr{F}(z(t_n)), \end{aligned} \tag{35}$$

where b_1, b_2, b_3, g_1, g_2 are positive parameters reflecting the kinetics of the involved hormones that also imply $b_i \neq b_j$ for $i \neq j$.

In biology, Hill functions [49] are habitually used to approximate monotonous nonlinearities and can be applied to capture the shape of $\mathscr{F}(\cdot)$, $\Phi(\cdot)$

$$\Phi(z) = k_1 + k_2 \frac{(z/h)^\nu}{1+(z/h)^\nu}, \quad \mathscr{F}(z) = k_3 + \frac{k_4}{1+(z/h)^\nu},$$

with some positive parameters $k_1, k_2, k_3, k_4, h, \nu$. System (35) is subject to the initial conditions $x(0^-) = x_0$, $y(t) = \varphi_2(t)$ with $-\tau \leqslant t \leqslant 0$, $z(0) = z_0$. Here, $x(0^+) = x(0^-) + \Phi(z_0)$, $\varphi_2(t)$ is a continuous initial function.

For the system at hand,

$$\mathbf{A}_0 = \begin{bmatrix} -b_1 & 0 & 0 \\ g_1 & -b_2 & 0 \\ 0 & 0 & -b_3 \end{bmatrix}, \quad \mathbf{A}_1 = \begin{bmatrix} 0 & 0 & 0 \\ 0 & 0 & 0 \\ 0 & g_2 & 0 \end{bmatrix}, \quad \mathbf{B}^T = \begin{bmatrix} 1 & 0 & 0 \end{bmatrix}, \quad \mathbf{C} = \begin{bmatrix} 0 & 0 & 1 \end{bmatrix}.$$

The linear part of the system is represented in the block matrix form of (9) and is therefore FD-reducible. Here,

$$U = \begin{bmatrix} -b_1 & 0 \\ g_1 & -b_2 \end{bmatrix}, \quad B_1 = \begin{bmatrix} 1 \\ 0 \end{bmatrix}, \quad V = [-b_3], \quad W = [0 \;\; 0], \quad \bar{W} = [0 \;\; g_2].$$

Introduce the numbers

$$\mu_i = \prod_{\substack{j=1\\ j\neq i}}^{3} \frac{1}{b_j - b_i}, \quad i = 1, 2, 3.$$

Applying formulas (21), (22), one gets

$$E_r^s(t) = \sum_{k=r}^{s} \lambda_k [0 \quad 0 \quad \tilde{K}(t - t_k - \tau)]^{\mathrm{T}}, \quad \text{where} \quad \tilde{K}(t) = g_1 g_2 \sum_{i=1}^{3} \mu_i \mathrm{e}^{-b_i t}.$$

The matrix exponential $\mathrm{e}^{\mathbf{D}t}$ is given by

$$\mathrm{e}^{\mathbf{D}t} = \begin{bmatrix} \mathrm{e}^{-b_1 t} & 0 & 0 \\ E_{21}(t) & \mathrm{e}^{-b_2 t} & 0 \\ E_{31}(t) & E_{32}(t) & \mathrm{e}^{-b_3 t} \end{bmatrix}$$

with

$$E_{21}(t) = -\frac{g_1}{b_1 - b_2}\left(\mathrm{e}^{-b_1 t} - \mathrm{e}^{-b_2 t}\right), \quad E_{32}(t) = -\frac{g_2 \mathrm{e}^{b_2 \tau}}{b_2 - b_3}\left(\mathrm{e}^{-b_2 t} - \mathrm{e}^{-b_3 t}\right),$$
$$E_{31}(t) = g_1 g_2 \left[\mu_1 \mathrm{e}^{b_1 \tau}\left(\mathrm{e}^{-b_1 t} - \mathrm{e}^{-b_3 t}\right) + \mu_2 \mathrm{e}^{b_2 \tau}\left(\mathrm{e}^{-b_2 t} - \mathrm{e}^{-b_3 t}\right)\right].$$

Hence, it applies that

$$\mathrm{e}^{\mathbf{D}T_n} E_r^s(t) = \mathrm{e}^{-b_3 T_n} E_r^s(t) \quad \text{for all} \quad t.$$

Thus, for this particular case, (23) and (26) are now written in explicit analytical form.

Introduce

$$\bar{X}_n = \begin{bmatrix} x_n \\ y_n \\ z_n \end{bmatrix}, \quad x_n = x(t_n^-), \quad y_n = y(t_n), \quad z_n = z(t_n),$$

where $\bar{X}_n$ obeys discrete time map (23) for those n that satisfy $t_n \geqslant \tau$. Consider the choice of initial values for map (23).

Proposition 17.1 *Let $g_2 \neq 0$ and choose arbitrarily the numbers x_0, y_0 and a sequence $z_0, \ldots, z_{m_0}$ such that*

$$\sum_{i=0}^{m_0-1} \Phi(z_i) \leqslant \tau, \quad \sum_{i=0}^{m_0} \Phi(z_i) > \tau. \tag{36}$$

(Notice that (7) implies $\tau/\Phi_2 < m_0 \leqslant \tau/\Phi_2$.) Define the numbers x_i, y_i, $i = 1, \ldots, m_0$ in the following manner

$$\begin{bmatrix} x_i \\ y_i \end{bmatrix} = e^{U\Phi(z_{i-1})} \begin{bmatrix} x_{i-1} + \mathscr{F}(z_{i-1}) \\ y_{i-1} \end{bmatrix}. \tag{37}$$

Then there exists a solution of (35) *such that*

$$t_0 = 0, \quad t_i = \sum_{j=0}^{i-1} \Phi(z_j), \quad i = 1, \ldots, m_0,$$

and

$$x(t_i^-) = x_i, \quad y(t_i) = y_i, \quad z(t_i) = z_i, \quad i = 0, \ldots, m_0.$$

Proof Inequalities (36) imply $t_{m_0-1} \leqslant \tau < t_{m_0}$. One has

$$z' = -b_3 z(t) + g_2 \varphi_2(t - \tau), \quad 0 \leqslant t \leqslant \tau. \tag{38}$$

Let $\xi(t)$ be any continuously differentiable function defined for $0 \leqslant t \leqslant \tau$. Since $g_2 \neq 0$, selecting

$$\varphi_2(t) = \frac{1}{g_2} \left(\xi'(t + \tau) + b_3 \xi(t + \tau) \right), \quad -\tau \leqslant t \leqslant 0,$$

yields the solution to (38) $z(t) \equiv \xi(t), 0 \leqslant t \leqslant \tau$. In particular, $\xi(t)$ can be chosen in such a way that $\varphi_2(0) = y_0, \xi(t_i) = z_i, i = 0, 1, \ldots, m_0 - 1$, and, moreover, $\xi(\tau)$ can be assigned to any desired value.

Since

$$\begin{bmatrix} x' \\ y' \end{bmatrix} = U \begin{bmatrix} x(t) \\ y(t) \end{bmatrix}, \quad t \neq t_i,$$

and $x(t_i^+) = x(t_i^-) + F(z_i)$, it follows

$$\begin{bmatrix} x(t) \\ y(t) \end{bmatrix} = e^{U(t-t_{i-1})} \begin{bmatrix} x(t_{i-1}^+) \\ y(t_{i-1}) \end{bmatrix}, \quad t_{i-1} < t < t_i, \quad i \leqslant m_0. \tag{39}$$

In particular, (37) is valid for $t = t_i^-$. One has

$$z' = -b_3 z(t) + g_2 y(t - \tau), \quad \tau \leqslant t \leqslant t_{m_0},$$

where $y(t)$ is obtained from (39) and, as it was mentioned previously, $z(\tau)$ can be assigned arbitrarily by the choice of $\varphi_2(t)$. Then, an arbitrary value z_{m_0} to $z(t_{m_0}^-)$ can be assigned by the choice of $z(\tau)$.

The proof is complete.

Remark 17.2 Proposition 17.1 can be extended to the general system expressed by (9). Then, the assumption of $g_2 \neq 0$ has to be replaced by controllability of the matrix pair $(V, \bar{W})$.

6.1 Bifurcation Analysis: Multi-Stability and Quasi-Periodicity

In what follows, the discrete map given by (23) was applied to the analysis of non-linear behaviors in (35) for sufficiently large values of time delay. On this way, new dynamical phenomena were discovered that had not so far been observed in the system in hand [37, 39]. The previously studied effects of multi-stability and quasi-periodicity are briefly illustrated below to provide a reference for comparison with further presented new results. Primarily, periodic solutions of (35) with exactly m jumps in the least period are considered. Their corresponding orbits are usually termed m-cycles. Parameters of m-cycles can be calculated from the fixed points of mth iterations of the Poincaré map.

In the oncoming analysis, the time delay τ $(0 < \tau < 205.0)$ and b_1 are used as bifurcation parameters. The remaining model parameters are kept constant: $b_2 = 0.15$, $b_3 = 0.2$, $k_1 = 60.0$, $k_2 = 40.0$, $k_3 = 3.0$, $k_4 = 2.0$, $g_1 = 2.0$, $g_2 = 0.5$, $h = 2.7$ and $\nu = 2$.

In order to obtain a more complete picture of the set of attracting states and the bifurcations they undergo, a careful numerical investigation of the map (23) was performed. Fig. 1 provides an overview of the bifurcation structure that can be observed in the dynamics of the system (23). In order to distinguish between branches of the bifurcation diagram, the notation $B_{m,i}$ is used, where m denotes the periodicity of the cycle, from which the branch stars and i numbers different cycles with the same m.

Depending on the time delay τ, a variety of different scenarios (Fig. 1) is observed.

(i) If $\tau < T_n$, map (23) can display a finite or infinite period doubling cascade. An example of such transition is shown in Fig. 1. One can see that when the time delay τ decreases (see the branch $B_{1,1}$), 1-cycle undergoes a period-doubling bifurcation at the point $\tau = \tau_d$, leading to the appearance of a stable 2-cycle. Here the dashed line represents an unstable 1-cycle.

(ii) If the values of the time delay parameter τ are chosen such that $\tau < T_n + T_{n-1}$, more complex bifurcation phenomena are possible. In particular (see Fig. 1), as the parameter τ decreases, the stable 1-cycle belonging to the branch $B_{1,2}$ undergoes a Neimark–Sacker bifurcation. As a result, a closed invariant curve corresponding to the quasi-periodic dynamics softly appears and the 1-cycle becomes an unstable focus. As shown in a recent publication [37], in the region of overlap between two branches $B_{1,1}$ and $B_{1,2}$, the system displays quasiperiodic dynamics, border collisions, multi-stability, overlapping resonance tongues of different periodicity, hidden attractors [42, 50], as well as period-doubling cascades and deterministic chaos.

(iii) If $\tau < T_n + T_{n-1} + T_{n-2}$, as illustrated in Fig. 1, the bifurcation diagram contains three branches $B_{1,1}$, $B_{1,2}$ and $B_{1,3}$. Around the region of overlap between the branches $B_{1,2}$ and $B_{1,3}$, the bifurcation pattern is qualitatively similar to the previous case $\tau < T_n + T_{n-1}$. A difference is that the system demonstrates a quasiperiodic period-doubling and a subcritical Neimark–Sacker bifurcation,

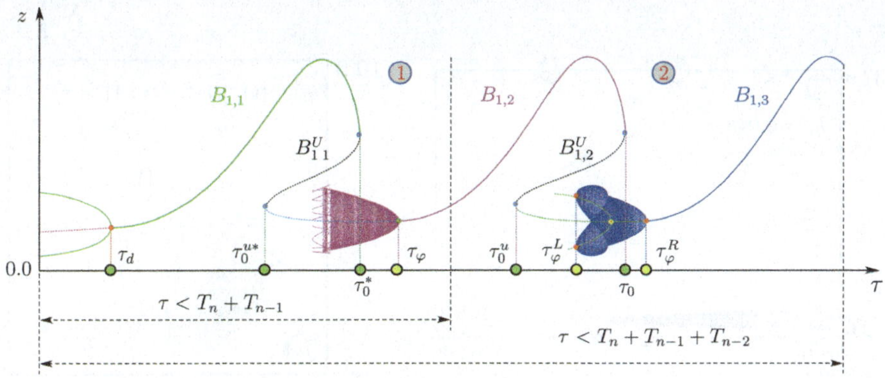

Fig. 1 A sketch illustrating the bifurcation structure for $0 < \tau < T_n + T_{n-1} + T_{n-2}$. If $\tau < T_n$ the bifurcation diagram contains a single branch $B_{1,1}$. When decreasing the time delay τ, the 1-cycle undergoes a period-doubling bifurcation at the point τ_d. If $\tau < T_n + T_{n-1}$ the bifurcation diagram contains two branches $B_{1,1}$ and $B_{1,2}$. As the parameter τ decreases the stable 1-cycle belonging to the branch $B_{1,2}$ undergoes a Neimark–Sacker bifurcation at τ_φ, leading to the birth of a stable closed invariant curve. If $\tau < T_n + T_{n-1} + T_{n-2}$, the bifurcation diagram contains three branches $B_{1,1}$, $B_{1,2}$ and $B_{1,3}$. These branches arise in a saddle-node bifurcation. Here $B^U_{1,1}$, $B^U_{1,2}$ are the unstable branches, corresponding to the saddle 1-cycles, and τ_0^{u*}, τ_0^u, τ_0^* and τ_0 are the saddle-node bifurcation points. τ_φ^L and τ_φ^R are the Neimark–Sacker bifurcation points for the branch $B_{1,3}$. Numbers 1 and 2 denote regions to be examined in what follows

leading to appearance of two closed invariant curves, one attracting and one repelling. These phenomena have not been investigated so far in the dynamics of time-delay impulsive model (35).

The bifurcation behavior for $\tau < T_n + T_{n-1}$ is analyzed below in more detail in order to illustrate the mechanisms of formation of the coexisting attractors (see Fig. 1). Fig. 2a displays the bifurcation diagram for the case $\tau < T_n + T_{n-1}$ and $b_1 = 0.048$.

The branches $B_{1,1}$ and $B_{1,2}$ start from the 1-cycles arising via saddle-node bifurcation. The dashed lines in Fig. 2 represent the unstable 1-cycles. The branch $B^U_{1,1}$ denotes the saddle 1-cycle. As the time delay τ decreases, the system enters the region of multi-stability through the saddle-node bifurcation point τ_0^*. On the part of this region that falls between the points τ_0^* and τ_φ, the stable node 1-cycle ($B_{1,1}$) coexists with the stable focus 1-cycle ($B_{1,2}$). The two-dimensional projection of the phase portrait in Fig. 2b depicts the coexistent stable focus 1-cycle F_1 ($B_{1,2}$) and stable node 1-cycle N_1 ($B_{1,1}$) for $\tau = 94.36$ and $b_1 = 0.048$. Here $W^U_\pm$ are the unstable manifolds of the saddle 1-cycle S_1. The stable manifold of the saddle 1-cycle S_1 ($B^U_{1,2}$) separates the basins of attraction of the coexisting stable 1-cycles F_1 and N_1.

Figure 2c presents a magnified part of the bifurcation diagram outlined by rectangle in Fig. 2a and shows the transition from a stable 1-cycle F_1 to the quasiperiodic dynamics. As the time delay τ is further decreased, the stable focus 1-cycle F_1 becomes an unstable focus in a Neimark–Sacker bifurcation at the point $\tau = \tau_\varphi$

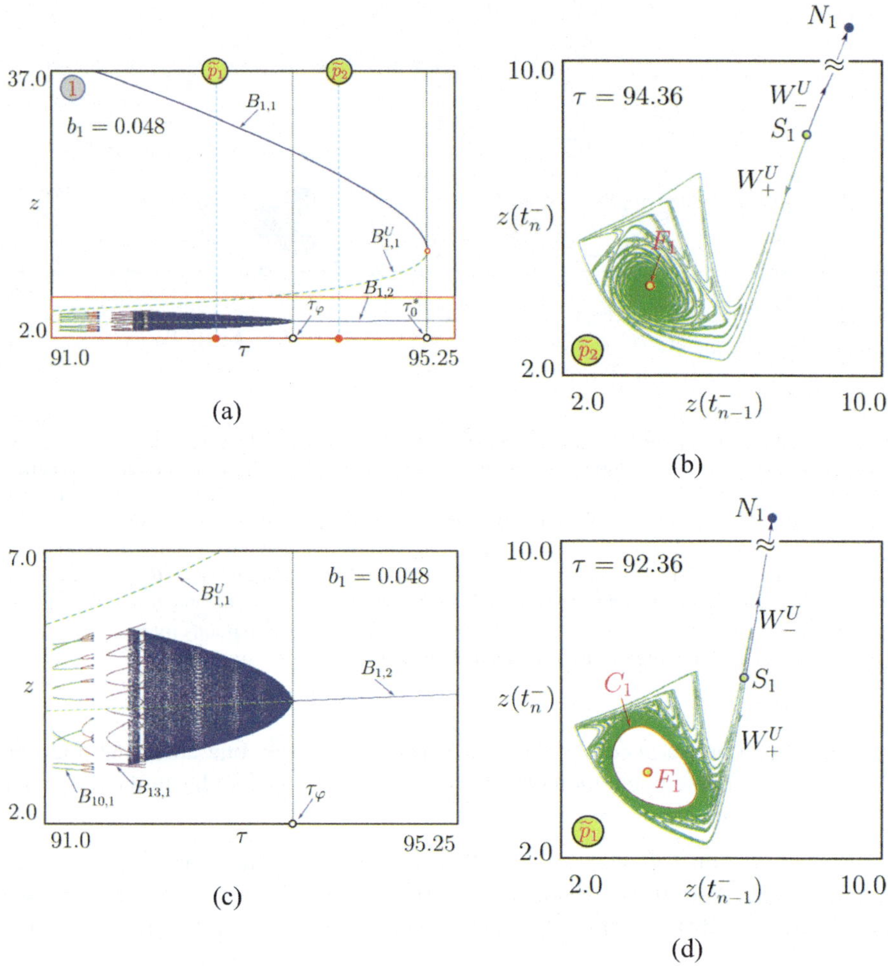

Fig. 2 **a** Bifurcation diagram for $\tau < T_n + T_{n-1}$ that is denoted by 1 in Fig. 1. $b_1 = 0.048$. Here, τ_φ is the point of Neimark–Sacker bifurcation and τ_0 is the point of saddle-node bifurcation, in which stable $(B_{1,1})$ and saddle $(B^U_{1,1})$ 1-cycles are born. **b** Phase portrait of the map for in the region $\tau_\varphi < \tau < \tau^*_0$, where the stable node 1-cycle N_1 coexists with a stable focus 1-cycle F_1: $\tau = 94.36$ and $b_1 = 0.048$. **c** Magnified part of the bifurcation diagram outlined by the rectangle in (b). **d** Phase portrait of the map after the Neimark–Sacker bifurcation. Stable node 1-cycle N_1 coexists with a closed invariant curve C_1: $\tau = 92.36$ and $b_1 = 0.048$.Z

where a pair of complex-conjugate multipliers leave the unit circle. The loss of stability for the 1-cycle is accompanied by the soft appearance of a stable closed invariant curve, associated with quasiperiodic dynamics. The green dashed line in Fig. 2c denotes the unstable focus 1-cycle after this bifurcation.

Figure 2d presents the phase portrait of the map after the Neimark–Sacker bifurcation. Here, the stable closed invariant curve C_1 corresponding to the quasi-periodic

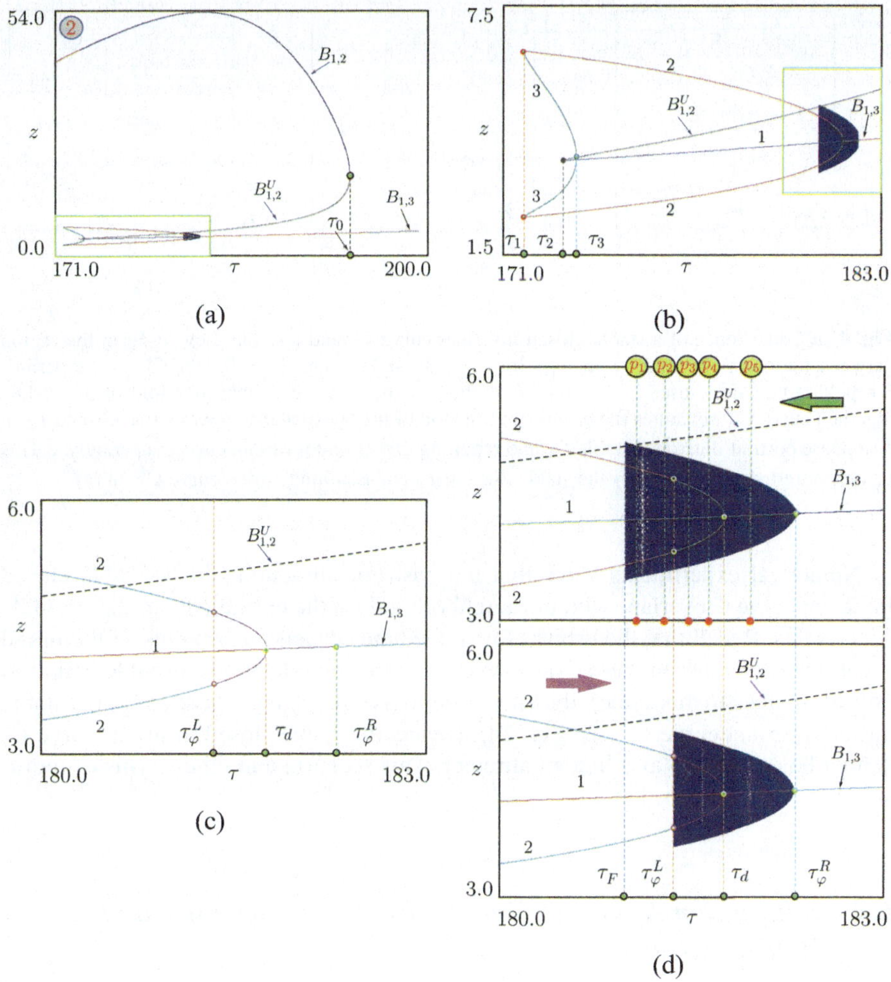

Fig. 3 **a** Bifurcation diagram for the large time delay $\tau < T_k + T_{k-1} + T_{k-2}$ and $b_1 = 0.048$. **b** Magnified part of the bifurcation diagram outlined by the rectangle in (a) showing the transition from a stable 1-cycle to a quasi-periodic attractor in a Neimark–Sacker bifurcation. This diagram presents also the bifurcational structure in the region of coexistence for stable 1- and 2-cycles. **c** and **d** Bifurcation diagrams illustrating the subcritical Neimark–Sacker bifurcation of the stable 2-cycle at the point τ^L_φ with subsequent transition to the unstable 1-cycle through a reverse period-doubling bifurcation transverse to the stable closed invariant curve at the point τ_d. Here, τ_φ is the point of supercritical Neimark–Sacker bifurcation, in which a stable closed invariant curve appears

motion coexists with a stable 1-cycle N_1. The stable manifold of the saddle 1-cycle S_1 delineates the basins of attraction for the stable 1-cycle N_1 and the stable closed invariant curve C_1.

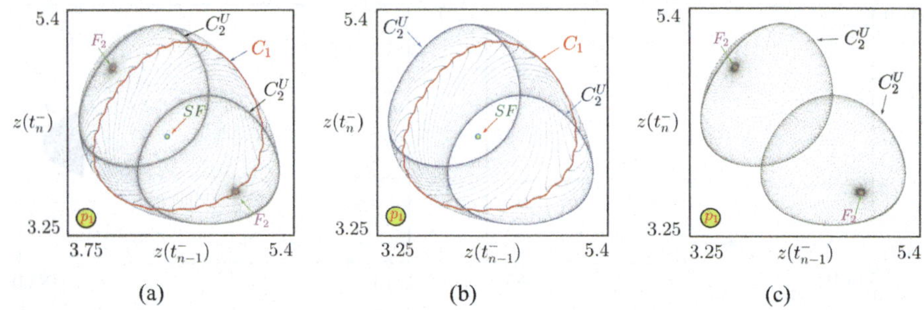

Fig. 4 **a** Coexistence of a stable closed invariant curve C_1 and a stable 2-cycle F_2 in the region between the points τ_F and τ_φ^L in Fig. 3d: $\tau = 180.983568$ and $b_1 = 0.048$. C_2^U is the period-2 repelling (saddle) closed curve and SF is the unstable 1-cycle. Stable manifold of the saddle invariant curve C_2^U separates the basins of attraction of the coexisting stable motions C_1 and F_2. **b** The phase portrait illustrates stable C_1 and repelling C_2^U closed invariant curves separately. **c** This phase portrait shows separately the stable 2-cycle F_2 and repelling closed curve C_2^U in (a)

Numerical experiments show that the basin of attraction for the stable closed invariant curve C overlaps with the neighborhoods of the unstable focus F_1 or saddle S_1 1-cycles. Recall that the orbits of map (23) are sequences of points. If the initial point is chosen close to the saddle 1- cycle S_1 (for example on the unstable manifold W_+^U of S_1 close to this cycle), then the system asymptotically converges to the stable closed invariant curve C_1 (see Fig. 2d). Hence, the stable closed invariant curve C_1 cannot be referred to as a hidden attractor. This scenario and other forms of multi-stability were described in more detail in Refs. [37, 39].

6.2 *Bifurcation Analysis: Crater Bifurcation Scenario and Hidden Attractors*

In this section, a number of new bifurcation phenomena that has not previously been revealed in the delay impulsive equations are examined. As mentioned above, operating in a region of large time delays leads to the more complex nonlinear phenomena that do not occur in the case $\tau < T_n + T_{n-1}$. The aim of this part is to exemplify such behaviors.

Figure 3a–d display the results of a bifurcation analysis for the slightly higher values of the time delay $\tau < T_n + T_{n-1} + T_{n-2}$ and $b_1 = 0.048$, wherein the bifurcation pattern is qualitatively similar to what is depicted in Fig. 2a. However, the nature of the quasiperiodic motion has certain peculiarities in comparison with Fig. 2a and c.

The bifurcation diagram in Fig. 3a contains three branches $B_{1,2}$, $B_{1,3}$ and $B_{1,2}^U$. The branch $B_{1,3}$ begins and ends at the saddle-node bifurcation points and $B_{1,2}^U$ presents a saddle 1-cycle, arising at the point τ_0.

A magnified part of the bifurcation diagram, outlined by the rectangle in Fig. 3a, is shown in Figs. 3b and c is the magnification of part of Fig. 3b. As the parameter τ decreases, the stable focus 1-cycle belonging to the branch $B_{1,3}$ undergoes a Neimark–Sacker bifurcation at the point τ_{φ}^{R} (Fig. 3c and d). As a result, a stable closed invariant curve softly appears from the stable focus 1-cycle. After this bifurcation, the 1-cycle becomes an unstable focus. With further decrease of the value of τ, the unstable focus 1-cycle is doubled via a period-doubling bifurcation transverse to the stable closed invariant curve at the point τ_d (Fig. 3c). This leads to the appearance of the unstable focus 2-cycle. The unstable 1-cycle after the transverse period-doubling bifurcation is denoted by 1 in Figs. 3b and c. As the time delay parameter τ passes the value τ_{φ}^{L}, the unstable focus 2-cycle becomes stable.

In the part of the bifurcation in Figs. 3b and c that falls to the left of the point τ_{φ}^{L}, the interval $\tau_1 < \tau < \tau_{\varphi}^{L}$ is observed, where the stable 2-cycle (denoted by 2)

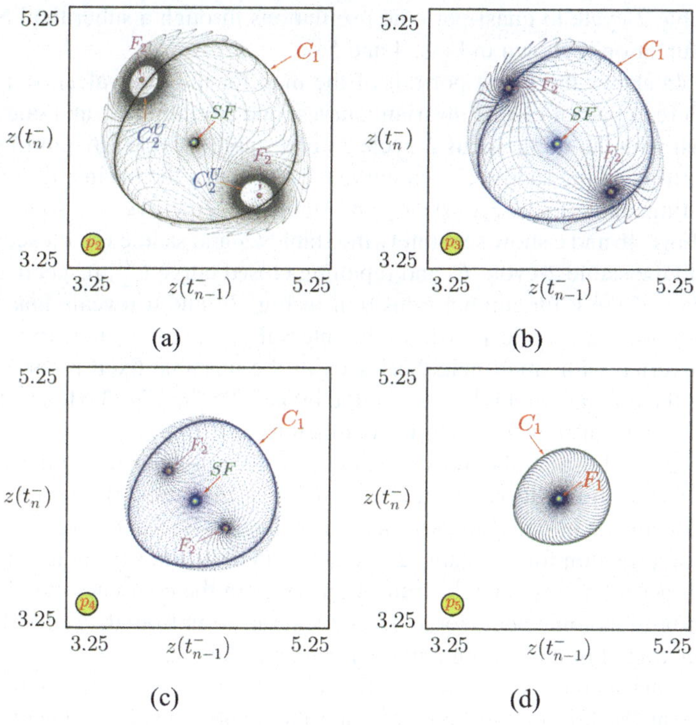

Fig. 5 Subcritical Neimark–Sacker bifurcation of the stable 2-cycle F_2. **a** Two-dimensional projection of the phase portrait before the Neimark–Sacker bifurcation. $\tau = 181.3122$. **b** Two-dimensional projection of the phase portrait after the subcritical Neimark–Sacker bifurcation. Here, F_2 is the unstable 2-cycle, C_1 is the stable closed curve and SF is the repelling 1-cycle: $\tau = 181.399$. **c** Two-dimensional projection of the phase portrait near the transverse period-doubling bifurcation point τ_d. $\tau = 181.6432$. **d** Stable closed invariant curve C_1 arising through a supercritical Neimark–Sacker bifurcation at the point τ_{φ}. Here, F_1 is the unstable focus 1-cycle: $b_1 = 182.002$

coexists with the stable 1-cycle belonging to the branch $B_{1,2}$. As illustrated in Fig. 3b, a pair of 2-cycles (a stable one marked as 2 and a saddle cycle marked by 3) appear in a saddle-node bifurcation at the point τ_1. When the parameter τ increases, a saddle 2-cycle undergoes a reverse period-doubling bifurcation at the point τ_3 that produces the saddle 1-cycle, associated with the branch $B^U_{1,2}$. Returning to Figs. 3b and c, to follow the course of the unstable 1-cycle (denoted by 1) produced in the period-doubling bifurcation at the point τ_d, one notes that the cycle initially continues to the left in the bifurcation diagram until it reaches τ_2 where it disappears in the saddle-node bifurcation.

Finally, if the time delay τ is increased over the value τ_1, the stable focus 2-cycle F_2 is destabilized via a subcritical Neimark–Sacker bifurcation at the point τ^L_φ (see Fig. 3d). Arrows in Fig. 3d denote scanning directions. Consider the characteristics of the subcritical Neimark–Sacker bifurcation shown in Fig. 3d in more detail in order to understand the mechanism of this transition.

A sequence of typical phase portraits illustrating the main stages of the transition from a stable 2-cycle to quasi-periodic oscillations through a subcritical Neimark–Sacker bifurcation is shown in Figs. 4 and 5.

Figure 4a shows the phase portrait of the map for the interval $\tau_F < \tau < \varphi^L$ in Fig. 3d. Here, a pair of closed invariant curves, one attracting C_1 and one repelling C^U_2, coexist with the stable focus 2-cycle F_2 and stable 1-cycle ($B_{1,2}$ in Fig. 3a. The stable manifold of the saddle closed curve C^U_2 separates the basins of attraction of the coexisting stable focus 2-cycle F_2 and stable closed curve C_1. To simplify the analysis, Figs. 4b and c show separately the stable C_1 and saddle C^U_2 closed invariant curves, and the stable 2-cycle F_2 and repelling closed curve C^U_2, respectively.

Detailing of the bifurcational transition in Fig. 3c and d reveals that the basin of attraction for the 2-cycle F_2 within the interval $\tau^L_\varphi < \tau < \tau_d$ (see Fig. 3c and d) does not overlap with small neighborhoods of the unstable fixed point SF and of the saddle fixed point, which belongs to the branch $B^U_{1,2}$ in Fig. 3. Hence, the stable 2-cycle F_2 can be attributed to a hidden attractor.

Figures 5a–d illustrate the transformations of the phase portrait in the subcritical Neimark–Sacker bifurcation. As the time delay τ increases, the repelling closed curve C^U_2 decreases in size (Fig. 5a). Near the Neimark–Sacker bifurcation point τ^L_φ, the basin of attraction for the stable 2-cycle F_2 can be arbitrarily small. In this case, even weak perturbations, such as truncation errors in the computer simulation, can lead to a non-monotonic behavior and a sudden transition from the stable 2-cycle F_2 to the attracting closed invariant curve C_1 can be expected.

With further increase of τ, the repelling closed curve C^U_2 merges with the focus 2-cycle F_2 at the bifurcation point τ^L_φ, and the stable 2-cycle F_2 becomes unstable (Fig. 5b). When crossing the bifurcation point τ^L_φ with increasing τ, the system displays an abrupt transition from the stable 2-cycle to a quasi-periodic attractor. This scenario may be considered as a modification of the so-called "*crater bifurcation*" [51, 52].

Figure 5a presents a phase portrait near the subcritical Neimark—Sacker bifurcation point τ^L_φ. Figure 5b shows the phase portrait after the system has passed the

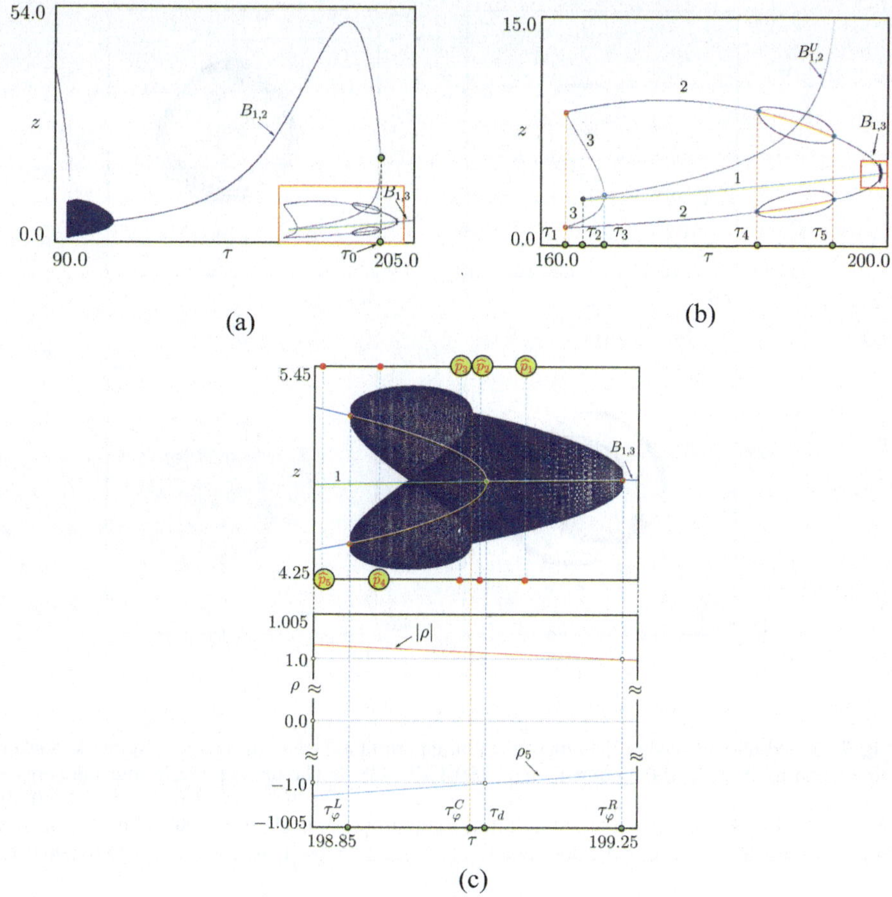

Fig. 6 **a** Bifurcation diagram for the large time delay $\tau < T_k + T_{k-1} + T_{k-2}$ and $b_1 = 0.055$. **b** Magnified part of the bifurcation diagram outlined by the rectangle in (a). **c** Magnified part of the bifurcation diagram that is outlined by the rectangle in (b), showing a quasi-periodic period-doubling. Multiplier diagram for the 1-cycle is shown in the lower panel. Here, $|\rho|$ is the absolute value of the complex-conjugate multipliers of the 1-cycle

Neimark–Sacker bifurcation at τ_φ^L. Here the unstable 2-cycle F_2 coexists with the stable closed curve C_1. Hereafter, the unstable 2-cycle F_2 undergoes a reverse period-doubling bifurcation transverse to the stable closed invariant curve C_1 at the point τ_d. Figures 5c and d show the phase portrait of the map before, and after the reverse period-doubling bifurcation, respectively. Here, F_1 is the unstable focus 1-cycle, surrounded by the stable closed curve C_1.

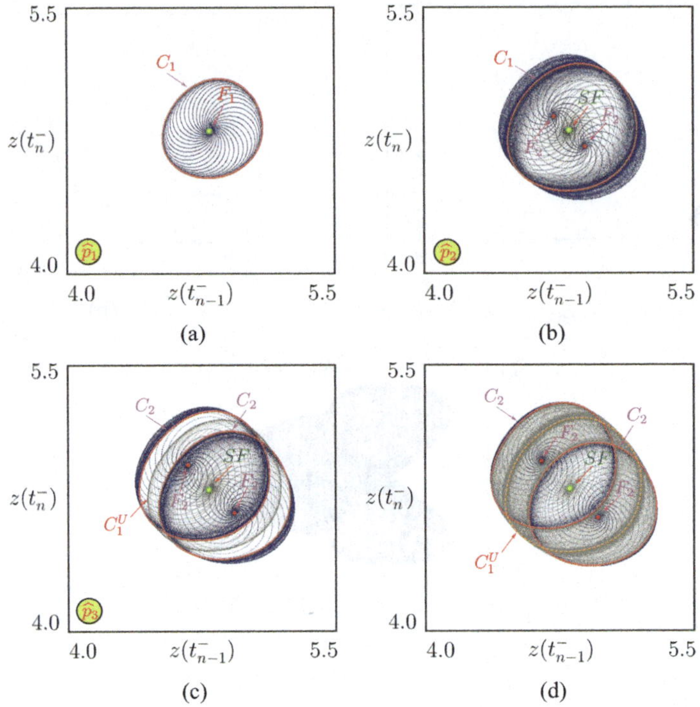

Fig. 7 **a** Two-dimensional projection of the phase portrait after the supercritical Neimark–Sacker bifurcation for 1-cycle at the point τ_φ. $\tau = 199.12480438$. **b** The phase portrait after the period-doubling bifurcation of the unstable 1-cycle at the point τ_d. $\tau = 199.05344288$. **c** The phase portrait after the quasi-periodic period-doubling bifurcation at the point τ_φ^C. $\tau = 199.04092332$). **d** The phase portrait for the for the fully developed stable period-2 closed curve C_2. $\tau = 199.03090767$

6.3 *Bifurcation Analysis: Quasi-Periodic Period-Doubling*

By making use of the discrete map of Theorem 17.2, one more example of complex dynamical behavior that is inherent in the "large delay" case is presented below.

Figure 6 depicts the results of the bifurcation analysis for $b_1 = 0.055$. Figures 6a and b exhibit a picture qualitatively similar to the case with $b_1 = 0.048$ (see Figs. 3a and b). But, unlike the previous case, the stable 2-cycle undergoes two period-doubling bifurcations: direct at the point τ_4 and reverse one at the point τ_5 (see Fig. 6b). Moreover, as the parameter b_1 increases from the value $b_1 = 0.048$, the subcritical Neimark–Sacker bifurcation for the 2-cycle transforms into the supercritical type, and the region of quasi-periodic dynamics moves from the inside to the outside of the region of multi-stability (see Figs. 3a and 6a).

Figure 6b illustrates the transitions from 1- and 2-cycles to quasi-periodic dynamics. As the time delay τ increases, the stable 2-cycle undergoes a supercritical Neimark–Sacker bifurcation at the point τ_φ^L giving birth to a stable period-2 closed

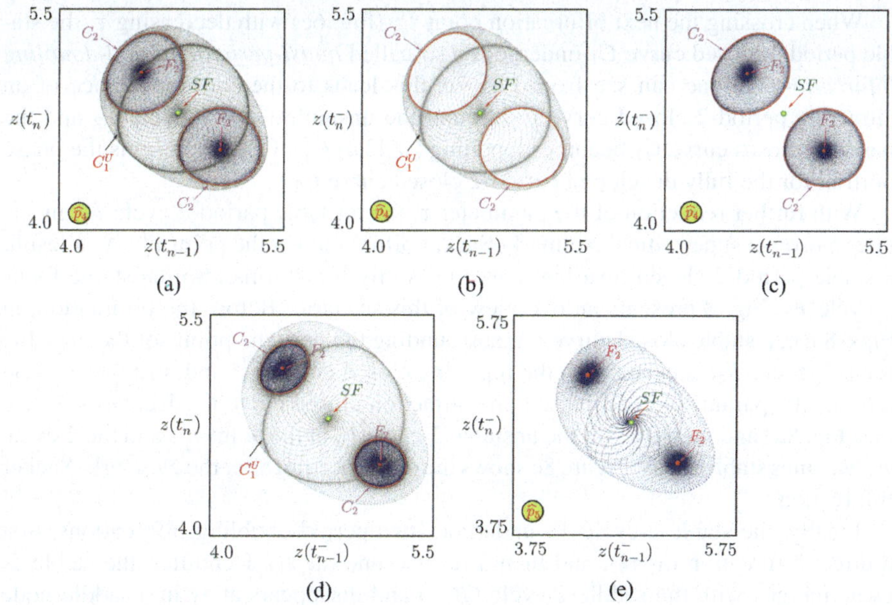

Fig. 8 Transition from the stable period-2 closed curve C_2 to the stable 2-cycle. **a** The phase portrait in the region between the points τ_φ^C and τ_φ^L. $\tau = 198.97269171$. Here, C_2 is the stable period-2 closed curve, C_1^U is the saddle period-1 closed curve, F_2 is the unstable 2-cycle, and SF is the unstable 1-cycle. **b** The phase portrait illustrates stable C_2 and repelling C^U closed invariant curves separately. **c** This phase portrait shows separately the unstable stable 2-cycle F_2 and stable closed curve C_2 in (a). **d** The phase portrait near the Neimark–Sacker bifurcation point τ_φ^L. $\tau = 198.92449$. **e** The phase portrait after the Neimark–Sacker bifurcation. Here, F_2 is the stable 2-cycle. $\tau = 198.88067$

invariant curve. Analogously, the stable 1-cycle is destabilized through a supercritical Neimark–Sacker bifurcation at the point τ_φ^R, leading to the appearance of a stable period-1 closed invariant curve. The diagram in the lower panel of Fig. 6b illustrates how a pair of complex-conjugated multipliers for the 1-cycle leave unit circle at a point τ_φ^R.

Examine now how a stable period-1 closed invariant curve is transformed into a stable period-2 closed curve. A sequence of typical phase portraits illustrating the main stages of this transition is shown in Fig. 7 (see also Fig. 6c).

At the starting point, the system displays an attracting period-1 closed invariant curve C_1 (Fig. 7a). As the value of parameter τ is decreased, the unstable focus 1-cycle F_1 undergoes a period-doubling bifurcation transverse the stable closed curve C_1 as one of the real multipliers ρ_5 leaves the unit circle through -1 at τ_d (see Fig. 6c) and the unstable period-2 cycle F_2 softly arises (Fig. 7b). The original unstable focus 1-cycle turns into the repelling focus 1-cycle SF. This is illustrated in Fig. 7a and b. The stable period-1 closed curve C_1 continues to exist after this bifurcation (Fig. 7c).

When crossing the next bifurcation point τ_{φ}^{C} (Fig. 6c) with decreasing τ, the stable period-1 closed curve C_1 undergoes a so-called *quasi-periodic period-doubling bifurcation*. As one can see from Fig. 7c, this leads to the soft appearance of an attracting period-2 closed curve C_2 around the unstable focus 2-cycle F_2 and the period-1 closes curve C_1 becomes repelling (saddle) C_1^U. Fig. 7d presents the phase portrait for the fully developed period-2 closed curve C_2.

With further reduction of the parameter τ, the unstable period-2 cycle F_2 undergoes a reverse supercritical Neimark–Sacker bifurcation at the point τ_{φ}^{L}. As a result, a stable period-2 closed invariant curve C_2 softly transforms into the stable focus 2-cycle F_2. Fig. 8 presents an overview of this scenario. Before the bifurcation, in Figs. 8a–c, a stable closed curve C_2, surrounding the periodic points of the unstable focus 2-cycle F_2, coexists with the repelling closed curve C_1^U and unstable 1-cycle SF. As the parameter τ decreased, the attraction closed curve C_2 decreases in size (see Fig. 8d) and merges with the unstable 2-cycle F_2 at the point τ_{φ}^{L}, and the 2-cycle F_2 becomes stable focus. Figure 8e shows the phase portrait after the Neimark–Sacker bifurcation.

Finally, the stable 2-cycle F_2 undergoes two period-doubling bifurcations, first a direct bifurcation (at τ_5), and then a reverse one (at τ_4). Hereafter, the stable 2-cycle merges with the saddle 2-cycle ($B_{1,1}^U$) and disappears at τ_1 in a saddle-node bifurcation (see Fig. 6b).

Moreover, in the region of multi-stability, the basins of attractions for the stable 2- or 4-cycles do not include the neighborhoods of the unstable 1-cycles (see Fig. 6b).

7 Conclusions

A specific class of delay impulsive equations with an arbitrary but finite value of the delay is considered. The previously proposed approach of reducing the analysis of an impulsive time-delay system to that of a discrete-time system is revisited and any restrictions on the duration of the time delay are removed. The hybrid dynamics are shown to be captured by a discrete impulse-to-impulse Poincaré map with memory that is proved to be instrumental in the analysis of complex nonlinear phenomena exhibited by the model.

Facilitated by the obtained discrete map, a detailed numerical bifurcation analysis of the system dynamics with respect to the time delay value reveals new complex nonlinear dynamics phenomena in the delay impulsive system, inherent to the case of large delay.

As an example of the complex dynamical structures that one can encounter, for slightly higher values of the time delay $\tau < T_n + T_{n-1} + T_{n-2}$, a structure is described, in which a pair of closed invariant curves, one attracting and one repelling, coexist with a stable focus 2-cycle. The stable manifold of the repelling closed curve separates the basins of attraction of the coexisting stable motions.

Under this condition, the stable focus 2-cycle qualifies as a hidden attractor since the basin of attraction for this stable motion does not intersect with small neighborhoods of the unstable focus or saddle 1-cycles.

As the time delay is increased, the repelling one decreases in size and merges with the stable focus 2-cycle at the subcritical Neimark–Sacker bifurcation point. This bifurcation leads to an abrupt transition from a stable 2-cycle to a stable closed invariant curve, associated with the quasi-periodic dynamics.

As another example, a bifurcation scenario is presented, in which a stable period-1 closed invariant curve undergoes a quasi-periodic period-doubling bifurcation, leading to the appearance of the attracting period-2 and repelling period-1 closed curves.

Acknowledgements A. N. Churilov was partly supported by the Government of Russian Federation, Grant 08-08. A. Medvedev was in part financed by Grant 2015-05256 from the Swedish Research Council.

References

1. V. Lakshmikantham, D. Bainov, P. Simeonov, *Theory of Impulsive Differential Equations* (World Scientific, Singapore, 1989)
2. D. Bainov, P. Simeonov, *Impulsive Differential Equations: Periodic Solutions and Applications, Harlow* (Longman, Harlow, 1993)
3. A.M. Samoilenko, N.A. Perestyuk, *Impulsive Differential Equations* (World Scientific, Singapore, 1995)
4. A.V. Anokhin, Linear impulsive systems for functional-differential equations. Dokl. Akad. Nauk SSSR **286**(5), 1037–1040 (1986)
5. K. Gopalsamy, B.G. Zhang, On delay differential equations with impulses. J. Math. Anal. Appl. **139**, 110–122 (1989)
6. D. Bainov, B. Covachev, I. Stamova, Estimates of the solutions of impulsive quasilinear functional differential equations. Ann. Fac. Sci. Toulouse Math. Ser. 5 **12**(2), 149–161 (1991)
7. D.D. Bainov, I. Stamova, Lipschitz stability of linear impulsive differential-difference equations. Note Matemat. **15**(2), 137–142 (1995)
8. A. Domoshnitsky, M. Drakhlin, Nonoscillation of first order impulse differential equations with delay. J. Math. Anal. Appl. **206**, 254–269 (1997)
9. L. Berezansky, E. Braverman, On oscillation of a second order impulsive linear delay differential equation. J. Math. Anal. Appl. **233**(2), 276–300 (1999)
10. G.H. Ballinger, Qualitative theory of impulsive delay differential equations. Ph.D. dissertation, University of Waterloo, Waterloo, ON, Canada (1999)
11. G. Ballinger, X. Liu, Existence, uniqueness and boundedness results for impulsive delay differential equations. Appl. Anal. **74**(1–2), 71–93 (2000)
12. X. Liu, X. Shen, Y. Zhang, Q. Wang, Stability criteria for impulsive systems with time delay and unstable system matrices. IEEE Trans. Circuits Syst. I: Reg. Papers **54**(10), 2288–2298 (2007)
13. I. Stamova, *Stability Analysis of Impulsive Functional Differential Equations* (Walter de Gruyter, Berlin, 2009)
14. R.P. Agarval, F. Karakoç, A survey on oscillation of impulsive delay differential equations. Comput. Math. Appl. **60**, 1648–1685 (2010)
15. G.T. Stamov, *Almost Periodic Solutions of Impulsive Differential Equations* (Springer, Berlin, 2012)

16. I. Stamova, G. Stamov, *Applied Impulsive Mathematical Models* (Springer, Berlin, 2016)
17. A.N. Churilov, A. Medvedev, A.I. Shepeljavyi, Mathematical model of non-basal testosterone regulation in the male by pulse modulated feedback. Automatica **45**(1), 78–85 (2009)
18. W. Smith, Hypothalamic regulation of pituitary secretion of lutheinizing hormone: II Feedback control of gonadotropin secretion. Bull. Math. Biol. **42**, 57–78 (1980)
19. W. Smith, Qualitative mathematical models of endocrine systems. Am. J. Physiol. **245**(4), R473–R477 (1983)
20. B.C. Goodwin, Oscillatory behavior in enzymatic control processes, in *Advances in Enzyme Regulation*, vol. 3, ed. by G. Weber (Pergamon, Oxford, 1995), pp. 425–438
21. J. Walker, J. Terry, K. Tsaneva-Atanasova, S. Armstrong, C. McArdle, S. Lightman, Encoding and decoding mechanisms of pulsatile hormone secretion. J. Neuroendocrinol. **22**, 1226–1238 (2009)
22. P. Mattsson, A. Medvedev, Modeling of testosterone regulation by pulse-modulated feedback. Adv. Exp. Med. Biol. **823**, 23–40 (2015)
23. Z.T. Zhusubaliyev, A.N. Churilov, A. Medvedev, Bifurcation phenomena in an impulsive model of non-basal testosterone regulation. Chaos **22**(1), 013121 (2012)
24. M. Cartwright, M. Husain, A model for the control of testosterone secretion. J. Theor. Biol. **123**, 239–250 (1986)
25. P. Das, A.B. Roy, A. Das, Stability and oscillations of a negative feedback delay model for the control of testosterone secretion. BioSystems **32**(1), 61–69 (1994)
26. S. Ruan, J. Wei, On the zeros of a third degree exponential polynomial with applications to a delayed model for the control of testosterone secretion. IMA J. Math. Appl. Medic. Biol. **18**(1), 41–52 (2001)
27. B. Mukhopadhyay, R. Bhattacharyya, A delayed mathematical model for testosterone secretion with feedback control mechanism. Int. J. Math. Math. Sciences **2004**(3), 105–115 (2004)
28. H. Ren, Stability analysis of a simplified model for the control of testosterone secretion. Discrete Contin. Dynam. Syst., Ser. B **4**(3), 729–738 (2004)
29. G. Enciso, E. Sontag, On the stability of a model of testosterone dynamics. J. Math. Biol. **49**, 627–634 (2004)
30. D.V. Efimov, A.L. Fradkov, Oscillatority conditions for nonlinear systems with delay. J. Appl. Math. **2007**, 72561 (2007)
31. A.N. Churilov, A. Medvedev, P. Mattsson, Periodical solutions in a time-delay model of endocrine regulation by pulse-modulated feedback. Proceedings of the 51st IEEE Conference on Decision and Control (IEEE, New York, 2012), pp. 362–367
32. A.N. Churilov, A. Medvedev, P. Mattsson, Finite-dimensional reducibility of time-delay systems under pulse-modulated feedback. Proceedings of the 52nd IEEE Conference on Decision and Control (IEEE, New York, 2013), pp. 2078–2083
33. A.N. Churilov, A. Medvedev, P. Mattsson, Periodical solutions in a pulse-modulated model of endocrine regulation with time-delay. IEEE Trans. Automat. Control **59**(3), 728–733 (2014)
34. W. Haddad, V. Chellaboina, S. Nersesov, *Impulsive and Hybrid Dynamical Systems: Stability, Dissipativity, and Control* (Princeton University Press, Princeton, 2006)
35. T. Parker, L. Chua, *Practical Numerical Algorithms for Chaotic Systems* (Springer, New York, 1989)
36. A.N. Churilov, A. Medvedev, An impulse-to-impulse discrete-time mapping for a time-delay impulsive system. Automatica **50**(8), 2187–2190 (2014)
37. Z.T. Zhusubaliyev, A.N. Churilov, A. Medvedev, Time delay induced multistability and complex dynamics in an impulsive model of endocrine regulation. Proceedings of the 13th European Control Conference (ECC) (IEEE, New York, 2014), pp. 2304–2309
38. A.N. Churilov, A. Medvedev, Z.T. Zhusubaliyev, Periodic modes and bistability in an impulsive Goodwin oscillator with large delay. Proceedings of the 19th World IFAC Congress (2014), pp. 3340–3345
39. Z.T. Zhusubaliyev, E. Mosekilde, A.N. Churilov, A. Medvedev, Multistability and hidden attractors in an impulsive Goodwin oscillator with time delay. Eur. Phys. J. Special Topics **224**, 1519–1539 (2015)

40. G.A. Leonov, N.V. Kuznetsov, V.I. Vagaitsev, Localization of hidden Chua's attractors. Phys. Lett. A **375**, 2230–2233 (2011)
41. G.A. Leonov, N.V. Kuznetsov, V.I. Vagaitsev, Hidden attractor in smooth Chua systems. Physica D **241**, 1482–1486 (2012)
42. G.A. Leonov, N.V. Kuznetsov, Hidden attractors in dynamical systems: From hidden oscillation in Hilbert-Kolmogorov, Aizerman and Kalman problems to hidden chaotic attractor in Chua circuits. Int. J. Bifur. Chaos **23**, 1330002 (2013)
43. D. Dudkowski, S. Jafari, T. Kapitaniak, N.V. Kuznetsov, G.A. Leonov, A. Prasad, Hidden attractors in dynamical systems. Phys. Rep. **637**(3), 1–50 (2016)
44. N.V. Kuznetsov, *"Hidden Attractors in Fundamental Problems and Engineering Models: A Short Survey, in AETA 2015: Recent Advances in Electrical Engineering and Related Sciences*, Lecture Notes in Electrical Engineering, vol. 371 (Springer, Cham, 2016), pp. 13–22
45. Z.T. Zhusubaliyev, E. Mosekilde, Multistability and hidden attractors in a multilevel DC/DC converter. Math. Comput. Simulation **109**, 32–45 (2015)
46. Z.T. Zhusubaliyev, E. Mosekilde, V. Rubanov, R. Nabokov, Multistability and hidden attractors in a relay system with hysteresis. Physica D **306**, 6–15 (2015)
47. H. Ahmed, R. Ushirobira, D. Efimov, On robustness of phase resetting to cell division under entrainment. J. Theor. Biol. **387**, 206213 (2015)
48. A.N. Churilov, A. Medvedev, Z.T. Zhusubaliyev, Impulsive Goodwin oscillator with large delay: periodic oscillations, bistability, and attractors. Nonlin. Anal. Hybrid Syst. **21**, 171–183 (2016)
49. J. Murray, *Mathematical Biology. I: An Introduction*, 3rd edn. (Springer, New York, 2002)
50. G.A. Leonov, N.V. Kuznetsov, T.N. Mokaev, Homoclinic orbits, and self-excited and hidden attractors in a Lorenz-like system describing convective fluid motion - Homoclinic orbits, and self-excited and hidden attractors. Eur. Phys. J. Special Topics **224**, 1421–1458 (2015)
51. S. Rasband, *Chaotic Dynamics of Nonlinear Systems* (Wiley, New York, 1990)
52. A. Agliari, Homoclinic connections and subcritical Neimark bifurcations in a duopoly model with adaptively adjusted productions. Chaos Soliton. Fract. **29**, 739–755 (2006)

Unconventional Algorithms and Hidden Chaotic Attractors

Ivan Zelinka

1 Introduction

Evolutionary algorithms (EAs) and deterministic chaos, which is a complex behavior produced by complex as well as simple dynamical systems, are tightly joined to create an interdisciplinary fusion of two interesting areas. This chapter discusses the use of EAs for numerical identification of the existence of the so-called hidden attractors (a full report is in [1]), which are part of the chaotic dynamics, as well as their synthesis [2, 3]. To understand this very specific topic, it is important to briefly explain the significance of hidden attractors for industrial technologies and also for the dynamics of evolutionary algorithms, as well as previous usages of EAs for chaotic dynamics control, identification and synthesis. In this section, we provide an overview, consequently of a) hidden attractors, b) EAs used for chaotic dynamics, and c) existence of chaos inside EAs and its impact on EAs performance, including its control.

Hidden attractors (see Fig. 4) constitute a special set of points that reflect the dynamics of the observed system, as reported in [4, 5]. In general and from a computational point of view, attractors can be regarded as self-excited or hidden attractors. Self-excited attractors can be localized numerically by a standard computational procedure, in which after a transient process a trajectory, starting from a point of unstable manifold in a neighborhood of an equilibrium, reaches a state of equilibrium or oscillation; therefore, one can easily identify it. In contrast, for a hidden attractor, the basin of attraction does not intersect with any small neighborhoods of the equilibriums. Normally, standard basins of attraction are solid part-sets that represent initial conditions, which lead a trajectory to the attractor, as demonstrated

I. Zelinka (✉)
Department of Computer Science, FEI, VSB Technical University of Ostrava, Tr. 17. Listopadu 15, Ostrava, Czech Republic
e-mail: zelinkaivan65@gmail.com

X. Wang et al. (eds.), *Chaotic Systems with Multistability and Hidden Attractors*, Emergence, Complexity and Computation 40,
https://doi.org/10.1007/978-3-030-75821-9_18

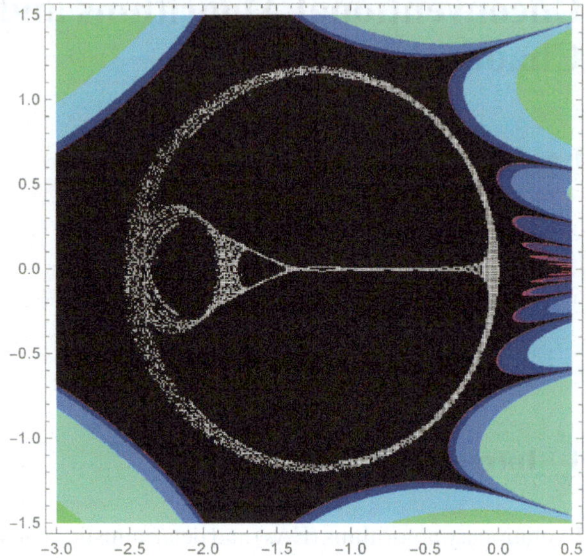

Fig. 1 A typical example of basin of attraction for the Burger system, see Eq. 1

in Fig. 1 and Eq. (1), or Fig. 2 and Eq. (2), while the basin of attraction of a hidden attractor can be quite tiny, see Fig. 6.

A hidden attractor can be a chaotic or a periodic solution—e.g. the case of coexistence of a single stationary point that is stable and a stable limit cycle (e.g., in the counterexamples to the Kalman and Aizerman conjecture) [4, 5]. On the contrary, classical attractors are self-excited attractors and, therefore, can be obtained and identified numerically by a standard computational procedure such as that for the Lorenz system. One can easily predict the existence of a self-excited attractor, while for a hidden attractor the main problem is how to predict its existence in the phase space. Thus, for localization of hidden attractors, it is important to develop special procedures, since there are no similar transient processes leading to such attractors. Few novel methods have been developed recently, for instance in [6, 7].

$$\begin{aligned} x_{n+1} &= Ax_n - y_n^2 \\ y_{n+1} &= By_n + x_n y_n \end{aligned} \tag{1}$$

and

$$\begin{aligned} x_{n+1} &= x_n^2 - y_n^2 + Ax_n + By_n \\ y_{n+1} &= 2x_n y_n + Cx_n + Dy_n \end{aligned} \tag{2}$$

If a hidden attractor is present in the system dynamics and if it is coincidentally reached, then a device (an airplane, el. circuit, etc.) starts to show quasi-cyclic behaviors that, depending on the kind of device, can cause real disasters. As an example,

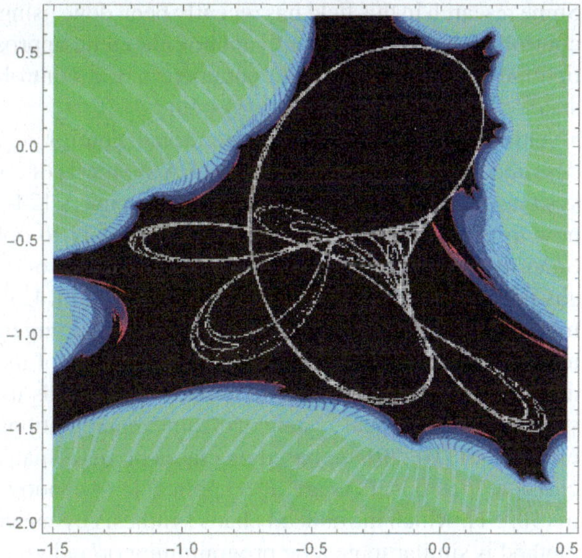

Fig. 2 A typical example of basin of attraction for the Burger system, see Eq. 2

we can use Gripen jet fighter crash[1] or F-22 raptor crash landing[2] caused by computer malfunction that led to oscillations (also called wind-up in control theory). Remember that even a computer is an electronic device containing nonlinear parts that can generate chaotic behaviors (e.g., Chua's circuit). Hidden attractors, as a part of deterministic chaos, can be studied [4, 5] just like the deterministic chaos itself, for example see [8–10]. The latest research papers, which discussed hidden attractor topics, are a very good source of information [11, 12]. In those papers, there are discussed topics like controlling hidden attractors [13] and/or its theoretical backgrounds, for example, in [6, 14–16] and [7].

In the last 15 years, it was demonstrated that evolutionary algorithms can be used successfully for deterministic chaos systems control, its identification and/or synthesis. Deterministic chaos control, see [10] (this handbook can serve as a very good reference to that topic and related areas), while control law synthesis is another area of EA applications. The control of chaotic systems has been an active area of research during the past decade. Numerous papers focusing on chaos control with EAs have been published. As an example, we can mention paper [17], where the basic ideas about chaos control, or about CML systems control, are presented. One usually expects some preliminary information to derive control law (for classical controllers). One of the first and important initial studies of EAs for control (including CML systems control) was reported in [18, 19] and [20], where the control law was based on the Pyragas method: Extended delay feedback control–ETDAS [21]. Other papers focus on the tuning of several parameters inside the control laws for a chaotic system.

---

[1] https://www.youtube.com/watch?v=jP-QMmzGL5I.

[2] https://www.youtube.com/watch?v=faB5bIdksi8.

Some research in this field has recently been done using EAs for optimization of local control of chaos based on a Lyapunov-theoretic approach [22, 23]. EA searched for optimal setting of adjustable parameters of a control method to reach the desired state or behavior of the chaotic system.

Other applications of unconventional control of chaotic systems by EAs are described in [2] and [24–26], where EA synthesis of control laws for discrete chaotic systems is discussed. Compared to that, papers [24–26] show the possibility and how to generate a whole control law (not only to optimize several parameters) in order to stabilize a chaotic system. The synthesis of control is inspired by the Pyragas delayed feedback control technique [27, 28]. Unlike the original OGY control method [17], it can be simply considered as a targeting and stabilizing algorithm together in one package. Another big advantage of the Pyragas method is the small number of accessible control parameters. Methods used in generating new chaotic systems from physical systems or from "manipulations" (e.g., control and parameter estimation) [10, 29] are based on classical mathematical analysis. Along with these classical methods, EAs are also applicable to chaotic system synthesis, as reported in [30, 31], which introduced chaos synthesis by means of a novel EA method. This method is similar to genetic programming or/and grammatical evolution. As shown in [30, 31], such an approach is able to synthesize new and "simple" chaotic systems based on some elements contained in a pre-chosen existing chaotic system and a properly defined cost function. The most significant results are carefully selected, visualized and reported in [31].

Chaos control or synthesis is not only a field for research, but also a research demonstration that chaos can also be observed inside evolutionary dynamics [31, 32], which posts a lot of questions that we will discuss below.

Chaotic systems were also used as pseudorandom number generators, like the logistic map [33], which is not the only example to use the logistic map. Another paper [34] discusses the use of the logistic map as a chaos-based true random number generator embedded in reconfigurable switched-capacitor hardware. Xing in [35] proposed an algorithm of generating a pseudorandom number generator and combined the coupled map lattice [10] via chaotic iterations. The authors also tested the algorithm in NIST 800-22 statistical test suits for its applicability in image encryption. In [36], the authors investigated interesting properties of chaotic systems in order to design a random bit generator (called CCCBG), in which two chaotic systems are cross-coupled with each other. For evaluation of the bit streams generated by the CCCBG, four basic tests are performed: monobit test, serial test, auto-correlation, and Poker test. Also, the most stringent tests of randomness: the NIST suite tests, have been used.

Several studies already dealt with the possibilities of integrating chaotic systems into the particle swarm optimization (PSO) algorithm and the performances of such algorithms, see [37, 38]. Papers [39, 40] extend the experiments of [38] and investigate the impact of using different chaotic maps on the behavior of the PSO algorithm, especially in terms of convergence speed and premature convergence risk. Three different chaotic systems (maps) are used and their impacts are compared in this study,

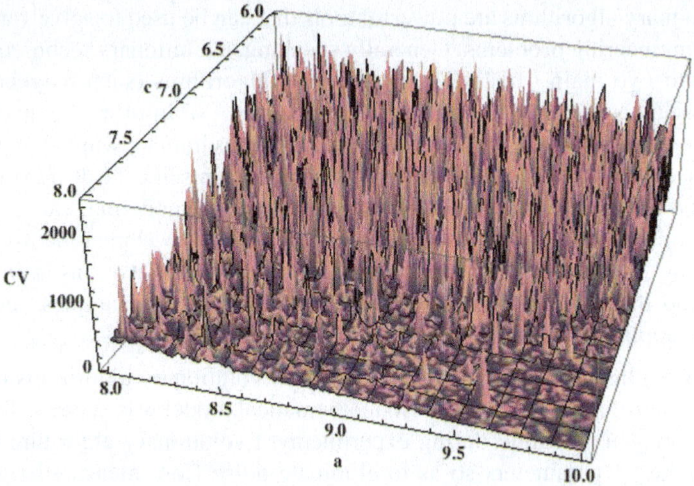

Fig. 3 A typical example of cost function surface from experiments based on EAs and chaos synchronization in [31]. The surface typically exhibits a lot of local extremes and is in general very complex

aiming to find a link between a specific chaotic system and a specific behavior of the PSO algorithm.

The above-mentioned information and references are only a fraction of existing research papers that discuss mutual fusion of EAs and chaotic dynamics. This chapter discusses the possibility on EA identification of hidden attractor existence and its synthesis, because, as stated in many papers and books (e.g. [8, 10]), basin of attraction (i.e., the set of "start" points that lead system dynamics to a hidden attractor) is very hardly identifiable. This is also discussed in papers [41] and [42], which present identification of basin of attraction for a hidden attractor (in fact the hidden attractor existence) by means of classical numerical algorithms. Identification by EAs is based on a suitable cost function definition that expresses the quality of the system state trajectory and that can be visualized as a surface. The global extreme on the surface then represents an optimal solution (or a set of start/initial points "leading" the trajectories to the hidden attractor). As an example of the complexity of such surfaces, see Fig. 3 and also [31]. From the geometry of such a surface, it is clear that EAs are the most promising tools to solve such kind of problems due to their capability to avoid local extremes.

2 Unconventional Algorithms—Motivation and Brief Introduction

Before we start to describe how EAs can be used for identification and synthesis, we explain at least basic principles of EAs (for a fully detailed introduction, see [31]).

Evolutionary algorithms are powerful tools that can be used to solve various very complex engineering problems. Generally speaking, evolutionary techniques can be divided into two main categories: evolutionary algorithms (such as genetic algorithms, particle swarming, ant colony optimization, ...). In principle, evolutionary techniques solve selected problems in the same way as human, which in general can be used successfully to a large set of engineering problems like the design of different devices and complex systems identification, control and modeling, etc.

Evolutionary techniques have been in existence for quite a long time, successfully solved many complex problems, showing their powerful applications in engineering practice and theory (see for example [31]). The following examples demonstrate successful applications of evolutionary techniques:

- **Real-time plasma reactor control.** Selected evolutionary algorithms have been used to control a plasma reactor. No mathematical model was needed. The reactor was running in real time during experiments. Evolutionary algorithm was used to estimate 14 parameters so as to eliminate noise from measured signals (see [43] and [44]). The most important aspect of this contribution is the demonstrated ability of evolutionary techniques for controlling black-box problems in real-time. Another very important contribution is the use of laboratory hardware equipment to calculate the fitness of just-in-time synthesized solutions without knowing a cost function.
- **Fingerprint identification.** Evolutionary algorithms have been used by computer scientists (e.g., Grasemann and Miikkulainen, Neural Networks Research Group from the University of Texas at Austin, USA). Genetic algorithm has been used to develop a program, which can digitally improve the quality of fingerprint images better than programs created by human programmers. During evolution, the best solutions of each generation in the algorithm were recorded and discovered. After 50 generations, the algorithm outperforms the comparable human program. The algorithm was then synthesized. In a typical FBI application, taking into consideration that the FBI has more than 50 million sets of fingerprints in its archives, so any algorithm needs to perform about 60,000 digital fingerprint image transactions every day, it is clear that evolutionary algorithms might soon help speed up such a time-consuming identification process.Results showed that the genetic algorithm was about 13% more accurate than the well-known 2D algorithm within the same running time.
- **Airplane optimization.** Evolutionary techniques are also widely used in airplane engineering [45]. There are numerous examples of wing optimization and optimal design of various mechanical parts of an airplane under investigation. An interesting approach is described in [46], where minimization of sonic boom on a supersonic aircraft was based on an evolutionary algorithm.
- **Antenna design.** In this application, evolutionary hardware has been used to design special antenna for NASA space mission (e.g., the Space Technology 5 Project (ST5)[3]). The ST5 space program is focused on the use of identical satellites to test new space technologies. It is part of the New Millennium Program (NMP). The

[3] http://nmp.jpl.nasa.gov/st5/.

NMP was created to identify, develop, build, and test innovative technologies and concepts to incorporate into future space missions. Despite numerous engineering designs, this particular design was deemed the best.

- **Flow shop scheduling—permutation-based combinatorial optimization.** A very important application in industry is scheduling, to which a number of manufacturing problems are associated. There are many such problems that cannot be solved by using conventional methods or cannot be solved in reasonable time (see [31]). Such problems can be successfully solved by evolutionary algorithms, however. A typical problem is the traveling salesman problem. Today, there exists the so-called ACO algorithm (see [47]), which is able to solve this hard problem with cardinality 10000! in a reasonable time yielding good results. The class of permutation-based combinatorial optimization problem is one of the famous optimization problems just like traveling salesman problem and vehicle routing problem. The most realistic and interesting are the shop scheduling problems for flow shop and job shop.
- **Chemical reactor design.** In chemical engineering, evolutionary optimization has been applied to system identification [48, 49], where a model of a process is built and then its parameters are identified by error minimization against experimental data. Evolutionary optimization has been widely applied to the evolution of neural network models for use in control applications (e.g., [50]). There has been increasing awareness of textbook knowledge and heuristics [51], which were commonly employed in the development of chemical reactors, were deemed responsible for the lack of innovation, quality, and efficiency that characterizes many industrial designs. In such examples, among many others, it has been proved that evolutionary techniques are highly effective for the application in chemical engineering.
- **Bioinformatics.** Bioinformatics applies information technology to the field of molecular biology. The term bioinformatics was coined by Hogeweg in 1978 for the study of informatic processes in biotic systems. Today, it entails the creation and advancement of databases, algorithms, computational and statistical techniques, as well as the theory to solve formal and practical problems arising from management and analysis of biological data. Evolutionary algorithms have been successfully applied in, for example, multi-objective optimization in modeling of protein structure prediction [52], evolutionary optimization of metabolic pathways in [53], and so on (see [54]).

There are many other examples worth of mentioning, where evolutionary algorithms find successful applications. Standard evolutionary algorithms exist since the famous seminal paper [55], however it is not completely true, for more see [31].

If the evolutionary principles are used for the purposes of complex calculations, see [31], the following EA principles are used:

1. Specification of the evolutionary parameters: For each algorithm, parameters that control the run of the algorithm or terminate it must be defined regularly, if the termination criterion defined in advance is fulfilled (for example, the number of cycles—generations). Part of this point is the definition of the cost function (objective function) or, as the case may be, what is called fitness–a modified

return value of the objective function). The objective function is usually a mathematical model of the problem, whose minimization or maximization leads to the solution of the problem. This function with possible limiting conditions is some kind of environmental equivalent in which the quality of current individuals is assessed.
2. Generation of the initial population (generally $N \times M$ matrix, where N is the number of parameters of an individual—D is used hereinafter in this publication—and M is the number of individuals in the population): Depending on the number of optimized arguments of the objective function and the user's criteria, the initial population of individuals is generated. An individual is a vector of numbers having such a number of components as the number of optimized parameters of the objective function. These components are set randomly and each individual thus represents one possible specific solution to the problem. The set of individuals is called population.
3. All the individuals are evaluated through a defined objective function, and each of them is assigned either a) a direct value of the return objective function, or b) a fitness value, which is a modified (usually normalized) value of the objective function.
4. Now parents are selected according to their quality (fitness, value of the objective function) or, as the case may be, also according to other criterions.
5. Descendants are created by crossbreeding the parents. The process of crossbreeding is different for each algorithm. Parts of parents are changed in classic genetic algorithms, in a differential evolution, crossbreeding is a certain vector operation, etc.
6. Every descendant is mutated. In other words, a new individual is changed by means of a suitable random process. This step is equivalent to the biological mutation of the genes of an individual.
7. Every new individual is evaluated in the same manner as in step 3.
8. The best individuals are selected.
9. The selected individuals fill a new population.
10. The old population is forgotten (eliminated, deleted, dies,..) and is replaced by a new population; step 4 represents further continuation.

Steps 4–10 are repeated until the number of evolution cycles specified before by the user is reached or until the required quality of the solution is achieved. The principle of the evolutionary algorithm outlined above is general and may more or less differ in specific cases. For more explanation, it is recommended to read chapters 1–3 in [31].

There are also exemptions that do not adhere to steps 1–10; in such a case, the corresponding algorithms are not denoted as evolutionary algorithms, but usually as algorithms that belong to EAs. Some evolutionists exclude them completely from the EA class. The ACO algorithm (*Ant Colony Optimization*), see [56], may be an example–it simulates the behavior of an ant colony and can solve extremely complicated combinatorial problems. It is based on the principles of cooperation of several individuals belonging to the same colony, which are ants in the present case.

As an example, we can mention two kinds of algorithms, i.e. differential evolution (DE), that has been derived from genetic algorithms and belongs today amongst the most powerful algorithms and SOMA, that is of comparable performance to modern algorithms including DE and represents swarm algorithms as well as PSO for example.

Differential Evolution [57] is a population-based optimization method that works on real-number coded individuals. For each individual $\mathbf{x}_{i,G}$ in the current generation G, differential evolution (DE) generates a new trial individual $\mathbf{x}'_{i,G}$ by adding the weighted difference between two randomly selected individuals $\mathbf{x}_{r1,G}$ and $\mathbf{x}_{r2,G}$ to a randomly selected third individual $\mathbf{x}_{r3,G}$. The resulting individual $\mathbf{x}'_{i,G}$ is crossed-over with the original individual $\mathbf{x}_{i,G}$. The fitness of the resulting individual, referred to as a perturbed vector $\mathbf{u}_{i,G+1}$, is then compared with the fitness of $\mathbf{x}_{i,G}$. If the fitness of $\mathbf{u}_{i,G+1}$ is greater than the fitness of $\mathbf{x}_{i,G}$, then $\mathbf{x}_{i,G}$ is replaced with $\mathbf{u}_{i,G+1}$; otherwise $\mathbf{x}_{i,G}$ remains in the population as $\mathbf{x}_{i,G+1}$. Differential Evolution is robust, fast, and effective with a global optimization ability. It does not require the objective function to be differentiable, and it works well even with noisy, epistatic and time-dependent objective functions.

Pseudocode for DE, especially for DERand1Bin, is:

$$
\begin{array}{l}
\text{1. Input}: D,\ G_{\max},\ NP \geqslant 4,\ F \in (0, 1+),\ CR \in [0, 1],\ \text{and initial bounds}: \mathbf{x}^{(lo)},\ \mathbf{x}^{(hi)}. \\
\text{2. Initialize}: \begin{cases} \forall i \leqslant NP \wedge \forall j \leqslant D: x_{i,j,G=0} = x_j^{(lo)} + rand_j[0,1] \cdot \left(x_j^{(hi)} - x_j^{(lo)}\right) \\ i = \{1, 2, \ldots, NP\},\ j = \{1, 2, \ldots, D\},\ G = 0,\ rand_j[0,1] \in [0,1] \end{cases} \\
\begin{cases}
\text{3. While } G < G_{\max} \\
\forall i \leqslant NP \begin{cases}
\text{4. Mutate and recombine}: \\
\text{4.1 } r_1, r_2, r_3 \in \{1, 2, \ldots, NP\},\ \text{randomly selected, except}: \\
\quad r_1 \neq r_2 \neq r_3 \neq i \\
\text{4.2 } j_{rand} \in \{1, 2, \ldots, D\},\ \text{randomly selected once each } i \\
\text{4.3 } \forall j \leqslant D,\ u_{j,i,G+1} = \begin{cases} x_{j,r_3,G} + F \cdot (x_{j,r_1,G} - x_{j,r_2,G}) \\ \text{if}(rand_j[0,1] < CR \vee j = j_{rand}) \\ x_{j,i,G} \text{otherwise} \end{cases} \\
\text{5. Select} \\
\mathbf{x}_{i,G+1} = \begin{cases} \mathbf{u}_{i,G+1} \text{ if } f(\mathbf{u}_{i,G+1}) \leqslant f(\mathbf{x}_{i,G}) \\ \mathbf{x}_{i,G} \text{ otherwise} \end{cases}
\end{cases} \\
G = G + 1
\end{cases}
\end{array}
\tag{3}
$$

SOMA is a stochastic optimization algorithm that is modeled based on the social behavior of competitive-cooperating individuals [58]. It was chosen because it has been proved that this algorithm has the ability to converge towards the global optimum [58]. SOMA works on a population of candidate solutions in loops, called migration loops. The population is initialized by uniform random distribution over the search space at the beginning of the search. In each loop, the population is evaluated and the solution with the best cost value becomes the Leader. Apart from the Leader, in one migration loop, all individuals will traverse the searched space in the direction of the leader. It ensures diversity amongst all the individuals and it also provides a means to restore lost information in a population. Mutation is different in SOMA as compared with other EAs. SOMA uses a parameter called PRT to achieve perturbations. The PRT vector defines the final movement of an active individual in the search space. The randomly generated binary perturbation vector controls the allowed dimensions for an individual. If an element of the perturbation vector is set to zero, then the individual is not allowed to change its position in the corresponding dimension. An individual will travel over a certain distance (called the $PathLength$) towards the leader in finite steps of the defined length. If the $PathLength$ is chosen to be greater than one, then the individual will overshoot the Leader. This path is perturbed randomly.

Pseudocode for SOMA is:

$$\text{Input}: N, Migrations, PopSize \geqslant 2, PRT \in [0,1], Step \in (0,1], \text{ MinDiv} \in (0,1], \text{ PathLength} \in (0,5], Specimen \text{ with uper and lower bound } x_j^{(hi)}, x_j^{(lo)}$$

$$\text{Inicialization}: \begin{cases} \forall i \leqslant PopSize \wedge \forall j \leqslant N : x_{i,j,Migrations=0} = x_j^{(lo)} + \\ \qquad rand_j\,[0,1] \cdot \left(x_j^{(hi)} - x_j^{(lo)}\right) \\ i = \{1, 2, \ldots, Migrations\},\ j = \{1, 2, \ldots, N\}, \\ \qquad Migrations = 0, rand_j[0,1] \in [0,1] \end{cases}$$

$$\begin{cases} \text{While } Migrations < Migrations_{\max} \\ \forall i \leqslant PopSize \begin{cases} \text{While} t \leqslant PathLength \\ if\ rnd_j < PRT\ pak\ PRTVector_j = 1\ else\ 0,\ j = 1, \ldots, N \\ x_{i,j}^{ML+1} = x_{i,j,start}^{ML} + (x_{L,j}^{ML} - x_{i,j,start}^{ML})\ t\ PRTVector_j \\ f\left(x_{i,j}^{ML+1}\right) = \text{ if } f\left(x_{i,j}^{ML}\right) \leqslant f\left(x_{i,j,start}^{ML}\right) \text{ else } f\left(x_{i,j,start}^{ML}\right) \\ t = t + Step \end{cases} \\ Migrations = Migrations + 1 \end{cases} \tag{4}$$

Both algorithms have been used in all our experiments mentioned here and also can be used in all of our research proposals as reported at the end.

3 Evolutionary Identification—Case Example

The performance and possibility of EAs use on hidden attractor identification have been fully described and demonstrated in [1]. In this chapter, we only briefly mentioned the principles and some results. For full information, it is recommended to read [1]. In this case study, DE and SOMA have been used in order to identify whether hidden attractor exists inside system or not, i.e. EA identification of hidden attractors did not reconstruct basin of attraction of hidden attractors but served as a binary classification whether hidden attractor is present or not.

3.1 Used Algorithm and Its Setting

All experiments reported in [1] were done with selected EAs in repeated numerical simulations. The used algorithms were differential evolution (DERand1Bin) and

[57], SOMA (AllToOne) [58]. In our future research, other algorithms, like genetic algorithms (GAs) [59], simulated annealing (SA) [60, 61], PSO [62] or ES [63, 64], are extensions of this study with more systems that contain hidden attractors. The setting of the algorithm is given in [1].

Both algorithms (i.e. SOMA + DE) have been applied 100 times in order to find the optimum of both case studies. The primary aim of this comparative study **is not to show** which algorithm is better or worse, but to show whether evolutionary algorithms are applicable to the identification of the basin of attraction related to a hidden attractor.

All experiments were done in Mathematica 10, on MacBook Pro, 2.8 GHz Intel Core 2 Duo.

3.2 Used System with Hidden Attractor

Typical hidden attractor, used in paper [1] comes from electronics and has been reported in [4, 65] or in special issue [5] and another research papers,[4] [11, 12]. It is Chua's attractor that can be observed, for example, in the electronic circuit of Chua's oscillator. Electronic circuits are among the most popular systems used to demonstrate deterministic chaos. Their popularity stems from the fact that electronic circuits are easy to set up and provide fast response to control inputs and settings. Typical representatives of electronic circuits with deterministic chaos is Chua's oscillator, see Fig. 4. The core of Chua's circuit is a nonlinear resistor, sometimes called Chua's diode [1, 66].

In Fig. 4 Chua's attractor visualized by the program Mathematica together with its hidden attractor. Chua's circuit can be described mathematically by Eq. (5), which can be used to simulate the behavior of the circuit:

$$
\begin{aligned}
&\text{Chua's system}\\
&x'(t) = \alpha(y(t) - x(t) - m_1 x(t) - 0.5(m_0 - m_1) sat(x))\\
&y'(t) = x(t) - y(t) + z(t)\\
&z'(t) = -\beta y(t) - \gamma z(t)\\
&\\
&\text{with saturation function}\\
&sat(x) = |x(t) + 1| - |x(t) - 1|\\
&\\
&\text{initial conditions}\\
&x(0) \in [-10, 10], y(0) \in [-10, 10], z(0) \in [-15, 15]\\
&\\
&\text{and parameter setting}\\
&\alpha = 8.4562, \beta = 12.0732, \gamma = 0.0052\\
&m_0 = -0.1768, m_1 = -1.1468
\end{aligned}
\tag{5}
$$

[4] http://www.math.spbu.ru/user/nk/.

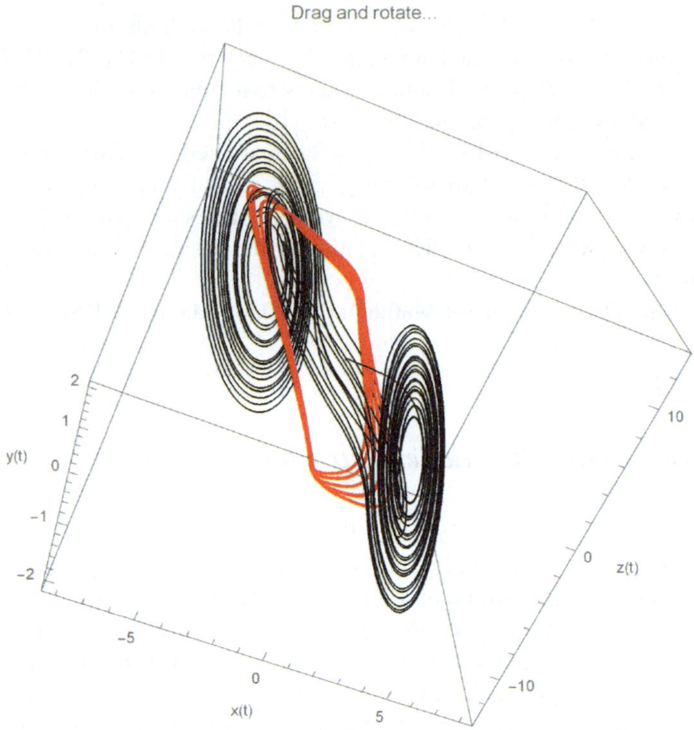

Fig. 4 Chua's attractor with a hidden attractor in red "circle" of points in bigger Chua's attractors, both were obtained for different parameter sets

If suitable initial conditions are set as described in (5), a chaotic attractor can be found in the system (Fig. 4). In our experiments Chua's and hidden attractor have been set according to [4] and [5]. It is reported in Eq. (5). Besides Chua's system, there are also discrete systems exhibiting HA, as reported in [6], however in [1] is reported Chua's system as a more real (i.e. electronic) system.

3.3 Cost Function and Its Visualization

The most important part of the experiments was the cost function definition, that is then used by evolutionary algorithms and is a core of whole process. With wrong cost function definition one would get definitely wrong results. So, its construction–setting up, is crucial for the EAs use. The cost function was based on Chua's circuit with setting for its hidden attractor. For experiments reported here few simplifications have been done, because in fact, in Chua's system there are 8 adjustable parameters: 3 for initial conditions ($x(0)$, $y(0)$ and $z(0)$) and 5 for parameter settings (m_0, m_1, α, β and

γ). So in total it is possible to search in the 8-dimensional space (compare complexity of simpler chaotic case in Fig. 3, [1]). For simplicity we fixed 5 parameters (so that hidden attractor would exist) and only 3 initial conditions were under investigation by EAs in order to find points belonging to the domain of its attraction. The cost function was set up as in Eq. (6). System (5) of three differential equations has been calculated for each evaluation of Eq. (6) in time $t \in [0, 200]$ and trajectories data from the last 50s has been used (to avoid possible initial trajectory transition with bigger amplitudes, as can be observed in [1]). The parameter Diam(eter) has been set empirically to be $\in [3, 11]$ and guarantees that if trajectory fulfills Eq. (6), then trajectory is trapped in hidden attractor. Otherwise, the returned value of Eq. (6) was usually extremely big, as 10^8, 10^{153}, etc., due to the trajectory escaping to the infinity. Thus, individuals (i.e. coordinates of the start trajectory) with CF=0 represented solutions.

$$\begin{aligned} &CF_{temp} = Max\left\{ \sum_{t=150s}^{200} \sqrt{x(t)^2 + y(t)^2 + z(t)^2} \right\} \\ &IF\ \mathrm{CF_{temp}} \in \mathrm{Diam\ then\ CF} = 0\ \mathrm{else\ CF} = \mathrm{CF_{temp}} \\ &Diam \in [3, 11] \end{aligned} \tag{6}$$

Dynamics of Eq. (5) according to Eq. (6), can be visualized in few different ways and as an example, few typical visualizations are depicted here. Because there were 3 variables (i.e. cost function surface is in 4-dimensional space), cost function surface can be visualized so that one variable is set to constant value and the remaining two are changed in the allowed interval. Typical examples of figures is then Fig. 6, for more see [1]. The black area represents the basin of attraction, i.e. the set of initial points whose trajectory leads to the standard Chua's attractor (see bigger attractor in Fig. 4). On the contrary, when parameters (m_0, m_1, α, β and γ) are set for hidden attractor existence, then basin of attraction is very small, as shown in Fig. 5, or even complex as Fig. 6 suggest. Both figures show, how basin of attraction changes its shape and structure, when for example $z(0)$ is changed (in fact they are xy slice of the attraction domain for a fixed $z(0)$.). It is also important to mention that graphic visualization of basins of attraction is only an approximation of the real one and depends on the setting of the graphical resolution which, of course, is limited on PCs. Confront with Fig. 6, where high resolution has to be set in numerical computation in order to made those few points visible.

It is logical to expect that the structure of the complete basin of attraction based on Eq. (6), will be more complex and changed as all 8 parameters of the Eq. (5) will be changed. Here, as in the initial study on this matter, we are concerned on EAs identification of initial conditions that lead to hidden attractor, when parameters $x(0)$ and $y(0)$ searched in interval $[-10, 10]$, for $z(0) = 8.7739$ and -13.4705, i.e. individual has "only" dimension 2. For more details, see [1].

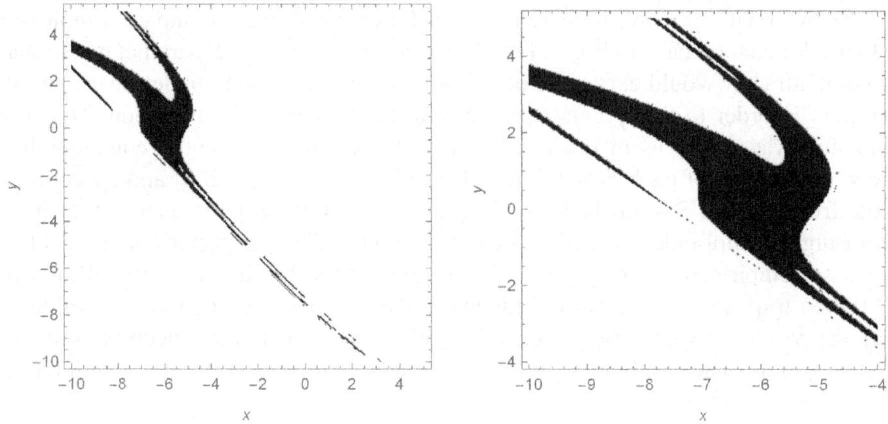

Fig. 5 Basin of attraction (black area) for Chua's hidden attractor with specific initial conditions in 2D ($z(0) = 8.7739$)

3.4 Results

The results of EAs identification of the basin of attraction come from experiments with SOMA and DE [1]. Both algorithms were repeated 100 times and for each individual, and its cost function evaluation was calculated by Eq. (5). All visualizations have been done by means of Mathematica 10. In both algorithms and all experiments were successfully located initial points that belong to the basin of attraction, i.e. we have successfully identified its existence within the system. As results in [1] shows then for algorithm SOMA (for example) has located basin of attraction after few migrations. The same can be stated for DE. Another interesting result was that both EAs, during their evolution, have identified not only one hidden attractor, as reported in [4], but also stable limit cycle in it, as is depicted in [1] where it is visible that despite the fact that both kinds of trajectories are very close at the start, they end up in different behavior as hidden attractor or limit cycle is. This just confirms that basin of attraction shall be really complex and tiny. For more details see [1].

Besides continuous dynamical systems like Chua's system, there are discrete dynamical systems that exhibit chaotic behavior, which are already under study in [6]. Such systems have different basins of attraction and of course attractors themselves, see Fig. 7, Eq. (7). The use of EAs on such systems is, in principle, the same and is at the moment under our investigation.

$$\begin{aligned} x_{n+1} &= y_n \\ y_{n+1} &= x_n - 0.3y_n^2 + 0.3x_n y_n + 1.09 \end{aligned} \tag{7}$$

Another possibility to use EAs on hidden attractors research is the so-called symbolic regression, i.e. evolutionary synthesis of complex structures, in this case of chaotic systems containing hidden attractors. In the next subsections we discuss

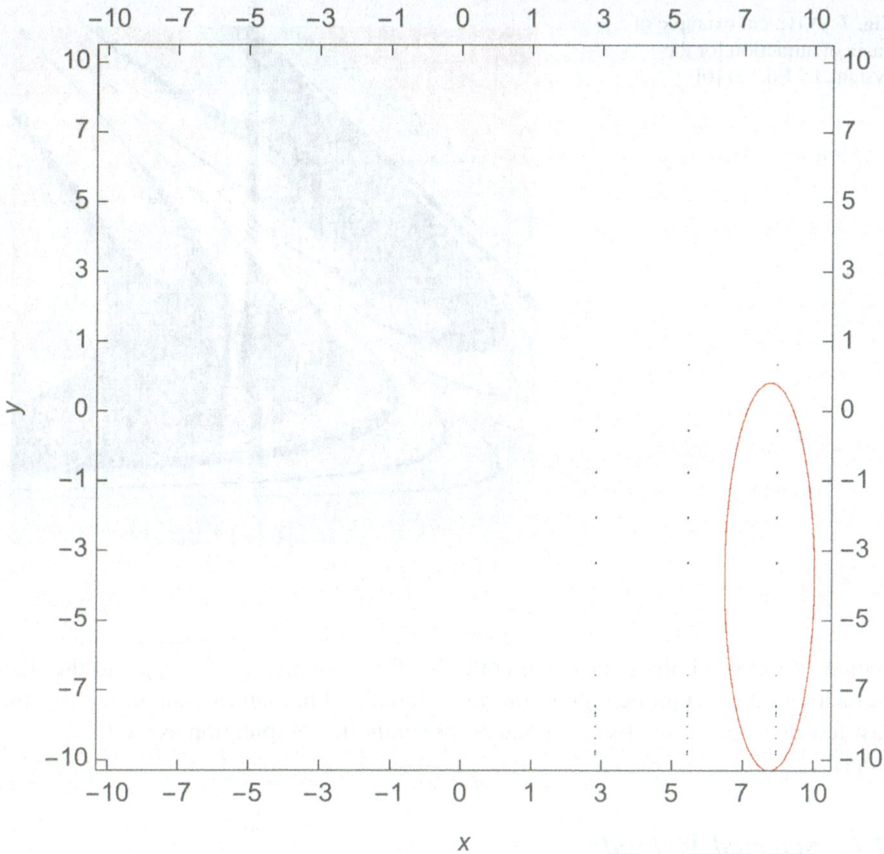

Fig. 6 Basin of attraction (black dots) for Chua's hidden attractor with specific initial conditions in 2D ($z(0) = -13.4705$). Black dots are scattered through space of possible solutions (see red oval capturing a few of them)

basic principles of symbolic regression and show its use on classical chaotic system synthesis use. Possibilities on synthesis of hidden attractors containing systems are then discussed in further section about research ideas—proposals at the end.

4 Evolutionary Synthesis

Evolutionary synthesis is, in fact, a process during which evolution is used as a tool that **creates** new structures based on user demands and basic object (i.e. building blocks) definition. It can be generally called a *symbolic regression*. The term "symbolic regression" represents a process during which measured data sets are fitted thereby a corresponding mathematical formula is obtained in an analytical way. An

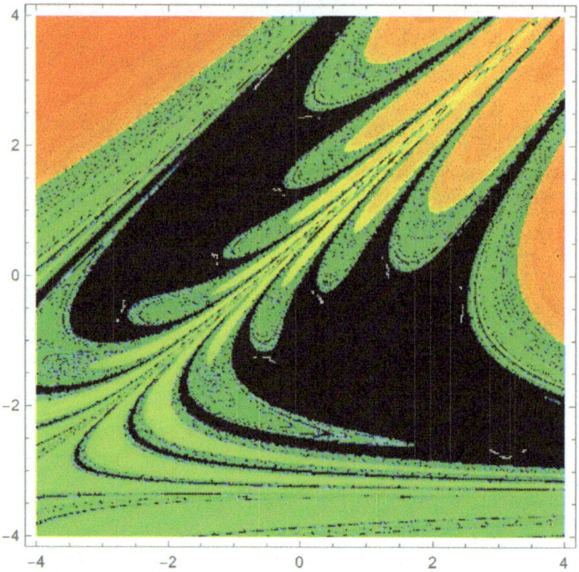

Fig. 7 A typical example of basin of attraction for the system, by Eq. (7) [6]

output of the symbolic expression could be, for example, $\sqrt[N]{x^2 + \frac{y^3}{k}}$, and the like. For a long time, symbolic regression was a domain of human calculations, but in the last few decades, it involves computers for symbolic computation as well.

4.1 Selected Methods

The initial idea of symbolic regression by means of a computer program was proposed in Genetic Programming (GP) [3, 67]. The other approaches are Grammatical Evolution (GE) developed in [68] and Analytic Programming (AP) in [2]. Other interesting investigations using symbolic regression were carried out in [69] on Artificial Immune Systems and Probabilistic Incremental Program Evolution (PIPE), which generates functional programs from an adaptive probability distribution over all possible programs. As an extension of GE to the other algorithms is also [70], where differential evolution was used with the GE. Symbolic regression is, generally speaking, a process which combines, evaluates and creates more complex structures based on some elementary and non-complex objects, in an evolutionary way. Such elementary objects are usually simple mathematical operators ($+$, $-$, $\times$, ...), simple functions (sin, cos, And, Not, ...), user-defined functions (simple commands for robots–MoveLeft, TurnRight, ...), etc. An output of symbolic regression is a more complex “object” (formula, function, command,...), solving a given problem like data fitting of the so-called Sextic and Quintic problem described in [2, 3], randomly synthesized function [2], Boolean problems of parity and symmetry solution (basically logical circuits synthesis) or synthesis of quite complex robot control command as in [2, 3]. Examples mentioned here are just a few samples from numerous repeated experiments done by AP, which are used to demonstrate how complex structures can be produced by symbolic regression in general for different problems.

The method described and demonstrated for experiments in this part is called Analytic Programming (AP), which has been compared to GP with very good results (see, for example, [2, 3, 31]).

4.2 *Chaotic Systems Synthesis*

The capability of the EAs to synthesize chaotic systems has been clearly demonstrated in [30] and [31] where an artificial (i.e. no real equivalent systems are in nature) chaotic systems have been synthesized. In this case, synthesis was focused only on "non-hidden" chaotic systems, however it is clear that it is obviously only a matter of little changes in our previous synthesis algorithm.

For an inspiration and motivation, consider how EAs were used for chaos synthesis by mentioning some main results. For full and detailed information it is recommended to read [30] and [31]. The synthesis was based on logistic equation structure and on the fact that the logistic equation is a well-known simplest system that can produce chaotic behavior. This equation is also well analyzed. It was expected that evolutionary search would be possible to synthesize the logistic equation, which was, in fact, a source of elements for GFS. Evolutionary synthesis of logistic equation was actually observed, as further discussed in [30] and [31]. Another reason behind the selection of the logistic equation is that results from designed experiments can be easily compared, verified and analyzed.

Basic set of objects used in symbolic regression are $\{\ x, A, +, -, \times\ /\}$. It is also important to note that experiments provided here, i.e., evolutionary synthesis of chaotic systems, are not restricted to one-dimensional chaotic maps but can be applied in principle to the synthesis of higher-dimensional and more complex chaotic systems. This declaration is based on many other successful complex examples accomplished by GP, GE and AP in the past.

In this investigation, a total of 1300 independent simulations were completed, 100 trials by each of the 13 algorithms. Each simulation was started at randomly selected initial conditions (i.e., each initial population was randomly generated).

During our experimentation we have got a lot of very interesting chaotic systems, for selected examples see Figs. 8, 9, 10 and 11 related to Eqs. (8)–(15). In [30] and [31], one can see full results and analysis of obtained systems from different point of views.

$$-\frac{A(A+2x)}{2\left(\frac{A\left(Ax+\frac{A}{x}\right)}{\left(A^2-A+1\right)(Ax+2x)(A(x-A)+A)}+x\right)} \tag{8}$$

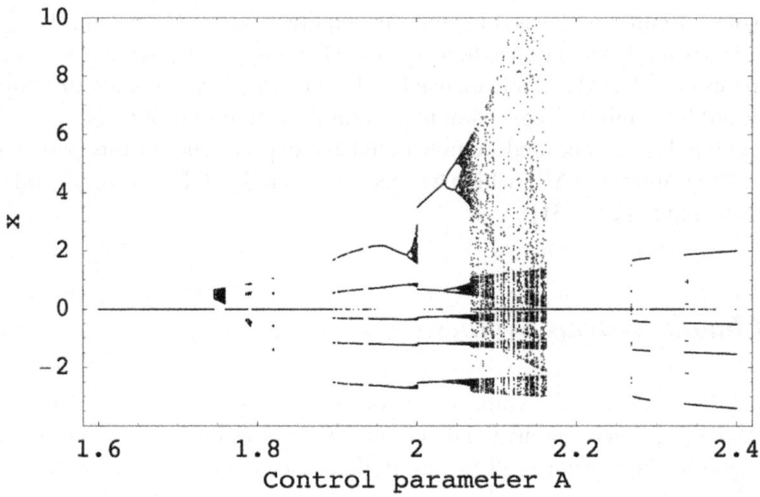

(i)

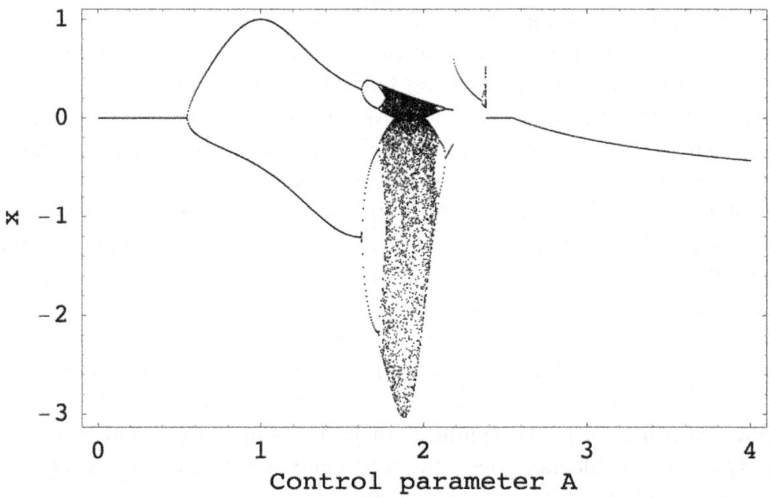

(ii)

Fig. 8 Engineering design: bifurcation diagram

$$\frac{A\left(-A^2+A+x\right)+A^2+A-x}{A^2(-x)+A\left(A^2+x\right)-\frac{A+x}{Ax}-A-2x} \tag{9}$$

$$-\frac{Ax\left(Ax-A^2(x-2A)\right)}{A\left(-A-x^2+x\right)-x} \tag{10}$$

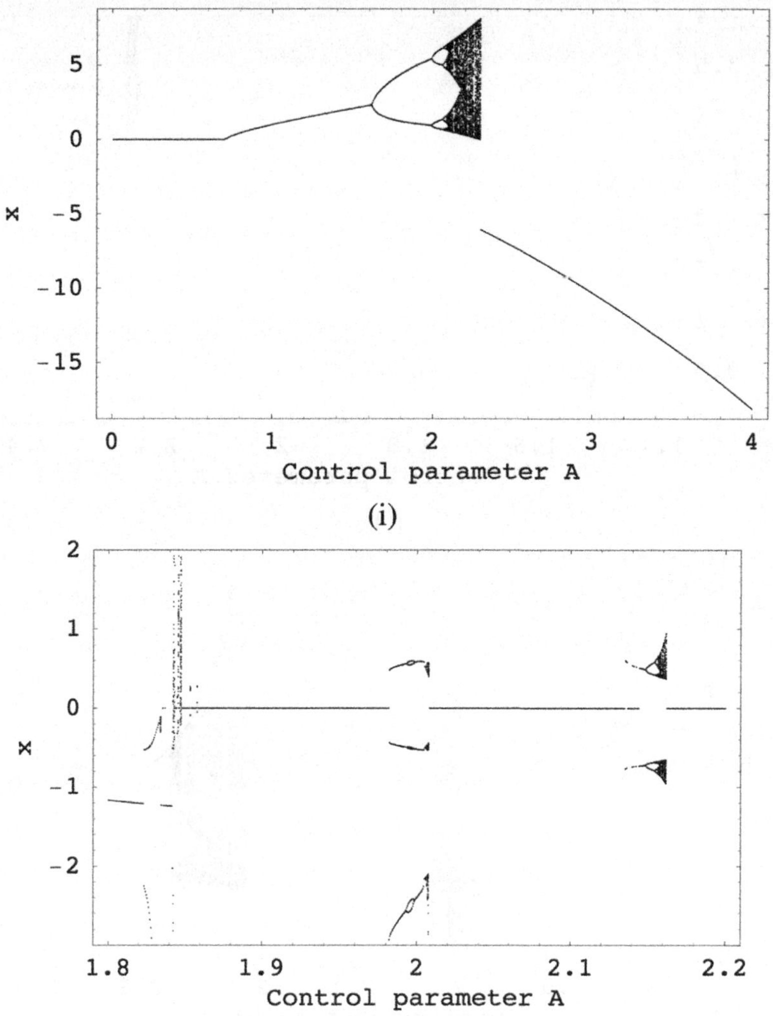

(ii)

Fig. 9 Engineering design: bifurcation diagram

$$\frac{Ax}{\left((A+x)\left(-A^2+A+x\right)+A+1\right)\left(A-\frac{A\left(\frac{x}{2\left(-\frac{A}{x}+A-x\right)(A+x)}+\frac{x+1}{A}\right)}{2x}\right)} \tag{11}$$

$$\frac{A\left(2A(x-A)+\frac{1}{A-x}+A\right)}{-\frac{1}{A^2x^2}+\frac{4x^3}{A}-x(x-A)-x} \tag{12}$$

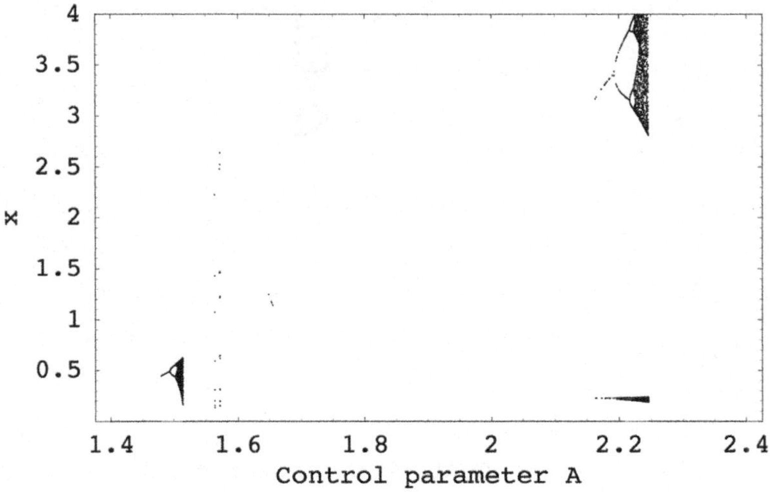

(i)

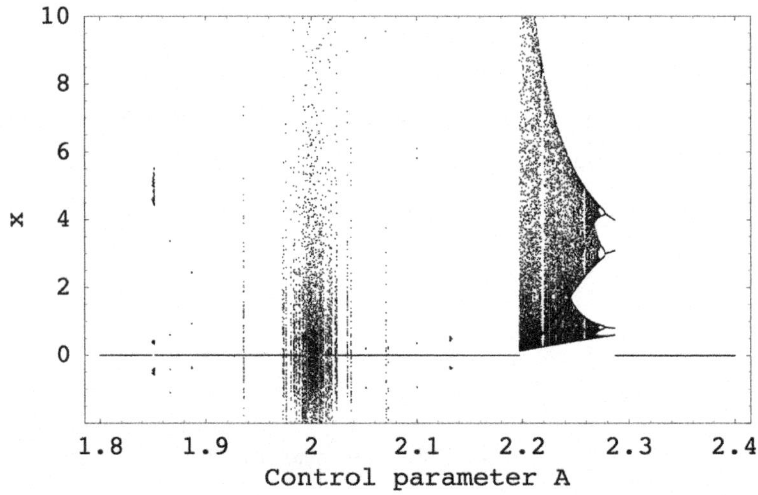

(ii)

Fig. 10 Engineering design: bifurcation diagram

$$-\frac{Ax^2}{-A^2+2A-x^3+x} \tag{13}$$

$$\frac{A}{x\left(\frac{A}{-A^2+Ax+A-x^2+x}+Ax^2-x\right)+\frac{A}{2x}+A+x} \tag{14}$$

$$\frac{A^2}{\left(Ax(A-x)+\frac{x}{A}+A-3x\right)\left(\left(\frac{x}{A}+A-x^2\right)\left(Ax^2+x\right)+\frac{Ax\left(\frac{x}{2A+2x}-Ax\right)}{\frac{A}{x(A+x)}-\frac{x}{A-x}}-A\right)-\frac{A}{A+x}-A} \tag{15}$$

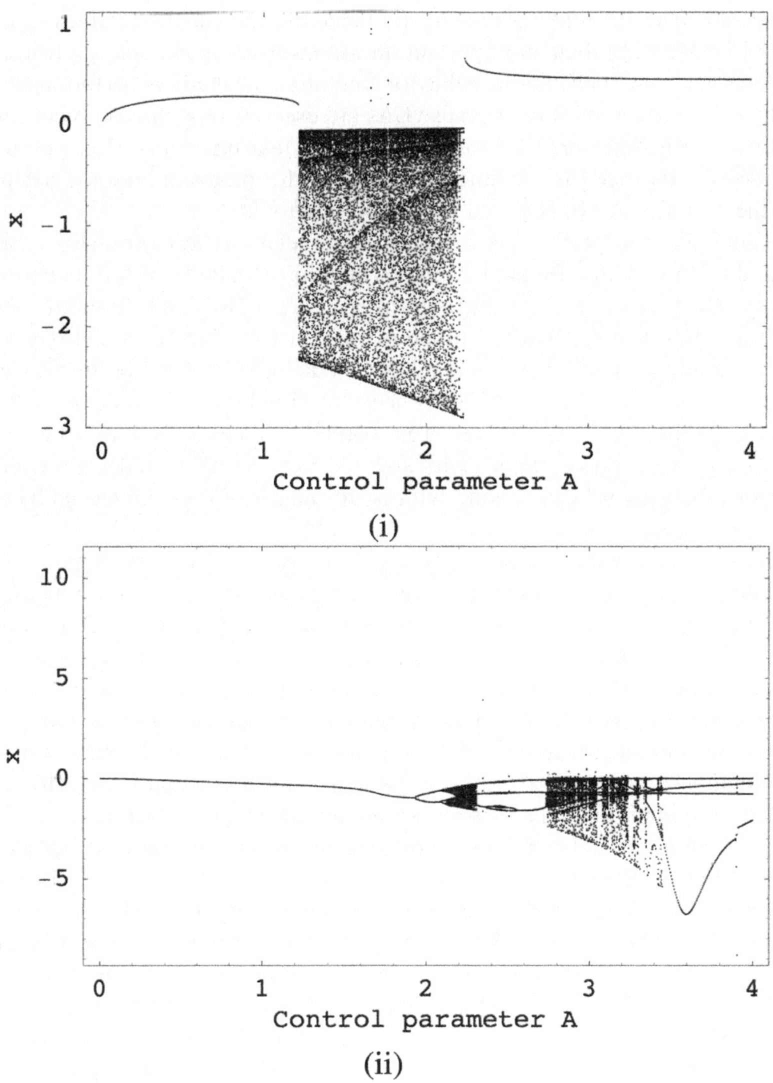

Fig. 11 Engineering design: bifurcation diagram

5 A Few Research Questions and Ideas

The above text has discussed the possibility on identification and possible synthesis of chaotic systems that contain hidden attractors. The topic of mutual intersection of hidden attractors and EAs can also be discussed from another point of view, which would generate a lot of interesting research questions. Let's discuss–propose a few of them.

In [32] and [31] it is demonstrated that inside EAs dynamics chaotic behavior can be observed. If in EAs there exist chaotic behavior, i.e. chaotic regimes (see Figs. 12 and 13, [31, 32]), then an important question for computer science researchers is whether and how such chaotic behavior can influence the EAs performance. This is quite an important question, because EAs are used on very hard problems, whose solutions by brute force or by classical methods can take much much longer time than our universe existence [31] or time, which is not, for practical reasons, acceptable. Thus, the performance of EAs is a very important topic.

It is not only about performance, but also about relations between chaos existing in EAs, and relation with different EA phases and the so-called stagnation, population diversity and speed of the algorithm convergence toward to global optimum. Another partial question whether hidden attractors exist in EAs dynamics and if yes, then again, what impact does it have on that? Does it disturb algorithm performance? Is it related to algorithm stagnation [71]? Concerning simple chaotic systems such as the logistic equation, they were derived from natural systems as predator-prey is, thus from systems based on swarm behavior and structure. Are then hidden attractors also inside swarm systems? Concerning with performance of EAs influenced by chaos, important research papers have already been published, such as [39, 72].

It is also joined with control of chaotic systems whose chaotic regimes are not acceptable in classical engineering and thus questions arise, like, can we successfully avoid by control being trapped in a hidden attractor. Or, if we can control hidden attractors is also important. Remember that all systems reported here and in many other papers are artificial and well known a priori. Thus, its control and analysis are very easy (due to easy model readability and knowledge) comparing it with possible hidden attractors present in black box of real-time systems. In the past, it has been clearly demonstrated that EAs are capable of such tasks, see [43, 44]. However, it is still an important topic that deserves more and deeper research. Can we identify hidden attractors in real-time black box systems? Can we control it in/out of the hidden attractor regime?

Back to swarm systems, there are also other interesting directions. A simpler one is about artificial neural networks (ANNs) with chaos inside it, see for example [73, 74]. During learning, there are different regimes during which information is stored in ANN. Existence of hidden attractor can then have fundamental impact on ANN capabilities and performance. Next, another interdisciplinary research joining EAs, chaos, hidden attractors and control is proposed in [1] and is under process by our research group[5]. The main idea is captured in Fig. 14, see [75, 76] and [77]. Here,

---

[5] navy.cs.vsb.cz.

swarm dynamics of selected algorithms is converted into complex networks that reflect its dynamics and thus by means of classic complex network analysis we can get information about EA dynamics and use it backward to control EAs performance. However, it can be translated further and complex networks can be converted into a CML (coupled map lattices) systems [10], which can also be controlled and analyzed but in a different way. In complex networks, we can also analyze the presence of chaos as well as in CML [10], thus a research space for hidden attractors to exist in such systems (i.e. in EAs, CN or CML) is provided.

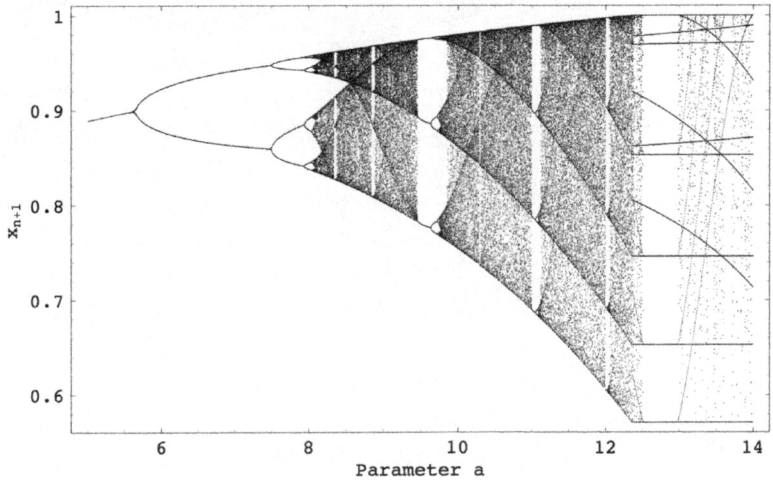

(i)

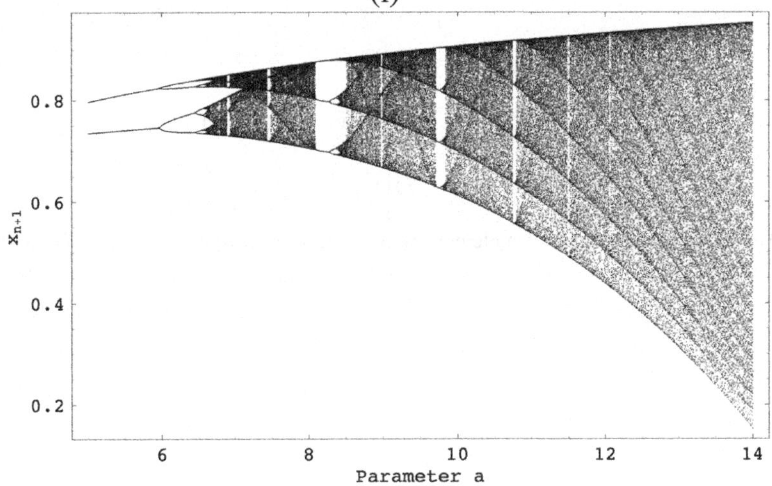

(ii)

Fig. 12 Bifurcation diagram of simple genetic algorithm, (i) version I and (ii) version II, see [31]

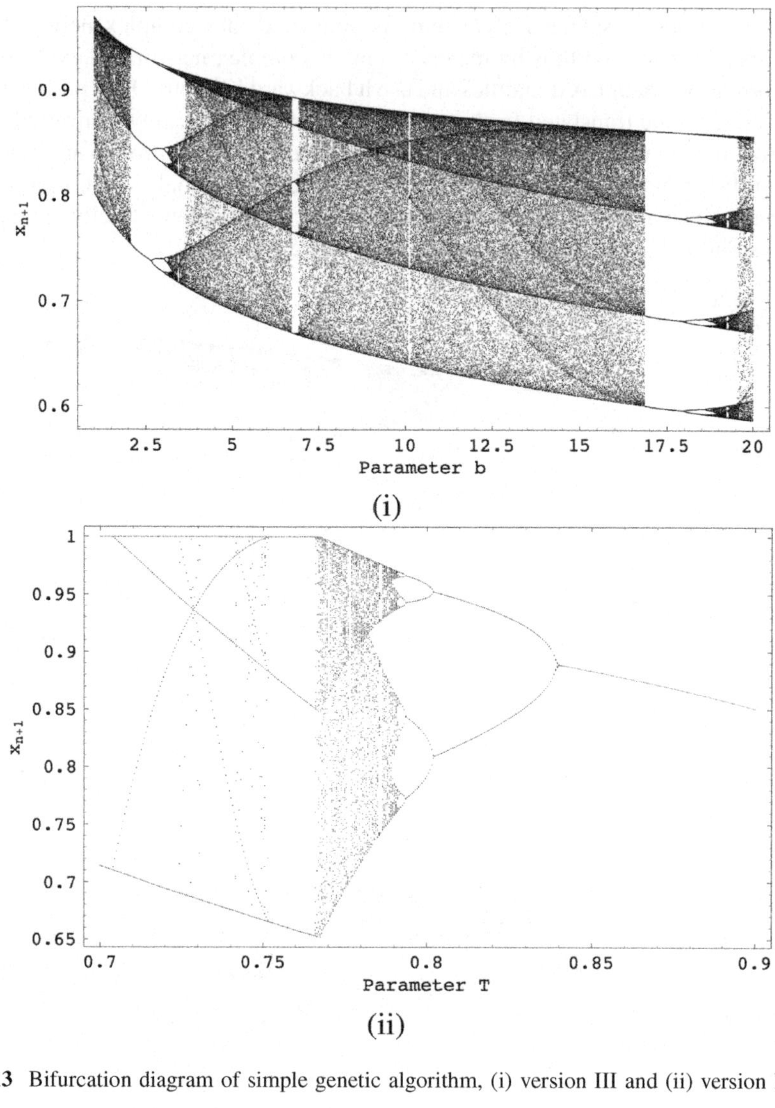

Fig. 13 Bifurcation diagram of simple genetic algorithm, (i) version III and (ii) version IV, see [31]

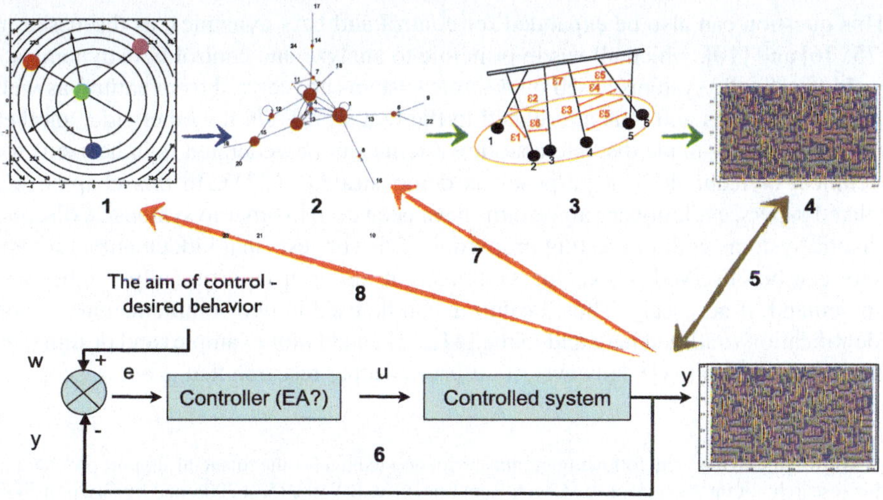

Fig. 14 Scheme of SEA dynamics conversion into complex network, CML system and its control

6 Conclusion

The main motivation of the research on the existence of hidden attractors is not only its theoretical importance, but also its impact on technological and industrial devices, as mentioned above in the Introduction section. The main interest here is whether EAs can be used on evolutionary identification of the basin of attraction that belongs to a hidden attractor, and its synthesis. Successful identification then can help to avoid critical moments and device malfunctions. For numerical identifications, there are already published representative papers like [41] and [42], based on classical (i.e. non-evolutionary) algorithms. On the other side, EAs are well known for their very good performance (including on black-box systems) and ability to avoid local extremes, e.g. Fig. 3, shall be very promising candidates to solve such tasks as well.

Based on results obtained from our initial experiments [1], it can be stated that EAs are applicable to such problems and their performances are very good. We have also discussed the possibility on evolutionary synthesis of chaotic systems and suggested its use on synthesis of chaotic systems that will (or will not) contain HA, based on the user's expectation.

In the last part, few research ideas have been suggested. All of them come from mutual intersection-fusion of nonlinear dynamics and evolutionary algorithms, which are regarded as nonlinear dynamical systems that also exhibit chaotic behaviors, with or without HA. Related and unanswered question from this research (and mutual intersections discussed in [31]) is whether one can observe hidden attractors also in evolutionary algorithms dynamics and what impact it has on EAs performance. The existence of chaos inside EAs has been "proven" and numerically demonstrated in [32] and thus possible existence of hidden attractors is theoretically almost surely.

This question can also be expanded for control and EAs dynamics, as discussed in [75, 76] and [10], which allows in principle to analyze and control EAs dynamics.

Many chaotic systems based on discrete systems are derived from natural as well as social systems and thus are related to the real world. On the other side, another set of artificially made discrete chaotic systems can be generated (synthesized) for technical or technological purposes as demonstrated in [31]. In this chapter and related papers, evolutionary algorithms have been used in order to synthesize discrete chaotic systems with interesting properties. This suggests that hidden attractors not only can be observed in existing systems, but also can be artificially synthesized on demand, if necessary. Thus, evolution can be used in two complementary ways: identification (outlined here and in the [41, 42] and [1] for example) and design (see an initial study in [31]). However, it deserves deeper research that is out of scope of this chapter.

Acknowledgements The following grants are acknowledged for the financial support provided to this research: Grant Agency of the Czech Republic–GACR P103/15/06700S and by Grant of SGS No. SP2016/175, VSB–Technical University of Ostrava.

References

1. I. Zelinka, Evolutionary identification of hidden chaotic attractors. Eng. Appl. Artif. Intell. **50**, 159–167 (2016)
2. I. Zelinka, D. Davendra, R. Jasek, R. Senkerik, Z. Oplatkova, *Analytical Programming – A Novel Approach for Evolutionary Synthesis of Symbolic Structures* (INTECH Open Access Publisher, 2011)
3. J.R. Koza, Human-competitive results produced by genetic programming. Genet. Program. Evolvable Mach. **11**(3–4), 251–284 (2010)
4. N.V. Kuznetsov, G.A. Leonov, V.I. Vagaitsev, Analytical-numerical method for attractor localization of generalized Chua's system. Periodic Control Syst. **4**(11), 29–33 (2010)
5. T. Kapitaniak, G.A. Leonov, Multistability: uncovering hidden attractors. Eur. Phys. J. Spec. Top. **224**(8), 1405–1408 (2015)
6. H. Jiang, Y. Liu, Z. Wei, L. Zhang, Hidden chaotic attractors in a class of two-dimensional maps. Nonlinear Dyn. 1–9 (2016)
7. D. Dudkowski, S. Jafari, T. Kapitaniak, N.V. Kuznetsov, G.A. Leonov, A. Prasad, Hidden attractors in dynamical systems. Physics Reports (2016)
8. R.C. Hilborn, S. Coppersmith, A.J. Mallinckrodt, S. Mckay, Chaos and nonlinear dynamics: an introduction for scientists and engineers. Am. J. Phys. **62**(9), 235 (1994)
9. H. Kantz, T. Schreiber, *Nonlinear Time Series Analysis* (Cambridge University Press, 2004)
10. E. Schöll, H.G. Schuster, *Handbook of Chaos Control* (Wiley, 2008)
11. G.A. Leonov, N.V. Kuznetsov, M.A. Kiseleva, E.P. Solovyeva, A.M. Zaretskiy, Hidden oscillations in mathematical model of drilling system actuated by induction motor with a wound rotor. Nonlinear Dyn. **77**(1–2), 277–288 (2014)
12. F.R. Tahir, S. Jafari, V. Pham, C. Volos, X. Wang, A novel no-equilibrium chaotic system with multiwing butterfly attractors. Int. J. Bifurc. Chaos **25**(04) (2015)
13. P.R. Sharma, M.D. Shrimali, A. Prasad, N.V. Kuznetsov, G.A. Leonov, Controlling dynamics of hidden attractors. Int. J. Bifurc. Chaos **25**(4) (2015)
14. G.A. Leonov, Iwcfta2012 keynote speech i – hidden attractors in dynamical systems: from hidden oscillation in hilbert-kolmogorov, aizerman and kalman problems to hidden chaotic attractor in chua circuits. Int. J. Bifurc. Chaos **23**(1), xv–xvii (2012)

15. M. Molaie, S. Jafari, J.C. Sprott, Simple chaotic flows with one stable equilibrium. Int. J. Bifurc. Chaos **23**(11), 151–167 (2013)
16. J.C. Sprott, S. Jafari, V.T. Pham, Z.S. Hosseini, A chaotic system with a single unstable node. Phys. Lett. A **379**(36), 2030–2036 (2015)
17. E. Ott, C. Grebogi, J.A. York, Controlling chaos. Phys. Rev. Lett. **64**(1), 2837 (1990)
18. I. Zelinka, R. Senkerik, E. Navratil, Investigation on evolutionary optimization of chaos control. Chaos, Solitons Fractals **40**(1), 111–129 (2009)
19. R. Senkerik, I. Zelinka, E. Navratil, Optimization of feedback control of chaos by evolutionary algorithms. IFAC Proc. **39**(8), 77–82 (2006)
20. I. Zelinka, R. Senkerik, E. Navratil, Investigation on realtime deterministic chaos control by means of evolutionary algorithms. IFAC Proc. **39**(8), 190–196 (2006)
21. K. Pyragas, Control of chaos via extended delay feedback. Phys. Lett. A **206**(5), 323–330 (1995)
22. H. Richter, K.J. Reinschke, Optimization of local control of chaos by an evolutionary algorithm. Physica D: Nonlinear Phenomena **144**(3), 309–334 (2000)
23. H. Richter, An evolutionary algorithm for controlling chaos: the use of multi-objective fitness functions, in *International Conference on Parallel Problem Solving from Nature* (Springer, 2002), pp. 308–317
24. Z. Oplatkova, R. Senkerik, I. Zelinka, J. Holoska, Synthesis of control law for chaotic henon system preliminary study, in *ECMS* (2010), pp. 277–282
25. Z. Oplatková, R. Šenkeřík, S. Bělašková, I. Zelinka, Synthesis of control rule for synthesized chaotic system by means of evolutionary techniques, *MENDEL 2010* (2010)
26. Z. Oplatková, R. Senkerik, I. Zelinka, J. Holoska, Synthesis of control law for chaotic logistic equation-preliminary study, in *2010 Fourth Asia International Conference on Mathematical/Analytical Modelling and Computer Simulation*, IEEE (2010), pp. 65–70
27. W. Just, Principles of time delayed feedback control, *Handbook of Chaos Control* (Wiley, 1999), pp. 21–41
28. K. Pyragas, Continuous control of chaos by self-controlling feedback. Phys. Lett. A **170**(6), 421–428 (1992)
29. G. Hu, F. Xie, J. Xiao, J. Yang, Z. Qu, Control of patterns and spatiotemporal chaos and its applications, *Handbook of Chaos Control* (Wiley-VCH, 1999), pp. 43–86
30. I. Zelinka, G. Chen, S. Celikovsky, Chaos synthesis by means of evolutionary algorithms. Int. J. Bifurc. Chaos **18**(04), 911–942 (2008)
31. I. Zelinka, S. Celikovskỳ, H. Richter, G. Chen, *Evolutionary Algorithms and Chaotic Systems*, vol. 267 (Springer, 2010)
32. A.H. Wright, A. Agapie, Cyclic and chaotic behavior in genetic algorithms, in *Proceedings of the 3rd Annual Conference on Genetic and Evolutionary Computation* (2001), pp. 718–724
33. K.J. Persohn, R.J. Povinelli, K.J. Persohn, R.J. Povinelli, Analyzing logistic map pseudorandom number generators for periodicity induced by finite precision floating-point representation. Chaos, Solitons Fractals **45**(3), 238–245 (2012)
34. M. Drutarovsky, P. Galajda, A robust chaos-based true random number generator embedded in reconfigurable switched-capacitor hardware, in *2007 17th International Conference on Radioelektronika*, IEEE (2007), pp. 1–6
35. X.-Y. Wang, X. Qin, A new pseudo-random number generator based on cml and chaotic iteration. Nonlinear Dyn. **70**(2), 1589–1592 (2012)
36. N.K. Pareek, V. Patidar, K.K. Sud, A random bit generator using chaotic maps. Int. J. Netw. Secur. **10**(1), 32–38 (2010)
37. E. Araujo, L.D.S. Coelho, Particle swarm approaches using lozi map chaotic sequences to fuzzy modelling of an experimental thermal-vacuum system. Appl. Soft Comput. **8**(4), 1354–1364 (2008)
38. B. Alatas, E. Akin, A.B. Ozer, Chaos embedded particle swarm optimization algorithms. Chaos, Solitons Fractals **40**(4), 1715–1734 (2009)
39. M. Pluhacek, R. Senkerik, D. Davendra, Z.K. Oplatkova, I. Zelinka, On the behavior and performance of chaos driven pso algorithm with inertia weight. Comput. Math. Appl. **66**(2), 122–134 (2013)

40. M. Pluhacek, R. Senkerik, I. Zelinka, "*Impact of Various Chaotic Maps on the Performance of Chaos Enhanced PSO Algorithm with Inertia Weight-an Initial Study, in Nostradamus: Modern Methods of Prediction, Modeling and Analysis of Nonlinear Systems* (Springer, 2013), pp. 153–166
41. N. Kuznetsov, O. Kuznetsova, G. Leonov, V. Vagaitsev, *Analytical-Numerical Localization of Hidden Attractor in Electrical Chua's Circuit* (Springer, Berlin-Heidelberg, 2013)
42. V.O. Bragin, V.I. Vagaitsev, N.V. Kuznetsov, G.A. Leonov, Algorithms for finding hidden oscillations in nonlinear systems: the aizerman and kalman problems and chua's circuits. Int. J. Comput. Syst. Sci. **50**(4), 511–543 (2011)
43. I. Zelinka, L. Nolle, Plasma reactor optimizing using differential evolution, *Differential Evolution: A Practical Approach to Global Optimization* (2005), pp. 499–512
44. L. Nolle, I. Zelinka, A.A. Hopgood, A. Goodyear, Comparison of an self-organizing migration algorithm with simulated annealing and differential evolution for automated waveform tuning. Adv. Eng. Softw. **36**(10), 645–653 (2005)
45. E.L. Houghton, P.W. Carpenter, *Aerodynamics for Engineering Students* (Butterworth-Heinemann, 2003)
46. C.L. Karr, R. Bowersox, V. Singh, Minimization of sonic boom on supersonic aircraft using an evolutionary algorithm, in *Genetic and Evolutionary Computation Conference* (Springer, 2003), pp. 2157–2167
47. M. Dorigo, M. Birattari, T. Stutzle, Ant colony optimization. IEEE Comput. Intell. Mag. **1**(4), 28–39 (2006)
48. Q.T. Pham, Dynamic optimization of chemical engineering processes by an evolutionary method. Comput. Chem. Eng. **22**(7), 1089–1097 (1998)
49. Q. Pham, S. Coulter, Modelling the chilling of pig carcasses using an evolutionary method," in *Proceedings of the International Congress of Refrigeration*, vol. 3 (1995), pp. 676–683
50. Y. Li, A. Häuβler, Artificial evolution of neural networks and its application to feedback control. Artif. Intell. Eng. **10**(2), 143–152 (1996)
51. O. Levenspiel, *Chemical Reaction Engineering: An Introduction to the Design of Chemical Reactors* (Wiley, 1962)
52. M. Judy, K. Ravichandran, K. Murugesan, A multi-objective evolutionary algorithm for protein structure prediction with immune operators. Comput. Methods Biomech. Biomed. Eng. **12**(4), 407–413 (2009)
53. O. Ebenhöh, R. Heinrich, Evolutionary optimization of metabolic pathways. theoretical reconstruction of the stoichiometry of atp and nadh producing systems. Bull. Math. Biol. **63**(1), 21–55 (2001)
54. G.B. Fogel, D.W. Corne, *Evolutionary Computation in Bioinformatics* (Morgan Kaufmann, 2002)
55. J.H. Holland, Adaptation in natural and artificial systems: an introductory analysis with applications to biology, control and artificial intelligence. Control Artif. Intell. Univ. Mich. Press **6**(2), 126–137 (1975)
56. M. Dorigo, M. Birattari, C. Blum, *Ant Colony Optimization and Swarm Intelligence*, vol. 5217(8) (Springer, 2004), pp. 767–771
57. V. Kenneth, *Price: An Introduction to Differential Evolution, New Ideas in Optimization* (McGraw-Hill, London, 1999)
58. I. Zelinka, Soma—self-organizing migrating algorithm, in *New Optimization Techniques in Engineering* (Springer, 2004), pp. 167–217
59. J.H. Holland, *Adaptation in Natural and Artificial Systems: An Introductory Analysis with Applications to Biology, Control, and Artificial Intelligence* (University of Michigan Press, 1975)
60. B.S. Kirkpatrick, C. Gelatt, D. Vecchi, Optimization by simulated annealing. Science **220**(4598), 671–80 (1983)
61. V. Černý, Thermodynamical approach to the traveling salesman problem: an efficient simulation algorithm. J. Optim. Theory Appl. **45**(1), 41–51 (1985)

62. R. Rajendra, D.K. Pratihar, Particle swarm optimization, in *International Conference on Biomedical Engineering & Informatics* (1942), pp. 129–132
63. H.P. Schwefel, Numerische optimierung von computer – modellen (2010)
64. B.I. Rechenberg, *Evolutionstrategie: Optimierung Technischer Systeme nach Prinzipien dier Biolischen Evolution* (Frommann-Holzboog, 2013)
65. N.V. Kuznetsov, G.A. Leonov, Hidden attractors in dynamical systems: systems with no equilibria, multistability and coexisting attractors. IFAC Proc. Vol. **47**(3), 5445–5454 (2014)
66. M.A. Van Wyk, W.-H. Steeb, *Chaos in Electronics*, vol. 2 (Springer Science & Business Media, 2013)
67. J.R. Koza, *Genetic Programming: A Paradigm for Genetically Breeding Populations of Computer Programs to Solve Problems* (Stanford University, 1990)
68. C. Ryan, J. Collins, M.O. Neill, Grammatical evolution: Evolving programs for an arbitrary language, in *European Conference on Genetic Programming* (Springer, 1998), pp. 83–96
69. C.G. Johnson, Artificial immune system programming for symbolic regression, in *European Conference on Genetic Programming* (Springer, 2003), pp. 345–353
70. M. O'Neill, A. Brabazon, Grammatical Differential Evolution, in *Proceedings of International Conference on Artificial Intelligence* (2006)
71. J. Lampinen, I. Zelinka, On stagnation of the differential evolution algorithm, in *Proceedings of MENDEL* (2000), pp. 76–83
72. I. Zelinka, M. Chadli, D. Davendra, R. Senkerik, M. Pluhacek, J. Lampinen, Do evolutionary algorithms indeed require random numbers? extended study, in *Nostradamus 2013: Prediction, Modeling and Analysis of Complex Systems* (Springer, 2013), pp. 61–75
73. K. Aihara, T. Takabe, M. Toyoda, Chaotic neural networks. Phys. Lett. A **144**(6–7), 333–340 (1990)
74. I. Tsuda, Dynamic link of memory—chaotic memory map in nonequilibrium neural networks. Neural Netw. **5**(2), 313–326 (1992)
75. M.T. Thai, P.M. Pardalos, *Handbook of Optimization in Complex Networks* (Springer, 2012)
76. I. Zelinka, D. Davendra, S. Roman, J. Roman, Do evolutionary algorithms dynamics create complex network structures? Complex Syst. **20**(2), 127 (2011)
77. I. Zelinka, A survey on evolutionary algorithms dynamics and its complexity-mutual relations, past, present and future. Swarm Evol. Comput. **25**, 2–14 (2015)

Hidden Attractors in a Dynamical System with a Sine Function

Christos Volos, Jamal-Odysseas Maaita, Viet-Thanh Pham, and Sajad Jafari

1 Introduction

In the last decades, chaotic systems and their applications have received significant attention [1–13]. When discovering chaotic systems, equilibrium points are studied seriously because of their important role [12, 14–16]. Conventional chaotic systems, especially 3-D chaotic systems, like Lorenz system [17], Rössler system [18], Arneodo system [19], Chen system [20] and Sprott systems (except case A) [21] have a finite or countable number of equilibrium points. Therefore, chaotic behavior in systems like those mentioned above can be verified by applying the well-known Šil'nikov criteria [22, 23].

Recently, great interest in dynamical systems with hidden attractors has been raised. The term "hidden attractor" refers to the fact that in this class of systems the attractor is not associated with an unstable equilibrium and thus often remains undiscovered because it may occur in a small region of parameter space with a small basin of attraction in the space of initial conditions [24–29]. Hidden attractors, in

C. Volos
Department of Physics, Aristotle University of Thessaloniki, 54124 Thessaloniki, Greece
e-mail: volos@physics.auth.gr

J.-O. Maaita (✉)
Department of Physics, Aristotle University of Thessaloniki, 54124 Thessaloniki, Greece
e-mail: jmaay@physics.auth.gr

V.-T. Pham
School of Electronics and Telecommunications, Hanoi University of Science and Technology, 01 Dai Co Viet, Hanoi, Vietnam
e-mail: pvt3010@gmail.com

S. Jafari
Biomedical Engineering Department, Amirkabir University of Technology, 15875-4413 Tehran, Iran
e-mail: sajadjafari@aut.ac.ir

X. Wang et al. (eds.), *Chaotic Systems with Multistability and Hidden Attractors*, Emergence, Complexity and Computation 40,
https://doi.org/10.1007/978-3-030-75821-9_19

contrast to self-excited attractors, cannot be computed by the standard procedure because their basins of attraction do not contain neighborhoods of any equilibria. In general, chaotic systems with hidden attractors belong to three main families: (i) systems with stable equilibria, (ii) systems with an infinite number of equilibrium points and (iii) systems without any equilibrium point. So, there has been an increasing interest in chaotic systems with the presence of hidden attractors recently.

In 2010, for the first time, a chaotic hidden attractor was discovered in the most well-known nonlinear circuit, Chua's circuit, which is described by a three-dimensional dynamical system [24, 30]. However, the problem of analyzing hidden oscillations arose for the first time in the second part of Hilbert's 16th problem (1900) for two-dimensional polynomial systems [31]. The first nontrivial results were obtained in Bautin's works [32], which were devoted to constructing nested limit cycles in quadratic systems, showing the necessity of studying hidden oscillations for solving this problem.

Later, in the middle of the 20th century, Kapranov [33] studied the qualitative behavior of Phase-Locked Loop (PLL) systems, which are used in telecommunications and computer architectures, and estimating system stability domains. In that work, Kapranov assumed that in PLL systems there were only self-excited oscillations. However, in 1961, Gubar [34] revealed a gap in Kapranov's work and analytically showed the possibility of the existence of hidden oscillations in a two-dimensional system of PLL. Thus, from a computational point of view, the system considered was globally stable which, however, has only a bounded domain of attraction.

Also, in the same period of time, the investigations of widely-known Markus-Yamabe [35] and Kalman [36] conjectures on absolute stability led to the finding of hidden oscillations in automatic control systems with a unique stable stationary point and a special nonlinearity belonging to the sector of linear stability [37–39].

Furthermore, systems with hidden attractors have received increasing attention due to their practical and theoretical importance in other scientific branches such as mechanics, fluids, electronic circuits, etc. [40–49]. Therefore, the study of these systems is an interesting topic of extremely significant importance.

In this chapter, a three-dimensional autonomous chaotic system with a sine function, which can display infinitely many equilibria, is presented for the first time. System's dynamics are discovered by analyzing equilibrium points, and by using well-known tools from nonlinear theory, such as phase portrait, Poincaré map, bifurcation diagram, maximum Lyapunov fvexponents, and the Kaplan-Yorke fractional dimension. Also, the design of an electronic circuit emulating the system, in which the sine function has been replaced with a Taylor series expansion, proves the feasibility of the proposed framework.

The chapter is organized as follows. In Sect. 2 related works are summarized. The mathematical model of the system with infinitely many equilibria is introduced in Sect. 3, while in Sect. 4 its dynamical properties are presented. In Sect. 5, the system with a Taylor series expansion of the sine function is studied. The circuital implementation of the system is reported in Sect. 6. Finally, some concluding remarks are given in Sect. 7.

2 Related Works

As it is mentioned, one of the classes of systems with hidden attractors is the class of systems with uncountably many equilibria [41, 50–55]. Today, there are two main research directions related to chaotic systems with an infinite number of equilibrium points, as illustrated in Fig. 1.

On one hand, motivated by the key work of Jafari and Sprott [56], which focuses on chaotic systems with a line of equilibrium points, five new chaotic flows with line equilibrium points and especially a complicated one with two infinite parallel lines of equilibrium points, were discovered by Li and Sprott [50]. By using signum functions and absolute-value functions, chaotic systems with a line or two perpendicular lines of equilibrium points were introduced in [57].

On the other hand, after the discovery of a new class of chaotic systems with circular equilibrium points [58], which concentrates on chaotic systems with a curve of equilibrium points, Kingni et al. [59] presented a 3D chaotic autonomous system with a circular curve of equilibrium points and its fractional-order form. Then, Gotthans et al. [53] reported another chaotic system with a circular curve of equilibrium points, where a 3D system was reported with a square curve of equilibrium points, constructed by modifying the system with a circular curve of equilibrium points in [53]. In addition, a system exhibiting a chaotic attractor with an ellipse curve of equilibrium points, a chaotic attractor with a square-shaped curve of equilibrium points, and a chaotic attractor with a rectangle-shaped curve of equilibrium points, were presented in Ref. [60].

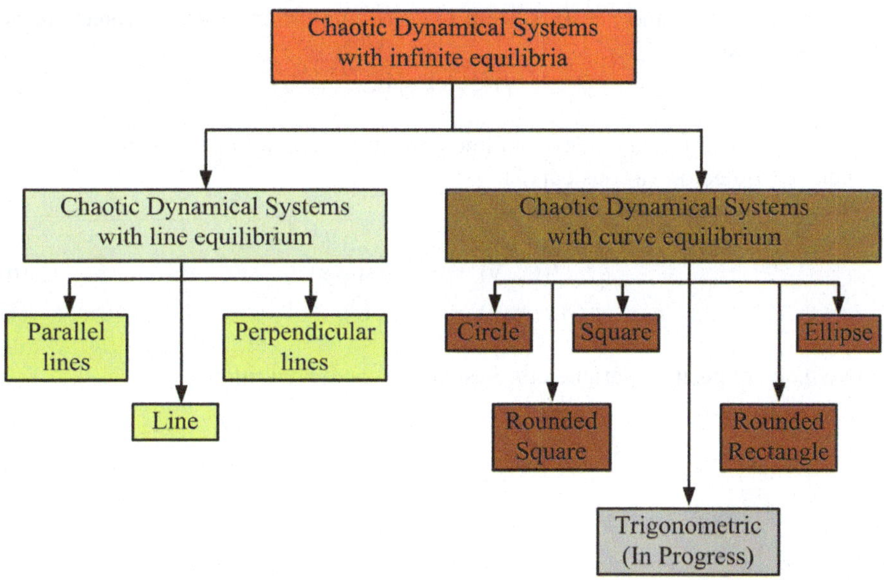

Fig. 1 Main research directions in constructing chaotic systems with infinitely many equilibria

The aim of this chapter is to open up a new research direction to the construction of chaotic systems with infinitely many equilibrium points, by using the well-known trigonometric functions. In this way, system's equilibria are located in trigonometric curves. In more detail, the sine function is chosen for use in a selected system, so as to make the system display hidden attractors.

3 Model of the Proposed System

Consider the following three-dimensional system:

$$\begin{aligned} \dot{x} &= -z, \\ \dot{y} &= xz^2 + a\ sgn(z), \\ \dot{z} &= f_1(x, y) + zf_2(x, y, z), \end{aligned} \tag{1}$$

where $x,\ y,\ z$ are state variables, a is a positive parameter, and $f_1,\ f_2$ are two nonlinear functions.

The equilibrium points of system (1) can be found by solving

$$-z = 0, \tag{2}$$

$$xz^2 + a\ sgn(z) = 0, \tag{3}$$

$$f_1(x, y) + zf_2(x, y, z) = 0. \tag{4}$$

Therefore, on the plane $z = 0$, system (1) has equilibrium points located on the following curve:

$$f_1(x, y) = 0. \tag{5}$$

So, by choosing different kinds of nonlinear function f_1, equilibrium points of system (1) can be located on various curves.

Select the following two functions $f_1,\ f_2$:

$$f_1(x, y) = x - b\sin(y), \tag{6}$$

$$f_2(x, y, z) = cy^2 - z^2, \tag{7}$$

where $b,\ c$ are positive parameters. System (1) can be rewritten as

$$\begin{aligned} \dot{x} &= -z, \\ \dot{y} &= xz^2 + a\ sgn(z), \\ \dot{z} &= x - b\sin(y) + z(cy^2 - z^2). \end{aligned} \tag{8}$$

From Eqs. (5) and (6), it is easy to see that system (8) has infinitely equilibria (E) located on a trigonometric curve, $E(b\sin(y^*), y^*, 0)$. As a result, the proposed system (8) can be considered as a chaotic system with hidden attractors [25, 39, 41].

When selecting $a = 0.28,\ b = 0.2,\ c = 2.5$ and initial conditions $(x_0, y_0, z_0) = (0.1, 0.1, 0.1)$, the Lyapunov exponents of system (8) are obtained as

$$L_1 = 0.09254,\ L_2 = 0,\ L_3 = -0.47397. \tag{9}$$

There is one positive Lyapunov exponent in the LE spectrum (9), thus the proposed system (8) exhibits chaotic behavior, as it is confirmed by the phase portraits in Figs. 2, 3 and 4 on various planes, and the strange attractor of the Poincaré map in Fig. 5.

In addition, since $L_1 + L_2 + L_3 = -0.38143 < 0$, it indicates that the system (8) is dissipative.

The Kaplan-Yorke fractional dimension, which presents the complexity of attractor is defined by

$$D_{KY} = j + \frac{1}{L_{j+1}} \sum_{i=1}^{j} L_i, \tag{10}$$

where j is the largest integer satisfying $\sum_{i=1}^{j} L_i \leq 0$ and $\sum_{i=1}^{j+1} L_i < 0$. The Kaplan-Yorke dimension of the proposed system (8) is calculated as

$$D_{KY} = 2 + \frac{L_1 + L_2}{|L_3|} = 2.1952, \tag{11}$$

which is fractional.

4 Dynamics and Properties of the Proposed System

4.1 *Dissipativity and Invariance*

By applying the general condition of dissipativity into system (8), we have

$$\text{div}\, \mathbf{u} = \frac{\partial \dot{x}}{\partial x} + \frac{\partial \dot{y}}{\partial y} + \frac{\partial \dot{z}}{\partial z}. \tag{12}$$

Therefore, system (8) is dissipative for $cy^2 - 3z^2 < 0$. The invariance of system (8) is indicated by the coordinate transformation $(x, y, z) \longrightarrow (-x, -y, -z)$.

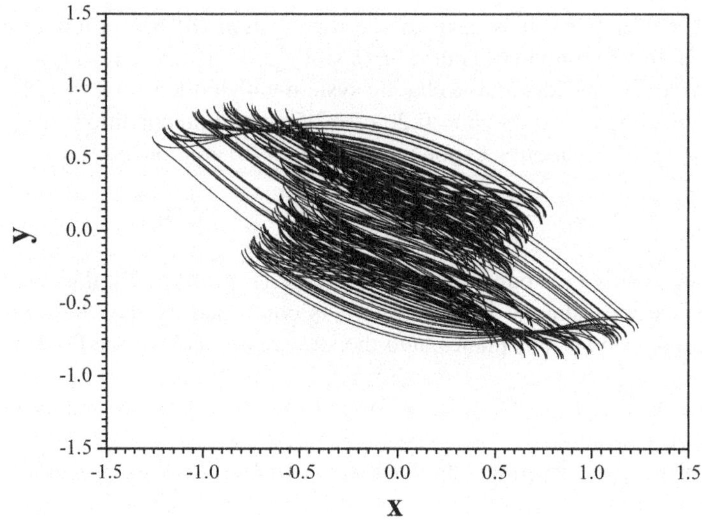

Fig. 2 2D projection of the chaotic system (8) on the (x, y)-plane, for $a = 0.28$, $b = 0.2$ and $c = 2.5$, with initial conditions $(x_0, y_0, z_0) = (0.1, 0.1, 0.1)$

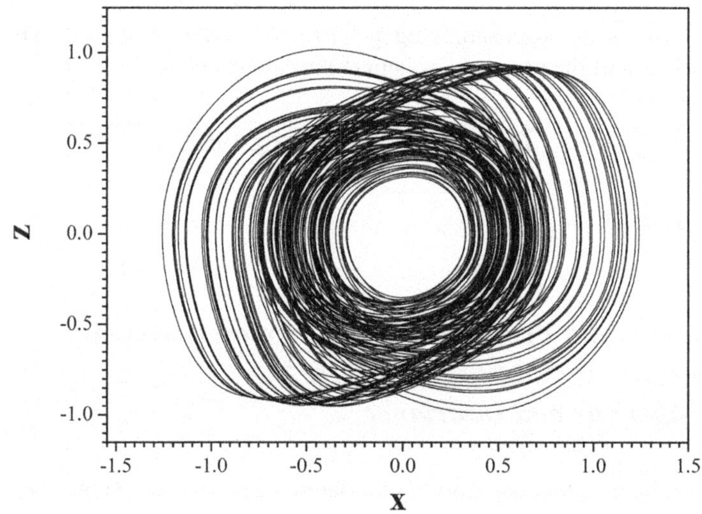

Fig. 3 2D projection of the chaotic system (8) on the (x, z)-plane, for $a = 0.28$, $b = 0.2$ and $c = 2.5$, with initial conditions $(x_0, y_0, z_0) = (0.1, 0.1, 0.1)$

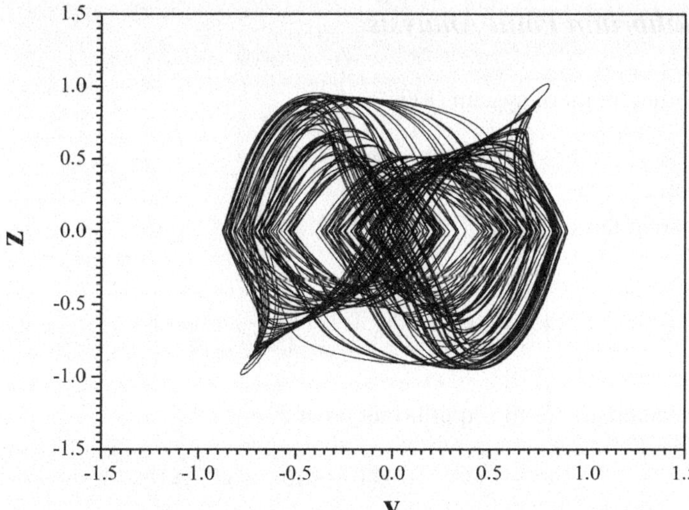

Fig. 4 2D projection of the chaotic system (8) on the (y, z)-plane, for $a = 0.28$, $b = 0.2$ and $c = 2.5$, with initial conditions $(x_0, y_0, z_0) = (0.1, 0.1, 0.1)$

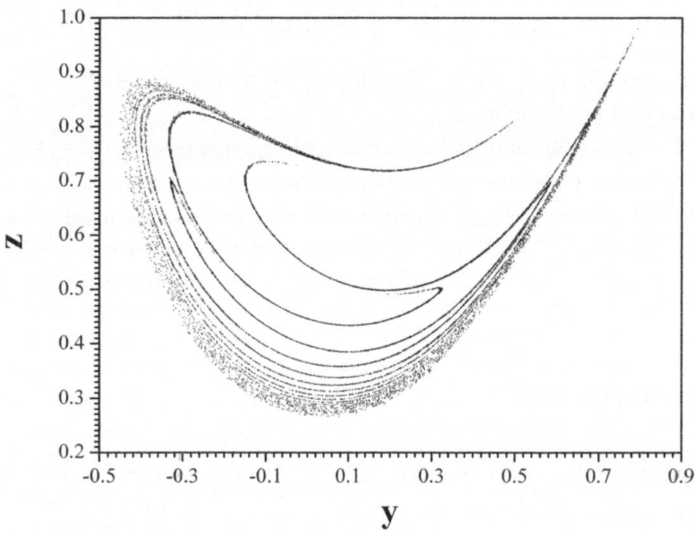

Fig. 5 Poincaré map of system (8) on the $y = 0$ plane, for $a = 0.28$, $b = 0.2$ and $c = 2.5$, with initial conditions $(x_0, y_0, z_0) = (0.1, 0.1, 0.1)$

4.2 Equilibrium Point Analysis

The equilibrium point of system (8) is

$$E = (b \sin y^*, y^*, 0). \tag{13}$$

The Jacobian of the system is

$$\begin{bmatrix} 0 & 0 & -1 \\ z^2 & 0 & 2xz + 2a\delta(z) \\ 1 & -b\cos y + 2czy & cy^2 - 3cz^2, \end{bmatrix}$$

and the characteristic for the equilibrium point E is

$$\lambda^3 - cy^{*2}\lambda^2 + \lambda(1 + 2ab\cos y^*) = 0. \tag{14}$$

So, the system always has one eigenvalue equal to zero ($\lambda_1 = 0$) and two other eigenvalues determined by the discriminant of the second-order part of the characteristic, given by

$$\Delta = c^2 y^{*4} - 8ab\cos y^* - 4\,. \tag{15}$$

When $\Delta = 0$, the system has a double real positive eigenvalue $\lambda = \frac{cy^{*2}}{2}$ and the equilibrium point is unstable.

When $\Delta > 0$, the system has a pair of real eigenvalues $\lambda = \frac{cy^{*2} \pm \sqrt{\Delta}}{2}$, where at least one of them is positive, so the equilibrium point is unstable.

When $\Delta < 0$, the system has a pair of complex conjugate eigenvalues $\lambda = \frac{cy^{*2} \pm \sqrt{\Delta}}{2}$ with positive real parts, and so the equilibrium point is unstable.

From the above analysis, we conclude that E is an unstable equilibrium point.

4.3 System Dynamics

In this subsection, the system's bifurcation diagram and Lyapunov exponent spectrum, which are classical tools from nonlinear theory, are used for the investigation of nonlinear system dynamics [13]. The bifurcation diagram and Lyapunov exponent spectrum of system (8) are reported in Figs. 6 and 7, respectively. As shown in Figs. 6 and 7, system (8) exhibits chaotic and periodic behaviors when changing the value of the bifurcation parameter c from 0.0 to 3.5. Also, the system exhibits the well-known route to chaos through the mechanism of period doublings, as the parameter c increases. In more details, when the value of $c < 0.928$, system (8) remains in the periodic state (as illustrated in Figs. 8, 9, for $c = 0.1$ and $c = 0.5$), where all the Lyapunov exponents are $LCE \leq 0$. When $c > 0.928$, it is easy to see that more complex behavior is merged (the system (8) has one positive Lyapunov exponent

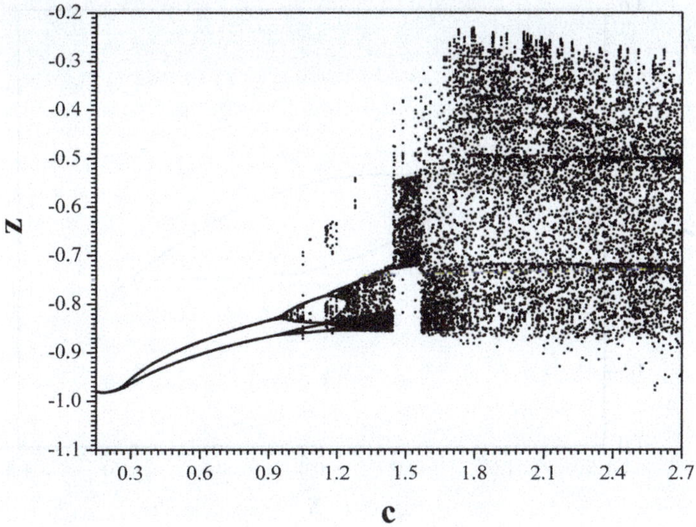

Fig. 6 Bifurcation diagram of system (8), when changing the value of parameter c, for $a = 0.28$ and $b = 0.2$, with initial conditions $(x_0, y_0, z_0) = (0.1, 0.1, 0.1)$

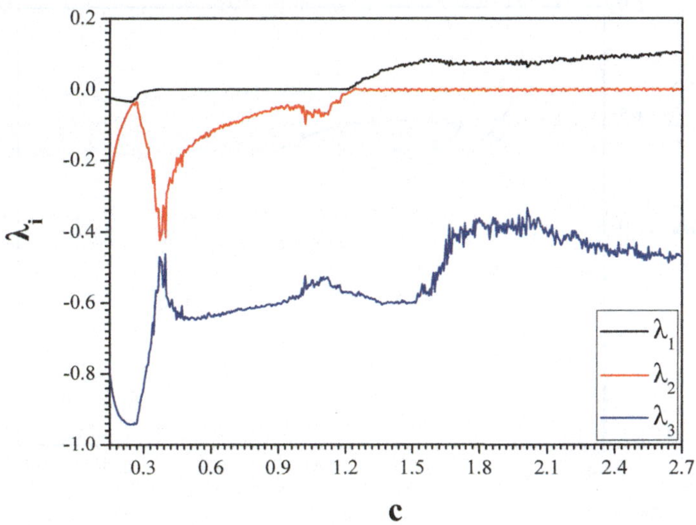

Fig. 7 Lyapunov exponents (λ_i) versus the parameter c, for $a = 0.28$ and $b = 0.2$, with initial conditions $(x_0, y_0, z_0) = (0.1, 0.1, 0.1)$

in Fig. 7). Figure 10 presents the chaotic attractor for $c = 1.4$, while for $c > 1.5$ an internal crisis (Ott, 1992) is observed and the chaotic attractor is enlarged, as shown in Figs. 2, 3 and 4, for $c > 2.5$.

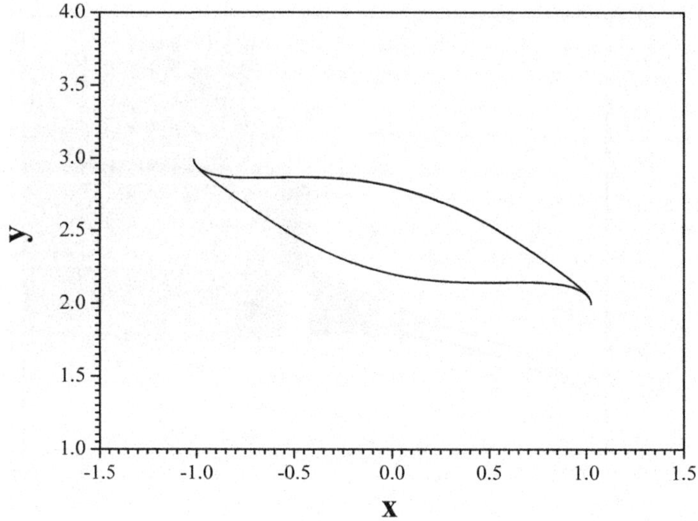

Fig. 8 Phase portrait of system (8), displayed on the $x - y$ plane, for $a = 0.28, b = 0.2$ and $c = 0.1$, with initial conditions $(x_0, y_0, z_0) = (0.1, 0.1, 0.1)$

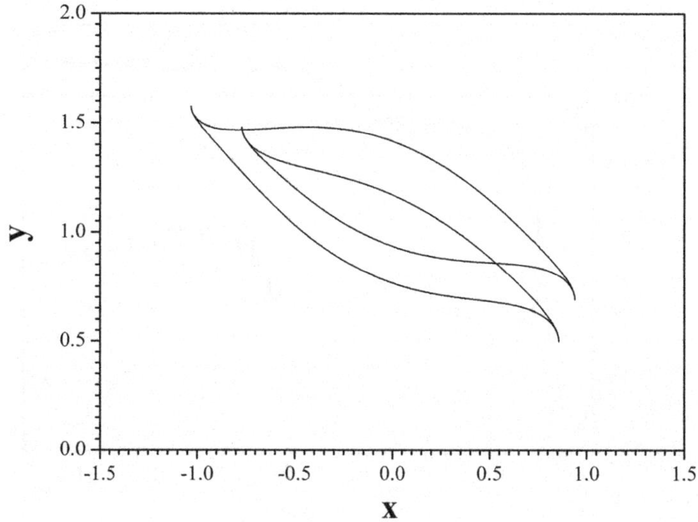

Fig. 9 Phase portrait of system (8), displayed on the $x - y$ plane, for $a = 0.28, b = 0.2$ and $c = 0.5$, with initial conditions $(x_0, y_0, z_0) = (0.1, 0.1, 0.1)$

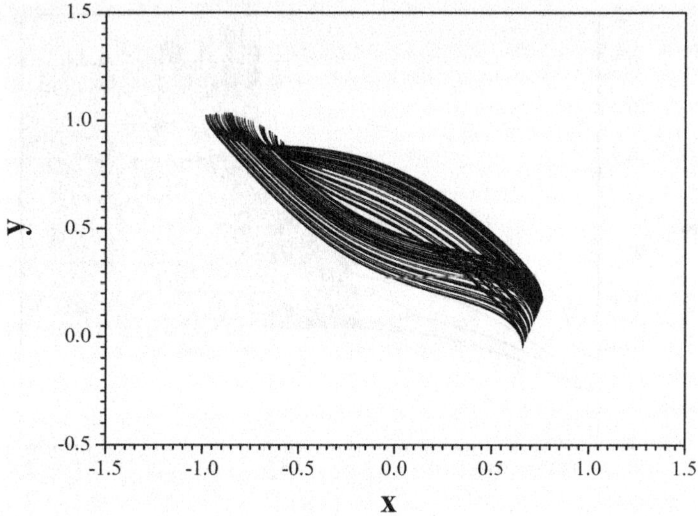

Fig. 10 Phase portrait of system (8), displayed on the $x - y$ plane, for $a = 0.28$, $b = 0.2$ and $c = 1.4$, with initial conditions $(x_0, y_0, z_0) = (0.1, 0.1, 0.1)$

4.4 Coexistence of Hidden Attractors

As it is mentioned, system (8) has rotational symmetry with respect to the origin of the axes as evidenced by their invariance under the transformation from (x, y, z) to $(-x, -y, -z)$. Therefore, any projection of the attractor has symmetry. As it has been known, symmetric systems generally have coexisting attractors [61–63]. Thus, such complex dynamics of system (8) is investigated here by plotting forward and backward continuations for the bifurcation parameter c, with $a = 0.25$ and $b = 0.2$. The continuations of system (8) indicate the presence of multi-stability in the system (Fig. 11). For example, there are coexisting chaotic attractors when $c = 1.4$, as shown in Fig. 12

5 Approximation of Sine Function with Taylor Series Expansion

The aim of this work is also to present an electronic circuit that emulates the proposed system (8) with hidden attractors and study its behavior. However, the absence of a chip that can produce the trigonometric functions is a big problem in this consideration. Two decades ago, a chip (AD639) was presented, which was a high-accuracy monolithic function converter that provides all the standard trigonometric functions and their inverses via pin-strapping. Nevertheless, this chip is not produced anymore today. This is the main reason that the trigonometric functions, mainly sine and

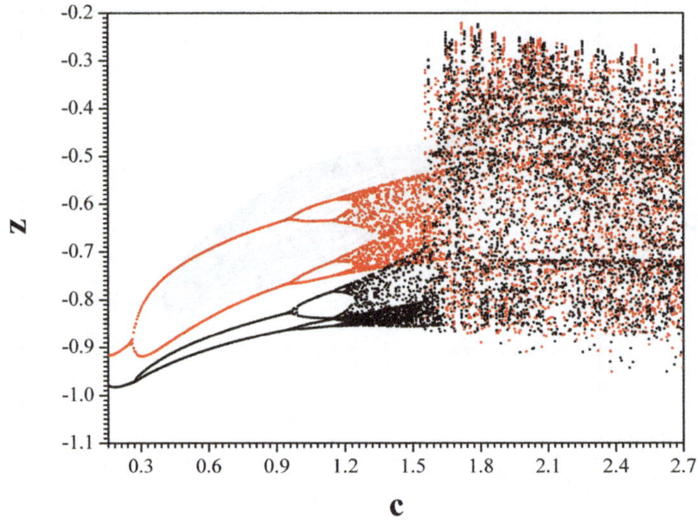

Fig. 11 Continuations of system (8) when varying c: forward continuation (black) and backward continuation (red), for $a = 0.25$ and $b = 0.2$

cosine, are not used as nonlinear functions in electronic circuits. As a consequence, other ways for producing such functions should be searched for.

The most appropriate way for implementing the sine and cosine functions is using the Taylor series expansions, which can be implemented easily with the analog multipliers (AD633). In mathematics, a Taylor series is a representation of a function as an infinite sum of terms that are calculated from the values of the function's derivatives at a single point. The concept of a Taylor series was formulated by the Scottish mathematician James Gregory and formally introduced by the English mathematician Brook Taylor in 1715 [64, 65]. If the Taylor series is centered at zero, then the series is also called a Maclaurin series, named after the Scottish mathematician Colin Maclaurin, who made extensive use of this special case of Taylor series in the 18th century.

A function can be approximated by using a finite number of terms of its Taylor series. Taylor's theorem gives quantitative estimates on the errors introduced by the use of such an approximation. The polynomial formed by taking some initial terms of the Taylor series is called a Taylor polynomial. The Taylor series of a function is the limit of that function's Taylor polynomials as the degree increases, provided that the limit exists. A function may not be equal to its Taylor series, even if its Taylor series converges at every point. A function that is equal to its Taylor series in an open interval (or a disc in the complex plane) is known as an analytic function in that interval.

In the case of the sine function, its Taylor series approximation is given by

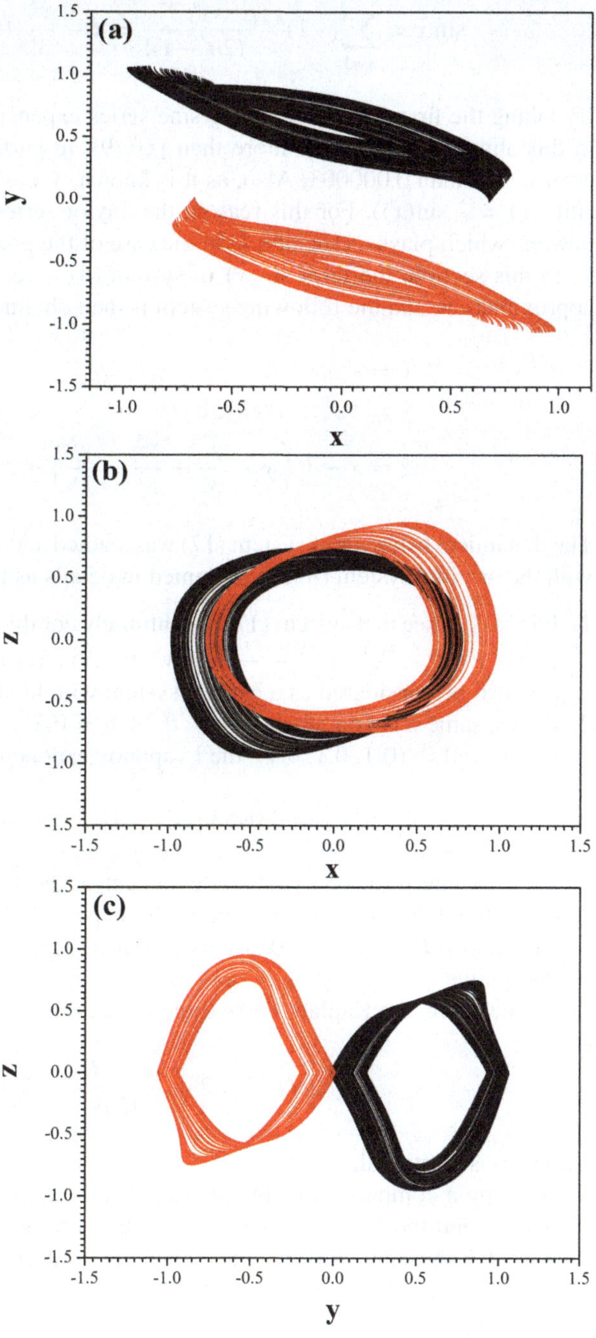

Fig. 12 Coexisting attractors of system (8), for $a = 0.25$, $b = 0.2$ and $c = 1.3$; the initial conditions $(x_0, y_0, z_0) = (0, 0.1, 0)$ (black) and the initial conditions $(x_0, y_0, z_0) = (0, -0.1, 0)$ (red) on (**a**) $x - y$ plane, (**b**) $x - z$ plane and (**c**) $y - z$ plane

$$\sin x = \sum_{n=1}^{\infty}(-1)^{n-1}\frac{x^{2n-1}}{(2n-1)!} = x - \frac{x^3}{3!} + \frac{x^5}{5!} - \frac{x^7}{7!} - \cdots \tag{16}$$

By taking the first four terms of the sine series expansion, as in Eq. (16), the error in this approximation is no more than $|x|^9/9!$. In particular, for $-1 < x < 1$, the error is less than 0.000003. Also, as it is known, $y = \sin x$ is an odd function (i.e., $\sin(-x) = -\sin(x)$). For this reason, the Taylor series of $y = \sin x$ has only odd powers, which plays a crucial role in the case of the proposed system (8).

In this section, the term $\sin(y)$ in system (8) is replaced with its Taylor series approximation, and the following system is then obtained:

$$\begin{aligned} \dot{x} &= -z, \\ \dot{y} &= xz^2 + asgn(z), \\ \dot{z} &= x - b\left(y - \frac{y^3}{3!} + \frac{y^5}{5!} - \frac{y^7}{7!}\right) + z(cy^2 - z^2). \end{aligned} \tag{17}$$

The dynamical behavior of system (17) was studied and the results of the comparison with the original system (8) are presented in details as follows.

1. It is easy to see that system (17) has infinitely equilibrium points (E) located on a curve, $E(y^* - \frac{(y^*)^3}{3!} + \frac{(y^*)^5}{5!} - \frac{(y^*)^7}{7!}, y^*, 0)$. As a result, the proposed system (17) can also be considered as a chaotic system with hidden attractors.
2. For the same set of parameters ($a = 0.28$, $b = 0.2$, $c = 2.5$) and initial conditions $(x_0, y_0, z_0) = (0.1, 0.1, 0.1)$, the Lyapunov exponents of the system (17) are:

$$L_1 = 0.09533,\ L_2 = 0,\ L_3 = -0.43701. \tag{18}$$

   There is one positive Lyapunov exponent in the Lyapunov exponent spectrum (18), thus the system (17) exhibits the expected chaotic behavior. In addition, since $L_1 + L_2 + L_3 = -0.34168 < 0$, it indicates that the system (17) is also dissipative.
   Furthermore, the Kaplan-Yorke dimension of the system (17) is

$$D_{KY} = 2 + \frac{L_1 + L_2}{|L_3|} = 2.1814 \tag{19}$$

   which is fractional.
3. Also, by a comparison of the phase portraits in Figs. 13, 14 and 15 on various planes, and the Poincaré map in Fig. 16, with the respective phase portraits in Figs. 2, 3 and 4 and the Poincaré map in Fig. 5, a very good agreement is obtained.

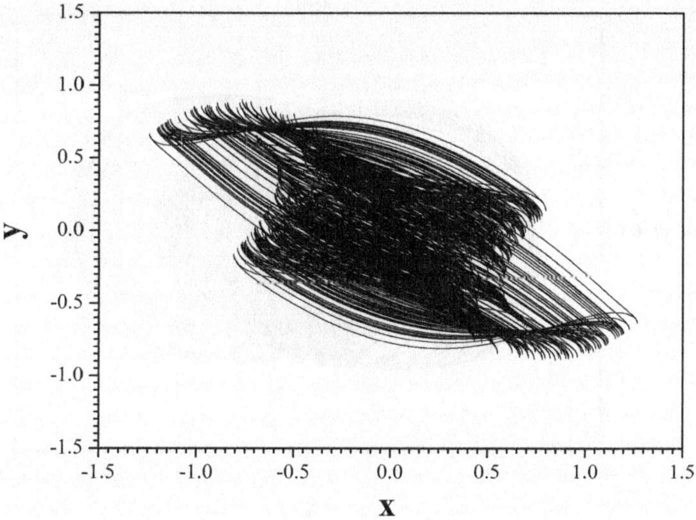

Fig. 13 2D projection of the chaotic system (17) on the (x, y)-plane, for $a = 0.28$, $b = 0.2$ and $c = 2.5$, with initial conditions $(x_0, y_0, z_0) = (0.1, 0.1, 0.1)$

4. By applying the general condition of dissipativity to system (17), the same results as in system (8) are concluded:

$$\text{div}\,\mathbf{u} = \frac{\partial \dot{x}}{\partial x} + \frac{\partial \dot{y}}{\partial y} + \frac{\partial \dot{z}}{\partial z} = cy^2 - 3z^2. \tag{20}$$

   Therefore, system (17) is also dissipative for $cy^2 - 3z^2 < 0$.
5. From the comparison of the bifurcation diagram of system (17) with the respective bifurcation diagram of system (8), a good agreement in their dynamics is concluded (see Fig. 17).
6. System (17) has also rotational symmetry with respect to the origin of the axes, as evidenced by their invariance under the transformation from (x, y, z) to $(-x, -y, -z)$. Therefore, any projection of the attractor has symmetry. By plotting forwards and backward continuations for the bifurcation parameter c, the presence of multi-stability in system (17) can be investigated (Fig. 18). For example, there are coexisting chaotic attractors when $c = 1.4$, as shown in Fig. 19.

From the study of dynamical behavior of system (17), a good agreement with the results produced by system (8), with the sine function, is obtained. So, system (17) with the Taylor series approximation of the sine function should be circuitry implementable as it will be presented in the next section.

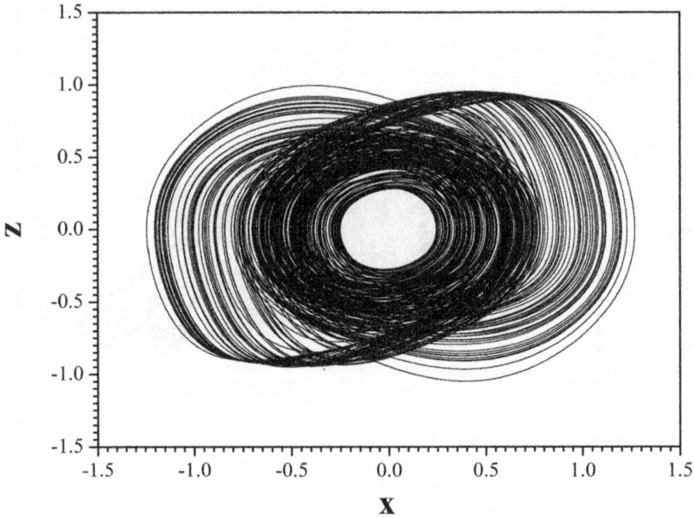

Fig. 14 2D projection of the chaotic system (17) on the (x, z)-plane, for $a = 0.28$, $b = 0.2$ and $c = 2.5$, with initial conditions $(x_0, y_0, z_0) = (0.1, 0.1, 0.1)$

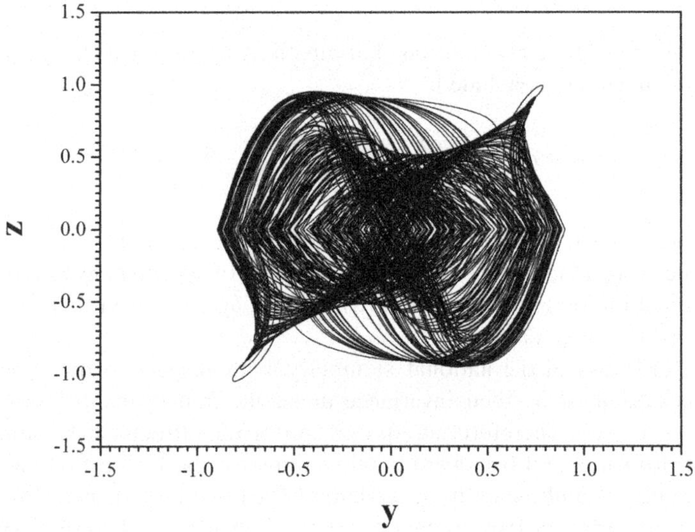

Fig. 15 2D projection of the chaotic system (17) on the (y, z)-plane, for $a = 0.28$, $b = 0.2$ and $c = 2.5$, with initial conditions $(x_0, y_0, z_0) = (0.1, 0.1, 0.1)$

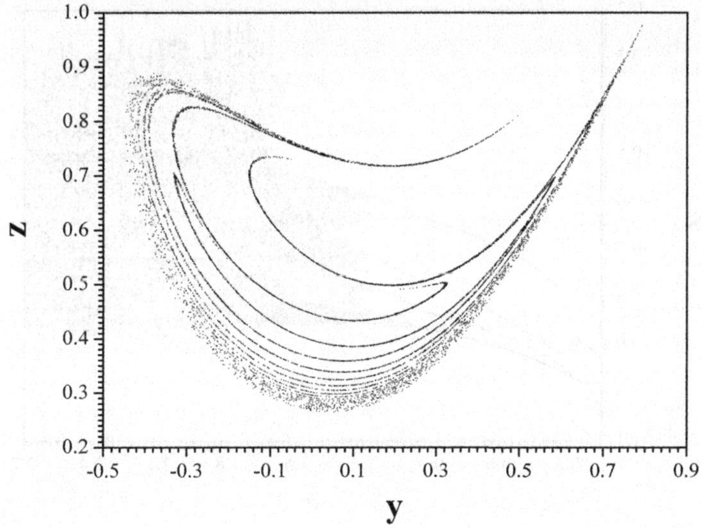

Fig. 16 Poincaré map of system (17) on the $y = 0$ plane, for $a = 0.28$, $b = 0.2$ and $c = 2.5$, with initial conditions $(x_0, y_0, z_0) = (0.1, 0.1, 0.1)$

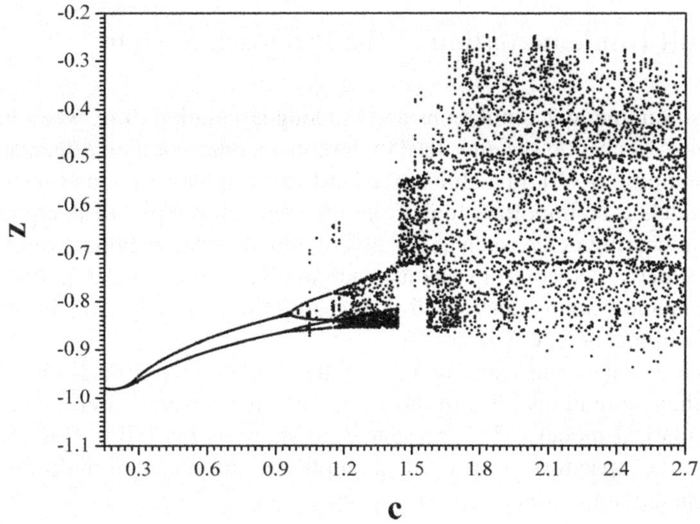

Fig. 17 Bifurcation diagram of system (17), with red dots, regarding the system (8), with black dots, when changing the value of parameter c, for $a = 0.28$ and $b = 0.2$, with initial conditions $(x_0, y_0, z_0) = (0.1, 0.1, 0.1)$

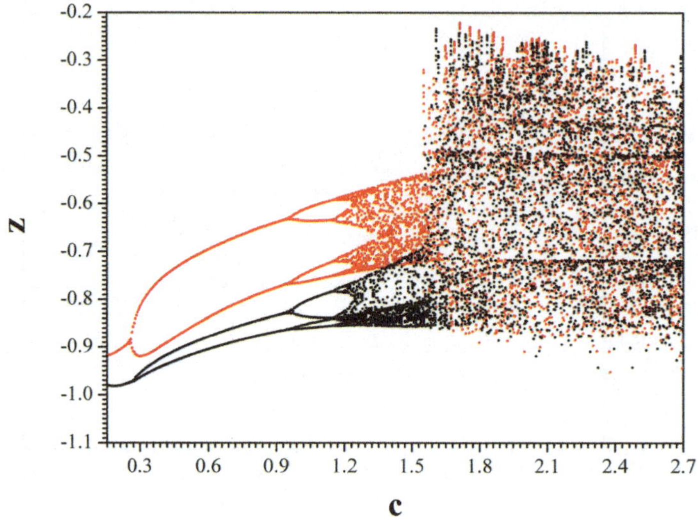

Fig. 18 Continuations of system (17) when varying c: forward continuation (black) and backward continuation (red), for $a = 0.28$ and $b = 0.2$

6 Circuit Implementation of the Proposed System

The errors in numerical simulations and the long simulation time, when investigating dynamics of chaotic systems, lead to the consideration of their effective physical implementation [66–69]. In addition, the hardware implementation of mathematical chaotic models is an important topic from the viewpoint of practical applications in various engineering fields, such as secure communications, robotics, signal generation, neural networks, control systems, encryption schemes, etc. [67, 70–75]. Nonlinear dynamical systems can be realized by using commercially available amplifiers, integrated circuits or FPGAs [76–78].

It is noted that, when using commercially available amplifiers, one should be aware of their limitations [79]. In this direction, an electronic circuit that emulates the mathematical model (17) is presented, to show its feasibility (Fig. 20). It has 20 resistors, 3 capacitors, 9 operational amplifiers and 9 analog multipliers. There are three integrators, which are created by the operational amplifiers ($U_1 - U_3$), five differential amplifiers ($U_4 - U_8$) and a comparator (U_9), implemented with the LF411. Also, there are nine signals multipliers (U_{10}, U_{18}) by using the AD633.

The state variables x, y, and z of system (17) are scaled up in order to avoid problems in the system realization, which are related with the endogenous scale factor of $10V$ of each of the multipliers that have been used. Therefore, the system (17) will be changed to

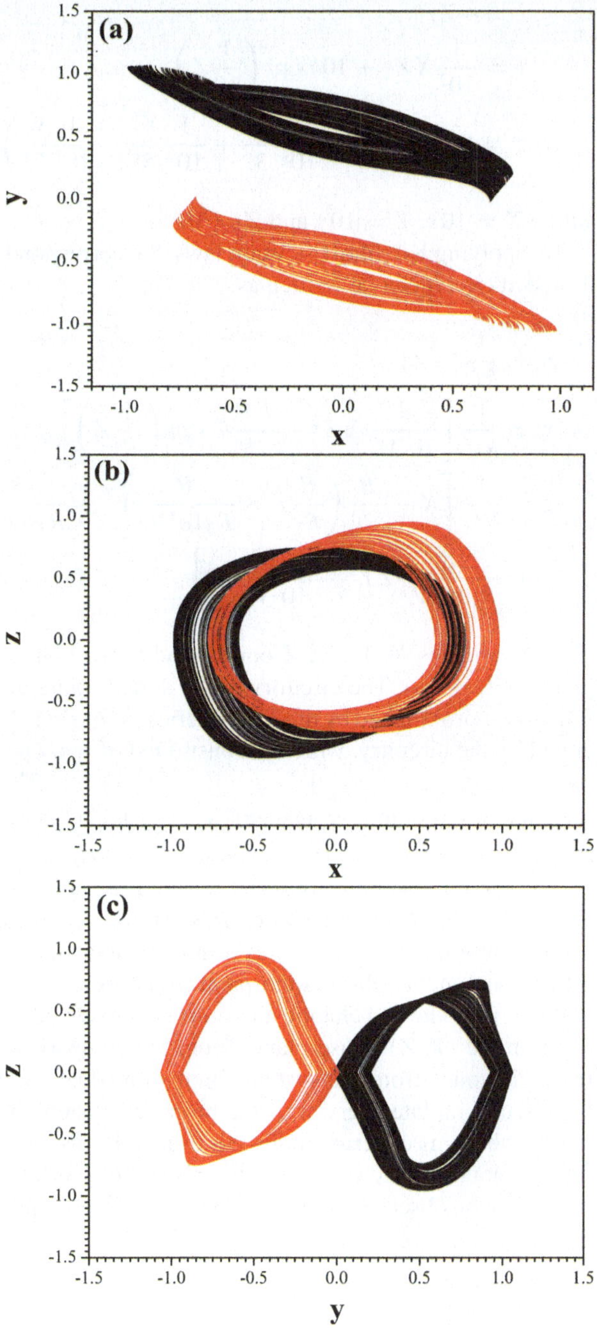

Fig. 19 Coexisting attractors of system (17), for $a = 0.25$, $b = 0.2$ and $c = 1.3$; the initial conditions $(x_0, y_0, z_0) = (0, 0.1, 0)$ (black) and the initial conditions $(x_0, y_0, z_0) = (0, -0.1, 0)$ (red) in (**a**) $x - y$ plane, (**b**) $x - z$ plane and (**c**) $y - z$ plane

$$
\begin{aligned}
\dot{x} &= -Z, \\
\dot{y} &= \frac{1}{10^2} X Z^2 + 10 a sgn\left(\frac{1}{10} Z\right), \\
\dot{z} &= X - b\left(\frac{Y}{10} - \frac{1}{10^3}\frac{Y^3}{3!} + \frac{1}{10^5}\frac{Y^5}{5!} - \frac{1}{10^7}\frac{y^7}{7!}\right) + \frac{1}{10^2} Z(cY^2 - Z^2),
\end{aligned} \tag{21}
$$

where $X = 10x$, $Y = 10y$ and $Z = 10z$.

By applying Kirchhoff's circuit laws, the corresponding circuital equations of the designed circuit can be written as

$$
\begin{aligned}
\dot{x} &= \frac{1}{RC}[-Z], \\
\dot{y} &= \frac{1}{RC}\left[\frac{1}{10^2 V^2} X Z^2 + \frac{R V_{sat}}{R_a} sgn\left(\frac{1}{10} Z\right)\right], \\
\dot{z} &= \frac{1}{RC}\left[X - \frac{R}{R_b}\left(\frac{R}{R_1} Y - \frac{R}{R_3 10^2 V^2} Y^3 + \frac{R}{R_5 10^4 V^4} Y^5 - \frac{R}{R_7 10^6 V^6} Y^7\right)\right. \\
&\quad \left. + \frac{R}{R_d 10^2 V^2} Z Y^2 - \frac{1}{10^2 V^2} Z^3\right],
\end{aligned} \tag{22}
$$

The variables X, Y and Z correspond to the voltages in the outputs of the integrators ($U_1 - U_3$). The circuitry in the dotted frame produces the signal of the sine function approximated by the Taylor series expansion. Also, the signal $sgn(Z)$ is generated by the circuitry, which is constituted of two operational amplifiers (U_7, U_8). Normalizing the differential equations of system (22) by using $t = \frac{t}{RC}$, it can be seen that this system is equivalent to the system (17), with $a = \frac{R V_{sat}}{10 R_a}$, $b = \frac{R}{10 R_b}$, and $c = \frac{R}{R_c}$.

The circuit components were selected as: $R = 10k\Omega$, $R_a = 58.4k\Omega$, $R_b = 5k\Omega$, $R_c = 4k\Omega$, $R_1{=}100k\Omega$, $R_3{=}60k\Omega$, $R_5{=}1.2M\Omega$, $R_7 = 50.4M\Omega$, $C = 10nF$, while the power supplies of all active devices are $\pm 15V_{DC}$. For the chosen set of components, the system parameters are: $a = 0.25$, $b = 0.2$ and $c = 1.3$. In Figs. 21, periodic and chaotic attractors are presented, which are obtained from Multisim on the (Y, Z) phase plane. From the comparison of these attractors with the respective ones from the system simulation (Figs. 2, 8 and 9), it can be seen that the circuit emulates very well the proposed system (8). Also, the phenomenon of multi-stability is experimentally confirmed. By using different initial conditions, as initial voltages at the capacitors C_1, C_2 and C_3, the coexisting attractors on various planes can be observed (Figs. 22, 23 and 24), as is expected by the simulation results presented in previous sections (Figs. 12, 19).

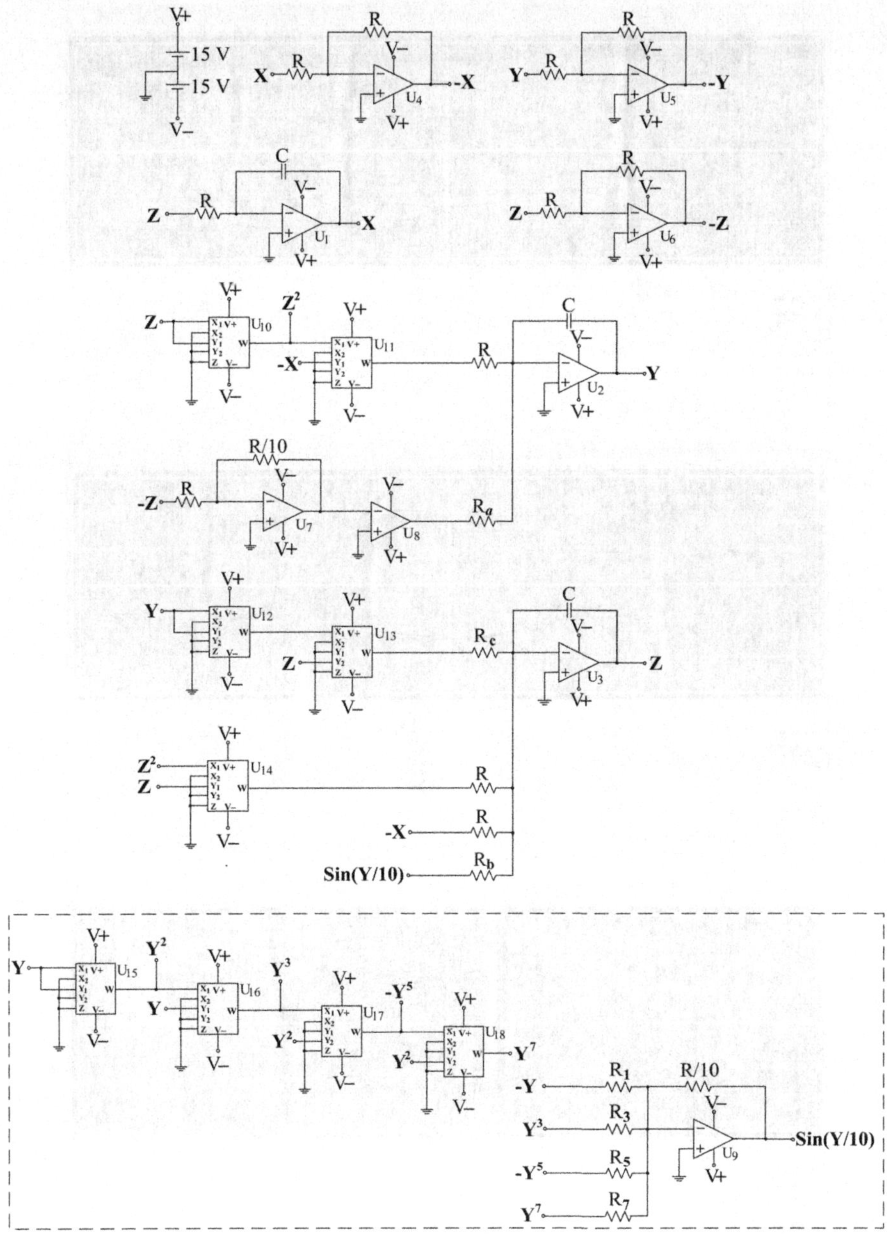

Fig. 20 Schematic of the master circuit

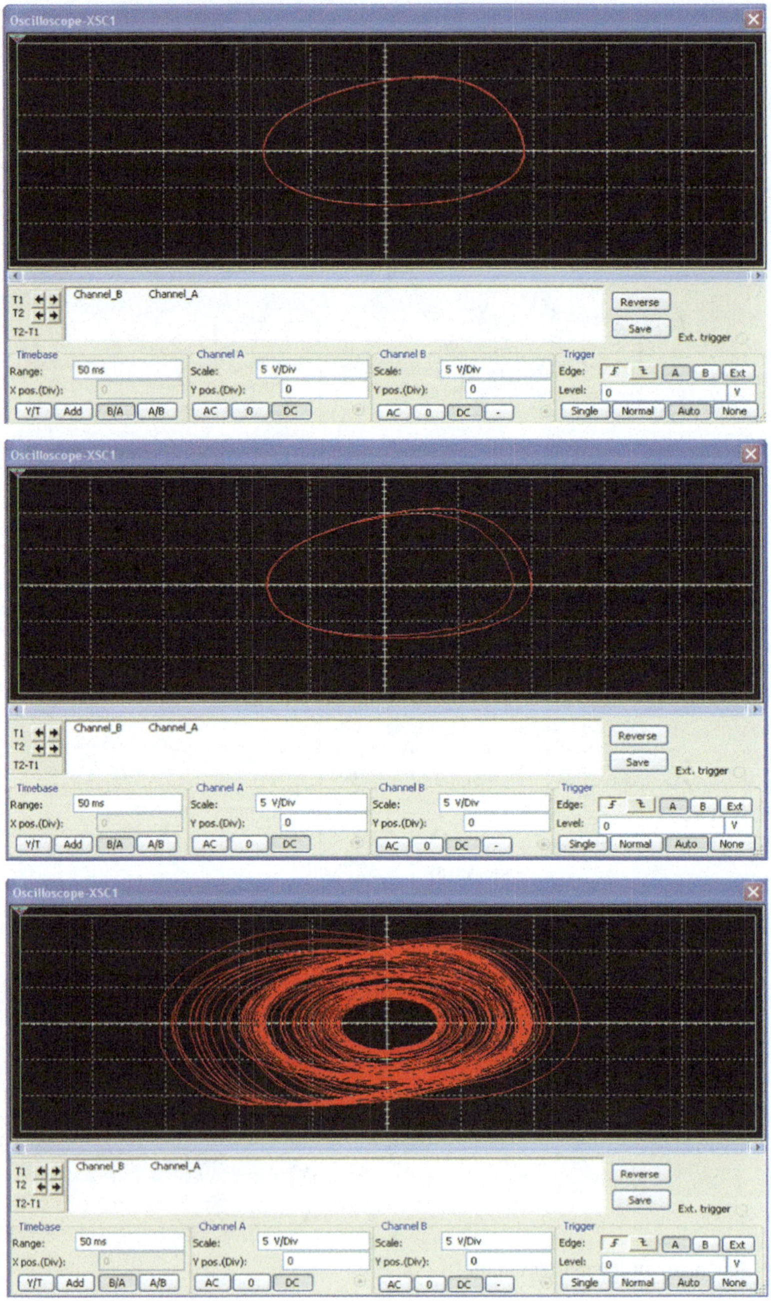

Fig. 21 Phase portraits from the designed circuit obtained by Multisim on the (Y, Z) phase plane, for $a = 0.28$, $b = 0.2$ and (**a**) $c = 0.1$, (**b**) $c = 0.5$, (**c**) $c = 2.5$

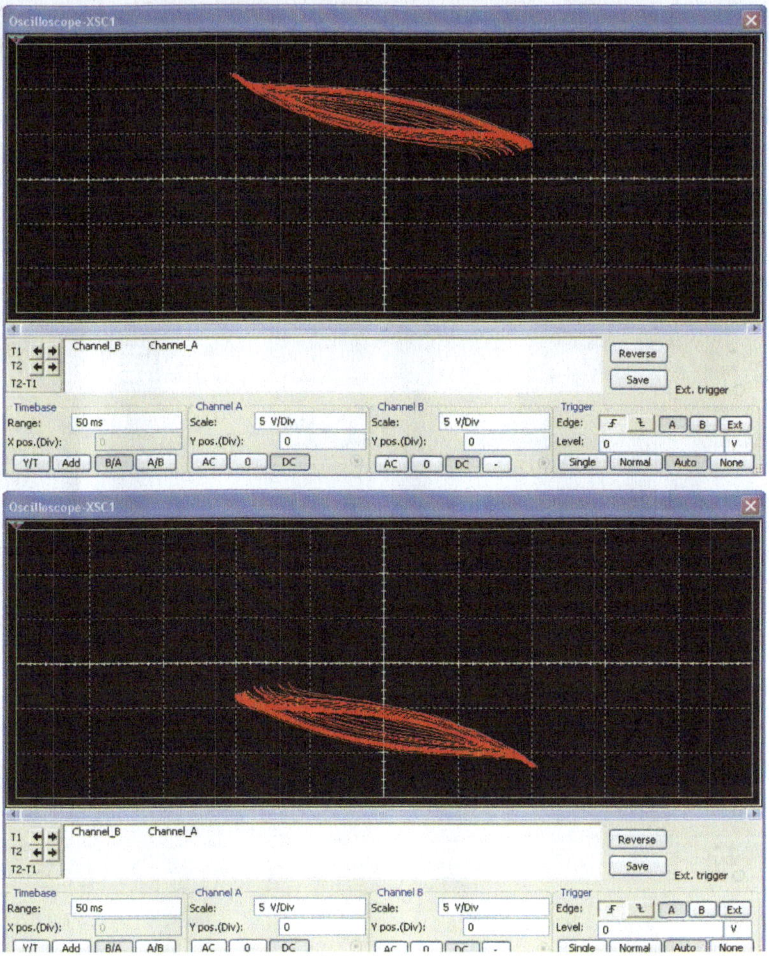

Fig. 22 Chaotic coexisting attractors of the designed circuit obtained from Multisim on the (X, Y) phase plane, for different initial conditions

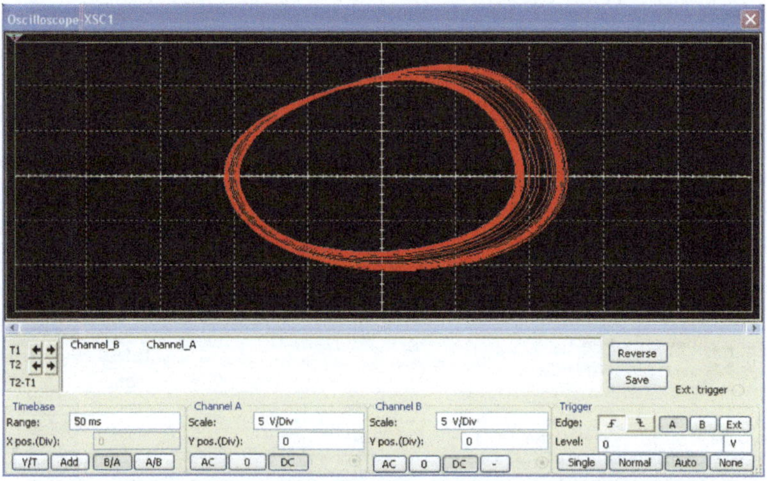

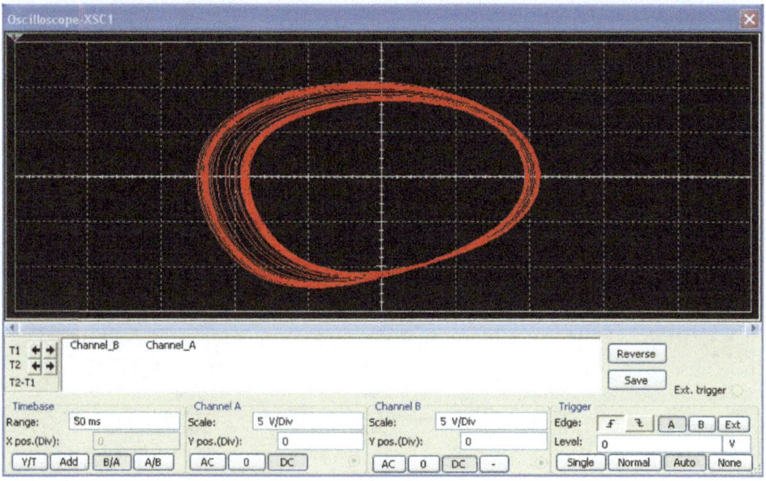

Fig. 23 Chaotic coexisting attractors of the designed circuit obtained from Multisim on the (X, Z) phase plane, for different initial conditions

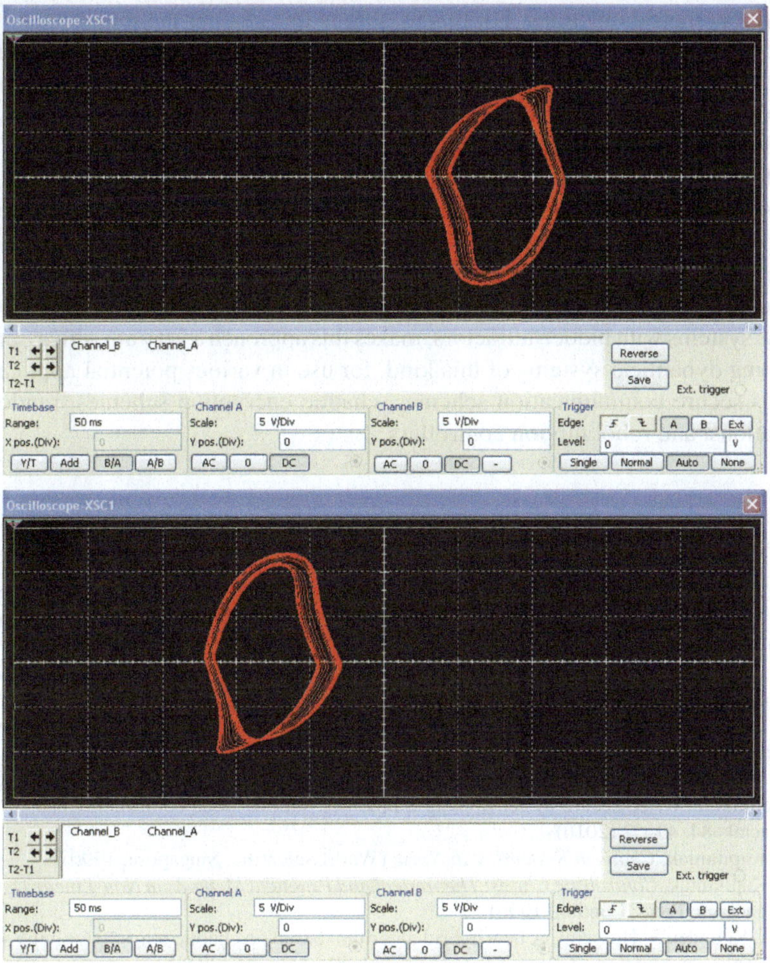

Fig. 24 Chaotic coexisting attractors of the designed circuit obtained from Multisim on the (Y, Z) phase plane, for different initial conditions

7 Conclusion

In this chapter, a new three-dimensional autonomous chaotic system with a sine function, which can display infinitely many equilibria, is presented. It is the first attempt to present a dynamical system with hidden attractors located onto a trigonometric curve. The proposed system has rich dynamics as confirmed by using well-known tools from nonlinear theory, such as phase portrait, Poincaré map, bifurcation diagram, maximum Lyapunov exponent, and the Kaplan-Yorke fractional dimension.

The absence from the market of a chip that can produce trigonometric functions is a drawback in the design and implementation of a circuitry realizing dynamical systems with trigonometric functions. For this reason, in this chapter the replacement of the sine function in the proposed system with its Taylor series expansion is presented. A comparative study of the two systems presents a good agreement from a dynamical point of view.

This investigation has driven us to design an analog circuit emulating the theoretical system with the Taylor series expansion and simulating its behavior by using Multisim. The results confirmed the usefulness of the proposed framework on circuitry design of sine function based on Taylor series expansion in realizing nonlinear dynamical systems.

The ease of the specific design method, for the use of trigonometric functions in chaotic systems with hidden attractors, makes this approach a very attractive option in designing dynamical systems of this kind, for use in various potential applications, such as secure communication schemes, chaotic encryption schemes, random bit generations, and robot motion controllers.

References

1. S. Behnia, Z. Pazhotan, N. Ezzati, A. Akhshani, Reconfigurable chaotic logic gates based on novel chaotic circuit. Chaos, Solitons Fractals **69**, 74–80 (2014)
2. M. Berezowski, D. Dubaj, Chaotic oscillations of coupled chemical reactors. Chaos, Solitons Fractals **78**(1), 22–25 (2015)
3. M.A. Chenaghlu, S. Jamali, N.N. Khasmakhi, A novel keyed parallel hashing scheme based on a new chaotic system. Chaos, Solitons Fractals **87**, 216–225 (2016)
4. A. Kajbaf, M.A. Akhaee, M. Sheikhan, Fast synchronization of non-identical chaotic modulation-based secure systems using a modified sliding mode controller. Chaos, Solitons Fractals **84**, 49–57 (2016)
5. T. Kapitaniak, *Chaos in Systems with Noise* (World Scientific, Singapore, 1990)
6. T. Kapitaniak, *Controlling Chaos: Theoretical and Practical Methods in Non-Linear Dynamics* (Academic Press, London, 1996)
7. J.A. Koupaei, S. Hosseini, A new hybrid algorithm based on chaotic maps for solving systems of nonlinear equations. Chaos, Solitons Fractals **81**, 233–245 (2015)
8. E. Lorenz, Deterministic non-periodic flow. J. Atmos. Sci. **20**, 130–141 (1963)
9. C. Morel, R. Vlad, J.Y. Morel, D. Petreus, Generating chaotic attractors on a surface. Math. Comput. Simula. **81**(11), 2549–2563 (2011)
10. I. Pan, S. Das, When Darwin meets Lorenz: Evolving new chaotic attractors through genetic programming. Chaos, Solitons Fractals **76**, 141–155 (2015)
11. L.M. Pecora, T.L. Carroll, Synchronization in chaotic signals. Phys. Rev. A **64**, 821–824 (1990)
12. J.C. Sprott, *Elegant Chaos: Algebraically Simple Chaotic Flows* (World Scientific, Singapore, 2010)
13. R.E. Mirollo, S.H. Strogatz, Synchronization of pulse-coupled biological oscillators. SIAM J. Appl. Math. **50**(6), 1645–1662 (1990)
14. J.H. Lu, G. Chen, Generating multiscroll chaotic attractors: Theories, methods and applications. Int. J. Bifurc. Chaos **16**(4), 775–858 (2006)
15. X. Wang, G. Chen, A chaotic system with only one stable equilibrium. Commun. Nonlinear Sci. Numer. Simul. **17**(3), 1264–1272 (2012)
16. X. Wang, G. Chen, Constructing a chaotic system with any number of equilibria. Nonlinear Dyn. **71**(3), 429–436 (2012)
17. E.N. Lorenz, Deterministic nonperiodic flow. J. Atmos. Sci. **20**(2), 130–141 (1963)

18. O.E. Rössler, An equation for continuous chaos. Phys. Lett. A **57**(5), 397–398 (1976)
19. A. Arneodo, P. Coullet, C. Tresser, Possible new strange attractors with spiral structure. Commun. Math. Phys. **79**(4), 573–579 (1981)
20. G. Chen, T. Ueta, Yet another chaotic attractor. Int. J. Bifurc. Chaos **9**, 1465–1466 (1999)
21. J.C. Sprott, Some simple chaotic flows. Phys. Rev. E **50**(2), R647 (1994)
22. L.P. Šilnikov, A case of the existence of a denumerable set of periodic motions. Dokl. Akad. Nauk SSSR **160**, 558–561 (1965)
23. L.P. Shilnikov, A.L. Shilnikov, D.V. Turaev, L.O. Chua, *Methods of Qualitative Theory in Nonlinear Dynamics* (World Scientific, Singapore, 1998)
24. N.V. Kuznetsov, G.A. Leonov, V.I. Vagaitsev, Analytical-numerical method for attractor localization of generalized Chua's system. Periodic Control Systems **4**(11), 29–33 (2010)
25. G.A. Leonov, N.V. Kuznetsov, V.I. Vagaitsev, Localization of hidden Chua's attractors. Phys. Lett. A **375**(23), 2230–2233 (2011)
26. G.A. Leonov, N.V. Kuznetsov, Analytical-numerical methods for investigation of hidden oscillations in nonlinear control systems. IFAC Proc. **44**(1), 2494–2505 (2011)
27. G. Leonov, N. Kuznetsov, V. Vagaitsev, Hidden attractor in smooth Chua systems. Physica D **241**(18), 1482–1486 (2012)
28. V.T. Pham, C. Volos, S. Jafari, X. Wang, S. Vaidyanathan, Hidden hyperchaotic attractor in a novel simple memristive neural network. Optoelectron. Adv. Mater.: Rapid Commun. **8**(11–12), 1157–1163 (2014)
29. V. Pham, S. Jafari, C. Volos, X. Wang, Is that really hidden? The presence of complex fixed-points in chaotic flows with no equilibria. Int. J. Bifurc. Chaos **24**(11), 1450146 (2014)
30. G. Leonov, N. Kuznetsov, V. Vagaitsev, Localization of hidden Chua's attractors. Phys. Lett. A **375**(23), 2230–2233 (2011)
31. D. Hilbert, Ueber flachen von constanter gaussscher krummung. Trans. Am. Math. Soc. **2**(1), 87–99 (1901)
32. N.N. Bautin, On the number of limit cycles appearing with variation of the coefficients from an equilibrium state of the type of a focus or a center. Mat. sb, **30**(72), 181–196 (1952)
33. M. Kapranov, Locking band for phase-locked loop [in Russian]. Radiofizika **2**(12), 37–52 (1956)
34. N.A. Gubar', Investigation of a piecewise linear dynamical system with three parameters. J. Appl. Math. Mech. **25**(6), 1519–1535 (1961)
35. L. Markus, H. Yamabe, Global stability criteria for differential systems. Osaka J. Math. **12**(2), 305–317 (1960)
36. R.E. Kalman, Physical and mathematical mechanisms of instability in nonlinear automatic control systems. Trans. ASME **79**(3), 553–563 (1957)
37. R. Fitts, Two counterexamples to Aizerman's conjecture. IEEE Trans. Autom. Control **11**(3), 553–556 (1966)
38. J. Bernat, J. Llibre, Counterexample to Kalman and Markus-Yamabe conjectures in larger than 3. Dyn. Contin. Discrete Impuls. Syst. **2**(3), 337–379 (1996)
39. G.A. Leonov, N.V. Kuznetsov, Hidden attractors in dynamical systems. From hidden oscillations in Hilbert-Kolmogorov, Aizerman, and Kalman problems to hidden chaotic attractors in Chua circuits. Int. J. Bifurc. Chaos **23**, 12330002 (2013)
40. T. Lauvdal, R.M. Murray, T.I. Fossen, Stabilization of integrator chains in the presence of magnitude and rate saturations: A gain scheduling approach. Proceedings of the 36th IEEE Conference on Decision and Control, vol. 4 (IEEE, New York, 1997), pp. 4004–4005
41. S. Jafari, J.C. Sprott, Elementary quadratic chaotic flows with no equilibria. Phys. Lett. A **377**(9), 699–702 (2013)
42. Z. Wei, Dynamical behaviors of a chaotic system with no equilibria. Phys. Lett. A **376**(2), 102–108 (2011)
43. S. Jafari, J. Sprott, Simple chaotic flows with a line equilibrium. Chaos, Solitons Fractals **57**, 79–84 (2013)
44. X. Wang, G. Chen, A chaotic system with only one stable equilibrium. Commun. Nonlinear Sci. Numer. Simul. **17**(3), 1264–1272 (2011)

45. X. Wang, G. Chen, Constructing a chaotic system with any number of equilibria. Nonlinear Dyn. **71**, 429–436 (2013)
46. G. Leonov, Fishing principle for homoclinic and heteroclinic trajectories. Nonlinear Dyn. **78**(4), 2751–2758 (2014)
47. T. Zhusubaliyev, E. Mosekilde, V. Rubanov, R. Nabokov, Multistability and hidden attractors in a relay system with hysteresis. Physica D **306**, 6–15 (2015)
48. Q. Xu, Y. Lin, B. Bao, M. Chen, Multiple attractors in a non-ideal active voltage-controlled memristor based Chua's circuit. Chaos, Solitons Fractals **83**, 186–200 (2016)
49. O.S. Ojoniyi, A.N. Njah, A 5D hyperchaotic Sprott B system with coexisting hidden attractors. Chaos, Solitons Fractals **87**, 172–181 (2016)
50. C. Li, J.C. Sprott, Chaotic flows with a single nonquadratic term. Phys. Lett. A **378**(3), 178–183 (2014)
51. C. Li, J.C. Sprott, Multistability in the Lorenz system: A broken butterfly. Int. J. Bifurc. Chaos **24**(10), 1450131 (2014)
52. T. Gotthans, J. Petržela, New class of chaotic systems with circular equilibrium. Nonlinear Dyn. **81**(3), 1143–1149 (2015)
53. T. Gotthans, J.C. Sprott, J. Petrzela, Simple chaotic flow with circle and square equilibrium. Int. J. Bifurc. Chaos **26**(8), 1650137 (2016)
54. V.-T. Pham, S. Vaidyanathan, C. Volos, S. Jafari, S. Kingni, A no-equilibrium hyperchaotic system with a cubic nonlinear term. Optik: Int. J. Light Electron Opt. **127**(6), 3259–3265 (2016)
55. V.T. Pham, S. Jafari, X. Wang, J. Ma, A chaotic system with different shapes of equilibria. Int. J. Bifurc. Chaos **26**(4), 1650069 (2016)
56. S. Jafari, J.C. Sprott, Simple chaotic flows with a line equilibrium. Chaos, Solitons Fractals **57**(4), 79–84 (2013)
57. C. Li, J.C. Sprott, Z. Yuan, H. Li, Constructing chaotic systems with total amplitude control. Int. J. Bifurc. Chaos **25**(10), 1530025 (2015)
58. T. Gotthans, J. Petržela, New class of chaotic systems with circular equilibrium. Nonlinear Dyn. **81**(3), 1–7 (2015)
59. V.T. Pham, S. Vaidyanathan, C. Volos, S. Jafari, S.T. Kingni, V.T. Pham, S. Vaidyanathan, C. Volos, S. Jafari, S.T. Kingni, A no-equilibrium hyperchaotic system with a cubic nonlinear term. Optik: Int. J. Light Electron Opt. **127**(6), 3259–3265 (2016)
60. V.T. Pham, C. Volos, S. Jafari, S. Vaidyanathan, T. Kapitaniak, X. Wang, A chaotic system with different families of hidden attractors. Int. J. Bifurc. Chaos **26**(8), 1650139 (2016)
61. J. Kengne, J. Chedjou, M. Kom, K. Kyamakya, V.K. Tamba, Regular oscillations, chaos, and multistability in a system of two coupled van der pol oscillators: numerical and experimental studies. Nonlinear Dyn. **76**(2), 1119–1132 (2014)
62. J. Kengne, Z. Njitacke, H. Fotsin, Dynamical analysis of a simple autonomous jerk system with multiple attractors. Nonlinear Dyn. **83**(1–2), 751–765 (2016)
63. E. Ott, C. Grebogi, J.A. Yorke, Controlling chaos. Phys. Rev. Lett. **64**, 1196–1199 (1990)
64. C.B. Boyer, U.C. Merzbach, *A History of Mathematics*, 3rd ed. (Wiley, New York, 2011)
65. T. Finney, *Elements of Calculus & Analytic Geometry: Solution Manual* (Addison-Wesley, Reading, 1989)
66. J.O. Maaita, C.K. Volos, I.N. Stouboulos, I.M. Kyprianidis, The dynamics of a cubic nonlinear system with no equilibrium point. Nonlinear Dyn. **2015**, 257923 (2015)
67. L. Fortuna, M. Frasca, M.G. Xibilia, *Chua's Circuit Implementations: Yesterday* (World Scientific, Today and Tomorrow, 2009)
68. S. Vaidyanathan, S. Sampath, C.K. Volos, V.T. Pham, An eight-term novel four-scroll chaotic system with cubic nonlinearity and its circuit simulation. J. Eng. Sci. Technol. Rev. **8**(2), 1–6 (2015)
69. F.R. Tahir, S. Jafari, V. Pham, C. Volos, X. Wang, A novel no-equilibrium chaotic system with multiwing butterfly attractors. Int. J. Bifurc. Chaos **25**(4), 1550056 (2015)
70. V.-T. Pham, S. Vaidyanathan, C. Volos, S. Jafari, Hidden attractors in a chaotic system with an exponential nonlinear term. Eur. Phys. J.: Special Topics **224**(8), 1507–1517 (2015)

71. S. Vaidyanathan, Analysis and adaptive synchronization of eight-term novel 3-D chaotic system with three quadratic nonlinearities. Eur. Phys. J.: Special Topics **223**, 1519–1529 (2014)
72. C.K. Volos, I.M. Kyprianidis, I.N. Stouboulos, Various synchronization phenomena in bidirectionally coupled double scroll circuits. Commun. Nonlinear Sci. Numer. Simul. **71**, 3356–3366 (2011)
73. C.K. Volos, I.M. Kyprianidis, I.N. Stouboulos, Experimental investigation on coverage performance of a chaotic autonomous mobile robot. Robot. Auton. Syst. **61**(12), 1314–1322 (2013)
74. C.K. Volos, I.M. Kyprianidis, I.N. Stouboulos, Image encryption process based on chaotic synchronization phenomena. Signal Process. **93**, 1328–1340 (2013)
75. C.K. Volos, I.M. Kyprianidis, I.N. Stouboulos, Text encryption scheme realized with a chaotic pseudo-random bit generator. J. Eng. Sci. Technol. Rev. **6**(4), 9–14 (2013)
76. G. Kaddoum, F. Gagnon, Design of a high-data-rate differential chaos-shift keying system. IEEE Trans. Circuits Syst. II: Express Briefs **59**(7), 448–452 (2012)
77. R. Trejo-Guerra, E. Tlelo-Cuautle, V.H. Carbajal-Gómez, G. Rodriguez-Gómez, A survey on the integrated design of chaotic oscillators. Appl. Math. Comput. **219**(10), 5113–5122 (2013)
78. D. Valli, B. Muthuswamy, S. Banerjee, M.R.K. Ariffin, A.W.A. Wahab, K. Ganesan, C.K. Subramaniam, J. Kurths, Synchronization in coupled Ikeda delay systems. experimental observations using field programmable gate arrays. Eur. Phys. J.: Special Topics **223**(8), 1465–1479 (2014)
79. E. Ortega-Torres, C. Sánchez-López, J. Mendoza-López, Frequency behavior of saturated nonlinear function series based on opamps. Rev. Mex. Fis. **59**(59), 504–510 (2013)

Dynamics of a 4D Hyperchaotic System with No or with Infinitely Many Isolated Equilibria

Qigui Yang and Yanhong Zhang

1 Introduction

The first four-dimensional (4D) hyperchaotic system was obtained by computer simulation in [1]. Due to the great potential in technological applications, such as secure communications, control systems, and neural networks, hyperchaos generation has become a focus research topic in the last four decades (see e.g. [1–6] as well as references therein). Hyperchaos is characterized as a chaotic system with at least two positive Lyapunov exponents, indicating that its dynamics are expended in two or more directions simultaneously. Thus, hyperchaotic systems have more complex dynamical behaviors and higher application values than common chaotic systems such as the Lorenz system [7], Chen system [8], Lü system [9], Yang system [10], and Hisch et al. [11], and so on. However, many hyperchaotic systems have been introduced by adding a feedback controller and other methods to a chaotic system with finitely many equilibria [12–14]. By now, it is still difficult to generate a hyperchaotic system because there is not a unified and effective method to do so.

The hyperchaos theory is still in its preliminary stage, and the complex dynamics of hyperchaotic systems have not been completely understood. Therefore, deeper studies of hyperchaotic systems are needed. As is well known, the lowest possible dimension of continuous-time hyperchaos is four, so 4D systems are the main interest in investigations. Some typical examples of 4D hyperchaotic systems include the hyperchaotic Rössler system [1], hyperchaotic Chua's circuit [15], and hyperchaotic modified Chua's circuit [16]. These examples stimulate us to further research on the properties of hyperchaos and some delicate characteristics of the new hyperchaotic

Q. Yang (✉) · Y. Zhang
School of Mathematical Sciences, South China University of Technology, Wushan District, Guangzhou 510640, Guangdong, People's Republic of China
e-mail: qgyang@scut.edu.cn

Y. Zhang
e-mail: 504832542@qq.com

X. Wang et al. (eds.), *Chaotic Systems with Multistability and Hidden Attractors*, Emergence, Complexity and Computation 40,
https://doi.org/10.1007/978-3-030-75821-9_20

system to be studied in this chapter, which has infinitely many isolated equilibria or without any equilibria.

In chaos theory, the number of equilibria is significant for understanding the complicated dynamics. To date, most hyperchaotic systems reported have a finitely many and a countable number of isolated equilibria. For 4D systems, only a few hyperchaotic systems with a line of equilibria or without equilibria are found [17]. Recently, Chen and Yang introduced a new Lorenz-type hyperchaotic system with a curve of equilibria [18]. To the best of our knowledge, little seems to be known about hyperchaotic systems with infinitely many isolated equilibria. Thus, it is natural and interesting to ask whether a 4D system with infinitely many isolated equilibria or without equilibria can reveal complex dynamics like hyperchaotic, chaotic, quasi-periodic and periodic attractors, and whether the coexistence of different attractors is possible. Positively, the answer to these questions will be discussed in the present paper.

Based on the number of equilibria and with the help of feedback control techniques, this chapter reports and analyzes a new 4D hyperchaotic system with infinitely many isolated equilibria or without equilibria. To further research on the new system, some complex dynamical behaviors such as the stability of hyperbolic or non-hyperbolic equilibria are rigorously studied via the center manifold theory. This new hyperchaotic system displays hyperchaotic, chaotic, quasi-periodic and periodic dynamics, which are investigated through numerical simulations, including phase portrait, Lyapunov exponent spectrum and Poincaré map, etc. Especially, with the exception of infinitely many isolated equilibria or without equilibria, there are several other coexisting attractors in the phase space of this new hyperchaotic system, including the coexistence of hyperchaotic attractor and periodic attractor, chaotic attractor and periodic attractor, quasi-periodic attractor and periodic attractor, and hidden hyperchaotic attractor and hidden periodic attractor.

2 New Hyperchaotic System

This section proposes a new 4D autonomous hyperchaotic system with infinitely many isolated equilibria or without equilibria, which has nine items with one nonlinear term, but it is very different from other 4D hyperchaotic systems.

2.1 *Form of the Hyperchaotic System*

Consider the following two-dimensional planar linear autonomous system:

$$\begin{cases} \dot{x} = x - y, \\ \dot{y} = cx + dy, \end{cases} \tag{1}$$

and the following one-dimensional autonomous system:

$$\dot{z} = -a, \tag{2}$$

where a, c and d are parameters. By using the solution of the first-order linear equations, one can obtain the solution set of system (1) and system (2). Coupling system (1) and system (2) together, one gets the following system:

$$\begin{cases} \dot{x} = x - y, \\ \dot{y} = cx + dy, \\ \dot{z} = -a + by, \end{cases} \tag{3}$$

where by is a coupled item, with b being the coupled divisor. It is easy to see that system (3) has no chaos. To obtain chaos, designing a controller for the second equation of system (3), one obtains a three-dimensional autonomous system with only one nonlinear term, described by

$$\begin{cases} \dot{x} = x - y, \\ \dot{y} = cx + dy - e \sin z, \\ \dot{z} = -a + by, \end{cases} \tag{4}$$

where $abe \neq 0$, $d < 0$, c and d are constant parameters, a, b and e are three control parameters, which together determine the chaotic behaviors and bifurcation of the system. When $(a, b, c, d, e) = (3, 5.9, 3, -1.5, 2)$, system (4) has infinitely many isolated equilibria:

$$E_{1k}\left(\frac{30}{59}, \frac{30}{59}, \arcsin\frac{45}{118} + 2k\pi\right), \quad E_{2k}\left(\frac{30}{59}, \frac{30}{59}, \pi - \arcsin\frac{45}{118} + 2k\pi\right)$$

where $k \in \mathbb{Z} = \{0, \pm 1, \pm 2, \ldots\}$. The three Lyapunov exponents are

$$\lambda_{LE_1} = 0.4562, \qquad \lambda_{LE_2} = 0.0000, \qquad \lambda_{LE_3} = -0.9763.$$

and the Lyapunov dimension of the corresponding chaotic attractor is $D_L = 2.467$. Moreover, numerical simulations indicate that system (4) indeed has a chaotic attractor when $(a, b, c, d, e) = (3, 5.9, 3, -1.5, 2)$, as depicted in Fig. 1.

It is well known that the lowest dimension of an autonomous system for the generation of hyperchaos is four, and hyperchaotic systems have more complex dynamical behaviors and higher application values than chaotic systems. Therefore, 4D hyperchaotic systems deserve more attention. Based on chaotic system (4) with infinitely many isolated equilibria, we further study the complicated dynamics of 4D hyperchaotic systems.

By adding a linear feedback controller to the first equation of system (4), a new 4D hyperchaotic system is obtained, as

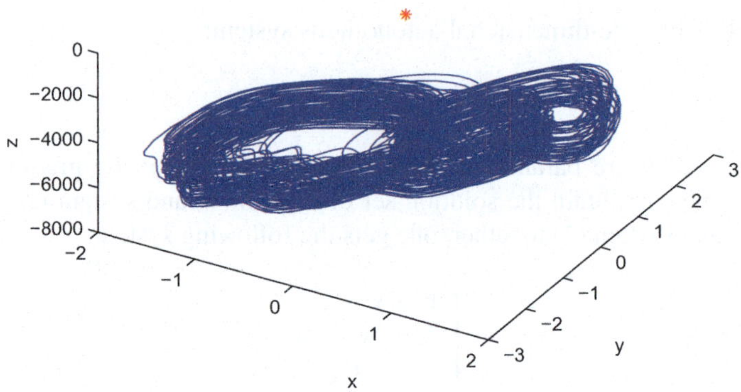

Fig. 1 Chaotic attractor of system (4): $(a, b, c, d, e) = (3, 5.9, 3, -1.5, 2)$

$$\begin{cases} \dot{x} = x - y + w, \\ \dot{y} = cx + dy - e \sin z, \\ \dot{z} = -a + by, \\ \dot{w} = -fx, \end{cases} \tag{5}$$

where the parameters a, b, c, d, e and f satisfy $abef \neq 0$ and $d < 0$. Under certain parameters conditions, system (5) has infinitely many isolated equilibria or without equilibria.

When $(a, b, c, d, e, f) = (3, 3, 5, -3, -15, 3.7)$, system (5) has a hyperchaotic attractor, with phase portraits shown in Fig. 2.

The corresponding four Lyapunov exponents of this hyperchaotic system are

$$\lambda_{LE_1} = 0.3593, \quad \lambda_{LE_2} = 0.1511, \quad \lambda_{LE_3} = 0.0000, \quad \lambda_{LE_4} = -2.5105.$$

The Lyapunov dimension is characterized by its Lyapunov exponents, i.e.

$$D_L = j + \frac{1}{|\lambda_{LE_{j+1}}|} \sum_{i=1}^{j} \lambda_{LE_i},$$

where j is an integer, which meets $\sum_{i=1}^{j} \lambda_{LE_i} \geqslant 0$ and $\sum_{i=1}^{j+1} \lambda_{LE_i} < 0$. Therefore, the Lyapunov dimension is $D_L = 3.2033$, which means that the hyperchaotic attractor has fractal dimension.

Moreover, the infinitely many isolated equilibria of system (5) are

$$E_{1k}\left(0, 1, \ \arcsin\frac{1}{5} + 2k\pi, 1\right), \quad E_{2k}\left(0, 1, \ \pi - \arcsin\frac{1}{5} + 2k\pi, 1\right), \quad k \in \mathbb{Z},$$

respectively. And, the corresponding eigenvalues of equilibria E_{1k}, E_{2k} are

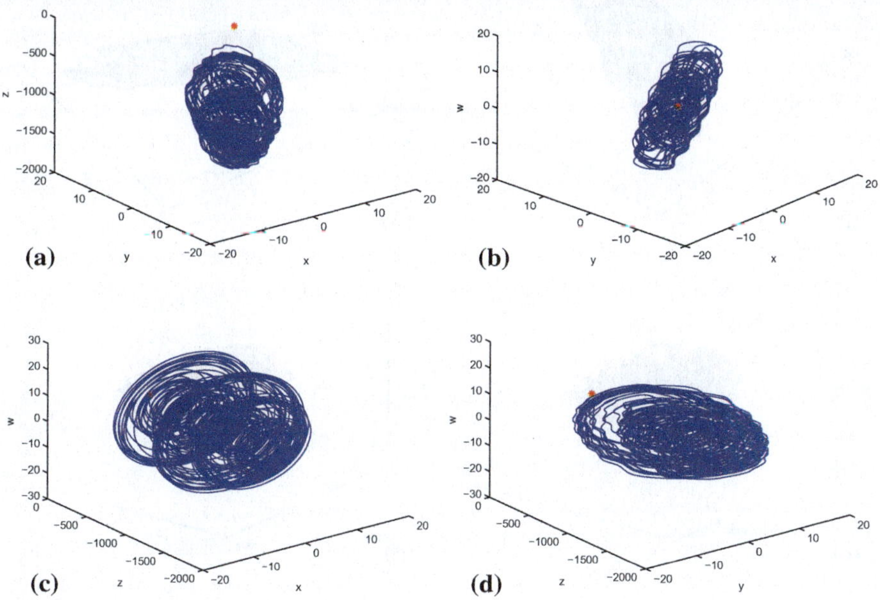

Fig. 2 Three-dimensional projections of the hyperchaotic attractor of system (5): $(a, b, c, d, e, f) = (3, 3, 5, -3, -15, 3.7)$

$$\lambda_1 = -7.9896, \quad \lambda_2 = 4.9091, \quad \lambda_3 = 0.5402 + 1.9666i, \quad \lambda_4 = 0.5402 - 1.9666i.$$

Thus, E_{1k} and E_{2k} are saddle-focus points, which have a one-dimensional stable manifold and a three-dimensional unstable manifold.

When $(a, b, c, d, e, f) = (11, 3, 5, -3, -10, 2.74)$, system (5) has no equilibria. But, system (5) has a hidden hyperchaotic attractor, and the phase portraits are shown in Fig. 3.

The corresponding Lyapunov exponents of the hidden hyperchaotic attractor are

$$\lambda_{LE_1} = 0.3002, \quad \lambda_{LE_2} = 0.0399, \quad \lambda_{LE_3} = 0.0000, \quad \lambda_{LE_4} = -2.3401.$$

and the Lyapunov dimension is $D_L = 3.1453$, which means that the hidden hyperchaotic attractor has a fractal dimension.

Remark 20.1 When the system (5) has infinitely many isolated equilibria or without equilibria, it can generate a hyperchaotic attractor or a hidden hyperchaotic attractor. To the best of our knowledge, this situation has not been discussed in other studies.

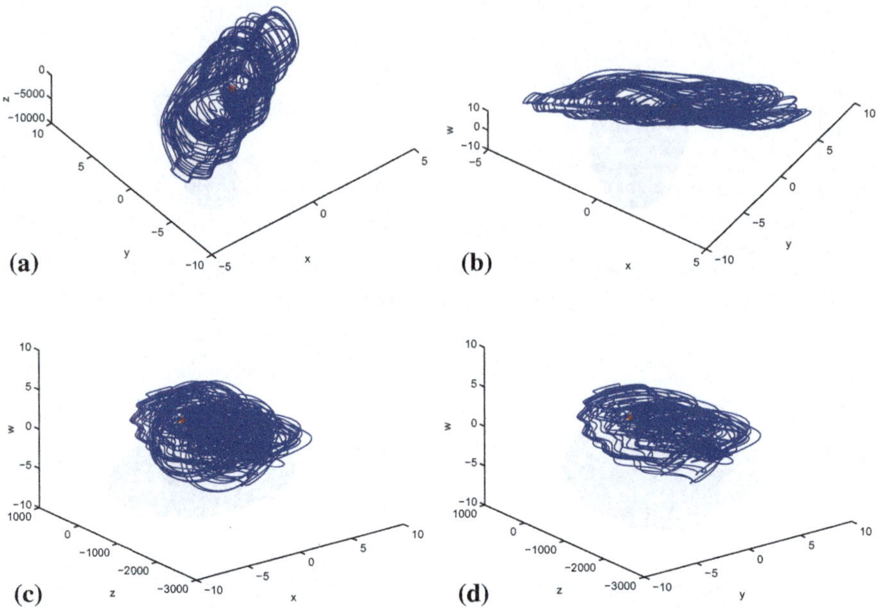

Fig. 3 Three-dimensional projections of the hidden hyperchaotic attractor of system (5): $(a, b, c, d, e, f) = (11, 3, 5, -3, -10, 2.74)$

2.2 *Dissipativity and Sensitive Dependence on the Initial Value*

This subsection examines the dynamical behaviors of the hyperchaotic system (5), including dissipativity and sensitive dependence on the initial value.

2.2.1 Dissipativity

The divergence of system (5) is

$$\nabla V = \frac{\partial \dot{x}}{\partial x} + \frac{\partial \dot{y}}{\partial y} + \frac{\partial \dot{z}}{\partial z} + \frac{\partial \dot{w}}{\partial w} = d + 1,$$

so the system will be dissipative if condition $d < -1$ is satisfied, and the exponential contraction rate: $\frac{dV}{dt} = (d+1)V$ for any values of a, b, c, e and f. That is, a volume element V_0 is contracted to $V_0 e^{(d+1)t}$ in time t, indicating that the volume of the phase space is shrinking to zero as $t \to \infty$ at an exponential rate: $(1 + d)$, which is independent of the state variables x, y, z and w. Therefore, when $d < -1$, all orbits of system (5) ultimately arrive to an attractor.

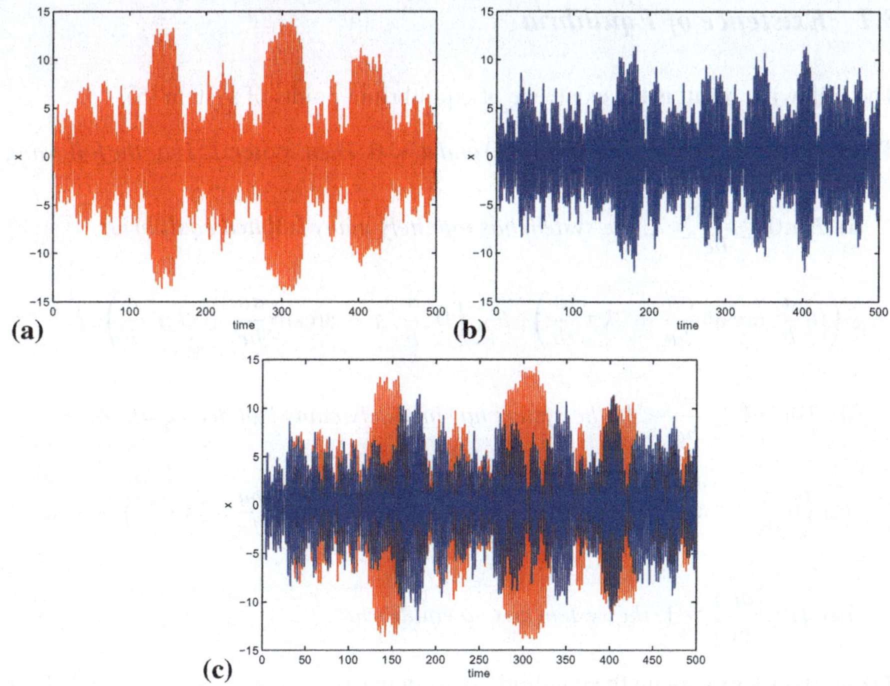

Fig. 4 Time series diagrams of system (5): $(a, b, c, d, e, f) = (3, 3, 5, -3, -15, 3.7)$. **a** $(x_0, y_0, z_0, w_0) = (0.1, 1.5, 3.6, -1)$; **b** $(x_0, y_0, z_0, w_0) = (0.101, 1.5, 3.6, -1)$; **c** red represents **a** and blue represents **b**

2.2.2 Sensitive Dependence on the Initial Value

An essential feature of chaotic dynamics is the sensitive dependence on initial conditions. Chaos has sensitive dependence on the initial conditions, which is known as "butterfly effect" and the self-similarity of the time scale. For its sensitive dependence on initial values, chaos can be applied to pattern recognition of the ones with extremely small differences. Figure 4 shows the time-sequence diagram about the abscissa under different initial values in two groups.

3 Local Dynamics of the Hyperchaotic System

This section studies the local dynamical behaviors of the hyperchaotic system (5), including the existence of equilibria and the stability of hyperbolic or non-hyperbolic equilibria.

3.1 Existence of Equilibria

The following result is the existence of equilibrium points of system (5).

Theorem 20.1 *Assume that $abef \neq 0$ and $d < 0$. Then, system (5) has the following properties:*

(i) *For $0 \leq \frac{ad}{be} \leq 1$, the system has infinitely many isolated equilibria:*

$$E_{1k}\left(0, \frac{a}{b}, \arcsin\frac{ad}{be} + 2k\pi, \frac{a}{b}\right), E_{2k}\left(0, \frac{a}{b}, \pi - \arcsin\frac{ad}{be} + 2k\pi, \frac{a}{b}\right), k \in \mathbb{Z};$$

(ii) *For $-1 \leq \frac{ad}{be} < 0$, the system has infinitely many isolated equilibria:*

$$E_{3k}\left(0, \frac{a}{b}, \pi + \arcsin\frac{ad}{be} + 2k\pi, \frac{a}{b}\right), E_{4k}\left(0, \frac{a}{b}, 2\pi - \arcsin\frac{ad}{be} + 2k\pi, \frac{a}{b}\right), k \in \mathbb{Z};$$

(iii) *For $\left|\frac{ad}{be}\right| > 1$, the system has no equilibria.*

Proof In order to obtain the equilibria of system (5), let $\dot{x} = \dot{y} = \dot{z} = \dot{w} = 0$. Then

$$\begin{cases} x - y + w = 0, \\ cx + dy - e\sin z = 0, \\ -a + by = 0, \\ -fx = 0, \end{cases} \tag{6}$$

implying that $x = 0$, $y = w = \frac{a}{b}$ and $\sin z = \frac{ad}{be}$. Thus, the following conclusions hold:

(i) When $0 \leq \frac{ad}{be} \leq 1$, one obtains $z = \arcsin\frac{ad}{be} + 2k\pi$ or $z = \pi - \arcsin\frac{ad}{be} + 2k\pi$, where $k \in \mathbb{Z}$, and the infinitely many isolated equilibria of system (5) are:

$$E_{1k}\left(0, \frac{a}{b}, \arcsin\frac{ad}{be} + 2k\pi, \frac{a}{b}\right), E_{2k}\left(0, \frac{a}{b}, \pi - \arcsin\frac{ad}{be} + 2k\pi, \frac{a}{b}\right).$$

(ii) When $-1 \leq \frac{ad}{be} < 0$, one has $z = \pi + \arcsin\frac{ad}{be} + 2k\pi$ or $z = 2\pi - \arcsin\frac{ad}{be} + 2k\pi$, where $k \in \mathbb{Z}$. The system (5) has infinitely many isolated equilibria:

$$E_{3k}\left(0, \frac{a}{b}, \pi + \arcsin\frac{ad}{be} + 2k\pi, \frac{a}{b}\right), E_{4k}\left(0, \frac{a}{b}, 2\pi - \arcsin\frac{ad}{be} + 2k\pi, \frac{a}{b}\right).$$

(iii) When $\left|\frac{ad}{be}\right| > 1$, which means that $|\sin z| > 1$, it is contradictory with $|\sin z| \leq 1$. So, (6) has no solutions, implying that system (5) has no equilibria.

3.2 *Stability of Equilibrium*

This subsection first shows the stability of the hyperbolic equilibria or non-hyperbolic equilibria E_{1k} and E_{4k} of system (5) in Theorem 20.1. It is similar to study the stability of equilibria E_{2k} and E_{3k}. For simplicity, denote $z^* \triangleq \arcsin \frac{ad}{be} + 2k\pi$, where $k \in \mathbb{Z}$, $T = \left|\frac{ad}{be}\right|$, and $m = \sqrt{b^2e^2 - a^2d^2}$.

3.2.1 The Hyperbolic Case with $T < 1$

The following theorem is the stability of infinitely many isolated equilibria E_{1k} and E_{4k} of system (5), when $T < 1$.

Theorem 20.2 *Let parameters a, b, d and e satisfy* $T < 1$ *i.e.* $a^2d^2 < b^2e^2$. *Then*

(i) *For* $be < 0$, E_{1k} *is local asymptotically stable, if and only if the following conditions hold:*

$$\begin{cases} d < -1, \quad f < 0, \quad fd - m - (1+d)(f+d+c-m) > 0, \\ (fd-m)^2 - fm(1+d)^2 - (fd-m)(1+d)(f+d+c-m) < 0. \end{cases} \tag{7}$$

Otherwise, equilibria E_{1k} *is non-hyperbolic or is unstable.*

(ii) *For* $be > 0$, E_{1k} *is local asymptotically stable, if and only if the following conditions hold:*

$$\begin{cases} d < -1, \quad f > 0, \quad fd + m - (1+d)(f+d+c+m) > 0, \\ (fd+m)^2 + fm(1+d)^2 - (fd+m)(1+d)(f+d+c+m) > 0. \end{cases} \tag{8}$$

Otherwise, equilibria E_{1k} *is non-hyperbolic or is unstable.*

(iii) *For* $be < 0$, E_{4k} *is local asymptotically stable, if and only if the conditions (7) hold. Otherwise, equilibria* E_{4k} *is non-hyperbolic or is unstable.*

(iv) *For* $be > 0$, E_{4k} *is local asymptotically stable, if and only if the condition (8) hold. Otherwise, equilibria* E_{4k} *is non-hyperbolic or is unstable.*

Proof Let $z^* = \arcsin \frac{ad}{be} + 2k\pi$, where $k \in \mathbb{Z}$. Then, the Jacobi matrix of system (5) at equilibria E_{1k} is

$$\mathbf{J} = \begin{bmatrix} 1 & -1 & 0 & 1 \\ c & d & -e\cos z & 0 \\ 0 & b & 0 & 0 \\ -f & 0 & 0 & 0 \end{bmatrix}. \tag{9}$$

(i) For $be < 0$, the corresponding characteristic equation is

$$\lambda^4 + a_1\lambda^3 + a_2\lambda^2 + a_3\lambda + a_4 = 0, \tag{10}$$

where $a_1 = -(1+d)$, $a_2 = f + d + c - m$, $a_3 = m - fd$ and $a_4 = -fm$. Compute the following determinants:

$$\Delta_1 \triangleq a_1, \quad \Delta_2 \triangleq \begin{vmatrix} a_1 & 1 \\ a_3 & a_2 \end{vmatrix}, \quad \Delta_3 \triangleq \begin{vmatrix} a_1 & 1 & 0 \\ a_3 & a_2 & a_1 \\ 0 & a_4 & a_3 \end{vmatrix}, \quad \Delta_4 \triangleq \begin{vmatrix} a_1 & 1 & 0 & 0 \\ a_3 & a_2 & a_1 & 1 \\ 0 & a_4 & a_3 & a_2 \\ 0 & 0 & 0 & a_4 \end{vmatrix}.$$

Under the condition (7), one gets

$$\Delta_1 > 0, \quad \Delta_2 > 0, \quad \Delta_3 > 0, \quad \Delta_4 > 0.$$

According to the Routh-Hurwitz criterion, all solutions of the characteristic polynomial (10) have negative real parts, therefore the equilibria E_{1k} are local asymptotically stable. Or else, if conditions (7) is not satisfied, some roots of the polynomial (10) have non-negative real parts, i.e. zero or positive real parts, then equilibria E_{1k} is non-hyperbolic or is unstable.

(ii) For $be > 0$, the corresponding characteristic equation is

$$\lambda^4 + a_1\lambda^3 + a_2\lambda^2 + a_3\lambda + a_4 = 0, \tag{11}$$

where $a_1 = -(1+d)$, $a_2 = f + d + c + m$, $a_3 = -m - fd$ and $a_4 = fm$. From the condition (8), it follows that

$$\Delta_1 > 0, \quad \Delta_2 > 0, \quad \Delta_3 > 0, \quad \Delta_4 > 0.$$

According to the Routh-Hurwitz criterion, similarly to the proof of case (i) above, one obtains conclusion (ii).

Similarly to proof of case (i), one can obtain the conclusions of (iii) and (iv). □

Theorem 20.3 *Let parameters a, b, d and e satisfy $T < 1$ i.e. $a^2d^2 < b^2e^2$. Then*

(i) *For $be < 0$, E_{2k} is local asymptotically stable, if and only if the following conditions (8) hold. Otherwise, equilibrium E_{2k} is non-hyperbolic or is unstable.*

(ii) *For $be > 0$, E_{2k} is local asymptotically stable, if and only if the conditions (7) hold. Otherwise, equilibrium E_{2k} is non-hyperbolic or is unstable.*

(iii) *For* $be < 0$*,* E_{3k} *is local asymptotically stable, if and only if the conditions (8) hold. Otherwise, equilibrium* E_{3k} *is non-hyperbolic or is unstable.*

(iv) *For* $be > 0$*,* E_{3k} *is local asymptotically stable, if and only if the condition (7) hold. Otherwise, equilibrium* E_{3k} *is non-hyperbolic or is unstable.*

Proof Similarly to proof of Theorem 20.2, Theorem 20.3 is verified.

3.2.2 The Non-hyperbolic Case with $T = 1$

It is easy to see that $T = 1$ implies $a^2d^2 = b^2e^2$. Therefore, the characteristic equation of system (5) at equilibria E_{1k} is

$$\lambda\left[\lambda^3 - (1+d)\lambda^2 + (f+d+c)\lambda - df\right] = 0,$$

implying that the equilibrium E_{1k} is non-hyperbolic.

Making a non-degenerate transformation $\mathscr{T}$:

$$x \to x, \quad y \to y + \frac{a}{b}, \quad z \to z + \arcsin\frac{ad}{be} + 2k\pi, \quad w \to w + \frac{a}{b}, \tag{12}$$

system (5) becomes

$$\begin{cases} \dot{x} = x - y + w, \\ \dot{y} = cx + dy + \dfrac{ad}{b} + \dfrac{m}{b}\sin z - \dfrac{ad}{b}\cos z, \\ \dot{z} = by, \\ \dot{w} = -fx. \end{cases} \tag{13}$$

Expanding the vector field of system (13) at equilibrium $E_0(0, 0, 0, 0)$, one has the following approximative system:

$$\begin{cases} \dot{x} = x - y + w, \\ \dot{y} = cx + dy + \dfrac{m}{b}z + \dfrac{ad}{2b}z^2 + O(z^3), \\ \dot{z} = by, \\ \dot{w} = -fx. \end{cases} \tag{14}$$

Since the Jacobi matrix and characteristic equation are the same as system (5) at equilibria E_{1k} and system (14) at the origin E_0, taking advantage of the characteristic equation of system (14) at E_0, one can get the characteristic equation of system (5) at E_{1k}. Note that the characteristic equation of system (14) at E_0 is (10). Thus, when $d = \dfrac{be}{a}$, system (5) is non-hyperbolic, and system (5) is transformed to

$$\begin{cases} \dot{x} = x - y + w, \\ \dot{y} = cx + \dfrac{be}{a}y + \dfrac{e}{2}z^2 + O(z^4), \\ \dot{z} = by, \\ \dot{w} = -fx. \end{cases} \tag{15}$$

Linearizing system (15) at E_0 now yields the Jacobi matrix

$$\mathbf{J1} = \begin{bmatrix} 1 & -1 & 0 & 1 \\ c & \dfrac{be}{a} & 0 & 0 \\ 0 & b & 0 & 0 \\ -f & 0 & 0 & 0 \end{bmatrix}.$$

The corresponding eigenvalues of $\mathbf{J1}$ are $\lambda_1 = 0$, λ_2, λ_3 and λ_4, where λ_2, λ_3 and λ_4 satisfy

$$g(\lambda) \triangleq \lambda^3 - \left(1 + \frac{be}{a}\right)\lambda^2 + \left(f + c + \frac{be}{a}\right)\lambda - \frac{be}{a}f = 0. \tag{16}$$

From $\lambda_1 = 0$, it follows that E_0 is non-hyperbolic. Moreover, E_0 is unstable if one of the other eigenvalues λ_2, λ_3 and λ_4 has positive real part.

Denote

$$\Omega_0 = \left\{(a, b, c, d, e, f) \,\middle|\, 1 + \frac{be}{a} < 0,\ \frac{bef}{a + be} - f - \frac{be}{a} < c,\ \frac{be}{a}f < 0\right\}.$$

By using the center manifold theorem, one obtains the following stability of the non-hyperbolic equilibria of system (5) for the case of $ad - be = 0$. When $ad + be = 0$, the discussion is similar to that of the case $ad - be = 0$.

Theorem 20.4 *Let $(a, b, c, d, e, f) \in \Omega_0$ and $d = \frac{be}{a}$. Then, system (5) has non-hyperbolic isolated equilibria E_{1k}, at which the linearized system has a zero eigenvalue. Furthermore, system (5) has a three-dimensional stable manifold, and the non-hyperbolic equilibrium E_{1k} is unstable.*

Proof To show the stability of equilibria E_{1k} of system (5), we only need to prove the stability of the equilibrium E_0 of system (15). Since λ_2, λ_3 and λ_4 satisfy (16), that is,

$$g(\lambda) = \lambda^3 + c_1\lambda^2 + c_2\lambda + c_3 = 0,$$

where $c_1 = -\left(1 + \frac{be}{a}\right)$, $c_2 = f + c + \frac{be}{a}$ and $c_3 = -\frac{bef}{a}$, one has

$$\Delta_1 \triangleq c_1 = -\left(1 + \frac{be}{a}\right),$$

$$\Delta_2 \triangleq \begin{vmatrix} c_1 & 1 \\ c_3 & c_2 \end{vmatrix} = \frac{bef}{a} - \left(1 + \frac{be}{a}\right)\left(f + \frac{be}{a} + c\right),$$

$$\Delta_3 \triangleq \begin{vmatrix} c_1 & 1 & 0 \\ c_3 & c_2 & c_1 \\ 0 & 0 & c_3 \end{vmatrix} = -\frac{bef}{a}\Delta_2.$$

The parameter condition Ω_0 means the Routh-Huiwitz conditions $\Delta_1 > 0$, $\Delta_2 > 0$ and $\Delta_3 > 0$. Therefore, all solutions of equation (16) have negative real parts.

To show the stability of equilibrium E_0 of system (15), the proof is divided into the following two cases.

Case I. $\Delta = -c_1^2 c_2^2 + 4c_1^3 c_3 + 4c_2^3 - 18c_1c_2c_3 + 27c_3^2 < 0$.

In this case, the equation $\lambda g(\lambda) = 0$ has eigenvalue $\lambda_1 = 0$, and three different negative real eigenvalues λ_2, λ_3 and λ_4 [19], with corresponding eigenvectors

$$\begin{aligned}
\xi_1 &= (0, 0, 1, 0)^T, \\
\xi_2 &= (\lambda_2(a\lambda_2 - be), ac\lambda_2, abc, -f(a\lambda_2 - be))^T, \\
\xi_3 &= (\lambda_3(a\lambda_3 - be), ac\lambda_3, abc, -f(a\lambda_3 - be))^T, \\
\xi_4 &= (\lambda_4(a\lambda_4 - be), ac\lambda_4, abc, -f(a\lambda_4 - be))^T.
\end{aligned}$$

Making the transformation

$$(x, y, z, w)^T = (\xi_1, \xi_2, \xi_3, \xi_4)(u, v, p, q)^T,$$

system (15) is transformed to

$$\begin{pmatrix} \dot{u} \\ \dot{v} \\ \dot{p} \\ \dot{q} \end{pmatrix} = \begin{pmatrix} 0 & 0 & 0 & 0 \\ 0 & \lambda_2 & 0 & 0 \\ 0 & 0 & \lambda_3 & 0 \\ 0 & 0 & 0 & \lambda_4 \end{pmatrix} \begin{pmatrix} u \\ v \\ p \\ q \end{pmatrix} + \begin{pmatrix} R_1 \\ R_2 \\ R_3 \\ R_4 \end{pmatrix} + O(||(u, v, p, q)||^4), \qquad (17)$$

where

$$\begin{cases}
R_1 = -\dfrac{1}{2}a(u + abc(p + q + v))^2, \\
R_2 = \dfrac{(u + abc(p + q + v))^2(be - a\lambda_3)(be - a\lambda_4)}{2a^2bc(\lambda_2 - \lambda_3)(\lambda_2 - \lambda_4)}, \\
R_3 = -\dfrac{(u + abc(p + q + v))^2(be - a\lambda_2)(be - a\lambda_4)}{2a^2bc(\lambda_2 - \lambda_3)(\lambda_3 - \lambda_4)}, \\
R_4 = -\dfrac{(u + abc(p + q + v))^2(be - a\lambda_2)(be - a\lambda_3)}{2a^2bc(\lambda_2 - \lambda_4)(-\lambda_3 + \lambda_4)}.
\end{cases}$$

According to the center manifold theory, the stability of equilibrium E_0 can be determined by studying the first-order ordinary differential equations on its center manifold, which can be represented as a graph over the variable u, as follows:

$$W^c(0) = \Big\{(u, v, p, q) \in \mathbb{R}^4 \Big| v = V(u),\ p = P(u),\ q = Q(u),\ |u| < \delta,$$
$$V(0) = P(0) = Q(0) = 0,\ DV(0) = DP(0) = DQ(0) = 0\Big\}$$

and δ sufficiently small.

Consider the center manifold

$$\begin{aligned} v &= V(u) = v_1u^2 + v_2u^3 + O(u^4),\\ p &= P(u) = p_1u^2 + p_2u^3 + O(u^4),\\ q &= Q(u) = q_1u^2 + q_2u^3 + O(u^4). \end{aligned} \tag{18}$$

Substituting expressions (18) into system (17), one obtains

$$v_1 = \frac{-(be - a\lambda_3)(be - a\lambda_4)}{2a^2bc\lambda_2(\lambda_2 - \lambda_3)(\lambda_2 - \lambda_4)},\quad v_2 = \frac{(be - a\lambda_3)(be - a\lambda_4)\varphi_1}{2a^3bc\lambda_2^2(\lambda_2 - \lambda_3)\lambda_3\lambda_4(\lambda_2 - \lambda_4)},$$
$$p_1 = \frac{(be - a\lambda_2)(be - a\lambda_4)}{2a^2bc\lambda_3(\lambda_2 - \lambda_3)(\lambda_3 - \lambda_4)},\quad p_2 = \frac{-(be - a\lambda_2)(be - a\lambda_4)\varphi_2}{2a^3bc\lambda_2(\lambda_2 - \lambda_3)\lambda_3^2\lambda_4(\lambda_3 - \lambda_4)},$$
$$q_1 = \frac{-(be - a\lambda_2)(be - a\lambda_3)}{2a^2bc\lambda_4(\lambda_2 - \lambda_4)(\lambda_3 - \lambda_4)},\quad q_2 = \frac{(be - a\lambda_2)(be - a\lambda_3)\varphi_3}{2a^3bc\lambda_2(\lambda_2 - \lambda_4)\lambda_3\lambda_4^2(\lambda_3 - \lambda_4)}.$$

Therefore,

$$\begin{aligned} v &= \frac{-(be - a\lambda_3)(be - a\lambda_4)}{2a^2bc\lambda_2(\lambda_2 - \lambda_3)(\lambda_2 - \lambda_4)}u^2 + \frac{(be - a\lambda_3)(be - a\lambda_4)\varphi_1}{2a^3bc\lambda_2^2(\lambda_2 - \lambda_3)\lambda_3\lambda_4(\lambda_2 - \lambda_4)}u^3 + O(u^4),\\ p &= \frac{(be - a\lambda_2)(be - a\lambda_4)}{2a^2bc\lambda_3(\lambda_2 - \lambda_3)(\lambda_3 - \lambda_4)}u^2 - \frac{(be - a\lambda_2)(be - a\lambda_4)\varphi_2}{2a^3bc\lambda_2(\lambda_2 - \lambda_3)\lambda_3^2\lambda_4(\lambda_3 - \lambda_4)}u^3 + O(u^4),\\ q &= \frac{-(be - a\lambda_2)(be - a\lambda_3)}{2a^2bc\lambda_4(\lambda_2 - \lambda_4)(\lambda_3 - \lambda_4)}u^2 + \frac{(be - a\lambda_2)(be - a\lambda_3)\varphi_3}{2a^3bc\lambda_2(\lambda_2 - \lambda_4)\lambda_3\lambda_4^2(\lambda_3 - \lambda_4)}u^3 + O(u^4). \end{aligned} \tag{19}$$

Substituting the expressions (19) into system (17), one gets the restricted vector field of system (17) on its center manifold

$$\dot{u} = -\frac{a}{2}u^2 + h_1u^3 + O(u^4), \tag{20}$$

where

$$h_1 = \frac{(be - a\lambda_3)(be - a\lambda_4)}{2\lambda_2(\lambda_2 - \lambda_3)(\lambda_2 - \lambda_4)} - \frac{(be - a\lambda_2)(be - a\lambda_4)}{2\lambda_3(\lambda_2 - \lambda_3)(\lambda_3 - \lambda_4)} - \frac{(be - a\lambda_2)(be - a\lambda_3)}{2\lambda_4(\lambda_2 - \lambda_4)(-\lambda_3 + \lambda_4)}.$$

From the assumption of $a \neq 0$ and (20), it follows that E_{1k} is unstable.

Case II. $\Delta = -c_1^2c_2^2 + 4c_1^3c_3 + 4c_2^3 - 18c_1c_2c_3 + 27c_3^2 > 0$.

In this case, the equation $\lambda g(\lambda) = 0$ has eigenvalues $\lambda_1 = 0$, negative real root λ_2, and conjugate complex roots with negative real parts λ_3 and λ_4 [19]. Let $\lambda_{3,4} = \beta \pm i\rho$. Then, the corresponding eigenvectors are

$$\xi_1 = (0, 0, 1, 0)^T,$$
$$\xi_2 = (\lambda_2(a\lambda_2 - be), ac\lambda_2, abc, -f(a\lambda_2 - be))^T,$$
$$\xi_3 = \left(\frac{a\beta - be}{ac}, 1, \frac{b\beta}{\beta^2 + \rho^2}, \frac{bef\beta - af(\beta^2 + \rho^2)}{ac(\beta^2 + \rho^2)}\right)^T,$$
$$\xi_4 = \left(\frac{\rho}{c}, 0, -\frac{b\rho}{\beta^2 + \rho^2}, -\frac{bef\rho}{ac(\beta^2 + \rho^2)}\right)^T.$$

Making the following transformation:

$$(x, y, z, w)^T = (\xi_1, \xi_2, \xi_3, \xi_4)(u, v, p, q)^T,$$

system (15) is changed to

$$\begin{pmatrix} \dot{u} \\ \dot{v} \\ \dot{p} \\ \dot{q} \end{pmatrix} = \begin{pmatrix} 0 & 0 & 0 & 0 \\ 0 & \lambda_2 & 0 & 0 \\ 0 & 0 & \beta & \rho \\ 0 & 0 & -\rho & \beta \end{pmatrix} \begin{pmatrix} u \\ v \\ p \\ q \end{pmatrix} + \begin{pmatrix} R_{11} \\ R_{22} \\ R_{33} \\ R_{44} \end{pmatrix} + O(||(u, v, p, q)||^4), \tag{21}$$

where

$$R_{11} = -\frac{a}{2(\beta^2 + \rho^2)^2}\varphi(u, v, p, q),$$
$$R_{22} = \frac{b^2e^2 - 2abe\beta + a^2(\beta^2 + \rho^2)}{2a^2bc(\beta^2 + \rho^2)^2(\beta^2 + \rho^2 - 2\beta\lambda_2 + \lambda_2^2)}\varphi(u, v, p, q),$$
$$R_{33} = -\frac{-abc(\beta^2 + \rho^2) + (b^2e^2 + a^2(\beta^2 + \rho^2))\lambda_2 - abe\lambda_2^2}{2ab(\beta^2 + \rho^2)^2(\beta^2 + \rho^2 - 2\beta\lambda_2 + \lambda_2^2)}\varphi(u, v, p, q),$$
$$R_{44} = \frac{(be - a\lambda_2)((be - a\beta)(\beta^2 + \rho^2) + (-be\beta + a(\beta^2 + \rho^2))\lambda_2)}{2ab\rho(\beta^2 + \rho^2)^2(\beta^2 + \rho^2 - 2\beta\lambda_2 + \lambda_2^2)}\varphi(u, v, p, q),$$

with

$$\varphi(u, v, p, q) = [u(\beta^2 + \rho^2) + b(p\beta - q\rho + acv(\beta^2 + \rho^2))]^2.$$

Repeating the same process as in Case I, one can show that the restricted vector field of system (21) on its center manifold becomes the following form:

$$\dot{u} = -\frac{a}{2}u^2 + \frac{b^2e^2 - 2abe\beta + a^2(\beta^2 + \rho^2) - a(be - 2a\beta)\lambda_2}{2(\beta^2 + \rho^2)\lambda_2}u^3 + O(u^4). \tag{22}$$

Therefore, it follows from $a \neq 0$ that E_{1k} is unstable.

Remark 20.2 Similarly to that in the discussion of Theorem 20.4, one can establish relevant stable conclusions for equilibria E_{2k}, E_{3k} and E_{4k} of system (5), which are omitted here for brevity.

4 Global Dynamics of the Hyperchaotic System

For a common chaotic or hyperchaotic system, its complicated dynamics are generally produced by the bifurcations at equilibria. However, for the hyperchaotic systems with infinitely many isolated equilibria or without equilibria, the equilibria of these systems are difficult to study to gain understanding of the complicated dynamics based on theoretical analysis. Thus, the methods of numerical analysis are particularly important for studying these systems. With the exception of the hyperchaotic attractor, many other complicated dynamics of system (5) are discovered by detailed numerical analysis. This section chooses some groups of fixed values of the parameters a, b, c, d and e, meanwhile varying the parameter f. With the help of phase portraits, Lyapunov exponents, Poincaré maps and bifurcation diagrams, it is possible to investigate the complicated dynamical characteristics of system (5) in detail.

4.1 *Complex Dynamics when $T > 1$*

When the parameters $a,\ b,\ d$ and e satisfying $T > 1$ implies that $a^2d^2 > b^2e^2$. It follows from Theorem 20.1 that system (5) has no equilibria. Fix $(a, b, c, d, e) = (11, 3, 5, -3, -10)$ and vary $f \in [2, 6]$, the corresponding Lyapunov exponent spectrum is shown in Fig. 5a, and the corresponding bifurcation diagram is given in Fig. 5b. Therefore, Fig. 5 show that system (5) may have hyperchaotic attractor, chaotic attractor and periodic attractor under proper parameters.

In particular, let $(a, b, c, d, e, f) = (11, 3, 5, -3, -10, 3.77)$. Then, the corresponding Lyapunov exponents are

$$\lambda_{LE_1} = 0.2752, \quad \lambda_{LE_2} = 0.0966, \quad \lambda_{LE_3} = -0.0002, \quad \lambda_{LE_4} = -2.3715.$$

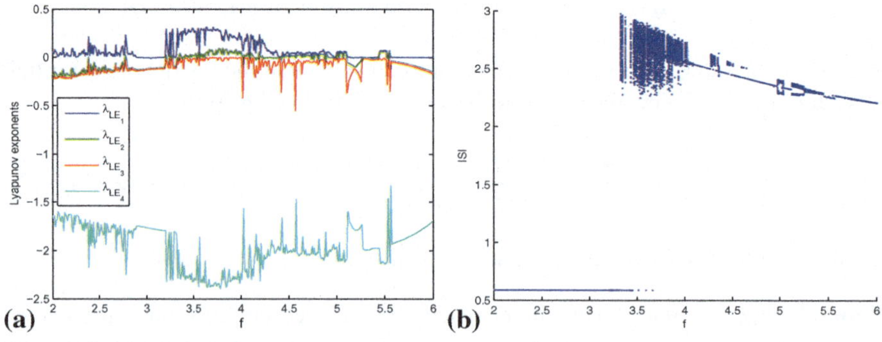

Fig. 5 System (5): $(a, b, c, d, e) = (11, 3, 5, -3, -10)$, $f \in [2, 6]$. **a** Lyapunov exponents; **b** bifurcation diagram

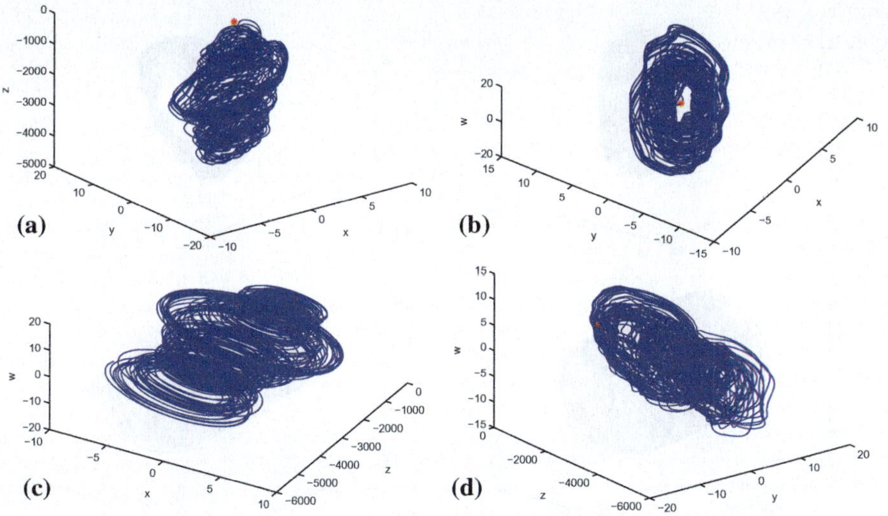

Fig. 6 Three-dimensional projections of hidden hyperchaotic attractor of system (5): $(a, b, c, d, e, f) = (11, 3, 5, -3, -10, 3.77)$

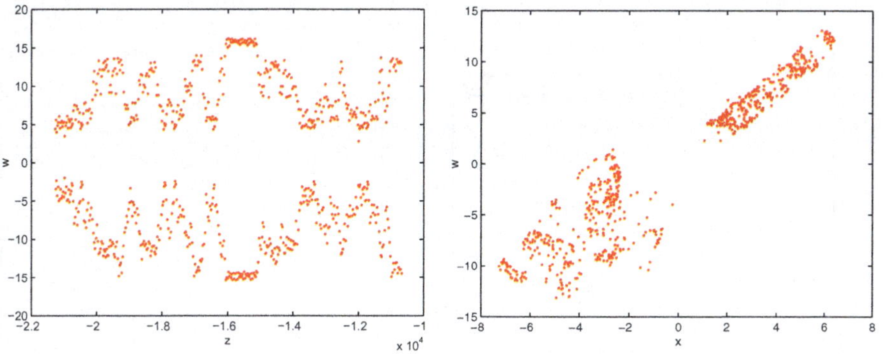

Fig. 7 Poincaré images of hidden hyperchaotic attractor of system (5): $(a, b, c, d, e, f) = (11, 3, 5, -3, -10, 3.77)$

and the Lyapunov dimension is $D_L = 3.1567$. Moreover, Fig. 6 shows the projections of this hidden attractor in the three-dimensional space, and Fig. 7 shows the Poincaré map in section $\{(x, y, z, w) \in \mathbf{R}^4 | y = 0\}$. Thus, system (5) has a hidden hyperchaotic attractor with two positive Lyapunov exponents.

Let $(a, b, c, d, e, f) = (11, 3, 5, -3, -10, 4.34)$. Then, the corresponding Lyapunov exponents are

$$\lambda_{LE_1} = 0.0650, \quad \lambda_{LE_2} = 0.0001, \quad \lambda_{LE_3} = -0.0424, \quad \lambda_{LE_4} = -2.0229.$$

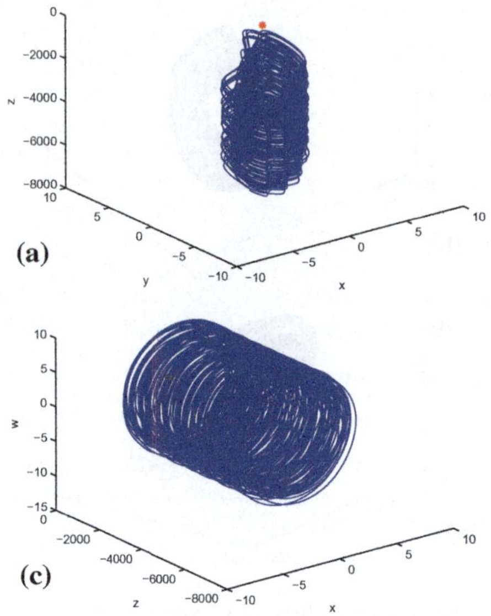

Fig. 8 Phase portraits of the hidden chaotic attractor of system (5): $(a, b, c, d, e, f) = (11, 3, 5, -3, -10, 4.34)$

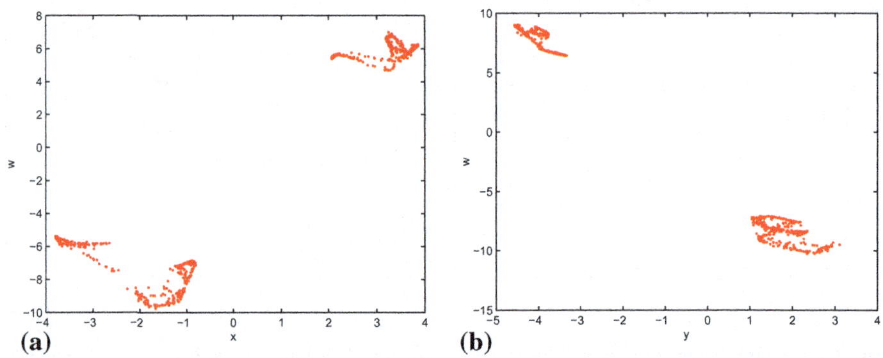

Fig. 9 Poincaré images of the hidden chaotic attractor of system (5): $(a, b, c, d, e, f) = (11, 3, 5, -3, -10, 4.34)$. **a** On section $\{(x, y, z, w) \in \mathbf{R}^4 | y = 0\}$; **b** on section $\{(x, y, z, w) \in \mathbf{R}^4 | x = 0\}$

The phase portraits of the hidden attractor are depicted in Fig. 8, and the Poincaré maps of the hidden attractor are depicted in Fig. 9. Thus, the numerical analysis verifies that system (5) has a hidden chaotic attractor.

Let $(a, b, c, d, e, f) = (11, 3, 5, -3, -10, 3.05)$. Then, the corresponding Lyapunov exponents are

$$\lambda_{LE_1} = 0.0001, \quad \lambda_{LE_2} = -0.1046, \quad \lambda_{LE_3} = -0.1229, \quad \lambda_{LE_4} = -1.7726,$$

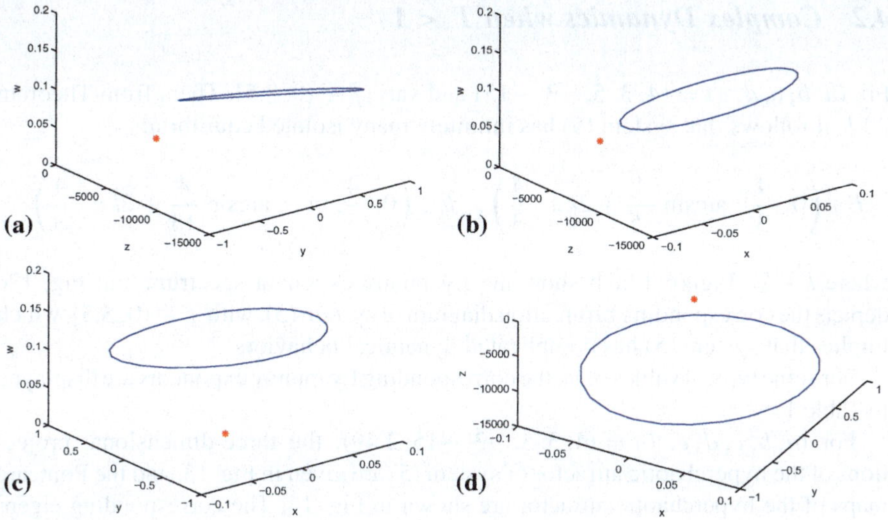

Fig. 10 Phase portraits of the hidden periodic attractor of system (5): $(a, b, c, d, e, f) = (11, 3, 5, -3, -10, 3.05)$

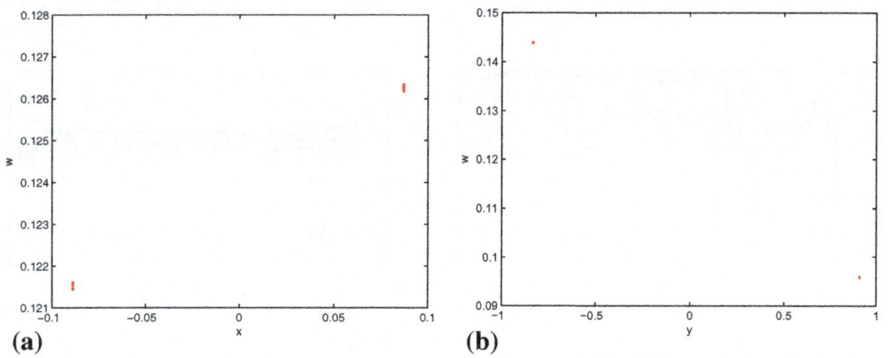

Fig. 11 Poincaré maps of the hidden periodic attractor of system (5): $(a, b, c, d, e, f) = (11, 3, 5, -3, -10, 3.05)$. **a** On section $\{(x, y, z, w) \in \mathbf{R}^4 | y = 0\}$; **b** on section $\{(x, y, z, w) \in \mathbf{R}^4 | x = -0.03\}$

which implies that system (5) has a hidden periodic attractor. The three-dimensional projections of the hidden periodic attractor are shown in Fig. 10, and the Poincaré maps of the hidden periodic attractor are shown in Fig. 11.

4.2 Complex Dynamics when $T < 1$

Fix $(a, b, c, d, e) = (4, 3, 5, -3, -15)$ and vary $f \in (0, 8.5]$. Then, from Theorem 20.1, it follows that system (5) has infinitely many isolated equilibria:

$$E_{1k}\left(0,\ \frac{4}{3},\ \arcsin\frac{4}{15} + 2k\pi,\ \frac{4}{3}\right),\quad E_{2k}\left(0,\ \frac{4}{3},\ \pi - \arcsin\frac{4}{15} + 2k\pi,\ \frac{4}{3}\right),$$

where $k \in \mathbb{Z}$. Figure 12a–b show the Lyapunov exponent spectrum and Fig. 12c depicts the corresponding bifurcation diagram of system (5), with $f \in (0, 8.5]$, which implies that system (5) has complicated dynamical behaviors.

For some typical values of f, the corresponding Lyapunov exponents are displayed in Table 1.

For $(a, b, c, d, e, f) = (4, 3, 5, -3, -15, 2.49)$, the three-dimensional projections of the hyperchaotic attractor of system (5) are given in Fig. 13, and the Poincaré maps of the hyperchaotic attractor are shown in Fig. 14. The corresponding eigenvalues at equilibria E_{1k} and E_{2k} are

$$\lambda_1 = -7.9289,\quad \lambda_2 = 4.8295,\quad \lambda_3 = 0.5497 + 1.5868i,\quad \lambda_4 = 0.5497 - 1.5868i.$$

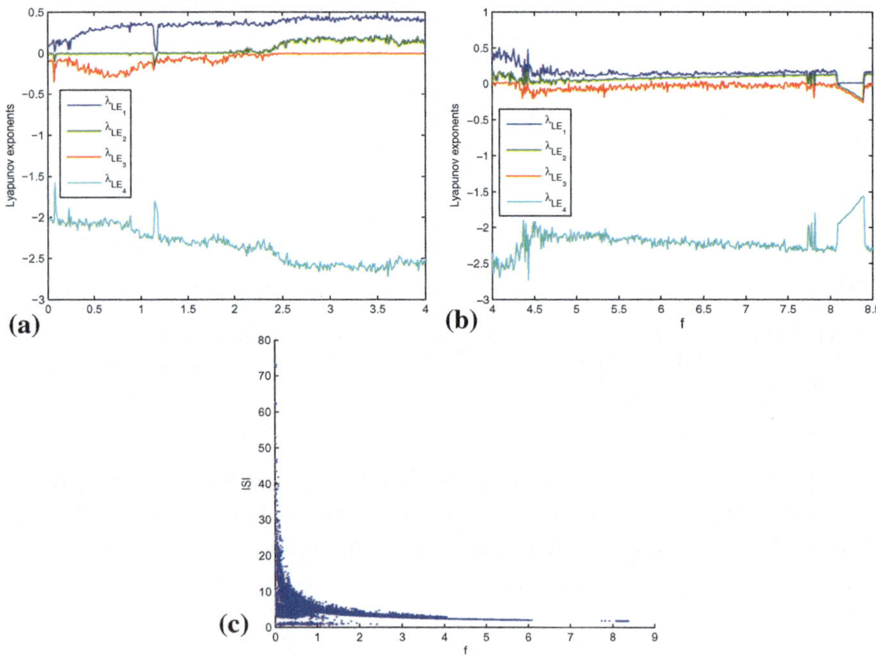

Fig. 12 System (5): $(a, b, c, d, e) = (4, 3, 5, -3, -15)$, $f \in (0, 8.5]$. **a–b** Lyapunov exponent spectrum; **c** bifurcation diagram

Table 1 Lyapunov exponents of system (5): $(a, b, c, d, e) = (4, 3, 5, -3, -15)$

f	λ_{LE_1}	λ_{LE_2}	λ_{LE_3}	λ_{LE_4}	Dynamical property
2.49	0.4131	0.1163	−0.0001	−2.5292	Hyperchaos
3.72	0.4180	0.1558	−0.0001	−2.5738	Hyperchaos
1.29	0.3726	0.0000	−0.0289	−2.3437	Chaos
0.62	0.3155	0.0000	−0.2818	−2.0337	Chaos
8.10	0.0003	−0.0345	−0.0425	−1.9233	Period

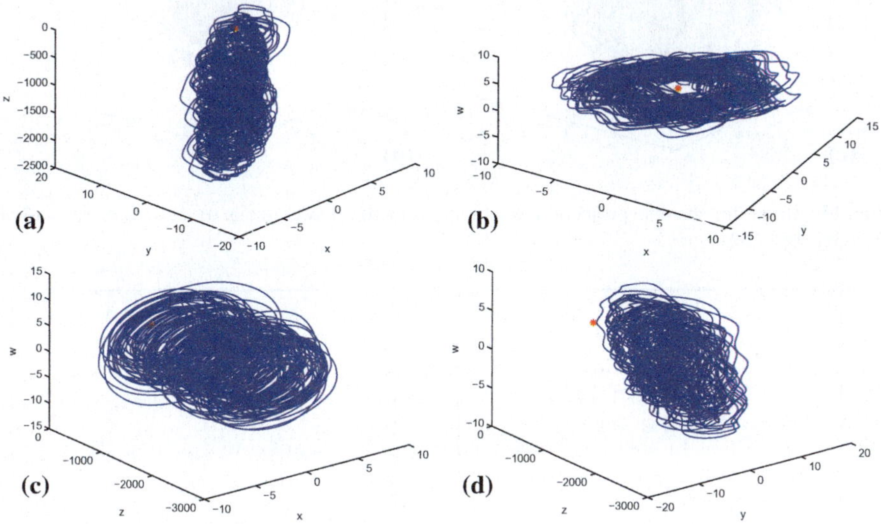

Fig. 13 Three-dimensional projections of the hyperchaotic attractor of system (5): $(a, b, c, d, e, f) = (4, 3, 5, -3, -15, 2.49)$

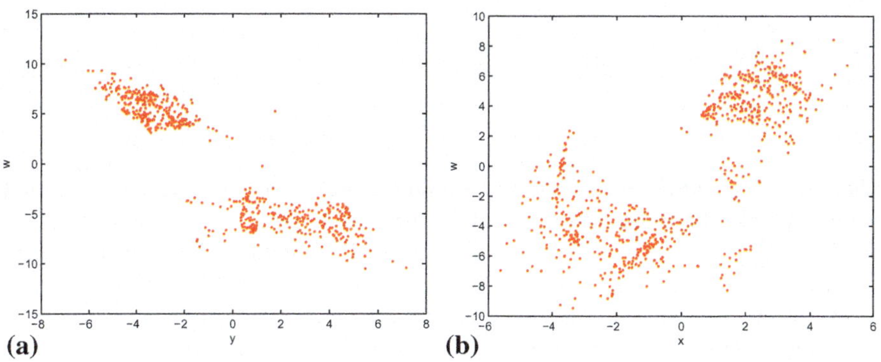

Fig. 14 Poincaré maps of the hyperchaotic attractor of system (5): $(a, b, c, d, e, f) = (4, 3, 5, -3, -15, 2.49)$. **a** On section $\{(x, y, z, w) \in \mathbf{R}^4 | x = 0\}$; **b** on section $\{(x, y, z, w) \in \mathbf{R}^4 | y = 0\}$

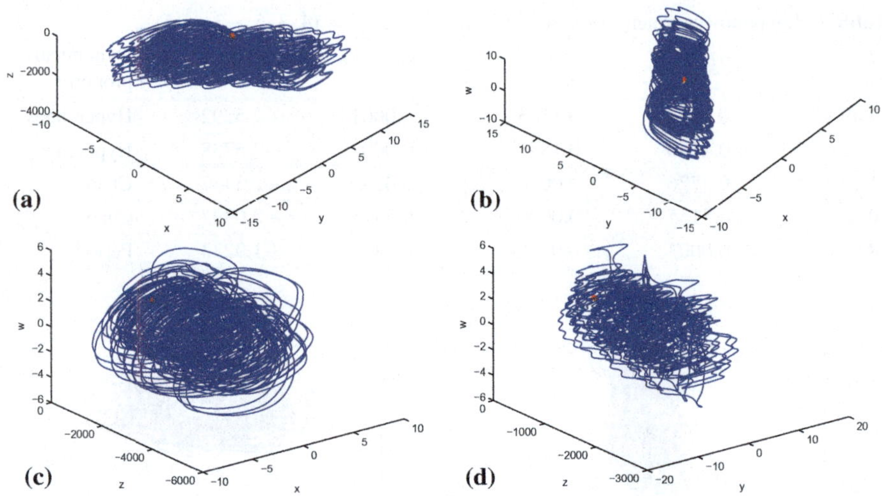

Fig. 15 Three-dimensional projections of chaotic attractor of system (5): $(a, b, c, d, e, f) = (4, 3, 5, -3, -15, 0.62)$

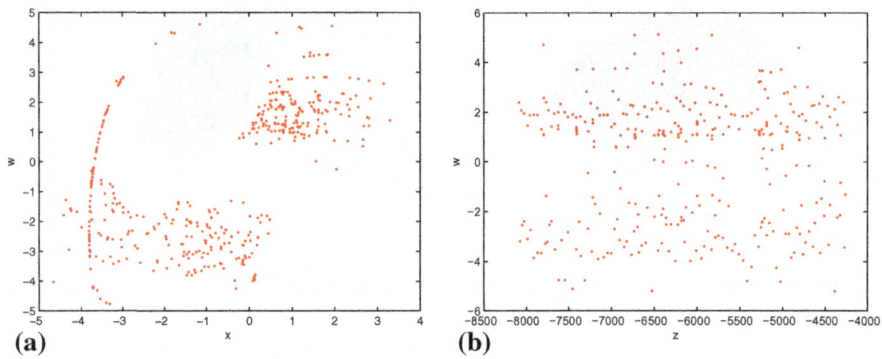

Fig. 16 Poincaré maps of chaotic attractor of system (5): $(a, b, c, d, e, f) = (4, 3, 5, -3, -15, 0.62)$. **a** On section $\{(x, y, z, w) \in \mathbf{R}^4 | y = 0\}$; **b** on section $\{(x, y, z, w) \in \mathbf{R}^4 | x = 0\}$

Therefore, E_{1k} and E_{2k} are saddle-focus points, which have a one-dimensional stable manifold and a three-dimensional unstable manifold.

For $(a, b, c, d, e, f) = (4, 3, 5, -3, -15, 0.62)$, system (5) has a chaotic attractor. The three-dimensional projections of the chaotic attractor of system (5) are shown in Fig. 15, and the Poincaré maps of chaotic attractor are shown in Fig. 16. The corresponding eigenvalues at equilibria E_{1k} and E_{2k} are

$$\lambda_1 = -7.9203, \quad \lambda_2 = 4.7856, \quad \lambda_3 = 0.5673 + 0.6226i, \quad \lambda_4 = 0.5673 - 0.6226i,$$

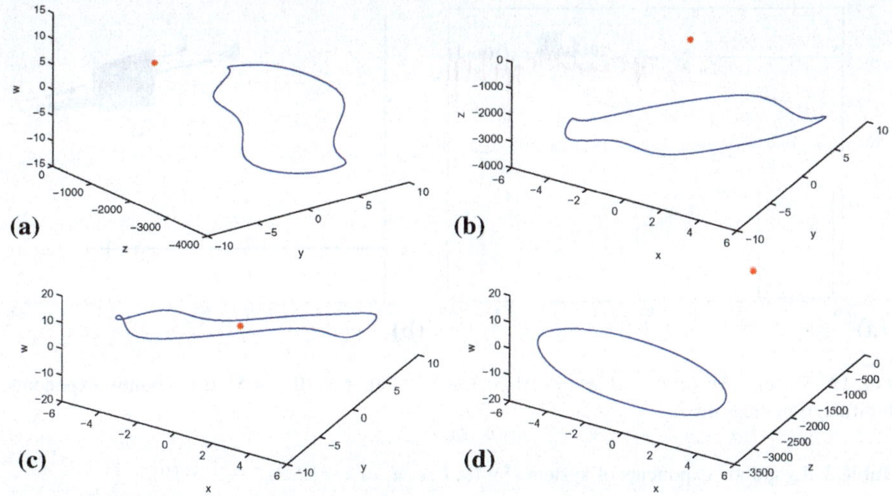

Fig. 17 Phase portraits of the periodic attractor of system (5): $(a, b, c, d, e, f) = (4, 3, 5, -3, -15, 8.10)$

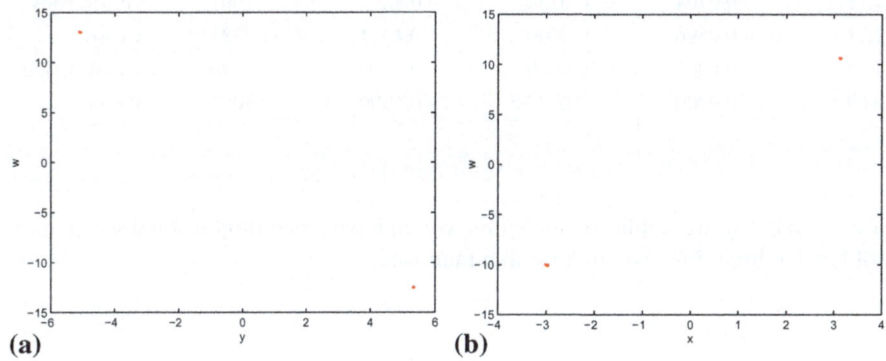

Fig. 18 Poincaré maps of the periodic attractor of system (5): $(a, b, c, d, e, f) = (4, 3, 5, -3, -15, 8.10)$. **a** On section $\{(x, y, z, w) \in \mathbf{R}^4 | x = 0\}$; **b** on section $\{(x, y, z, w) \in \mathbf{R}^4 | y = -0.3\}$

implying that E_{1k} and E_{2k} are saddle-focus points with a one-dimensional stable manifold and a three-dimensional unstable manifold.

For $(a, b, c, d, e, f) = (4, 3, 5, -3, -15, 8.10)$, system (5) has a periodic attractor with three-dimensional projections shown in Fig. 17. The Poincaré maps of the periodic attractor are shown in Fig. 18. The corresponding eigenvalues at the equilibria E_{1k}, E_{2k} are

$$\lambda_1 = -7.9523, \quad \lambda_2 = 4.9188, \quad \lambda_3 = 0.5168 + 2.5919i, \quad \lambda_4 = 0.5168 - 2.5919i.$$

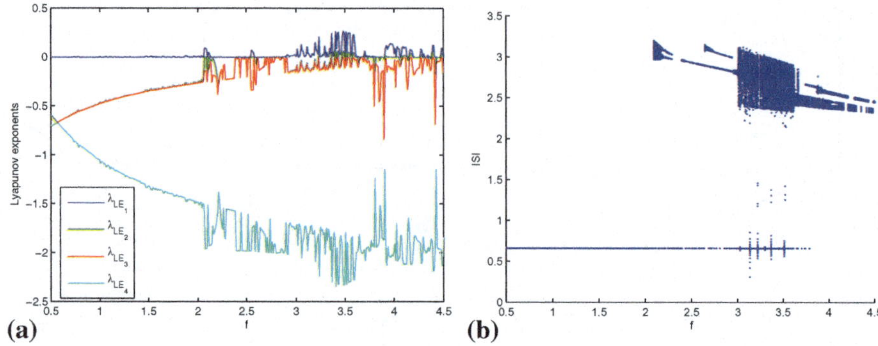

Fig. 19 System (5): $(a, b, c, d, e) = (10, 3, 5, -3, -10)$, $f \in [0.5, 4.5]$. **a** Lyapunov exponents, **b** bifurcation diagram

Table 2 Lyapunov exponents of system (5): $(a, b, c, d, e) = (10, 3, 5, -3, -10)$

f	λ_{LE_1}	λ_{LE_2}	λ_{LE_3}	λ_{LE_4}	Dynamical property
4.40	0.1270	0.0590	−0.0001	−2.1857	Hyperchaos
4.18	0.0704	0.0298	0.0002	−2.1001	Hyperchaos
2.11	0.0596	0.0000	−0.1311	−1.9283	Chaos
4.13	0.0001	0.0000	−0.2105	−1.7896	Quasi-period
0.69	0.0000	−0.5995	−0.5999	−0.8003	Period

So, E_{1k} and E_{2k} are saddle-focus points, which have a one-dimensional stable manifold and a three-dimensional unstable manifold.

4.3 Complex Dynamics when $T = 1$

When $(a, b, c, d, e) = (10, 3, 5, -3, -15)$ and $f \in [0.5, 4.5]$, it follows from Theorem 20.1 that system (5) has equilibria $E_k\left(0, \frac{10}{3}, \left(2k + \frac{1}{2}\right)\pi, \frac{10}{3}\right)$, where $k \in \mathbb{Z}$. Figure 19a displays the Lyapunov exponent spectrum of system (5), and Fig. 19b displays the corresponding bifurcation diagram.

These computation and simulation results demonstrate that system (5) may have complex dynamics such as periodic, quasi-periodic, chaotic and hyperchaotic attractors. For some typical values of f, the corresponding Lyapunov exponents are shown in Table 2.

In particular, for $(a, b, c, d, e, f) = (10, 3, 5, -3, -10, 4.40)$, system (5) has a hyperchaotic attractor with the three-dimensional projection shown in Fig. 20a. The corresponding eigenvalues at equilibrium E_k are

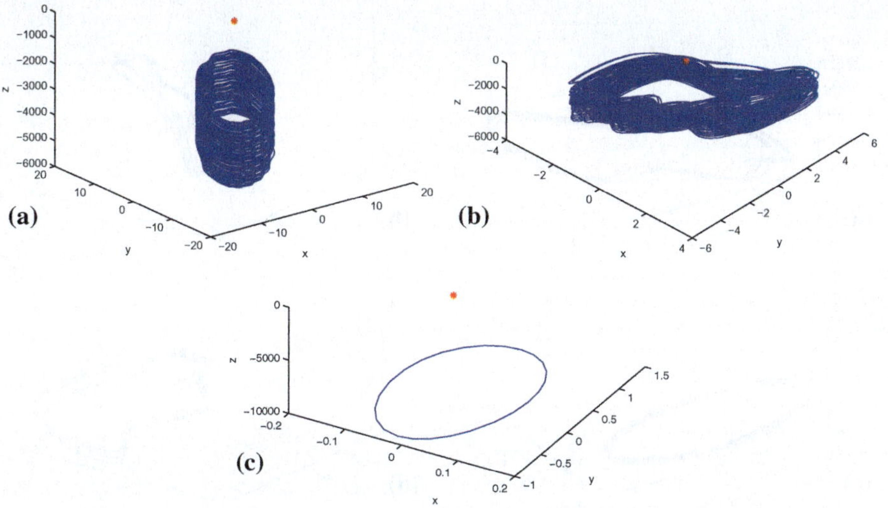

Fig. 20 Phase portraits of system (5): $(a, b, c, d, e) = (10, 3, 5, -3, -10)$. **a** Hyperchaotic attractor, **b** chaotic attractor, **c** periodic attractor

$$\lambda_1 = 0, \quad \lambda_2 = -2.0379, \quad \lambda_3 = 0.0190 + 2.5450i, \quad \lambda_4 = 0.0190 - 2.5450i,$$

implying that E_k is a saddle-focus point, which is also unstable.

For $(a, b, c, d, e, f) = (10, 3, 5, -3, -10, 2.11)$, system (5) has a chaotic attractor, with the three-dimensional projection shown in Fig. 20b. The corresponding eigenvalues at equilibrium E_k are

$$\lambda_1 = 0, \ \lambda_2 = -1.7345, \ \lambda_3 = -0.1328 + 1.9057i, \ \lambda_4 = -0.1328 - 1.9057i.$$

It can be verified that E_k is unstable by Theorem 20.4.

For $(a, b, c, d, e, f) = (10, 3, 5, -3, -10, 0.69)$, system (5) has a periodic attractor, with the three-dimensional projection shown in Fig. 20c, and the corresponding eigenvalues at equilibrium E_k are

$$\lambda_1 = 0, \ \lambda_2 = -1.1973, \ \lambda_3 = -0.4014 + 1.2521i, \ \lambda_4 = -0.4014 - 1.2521i.$$

From Theorem 20.4, it follows that E_k is unstable.

For $(a, b, c, d, e, f) = (10, 3, 5, -3, -10, 4.13)$, system (5) has a quasi-periodic attractor, with the three-dimensional projections and the Poincaré maps displayed in Fig. 21 and Fig. 22, respectively. The corresponding eigenvalues at equilibrium E_k are

$$\lambda_1 = 0, \ \lambda_2 = -2.0128, \ \lambda_3 = 0.0064 + 2.4811i, \ \lambda_4 = 0.0064 - 2.4811i,$$

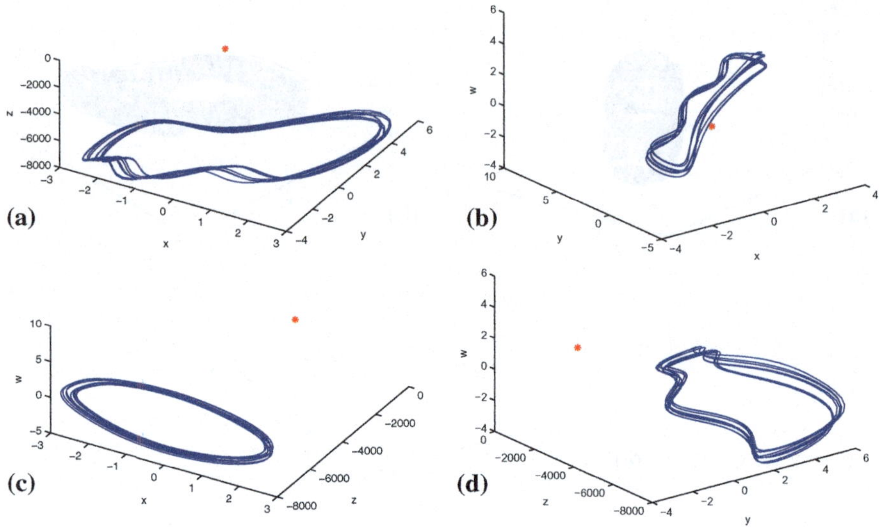

Fig. 21 Phase portraits of the quasi-periodic attractor of system (5): $(a, b, c, d, e, f) = (10, 3, 5, -3, -10, 4.13)$

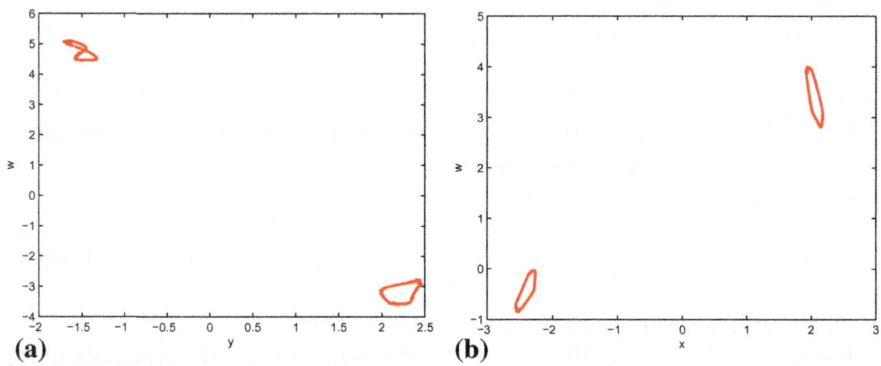

Fig. 22 Poincaré maps of the quasi-periodic attractor of system (5): $(a, b, c, d, e, f) = (10, 3, 5, -3, -10, 4.13)$. **a** On section $\{(x, y, z, w) \in \mathbf{R}^4 | x = 0\}$; **b** on section $\{(x, y, z, w) \in \mathbf{R}^4 | y = 0\}$

implying that E_k is a saddle-focus point, which is also unstable.

4.4 Coexisting Attractors

According to the analysis of Sects. 4.1, 4.2, 4.3, in addition to infinitely many isolated equilibria or no equilibria, system (5) can exhibit a lot of complex dynamics, like hyperchaotic, chaotic, periodic, and quasi-periodic motions. Now, in this subsection, under some proper parameter conditions, several coexisting attractors of system (5) are discussed.

The coexisting solutions of system (5), which satisfy different initial conditions, may possess very different dynamical behaviors. The following Sects. 4.4.1, 4.4.2, 4.4.3 show the coexisting attractors of system (5) with $T = 1$, but Sect. 4.4.4 shows the coexisting hidden attractors with $T > 1$.

4.4.1 Hyperchaotic Attractor and Periodic Attractor

Let $(a, b, c, d, e, f) = (10, 3, 5, -3, -10, 4.40)$ and choose the initial value (0.1, 1.5, 3.6, −1). Then, the corresponding Lyapunov exponents are

$$\lambda_{LE_1} = 0.1270, \quad \lambda_{LE_2} = 0.0590, \quad \lambda_{LE_3} = -0.0001, \quad \lambda_{LE_4} = -2.1857,$$

which demonstrates that the attractor is hyperchaotic. The corresponding Lyapunov dimension is $D_L = 3.0851$. The projection of this attractor onto the x-y-w space is depicted in Fig. 23a, and the Poincaré maps on the y-w section are depicted in Fig. 24a.

Under the same parameters, and choosing the initial condition (−1, −1, −1, −1.05), which implies that system (5) has a periodic attractor, the corresponding Lyapunov exponents are

$$\lambda_{LE_1} = 0.0001, \quad \lambda_{LE_2} = -0.0567, \quad \lambda_{LE_3} = -0.0573, \quad \lambda_{LE_4} = -1.8857.$$

The projection of the periodic attractor onto the three-dimension space is shown in Fig. 23b, and the Poincaré maps on the y-w section are depicted in Fig. 24b.

Therefore, for parameters $(a, b, c, d, e, f) = (10, 3, 5, -3, -10, 4.40)$, system (5) can exhibit complex dynamics with coexisting hyperchaotic attractor and periodic attractor, as shown in Fig. 23c.

4.4.2 Chaotic Attractor and Periodic Attractor

Fixing $(a, b, c, d, e, f) = (10, 3, 5, -3, -10, 2.11)$ and choosing the initial value (0.1, 1.5, 3.6, −1), system (5) has a chaotic attractor with projection into the x-y-w space shown in Fig. 25a and the Poincaré maps on the section $\{(x, y, z, w) \in \mathbf{R}^4 \mid x = 0\}$ displayed in Fig. 26a. Moreover, the corresponding Lyapunov exponents are

$$\lambda_{LE_1} = 0.0596, \quad \lambda_{LE_2} = 0.0000, \quad \lambda_{LE_3} = -0.1311, \quad \lambda_{LE_4} = -1.9283.$$

Under the same parameters, for the initial value (−1, −1, −1, −1.05), the corresponding Lyapunov exponents are

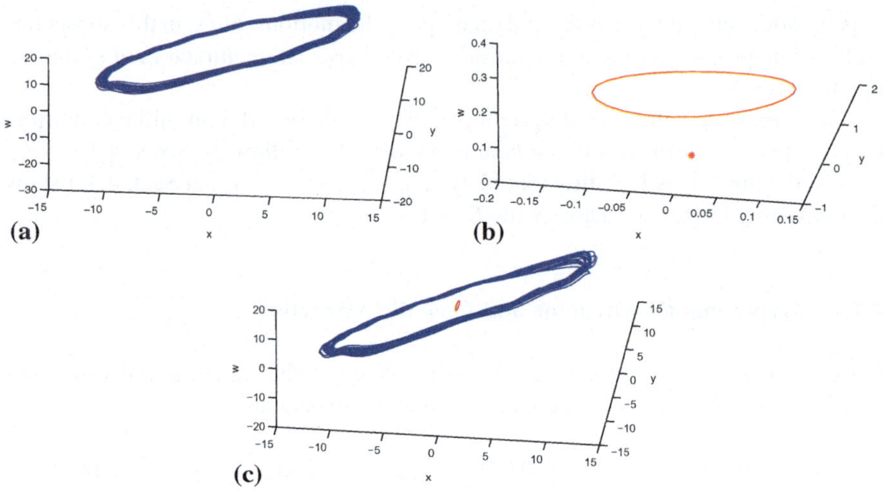

Fig. 23 Projections of coexisting attractors of system (5): $(a, b, c, d, e, f) = (10, 3, 5, -3, -10, 4.40)$. **a** Hyperchaotic attractor; **b** periodic attractor; **c** coexistence

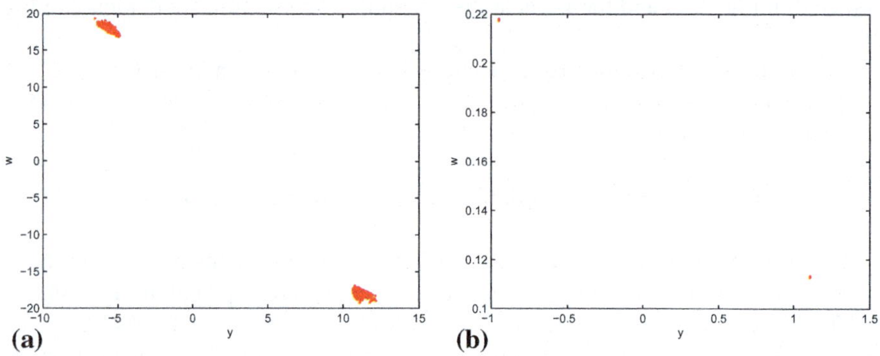

Fig. 24 Poincaré maps of the coexisting attractors of system (5): $(a, b, c, d, e, f) = (10, 3, 5, -3, -10, 4.40)$. **a** Hyperchaotic attractor; **b** periodic attractor

$$\lambda_{LE_1} = -0.0002, \quad \lambda_{LE_2} = -0.2478, \quad \lambda_{LE_3} = -0.2480, \quad \lambda_{LE_4} = -1.5038,$$

which implies that system (5) has a periodic attractor. The projection of this periodic attractor onto the x-y-w space is depicted in Fig. 25b, and the Poincaré maps on section $\{(x, y, z, w) \in \mathbf{R}^4 \mid x = 0\}$ are displayed in Fig. 26b.

Therefore, for the parameters $(a, b, c, d, e, f) = (10, 3, 5, -3, -10, 2.11)$, there coexist chaotic attractor and periodic attractor of system (5), as shown in Fig. 25c.

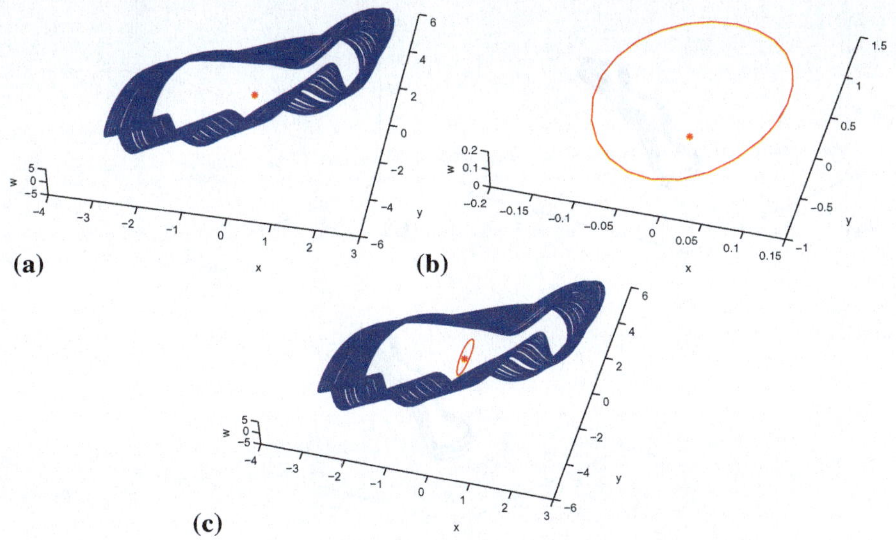

Fig. 25 Projections of coexisting attractors of system (5): $(a, b, c, d, e, f) = (10, 3, 5, -3, -10, 2.11)$. **a** Chaotic attractor; **b** periodic attractor; **c** coexisting attractors

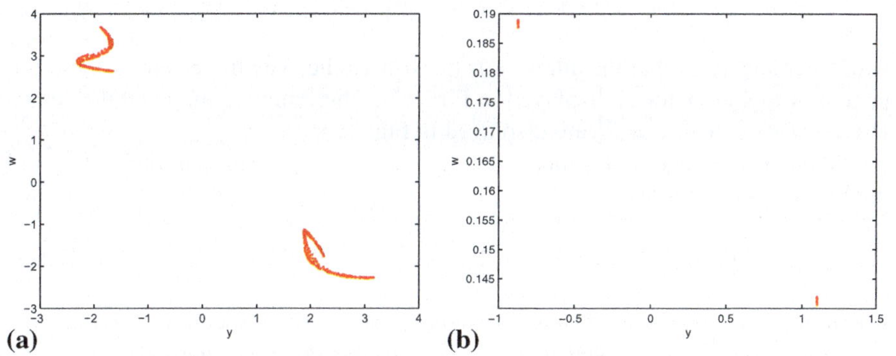

Fig. 26 Poincaré maps of coexisting attractors of system (5): $(a, b, c, d, e, f) = (10, 3, 5, -3, -10, 2.11)$. **a** Chaotic attractor; **b** periodic attractor

4.4.3 Quasi-periodic Attractor and Periodic Attractor

Let $(a, b, c, d, e, f) = (10, 3, 5, -3, -10, 4.13)$, and choose the initial value $(0.1, 1.5, 3.6, -1)$. Then, the corresponding Lyapunov exponents are

$$\lambda_{LE_1} = 0.0001, \quad \lambda_{LE_2} = 0.0000, \quad \lambda_{LE_3} = -0.2105, \quad \lambda_{LE_4} = -1.7896,$$

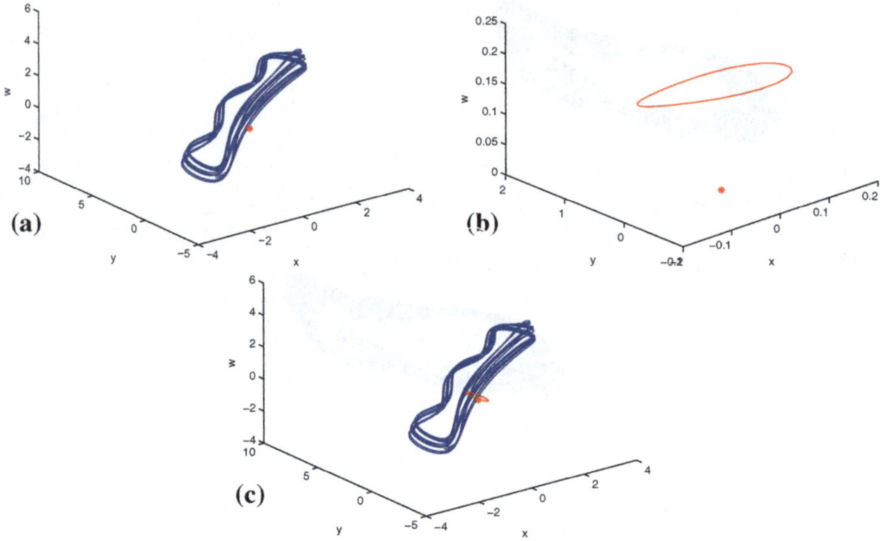

Fig. 27 Projections of coexisting attractors of system (5): $(a, b, c, d, e, f) = (10, 3, 5, -3, -10, 4.13)$. **a** Quasi-periodic attractor; **b** periodic attractor; **c** coexisting attractors

which demonstrates that the attractor is quasi-periodic. The three-dimensional projection of this attractor is displayed in Fig. 27a. The Poincaré maps on the section $\{(x, y, z, w) \in \mathbf{R}^4 \mid x = 0\}$ are displayed in Fig. 28a.

Choosing another initial value, $(-1, -1, -1, -1.05)$, one obtains the corresponding Lyapunov exponents

$$\lambda_{LE_1} = -0.0001, \quad \lambda_{LE_2} = -0.0722, \quad \lambda_{LE_3} = -0.0726, \quad \lambda_{LE_4} = -1.8548,$$

which implies that system (5) has a periodic attractor. The projection onto the three-dimension space is displayed in Fig. 27b, and the Poincaré maps on the section $\{(x, y, z, w) \in \mathbf{R}^4 \mid x = 0\}$ are displayed in Fig. 28b.

Thus, under the same parameters $(a, b, c, d, e, f) = (10, 3, 5, -3, -10, 4.13)$, system (5) coexists two different types of attractors: quasi-periodic attractor and periodic attractor. The coexisting attractors are further depicted in Fig. 27c.

4.4.4 Hidden Hyperchaotic Attractor and Hidden Periodic Attractor

Let $(a, b, c, d, e, f) = (11, 3, 5, -3, -10, 3.77)$. Then, system (5) has no equilibria. For the initial value $(0.1, 1.5, 3.6, -1)$, the corresponding Lyapunov exponents are

$$\lambda_{LE_1} = 0.2752, \quad \lambda_{LE_2} = 0.0966, \quad \lambda_{LE_3} = -0.0002, \quad \lambda_{LE_4} = -2.3715,$$

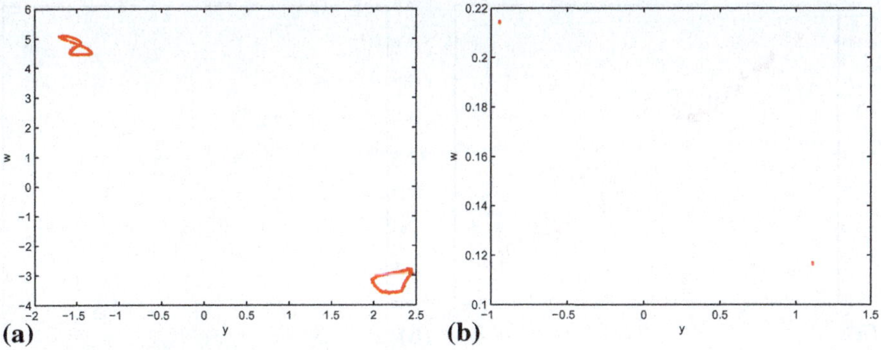

Fig. 28 Poincaré maps of coexisting attractors of system (5): $(a, b, c, d, e, f) = (10, 3, 5, -3, -10, 4.13)$. **a** Quasi-periodic attractor; **b** periodic attractor

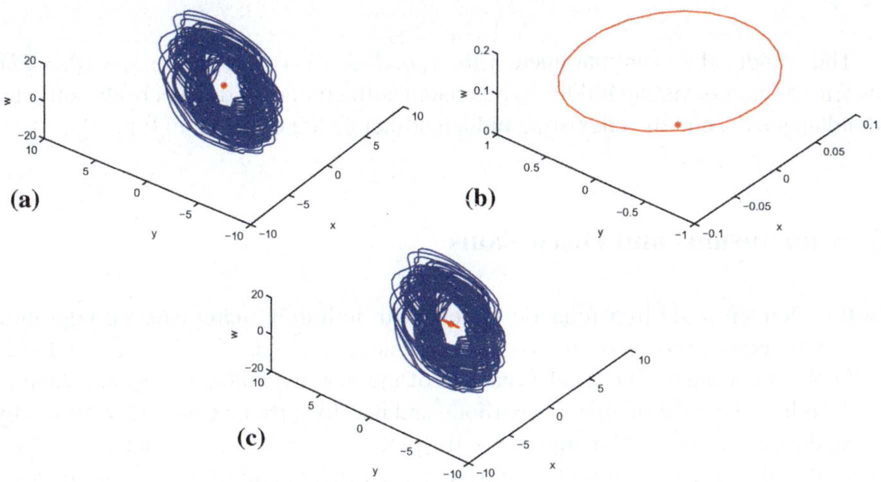

Fig. 29 Projections of coexisting hidden attractors of system (5): $(a, b, c, d, e, f) = (11, 3, 5, -3, -10, 3.77)$. **a** Hidden hyperchaotic attractor; **b** hidden periodic attractor; **c** coexisting attractors

and the Lyapunov dimension is $D_L = 3.1567$. This means that system (5) has a hidden hyperchaotic attractor, with the projection onto the x-y-w space depicted in Fig. 29a. The Poincaré maps on the y-w section are depicted in Fig. 30a.

For the initial condition $(0.1, -1.2, -3, -1)$, system (5) has a hidden periodic attractor, and the corresponding Lyapunov exponents are

$$\lambda_{LE_1} = -0.0002, \quad \lambda_{LE_2} = -0.0629, \quad \lambda_{LE_3} = -0.0668, \quad \lambda_{LE_4} = -1.8699.$$

The projection of the hidden periodic attractor onto the three-dimension space is shown in Fig. 29b, and the Poincaré maps on the y-w section are depicted in Fig. 30b.

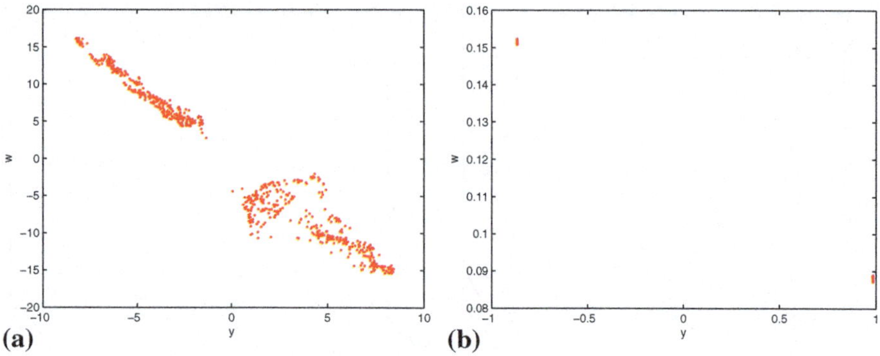

Fig. 30 Poincaré maps of coexisting attractors of system (5): $(a, b, c, d, e, f) = (11, 3, 5, -3, -10, 3.77)$. **a** Hidden hyperchaotic attractor; **b** hidden periodic attractor

Thus, under the same parameters $(a, b, c, d, e, f) = (11, 3, 5, -3, -10, 3.77)$, system (5) has coexisting hidden hyperchaotic attractor and hidden periodic attractor simultaneously, and the coexisting hidden attractors are depicted in Fig. 29c.

5 Conclusions and Discussions

In this chapter, a 4D hyperchaotic system with infinitely many isolated equilibria or without equilibria has been constructed by adding a linear feedback controller to a 3D chaotic system. The local dynamics of the new hyperchaotic system, such as the stability of equilibria in the hyperbolic and non-hyperbolic case, are analyzed by using the center manifold theorem. For the global dynamics, the numerical analysis show that the new system with infinitely many isolated equilibria or without equilibria can reveal hyperchaotic, chaotic, quasi-periodic and periodic dynamics. The corresponding hyperchaotic attractor and other attractors are numerically verified through investigating phase portraits, bifurcations, Lyapunov exponents, Poincaré sections, and fractional dimensions. The system can generate hyperchaotic attractor or hidden hyperchaotic attractor when system has infinitely many isolated equilibria or has no equilibria, under some proper parameter values. In particular, there are four kinds of coexistence of differential attractors coexisting for some parameters: (i) hyperchaotic attractor and periodic attractor, (ii) chaotic attractor and periodic attractor, (iii) quasi-periodic attractor and periodic attractor, and (iv) hidden hyperchaotic attractor and hidden periodic attractor.

In order to better understand four-dimensional hyperchaotic systems, there are a few problems needed to be further investigated:

(i) What is the hyperchaotic formation mechanism of four-dimensional autonomous systems with no equilibria?

(ii) What is the mechanism of hyperchaotic systems with infinitely many isolated equilibria?

(iii) How to change a 4D autonomous system from having no equilibria to having infinitely many isolated equilibria?

(iv) How to construct the coexisting mechanism of periodic attractor, quasi-periodic attractor, chaotic attractor and hyperchaotic attractor of a general hyperchaotic system without equilibria?

These problems in the study of a hyperchaotic system with infinitely many isolated equilibria or without equilibria may make the hyperchaos theories richer for more practical applications. It is hoped that the investigation of this chapter will shed some lights onto more systematic studies of 4D hyperchaotic systems.

Acknowledgements This work was supported by the National Natural Science Foundation of China (No. 12071151), Natural Science Foundation of Guangdong Province (No. 2021A1515010052) and the key project of the graduate textbooks of South China University of Technology.

References

1. O.E. Rössler, An equation for hyperchaos. Phys. Lett. A **71**, 155–157 (1979)
2. Y. Li, G. Chen, W. Tang, Controlling a unified chaotic system to hyperchaotic. IEEE Trans. Circuits Syst.-II **52**, 204–207 (2005)
3. G. Qi, M.A. Wyk, B.J. Wyk, G. Chen, On a new hyperchaotic system. Phys. Lett. A **372**, 124–136 (2008)
4. X. Wang, M. Wang, A hyperchaos generated from Lorzen system. Physica A **387**, 3751–3758 (2008)
5. Q. Yang, Y. Liu, A hyperchaotic system from a chaotic system with one saddle and two stable node-foci. J. Math. Anal. Appl. **360**(1), 293–306 (2009)
6. Q. Yang, K. Zhang, G. Chen, Hyperchaotic attractors from a linearly controlled Lorenz system. Nonlinear Anal. Real World Appl. **10**, 1601–1617 (2009)
7. E. Lorenz, Deterministic non-periodic flow. J. Atmos. Sci. **20**, 130–141 (1963)
8. G. Chen, T. Ueta, Yet another chaotic attractor. Int. J. Bifurc. Chaos **9**, 1465–1466 (1999)
9. J. Lu, G. Chen, A new chaotic attractor coined. Int. J. Bifurc. Chaos **12**, 659–661 (2002)
10. Q. Yang, G. Chen, A chaotic system with one saddle and two stable node-foci. Int. J. Bifurc. Chaos **18**, 1393–1414 (2008)
11. M. Hisch, S. Smale, R. Devaney, *Differential Equations, Dynamical Systems, and an Introduction to Chaos* (Elsevier Academic Press, New York, 2007)
12. Y. Chen, Q. Yang, Dynamics of a hyperchaotic Lorenz-type system. Nonlinear Dyn. **77**, 569–581 (2014)
13. Z. Chen, Y. Yang, G. Qi, A novel hyperchaos system only with one equilibrium. Phys. Lett. A **360**, 696–701 (2007)
14. W. Xue, G. Qi, J. Mu, H. Jia, Y. Guo, Hopf bifurcation analysis and circuit implementation for a novel four-wing hyperchaotic system. Chin. Phys. B **8**(080504) (2013)
15. T. Kapitaniak, L. Chua, Hyperchaotic attractor of unidirectionally coupled Chua's circuit. Int. J. Bifurc. Chaos **4**, 477–486 (1994)
16. K. Thamilmaran, M. Lakshmanan, A. Venkatesan, A hyperchaos in a modified canonical Chus circuit. Int. J. Bifurc. Chaos **14**, 221–243 (2004)
17. Z. Wei, R. Wang, A. Liu, A new find of the existence of hidden hyperchaotic attractors with no equilibria. Math. Compu. Simu. **100**, 13–23 (2014)

18. Y. Chen, Q. Yang, A new Lorenz-type hyperchaotic systems with a curve of equilibria. Math. Compu. Simu. **112**, 40–55 (2015)
19. L. Šil'nikov, A. Šil'nikov, D. Turaev, L. Chua, *Methods of Qualitative Theory in Nonlinear Dynamics, Part II* (World Scientific Publishing, Singapore, 2001)

Singular Cycles and Chaos in Piecewise-Affine Systems

Xiao-Song Yang, Lei Wang, and Tiantian Wu

1 Introduction to Piecewise-Smooth Systems and Piecewise-Affine Systems

There are a lot of interesting phenomena in nature, society and engineering that are closely related to the description of piecewise-smooth dynamical systems [1–6]. Among these are population dynamics, such as food chain dynamics and generic regulation networks in biology, switching power converters and relay systems, different types of pulse-width modulated control systems, mechanical systems, including legged robots, digital electronics and economic business cycles, to name just a few.

Piecewise-smooth dynamical systems can exhibit rich dynamics and bifurcations similar to smooth dynamical systems, and more interestingly display some unique features like border-collision, discontinuous bifurcation and periodic adding due to non-smoothness and discontinuity [1]. Therefore, investigation of various dynamics and bifurcations in piecewise-smooth dynamical systems is of great significance from both practical and theoretical points of view.

The first goal in studying piecewise-smooth dynamical systems is clearly on the piecewise-linear systems, or more generally, piecewise-affine systems. There are

X.-S. Yang (✉)
Hubei Key Laboratory of Engineering Modeling and Scientific Computing, School of Mathematics and Statistics, Huazhong University of Science and Technology, Wuhan 430074, People's Republic of China
e-mail: yangxs@hust.edu.cn

L. Wang
School of Artificial Intelligence and Big Data, Sino-German Institute of Applied Mathematics, Hefei University, Hefei 230601, China
e-mail: wanglei@hfuu.edu.cn

T. Wu
School of Mathematics and Statistics, Shandong Normal University, Jinan 250014, China
e-mail: wutiantian1021@126.com

X. Wang et al. (eds.), *Chaotic Systems with Multistability and Hidden Attractors*, Emergence, Complexity and Computation 40,
https://doi.org/10.1007/978-3-030-75821-9_21

two reasons for studying this class of piecewise-smooth dynamical systems. First, the piecewise-affine systems can display almost all of dynamical behaviors observed in smooth systems and are of fundamental theoretical significance in understanding general piecewise-smooth dynamical systems. Second, piecewise-affine systems can be investigated analytically by means of quantitatively and qualitatively approaches because of their piecewise nature.

Even from the smooth dynamical system perspective, it is worthy of making efforts to investigate the dynamics of piecewise-affine systems, especially complex dynamics such as chaos. An intensive investigation on the dynamics of piecewise-affine systems can also give some deep insights into the dynamics of smooth dynamical systems. Because singular cycles such as homoclinic orbits and heteroclinic cycles are of great importance in the study of nonlinear dynamical systems, due to their close connections with global bifurcations and chaotic behaviors of dynamical systems in terms of the Šil'nikov theory, in this chapter we focus on piecewise-affine systems and present some recent progress in the investigation of the existence of homoclinic orbits and heteroclinic cycles, and an extension of the well-known Šil'nikov theory to studying chaos in piecewise-affine systems.

This chapter is organized as follows. In the rest of this section, we review some notions and notations of piecewise-affine systems. In Sect. 2, we present recent results on the existence of homoclinic orbits and heteroclinic cycles in several subclasses of piecewise-affine systems. Section 3 offers a Šil'nikov-type theory for piecewise-affine systems. In Sect. 4, we present a gallery of chaotic piecewise-affine systems with computer simulations. In the last section, we give a brief summary and propose several topics that are of great value and common interest in the opinion of the authors.

Now, let us prepare some preliminaries of piecewise-affine systems.

Consider the following n-dimensional piecewise-affine systems with two zones:

$$\dot{\mathbf{x}} = \begin{cases} \mathbf{Ax} + \mathbf{a}, & \mathbf{c}^{\mathsf{T}}\mathbf{x} < d \\ \mathbf{Bx} + \mathbf{b}, & \mathbf{c}^{\mathsf{T}}\mathbf{x} \geq d, \end{cases} \tag{1}$$

where $\mathbf{x} = (x_1, \ldots, x_n)^{\mathsf{T}} \in \mathbb{R}^n$ is a **vector of state variables.** $\mathbf{a} = (a_1, \ldots, a_n)^{\mathsf{T}}, \mathbf{b} = (b_1, \ldots, b_n)^{\mathsf{T}}$ and $\mathbf{c} = (c_1, \ldots, c_n)^{\mathsf{T}}$ are all constant vectors in $\mathbb{R}^n$, d is a constant, and $\mathbf{A}$ and $\mathbf{B}$ are invertible matrices.

Let

$$\Sigma = \{\mathbf{x} \in \mathbb{R}^n \mid \mathbf{c}^{\mathsf{T}}\mathbf{x} = d\}.$$

Here, Σ is usually called a **discontinuity set**, **discontinuity boundary** or, sometimes, a **switching manifold** [1, 3].

Let

$$\Sigma^- = \{\mathbf{x} \in \mathbb{R}^n \mid \mathbf{c}^{\mathsf{T}}\mathbf{x} < d\}, \quad \Sigma^+ = \{\mathbf{x} \in \mathbb{R}^n \mid \mathbf{c}^{\mathsf{T}}\mathbf{x} > d\}.$$

In system (1), let the equilibrium point of the subsystem

$$\dot{\mathbf{x}} = \mathbf{A}\mathbf{x} + \mathbf{a} \tag{2}$$

be denoted by $\mathbf{p}$ with $\mathbf{p} = -\mathbf{A}^{-1}\mathbf{a} \in \Sigma^-$, and the equilibrium point of subsystem

$$\dot{\mathbf{x}} = \mathbf{B}\mathbf{x} + \mathbf{b} \tag{3}$$

be denoted by $\mathbf{q}$ with $\mathbf{q} = -\mathbf{B}^{-1}\mathbf{b}$.

Let

$$\mathbf{p} = (x_{1\mathbf{p}}, \ldots, x_{n\mathbf{p}})^{\mathsf{T}}, \quad \mathbf{q} = (x_{1\mathbf{q}}, \ldots, x_{n\mathbf{q}})^{\mathsf{T}}.$$

Without loss of generality, assume that

$$\mathbf{A} = \mathbf{P}\mathbf{J}_{\mathbf{A}}\mathbf{P}^{-1}, \tag{4}$$

$$\mathbf{B} = \mathbf{Q}\mathbf{J}_{\mathbf{B}}\mathbf{Q}^{-1}, \tag{5}$$

where $\mathbf{J}_{\mathbf{A}}$ and $\mathbf{J}_{\mathbf{B}}$ are the Jordan normal forms of matrices $\mathbf{A}$ and $\mathbf{B}$, the invertible matrices $\mathbf{P}$ and $\mathbf{Q}$ are given as follows:

$$\mathbf{P} = \left(\zeta_1, \ldots, \zeta_n\right),$$

with $\zeta_1, \ldots, \zeta_3 \in R^n$ being generalized eigenvectors of matrix $\mathbf{A}$, and

$$\mathbf{Q} = \left(\xi_1, \ldots, \xi_n\right),$$

with $\xi_1, \ldots, \xi_3 \in R^3$ being generalized eigenvectors of matrix $\mathbf{B}$.

Denote by $\Phi_A(t, \mathbf{x}_0)$ and $\Phi_B(t, \mathbf{y}_0)$ the solutions of system (2) and (3), with initial conditions $\Phi_A(0, \mathbf{x}_0) = \mathbf{x}_0$ and $\Phi_B(0, \mathbf{y}_0) = \mathbf{y}_0$, respectively. We obtain that

$$\Phi_A(t, \mathbf{x}_0) = e^{\mathbf{A}t}(\mathbf{x}_0 - \mathbf{p}) + \mathbf{p}, \tag{6}$$

$$\Phi_B(t, \mathbf{y}_0) = e^{\mathbf{B}t}(\mathbf{y}_0 - \mathbf{q}) + \mathbf{q}. \tag{7}$$

For convenience, let

$$\zeta_i = \begin{pmatrix} \zeta_{i1} \\ \vdots \\ \zeta_{in} \end{pmatrix}, \quad \xi_i = \begin{pmatrix} \xi_{i1} \\ \vdots \\ \xi_{in} \end{pmatrix}, \quad \text{for} \quad i = 1, \ldots, n.$$

2 Existence of Homoclinic Orbits and Heteroclinic Cycles

In this section, we provide several results about the existence of homoclinic orbits and heteroclinic cycles that cross the switching manifold transversally at two points in 3-dimensional or 4-dimensional piecewise-affine systems.

2.1 Existence of Homoclinic Orbits

2.1.1 Existence of Homoclinic Orbits to a Saddle-Focus in 3-Dimensional Piecewise-Affine Systems

Without loss of generality, this section considers system (1) with $n = 3$, $\mathbf{c} = (-1, 0, 0)^\mathsf{T}$ and $d = 0$, as follows:

$$\dot{\mathbf{x}} = \begin{cases} \mathbf{Ax} + \mathbf{a}, & x_1 > 0, \\ \mathbf{Bx} + \mathbf{b}, & x_1 \leq 0, \end{cases} \quad \mathbf{x} \in \mathbb{R}^3, \tag{8}$$

where the eigenvalues of $\mathbf{A}$ are $\alpha \pm i\beta,\ \lambda$, with $\alpha,\ \beta > 0,\ \lambda < 0$ and the eigenvalues of $\mathbf{B}$ are $\lambda_1,\ \lambda_2,\ \lambda_3$ with $\lambda_i \neq 0$, for $i = 1, 2, 3$.

In (4) and (5), suppose that

$$\mathbf{J_A} = \begin{pmatrix} \alpha & -\beta & 0 \\ \beta & \alpha & 0 \\ 0 & 0 & \lambda \end{pmatrix},$$

and $\mathbf{J_B}$ has the following three forms:

$$\mathbf{J}_1 = \begin{pmatrix} \lambda_1 & 0 & 0 \\ 0 & \lambda_2 & 0 \\ 0 & 0 & \lambda_3 \end{pmatrix}, \ \mathbf{J}_2 = \begin{pmatrix} \lambda_1 & 1 & 0 \\ 0 & \lambda_1 & 0 \\ 0 & 0 & \lambda_2 \end{pmatrix}, \ \mathbf{J}_3 = \begin{pmatrix} \lambda_1 & 1 & 0 \\ 0 & \lambda_1 & 1 \\ 0 & 0 & \lambda_1 \end{pmatrix}.$$

Moreover, for

$$\mathbf{x}_0 = \mathbf{p} + (\zeta_1, \zeta_2, \zeta_3) \begin{pmatrix} x_1 \\ x_2 \\ x_3 \end{pmatrix}, \quad \mathbf{y}_0 = \mathbf{q} + (\xi_1, \xi_2, \xi_3) \begin{pmatrix} x_1 \\ x_2 \\ x_3 \end{pmatrix},$$

from (6) and (7), we have

$$\Phi_A(t, \mathbf{x}_0) = \mathbf{p} + (\zeta_1, \zeta_2, \zeta_3) \begin{pmatrix} e^{\alpha t}(x_1 \cos(\beta t) - x_2 \sin(\beta t)) \\ e^{\alpha t}(x_1 \sin(\beta t) + x_2 \cos(\beta t)) \\ e^{\lambda t} x_3 \end{pmatrix}, \tag{9}$$

and

$$\Phi_B(t, \mathbf{y}_0) = \mathbf{q} + (\xi_1, \xi_2, \xi_3)e^{\mathbf{J_B t}} \begin{pmatrix} x_1 \\ x_2 \\ x_3 \end{pmatrix}, \tag{10}$$

where $e^{\mathbf{J_B t}}$ has the following three forms:

$$e^{\mathbf{J}_1 t} = \begin{pmatrix} e^{\lambda_1 t} & 0 & 0 \\ 0 & e^{\lambda_2 t} & 0 \\ 0 & 0 & e^{\lambda_3 t} \end{pmatrix}, \tag{11}$$

$$e^{\mathbf{J}_2 t} = \begin{pmatrix} e^{\lambda_1 t} & te^{\lambda_1 t} & 0 \\ 0 & e^{\lambda_1 t} & 0 \\ 0 & 0 & e^{\lambda_2 t} \end{pmatrix}, \tag{12}$$

$$e^{\mathbf{J}_3 t} = \begin{pmatrix} e^{\lambda_1 t} & te^{\lambda_1 t} & \frac{t^2}{2}e^{\lambda_1 t} \\ 0 & e^{\lambda_1 t} & te^{\lambda_1 t} \\ 0 & 0 & e^{\lambda_1 t} \end{pmatrix}. \tag{13}$$

From the expression of $\mathbf{A}$ in (4), the stable subspace $E^s(\mathbf{p})$ and the unstable subspace $E^u(\mathbf{p})$ are shown as follows:

$$E^s(\mathbf{p}) = \{\mathbf{p} + k\zeta_3 \mid k \in R\}, \tag{14}$$

$$E^u(\mathbf{p}) = \{\mathbf{p} + k_1\zeta_1 + k_2\zeta_2 \mid k_1, k_2 \in R\}.$$

Suppose that $\mathbf{p}_1 = E^s(\mathbf{p}) \cap \Sigma,\ L = E^u(\mathbf{p}) \cap \Sigma \neq \emptyset$, then we get

$$\mathbf{p}_1 = \left(0, x_{2\mathbf{p}} - \frac{\zeta_{32}}{\zeta_{31}}x_{1\mathbf{p}}, x_{3\mathbf{p}} - \frac{\zeta_{33}}{\zeta_{31}}x_{1\mathbf{p}}\right)^{\mathsf{T}},$$

$$L = \{\mathbf{x} = \mathbf{p} + k_1\zeta_1 + k_2\zeta_2 \mid x_{1\mathbf{p}} + k_1\zeta_{11} + k_2\zeta_{21} = 0, k_1, k_2 \in R\}.$$

For $\mathbf{p}_1$ and a point $\mathbf{p}_2 \in L$, let

$$\mathbf{p}_1 - \mathbf{q} = \mathbf{Q}(\rho_1, \rho_2, \rho_3)^{\mathsf{T}}, \tag{15}$$

$$\mathbf{p}_2 - \mathbf{q} = \mathbf{Q}(\sigma_1, \sigma_2, \sigma_3)^{\mathsf{T}}. \tag{16}$$

For the existence of a homoclinic orbit of system (1), satisfying all the above assumptions with $n = 3$, we have the following theorem [7].

Theorem 21.1 *System (8) has a homoclinic orbit to the equilibrium point* $\mathbf{p}$, *which crosses* Σ *transversally at two points, if and only if there exist real numbers* k_1, k_2 *and* $T > 0$ *such that the following conditions hold:*

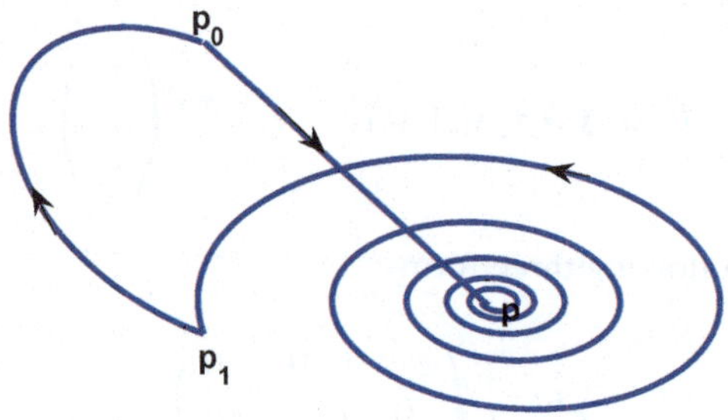

Fig. 1 Illustration of a homoclinic orbit

(I) $\mathbf{p}_2 = \mathbf{p} + k_1\zeta_1 + k_2\zeta_2 \in L$.

(II) From the expressions of $\mathbf{p}_1$ *and* $\mathbf{p}_2$ *in (15) and (16), respectively, it follows that*

$$e^{\mathbf{J_B}T}(\sigma_1, \sigma_2, \sigma_3)^\mathsf{T} = (\rho_1, \rho_2, \rho_3)^\mathsf{T},$$

$$(\xi_{11}, \xi_{21}, \xi_{31})\mathbf{J}(\sigma_1, \sigma_2, \sigma_3)^\mathsf{T} < 0, (\xi_{11}, \xi_{21}, \xi_{31})\mathbf{J}(\rho_1, \rho_2, \rho_3)^\mathsf{T} > 0.$$

(III)

$$\alpha x_{1\mathbf{p}} - \beta(k_1\zeta_{21} - k_2\zeta_{11}) > 0,$$

$$\sqrt{x_{1\mathbf{p}}^2 + (k_1\zeta_{21} - k_2\zeta_{11})^2}\frac{\beta}{\sqrt{\alpha^2+\beta^2}}e^{-\alpha T'} < x_{1\mathbf{p}},$$

where

$$T' = \frac{\pi}{\beta} + \frac{1}{\beta}\arctan\left(\frac{\beta}{\alpha}\right) + \frac{1}{\beta}\arcsin\left(\frac{x_{1\mathbf{p}}}{\sqrt{x_{1\mathbf{p}}^2 + (k_1\zeta_{21} - k_2\zeta_{22})^2}}\right).$$

In addition, the homoclinic orbit crosses Σ *transversally at* $\mathbf{p}_1$ *and* $\mathbf{p}_2$ *(Fig. 1).*

Proof of Theorem 21.1 If system (8) has a homoclinic orbit to $\mathbf{p}$ and the homoclinic orbit crosses Σ transversally at two points, then from the definition of a homoclinic orbit, one point is $\mathbf{p}_1 = E^s(\mathbf{p}) \cap \Sigma$ and the other one is located in line L.

System (8) has a homoclinic orbit to $\mathbf{p}$ and the homoclinic orbit crosses Σ transversally at $\mathbf{p}_1$ and $\mathbf{p}_2$ if and only if system (8) satisfies the following conditions:

(1) $\mathbf{p}_2 \in L$.
(2) The positive orbit of $\mathbf{p}_0$ satisfies $\{\Phi_A(t, \mathbf{p}_1) \mid t > 0\} \subset \Sigma^-$.
(3) The negative orbit of $\mathbf{p}_1$ satisfies $\{\Phi_A(t, \mathbf{p}_2) \mid t < 0\} \subset \Sigma^-$.
(4) For a constant $T > 0$ $\{\Phi_B(t, \mathbf{p}_2) \mid t \in (0, T)\} \subset \Sigma^+$ *and* $\Phi_B(T, \mathbf{p}_2) = \mathbf{p}_1$.
(5)

$$(1, 0, 0)^\mathsf{T}(\mathbf{A}\mathbf{p}_1 + \mathbf{a}) \cdot (1, 0, 0)^\mathsf{T}(\mathbf{B}\mathbf{p}_1 + \mathbf{b}) > 0,$$

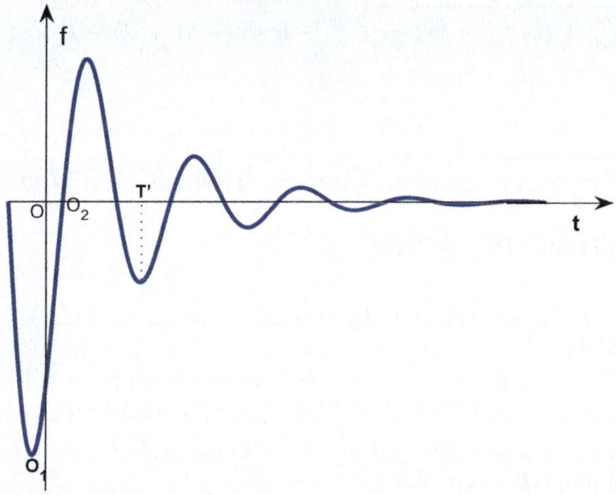

Fig. 2 Graph of $f(t)$

$$(1,0,0)^{\mathsf{T}}(\mathbf{A}\mathbf{p}_2+\mathbf{a})\cdot(1,0,0)^{\mathsf{T}}(\mathbf{B}\mathbf{p}_2+\mathbf{b})>0.$$

Notice that condition (I) in Theorem 21.1 is condition (1) here.

Consider the expression of $E^s(\mathbf{p})$ in (14). Then, $\{\Phi_A(t,\mathbf{p}_1)\mid t>0\}$ is a line connecting $\mathbf{p}_1$ and $\mathbf{p}$. From $\mathbf{p}_1\in\Sigma$, $\mathbf{p}\in\Sigma^-$, we have

$$\{\Phi_A(t,\mathbf{p}_1)\mid t>0\}\subset\Sigma^-,$$

which means that condition (2) holds.

Let

$$f(t)=(1,0,0)^{\mathsf{T}}(\Phi_A(-t,\mathbf{p}_2)-\mathbf{p}).$$

Then, $\{\Phi_A(t,\mathbf{p}_2)\mid t<0\}\subset\Sigma^-$ if and only if $f(t)>-x_{1\mathbf{p}}$ holds for $t>0$.

From (9) and condition (I), we get

$$f(t)=\sqrt{x_{1\mathbf{p}}^2+(k_1\zeta_{21}-k_2\zeta_{22})^2}e^{-\alpha t}\sin(-\beta t+\theta), \tag{17}$$

where

$$\sin\theta=\frac{-x_{1\mathbf{p}}}{\sqrt{x_{1\mathbf{p}}^2+(k_1\zeta_{21}-k_2\zeta_{22})^2}},\quad \cos\theta=\frac{k_1\zeta_{21}-k_2\zeta_{11}}{\sqrt{x_{1\mathbf{p}}^2+(k_1\zeta_{21}-k_2\zeta_{11})^2}}.$$

From (17), we have

$$f'(t) = \sqrt{x_{1\mathbf{p}}^2 + (k_1\zeta_{21} - k_2\zeta_{22})^2} e^{-\alpha t}[-\alpha \sin(-\beta t + \theta) - \beta \cos(-\beta t + \theta)], \tag{18}$$

and

$$f''(t) = \sqrt{x_{1\mathbf{p}}^2 + (k_1\zeta_{21} - k_2\zeta_{22})^2} e^{-\alpha t}[(\alpha^2 - \beta^2) \sin(-\beta t + \theta) + 2\alpha\beta \cos(-\beta t + \theta)].$$

Considering (17) and (18), we have

$$f(0) = -x_{1\mathbf{p}}, \quad f'(0) = \alpha x_{1\mathbf{p}} - \beta(k_1\zeta_{21} - k_2\zeta_{11}).$$

Since $f(0) = -x_{1\mathbf{p}}$, for $f(t) > -x_{1\mathbf{p}}$ ($t > 0$), we must have $f'(0) > 0$, which is the first inequality in condition (III) in Theorem 21.1. In addition, combining these with Fig. 2, $f(t) > -x_{1\mathbf{p}}$ holds for $t > 0$ if and only if $f(T') > -x_{1\mathbf{p}}$ holds for the local minimum point $T' \in (0, \frac{2\pi}{\beta})$.

From (18), we have

$$f'(t) = 0 \;\Leftrightarrow\; \tan(-\beta t + \theta) = \frac{-\beta}{\alpha}.$$

Equivalently, we have either

$$\sin(-\beta t + \theta) = \frac{\beta}{\sqrt{\alpha^2 + \beta^2}}, \quad \cos(-\beta t + \theta) = \frac{-\alpha}{\sqrt{\alpha^2 + \beta^2}}, \tag{19}$$

or

$$\sin(-\beta t + \theta) = \frac{-\beta}{\sqrt{\alpha^2 + \beta^2}}, \quad \cos(-\beta t + \theta) = \frac{\alpha}{\sqrt{\alpha^2 + \beta^2}}. \tag{20}$$

For t satisfying (19), we have $f''(t) < 0$, i.e. t is a local maximum point of f. For t satisfying (20), we have $f''(t) > 0$, i.e. t is a local minimum point of f, and

$$f(t) = \sqrt{x_{1\mathbf{p}}^2 + (k_1\zeta_{21} - k_2\zeta_{22})^2} e^{-\alpha t} \frac{-\beta}{\sqrt{\alpha^2 + \beta^2}}.$$

Moreover, combining with Fig. 2, the local minimum point T' of f in $(0, \frac{2\pi}{\beta})$ is

$$T' = \frac{\pi}{\beta} + \frac{1}{\beta} \arctan\left(\frac{\beta}{\alpha}\right) + \frac{1}{\beta} \arcsin\left(\frac{x_{1\mathbf{p}}}{\sqrt{x_{1\mathbf{p}}^2 + (k_1\zeta_{21} - k_2\zeta_{22})^2}}\right).$$

and

$$f(T') = \sqrt{x_{1\mathbf{p}}^2 + (k_1\zeta_{21} - k_2\zeta_{22})^2} e^{-\alpha T'} \frac{-\beta}{\sqrt{\alpha^2 + \beta^2}}. \tag{21}$$

Considering (21), $\{\Phi_A(t, \mathbf{p}_2) \mid t < 0\} \subset \Sigma^-$ if and only if condition (III) in Theorem 21.1 holds. That is, condition (3) holds if and only if condition (III) in Theorem 21.1 holds.

Next, consider the verification of condition (4). If system (8) has a homoclinic orbit to the equilibrium point $\mathbf{p}$, then from the above discussions, we must have $\Phi_B(T, \mathbf{p}_2) = \mathbf{p}_1$ for some constant $T > 0$ and $\{\Phi_B(t, \mathbf{p}_2) \mid t \in (0, T)\} \subset \Sigma^+$.

Assume that for a constant $T > 0$, $\Phi_B(T, \mathbf{p}_2) = \mathbf{p}_1$. From (9), it is the equality of condition (II) in Theorem 21.1. In the following, we prove that

$$\{\Phi_B(t, \mathbf{p}_2) \mid t \in (0, T)\} \subset \Sigma^+$$

if and only if condition (II) in Theorem 21.1 holds.

To do so, consider (9) and let

$$g_i(t) = (1, 0, 0)^{\mathsf{T}} \Phi_B(t, \mathbf{p}_1) \quad for \quad \mathbf{J_B} = \mathbf{J_i}, i = 1, 2, 3.$$

Then, for $\mathbf{J_B} = \mathbf{J_i}$,

$$\{\Phi_B(t, \mathbf{p}_2) \mid t \in (0, T)\} \subset \Sigma^+$$

if and only if $g_i(t) < 0$ for $t \in (0, T)$, $i = 1, 2, 3$.

In view of condition (I) in Theorem 21.1 and (9), we get

$$g_i(0) = g_i(T) = 0, \quad g_i'(0) = (\xi_{11}, \xi_{21}, \xi_{31})^{\mathsf{T}} \mathbf{J_i} \begin{pmatrix} \sigma_1 \\ \sigma_2 \\ \sigma_3 \end{pmatrix},$$

$$g_i'(T) = (\xi_{11}, \xi_{21}, \xi_{31})^{\mathsf{T}} \mathbf{J_i} \begin{pmatrix} \rho_1 \\ \rho_2 \\ \rho_3 \end{pmatrix}, \quad i = 1, 2, 3.$$

Since $g_i(0) = g_i(T) = 0$, to have $g_i(t) < 0$ for $t \in (0, T)$, we must have $g_i'(0) < 0$, $g_i'(T) > 0$, i.e. the inequalities in condition (II) in Theorem 21.1.

Now, we prove that $g_i(t) < 0$ holds for $t \in (0, T)$ if condition (II) in Theorem 21.1 holds. We only study $g_1(t)$. Functions $g_2(t)$ and $g_3(t)$ can be studied by the same method.

From (9), we get

$$g_1(t) = x_{1\mathbf{q}} + \xi_{11}\sigma_1 e^{\lambda_1 t} + \xi_{21}\sigma_2 e^{\lambda_2 t} + \xi_{31}\sigma_3 e^{\lambda_3 t},$$

$$\begin{aligned} g_1'(t) &= \xi_{11}\sigma_1\lambda_1 e^{\lambda_1 t} + \xi_{21}\sigma_2\lambda_2 e^{\lambda_2 t} + \xi_{31}\sigma_3\lambda_3 e^{\lambda_3 t} \\ &= e^{\lambda_1 t}(\xi_{11}\sigma_1\lambda_1 + \xi_{21}\sigma_2\lambda_2 e^{(\lambda_2-\lambda_1)t} + \xi_{31}\sigma_3\lambda_3 e^{(\lambda_3-\lambda_1)t}). \end{aligned}$$

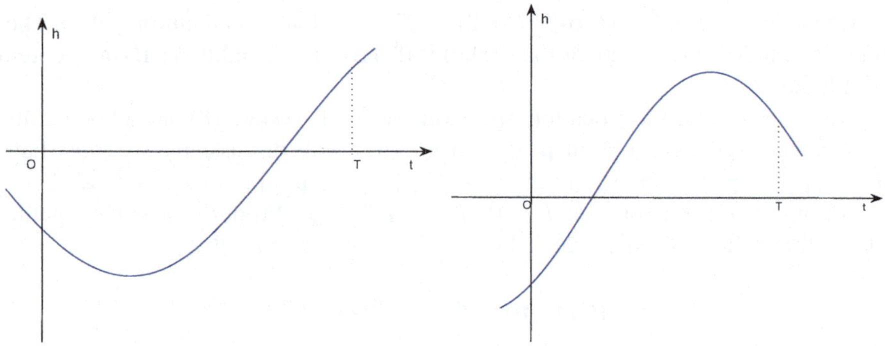

Fig. 3 The figure of h

Let

$$h(t) = \xi_{11}\sigma_1\lambda_1 + \xi_{21}\sigma_2\lambda_2 e^{(\lambda_2-\lambda_1)t} + \xi_{31}\sigma_3\lambda_3 e^{(\lambda_3-\lambda_1)t}.$$

Then,

$$g_1'(t) = e^{\lambda_1 t}h(t), \quad h(0) < 0, \quad h(T) > 0,$$

and

$$h'(t) = e^{(\lambda_2-\lambda_1)t}[\xi_{21}\sigma_2\lambda_2(\lambda_2-\lambda_1) + \xi_{31}\sigma_3\lambda_3(\lambda_3-\lambda_1)e^{(\lambda_3-\lambda_2)t}]. \tag{22}$$

Since $h(0)h(T) < 0$, $h(t) = 0$ has a solution in $(0, T)$ from the Mean Value Theorem and $h'(t)$ is not equal to zero. In view of (22), $h'(t) = 0$ has one solution at most.

If $h'(t) = 0$ has no solutions in $(0, T)$, then from $h(0) < 0$ and $h(T) > 0$, we conclude that $h'(t) > 0$ and $h(t) = 0$ has a unique solution in $(0, T)$; that is, $g_1'(t) = 0$ has a unique solution in $(0, T)$.

If the unique solution t' of $h'(t) = 0$ is in $(0, T)$, then we have either

$$h'(t) < 0 \quad \text{for} \quad t < t' \quad \text{and} \quad h'(t) > 0 \quad \text{for} \quad t > t', \tag{23}$$

or

$$h'(t) > 0 \quad \text{for} \quad t < t' \quad \text{and} \quad h'(t) < 0 \quad \text{for} \quad t > t'. \tag{24}$$

For $h(0) < 0$, $h(T) > 0$, the figure of $h(t)$ satisfying (23) is shown in Fig. 3 and the figure of $h(t)$ satisfying (24) is shown in Fig. 4.

Combining the above with Fig. 3, $h(t) = 0$ has a unique solution in $(0, T)$; that is, $g_1'(t) = 0$ has a unique solution in $(0, T)$.

In conclusion, $g_1'(t) = 0$ has a unique solution in $(0, T)$. Moreover, from $g_1(0) = g_1(T) = 0$, $g_1'(0) < 0$ and $g_1'(T) > 0$, we have $g_1(t) < 0$ for $t \in (0, T)$. The figure of $g_1(t)$ is shown as Fig. 4.

From the above discussions, for a constant $T > 0$, we have

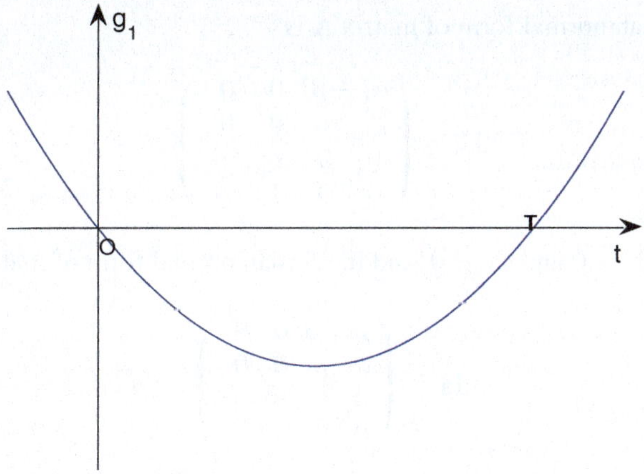

Fig. 4 The figure of g_1

$$\Phi_B(T, \mathbf{p}_1) = \mathbf{p}_0 \text{ and } \{\Phi_B(t, \mathbf{p}_1) \mid t \in (0, T)\} \subset \Sigma^+$$

if and only if condition (II) in Theorem 21.1 holds. Then, condition (4) holds.

Moreover, from inequalities in condition (II) and the first inequality in condition (III) in Theorem 21.1, we have

$$(1, 0, 0)^{\mathsf{T}}(\mathbf{A}\mathbf{p}_1 + \mathbf{a}) = -\lambda x_{1\mathbf{p}} > 0,$$

$$(1, 0, 0)^{\mathsf{T}}(\mathbf{A}\mathbf{p}_2 + \mathbf{a}) = -\alpha_1 x_{1\mathbf{p}} + \beta_1(k_1\zeta_{21} - k_2\zeta_{11}) < 0,$$

$$(1, 0, 0)^{\mathsf{T}}(\mathbf{B}\mathbf{p}_1 + \mathbf{b}) = (\xi_{11}, \xi_{21}, \xi_{31})\mathbf{J}(\rho_1, \rho_2, \rho_3)^{\mathsf{T}} > 0,$$

$$(1, 0, 0)^{\mathsf{T}}(\mathbf{B}\mathbf{p}_2 + \mathbf{b}) = (\xi_{11}, \xi_{21}, \xi_{31})\mathbf{J}(\sigma_1, \sigma_2, \sigma_3)^{\mathsf{T}} < 0.$$

Then, condition (5) holds.

In conclusion, conditions (1)–(5) hold if and only if conditions in Theorem 21.1 hold. The proof of Theorem 21.1 is thus completed.

2.1.2 Existence of Homoclinic Orbits to an Equilibrium Point with Two Pairs of Complex Eigenvalues in 4-Dimensional Piecewise Affine Systems

In this subsection, consider system (1) with $n = 4$, $\mathbf{c} = (-1, 0, 0, 0)^{\mathsf{T}}$ and $d = 0$, as follows:

$$\dot{\mathbf{x}} = \begin{cases} \mathbf{A}\mathbf{x} + \mathbf{a}, & x_1 > 0, \\ \mathbf{B}\mathbf{x} + \mathbf{b}, & x_1 \leq 0, \end{cases} \quad \mathbf{x} \in \mathbb{R}^4, \tag{25}$$

where the Jordan normal form of matrix **A** is

$$\mathbf{J_A} = \begin{pmatrix} \alpha_1 & -\beta_1 & 0 & 0 \\ \beta_1 & \alpha_1 & 0 & 0 \\ 0 & 0 & \alpha_2 & -\beta_2 \\ 0 & 0 & \beta_2 & \alpha_2 \end{pmatrix},$$

with $\alpha_2,\ \beta_1,\ \beta_2 > 0$ and $\alpha_1 < 0$, and the Jordan normal form of matrix **B** is

$$\mathbf{J_B} = \begin{pmatrix} \rho & -\sigma & 0 & 0 \\ \sigma & \rho & 0 & 0 \\ 0 & 0 & \rho & -\sigma \\ 0 & 0 & \sigma & \rho \end{pmatrix},$$

with $\sigma > 0$ and $\rho \in R$.

Then, the stable subspace $E^s(\mathbf{p})$ and unstable subspace $E^u(\mathbf{p})$ are given, respectively, as follows:

$$E^s(\mathbf{p}) = \{\mathbf{p} + k_1\zeta_1 + k_2\zeta_2 \mid k_1, k_2 \in \mathbb{R}\},$$

$$E^u(\mathbf{p}) = \{\mathbf{p} + k_3\zeta_3 + k_4\zeta_4 \mid k_3, k_4 \in \mathbb{R}\}.$$

Assume that

$$L_1 = E^s(\mathbf{p}) \cap \Sigma \neq \emptyset, \quad L_2 = E^u(\mathbf{q}) \cap \Sigma \neq \emptyset.$$

Then, we have

$$L_1 = \{\mathbf{p} + k_1\zeta_1 + k_2\zeta_2 \mid k_1\zeta_{11} + k_2\zeta_{21} = -x_{1\mathbf{p}}\},$$

$$L_2 = \{\mathbf{p} + k_3\zeta_3 + k_4\zeta_4 \mid k_3\zeta_{31} + k_4\zeta_{41} = -x_{1\mathbf{p}}\}.$$

For points $\mathbf{p}_1 \in L_1$ and $\mathbf{p}_2 \in L_2$, let

$$\mathbf{p}_1 - \mathbf{q} = \mathbf{Q}(l_1 \ \ l_2 \ \ l_3 \ \ l_4)^\top,$$

and

$$\mathbf{p}_2 - \mathbf{q} = \mathbf{Q}(l_1' \ \ l_2' \ \ l_3' \ \ l_4')^\top,$$

where the invertible matrix **Q** is defined in (5).

Similarly to the discussions on system (1) with $n = 3$ (system (8)) in Sect. 2.1.1, for the existence of a homoclinic orbit of system (25), satisfying all the above assumptions, we have the following theorem [8].

Theorem 21.2 *System (25) has a homoclinic orbit to the equilibrium point* $\mathbf{p}$*, which crosses* Σ *transversally at two points, if and only if there exist real numbers* k_i, $i = 1, 2, 3, 4$, *such that the following conditions hold:*

(I)

$$\mathbf{p}_1 = \mathbf{p} + k_1\zeta_1 + k_2\zeta_2 \in L_1,$$

$$\mathbf{p}_2 = \mathbf{p} + k_3\zeta_3 + k_4\zeta_4 \in L_2.$$

(II) In (11) and (12)

$$-x_{1\mathbf{q}}\rho + \sigma(\xi_{21}l_1 - \xi_{11}l_2 + \xi_{41}l_3 - \xi_{31}l_4) > 0,$$

$$-x_{1\mathbf{q}}\rho + \sigma(\xi_{21}l_1' - \xi_{11}l_2' + \xi_{41}l_3' - \xi_{31}l_4') < 0.$$

(III) If $x_{1\mathbf{q}} > 0$, *there is a constant* $0 < T_0 < \frac{\pi}{\sigma}$ *such that* $\Phi_B(T_0, \mathbf{p}_2) = \mathbf{p}_1$.
If $x_{1\mathbf{q}} = 0$, *we have* $\Phi_B(T_0, \mathbf{p}_2) = \mathbf{p}_1$ *for* $T_0 = \frac{\pi}{\sigma}$.
If $x_{1\mathbf{q}} < 0$, *there is a constant* $\frac{\pi}{\sigma} < T_0 < \frac{2\pi}{\sigma}$ *such that* $\Phi_B(T_0, \mathbf{p}_2) = \mathbf{p}_1$.

(IV)

$$-\alpha_1 x_{1\mathbf{p}} + \beta_1(k_1\zeta_{21} - k_2\zeta_{11}) > 0,$$

$$\alpha_2 x_{1\mathbf{p}} + \beta_2(k_4\zeta_{31} - k_3\zeta_{41}) > 0,$$

$$\sqrt{x_{1\mathbf{p}}^2 + (k_1\zeta_{21} - k_2\zeta_{11})^2}\frac{\beta_1}{\sqrt{\alpha_1{}^2 + \beta_1{}^2}}e^{\alpha_1 T_1} < x_{1\mathbf{p}},$$

$$\sqrt{x_{1\mathbf{p}}^2 + (k_4\zeta_{31} - k_3\zeta_{41})^2}\frac{\beta_2}{\sqrt{\alpha_2{}^2 + \beta_2{}^2}}e^{-\alpha_2 T_2} < x_{1\mathbf{p}},$$

where

$$T_1 = \frac{\pi}{\beta_1} + \frac{1}{\beta_1}\arctan\left(\frac{-\beta_1}{\alpha_1}\right) + \frac{1}{\beta_1}\arcsin\left(\frac{x_{1\mathbf{p}}}{\sqrt{x_{1\mathbf{p}}^2 + (k_1\zeta_{21} - k_2\zeta_{11})^2}}\right),$$

$$T_2 = \frac{\pi}{\beta_2} + \frac{1}{\beta_2}\arctan\left(\frac{\beta_2}{\alpha_2}\right) + \frac{1}{\beta_2}\arcsin\left(\frac{x_{1\mathbf{p}}}{\sqrt{x_{1\mathbf{p}}^2 + (k_3\zeta_{41} - k_4\zeta_{31})^2}}\right).$$

In addition, the homoclinic orbit crosses Σ *transversally at* $\mathbf{p}_1$ *and* $\mathbf{p}_2$.

Proof The method for the proof of Theorem 21.2 is similar to the proof of Theorem 21.1.

For more details about the proof of Theorem 21.2, we refer to [8].

2.1.3 Existence of Homoclinic Orbits to an Equilibrium Point with a Pair of Complex Eigenvalues and Two Real Eigenvalues in 4-Dimensional Piecewise-Affine Systems

This subsection considers system (1), with $n = 4$, $\mathbf{a} = \mathbf{b} = \mathbf{0}$, and $\mathbf{c} = (0, 1, 0, 1)^\mathsf{T}$, as follows:

$$\dot{\mathbf{x}} = \begin{cases} \mathbf{Ax}, & x_2 + x_4 \le 1, \\ \mathbf{Bx}, & x_2 + x_4 > 1, \end{cases} \quad \mathbf{x} = (x_1\ x_2\ x_3\ x_4)^\mathsf{T} \in \mathbb{R}^4, \tag{26}$$

where

$$\mathbf{A} = \begin{pmatrix} \alpha & -\beta & 0 & 0 \\ \beta & \alpha & 0 & 0 \\ 0 & 0 & \lambda & 0 \\ 0 & 0 & 0 & \rho \end{pmatrix}, \quad \mathbf{B} = \begin{pmatrix} \gamma & 0 & 0 & 0 \\ 0 & 0 & 0 & \delta \\ 0 & 0 & \mu & 0 \\ 0 & -\delta & 0 & 0 \end{pmatrix}$$

with

$$\beta > 0, \ \lambda\alpha\rho \neq 0, \ \gamma\delta\mu \neq 0.$$

The following theorem [9] provides some sufficient conditions on the existence of homoclinic orbits in system (26).

Theorem 21.3 *Suppose that $\alpha\rho < 0$, $\delta\rho > 0$ in system (26). Then, there exists a homoclinic orbit Γ connecting the origin to itself and intersecting the discontinuity manifold $\Sigma : x_2 + x_4 = 1$ at two points $\mathbf{p}_0 = (0, 1, 0, 0)^\mathsf{T}$ and $\mathbf{q}_0 = (0, 0, 0, 1)^\mathsf{T}$.*

Proof By direct calculations, we get

$$\Phi_A(t, \mathbf{x}_0) = \begin{pmatrix} e^{\alpha t}\cos\beta t & -e^{\alpha t}\sin\beta t & 0 & 0 \\ e^{\alpha t}\sin\beta t & e^{\alpha t}\cos\beta t & 0 & 0 \\ 0 & 0 & e^{\lambda t} & 0 \\ 0 & 0 & 0 & e^{\rho t} \end{pmatrix} \cdot \mathbf{x}_0, \tag{27}$$

and

$$\Phi_B(t, \mathbf{x}_0) = \begin{pmatrix} e^{\gamma t} & 0 & 0 & 0 \\ 0 & \cos\delta t & 0 & \sin\delta t \\ 0 & 0 & e^{\mu t} & 0 \\ 0 & -\sin\delta t & 0 & \cos\delta t \end{pmatrix} \cdot \mathbf{x}_0. \tag{28}$$

Note that, when $\alpha < 0,\ \rho > 0,\ \delta > 0\,(\alpha > 0,\ \rho < 0,\ \delta < 0)$, by (27) we obtain $\Phi_A(t, \mathbf{p}_0) \to \mathbf{0}$ as $t \to +\infty$ $(t \to -\infty)$ and $\Phi_A(t, \mathbf{q}_0) \to \mathbf{0}$ as $t \to -\infty$ $(t \to +\infty)$. In addition to this, by (27), we also have $e^{\alpha t}\cos\beta t < 1$ $(e^{\rho t} < 1)$ when $t > 0$ and $e^{\rho t} < 1$ $(e^{\alpha t}\cos\beta t < 1)$ when $t < 0$, i.e., $\Phi_A(t, \mathbf{p}_0), t \in (0, +\infty) (t \in (-\infty, 0))$ and $\Phi_A(t, \mathbf{q}_0), t \in (-\infty, 0) (t \in (0, +\infty))$ will stay in the action area of the half system $\dot{\mathbf{x}} = \mathbf{Ax}$.

Moreover, by (28), we have $\Phi_B(\pi/2\delta, \mathbf{q}_0) = \mathbf{p}_0$ and $\sin\delta t + \cos\delta t > 1$ when $\delta > 0,\ t \in (0, \frac{\pi}{2\delta})$, and $\cos\delta t - \sin\delta t > 1$ when $\delta < 0,\ t \in (0, -\frac{\pi}{2\delta})$, i.e., the trajectory of the half system $\dot{\mathbf{x}} = \mathbf{B}\mathbf{x}$ connecting the points $\mathbf{p}_0$ and $\mathbf{q}_0$ always stays in its action area.

So, the statements are proved.

2.2 *Existence of Heteroclinic Cycles*

2.2.1 Existence of Heteroclinic Cycles to Two Saddle-Focus in 3-Dimensional Piecewise-Affine System

Consider system (1) with $n = 3$, $d = 1$ and $\mathbf{c} = (2, 0, 1)^{\mathsf{T}}$, as follows:

$$\dot{\mathbf{x}} = \begin{cases} \mathbf{A}\mathbf{x} + \mathbf{a}, & 2x_1 + x_3 < 1, \\ \mathbf{B}\mathbf{x} + \mathbf{b}, & 2x_1 + x_3 \geq 1, \end{cases} \quad \mathbf{x} = (x_1, x_2, x_3)^{\mathsf{T}} \in \mathbb{R}^3, \tag{29}$$

where the Jordan normal forms of $\mathbf{A}$ and $\mathbf{B}$ are

$$\mathbf{J_A} = \begin{pmatrix} \alpha_A & -\beta_A & 0 \\ \beta_A & \alpha_A & 0 \\ 0 & 0 & \lambda_A \end{pmatrix}, \quad \mathbf{J_B} = \begin{pmatrix} \alpha_B & -\beta_B & 0 \\ \beta_B & \alpha_B & 0 \\ 0 & 0 & \lambda_B \end{pmatrix},$$

respectively, with $\lambda_A,\ \lambda_B < 0,\ \alpha_A,\ \alpha_B,\ \beta_A,\ \beta_B$.

The stable manifolds $W^s(\mathbf{p}),\ W^s(\mathbf{q})$ and the unstable manifolds $W^u(\mathbf{p}),\ W^u(\mathbf{q})$ are shown as follows:

$$W^s(\mathbf{p}) = \{\mathbf{p} + \mathbf{x} \mid \mathbf{x} \in span\{\zeta_3\}\},$$

$$W^s(\mathbf{q}) = \{\mathbf{q} + \mathbf{x} \mid \mathbf{x} \in span\{\xi_3\}\},$$

$$W^u(\mathbf{p}) = \{\mathbf{p} + \mathbf{x} \mid \mathbf{x} \in span\{\zeta_1, \zeta_2\}\},$$

$$W^u(\mathbf{q}) = \{\mathbf{q} + \mathbf{x} \mid \mathbf{x} \in span\{\xi_1, \xi_2\}\}.$$

Suppose that the intersections $W^u(\mathbf{p}) \cap \Sigma$, $W^s(\mathbf{p}) \cap \Sigma$, $W^u(\mathbf{q}) \cap \Sigma$ and $W^s(\mathbf{q}) \cap \Sigma$ are not empty. Then, $W^u(\mathbf{p}) \cap \Sigma$ is a line and $W^s(\mathbf{p}) \cap \Sigma$ is a point, $W^u(\mathbf{q}) \cap \Sigma$ is a line and $W^s(\mathbf{q}) \cap \Sigma$ is a point. Let $\mathbf{p}_0 = W^s(\mathbf{p}) \cap \Sigma$ and $\mathbf{q}_0 = W^s(\mathbf{q}) \cap \Sigma$. Then, we have

$$\mathbf{p}_0 = \mathbf{p} + k\zeta_3, \quad \mathbf{q}_0 = \mathbf{q} + l\xi_3,$$

with $k = \frac{1-2x_{1\mathbf{p}}-x_{3\mathbf{p}}}{2\zeta_{31}+\zeta_{33}}$ and $l = \frac{1-2x_{1\mathbf{q}}-x_{3\mathbf{q}}}{2\xi_{31}+\xi_{33}}$.

For the existence of a heteroclinic cycle in system (29), we have the following theorem [10].

Theorem 21.4 *Suppose that the following conditions hold:*

(I) There exist constant numbers $k_{10}, k_{20}, l_{10}, l_{20} \in R$, such that

$$\mathbf{p}_0 = \mathbf{q} + k_{10}\xi_1 + k_{20}\xi_2,$$

$$\mathbf{q}_0 = \mathbf{p} + l_{10}\zeta_1 + l_{20}\zeta_2,$$

(II)

$$\alpha_A\rho_1 + \beta_A\sigma_1 > 0, \quad \frac{\beta_A\sqrt{\rho_1^2+\sigma_1^2}}{\sqrt{\alpha_A^2+\beta_A^2}} e^{\frac{-T\alpha_A}{\beta_A}} < 1 - 2x_{1\mathbf{p}} - x_{3\mathbf{p}},$$

$$\alpha_B\rho_2 + \beta_B\sigma_2 < 0, \quad \frac{\beta_B\sqrt{\rho_2^2+\sigma_2^2}}{\sqrt{\alpha_B^2+\beta_B^2}} e^{\frac{-\bar{T}\alpha_B}{\beta_B}} < 2x_{1\mathbf{q}} + x_{3\mathbf{q}} - 1,$$

where

$$\rho_1 = 2\zeta_{11}l_{10} + 2\zeta_{21}l_{20} + \zeta_{13}l_{10} + \zeta_{23}l_{20},$$

$$\sigma_1 = -2\zeta_{11}l_{20} + 2\zeta_{21}l_{10} - \zeta_{13}l_{20} + \zeta_{23}l_{10},$$

$$\rho_2 = 2\xi_{11}k_{10} + 2\xi_{21}k_{20} + \xi_{13}k_{10} + \xi_{23}k_{20},$$

$$\sigma_2 = -2\xi_{11}k_{20} + 2\xi_{21}k_{10} - \xi_{13}k_{20} + \xi_{23}k_{10},$$

$$T = \frac{1}{\beta_A}\left[\arcsin\frac{\rho_1}{\sqrt{\rho_1^2+\sigma_1^2}} + \arctan\frac{\beta_A}{\alpha_A} + \pi\right],$$

$$\bar{T} = \frac{1}{\beta_B}\left[\arcsin\frac{\rho_2}{\sqrt{\rho_2^2+\sigma_2^2}} + \arctan\frac{\beta_B}{\alpha_B} + \pi\right].$$

Then, system (29) has a heteroclinic cycle connecting the equilibrium points $\mathbf{p}$ *and* $\mathbf{q}$. *In addition, the heteroclinic cycle crosses Σ transversally at* $\mathbf{p}_0$ *and* $\mathbf{q}_0$.

Proof The method for the proof of this theorem is also similar to the proof of Theorem 21.1. We refer to [10] for a detailed proof.

2.2.2 The Existence of Bifocal Heteroclinic Cycles in a Class of Four-Dimensional Piecewise-Affine Systems

Consider system (1), with $n = 4$, $d = 0$ and $\mathbf{c} = (1, 0, 0, 0)^\mathsf{T}$, as follows:

$$\dot{\mathbf{x}} = \begin{cases} \mathbf{A}\mathbf{x} + \mathbf{a}, & x_1 > 0, \\ \mathbf{B}\mathbf{x} + \mathbf{b}, & x_1 \le 0, \end{cases} \qquad \mathbf{x} = (x_1, x_2, x_2, x_3)^{\mathsf{T}} \in \mathbb{R}^n, \tag{30}$$

with the Jordan normal forms of matrices **A** and **B** being

$$\mathbf{J_A} = \begin{pmatrix} \alpha_1 & -\beta_1 & 0 & 0 \\ \beta_1 & \alpha_1 & 0 & 0 \\ 0 & 0 & \alpha_2 & -\beta_2 \\ 0 & 0 & \beta_2 & \alpha_2 \end{pmatrix}, \quad \mathbf{J_B} = \begin{pmatrix} \rho_1 & -\sigma_1 & 0 & 0 \\ \sigma_1 & \rho_1 & 0 & 0 \\ 0 & 0 & \rho_2 & -\sigma_2 \\ 0 & 0 & \sigma_2 & \rho_2 \end{pmatrix},$$

respectively. Then, the stable manifolds $W^s(\mathbf{p})$, $W^s(\mathbf{q})$ and unstable manifolds $W^u(\mathbf{p})$, $W^u(\mathbf{q})$ are shown as follows:

$$W^s(\mathbf{p}) = \{\mathbf{p} + k_1\zeta_1 + k_2\zeta_2 \mid k_1, k_2 \in R\},$$

$$W^u(\mathbf{p}) = \{\mathbf{p} + k_3\zeta_3 + k_4\zeta_4 \mid k_3, k_4 \in R\},$$

$$W^s(\mathbf{q}) = \{\mathbf{q} + l_1\xi_1 + l_2\xi_2 \mid l_1, l_2 \in R\},$$

$$W^u(\mathbf{q}) = \{\mathbf{q} + l_3\xi_3 + l_4\xi_4 \mid l_3, l_4 \in R\}.$$

Suppose that

$$L_1 = W^s(\mathbf{p}) \cap \Sigma, \quad L_2 = W^s(\mathbf{q}) \cap \Sigma,$$

$$L_3 = W^u(\mathbf{p}) \cap \Sigma, \quad L_4 = W^u(\mathbf{q}) \cap \Sigma,$$

$$L_1 \cap L_4 \neq \emptyset, \quad L_2 \cap L_3 \neq \emptyset.$$

Thus, L_i is clearly a line for $i = 1, 2, 3, 4$.

For the existence of a heteroclinic cycle, we have the following theorem [11].

Theorem 21.5 *System (30) has a heteroclinic cycle connecting equilibrium points* **p** *and* **q**, *which crosses Σ transversally at two points, if and only if there exist constant real numbers k_i, l_i, $i = 1, 2, 3, 4$, such that the following conditions hold:*

(I)

$$\mathbf{p}_1 = \mathbf{p} + k_1\zeta_1 + k_2\zeta_2 = \mathbf{q} + l_3\xi_3 + l_4\xi_4 \in \Sigma,$$

$$\mathbf{p}_2 = \mathbf{p} + k_3\zeta_3 + k_4\zeta_4 = \mathbf{q} + l_1\xi_1 + l_2\xi_2 \in \Sigma.$$

(II)

$$-\alpha_1 x_{1\mathbf{p}} + \beta_1(k_1\zeta_{21} - k_2\zeta_{11}) > 0,$$

$$\alpha_2 x_{1\mathbf{p}} + \beta_2(k_4\zeta_{31} - k_3\zeta_{41}) > 0,$$

$$-\rho_1 x_{\mathbf{1q}} + \sigma_1(l_1\xi_{21} - l_2\xi_{11}) < 0,$$

$$\rho_2 x_{\mathbf{1q}} + \sigma_2(l_4\xi_{31} - l_3\xi_{41}) < 0,$$

$$\sqrt{x_{\mathbf{1p}}^2 + (k_1\zeta_{21} - k_2\zeta_{11})^2}\frac{\beta_1}{\sqrt{{\alpha_1}^2 + {\beta_1}^2}}e^{\alpha_1 T_1} < x_{\mathbf{1p}},$$

$$\sqrt{x_{\mathbf{1p}}^2 + (k_4\zeta_{31} - k_3\zeta_{41})^2}\frac{\beta_2}{\sqrt{{\alpha_2}^2 + {\beta_2}^2}}e^{-\alpha_2 T_2} < x_{\mathbf{1p}},$$

$$\sqrt{x_{\mathbf{1q}}^2 + (l_1\xi_{21} - l_2\xi_{11})^2}\frac{\rho_1}{\sqrt{{\rho_1}^2 + {\sigma_1}^2}}e^{\rho_1 T_3} < -x_{\mathbf{1q}},$$

$$\sqrt{x_{\mathbf{1q}}^2 + (l_4\xi_{31} - l_3\xi_{41})^2}\frac{\rho_2}{\sqrt{{\rho_2}^2 + {\sigma_2}^2}}e^{-\rho_2 T_4} < -x_{\mathbf{1q}},$$

where

$$T_1 = \frac{\pi}{\beta_1} + \frac{1}{\beta_1}\arctan\left(\frac{-\beta_1}{\alpha_1}\right) + \frac{1}{\beta_1}\arcsin\left(\frac{x_{\mathbf{1p}}}{\sqrt{x_{\mathbf{1p}}^2 + (k_1\zeta_{21} - k_2\zeta_{11})^2}}\right),$$

$$T_2 = \frac{\pi}{\beta_2} + \frac{1}{\beta_2}\arctan\left(\frac{\beta_2}{\alpha_2}\right) + \frac{1}{\beta_2}\arcsin\left(\frac{x_{\mathbf{1p}}}{\sqrt{x_{\mathbf{1p}}^2 + (k_3\zeta_{41} - k_4\zeta_{31})^2}}\right),$$

$$T_3 = \frac{\pi}{\sigma_1} + \frac{1}{\sigma_1}\arctan\left(\frac{-\sigma_1}{\rho_1}\right) + \frac{1}{\sigma_1}\arcsin\left(\frac{-x_{\mathbf{1q}}}{\sqrt{x_{\mathbf{1q}}^2 + (l_1\xi_{21} - l_2\xi_{11})^2}}\right),$$

$$T_4 = \frac{\pi}{\sigma_2} + \frac{1}{\sigma_2}\arctan\left(\frac{\sigma_2}{\rho_2}\right) + \frac{1}{\sigma_2}\arcsin\left(\frac{-x_{\mathbf{1q}}}{\sqrt{x_{\mathbf{1q}}^2 + (l_3\xi_{41} - l_4\zeta_{31})^2}}\right).$$

In addition, the heteroclinic cycle crosses Σ transversally at $\mathbf{p}_1$ and $\mathbf{p}_2$.

Proof The proof of this theorem is similar to that for Theorems 21.1–21.4, but more complex and skillful, referring to [11] for details.

In Theorems 21.1–21.5, we study the existence of homoclinic orbits or heteroclinic cycles in various 3-dimensional or 4-dimensional piecewise-affine systems. The conditions in these theorems are a bit complicated but straightforward. The methods for the proofs of these theorems are based on conventional analyses. In fact, from these proofs, it is not hard to see that the key of these results is to ensure that the space positional relations among the stable manifolds of subsystems, the unstable manifolds of subsystems and the switching manifold, all satisfy some conditions. Notice that the stable (unstable) manifolds and the switching manifold are all lower-dimensional than the state variables of the systems.

Along the same way, in the next subsection, we will further study the existence of more types of heteroclinic cycles in some 3-dimensional piecewise-affine systems based on the plane positional relations between a trajectory of a plane linear system and a fixed line. Some more concise necessary and sufficient conditions for the existence of three types of heteroclinic cycles will be obtained.

2.2.3 The Existence of Three Types of Heteroclinic Cycles in a Class of 3-Dimensional Piecewise-Affine Systems Based on Two Results from Planar Linear Systems

All contents of this subsection are taken from [12].

We firstly introduce two results on the planar linear system given by

$$\dot{\mathbf{x}} = \mathrm{A}_0\mathbf{x}, \tag{31}$$

where $\mathbf{x} \in \mathbb{R}^2$. For $\mathbf{x}_0 \in \mathbb{R}^2$, we denote the positive semi-orbit of $\mathbf{x}_0$ by $O_+(\mathbf{x}_0)$, i.e.,

$$O_+(\mathbf{x}_0) = \{\Phi(t, \mathbf{x}_0)|t > 0\} = \{\exp(\mathrm{A}_0 t)\mathbf{x}_0|t > 0\},$$

where $\Phi(t, \cdot)$ denotes the flow generated by (31).

Let $L = \{\mathbf{x} \in \mathbb{R}^2|\mathbf{k}^T\mathbf{x} = c\}$, with $0 < c \in \mathbb{R}$ and $\mathbf{0} \neq \mathbf{k} = (k_1, k_2)^T \in \mathbb{R}^2$. Then, L is a straight line not passing through the origin, which divides the plane into three disjoint subsets, L, L_+ and L_-, where

$$L_+ = \{\mathbf{x} \in \mathbb{R}^2|\mathbf{k}^T\mathbf{x} > c\},\ L_- = \{\mathbf{x} \in \mathbb{R}^2|\mathbf{k}^T\mathbf{x} < c\}.$$

Obviously, the origin is in L_-.

Lemma 1 *For system (31), suppose that the eigenvalues of* A_0 *are given by* $\mu_{1,2} < 0$ *and* $\boldsymbol{x}_0 \in L$. *Then,*

$$O_+(\boldsymbol{x}_0) \subset L_- \text{ if and only if } \boldsymbol{k}^T\mathrm{A}_0\boldsymbol{x}_0 \leq 0.$$

Proof Here, we only present the proof of Lemma 1 under the assumption that A_0 is given by $\mathrm{N} : \begin{pmatrix} \mu_1 & 0 \\ 0 & \mu_2 \end{pmatrix} \mathrm{N}^{-1}$ with N being an invertible matrix.

If A_0 is given by $\mathrm{N}\begin{pmatrix} \mu_1 & 1 \\ 0 & \mu_2 \end{pmatrix} \mathrm{N}^{-1}$ with $\mu_1 = \mu_2$, the proof is similar.

(a) *Proof of necessity*

Let $g(t) = \mathbf{k}^T\exp(\mathrm{A}_0 t)\mathbf{x}_0$. Obviously, g is a smooth function with

$$g'(t) = \mathbf{k}^T\exp(\mathrm{A}_0 t)\mathrm{A}_0\mathbf{x}_0. \tag{32}$$

Moreover, $g(0) = c$ and $g(t) < c(t > 0)$ imply $g'(0) = \mathbf{k}^T \mathrm{A}_0 \mathbf{x}_0 \leq 0$.
(b) *Proof of sufficiency*
From (32), we have $g'(t) = \mathbf{k}^T \mathrm{N} \begin{pmatrix} e^{\mu_1 t} & 0 \\ 0 & e^{\mu_2 t} \end{pmatrix} \mathrm{N}^{-1} \mathrm{A}_0 \mathbf{x}_0$, which is actually a linear combination of $e^{\mu_1 t}$ and $e^{\mu_2 t}$; that is,

$$g'(t) = \sigma_1 e^{\mu_1 t} + \sigma_2 e^{\mu_2 t}, \tag{33}$$

where $\sigma_{1,2} \in \mathbb{R}$. Obviously, $g(t) \to 0$ and $g'(t) \to 0$ as $t \to +\infty$.

Note that σ_1 and σ_2 are not both zero. In fact, if σ_1 and σ_2 are both zero, i.e., $g'(t) \equiv 0$, then $\mathrm{N} \begin{pmatrix} e^{\mu_1 t} & 0 \\ 0 & e^{\mu_2 t} \end{pmatrix} \mathrm{N}^{-1} \mathrm{A}_0 \mathbf{x}_0 \in L$ for any $t \in \mathbb{R}$, which is contradictory to the fact that $\mathrm{N} \begin{pmatrix} e^{\mu_1 t} & 0 \\ 0 & e^{\mu_2 t} \end{pmatrix} \mathrm{N}^{-1} \mathrm{A}_0 \mathbf{x}_0 \to 0$ as $t \to +\infty$.

(i) $\mathbf{k}^T A_0 \mathbf{x}_0 = 0$. By (33), we know that 0 is the unique value such that $g'(t) = 0$. We affirm here that $g(t) < c$ for all $0 < t < +\infty$. In fact, if there exists $t_0 \in (0, +\infty)$ such that $g(t_0) \geq c$, then, since $g(0) = c$ and $g(t) \to 0$ $(t \to +\infty)$, there exists $t_1 \in (0, +\infty)$ satisfying $g'(t_1) = 0$, which is impossible.

(ii) $\mathbf{k}^T A_0 \mathbf{x}_0 < 0$. Then, $g'(0) < 0$. From (33), there exists a unique $t = t_* \in \mathbb{R}$ such that $g'(t_*) = 0$. Thus, $t_* \neq 0$ and $g'(t)$ has a fixed sign in interval $(-\infty, t_*)$ or $(t_*, +\infty)$.

If $t_* < 0$, then, since $g'(0) < 0$, we have $g'(t) < 0$ for $t \in (t_*, +\infty)$, which implies that $g(t) = \mathbf{k}^T \exp(\mathrm{A}_0 t) \mathbf{x}_0 < g(0) = c$ for $t \in (0, +\infty)$.

If $t_* > 0$, then, since $g'(0) < 0$, we have $g'(t) < 0$ for $t \in (-\infty, t_*)$. Therefore, for any $t \in (0, t_*]$, we have $g(t) < g(0) = c$.

We affirm here that, in this case, $g(t) < c$ for all $t_* < t < +\infty$. In fact, if there exists $t_2 \in (t_*, +\infty)$ such that $g(t_2) > c$, then, since $g(t_*) < c$ and $g(t) \to 0$ $(t \to +\infty)$, there exists $t_3 \in (t_*, +\infty)$ satisfying $g'(t_3) = 0$, which is impossible. □

Remark 1 From the proof of Lemma 1, for any $\mathbf{x}_0 \in L$, if $\mathbf{k}^T \mathrm{A}_0 \mathbf{x}_0 < 0$, then $O(\mathbf{x}_0)$ intersects L transversally at $\mathbf{x}_0$. Here, $O(\mathbf{x}_0)$ means the whole orbit with initial condition $\mathbf{x}_0$.

Lemma 2 *For system (31), suppose that the eigenvalues of* A_0 *are given by*

$$\alpha \pm \beta j$$

with $\alpha < 0$, $\beta > 0$ *and* $j = \sqrt{-1}$. *Let*

$$\boldsymbol{x}_* = \frac{c}{\boldsymbol{k}^T \mathrm{A}_0^{-1} \boldsymbol{k}^{\perp}} \mathrm{A}_0^{-1} \boldsymbol{k}^{\perp}, \tag{34}$$

with $\boldsymbol{k}^{\perp} = (-k_2, k_1)^T$.

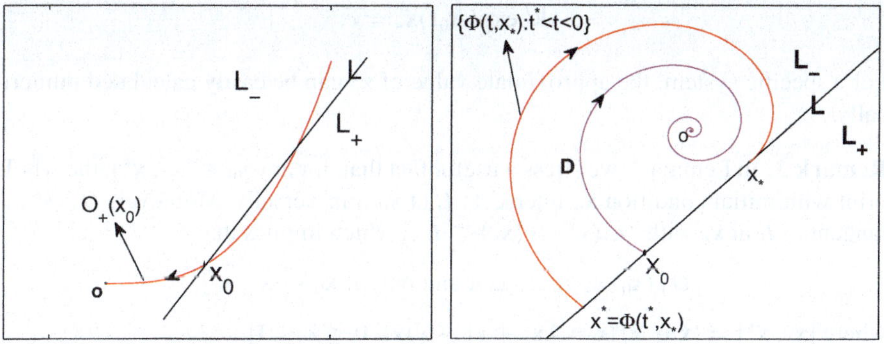

Fig. 5 Illustrations for Lemmas 1 and 2

Obviously, $\mathbf{x}_* \in L$. *Moreover, denote by* $\boldsymbol{x}^*$ *the first intersection of the flow with initial condition* $\mathbf{x}_*$ *under negative flight time with straight line* L *(see Fig. 5b). Then, for* $\boldsymbol{x}_0 \in L$,

$$O_+(\boldsymbol{x}_0) \subset L_- \text{ if and only if } \boldsymbol{x}_0 \in (\boldsymbol{x}_*, \boldsymbol{x}^*) \cup \{\mathbf{x}_*\},$$

where $(\boldsymbol{x}_*, \boldsymbol{x}^*)$ *denotes* $\{\boldsymbol{x} \in \mathbb{R}^2 | \boldsymbol{x} = \lambda \boldsymbol{x}_* + (1-\lambda)\boldsymbol{x}^*, 0 < \lambda < 1\}$.

Proof
(a) *Proof of sufficiency*
Simple calculations show that $\mathbf{x}_* = \frac{c}{\mathbf{k}^T \mathrm{A}_0^{-1} \mathbf{k}^\perp} \mathrm{A}_0^{-1} \mathbf{k}^\perp$ is the unique point at which the sector field is tangent to L. Obviously the origin O is a stable focus belonging to L_-. It is easy to see that $\{\Phi(t, x_*) | t^* < t < 0\} \subset L_-$, where t^* is the negative flight time from $\mathbf{x}_*$ to $\mathbf{x}^*$ (Fig. 5b).

Now, consider the region D surrounded by the union set of $\{\mathbf{x}^*\}$, $\{\Phi(t, \mathbf{x}_*) | t^* < t < 0\}$, $\{\mathbf{x}_*\}$ and $(\mathbf{x}_*, \mathbf{x}^*)$. Obviously, the interior of D belongs to L_-. Moreover, for any $\mathbf{x}_0 \in (\mathbf{x}^*, \mathbf{x}_*)$, the vector field at $\mathbf{x}_0$ must be transversal to the direction of L and point to the interior of D. If not, then there exists $\mathbf{x}_1 \in (\mathbf{x}^*, \mathbf{x}_*)$, at which the sector field points to L^+. Then, there exists $\mathbf{x}_2 \in (\mathbf{x}_1, \mathbf{x}^*) \subset (\mathbf{x}_*, \mathbf{x}^*)$, at which the sector field is tangent to L. But this is a contradiction with the fact that $\mathbf{x}_*$ is the unique point, at which the sector field is tangent to L.

From the above discussions, it is easy to see that D is positively invariant. Thus, $O_+(x_0) \subset L_-$ for any $\mathbf{x}_0 \in (\mathbf{x}_*, \mathbf{x}^*) \cup \{\mathbf{x}_*\}$.
(b) *Proof of necessity*
By the detailed proof of sufficiency, the proof of necessity is trivial, which is omitted.
□

Remark 2 In Lemma 2, although it is impossible to give an explicit formula for $\mathbf{x}^*$, for specific systems, we can first calculate $\mathbf{x}_*$ by (34) explicitly. Then, it can be seen that $\mathbf{x}^*$ is equivalent to $\exp(\mathrm{A}_0 t_*)\mathbf{x}_*$, where t_* is the largest negative solution of the following implicit equation of t:

$$\mathbf{k}^T \exp(A_0 t)\mathbf{x}_* = c.$$

For a specific system, the approximate value of $\mathbf{x}^*$ can be easily calculated numerically.

Remark 3 In Lemma 2, we stress a useful fact that, for any $\mathbf{x}_0 \in (\mathbf{x}_*, \mathbf{x}^*)$, the whole orbit with initial condition $\mathbf{x}_0$ intersects L at $\mathbf{x}_0$ transversally. Moreover, $O_+(\mathbf{x}^*)$ is tangent to L at $\mathbf{x}_*$ with $O_+(\mathbf{x}^*) - \{\mathbf{x}_*\} \subset L_-$, which implies that

$$O_+(\mathbf{x}_0) \subset L_- \cup L \text{ if and only if } \mathbf{x}_0 \in [\mathbf{x}_*, \mathbf{x}^*],$$

where $[\mathbf{x}_*, \mathbf{x}^*] = \{\mathbf{x} \in \mathbb{R}^2 | \mathbf{x} = \lambda\mathbf{x}_* + (1-\lambda)\mathbf{x}^*, 0 \leq \lambda \leq 1\}$.

Now, consider system (1) with $n = 3$, $\mathbf{c} = (c_1, c_2, 1)^\mathsf{T}$, as follows [12]:

$$\dot{\mathbf{x}} = \begin{cases} \mathbf{Ax} + \mathbf{a}, & \mathbf{c}^T\mathbf{x} \leq d \\ \mathbf{Bx} + \mathbf{b}, & \mathbf{c}^T\mathbf{x} > d \end{cases}, \mathbf{x} = (x_1, x_2, x_3)^\mathsf{T} \in \mathbb{R}^n, \tag{35}$$

with the invertible matrix

$$\mathbf{A} = \begin{pmatrix} a_{11} & a_{12} & 0 \\ a_{21} & a_{22} & 0 \\ 0 & 0 & \lambda_A \end{pmatrix}, \quad \mathbf{B} = \begin{pmatrix} b_{11} & b_{12} & 0 \\ b_{21} & b_{22} & 0 \\ 0 & 0 & \lambda_B \end{pmatrix}.$$

To investigate the existence of heteroclinic cycles in system (35), it should be clear that we only need to study the following three possibilities for the eigenvalues associated with A and B.

(E1) the eigenvalues of A: $\lambda_A, \lambda_1, \lambda_2$ with $\lambda_A > 0, \lambda_{1,2} < 0$; the eigenvalues of B: λ_B, μ_1, μ_2 with $\lambda_B > 0, \mu_{1,2} < 0$.

(E2) the eigenvalues of A: $\lambda_A, \alpha \pm \beta j$ with $\lambda_A > 0, \alpha < 0, \beta > 0$; the eigenvalues of B: λ_B, μ_1, μ_2 with $\lambda_B > 0, \mu_{1,2} < 0$.

(E3) the eigenvalues of A: $\lambda_A, \alpha \pm \beta j$ with $\lambda_A > 0, \alpha < 0, \beta > 0$; the eigenvalues of B: $\lambda_B, \rho \pm \omega j$ with $\lambda_B > 0, \rho < 0, \omega > 0$.

The heteroclinic cycles corresponding to the above (E1), (E2) and (E3) can be illustrated by Fig. 6a–c, respectively.

All other possibilities for the eigenvalues of A and B follow from the above three cases via time reversal or by exchanging the positions of A and B.

Now, we state the main results of this subsection.

Theorem 21.6 *For system (35) under the eigenvalues condition (E1), suppose the following conditions hold:*

(i) $\mathbf{c}^T\boldsymbol{p} < d, \mathbf{c}^T\boldsymbol{q} > d;$
(ii) $\mathbf{c}^T\mathbf{p}_0 = \mathbf{c}^T\mathbf{q}_0 = d;$
(iii) $\mathbf{c}^T\mathbf{A}(\boldsymbol{q}_0 - \boldsymbol{p}) \leq 0, \mathbf{c}^T\mathbf{B}(\boldsymbol{p}_0 - \boldsymbol{q}) \geq 0.$

Then, there exists a heteroclinic cycle Γ connecting $\mathbf{p}$ *and* $\mathbf{q}$*, which intersects the switching plane Σ at only two points,* $\mathbf{p}_0$ *and* $\mathbf{q}_0$*, see Fig. 6a.*

Moreover, assume that the above condition (iii) is substituted by the following more rigorous condition:

(iii') $c^T\mathbf{A}(q_0 - p) < 0$, $c^T\mathbf{B}(p_0 - q) > 0$.

Then, Γ *crosses* Σ *transversally at* $\mathbf{p}_0$ *and* $\mathbf{q}_0$.

Theorem 21.7 *For system (35) with the eigenvalues condition (E2), suppose the following conditions hold:*

(i) $c^T p < d$, $c^T q > d$;
(ii) $c^T p_0 = c^T q_0 = d$;
(iii) $c^T\mathbf{B}(p_0 - q) \geq 0$, $q_0 \in [x_*, x^*]$;

where $x_* = \frac{d - c^T p}{c^T \mathbf{A}^{-1} c^{\perp}} \mathbf{A}^{-1} c^{\perp} + p$, *with* $c^{\perp} = (c_2, -c_1, 0)^T$, $[x_*, x^*] = \{x \in \mathbb{R}^2 | x = \lambda x_* + (1 - \lambda)x^*, 0 \leq \lambda \leq 1\}$, *with* x^* *denoting the first intersection of the flow of system (35), with initial condition* x_*, *with straight line* $L = \{x \in \mathbb{R}^3 | c^T x = d\} \cap \{x \in \mathbb{R}^3 | z = z_p\}$ *under negative flight time. Then, there exists a heteroclinic cycle* Γ *connecting* $\mathbf{p}$ *and* $\mathbf{q}$ *together, see Fig. 6b.*

Moreover, assume that the above condition (iii) is substituted by the following more rigorous condition:

(iii') $c^T\mathbf{B}(p_0 - q) > 0$, $q_0 \in (x_*, x^*)$;

Then, Γ *intersects the switching plane* Σ *at only two points,* p_0 *and* q_0, *transversally.*

Theorem 21.8 *For system (35) with the eigenvalues condition (E3), suppose the following conditions hold:*

(i) $c^T p < d$, $c^T q > d$;
(ii) $c^T p_0 = c^T q_0 = d$;
(iii) $p_0 \in [\hat{x}_*, \hat{x}^*]$, $q_0 \in [\check{x}_*, \check{x}^*]$;

where

$$\hat{x}_* = \frac{d - c^T q}{c^T \mathbf{A}^{-1} c^{\perp}} \mathbf{A}^{-1} c^{\perp} + q \text{ and } \check{x}_* = \frac{d - c^T p}{c^T \mathbf{A}^{-1} c^{\perp}} \mathbf{A}^{-1} c^{\perp} + p,$$

with $c^{\perp} = (c_2, -c_1, 0)^T$, $\hat{x}^*$ *denotes the first intersection of the flow of system (35), with initial condition* $\hat{x}_*$, *under negative flight time with straight line* $L = \Sigma \cap \{x \in \mathbb{R}^3 | z = z_q\}$, *and* $\check{x}^*$ *denotes the first intersection of the flow of system (35), with initial condition* $\check{x}_*$, *under negative flight time with straight line* $L = \Sigma \cap \{x \in \mathbb{R}^3 | z = z_p\}$. *Then, there exists a heteroclinic cycle* Γ *connecting* $\mathbf{p}$ *and* $\mathbf{q}$, *see Fig. 6c.*

Moreover, assume that the above condition (iii) is substituted by the following more rigorous condition:

(iii') $p_0 \in (\hat{x}_*, \hat{x}^*)$, $q_0 \in (\check{x}_*, \check{x}^*)$;

Then, Γ *intersects the switching plane* Σ *at only two points,* p_0 *and* q_0, *transversally.*

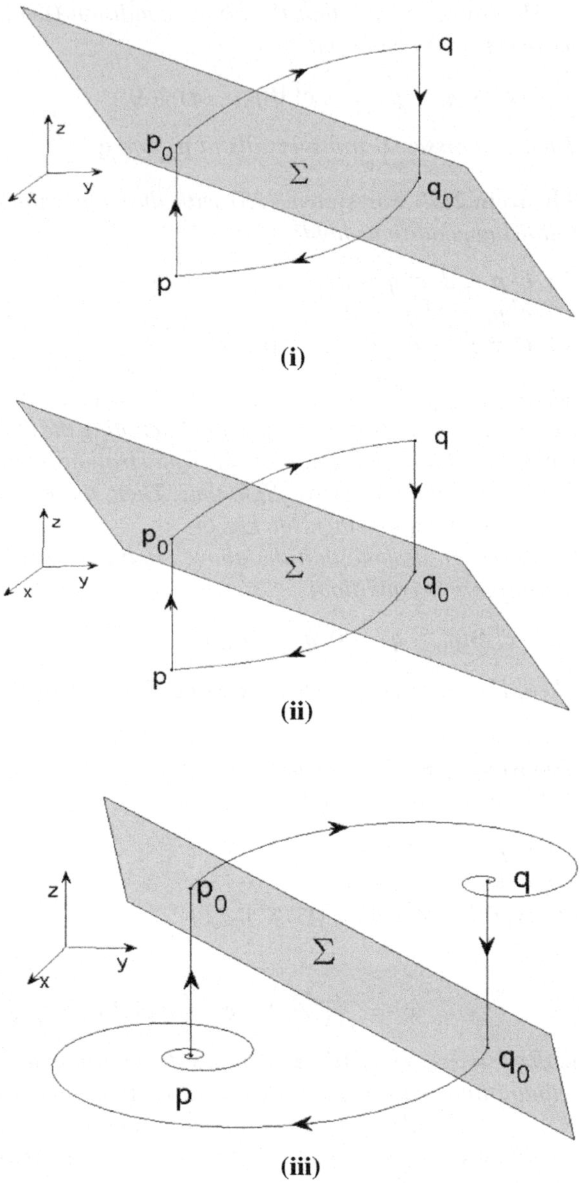

Fig. 6 (i) Heteroclinic cycle to two saddle points with purely real eigenvalues, (ii) Heteroclinic cycle to two saddle points, one is saddle point with purely real eigenvalues and the other is a saddle-focus, (iii) Heteroclinic cycle to two saddle-foci

Remark 4 From the proof to be presented in the next subsection, the heteroclinic cycle Γ in Theorems 21.1–21.3 can all be expressed explicitly by the following formula:

$$\Gamma = \{\exp(\mathbf{A}t)\mathbf{p}_0 | t \leq 0\} \cup \{\exp(\mathbf{B}t)\mathbf{p}_0 | t > 0\} \cup \{\exp(\mathbf{B}t)\mathbf{q}_0 | t < 0\} \cup \{\exp(\mathbf{A}t)\mathbf{q}_0 | t \geq 0\}.$$

Remark 5 The above results can be easily generalized to 3-dimensional piecewise-affine systems with more saddle points, which belong to different separated zones.

In the following, we present an analytical proof for Theorem 21.7, based on Lemmas 1 and 2. The proofs for Theorems 21.6 and 21.8 are similar. For simplicity, we omit them here.

Proof of Theorem 21.7 From condition (i), it is easy to see that $\mathbf{p}$ and $\mathbf{q}$ are the only two saddle points of system (35), with

$$\mathbf{p} \in \Sigma^- \text{ and } \mathbf{q} \in \Sigma^+.$$

Note that, if considering only the subsystem

$$\dot{\mathbf{x}} = \mathbf{A}\mathbf{x} + \mathbf{a}, \mathbf{x} \in \mathbb{R}^3, \tag{36}$$

then $\mathbf{p}$ is a saddle-focus of system (36), with a 1-dimensional unstable manifold and a 2-dimensional stable manifold described by

$$W^u(\mathbf{p}) = \{\mathbf{x} \in \mathbb{R}^3 | \mathbf{x} = \mathbf{p} + k(0, 0, 1)^T, k \in \mathbb{R}\} \text{ and } W^s(\mathbf{p}) = \{\mathbf{x} \in \mathbb{R}^3 | z = z_p\},$$

respectively. Similarly, if considering only the subsystem

$$\dot{\mathbf{x}} = \mathbf{B}\mathbf{x} + \mathbf{b}, \mathbf{x} \in \mathbb{R}^3, \tag{37}$$

then $\mathbf{q}$ is a saddle point of system (37), with a 1-dimensional unstable manifold and a 2-dimensional stable manifold described by

$$W^u(\mathbf{q}) = \{\mathbf{x} \in \mathbb{R}^3 | \mathbf{x} = \mathbf{q} + k(0, 0, 1)^T, k \in \mathbb{R}\} \text{ and } W^s(\mathbf{q}) = \{\mathbf{x} \in \mathbb{R}^3 | z = z_q\},$$

respectively.

From (ii), we have that

$$\mathbf{p}_0 \in W^u(\mathbf{p}) \cap \Sigma \cap W^s(\mathbf{q}), \;\; \mathbf{q}_0 \in W^u(\mathbf{q}) \cap \Sigma \cap W^s(\mathbf{p}).$$

Now, we denote the flows of systems (36) and (37) with initial condition $\mathbf{x}_0$ by $\Phi_A(t, \mathbf{x}_0)$ and $\Phi_B(t, \mathbf{x}_0)$, respectively. From (9), it is obvious that, as the flows of (36) and (37),

$$\Phi_A(t, \mathbf{p}_0) \to \mathbf{p}, t \to -\infty; \; \Phi_B(t, \mathbf{p}_0) \to \mathbf{q}, t \to +\infty;$$
$$\Phi_B(t, \mathbf{q}_0) \to \mathbf{q}, t \to -\infty; \; \Phi_A(t, \mathbf{q}_0) \to \mathbf{p}, t \to +\infty.$$

Obviously, to accomplish the first part of the proof of the theorem, it suffices to show that

$$\{\Phi_A(t, \mathbf{p}_0)|t < 0\} \subset \Sigma_-, \tag{38}$$

$$\{\Phi_B(t, \mathbf{p}_0)|t > 0\} \subset \Sigma_+, \tag{39}$$

$$\{\Phi_B(t, \mathbf{q}_0)|t < 0\} \subset \Sigma_+, \tag{40}$$

$$\{\Phi_A(t, \mathbf{q}_0)|t > 0\} \subset \Sigma_- \cup \Sigma. \tag{41}$$

Since $\mathbf{p} \in \Sigma_-$ and $\mathbf{p_0} \in W^u(\mathbf{p}) \cap \Sigma$, we know that as the negative semi-orbit of $\mathbf{p_0}$, $\{\Phi_A(t, \mathbf{p}_0)|t < 0\}$ is a straight line connecting $\mathbf{p}$ and $\mathbf{p_0}$, which belongs to Σ_- entirely. Thus, (38) holds. Similarly, (40) holds.

Now, for system (35), let $\mathbf{y} = \mathbf{x} - \mathbf{q}$. Then, system (35) is transformed to the following system:

$$\dot{\mathbf{y}} = \begin{cases} \mathbf{Ay} + \mathbf{Aq} + \mathbf{a}, & \text{if } \mathbf{c}^T\mathbf{y} \le d - \mathbf{c}^T\mathbf{q} \\ \mathbf{By}, & \text{if } \mathbf{c}^T\mathbf{y} > d - \mathbf{c}^T\mathbf{q}. \end{cases}$$

Thus,

$$\{\Phi_B(t, \mathbf{p}_0)|t > 0\} \subset \Sigma_+$$

if and only if

$$\{\tilde{\Phi}_B(t, \mathbf{p}_0 - \mathbf{q})|t > 0\} \subset \{\mathbf{y} \in \mathbb{R}^3|\mathbf{c}^T\mathbf{y} > d - \mathbf{c}^T\mathbf{q}\},$$

where $\tilde{\Phi}_B(t, \cdot)$ means the flow generated by the system $\dot{\mathbf{y}} = \mathbf{By}$ ($\mathbf{y} = (y_1, y_2, y_3)^T \in \mathbb{R}^3$). Furthermore, let $\mathbf{B}_0 = \begin{pmatrix} b_{11} & b_{12} \\ b_{21} & b_{22} \end{pmatrix}$. Then, $\mathbf{B} = \begin{pmatrix} \mathbf{B}_0 & \\ & \lambda_B \end{pmatrix}$. Elementary analysis indicates that

$$\{\tilde{\Phi}_B(t, \mathbf{p}_0 - \mathbf{q})|t > 0\} \subset \{\mathbf{y} \in \mathbb{R}^3|\mathbf{c}^T\mathbf{y} > d - \mathbf{c}^T\mathbf{q}\}$$

if and only if

$$\{\exp(\mathbf{B}_0 t)(x_{p_0} - x_q, y_{p_0} - y_q)^T|t > 0\} \subset \{(y_1, y_2)^T \in \mathbb{R}^2| - c_1y_1 - c_2y_2 < \mathbf{c}^T\mathbf{q} - d\}. \tag{42}$$

From (i), we know that $\mathbf{c}^T\mathbf{q} - d > 0$. In addition, by (iii) we obtain $\mathbf{c}^T\mathbf{B}(\mathbf{p}_0 - \mathbf{q}) \ge 0$, which implies that $(-c_1, -c_2)\mathbf{B}_0(x_{p_0} - x_q, y_{p_0} - y_q)^T < 0$. By Lemma 1, we know that (42) holds. Hence, (39) holds.

Analogously, in order to prove (41), let

$$\mathbf{z} = \mathbf{x} - \mathbf{p}. \tag{43}$$

Substituting (43) into (35) gives the following system:

$$\dot{\mathbf{z}} = \begin{cases} \mathbf{A}\mathbf{z}, & \text{if } \mathbf{c}^T\mathbf{z} \le d - \mathbf{c}^T\mathbf{p} \\ \mathbf{B}\mathbf{z} + \mathbf{B}\mathbf{p} + \mathbf{b}, & \text{if } \mathbf{c}^T\mathbf{z} > d - \mathbf{c}^T\mathbf{p}. \end{cases}$$

Obviously,

$$\{\Phi_A(t, \mathbf{q}_0) | t > 0\} \subset \Sigma_- \cup \Sigma$$

if and only if

$$\{\tilde{\Phi}_A(t, \mathbf{q}_0 - \mathbf{p}) | t > 0\} \subset \{\mathbf{z} \in \mathbb{R}^3 | \mathbf{c}^T\mathbf{z} \le d - \mathbf{c}^T\mathbf{p}\},$$

where $\tilde{\Phi}_A(t, \cdot)$ denotes the flow generated by the system $\dot{\mathbf{z}} = \mathbf{A}\mathbf{z}$ ($\mathbf{z} = (z_1, z_2, z_3)^T \in \mathbb{R}^3$). Thus,

$$\{\tilde{\Phi}_A(t, \mathbf{q}_0 - \mathbf{p}) | t > 0\} \subset \{\mathbf{z} \in \mathbb{R}^3 | \mathbf{c}^T\mathbf{z} \le d - \mathbf{c}^T\mathbf{p}\}$$

if and only if

$$\{\exp(\mathbf{A}_0 t)(x_{q_0} - x_p, y_{q_0} - y_p)^T | t > 0\} \subset \{(z_1, z_2)^T \in \mathbb{R}^2 | c_1 z_1 + c_2 z_2 \le d - \mathbf{c}^T\mathbf{p}\}, \tag{44}$$

with $\mathbf{A}_0 = \begin{pmatrix} a_{11} & a_{12} \\ a_{21} & a_{22} \end{pmatrix}$.

From (iii), we know that $\mathbf{q}_0 \in [\mathbf{x}_*, \mathbf{x}^*]$. Let $\hat{\mathbf{z}} = (z_1, z_2)^T$, $\hat{\mathbf{c}} = (c_1, c_2)^T$ and $\hat{\mathbf{z}}_* = \frac{d - \mathbf{c}^T\mathbf{p}}{\hat{\mathbf{c}}^T \mathbf{A}_0^{-1} \hat{\mathbf{c}}^\perp} \mathbf{A}_0^{-1} \hat{\mathbf{c}}^\perp$. Then, $\mathbf{q}_0 \in [\mathbf{x}_*, \mathbf{x}^*]$ if and only if $(x_{q_0} - x_p, y_{q_0} - y_p)^T \in [\hat{\mathbf{z}}_*, \hat{\mathbf{z}}^*]$, where $\hat{\mathbf{z}}^*$ means the first intersection of the flow of the plane system $\dot{\hat{\mathbf{z}}} = \mathbf{A}_0\hat{\mathbf{z}}$, with initial condition $\hat{\mathbf{z}}_*$, under negative flight time with plane straight line $\hat{L} = \{\hat{\mathbf{z}} \in \mathbb{R}^2 | \hat{\mathbf{c}}^T\hat{\mathbf{z}} = d - \mathbf{c}^T\mathbf{p}\}$. Note that $d - \mathbf{c}^T\mathbf{p} > 0$. By Lemma 2 and Remark 3, we know that (44) holds, which shows that (41) holds.

From the above discussions, if (i), (ii) and (iii) are all satisfied then there exists a heteroclinic cycle Γ connecting $\mathbf{p}$ and $\mathbf{q}$, which is the union set of $\{\Phi_A(t, \mathbf{p}_0) | t \le 0\}$, $\{\Phi_B(t, \mathbf{p}_0) | t > 0\}$, $\{\Phi_B(t, \mathbf{q}_0) | t < 0\}$ and $\{\Phi_A(t, \mathbf{q}_0) | t \ge 0\}$. Moreover, if we substitute condition (iii) for condition (iii′), then by Remarks 1 and 3, it is easy to see that Γ intersects Σ at $\mathbf{p}_0$ and $\mathbf{q}_0$ transversally with $\mathbf{p}_0$ and $\mathbf{q}_0$ being the only two intersections of Γ and Σ. Thus, Theorem 21.7 holds.

3 Šil'nikov-Type Theory for Piecewise-Affine Systems

Based on the ideas of Šil'nikov-type theory for smooth systems [13–15], we can get some similar Šil'nikov-type theory for piecewise-affine systems.

Now, we list some similar Šil'nikov theorems for the 3-dimensional or 4-dimensional piecewise-affine systems with homoclinic orbits or heteroclinic cycles discussed in the above section, without detailed proofs.

Theorem 21.9 *Suppose that system (8) has a homoclinic orbit to* **p**, *which crosses the switching manifold at two points, and*

$$\lambda_A < -\alpha_A < 0.$$

Then, system (8) has infinitely many chaotic invariant sets [7].

Theorem 21.10 *Suppose that system (26) satisfies*

$$\alpha\rho < 0, \ \ |\alpha| < |\rho| \ \text{and} \ \lambda < 0, \ \ \delta\rho > 0.$$

Then, there exist chaotic invariant sets [9].

Theorem 21.11 *Suppose that system (29) has a heteroclinic cycle connecting* **p** *and* **q**, *and intersecting the switching manifold transversally at two points, if*

$$\alpha_A\alpha_B - \lambda_A\lambda_B < 0.$$

Then, system (29) has infinitely many chaotic invariant sets [10].

Theorem 21.12 *For system (35) with condition (E2), suppose that there exists a heteroclinic cycle connecting* **p** *and* **q**, *and intersecting the switching manifold transversally at two points, if*

$$\alpha\mu_1 - \lambda_A\lambda_B < 1,$$

then system (35) has infinitely many chaotic invariant sets [12].

The main ideas for proving Theorems 21.9–21.12 are to define a Poincaré map on an appropriate cross-section located at a sufficiently small neighborhood of a homoclinic orbit or heteroclinic cycle, and then study the existence of horseshoes of the Poincaré map by the topological horseshoe theory. The methods originate from [13–17].

For readers' convenience, in the following we give a brief proof of Theorem 21.10, for illustration. For more details about the proofs of Theorems 21.9–21.12, we refer to [7, 9, 10, 12].

Proof of Theorem 21.10

Step one: Choice of a cross-section.

We known by Theorem 21.1 that, when the conditions in Theorem 21.10 are satisfied, there exists a homoclinic orbit Γ that intersects Σ at two points, $\mathbf{p} = (0, 1, 0, 0)$ and $\mathbf{q} = (0, 0, 0, 1)$. We only consider the case when $\alpha > 0, \ \rho < 0$ in detail; the other case, i.e., $\alpha < 0, \ \rho > 0$, can be discussed in the same manner.

Let

$$\Pi_0 = \left\{x_1 = 0, \varepsilon \cdot e^{\frac{-\alpha \cdot 2\pi}{\beta}} \le x_2 < \varepsilon, 0 \le x_3 < \varepsilon, 0 < x_4 \le \varepsilon\right\}, \quad \Pi_1 = \{x_4 = \varepsilon\},$$

where ε is a sufficiently small constant. Denote

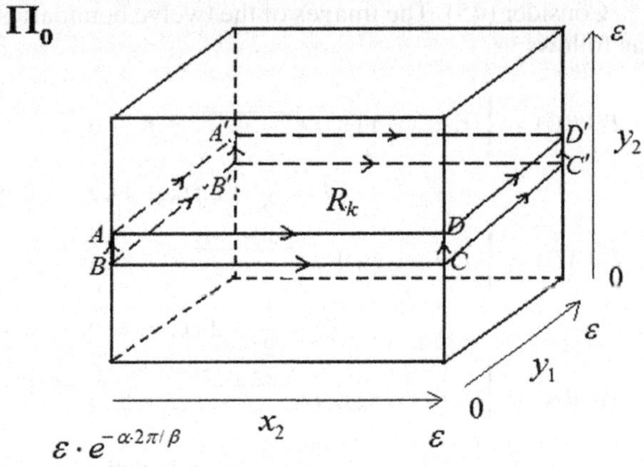

Fig. 7 Geometric structures of Π_0 and R_k

$$R_k = \left\{x_1 = 0,\ \varepsilon \cdot e^{\frac{-\alpha \cdot 2\pi}{\beta}} \le x_2 < \varepsilon,\ 0 \le x_3 < \varepsilon, \varepsilon \cdot e^{\frac{\rho \cdot 2\pi(k+1)}{\beta}} < x_4 \le \varepsilon \cdot e^{\frac{\rho \cdot 2\pi k}{\beta}}\right\}.$$

Then, $\Pi_0 = \bigcup_{k=0}^{\infty} R_k$.

By introducing the polar coordinates to the first two elements, i.e.,

$$x_1 = r \cdot \cos\theta,\ x_2 = r \cdot \sin\theta,$$

we can write R_k as

$$R_k = \{\varepsilon \cdot e^{\frac{-\alpha \cdot 2\pi}{\beta}} \le r < \varepsilon, \theta = \frac{\pi}{2},\ 0 \le x_3 < \varepsilon, \varepsilon \cdot e^{\frac{\rho \cdot 2\pi(k+1)}{\beta}} < x_4 \le \varepsilon \cdot e^{\frac{\rho \cdot 2\pi k}{\beta}}\}.$$

Step two: Construction of a Poincaré map.

Now, define the map $P_0 : \Pi_0 \to \Pi_1$. For a point $\mathbf{x}_0 = (r_0, \theta_0, x_{30}, x_{40}) \in \Pi_0$, $P_0(\mathbf{x}_0)$ is defined as the first intersection of $\Phi_A(t, \mathbf{x}_0)$ with Π_1. The time T for the flow from Π_0 to Π_1 is determined by $x_{40} \cdot e^{\rho T} = \varepsilon$. Then, we have

$$T = \frac{1}{\rho} \cdot \ln \frac{\varepsilon}{x_{40}}.$$

Using the polar coordinates $x_{10} = r_0 \cos\theta_0,\ x_{20} = r_0 \sin\theta_0$, and note that $\theta_0 = \pi/2$, the first section map $P_0 : \Pi_0 \to \Pi_1$ is given by

$$P_0 : \begin{pmatrix} r_0 \\ \frac{\pi}{2} \\ x_{30} \\ x_{40} \end{pmatrix} \longmapsto \begin{pmatrix} r_0 \cdot (\frac{\varepsilon}{x_{40}})^{\frac{\alpha}{\rho}} \\ \frac{\pi}{2} + \frac{\beta}{\rho} \cdot \ln \frac{\varepsilon}{x_{40}} \\ x_{30} \cdot (\frac{\varepsilon}{x_{40}})^{\frac{\lambda}{\rho}} \\ \varepsilon \end{pmatrix}. \tag{45}$$

Consider (45). The images of the twelve boundaries of R_k under P_0 are expressed as follows:

$$P_0(BC)=\Big\{(r,\theta,x_3,x_4)|\ \varepsilon e^{\frac{-\alpha\cdot 2\pi(k+2)}{\beta}}\le r\le \varepsilon e^{\frac{-\alpha\cdot 2\pi(k+1)}{\beta}},\ \theta=\frac{\pi}{2}-2\pi(k+1), x_3=0, x_4=\varepsilon\Big\};$$

$$P_0(AD)=\Big\{(r,\theta,x_3,x_4)|\ \varepsilon e^{\frac{-\alpha\cdot 2\pi(k+1)}{\beta}}\le r\le \varepsilon e^{\frac{-\alpha\cdot 2\pi k}{\beta}},\ \theta=\frac{\pi}{2}-2\pi k, x_3=0, x_4=\varepsilon\Big\};$$

$$P_0(BA)=\Big\{(r,\theta,x_3,x_4)|\ \varepsilon e^{\frac{-\alpha\cdot 2\pi(k+2)}{\beta}}\le r\le \varepsilon e^{\frac{-\alpha\cdot 2\pi(k+1)}{\beta}},\ \frac{\pi}{2}-2\pi(k+1)\le\theta\le\frac{\pi}{2}-2\pi k, x_3=0, x_4=\varepsilon\Big\};$$

$$P_0(CD)=\Big\{(r,\theta,x_3,x_4)|\ \varepsilon e^{\frac{-\alpha\cdot 2\pi(k+1)}{\beta}}\le r\le \varepsilon e^{\frac{-\alpha\cdot 2\pi k}{\beta}},\ \frac{\pi}{2}-2\pi(k+1)\le\theta\le\frac{\pi}{2}-2\pi k, x_3=0, x_4=\varepsilon\Big\};$$

$$P_0(B'C')=\Big\{(r,\theta,x_3,x_4)|\ \varepsilon e^{\frac{-\alpha\cdot 2\pi(k+2)}{\beta}}\le r\le \varepsilon e^{\frac{-\alpha\cdot 2\pi(k+1)}{\beta}},\ \theta=\frac{\pi}{2}-2\pi(k+1), x_3=\varepsilon\cdot e^{\frac{-\lambda\cdot 2\pi(k+1)}{\beta}}, x_4=\varepsilon\Big\};$$

$$P_0(A'D')=\Big\{(r,\theta,x_3,x_4)|\ \varepsilon e^{\frac{-\alpha\cdot 2\pi(k+1)}{\beta}}\le r\le \varepsilon e^{\frac{-\alpha\cdot 2\pi k}{\beta}},\ \theta=\frac{\pi}{2}-2\pi k, x_3=\varepsilon\cdot e^{\frac{-\lambda\cdot 2\pi k}{\beta}}, x_4=\varepsilon\Big\};$$

$$P_0(B'A')=\Big\{(r,\theta,x_3,x_4)|\ \varepsilon e^{\frac{-\alpha\cdot 2\pi(k+2)}{\beta}}\le r\le \varepsilon e^{\frac{-\alpha\cdot 2\pi(k+1)}{\beta}},\ \frac{\pi}{2}-2\pi(k+1)\le\theta\le\frac{\pi}{2}-2\pi k,\ \varepsilon e^{\frac{-\lambda\cdot 2\pi k}{\beta}}\le x_3\le \varepsilon e^{\frac{-\lambda\cdot 2\pi(k+1)}{\beta}}, x_4=\varepsilon\Big\};$$

$$P_0(C'D')=\Big\{(r,\theta,x_3,x_4)|\ \varepsilon e^{\frac{-\alpha\cdot 2\pi(k+1)}{\beta}}\le r\le \varepsilon e^{\frac{-\alpha\cdot 2\pi k}{\beta}},\ \frac{\pi}{2}-2\pi(k+1)\le\theta\le\frac{\pi}{2}-2\pi k,\ \varepsilon e^{\frac{-\lambda\cdot 2\pi k}{\beta}}\le x_3\le \varepsilon e^{\frac{-\lambda\cdot 2\pi(k+1)}{\beta}}, x_4=\varepsilon\Big\};$$

$$P_0(AA')=\Big\{(r,\theta,x_3,x_4)|\ r=\varepsilon e^{\frac{-\alpha\cdot 2\pi(k+1)}{\beta}}, \theta=\frac{\pi}{2}-2\pi k,\ 0\le x_3\le \varepsilon e^{\frac{-\lambda\cdot 2\pi k}{\beta}}, x_4=\varepsilon\Big\};$$

$$P_0(BB')=\Big\{(r,\theta,x_3,x_4)|\ r=\varepsilon e^{\frac{-\alpha\cdot 2\pi(k+2)}{\beta}}, \theta=\frac{\pi}{2}-2\pi(k+1),$$

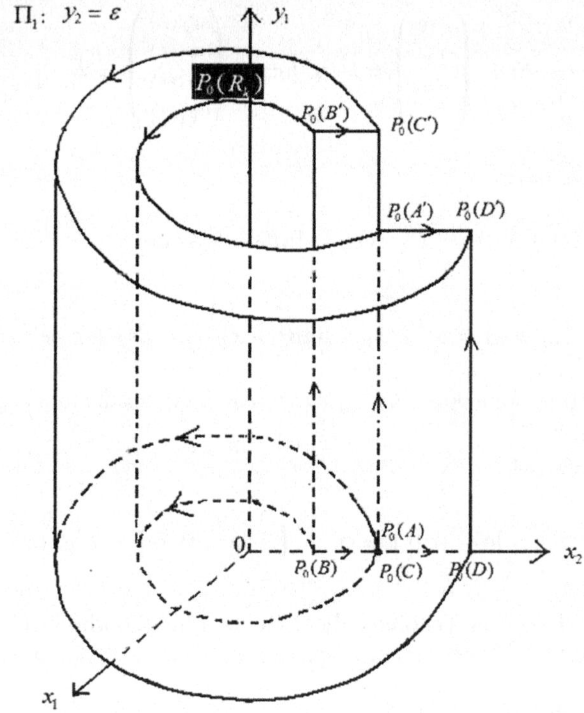

Fig. 8 The image of $P_0(R_k)$

$$0 \le x_3 \le \varepsilon e^{\frac{-\lambda \cdot 2\pi (k+1)}{\beta}}, x_4 = \varepsilon \Big\};$$

$$P_0(DD') = \Big\{ (r, \theta, x_3, x_4) |\ r = \varepsilon e^{\frac{-\alpha \cdot 2\pi k}{\beta}}, \theta = \frac{\pi}{2} - 2\pi k, 0 \le x_3 \le \varepsilon \cdot e^{\frac{-\lambda \cdot 2\pi k}{\beta}}, x_4 = \varepsilon \Big\};$$

$$P_0(CC') = \Big\{ (r, \theta, x_3, x_4) |\ r = \varepsilon e^{\frac{-\alpha \cdot 2\pi (k+1)}{\beta}}, \theta = \frac{\pi}{2} - 2\pi (k+1),$$

$$0 \le x_3 \le \varepsilon \cdot e^{\frac{-\lambda \cdot 2\pi (k+1)}{\beta}}, x_4 = \varepsilon \Big\}.$$

The geometry of $P_0(R_k) \subset \Pi_1$ is shown in Fig. 8. Note that the arrows in Figs. 7 and 8 represent corresponding direction relationships among the twelve boundaries and their images.

Denote $\mathbf{q}^* = \Gamma \cap \Pi_1$. Then, $\mathbf{q}^* = (0, 0, 0, \varepsilon)$ by (27). For a point $\mathbf{x}_0$ in a small neighborhood $U(\mathbf{q}^*)$ of $\mathbf{q}^*$, the map P_1 is defined as the first intersection of $\Phi_A(t, \mathbf{x}_0)$ $(t < 0)$ with Σ, i.e.,

$$P_1 : \begin{pmatrix} x_{10} \\ x_{20} \\ x_{30} \\ \varepsilon \end{pmatrix} \in U(q^*) \longmapsto \begin{pmatrix} x_1(t) \\ x_2(t) \\ x_3(t) \\ x_4(t) \end{pmatrix} \in \Sigma.$$

From (27), we have

$$x_2(t) + x_4(t) = e^{\alpha t} \cdot (x_{10} \cdot \sin\beta t + x_{20} \cdot \cos\beta t) + \varepsilon \cdot e^{\rho t}.$$

Let

$$F(t; x_{10}, x_{20}) = e^{\alpha t} \cdot (x_{10} \cdot \sin\beta t + x_{20} \cdot \cos\beta t) + \varepsilon \cdot e^{\rho t} - 1.$$

Then, the flight time for a point in $U(\mathbf{q}^*)$ to reach Σ equals the first negative solution of $F(t; x_{10}, x_{20}) = 0$.

In addition, we get

$$F\left(\frac{1}{\rho}\ln\frac{1}{\varepsilon}; 0, 0\right) = 0, \ F_t\left(\frac{1}{\rho}\ln\frac{1}{\varepsilon}; 0, 0\right) = \rho \neq 0.$$

By the Implicit Function Theorem, there exists a neighborhood $U \subset R^2$ of $(x_{10}, x_{20}) = (0, 0)$ and a C^1 function $t(x_{10}, x_{20})$ defined on U, such that

$$t(0, 0) = \frac{1}{\rho}\ln\frac{1}{\varepsilon}, \ t(x_{10}, x_{20}) = t_0 + a \cdot x_{10} + b \cdot x_{20} + O(\|x_{10}, x_{20}\|^2),$$

with $t_0 = t(0, 0)$ and

$$a = \frac{\partial t}{\partial x_{10}}(0, 0) = -\frac{1}{\rho} \cdot \left(\frac{1}{\varepsilon}\right)^{\alpha/\rho} \cdot \sin\left(\frac{\beta}{\rho}\ln\frac{1}{\varepsilon}\right),$$

$$b = \frac{\partial t}{\partial x_{20}}(0, 0) = -\frac{1}{\rho} \cdot \left(\frac{1}{\varepsilon}\right)^{\alpha/\rho} \cdot \cos\left(\frac{\beta}{\rho}\ln\frac{1}{\varepsilon}\right).$$

Substituting

$$e^{\alpha t} = e^{\alpha t_0} \cdot e^{\alpha(t-t_0)} = e^{\alpha t_0} \cdot \left[1 + \alpha(t - t_0) + \frac{\alpha^2}{2}(t - t_0)^2 + O((t - t_0)^3)\right],$$

$$\sin\beta t = \sin\beta t_0 + \beta\cos\beta t_0 \cdot (t - t_0) - \frac{\beta^2}{2}\sin\beta t_0 \cdot (t - t_0)^2 + O((t - t_0)^3),$$

$$\cos\beta t = \cos\beta t_0 - \beta\sin\beta t_0 \cdot (t - t_0) - \frac{\beta^2}{2}\cos\beta t_0 \cdot (t - t_0)^2 + O((t - t_0)^3),$$

into Eq. (27), we get

$$P_1:\begin{pmatrix}x_{10}\\x_{20}\\x_{30}\\\varepsilon\end{pmatrix}\mapsto\begin{pmatrix}e^{\alpha t_0}(\cos\beta t_0\cdot x_{10}-\sin\beta t_0\cdot x_{20})+O(\|x_{10},x_{20}\|^2)\\e^{\alpha t_0}(\sin\beta t_0\cdot x_{10}+\cos\beta t_0\cdot x_{20})+O(\|x_{10},x_{20}\|^2)\\x_{30}e^{\lambda t_0}[1+\lambda(ax_{10}+bx_{20})]+O(\|x_{10},x_{20}\|^2)\\\varepsilon e^{\rho t_0}[1+\rho(ax_{10}+bx_{20})]+O(\|x_{10},x_{20}\|^2)\end{pmatrix}.$$

Note that, when ε is sufficiently small, we have $P_0(R_k)\subset U(\mathbf{q}^*)$, $k=0,1,2,\ldots$. For ε sufficiently small, the expression of the second section map $P_1: P_0(R_k)\to\Sigma$ is

$$P_1:\begin{pmatrix}r_0\\\theta_0\\x_{30}\\\varepsilon\end{pmatrix}\mapsto\begin{pmatrix}r_0\cdot\varepsilon^{-\frac{\alpha}{\rho}}\\\theta_0-2\pi\cdot N_0\\x_{30}\cdot\varepsilon^{-\frac{\lambda}{\rho}}\\1-x_2\end{pmatrix}+O(r_0^2),$$

where the first two elements are given by the polar coordinates, N_0 is some positive integer (which can be realized by choosing ε properly), and

$$x_2=r_0\cdot\varepsilon^{-\alpha/\rho}\cdot\sin\left(\theta_0+\frac{\beta}{\rho}\cdot\ln\frac{1}{\varepsilon}\right),\quad\frac{\beta}{\rho}\cdot\ln\frac{1}{\varepsilon}=-2\pi\cdot N_0.$$

Moreover, define the section map from Σ to Σ as follows. Let $(x_1(t),x_2(t),x_3(t),x_4(t))^T=\Phi_B(t,\mathbf{x}_0)$. Then, (28) implies

$$x_2(t)=\cos\delta t\cdot x_{20}+\sin\delta t\cdot(1-x_{20}),\ x_4(t)=-\sin\delta t\cdot x_{20}+\cos\delta t\cdot(1-x_{20}).$$

Let $T<0$ be the first return time for the flow $\Phi_B(\cdot,\mathbf{x}_0)$ to reach Σ, and

$$F(t;x_{20})=x_2(t)+x_4(t)-1=\cos\delta t+(1-2x_{20})\cdot\sin\delta t-1.$$

Then, T is the first negative solution of $F(t;x_{20})=0$.

Note that $\frac{\pi}{2\delta}<t<0$. We have

$$F\left(\frac{\pi}{2\delta};0\right)=0,\ F_t\left(\frac{\pi}{2\delta};0\right)=-\delta\neq 0,\ F(t;0)\neq 0.$$

So, when x_{20} is sufficiently small, T can be expressed as a continuous function of x_{20}, i.e.,

$$T(x_{20})=T(0)+T'(0)\cdot x_{20}+O((x_{20})^2)=\frac{\pi}{2\delta}-\frac{2}{\delta}\cdot x_{20}+O((x_{20})^2).$$

From (28), there exists a sufficiently small neighborhood $V(\mathbf{q})\subset\Sigma$ of $\mathbf{q}$, such that

$$P_2 : V(\mathbf{q}) \longrightarrow \Sigma$$

$$\begin{pmatrix} x_{10} \\ x_{20} \\ x_{30} \\ 1 - x_{20} \end{pmatrix} \longmapsto \begin{pmatrix} x_{10} \cdot e^{\frac{\gamma\pi}{2\delta}} \\ 1 - x_{20} \\ x_{30} \cdot e^{\frac{\mu\pi}{2\delta}} \\ x_{20} \end{pmatrix} + O(x_{20}^2).$$

Now, define a map P_3 from Π_0 to Σ under the flow of system (26). For a point $\mathbf{x_0} = (0, x_{20}, x_{30}, x_{40})^T \in \Pi_0$, define $P_3(\mathbf{x_0})$ as the first intersection point of $\Phi_A(t, \mathbf{x}_0)$ $(t > 0)$ with Σ. By using a similar method to define P_1, for a sufficiently small constant ε, we have

$$P_3 : \Pi_0 \longrightarrow \Sigma$$

$$\begin{pmatrix} 0 \\ x_{20} \\ y_{10} \\ y_{20} \end{pmatrix} \mapsto \begin{pmatrix} \frac{\beta}{\alpha}(-1 + e^{\alpha t_0} \cdot x_{20} + e^{\rho t_0} \cdot y_{20}) \\ 1 - e^{\rho t_0} \cdot y_{20} \\ y_{10} \cdot e^{\lambda t_0} \\ y_{20} \cdot e^{\rho t_0} \end{pmatrix} + O(\|x_{20} - \Delta, y_{20}\|^2), \tag{46}$$

where $\Delta = e^{-\alpha t_0}$ and $t_0 = 2\pi m_0/\beta$ with m_0 satisfying

$$\varepsilon \cdot e^{\frac{-3\alpha\pi}{2\beta}} \leq e^{\frac{-\alpha 2\pi \cdot m_0}{\beta}} \leq \varepsilon \cdot e^{\frac{-\alpha\pi}{\beta}}. \tag{47}$$

Note that all the images of the boundaries of R_k under P_3 are on Σ. So, these images can be determined by their first three elements. By (46) and omitting the higher-order terms, we obtain

$$\begin{aligned}
P_3(BC) &= \left\{ v_2 \leq x_1 \leq v_1,\ x_2 = 1 - \varepsilon \cdot e^{\frac{\rho \cdot 2\pi \cdot (m_0 + k + 1)}{\beta}},\ x_3 = 0 \right\}, \\
P_3(AD) &= \left\{ v_4 \leq x_1 \leq v_3,\ x_2 = 1 - \varepsilon \cdot e^{\frac{\rho \cdot 2\pi \cdot (m_0 + k)}{\beta}},\ x_3 = 0 \right\}, \\
P_3(BA) &= \left\{ v_2 \leq x_1 \leq v_4,\ 1 - \varepsilon \cdot e^{\frac{\rho \cdot 2\pi \cdot (m_0 + k)}{\beta}} \leq x_2 \leq 1 - \varepsilon \cdot e^{\frac{\rho \cdot 2\pi \cdot (m_0 + k + 1)}{\beta}},\ x_3 = 0 \right\}, \\
P_3(CD) &= \left\{ v_1 \leq x_1 \leq v_3,\ 1 - \varepsilon \cdot e^{\frac{\rho \cdot 2\pi \cdot (m_0 + k)}{\beta}} \leq x_2 \leq 1 - \varepsilon \cdot e^{\frac{\rho \cdot 2\pi \cdot (m_0 + k + 1)}{\beta}},\ x_3 = 0 \right\}, \\
P_3(B'C') &= \left\{ v_2 \leq x_1 \leq v_1,\ x_2 = 1 - \varepsilon \cdot e^{\frac{\rho \cdot 2\pi \cdot (m_0 + k + 1)}{\beta}}, x_3 = \varepsilon \cdot e^{\frac{\lambda \cdot 2\pi m_0}{\beta}} \right\}, \\
P_3(A'D') &= \left\{ v_4 \leq x_1 \leq v_3,\ x_2 = 1 - \varepsilon \cdot e^{\frac{\rho \cdot 2\pi \cdot (m_0 + k)}{\beta}}, x_3 = \varepsilon \cdot e^{\frac{\lambda \cdot 2\pi m_0}{\beta}} \right\}, \\
P_3(B'A') &= \left\{ v_2 \leq x_1 \leq v_4,\ 1 - \varepsilon \cdot e^{\frac{\rho \cdot 2\pi \cdot (m_0 + k)}{\beta}} \leq x_2 \leq 1 - \varepsilon \cdot e^{\frac{\rho \cdot 2\pi \cdot (m_0 + k + 1)}{\beta}}, \right. \\
&\qquad \left. x_3 = \varepsilon \cdot e^{\frac{\lambda \cdot 2\pi m_0}{\beta}} \right\}, \\
P_3(C'D') &= \left\{ v_1 \leq x_1 \leq v_3,\ 1 - \varepsilon \cdot e^{\frac{\rho \cdot 2\pi \cdot (m_0 + k)}{\beta}} \leq x_2 \leq 1 - \varepsilon \cdot e^{\frac{\rho \cdot 2\pi \cdot (m_0 + k + 1)}{\beta}}, \right.
\end{aligned}$$

$$x_3 = \varepsilon \cdot e^{\frac{\lambda \cdot 2\pi m_0}{\beta}} \bigg\},$$

$$P_3(A'A') = \left\{x_1 = v_4,\ x_2 = 1 - \varepsilon \cdot e^{\frac{\rho \cdot 2\pi \cdot (m_0+k)}{\beta}},\ 0 \le x_3 \le \varepsilon \cdot e^{\frac{\lambda \cdot 2\pi m_0}{\beta}} \right\},$$

$$P_3(B'B') = \left\{x_1 = v_2,\ x_2 = 1 - \varepsilon \cdot e^{\frac{\rho \cdot 2\pi \cdot (m_0+k+1)}{\beta}},\ 0 \le x_3 \le \varepsilon \cdot e^{\frac{\lambda \cdot 2\pi m_0}{\beta}} \right\},$$

$$P_3(D'D') = \left\{x_1 = v_3,\ x_2 = 1 - \varepsilon \cdot e^{\frac{\rho \cdot 2\pi \cdot (m_0+k)}{\beta}},\ 0 \le x_3 \le \varepsilon \cdot e^{\frac{\lambda \cdot 2\pi m_0}{\beta}} \right\},$$

$$P_3(C'C') = \left\{x_1 = v_1,\ x_2 = 1 - \varepsilon \cdot e^{\frac{\rho \cdot 2\pi \cdot (m_0+k+1)}{\beta}},\ 0 \le x_3 \le \varepsilon \cdot e^{\frac{\lambda \cdot 2\pi m_0}{\beta}} \right\},$$

where

$$v_1 = -\frac{\beta}{\alpha} + \frac{\beta}{\alpha} \cdot \varepsilon \cdot e^{\frac{\alpha \cdot 2\pi \cdot m_0}{\beta}} + \frac{\beta}{\alpha} \cdot \varepsilon \cdot e^{\frac{\rho \cdot 2\pi \cdot (m_0+k+1)}{\beta}},$$

$$v_2 = -\frac{\beta}{\alpha} + \frac{\beta}{\alpha} \cdot \varepsilon \cdot e^{\frac{\alpha \cdot 2\pi \cdot (m_0-1)}{\beta}} + \frac{\beta}{\alpha} \cdot \varepsilon \cdot e^{\frac{\rho \cdot 2\pi \cdot (m_0+k+1)}{\beta}},$$

$$v_3 = -\frac{\beta}{\alpha} + \frac{\beta}{\alpha} \cdot \varepsilon \cdot e^{\frac{\alpha \cdot 2\pi \cdot m_0}{\beta}} + \frac{\beta}{\alpha} \cdot \varepsilon \cdot e^{\frac{\rho \cdot 2\pi \cdot (m_0+k)}{\beta}},$$

$$v_4 = -\frac{\beta}{\alpha} + \frac{\beta}{\alpha} \cdot \varepsilon \cdot e^{\frac{\alpha \cdot 2\pi \cdot (m_0-1)}{\beta}} + \frac{\beta}{\alpha} \cdot \varepsilon \cdot e^{\frac{\rho \cdot 2\pi \cdot (m_0+k)}{\beta}}.$$

When $\varepsilon > 0$ is sufficiently small, by (47), we obtain

$$v_2 < v_4 < 0 < v_1 < v_3.$$

Then, we can get the geometry of $P_3(R_k)$ as shown in Fig. 9.

Denote $P_2 \circ P_1 \circ P_0 \equiv P^*$ and $P^* \circ P_3^{-1} \equiv f$. Then,

$$f \circ P_3 = P_3 \circ P.$$

Remark 6 To study the existence of chaos in system (26), we can investigate the existence of chaotic invariant set of the full Poincaré map P, or, equivalently, of the map f.

Step Three: the geometric structure of the Poincaré map.

Denote by P_i^l the linear parts of the maps P_i, $i = 1, 2, 3$, and let

$$P^l = (P_3^l)^{-1} \circ P_2^l \circ P_1^l \circ P_0.$$

Since ε is sufficiently small, we can omit the higher-order terms and take $P = P^l$.

Note that $P^*(R_k)$ and $P_3(R_k)$ are both contained in Σ. So, the geometries of $P^*(R_k)$ and $P_3(R_k)$ can be determined by their first three coordinates x_1, x_2, x_3. In the sequel, we will prove the results in the x_1-x_2-x_3 space.

Denote by $P^*(R_k)|_{x_1x_2}$ and $P_3(R_k)|_{x_1x_2}$ the projections of $P^*(R_k)$ and $P_3(R_k)$ onto the x_1-x_2 plane, respectively. Then, we have the following result.

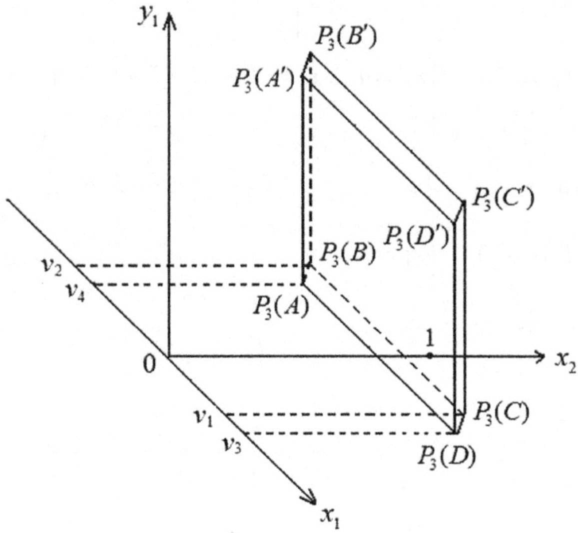

Fig. 9 The graph of $P_3(R_k)$

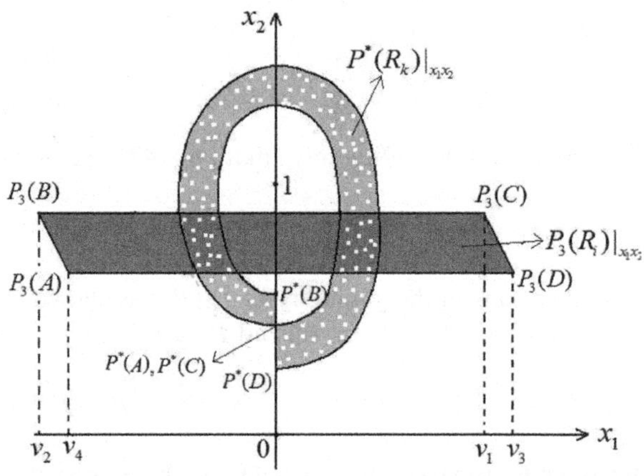

Fig. 10 Projections of $P^*(R_k)$ and $P_3(R_k)$ onto the x_1-x_2 plane

Lemma 3 *Suppose that $\alpha\rho < 0$, $\rho\delta > 0$, $|\alpha| < |\rho|$ and $\lambda < 0$ in system (26). Consider R_k for a fixed k, large enough. Then, the inner boundary of $P^*(R_k)|_{x_1x_2}$ will intersect the lower boundary of $P_3(R_i)|_{x_1x_2}$ at two points for $i \geq k$ (Fig. 10).*

Lemma 4 *Suppose that $\rho\alpha < 0$, $|\alpha| < |\rho|$, $\lambda < 0$, $\rho\delta > 0$ in system (26). For sufficiently large k, the Poincaré map P possesses an invariant Cantor set $\Sigma \subset R_k$, on which it is topologically conjugate to a full shift on two symbols (Fig. 11).*

From the above results, Theorem 21.10 is proved.

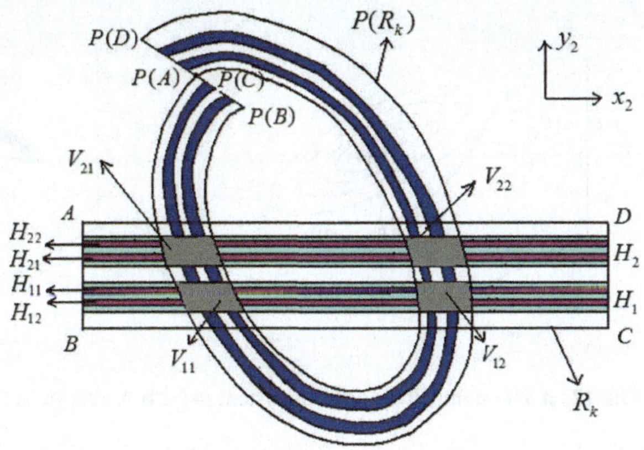

Fig. 11 Geometries of $P(R_k)$ and R_k on the $x_2 - x_4$ plane

4 A Gallery of Chaotic Piecewise-Affine Systems

In this section, we give some precise examples to illustrate the applications of the main results in Sects. 2 and 3.

Example 1 ([7]) Consider the system

$$\dot{\mathbf{x}} = \begin{cases} \mathbf{A}\mathbf{x} + \mathbf{a}, & x_1 > 0, \\ \mathbf{B}\mathbf{x} + \mathbf{b}, & x_1 \le 0, \end{cases} \tag{48}$$

where

$$\mathbf{A} = \begin{pmatrix} \frac{127}{16} & -\frac{2313}{144} & \frac{67}{16} \\ -\frac{353}{24} & -\frac{887}{72} & -\frac{47}{72} \\ -\frac{3743}{48} & \frac{5227}{144} & -\frac{2105}{144} \end{pmatrix}, \quad \mathbf{B} = \begin{pmatrix} \frac{5}{4} & -\frac{3}{4} & 0 \\ \frac{3}{4} & \frac{5}{4} & 0 \\ -3 & 1 & -1 \end{pmatrix},$$

$$\mathbf{a} = \left(\frac{565}{4}, 140, -\frac{2575}{12}\right)^{\mathsf{T}}, \quad \mathbf{b} = \left(\frac{23}{4}, \frac{17}{4}, 18\right)^{\mathsf{T}}.$$

By computation, the equilibrium point of system $\dot{\mathbf{x}} = \mathbf{A}\mathbf{x} + \mathbf{a}$ is $\mathbf{p} = (2, 9, -3)^{\mathsf{T}}$, and the equilibrium point of system $\dot{\mathbf{x}} = \mathbf{B}\mathbf{x} + \mathbf{b}$ is $\mathbf{q} = (-4, 1, -5)^{\mathsf{T}}$.

In addition, we get $\mathbf{A} = \mathbf{P}\mathbf{J}_{\mathbf{A}}\mathbf{P}^{-1}$ and $\mathbf{B} = \mathbf{Q}\mathbf{J}_{\mathbf{B}}\mathbf{Q}^{-1}$, with

$$\mathbf{J}_{\mathbf{A}} = \begin{pmatrix} 0.5 & -10 & 0 \\ 10 & 0.5 & 0 \\ 0 & 0 & -20 \end{pmatrix}, \quad \mathbf{P} = \begin{pmatrix} 1 & 1 & 2 \\ -\frac{1}{3} & -1 & 4 \\ -\frac{2}{3} & -8 & 2 \end{pmatrix},$$

$$\mathbf{J}_{\mathbf{A}} = \begin{pmatrix} -1 & 0 & 0 \\ 0 & -1 & 0 \\ 0 & 0 & 1 \end{pmatrix}, \quad \mathbf{Q} = \begin{pmatrix} 1 & 1 & 1 \\ 3 & 3 & \frac{1}{3} \\ 6 & 3 & -\frac{4}{3} \end{pmatrix}.$$

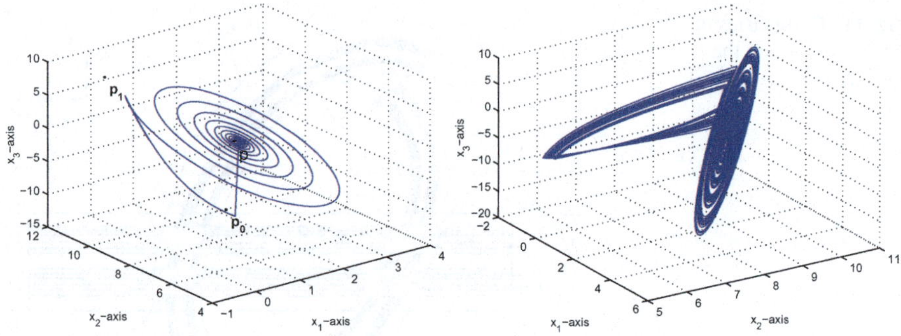

Fig. 12 **a** The homoclinic orbit of system (48); **b** A chaotic set of system

Moreover, we obtain that

$$\mathbf{p}_0 = (0, 5, -5), \quad \rho_1 = \frac{1}{3}, \quad \rho_2 = \frac{2}{3}, \quad \rho_3 = 3.$$

Considering conditions in Theorem 21.1, let $k_1 = k_2 = -1, \; T = \ln 3$. Then, we have

$$\mathbf{p}_1 = \left(0, \frac{31}{3}, \frac{17}{3}\right)^{\mathsf{T}} = \mathbf{p} + k_1\zeta_1 + k_2\zeta_2,$$

$$\sigma_1 = \sigma_3 = 1, \quad \sigma_2 = 2, \quad e^{\mathbf{J}T}(\sigma_1, \sigma_2, \sigma_3)^{\mathsf{T}} = (\rho_1, \rho_2, \rho_3)^{\mathsf{T}}.$$

$$(\xi_{11}, \xi_{21}, \xi_{31})\mathbf{J}(\sigma_1, \sigma_2, \sigma_3)^{\mathsf{T}} = -2 < 0,$$

$$(\xi_{11}, \xi_{21}, \xi_{31})\mathbf{J}(\rho_1, \rho_2, \rho_3)^{\mathsf{T}} = 2 > 0.$$

$$\alpha_1 x_{1\mathbf{p}} - \beta_1(k_1\zeta_{21} - k_2\zeta_{11}) = 1 > 0,$$

$$\sqrt{x_{1\mathbf{p}}^2 + (k_1\zeta_{21} - k_2\zeta_{11})^2}\frac{\beta_1}{\sqrt{\alpha_1{}^2 + \beta_1{}^2}}e^{-\alpha T'} < 2e^{-0.05\pi} < 2.$$

Thus, system (48) satisfies the conditions in Theorems 21.1 and 21.9. Hence, the system has a homoclinic orbit to equilibrium point $\mathbf{p}$, as shown in Fig. 12a.

In addition, from $\alpha_1 + \lambda = -19.5 < 0$, system (48) satisfies the conditions in Theorem 21.2. Thus, system (48) has infinitely many chaotic invariant sets. A chaotic invariant set is shown in Fig. 12b.

Moreover, from computer simulations, the invariant set appears to be a chaotic attractor.

Example 2 ([9]) In system (26), let $\gamma = -3, \; \mu = -1$, and in system (26), let

$$\alpha = 1, \; \rho = -1.5, \; \lambda = \delta = -1, \; \beta = 1\,.$$

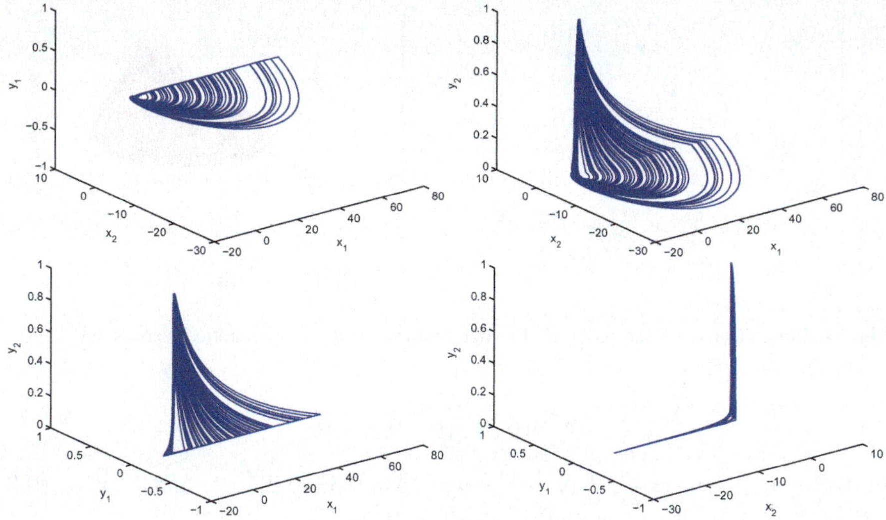

Fig. 13 Chaotic attractor in Example 2

Then, the conditions in Theorems 21.3 and 21.10 are satisfied. Hence, there exists a chaotic invariant set. In fact, by choosing randomly a number of initial states in a small neighborhood of the homoclinic orbit and computing their trajectories for a long time, we see that the chaotic invariant set is chaotic attractor as shown in Fig. 13.

Example 3 ([12]) For system (35), let

$$A = \begin{pmatrix} -0.3 & 4 & 0 \\ -4 & -0.3 & 0 \\ 0 & 0 & 10 \end{pmatrix}, \quad B = \begin{pmatrix} -0.2 & -5 & 0 \\ 0 & -0.3 & 0 \\ 0 & 0 & 15 \end{pmatrix},$$

with $\mathbf{a} = (-1.08, 1.69, -2)^T$, $\mathbf{b} = (1.66, 0.09, -9)^T$, $\mathbf{c} = (1, 0, 1)^T$ and $d = 1$. Then, the eigenvalues of A and B are

$$\lambda_A = 10, \alpha \pm \beta j = -0.3 \pm 4j,$$

and

$$\lambda_B = 15, \mu_1 = -0.2, \mu_2 = -0.3,$$

respectively. Additionally, $\mathbf{p} = (0.4, 0.3, 0.2)^T$, $\mathbf{q} = (0.8, 0.3, 0.6)^T$, $\mathbf{p}_0 = (0.4, 0.3, 0.6)^T$ and $\mathbf{q}_0 = (0.8, 0.3, 0.2)^T$. Hence,

$$\mathbf{c}^T\mathbf{p} = 0.6 < 1 = d, \mathbf{c}^T\mathbf{q} = 1.4 > 1 = d,$$

$$\mathbf{c}^T\mathbf{p}_0 = 1 = d, \mathbf{c}^T\mathbf{q}_0 = 1 = d,$$

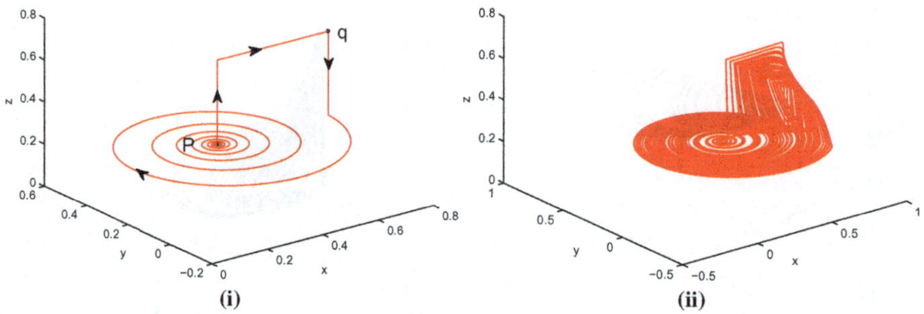

Fig. 14 Illustrations of Example 3: (i) The heteroclinic cycle, (ii) A chaotic invariant set

$$\mathbf{c}^T \mathrm{B}(\mathbf{p}_0 - \mathbf{q}) = 0.8 > 0.$$

Furthermore, $\mathbf{x}_* = \frac{d - \mathbf{c}^T \mathbf{p}}{\mathbf{c}\mathrm{A}^{-1}\mathbf{c}^{\perp}} \mathrm{A}^{-1}\mathbf{c}^{\perp} + \mathbf{p} = (0.8, 0.33, 0.2)^T$, and $\mathbf{x}^* = (0.8, -0.147126, 0.2)^T$ by numerical calculations. Thus,

$$\mathbf{q}_0 \in (\mathbf{x}_*, \mathbf{x}^*).$$

Therefore, all conditions in Theorem 21.7 are satisfied, so there exists a heteroclinic cycle connecting $\mathbf{p}$ and $\mathbf{q}$, and crossing Σ transversally at $\mathbf{p}_0$ and $\mathbf{q}_0$, as exhibited in Fig. 14a. In addition,

$$\frac{\alpha \mu_1}{\lambda_A \lambda_B} = 0.0004 < 1.$$

Therefore, by Theorem 21.12, there exists a chaotic invariant set, as shown in Fig. 14b.

5 Summary

In this chapter, we have reviewed the recent results on the existence of homoclinic orbits and heteroclinic cycles in piecewise-affine systems, and shown how to extend the well-known Šil'nikov theory to study chaos in this class of piecewise-smooth systems. As illustrations of the results, we have demonstrated a few examples that have homoclinic cycles or heteroclinic cycles.

There are numerous topics to be studied within the context of piecewise-affine systems that are of general interest. Here, we just mention a few of them:

(1) After establishing the existence of homoclinic orbits and heteroclinic cycles, a natural step is to prove the existence of chaotic invariant sets under some mild conditions. For dimensions greater than three, this is a hard and tedious job. It is expected to make some progress in the near future.

(2) It is also of significance to investigate global bifurcations near homoclinic orbits and heteroclinic cycles. As commented at the beginning of this chapter, detailed

investigations on the dynamics of piecewise-affine systems will also provide deep insights into the dynamics of smooth dynamical systems.

(3) The existence of periodic orbits, especially limit cycles, in piecewise-affine systems is a good topic. Since periodic phenomena are ubiquitous in natural and man-made systems, much work should be carried out on this topic. For two-dimensional piecewise-affine systems, there exist many publications on periodic orbits, as for higher-dimensional cases, much more work remains to be done to find methods and theories on the existence of periodic orbits.

Acknowledgements This study was supported by the National Natural Science Foundation of China (Grant numbers 11702077,11801329), the Fostering Master's Degree Empowerment Point Project of Hefei University (Grant number 2018xs03), the Talent Fund of Hefei University (Grant number 18-19RC58) and Natural Science Foundation of Shandong province (Grant number ZR2018BA002).

References

1. M.D. Bernardo, C.J. Budd, A.R. Champneys, P. Kowalczyk, *Piecewise-Smooth Dynamical Systems: Theory and Applications* (Springer, Berlin, 2008)
2. B. Brogliato, *Nonsmooth Mechanics: Models, Dynamics and Control* (Springer, Berlin, 2000)
3. A. Filippov, *Differential Equations with Discontinuous Right-Hand Side* (Springer, Berlin, 1988)
4. M. Kunze, *Non-smooth Dynamical Systems*. Lecture Notes in Mathematics, vol. 1744 (2000), pp. 97–100
5. R.I. Leine, H. Nijmeijer, *Dynamics and Bifurcations of Non-smooth Mechanical Systems* (Springer, Berlin, 2004)
6. C.K. Tse, M.D. Bernardo, Complex behavior in switching power converters. Proc. IEEE **90**(5), 768–781 (2002)
7. T. Wu, X.S. Yang, A new class of 3-dimensional piecewise affine systems with homoclinic orbits. Discret. Contin. Dyn. Syst. **36**(9), 5119–5129 (2016)
8. T. Wu, X.S. Yang, Construction of a class of four-dimensional piecewise affine systems with homoclinic orbits. Int. J. Bifurc. Chaos **26**, 1650099 (2016)
9. S. Huan, X. Yang, Existence of chaotic invariant set in a class of 4-dimensional piecewise linear dynamical systems. Int. J. Bifurc. Chaos **24**(12), 1450158 (2014)
10. T. Wu, L. Wang, X.S. Yang, Chaos generator design with piecewise affine systems. Nonlinear Dyn. **84**(2), 817–832 (2016)
11. T. Wu, X.S. Yang, On the existence of bifocal heteroclinic cycles in a class of four-dimensional piecewise affine systems. Chaos **26**(5), 454–466 (2016)
12. L. Wang, X.S. Yang, Heteroclinic cycles in a class of 3-dimensional piecewise affine systems. Nonlinear Anal.: Hybrid Syst. **23**, 44–60 (2017)
13. S. Wiggins, *Introduction to Applied Nonlinear Dynamical Systems and Chaos*, 2nd edn. (Springer, Berlin, 2003)
14. L. Shil'nikov, A. Shil'nikov, D. Turaev, L. Chua, *Methods of Qualitative Theory in Nonlinear Dynamics (Part II)* (World Scientific, Singapore, 2001)
15. C. Tresser, About some theorems by L.P. Shil'nikov. Ann. Inst. H. Poincaré **40**, 440–461 (1984)
16. S. Huan, Q. Li, X.S. Yang, Chaos in three-dimensional hybrid systems and design of chaos generators. Nonlinear Dyn. **69**(4), 1915–1927 (2012)
17. S. Ibánez, A. Rodrigues, On the dynamics near a homoclinic network to a bifocus: switching and horseshoes. Int. J. Bifurc. Chaos **25**, 1530030 (2015)

A New Chaotic System with Equilibria Located on a Line and Its Circuit Implementation

Fahimeh Nazarimehr, Mohammad-Ali Jafari, Sajad Jafari, Viet-Thanh Pham, Xiong Wang, and Guanrong Chen

Abstract In this chapter, a new three-dimensional chaotic system with infinitely many equilibria located on a line is proposed. Investigation of dynamical properties of the new system shows its various complex dynamical behaviours. Circuit implementation verifies the feasibility of the system for engineering applications.

F. Nazarimehr (✉) · S. Jafari
Biomedical Engineering Department, Amirkabir University of Technology,
Tehran 15875-4413, Iran
e-mail: fahimenazarimehr@gmail.com

S. Jafari
e-mail: sajadjafari@aut.ac.ir

M.-A. Jafari · S. Jafari
Health Technology Research Institute, Amirkabir University of Technology, No. 350, Hafez Ave, Valiasr Square, 159163-4311 Tehran, Iran
e-mail: mohammadalijafari86@gmail.com

V.-T. Pham
School of Electronics and Telecommunications, Hanoi University of Science and Technology,
01 Dai Co Viet, Hanoi 10000, Vietnam
e-mail: pvt3010@gmail.com

X. Wang
Institute for Advanced Study, Shenzhen University, Shenzhen, Guangdong 518060, People's Republic of China
e-mail: wangxiong8686@szu.edu.cn

G. Chen
IDepartment of Electrical Engineering, City University of Hong Kong,
Hong Kong SAR 999077, People's Republic of China
e-mail: eegchen@cityu.edu.hk

X. Wang et al. (eds.), *Chaotic Systems with Multistability and Hidden Attractors*,
Emergence, Complexity and Computation 40,
https://doi.org/10.1007/978-3-030-75821-9_22

1 Introduction

Most known chaotic flows such as the Lorenz [1], Rössler [2], Chen [3], Sprott (cases B–S) [4] attractors are excited by unstable equilibria. Thus, they can be found by putting initial conditions in the vicinity of the saddle points [5]. Recently, a new category, which is called "hidden attractors" amuses a lot of attention [5–16]. Chaotic flows of this category are not associated with any saddle point, including the cases without any equilibrium points, with only stable equilibria, or with an infinite number of equilibrium points [17–32]. The attractors of this category are called hidden because of the difficulty in finding them without any systematic way. On the other hand, because of their unexpected and potentially disastrous responses to perturbations in such structures like a bridge or an aircraft wing, hidden attractors are important in engineering applications.

In this chapter, we introduce a new chaotic flow with a line of equilibria that has a different structure from those reported previously [30, 31]. In the next section, we describe the new chaotic system. In Sect. 3, we investigate the behavioral properties of the system from the viewpoint of bifurcation analysis and Lyapunov exponents. We propose circuit implementation of the system in Sect. 4. Finally, we conclude the study in Sect. 5.

2 Model of Simple Chaotic Flows with a Line of Equilibria

Consider the following system with six terms:

$$\begin{aligned} \dot{x} &= y \\ \dot{y} &= 0.4xz \\ \dot{z} &= 0.3y - 0.1z - 1.4y^2 + kxy. \end{aligned} \tag{1}$$

which is a three-dimensional flow with three state variables, $x,\ y,\ z$ and one parameter, k. The divergence of the system (1) is

$$\nabla V = \frac{\partial \dot{x}}{\partial x} + \frac{\partial \dot{y}}{\partial y} + \frac{\partial \dot{z}}{\partial z} = -0.1. \tag{2}$$

Thus, this system is dissipative with $\nabla V < 0$. In order to find the equilibrium points of the system, the right-hand sides of these equations are set to zeros:

$$\begin{aligned} &y = 0 \\ &0.4xz = 0 \\ &0.3y - 0.1z - 1.4y^2 + kxy = 0 \rightarrow z = 0. \end{aligned} \tag{3}$$

According to Eq. (3), the system of Eq. (1) has an infinite number of equilibria located on the line $E^* : y = 0\ \&\ z = 0$.

The Jacobian matrix of system (1) along the line of equilibria E^* is

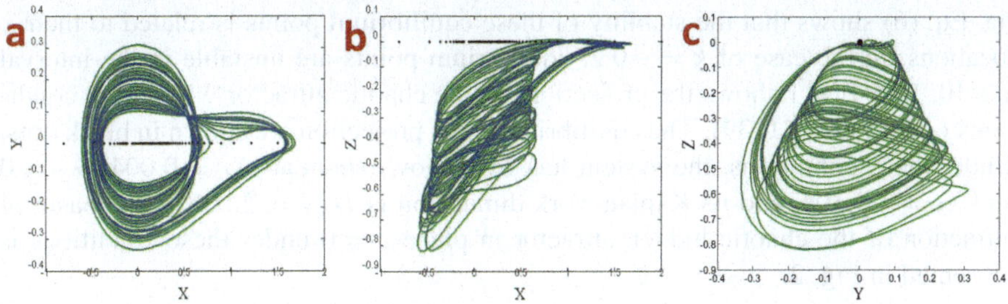

Fig. 1 The strange attractor's projections for system (1) with parameter $k = -0.2$ and initial conditions $(-1.53, 0.33, 0.39)$. Black dots show the line of equilibria

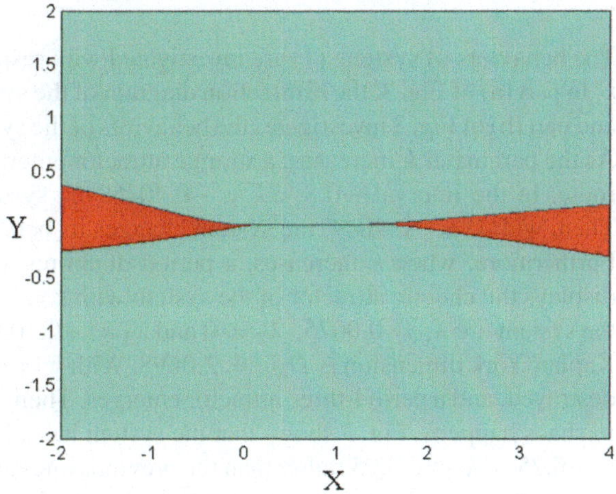

Fig. 2 Basins of attraction for system (1) with parameter $k = -0.2$ in the plane $z = 0$. Initial conditions in the light blue region lead to a chaotic attractor, and the red leads to the equilibrium points

$$J_{E^*} = \begin{bmatrix} 0 & 1 & 0 \\ 0.4z^* & 0 & 0.4x^* \\ ky^* & 0.3 - 2.8y^* + kx^* & -0.1 \end{bmatrix} = \begin{bmatrix} 0 & 1 & 0 \\ 0 & 0 & 0.4x \\ 0 & 0.3 + kx & -0.1 \end{bmatrix}. \tag{4}$$

Its characteristic equation, given by evaluating $|\lambda I - J| = 0$, is

$$\lambda \times (\lambda^2 + 0.1\lambda - 0.12x - 0.4kx^2) = 0. \tag{5}$$

Thus, the system has three eigenvalues:

$$\begin{aligned} \lambda &= 0, \\ \lambda &= \tfrac{-0.1 \pm \sqrt{0.01 + 0.48x + 1.6kx^2}}{2}. \end{aligned} \tag{6}$$

It is clear that the sign of the real part of the two nonzero eigenvalues is dependent on the x position of the equilibrium points on the line E^*.

Considering system (1) in terms of the parameter $k = -0.2$ shows that the system has a chaotic hidden attractor, which could not be found using the line of equilibria. Eq. (6) shows that the stability of these equilibrium points is related to their x locations. In the case of $k = -0.2$, equilibrium points are unstable in the interval $x \in [0, 1.5]$. Fig. 1 shows the projections of the chaotic attractor with initial conditions $(-1.53, 0.33, 0.39)$. The equilibria in each projection are shown in black dots. Under these conditions, the system has Lyapunov exponents $\lambda_1 = 0.004$, $\lambda_2 = 0$ and $\lambda_3 = -0.104$, and its Kaplan-York dimension is $D_{KY} = 2.0385$. The basin of attraction of the chaotic hidden attractor in plane $z = 0$ under these conditions is presented in Fig. 2.

3 Bifurcation Analysis

The behaviors of system (1) are investigated with respect to the changing parameter k. In part (a) of Fig. 3, the bifurcation diagram of the system is shown by increasing k, and part (b) of Fig. 3 investigates the behaviors of the system via Lyapunov exponents. As the parameter k increases, a strange attractor emerged through a period-doubling route. In the interval $-0.8 < k < -0.4045$, the system has an equilibrium point. Then, with $k = -0.4044$, the system changes its behavior and creates a limit cycle. Furthermore, when k increases, a period-doubling route to chaos happens. Fig. 4 displays the chaotic attractor of the system with $k = -0.27$. Lyapunov exponents of the system are $\lambda_1 = 0.0075$, $\lambda_2 = 0$ and $\lambda_3 = -0.1075$ under this condition, and its Kaplan-York dimension is $D_{KY} = 2.0698$. With $k = -0.268$, the chaotic attractor is destroyed, and a period-three attractor emerges. Then, another period-doubling route to chaos happens. Fig. 3 shows that the system has a little different chaotic behavior in $-0.25 < k < -0.05$ rather than the previous ones and walks through two separate regions. Comparing the attractors in Figs. 1 and 4 highlights the differences of these two chaotic attractors.

4 Circuit Implementation

Physical realizations of chaotic models have been applied in various engineering areas. In this section, an electronic circuit is introduced to realize the theoretical system (1). By using common electronic components, the circuit is designed as presented in Fig. 5. In Fig. 5, the voltages of capacitors are denoted as X, Y, Z.

By applying Kirchhoff's circuit laws to the circuit, its circuital equations are obtained as follows:

$$\begin{aligned} \dot{X} &= \frac{1}{R_1C_1}Y \\ \dot{Y} &= \frac{1}{R_2C_2 10V}XZ \\ \dot{Z} &= \frac{1}{R_3C_3}Y - \frac{1}{R_4C_3}Z - \frac{1}{R_5C_3 10V}Y^2 + \frac{1}{R_6C_3 10V}XY \end{aligned} \tag{7}$$

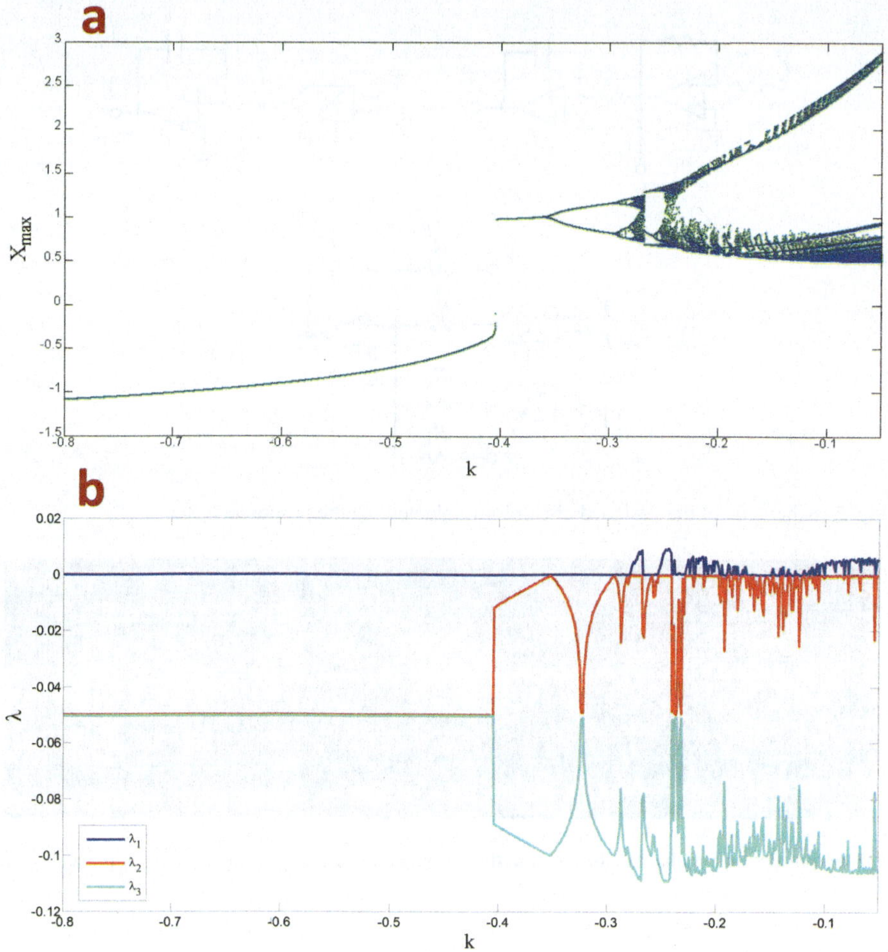

Fig. 3 Bifurcation diagram and Lyapunov exponents of system (1). **a** Bifurcation diagram with respect to parameter k and initial conditions $(-1.53, 0.33, 0.39)$. **b** Lyapunov exponents with respect to parameter k

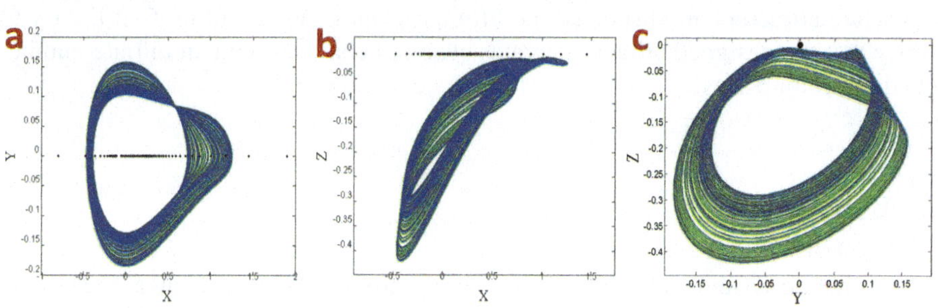

Fig. 4 Projections of the strange attractor in system (1) with parameter $k = -0.27$ and initial conditions $(-1.53, 0.33, 0.39)$. Black dots show the line of equibria

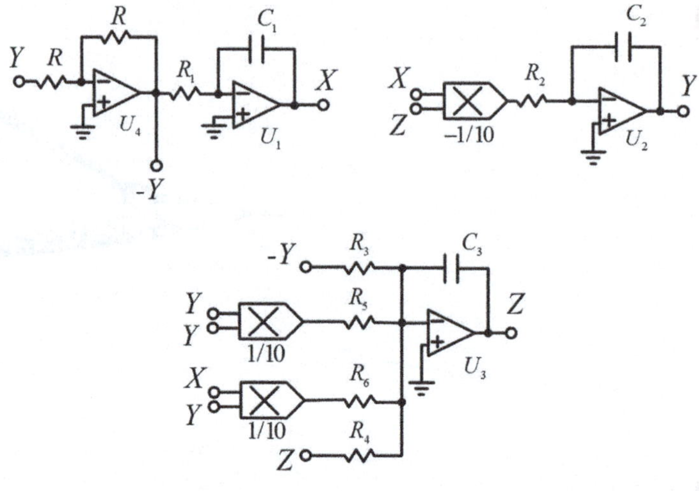

Fig. 5 Schematic of the designed circuit for implementing the proposed system (1)

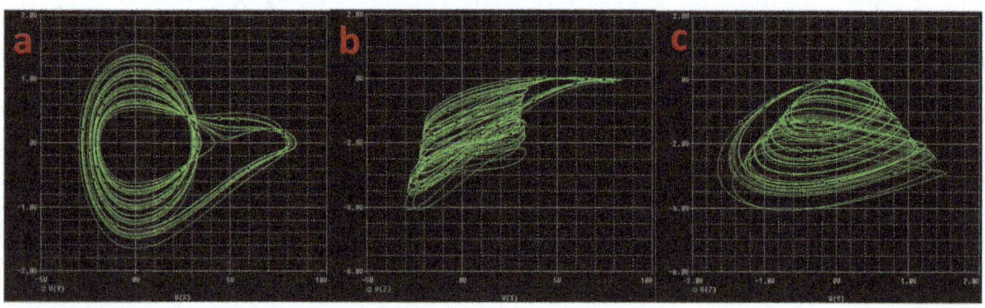

Fig. 6 Projections of chaotic attractors in the designed circuit. **a** X–Y plane. **b** X–Z plane. **c** Y–Z plane

The circuit has been implemented in the electronic simulation package Cadence OrCAD, with $R_1 = R = 21k\Omega,\ R_2 = 26.25k\Omega,\ R_3 = 70k\Omega,\ R_4 = 210k\Omega,\ R_5 = 7.5k\Omega,\ R_6 = 52.5k\Omega$, and $C_1 = C_2 = C_3 = C = 2.2nF$. The obtained projections of chaotic attractors are displayed on different planes shown in Fig. 6. It is easy to verify that the designed circuit generates chaotic behaviors with an infinite number of equilibrium points.

5 Conclusion

A new chaotic system with a line of equilibria is proposed in this chapter. The system is a simple one with six terms and one parameter. Bifurcation analysis and circuit implementation of the proposed system are investigated. Bifurcation analysis of the simple system shows its ability to have complex behaviors. Circuit Implementation of the system demonstrates its feasibility and a good match between numerical and circuit implementation results, illustrating the ability of the system for real applications.

References

1. E.N. Lorenz, Deterministic nonperiodic flow. J. Atmos. Sci. **20**(2), 130–141 (1963)
2. O.E. Rössler, An equation for continuous chaos. Phys. Lett. A **57**(5), 397–398 (1976)
3. G. Chen, T. Ueta, Yet another chaotic attractor. Int. J. Bifurc. Chaos **9**(07), 1465–1466 (1999)
4. J. Sprott, Some simple chaotic flows. Phys. Rev. E **50**(2), R647 (1994)
5. G. Leonov, N. Kuznetsov, V. Vagaitsev, Localization of hidden Chua's attractors. Phys. Lett. A **375**(23), 2230–2233 (2011)
6. G. Leonov, N. Kuznetsov, Hidden oscillations in dynamical systems. 16 Hilbert's problem, Aizerman's and Kalman's conjectures, hidden attractors in Chua's circuits. J. Math. Sci. **201**(5), 645–662 (2014)
7. G. Leonov, N. Kuznetsov, M. Kiseleva, E. Solovyeva, A. Zaretskiy, Hidden oscillations in mathematical model of drilling system actuated by induction motor with a wound rotor. Nonlinear Dyn. **77**(1–2), 277–288 (2014)
8. G. Leonov, N. Kuznetsov, Analytical-numerical methods for investigation of hidden oscillations in nonlinear control systems. IFAC Proc. **18**(1), 2494–2505 (2011)
9. G. Leonov, N. Kuznetsov, T. Mokaev, Hidden attractor and Homoclinic orbit in Lorenz-like system describing convective fluid motion in rotating cavity. Commun. Nonlinear Sci. Numer. Simul. **28**(1), 166–174 (2015)
10. G. Leonov, N. Kuznetsov, V. Vagaitsev, Hidden attractor in smooth Chua systems. Physica D **241**(18), 1482–1486 (2012)
11. G. Leonov, N. Kuznetsov, Hidden attractors in dynamical systems. from hidden oscillations in Hilbert–Kolmogorov, Aizerman, and Kalman problems to hidden chaotic attractor in Chua circuits. Int. J. Bifurc. Chaos **23**(1), 1330002 (2013)
12. G. Leonov, N. Kuznetsov, Prediction of hidden oscillations existence in nonlinear dynamical systems: analytics and simulation, in *Nostradamus 2013: Prediction, Modeling and Analysis of Complex Systems* (Springer, Berlin, 2013), pp. 5–13
13. V. Bragin, V. Vagaitsev, N. Kuznetsov, G. Leonov, Algorithms for finding hidden oscillations in nonlinear systems. the Aizerman and Kalman conjectures and Chua's circuits. J. Comput. Syst. Sci. Int. **50**(4), 511–543 (2011)
14. N. Kuznetsov, G. Leonov, S. Seledzhi, Hidden oscillations in nonlinear control systems. IFAC Proc. **18**(1), 2506–2510 (2011)
15. N. Kuznetsov, G. Leonov, V. Vagaitsev, Analytical-numerical method for attractor localization of generalized Chua's system. PSYCO, 2010, pp. 29–33
16. G. Leonov, Hidden oscillation in dynamical systems, in *From Physics to Control Through an Emergent View*, ed. by L. Fortuna, A. Fradkov, M. Frasca (World Scientific, Singapore, 2010)
17. S. Jafari, J. Sprott, S.M.R.H. Golpayegani, Elementary quadratic chaotic flows with no equilibria. Phys. Lett. A **377**(9), 699–702 (2013)
18. S. Jafari, J. Sprott, F. Nazarimehr, Recent new examples of hidden attractors. European Physical Journal: Special Topics **224**(8), 1469–1476 (2015)
19. S. Kingni, S. Jafari, H. Simo, P. Woafo, Three-dimensional chaotic autonomous system with only one stable equilibrium: Analysis, circuit design, parameter estimation, control, synchronization and its fractional-order form. Eur. Phys. J. Plus **129**(5), 1–16 (2014)
20. S. Lao, Y. Shekofteh, S. Jafari, J. Sprott, Cost function based on Gaussian mixture model for parameter estimation of a chaotic circuit with a hidden attractor. Int. J. Bifurc. Chaos **24**(01), 1450010 (2014)

21. M. Molaie, S. Jafari, J. Sprott, S.M.R.H. Golpayegani, Simple chaotic flows with one stable equilibrium. Int. J. Bifurc. Chaos **23**(11), 1350188 (2013)
22. V.-T. Pham, S. Jafari, C. Volos, X. Wang, S.M.R.H. Golpayegani, Is that really hidden? the presence of complex fixed-points in chaotic flows with no equilibria. Int. J. Bifurc. Chaos **24**(11), 1450146 (2014)
23. V.-T. Pham, S. Vaidyanathan, C. Volos, S. Jafari, Hidden attractors in a chaotic system with an exponential nonlinear term. Eur. Phys. J.: Special Topics **224**(8), 1507–1517 (2015)
24. V.-T. Pham, C. Volos, S. Jafari, X. Wang, S. Vaidyanathan, Hidden hyperchaotic attractor in a novel simple memristive neural network. Optoelectron. Adv. Mater. Rapid Commun. **8**, 1157–1163 (2014)
25. V.-T. Pham, C. Volos, S. Jafari, Z. Wei, X. Wang, Constructing a novel no-equilibrium chaotic system. Int. J. Bifurc. Chaos **24**(05), 1450073 (2014)
26. M. Shahzad, V.-T. Pham, M. Ahmad, S. Jafari, F. Hadaeghi, Synchronization and circuit design of a chaotic system with coexisting hidden attractors. Eur. Phys. J.: Special Topics **224**(8), 1637–1652 (2015)
27. F.R. Tahir, S. Jafari, V.-T. Pham, C. Volos, X. Wang, A novel no-equilibrium chaotic system with multiwing butterfly attractors. Int. J. Bifurc. Chaos **25**(04), 1550056 (2015)
28. V.-T. Pham, S. Vaidyanathan, C. Volos, S. Jafari, S. Kingni, A no-equilibrium hyperchaotic system with a cubic nonlinear term. Optik: Int. J. Light Electron Opt. **127**(6), 3259–3265 (2016)
29. S. Goudarzi, S. Jafari, M.H. Moradi, J. Sprott, Narx prediction of some rare chaotic flows: Recurrent fuzzy functions approach. Phys. Lett. A **380**(5), 696–706 (2016)
30. S. Jafari, J. Sprott, Simple chaotic flows with a line equilibrium. Chaos, Solitons Fractals **57**, 79–84 (2013)
31. S. Jafari, J. Sprott, Erratum to:“Simple chaotic flows with a line equilibrium [Chaos Solitons and Fractals 57 (2013) 79–84]". Chaos, Solitons Fractals **77**, 341–342 (2015)
32. S. Jafari, J. Sprott, M. Molaie, A simple chaotic flow with a plane of equilibria. Int J. Bifurc. Chaos **26**(06), 1650098 (2016)

A Comprehensive Analysis on the Wang-Chen System: A Challenging Case for the Šil'nikov Theory

Atiyeh Bayani, Mohammad-Ali Jafari, Sajad Jafari, and Viet-Thanh Pham

1 Introduction

Discovering and analyzing new chaotic systems with desired properties have received more and more interest and attention because of the potential applications of such systems in biology [1–8], economics [9, 10], ecosystems [11, 12] and electronic circuits [13, 14], etc. Meanwhile, having preferred number(s) of equilibria has attracted even more interest due to the special features of such dynamical systems. Some of such systems are chaotic with no equilibrium [15–23], only one stable equilibrium [24–27], a curve of equilibria [28–32], or a plane of equilibria [33].

In this chapter, an interesting three-dimensional chaotic system, known as the Wang-Chen system [23], is analyzed, which has no equilibrium in some ranges of its key parameter. This system is clearly in the class of systems with hidden attractors [34, 35]. It is well known that some complex dynamical systems have coexisting attractors. This property of systems is called multistability [36–38]. As the Wang-Chen system does not have homoclinic and heteroclinic orbits, the Šil'nikov criterion

A. Bayani (✉) · S. Jafari
Biomedical Engineering Department, Amirkabir University of Technology, Tehran 15875-4413, Iran
e-mail: atiyeh.bayani@yahoo.com

S. Jafari
e-mail: sajadjafari@aut.ac.ir

M.-A. Jafari · S. Jafari
Health Technology Research Institute, Amirkabir University of Technology, No. 350, Hafez Ave, Valiasr Square, Tehran 159163-4311, Iran
e-mail: mohammadalijafari86@gmail.com

V.-T. Pham
V.-T. Pham School of Electronics and Telecommunications, Hanoi University of Science and Technology,01 Dai Co Viet, Hanoi 10000, Vietnam
e-mail: pvt3010@gmail.com

X. Wang et al. (eds.), *Chaotic Systems with Multistability and Hidden Attractors*, Emergence, Complexity and Computation 40,
https://doi.org/10.1007/978-3-030-75821-9_23

is not applicable to analyze this system [39]. We thus use Lyapunov exponents (LEs), bifurcation diagram and basin of attraction to evaluate this system. Furthermore, stability analysis of the system is performed in parameter ranges where it has equilibria, so as to characterize its statistical properties. Furthermore, an electronic implementation of the system is presented in Sect. 3. Finally, concluding remarks are given in Sect. 4.

2 A Chaotic System

A 3D chaotic autonomous system, the Wang-Chen system [23], is described by

$$\begin{aligned} \dot{x} &= y \\ \dot{y} &= z \\ \dot{z} &= -y + 3y^2 - x^2 - xy + a \end{aligned} \tag{1}$$

where a is a constant parameter.

When $a > 0$, this system has two symmetrical equilibria: $(\sqrt{a}, 0, 0)$ and $(-\sqrt{a}, 0, 0)$, with the Jacobian matrix given by

$$J = \begin{bmatrix} 0 & 1 & 0 \\ 0 & 0 & 1 \\ -2x - z & -1 + 6y & -x \end{bmatrix} = \begin{bmatrix} 0 & 1 & 0 \\ 0 & 0 & 1 \\ \mp 2\sqrt{a} & -1 & \mp\sqrt{a} \end{bmatrix} \tag{2}$$

The characteristic equation is

$$\lambda^3 \pm \sqrt{a}\lambda^2 + \lambda \pm 2\sqrt{a} = 0\ . \tag{3}$$

with eigenvalues $(\sqrt{a}, 0, 0)$ and $(-\sqrt{a}, 0, 0)$ as follows:

$$\begin{aligned} \lambda_{1(\sqrt{a},0,0)} &= \frac{\sqrt[3]{2}(a-3)}{3\sqrt[3]{-2a^{3/2}+\sqrt{216a^2+1917a+108}-45\sqrt{a}}} \\ &+ \frac{\sqrt[3]{-2a^{3/2}+\sqrt{216a^2+1917a+108}-45\sqrt{a}}}{3\sqrt[3]{2}} - \frac{\sqrt{a}}{3} \\ \lambda_{2,3(\sqrt{a},0,0)} &= -\frac{(1\pm i\sqrt{3})(a-3)}{3\times 2^{2/3}\sqrt[3]{-2a^{3/2}+\sqrt{216a^2+1917a+108}-45\sqrt{a}}} \\ &+ \frac{(1\mp i\sqrt{3})\sqrt[3]{-2a^{3/2}+\sqrt{216a^2+1917a+108}-45\sqrt{a}}}{6\sqrt[3]{2}} - \frac{\sqrt{a}}{3} \end{aligned} \tag{4}$$

and

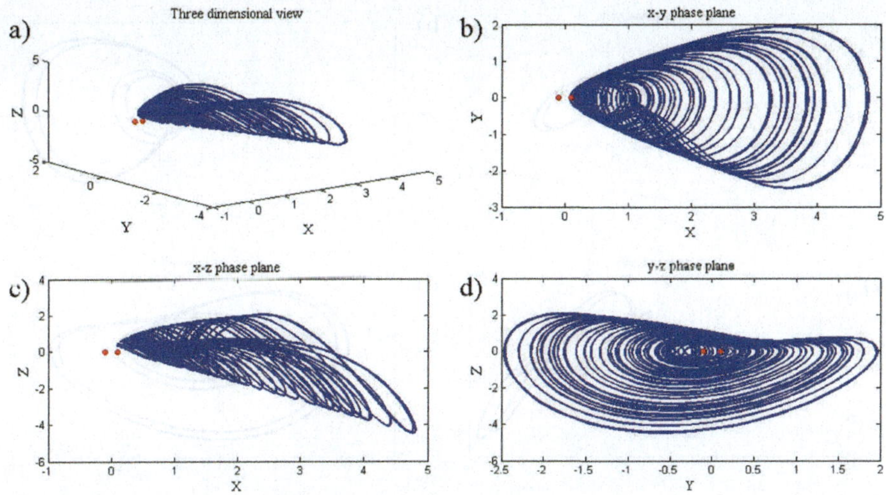

Fig. 1 The chaotic attractor of system (1), which has two symmetrical equilibria when $a = 0.01$. The 3D view (**a**), and projections on planes x–y (**b**), $x - z$ (**c**), $y - z$ (**d**). Red dots are the equilibria

$$\begin{aligned}
\lambda_{1(-\sqrt{a},0,0)} &= \frac{\sqrt[3]{2}(a-3)}{3\sqrt[3]{2a^{3/2}+\sqrt{216a^2+1917a+108}+45\sqrt{a}}} \\
&+\frac{\sqrt[3]{2a^{3/2}+\sqrt{216a^2+1917a+108}+45\sqrt{a}}}{3\sqrt[3]{2}}+\frac{\sqrt{a}}{3} \\
\lambda_{2,3(-\sqrt{a},0,0)} &= -\frac{(1\pm i\sqrt{3})(a-3)}{3\times 2^{2/3}\sqrt[3]{2a^{3/2}+\sqrt{216a^2+1917a+108}+45\sqrt{a}}} \\
&+\frac{(1\mp i\sqrt{3})\sqrt[3]{2a^{3/2}+\sqrt{216a^2+1917a+108}+45\sqrt{a}}}{6\sqrt[3]{2}}+\frac{\sqrt{a}}{3}
\end{aligned} \tag{5}$$

Considering these eigenvalues, both equilibria are spiral repellers for $a > 0$, at which these equilibria exist. In these ranges, the system has chaotic and periodic dynamics as shown in Figs. 1 and 2, respectively.

When $a = 0$, these two symmetrical equilibria merge into one, the origin (0, 0, 0).

When $a < 0$, there is no equilibrium in this system, but interestingly the system can generate a chaotic attractor, as shown in Fig. 3.

2.1 Lyapunov Exponents

To evaluate the chaoticity of the Wang-Chen system, LEs are calculated as functions of the key parameter a.

In Fig. 4a, LEs are shown when parameter a is changed from -0.078 to 5.0. To find multistability of this system, consider the Largest Lyapunov Exponent (LLE). Other LEs are also plotted as the parameter a is changed from 5.0 to -0.078, as shown in Fig. 5a.

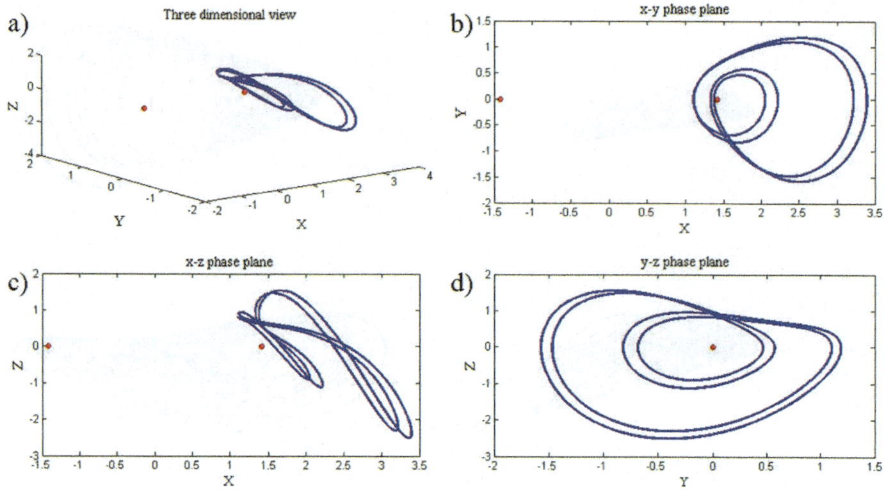

Fig. 2 The periodic attractor of system (1), which has two symmetrical equilibria when $a = 2$. The 3D view (**a**), and projections on planes $x - y$ (**b**), $x - z$ (**c**) and $y - z$ (**d**). Red dots are the equilibria

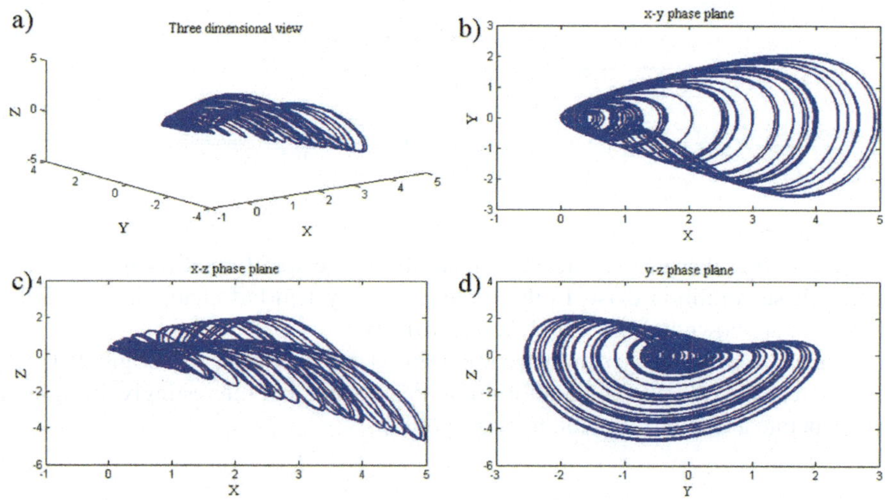

Fig. 3 The chaotic attractor of system (1) when $a = -0.05$. The 3D view (**a**) and projections on planes $x - y$ (**b**), $x - z$ (**c**) and $y - z$ (**d**)

Comparing Figs. 4a and 5a, for some values of the parameter ($a = 0.196 - 0.244$), the signs of LLE in these figures are opposite, which shows that a chaotic attractor (sign of LLE is positive) coexists with another attractor (sign of LLE is negative) with these parameter values. As this system does not have any equilibria within this range, the other attractor would be a limit cycle or a torus.

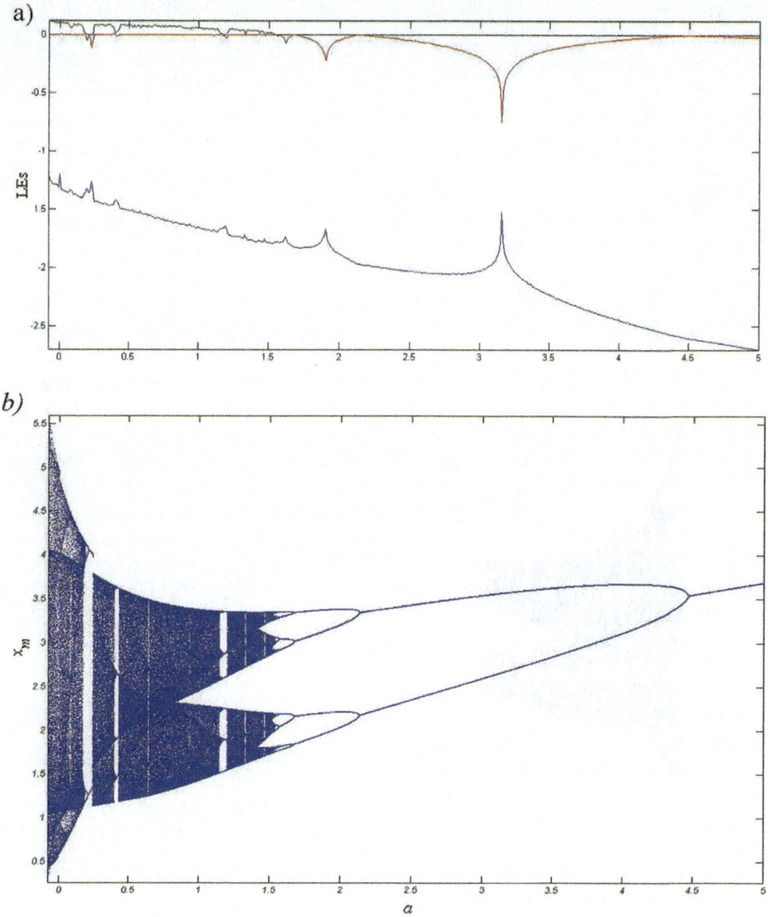

Fig. 4 **a** Lyaponuv exponents (LEs). **b** Bifurcation diagram are shown as functions of the parameter a. The parameter a increases from -0.078 to 5.0

Furthermore, other available methods like bifurcation diagram, basin of attraction and phase portrait, which can determine the types of a multistable attractor, are presented in the following.

2.2 Bifurcation Analysis

To illustrate the qualitative changes of the system versus the key parameter a, bifurcation diagram of the system is presented in Fig. 4b. Regarding Fig. 4b, the system has a period-doubling bifurcation route to chaos and some periodic windows exist

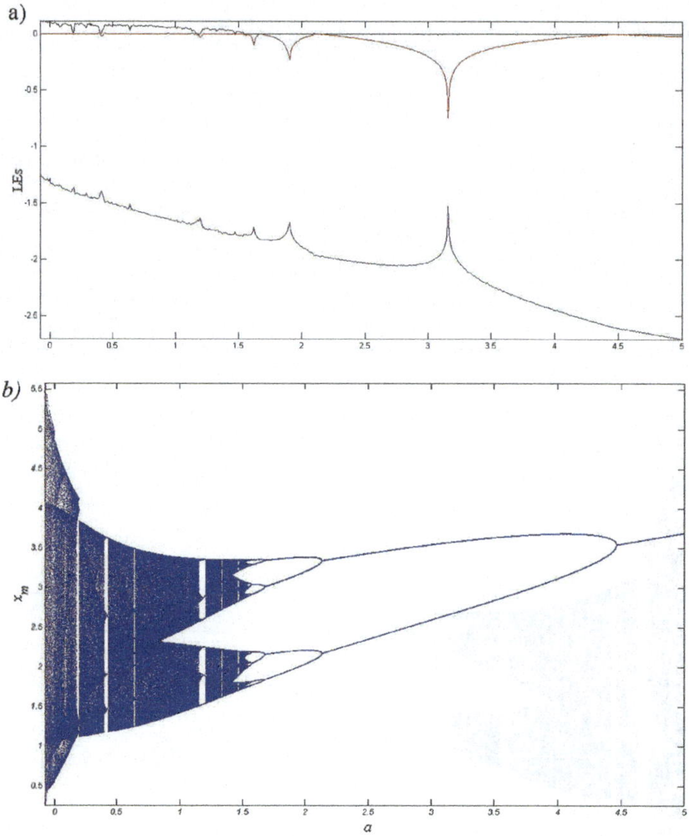

Fig. 5 **a** Lyaponuv exponents (LEs). **b** Bifurcation diagram, as functions of the parameter a. The parameter a decreases from 5.0 to -0.078

inside the chaotic regions. Like most chaotic systems, different selections of initial conditions may result in different attractors from the system, so other bifurcations may happen. It is necessary to select, when $a = -0.078$, the initial condition $(2.1428, -1.7663, 1.4481)$. The initial condition for each parameter value is the final value of x, y and z, corresponding to its previous parameter value, which is reached after $t = 2000s$ in simulations.

To examine the existence of multistability, a bifurcation diagram of the system is shown as the parameter decreases from 5.0 to -0.078. To compare Fig. 4b with Fig. 5b, these figures are plotted together in Fig. 6. Multistability is seen when $a = 0.196 - 0.244$, where a chaotic attractor coexists with a limit cycle.

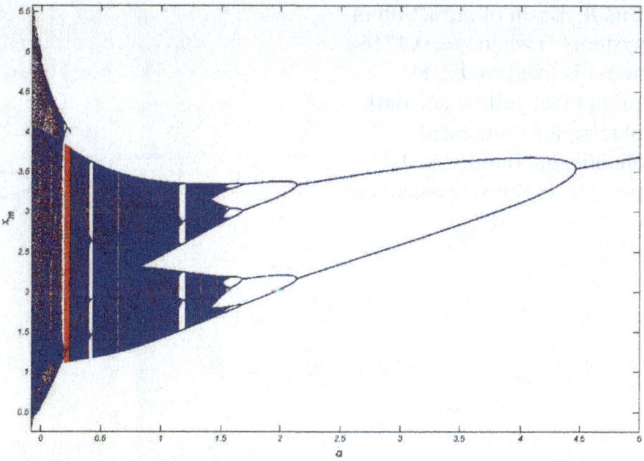

Fig. 6 Bifurcation diagram of the system as a increases from -0.078 to 5.0 (blue) and decreases from 5.0 to -0.078. Multistability is seen when $a = 0.196$ to 0.244, where a chaotic attractor coexists with a limit cycle

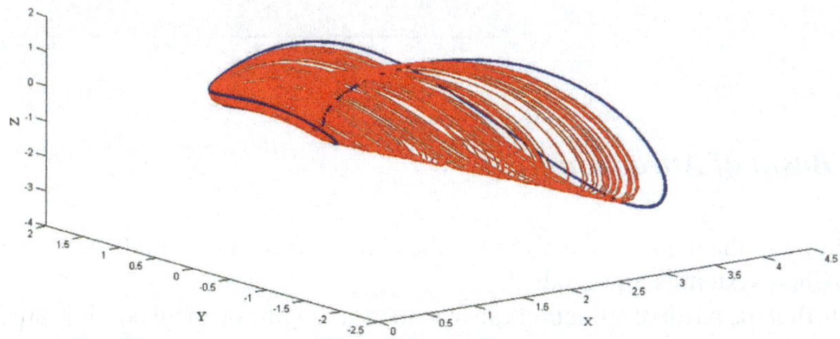

Fig. 7 When $a = 0.2180$, with initial conditions (3.022,1.196,1.643) and (1.276, -0.190, 0.471), a limit cycle and a chaotic attractor appears, respectively

2.3 *Coexisting Attractors*

Complex dynamical systems usually have some coexisting attractors, which are distinguished from one another by appropriate selections of initial conditions. This property of complex dynamical systems may occur between different or even the same kind of systems. For example, when the control parameter is fixed, a limit cycle may coexist with a strange attractor in a system, if they have their distinct basins of attraction.

In the Wang-Chen system, as shown in Fig. 7, when $a = 0.2180$, for initial conditions (3.022,1.196,1.643) and (1.276, -0.190, 0.471), a limit cycle and a chaotic attractor appears, respectively.

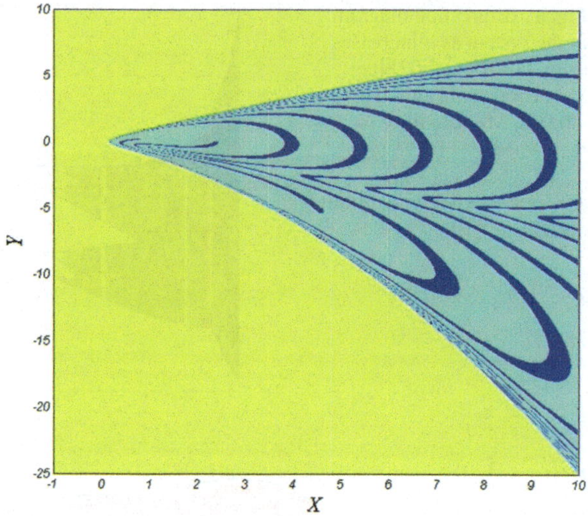

Fig. 8 Basin of attraction in system (1) when $z = 0.47160$ and a is fixed to 0.2180. Light blue, yellow and dark blue regions represent chaotic, unbounded and periodic regions, respectively

2.4 *Basin of Attraction*

In addition to the dynamics analyzed in the above two sections, multistability of the Wang-Chen system is now studied.

Note that the basin of attraction can lead to the recognition of all possible attractors coexisting with each other. To identify initial condition regions, parameter a is fixed to 0.2180. The basin of attraction of each attractor is shown in Fig. 8 when $z = 0.47160$, and in Fig. 9 when $z = 0$. In Fig. 9, one can see the equilibria of this system on the x-y plane. As mentioned in Sect. 2, both of these equilibria are unstable.

As shown by Fig. 9, $(-\sqrt{a}, 0, 0)$ is in the basin of an unbounded attractor and, interestingly, $(\sqrt{a}, 0, 0)$ is in the basins of both periodic and chaotic attractors. So, both periodic and chaotic attractors are self-excited via this parameter. Also, the basin of attraction of the system is considered when the system does not have any number of equilibrium (when $a = -0.05$), as shown in Fig. 10. As Fig. 10 shows, with this parameter, the system has a chaotic attractor and an unbounded orbit. Because there is no equilibrium with this parameter value, the chaotic attractor is a hidden attractor.

In general, considering the basin of attraction of the Wang-Chen system would be the most effective way to identify distinct attractors. But, in the case that the same type of attractors coexists with each other, it is useful to consider bifurcation diagrams beside the basin of attraction.

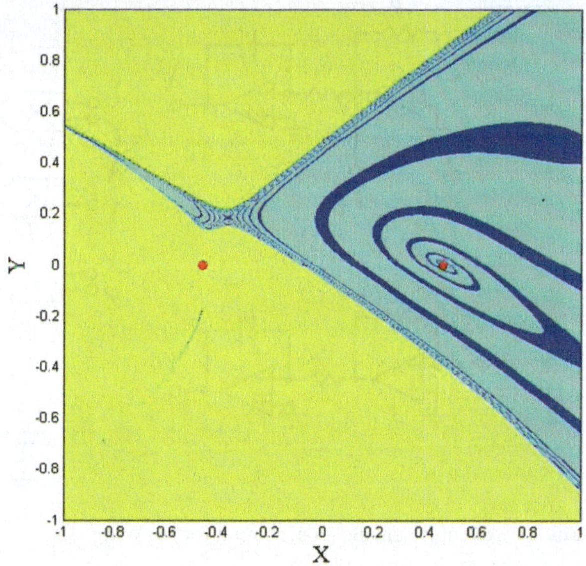

Fig. 9 Basin of attraction in system (1) when $z = 0$ and a is fixed to 0.2180. Light blue, yellow and dark blue regions represent chaotic, unbounded and periodic regions, respectively. Both equilibria of the systems are shown in red

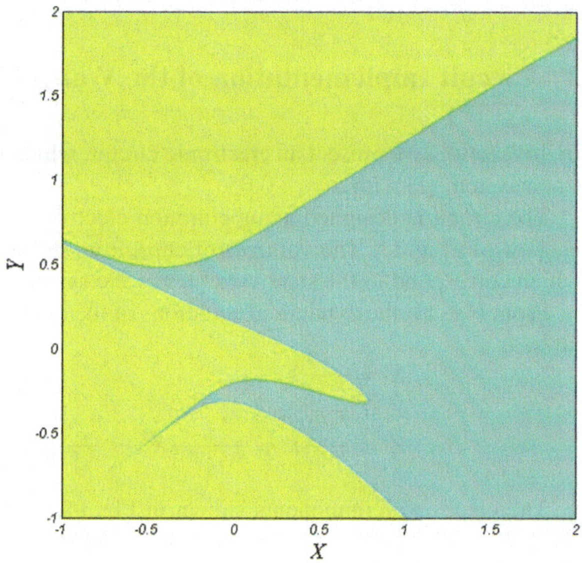

Fig. 10 Basin of attraction in system (1) when $z = 0$ and a is fixed to -0.05. Light blue and yellow represent chaotic and unbounded regions, respectively

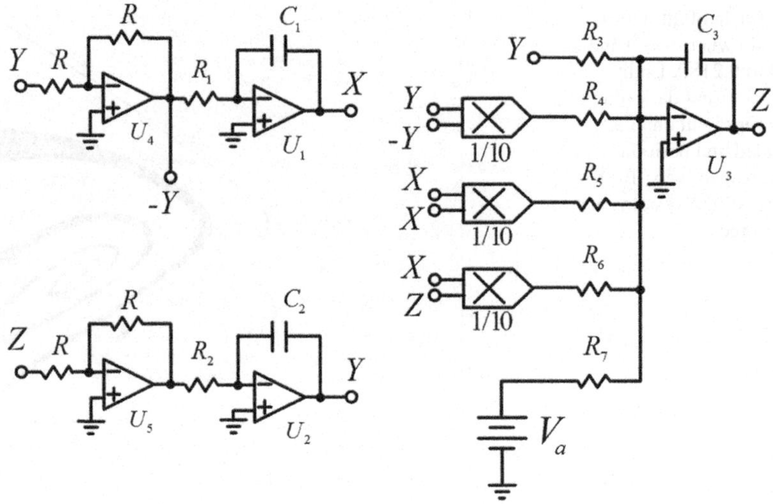

Fig. 11 An equivalent electronic circuit for the Wang-Chen system (1)

3 Circuit Implementation of the Wang-Chen System

In this section, we present an electronic circuit, which realizes the theoretical Wang-Chen system (1) with $a = -0.05$.

The circuit is designed using common electronic components and its schematic is shown in Fig. 11. The voltages of capacitors in Fig. 11 are denoted as X, Y, Z, which correspond to the state variables of the system (1).

From Fig. 11, the dynamical equations of the circuit are derived from Kirchhoff's laws, as

$$\begin{aligned} \dot{X} &= \tfrac{1}{R_1C_1}Y \\ \dot{Y} &= \tfrac{1}{R_2C_2 10V}Z \\ \dot{Z} &= \tfrac{-1}{R_3C_3}Y + \tfrac{1}{R_4C_3 10V}X^2 - \tfrac{1}{R_6C_3 10V}XZ + \tfrac{1}{R_7C_3}V_a. \end{aligned} \tag{6}$$

The electronic components shown in Fig. 11 are selected as: $R_1 = R_2 = R_3 = R_7 = R = 30\,k\Omega$, $R_4 = 1\,k\Omega$, $R_5 = R_6 = 3\,k\Omega$, $V_a = 0.05\,V_{DC}$, and $C_1 = C_2 = C_3 = C = 2\,nF$. The designed circuit is implemented by using PSpice. Fig. 12 illustrates the PSpice results of the designed circuit.

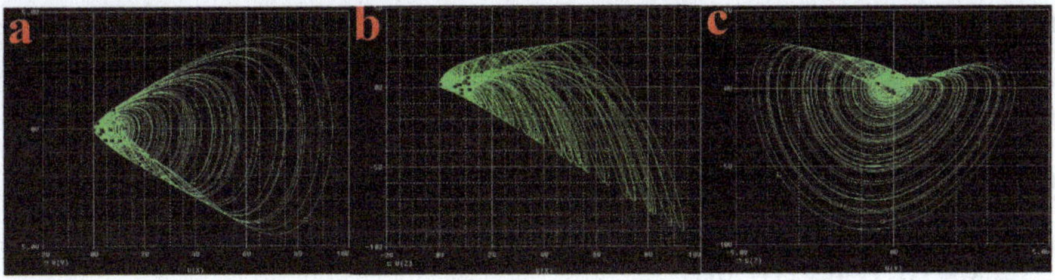

Fig. 12 Phase portraits from the designed circuit. **a** $x - y$ plane. **b** $x - z$ plane. **c** $y - z$ plane

4 Conclusion

In this chapter, we have analyzed the simple chaotic Wang-Chen system, which has no equilibrium, as a feature that gives interesting properties to unusual chaotic systems. This system belongs to a new category of chaotic systems with hidden attractors.

As the Šil'nikov criterion is not applicable to such systems, having a positive largest Lyapunov exponent with bifurcation diagrams offers a basic method for analysis and verification of the existence of chaos and multistabilties.

References

1. H. Degn, A.V. Holden, L.F. Olsen, *Chaos in Biological Systems*, vol. 138 (Springer Science & Business Media, Berlin, 2013)
2. M. Molaie, R. Falahian, S. Gharibzadeh, S. Jafari, J.C. Sprott, Artificial neural networks: Powerful tools for modeling chaotic behavior in the nervous system. Front. Comput. Neurosci. **8**, 40 (2014)
3. R. Falahian, M.M. Dastjerdi, M. Molaie, S. Jafari, S. Gharibzadeh, Artificial neural network-based modeling of brain response to flicker light. Nonlinear Dyn. **81**(4), 1951–1967 (2015)
4. H. Kantz, T. Schreiber, *Nonlinear Time Series Analysis*, vol. 7 (Cambridge University Press, Cambridge, 2004)
5. F. Hadaeghi, M.R.H. Golpayegani, G. Murray, Towards a complex system understanding of bipolar disorder: A map based model of a complex winnerless competition. J. Theor. Biol. **376**, 74–81 (2015)
6. F. Hadaeghi, M.R. Hashemi Golpayegani, K. Moradi, Does 'crisis-induced intermittency' explain bipolar disorder dynamics? Front. Comput. Neurosci. **7**, 116 (2013)
7. G. Baghdadi, S. Jafari, J. Sprott, F. Towhidkhah, M.H. Golpayegani, A chaotic model of sustaining attention problem in attention deficit disorder. Commun. Nonlinear Sci. Numer. Simul. **20**(1), 174–185 (2015)
8. S. Jafari, G. Baghdadi, S.R.H. Golpayegani, F. Towhidkhah, S. Gharibzadeh, Is attention deficit hyperactivity disorder a kind of intermittent chaos? J. Neuropsychiatry Clin. Neurosci. **25**(2), E02–E02 (2013)
9. L. Chen, G. Chen, Controlling chaos in an economic model. Phys. A **374**(1), 349–358 (2007)
10. R. Donangelo, H. Fort, V. Dakos, M. Scheffer, E.H. Van Nes, Early warnings for catastrophic shifts in ecosystems: Comparison between spatial and temporal indicators. Int. J. Bifurc. Chaos **20**(02), 315–321 (2010)

11. E. Benincà, J. Huisman, R. Heerkloss, K.D. Jöhnk, P. Branco, E.H. Van Nes, M. Scheffer, S.P. Ellner, Chaos in a long-term experiment with a plankton community. Nature **451**(7180), 822–825 (2008)
12. D. Patil, B.R. Hunt, E. Kalnay, J.A. Yorke, E. Ott, Local low dimensionality of atmospheric dynamics. Phys. Rev. Lett. **86**(26), 5878 (2001)
13. T. Kapitaniak, L. Kocarev, L.O. Chua, Controlling chaos without feedback and control signals. Int. J. Bifurc. Chaos **3**(02), 459–468 (1993)
14. G. Leonov, N. Kuznetsov, V. Vagaitsev, Localization of hidden Chua's attractors. Phys. Lett. A **375**(23), 2230–2233 (2011)
15. S. Jafari, J. Sprott, S.M.R.H. Golpayegani, Elementary quadratic chaotic flows with no equilibria. Phys. Lett. A **377**(9), 699–702 (2013)
16. S. Jafari, J.C. Sprott, V.-T. Pham, S.M.R.H. Golpayegani, A.H. Jafari, A new cost function for parameter estimation of chaotic systems using return maps as fingerprints. Int. J. Bifurc. Chaos **24**(10), 1450134 (2014)
17. V.-T. Pham, C. Volos, S. Jafari, Z. Wei, X. Wang, Constructing a novel no-equilibrium chaotic system. Int. J. Bifurc. Chaos **24**(05), 1450073 (2014)
18. F.R. Tahir, S. Jafari, V.-T. Pham, C. Volos, X. Wang, A novel no-equilibrium chaotic system with multiwing butterfly attractors. Int. J. Bifurc. Chaos **25**(04), 1550056 (2015)
19. S. Jafari, V.-T. Pham, T. Kapitaniak, Multiscroll chaotic sea obtained from a simple 3d system without equilibrium. Int. J. Bifurc. Chaos **26**(02), 1650031 (2016)
20. V.-T. Pham, S. Vaidyanathan, C. Volos, S. Jafari, S.T. Kingni, A no-equilibrium hyperchaotic system with a cubic nonlinear term. Optik-Int. J. Light Electron Opt. **127**(6), 3259–3265 (2016)
21. V.-T. Pham, S. Vaidyanathan, C. Volos, S. Jafari, N. Kuznetsov, T. Hoang, A novel memristive time-delay chaotic system without equilibrium points. Eur. Phys. J. Special Topics **225**(1), 127–136 (2016)
22. Z. Wei, Dynamical behaviors of a chaotic system with no equilibria. Phys. Lett. A **376**(2), 102–108 (2011)
23. X. Wang, G. Chen, Constructing a chaotic system with any number of equilibria. Nonlinear Dyn. **71**(3), 429–436 (2013)
24. M. Molaie, S. Jafari, J.C. Sprott, S.M.R.H. Golpayegani, Simple chaotic flows with one stable equilibrium. Int. J. Bifurc. Chaos **23**(11), 1350188 (2013)
25. S. Kingni, S. Jafari, H. Simo, P. Woafo, Three-dimensional chaotic autonomous system with only one stable equilibrium: Analysis, circuit design, parameter estimation, control, synchronization and its fractional-order form. Eur. Phys. J. Plus **129**(5), 1–16 (2014)
26. S.-K. Lao, Y. Shekofteh, S. Jafari, J.C. Sprott, Cost function based on Gaussian mixture model for parameter estimation of a chaotic circuit with a hidden attractor. Int. J. Bifurc. Chaos **24**(01), 1450010 (2014)
27. V.-T. Pham, C. Volos, S. Jafari, X. Wang, Generating a novel hyperchaotic system out of equilibrium. Optoelectron. Adv. Mater. Rapid Commun. **8**(5–6), 535–539 (2014)
28. S. Jafari, J. Sprott, Simple chaotic flows with a line equilibrium. Chaos, Solitons Fractals **57**, 79–84 (2013)
29. S.T. Kingni, V.-T. Pham, S. Jafari, G.R. Kol, P. Woafo, Three-dimensional chaotic autonomous system with a circular equilibrium: Analysis, circuit implementation and its fractional-order form. Circuits Syst. Sig. Process. **35**(6), 1933–1948 (2016)
30. V.-T. Pham, S. Jafari, C. Volos, S. Vaidyanathan, T. Kapitaniak, A chaotic system with infinite equilibria located on a piecewise linear curve. Optik: Int. J. Light Electron Opt. **127**(20), 9111–9117 (2016)
31. V.-T. Pham, S. Jafari, X. Wang, J. Ma, A chaotic system with different shapes of equilibria. Int. J. Bifurc. Chaos **26**(04), 1650069 (2016)
32. T. Gotthans, J. Petržela, New class of chaotic systems with circular equilibrium. Nonlinear Dyn. **81**(3), 1143–1149 (2015)
33. S. Jafari, J. Sprott, M. Molaie, A simple chaotic flow with a plane of equilibria. Int. J. Bifurc. Chaos **26**(06), 1650098 (2016)

34. G.A. Leonov, N.V. Kuznetsov, Hidden attractors in dynamical systems. from hidden oscillations in Hilbert–Kolmogorov, Aizerman, and Kalman problems to hidden chaotic attractor in Chua circuits. Int. J. Bifurc. Chaos **23**(1), 1330002 (2013)
35. D. Dudkowski, S. Jafari, T. Kapitaniak, N.V. Kuznetsov, G.A. Leonov, A. Prasad, Hidden attractors in dynamical systems. Physics Reports (2016)
36. P. Sharma, M. Shrimali, A. Prasad, N. Kuznetsov, G. Leonov, Control of multistability in hidden attractors. Eur. Phys. J.: Special Topics **224**(8), 1485–1491 (2015)
37. N. Kuznetsov, G. Leonov, Hidden attractors in dynamical systems: systems with no equilibria, multistability and coexisting attractors. IFAC Proc. **47**(3), 5445–5454 (2014)
38. T. Kapitaniak, G. Leonov, Multistability: Uncovering hidden attractors. Eur. Phys. J.: Special Topics **224**(8), 1405–1408 (2015)
39. L.P. Šil'nikov, A.L. Šil'nikov, D.V. Turaev, L.O. Chua, *Methods of Qualitative Theory in Nonlinear Dynamics* (World Scientific, Singapore, 1998)

A New 3D Chaotic System with only Quadratic Nonlinearities: Analysis and Circuit Implantation

Seyede Sanaz Hosseini, Mohammad-Ali Jafari, Sajad Jafari, Viet-Thanh Pham, and Xiong Wang

1 Introduction

Chaos is an interesting nonlinear behavior of many natural and artificial systems. Many researchers are studying chaos theory and chaotic systems, and various applications of chaos have been introduced in such as biological systems, communications, information encryption, electronic circuits, lasers, etc. [1–4]. Thus, researchers try to find and introduce new chaotic systems for their potential chaos-based applications [5–10]. For example, recently chaotic systems with especial properties like having hidden attractors [11–18], without any equilibrium points [7, 8, 10, 19–24], with only stable equilibria [25–28], or with curves of equilibria [29–33], have attracted lots of interest.

S. S. Hosseini (✉) · S. Jafari
Biomedical Engineering Department, Amirkabir University of Technology, Tehran 15875-4413, Iran
e-mail: Hosseinisanaz92@gmail.com

S. Jafari
e-mail: sajadjafari@aut.ac.ir

M.-A. Jafari
Biomedical Engineering Department, Qazvin Islamic Azad University, Qazvin, Iran
e-mail: mohammadalijafari86@gmail.com

V.-T. Pham
School of Electronics and Telecommunications, Hanoi University of Science and Technology, 01 Dai Co Viet, Hanoi, Vietnam
e-mail: pvt3010@gmail.com

X. Wang
Institute for Advanced Study, Shenzhen University, Shenzhen, Guangdong 518060, People's Republic of China
e-mail: wangxiong8686@szu.edu.cn

X. Wang et al. (eds.), *Chaotic Systems with Multistability and Hidden Attractors*, Emergence, Complexity and Computation 40,
https://doi.org/10.1007/978-3-030-75821-9_24

In this chapter, we introduce a new three-dimensional chaotic system and analyze it. The unique feature of this system is that the nonlinear terms are only quadratic, while there is no linear term. The chapter is organized as follows: in the next section, the model and its stability analysis are presented. In the third section, bifurcation analysis and Lyapunov exponent diagrams with respect to one of the parameters are given. In the fourth section, a circuit implementation of the proposed system is presented. Finally, the chapter is summarized in the last section.

2 The Model and Equilibria Analysis

The system we introduce in this chapter is described by

$$\begin{aligned} \dot{x} &= ax^2 - 2y^2 - 0.4, \\ \dot{y} &= -x^2 + 0.2y^2 + z^2 + 2, \\ \dot{z} &= -1.4x^2 + 1.6y^2 + 5, \end{aligned} \tag{1}$$

where a is the bifurcation parameter. In the above equations, the terms are only quadratic with a constant term in each equation.

The equilibrium points of the system (1) are obtained by solving $\dot{x} = 0$, $\dot{y} = 0$ and $\dot{z} = 0$, namely

$$ax^2 - 2y^2 - 0.4 = 0, \tag{2}$$

$$-x^2 + 0.2y^2 + z^2 + 2 = 0, \tag{3}$$

$$-1.4x^2 + 1.6y^2 + 5 = 0. \tag{4}$$

So, the equilibrium points are

$$x^* = \pm\sqrt{\frac{5.85}{1.75 - a}}, \tag{5}$$

$$y^* = \pm\sqrt{\frac{3.125a - 0.35}{1.75 - a}}, \tag{6}$$

$$z^* = \pm\sqrt{\frac{1.375a + 2.42}{1.75 - a}}. \tag{7}$$

The equilibrium points are located on the vertexes of a cube in the space of (x, y, z) for each value of a.

Now, consider system (1) with $a = 1$. In this case, the system becomes

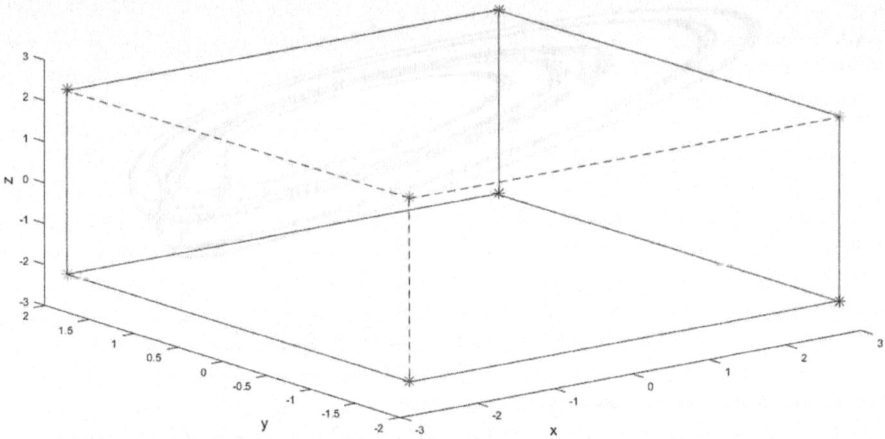

Fig. 1 Equilibrium points of the system are placed at the vertexes of a cube [represented points are the equilibrium points of system (8)]

$$\begin{aligned} \dot{x} &= x^2 - 2y^2 - 0.4, \\ \dot{y} &= -x^2 + 0.2y^2 + z^2 + 2, \\ \dot{z} &= -1.4x^2 + 1.6y^2 + 5. \end{aligned} \tag{8}$$

The eight equilibrium points of the system (8) are $(\sqrt{7.8}, \sqrt{3.7}, \sqrt{5.06})$, $(-\sqrt{7.8}, \sqrt{3.7}, \sqrt{5.06})$, $(\sqrt{7.8}, -\sqrt{3.7}, \sqrt{5.06})$, $(\sqrt{7.8}, \sqrt{3.7}, -\sqrt{5.06})$, $(-\sqrt{7.8}, -\sqrt{3.7}, \sqrt{5.06})$, $(-\sqrt{7.8}, \sqrt{3.7}, -\sqrt{5.06})$, $(\sqrt{7.8}, -\sqrt{3.7}, -\sqrt{5.06})$ and $(-\sqrt{7.8}, -\sqrt{3.7}, -\sqrt{5.06})$.

As can be seen, the equilibrium points are placed at the vertexes of a cube in the space of (x, y, z).

The Jacobian matrix of the system (8) at the equilibrium point E^* is

$$J_{E^*} = \begin{bmatrix} 2x^* & -4y^* & 0 \\ -2x^* & 0.4y^* & 2z^* \\ -2.8x^* & 3.2y^* & 0 \end{bmatrix}, \tag{9}$$

where $E^* = (x^*, y^*, z^*)$.

The characteristic equation of matrix A in variable λ is defined by $det(A - \lambda I) = 0$, where I is the identity matrix. Thus, for the Jacobian matrix 9, the characteristic equation is

$$\lambda^3 - (2x^* + 0.4y^*)\lambda^2 - (6.4y^*z^* + 7.2x^*y^*)\lambda - 9.6x^*y^*z^* = 0. \tag{10}$$

By calculating the eigenvalues of the characteristic Eq. (10), it is found that all of them have at least one positive real root or a complex root with positive real part. Thus, they are all unstable.

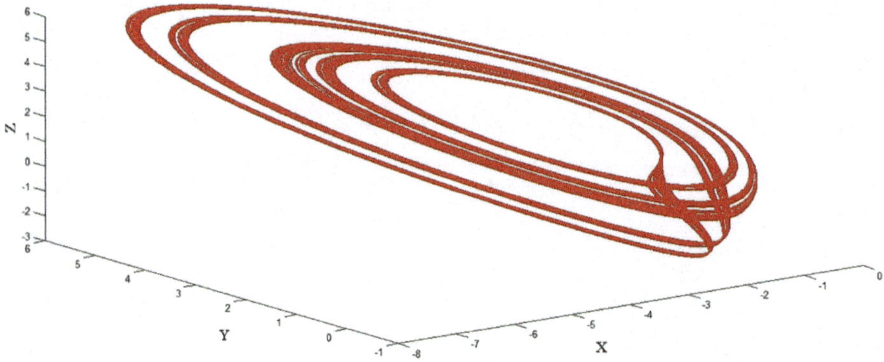

Fig. 2 Strange attractor of system (8)

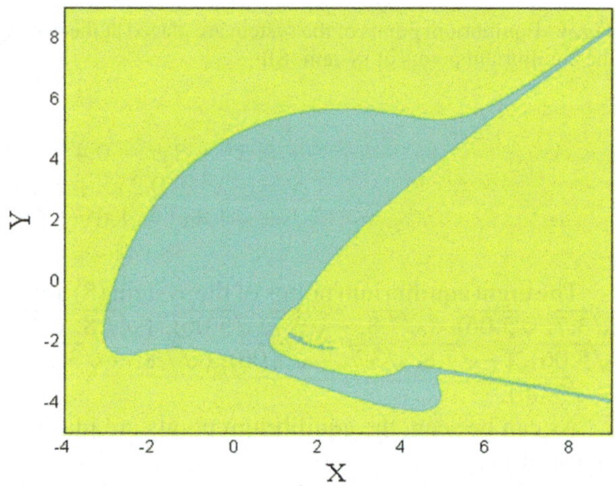

Fig. 3 Basin of attraction of system (8) in the plane $z = 0$, where the yellow region indicates unbounded area and the cyan region leads to a chaotic attractor

Figure 2 presents the trajectory of the system (8) with initial condition $(x_0, y_0, z_0) = (-1.8387, 2.4497, 3.2453)$.

The basin of attraction of system (8) in the plane $z = 0$ is shown in Fig. 3.

3 Bifurcation Analysis and Dynamics

In order to have physical meaningful equilibrium points, the radicands in (5), (6) and (7) should be positive:

$$\frac{5.85}{1.75 - a} \geqslant 0, \tag{11}$$

$$\frac{3.125a - 0.35}{1.75 - a} \geqslant 0, \tag{12}$$

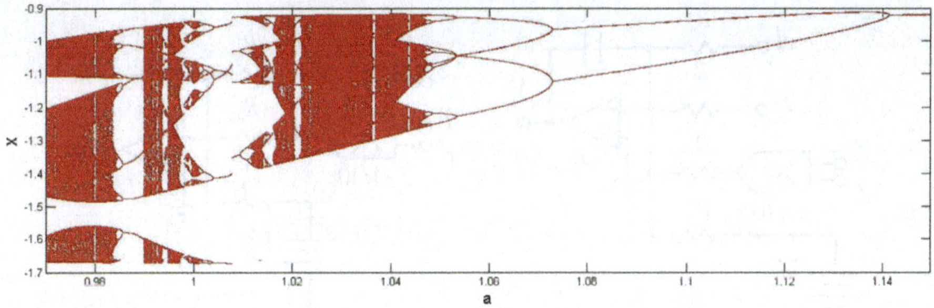

Fig. 4 Bifurcation diagram of the variable x with respect to parameter a

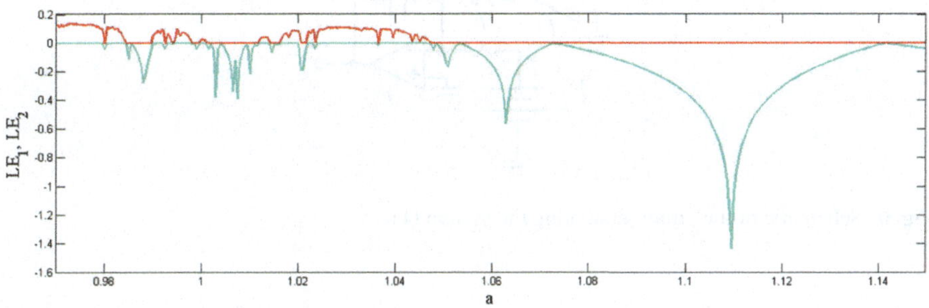

Fig. 5 Lyapunov exponent diagram

$$\frac{1.375a + 2.42}{1.75 - a} \geqslant 0, \tag{13}$$

which led to $0.112 < a < 1.75$.

Now, consider a between 0.97 and 1.15, which is in the allowed interval. In Fig. 4, the bifurcation diagram of the variable x with parameter a within the interval of (0.97, 1.15) is presented.

The Lyapunov exponent diagram corresponding to Fig. 4 is represented in Fig. 5. Notice that only two first Lyapunov exponents are plotted in this figure, while the third Lyapunov exponent is always negative.

4 Circuit Implementation

As has been reported in the literature, electronic realizations of theoretical chaotic models are useful for practical engineering applications. In this section, we introduce an electronic circuit for realizing the system (1). By using common electronic components such as resistors, capacitors, operational amplifiers and multipliers, a circuit is designed as represented in Fig. 6, in which the voltages of the three capacitors are denoted as $X,\ Y,\ Z$.

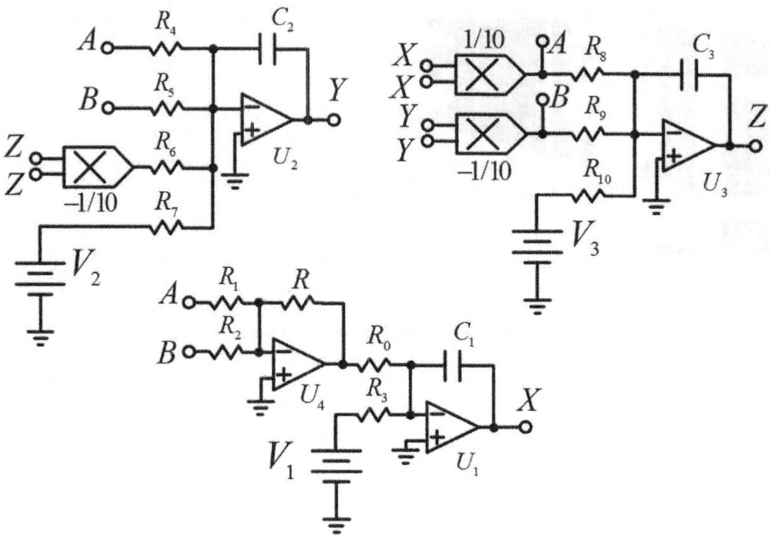

Fig. 6 Schematic of the circuit emulating the system (1)

By applying Kirchhoff's circuit laws to the designed circuit shown in Fig. 6, its circuital equations are obtained as follows:

$$\begin{aligned} \dot{X} &= \frac{R}{R_1 R_0 C_1 10V} X^2 - \frac{R}{R_2 R_0 C_1 10V} Y^2 - \frac{1}{R_3 C_1} V_1, \\ \dot{Y} &= -\frac{1}{R_4 C_2 10V} X^2 + \frac{1}{R_5 C_2 10V} Y^2 + \frac{1}{R_6 C_2 10V} Z^2 - \frac{1}{R_7 C_2} V_2, \\ \dot{Z} &= -\frac{1}{R_8 C_3 10V} X^2 + \frac{1}{R_9 C_3 10V} Y^2 - \frac{1}{R_{10} C_3} V_3. \end{aligned} \tag{14}$$

We have implemented the circuit using the electronic simulation package Cadence OrCAD. The values of electronic components shown in Fig. 6 are $R_1 = R_3 = R_7 = R_{10} = R = 56\,k\Omega$, $R_2 = R_5 = 28\,k\Omega$, $R_4 = R_6 = 5.6\,k\Omega$, $R_8 = 4\,k\Omega$, $R_9 = 3.5\,k\Omega$, $V_1 = 0.4\,V_{DC}$, $V_2 = -2\,V_{DC}$, $V_3 = -5\,V_{DC}$, and $C_1 = C_2 = C_3 = C = 2\,nF$.

The obtained projections of the chaotic attractors are illustrated in Fig. 7. It is straightforward to verify that the designed circuit indeed displays chaotic signals.

5 Conclusion

In the presented chapter, we have introduced a new three-dimensional chaotic system with eight equilibrium points located at the vertexes of a cube. Nonlinear dynamics of the system have been studied and the circuit implementation of this system has been realized. The main goal of introducing chaotic systems is for their potential

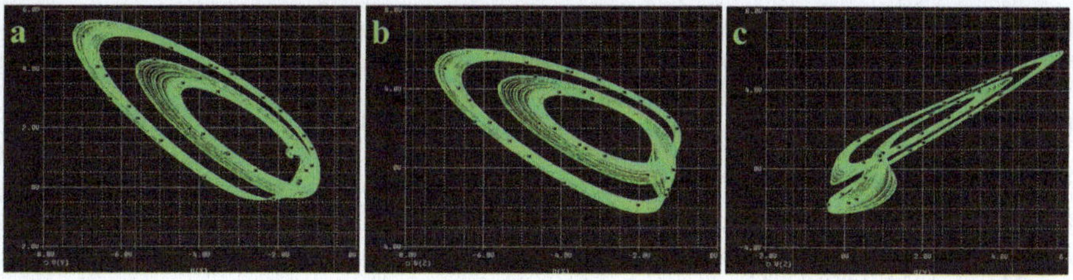

Fig. 7 Projections of chaotic attractors in the designed circuit. **a** $X - Y$ plane. **b** $X - Z$ plane. **c** $Y - Z$ plane

chaos-based applications, especially in information encoding. For this reason, the advantages of the proposed system in this chapter will be explored in the future.

References

1. H. Gao, Y. Zhang, S. Liang, D. Li, A new chaotic algorithm for image encryption. Chaos, Solitons Fractals **29**(2), 393–399 (2006)
2. M.J. Ogorzalek, *Chaos and Complexity in Nonlinear Electronic Circuits* (World Scientific, Singapore, 1997)
3. C.K. Volos, I.M. Kyprianidis, I.N. Stouboulos, Image encryption process based on chaotic synchronization phenomena. Signal Process. **93**(5), 1328–1340 (2013)
4. V.-T. Pham, S. Jafari, C. Volos, A. Giakoumis, S. Vaidyanathan, T. Kapitaniak, A chaotic system with equilibria located on the rounded square loop and its circuit implementation. IEEE Trans. Circuits Syst. II Express Briefs **63**(9), 878–882 (2016)
5. J. Chen, X. Zhang, J. Peng, Time-delayed chaotic circuit design using all-pass filter. IEEE Trans. Circuits Syst. I Regul. Pap. **61**(10), 2897–2903 (2014)
6. C. Shen, S. Yu, J. Lü, G. Chen, A systematic methodology for constructing hyperchaotic systems with multiple positive lyapunov exponents and circuit implementation. IEEE Trans. Circuits Syst. I Regul. Pap. **61**(3), 854–864 (2014)
7. V.-T. Pham, S. Vaidyanathan, C. Volos, S. Jafari, S.T. Kingni, A no-equilibrium hyperchaotic system with a cubic nonlinear term. Optik: Int. J. Light Electron Opt. **127**(6), 3259–3265 (2016)
8. F.R. Tahir, S. Jafari, V.-T. Pham, C. Volos, X. Wang, A novel no-equilibrium chaotic system with multiwing butterfly attractors. International Journal of Bifurcation and Chaos **25**(04), 1550056 (2015)
9. V.-T. Pham, C. Volos, S. Jafari, X. Wang, Generating a novel hyperchaotic system out of equilibrium. Optoelectronics and Advanced Materieals: Rapid Communication **8**(5–6), 535–539 (2014)
10. V.-T. Pham, C. Volos, S. Jafari, Z. Wei, X. Wang, Constructing a novel no-equilibrium chaotic system. Int. J. Bifurc. Chaos **24**(05), 1450073 (2014)
11. G. Leonov, N. Kuznetsov, V. Vagaitsev, Localization of hidden chua's attractors. Phys. Lett. A **375**(23), 2230–2233 (2011)
12. G. Leonov, N. Kuznetsov, V. Vagaitsev, Hidden attractor in smooth Chua systems. Physica D **241**(18), 1482–1486 (2012)
13. G.A. Leonov, N.V. Kuznetsov, Hidden attractors in dynamical systems. from hidden oscillations in Hilbert-Kolmogorov, Aizerman, and Kalman problems to hidden chaotic attractor in Chua circuits. Int. J. Bifurc. Chaos **23**(1), 1330002 (2013)

14. G. Leonov, N. Kuznetsov, M. Kiseleva, E. Solovyeva, A. Zaretskiy, Hidden oscillations in mathematical model of drilling system actuated by induction motor with a wound rotor. Nonlinear Dyn. **77**(1–2), 277–288 (2014)
15. G. Leonov, N. Kuznetsov, T. Mokaev, Hidden attractor and homoclinic orbit in Lorenz-like system describing convective fluid motion in rotating cavity. Commun. Nonlinear Sci. Numer. Simul. **28**(1), 166–174 (2015)
16. G. Leonov, N. Kuznetsov, T. Mokaev, Homoclinic orbits, and self-excited and hidden attractors in a Lorenz-like system describing convective fluid motion. Eur. Phys. Jo.: Special Topics **224**(8), 1421–1458 (2015)
17. P. Sharma, M. Shrimali, A. Prasad, N. Kuznetsov, G. Leonov, Control of multistability in hidden attractors. Eur. Phys. J.: Special Topics **224**(8), 1485–1491 (2015)
18. P.R. Sharma, M.D. Shrimali, A. Prasad, N.V. Kuznetsov, G.A. Leonov, Controlling dynamics of hidden attractors. Int. J. Bifurc. Chaos **25**(04), 1550061 (2015)
19. S. Jafari, J. Sprott, S.M.R.H. Golpayegani, Elementary quadratic chaotic flows with no equilibria. Phys. Lett. A **377**(9), 699–702 (2013)
20. S. Jafari, J.C. Sprott, V.-T. Pham, S.M.R.H. Golpayegani, A.H. Jafari, A new cost function for parameter estimation of chaotic systems using return maps as fingerprints. Int. J. Bifurc. Chaos **24**(10), 1450134 (2014)
21. S. Jafari, V.-T. Pham, T. Kapitaniak, Multiscroll chaotic sea obtained from a simple 3d system without equilibrium. Int. J. Bifurc. Chaos **26**(02), 1650031 (2016)
22. V.-T. Pham, S. Vaidyanathan, C. Volos, S. Jafari, N. Kuznetsov, T. Hoang, A novel memristive time-delay chaotic system without equilibrium points. Eur. Phys. J.: Special Topics **225**(1), 127–136 (2016)
23. Z. Wei, Dynamical behaviors of a chaotic system with no equilibria. Phys. Lett. A **376**(2), 102–108 (2011)
24. X. Wang, G. Chen, Constructing a chaotic system with any number of equilibria. Nonlinear Dyn. **71**(3), 429–436 (2013)
25. M. Molaie, S. Jafari, J.C. Sprott, S.M.R.H. Golpayegani, Simple chaotic flows with one stable equilibrium. Int. J. Bifurc. Chaos **23**(11), 1350188 (2013)
26. S. Kingni, S. Jafari, H. Simo, P. Woafo, Three-dimensional chaotic autonomous system with only one stable equilibrium: Analysis, circuit design, parameter estimation, control, synchronization and its fractional-order form. Eur. Phys. J. Plus **129**(5), 1–16 (2014)
27. S.-K. Lao, Y. Shekofteh, S. Jafari, J.C. Sprott, Cost function based on Gaussian mixture model for parameter estimation of a chaotic circuit with a hidden attractor. Int. J. Bifurc. Chaos **24**(01), 1450010 (2014)
28. V.-T. Pham, C. Volos, S. Jafari, X. Wang, Generating a novel hyperchaotic system out of equilibrium. Optoelectronics and Advanced Materiels: Rapid Communication **8**(5–6), 535–539 (2014)
29. S. Jafari, J. Sprott, Simple chaotic flows with a line equilibrium. Chaos, Solitons Fractals **57**, 79–84 (2013)
30. S.T. Kingni, V.-T. Pham, S. Jafari, G.R. Kol, P. Woafo, Three-dimensional chaotic autonomous system with a circular equilibrium: Analysis, circuit implementation and its fractional-order form. Circuits Systems Signal Process. **35**(6), 1933–1948 (2016)
31. V.-T. Pham, S. Jafari, C. Volos, S. Vaidyanathan, T. Kapitaniak, A chaotic system with infinite equilibria located on a piecewise linear curve. Optik: Int. J. Light Electron Opt. **127**(20), 9111–9117 (2016)
32. V.-T. Pham, S. Jafari, X. Wang, J. Ma, A chaotic system with different shapes of equilibria. Int. J. Bifurc. Chaos **26**(04), 1650069 (2016)
33. T. Gotthans, J. Petržela, New class of chaotic systems with circular equilibrium. Nonlinear Dyn. **81**(3), 1143–1149 (2015)

Globally Attracting Hidden Attractors

Julien Clinton Sprott

Abstract The many examples in the previous chapters should leave no doubt that hidden attractors are common in nonlinear dynamical systems. Remarkably, hidden attractors can have basins that fill the entire space with every initial condition on the attractor. Two such examples are shown here.

1 Introduction

The many examples in the previous chapters should leave no doubt that hidden attractors are common in nonlinear dynamical systems. Previous authors have echoed the claim that they are hard to find because there is no systematic method to identify initial conditions in their basin of attraction.

Thus it is fitting to temper those claims with some examples of hidden attractors that are *globally* attracting. Not only is every initial condition in their basin of attraction, but every initial condition lies on the attractor, and thus they could hardly be less hidden. Furthermore, such attractors have been known and studied long before the recent hoopla about hidden attractors, and they have other remarkable properties to be recounted here.

2 Conservative Nosé–Hoover System

The interest in chaotic systems whose orbit visits the entire state space (called *ergodic*) arose long ago from a quest among molecular dynamicists to find a simple

J. C. Sprott (✉)
University of Wisconsin-Madison, Madison, WI, USA
e-mail: sprott@physics.wisc.edu

X. Wang et al. (eds.), *Chaotic Systems with Multistability and Hidden Attractors*,
Emergence, Complexity and Computation 40,
https://doi.org/10.1007/978-3-030-75821-9_25

dynamical system that would model the behavior of a harmonic oscillator in thermal equilibrium with a heat bath at a constant temperature. Prior to the modern chaos era, it had been assumed that any such model would need many variables. The first breakthrough came in 1984 when Shuichi Nosé found a Hamiltonian system with four variables consistent with the necessary condition that the probability distribution functions of position and momentum should be Gaussian [1] as expected for Gibbs' canonical ensemble [2].

The following year, Bill Hoover showed that Nosé system could be reduced to a three-dimensional form with the same properties, now known as the Nosé–Hoover system [3]:

$$\begin{cases} \dot{x} = y \\ \dot{y} = -x - zy \\ \dot{z} = y^2 - 1. \end{cases} \tag{1}$$

This system was independently discovered in a search for three-dimensional chaotic flows with five terms and two quadratic nonlinearities, and thus it is also known as the Sprott A system [4, 5] and has been widely studied. It is the simplest in a large class of systems with similar properties [6].

Absent the zy term, this system is a simple harmonic oscillator with x playing the role of position, and y is its canonically conjugate momentum. The zy term represents a nonlinear damping (for positive z) or anti-damping (for negative z) with z controlled by the $\dot{z}$ equation such that the damping averages to zero when the mean square momentum $\langle y^2 \rangle$ is unity. Thus z acts as a thermostat, controlling the average energy of the chaotic oscillator, but allowing it to fluctuate as desired to model an oscillator in equilibrium with a heat bath [7].

System (1) is unusual because it has no equilibrium points, but neither does it have an attractor because it is derived from a Hamiltonian [8] in which the fourth variable is a slave of the other three and thus does not influence the dynamics. Hence the system is conservative with the missing energy in the hidden variable, and the oscillator is isothermal rather than isoenergetic. Such systems are called nonuniformly conservative [9], and they share many of the properties of conventional conservative systems.

The third variable allows the system to oscillate chaotically with a chaotic sea whose Lyapunov exponents are (0.0139, 0, –0.0139) and that stretches to infinity in all three dimensions, but that encloses an intricate set of nested and intertwined invariant tori on which the orbits are quasiperiodic with Lyapunov exponents of (0, 0, 0). All orbits, both in the chaotic sea and on the tori, repeatedly cross the $z = 0$ plane, which allows the dynamics to be completely characterized by examining a cross section of the orbit in that plane. In particular, quasiperiodic orbits embedded anywhere in the chaotic sea will appear as 'holes' in that cross section of the flow. The system is time-reversal invariant under the transformation $(x; y; z; t) \rightarrow (x; -y; -z; -t)$ as expected for a conservative system.

3 Dissipative Nosé–Hoover System

There are many ways to add dissipation to a thermostatted oscillator without introducing equilibrium points and thus producing a hidden attractor. For example, the constant term in the $\dot{z}$ equation of system (1) can be replaced by a function $f(x)$ that is everywhere positive:

$$\begin{cases} \dot{x} = y \\ \dot{y} = -x - zy \\ \dot{z} = y^2 - f(x). \end{cases} \tag{2}$$

Physically, this corresponds to a harmonic oscillator in a heat bath with a one-dimensional temperature gradient given by $\frac{df}{dx}$.

This system with $f(x) = 1 + \varepsilon \tanh(x)$ (corresponding to a temperature that varies from $1-\varepsilon$ at $x = -\infty$ to $1+\varepsilon$ at $x = \infty$ with a maximum gradient of ε at $x = 0$) was originally proposed and studied by Posch and Hoover in 1997 [10]. For $\varepsilon = 0.38$ it has a hidden chaotic attractor that extends to infinity in all three dimensions and encloses a region in the vicinity of the origin with conservative tori and quasiperiodic orbits as shown in Fig. 1 [11]. The chaotic attractor fills the entirety of its basin of attraction, but with a highly nonuniform measure.

The attractor has Lyapunov exponents of (0.0019, 0, –0.0020) and a Kaplan–Yorke dimension of 2.945. Thus it differs markedly from essentially all the other chaotic attractors in this book for three-dimensional autonomous flows whose Kaplan–Yorke dimensions are only slightly greater than 2.0. In fact, the attractor is multifractal with a capacity dimension of exactly 3.0, and it stretches to infinity in all three dimensions but with a rapidly decreasing measure. Furthermore, as ε is decreased, the Kaplan–Yorke dimension further increases until it reaches a value of 3.0 for $\varepsilon = 0$, where the standard Nosé–Hoover system (1) with a chaotic sea is recovered.

As a consequence, its basin of attraction fills the entire space except for a finite region in the vicinity of the origin wherein tori with quasiperiodic orbits reside. This is an example of a *Class 1b* basin of attraction [12]. Although it is not a global attractor, a randomly chosen initial condition not too close to the origin is overwhelmingly likely to lie in the basin, and it will lie on the attractor, although usually in a region rarely visited by the orbit.

Remarkably, this dissipative system is time-reversal invariant, just like its conservative counterpart. The system has a repellor that overlaps the attractor, and that becomes an attractor when time is reversed. The attractor-repellor pair as shown by a portion of the orbits in Fig. 2 have only an imperceptible shift in the z-direction of $\langle z \rangle \approx \pm 1.2 \times 10^{-4}$ with $\langle x \rangle \approx -0.6855$ and $\langle y \rangle = 0$.

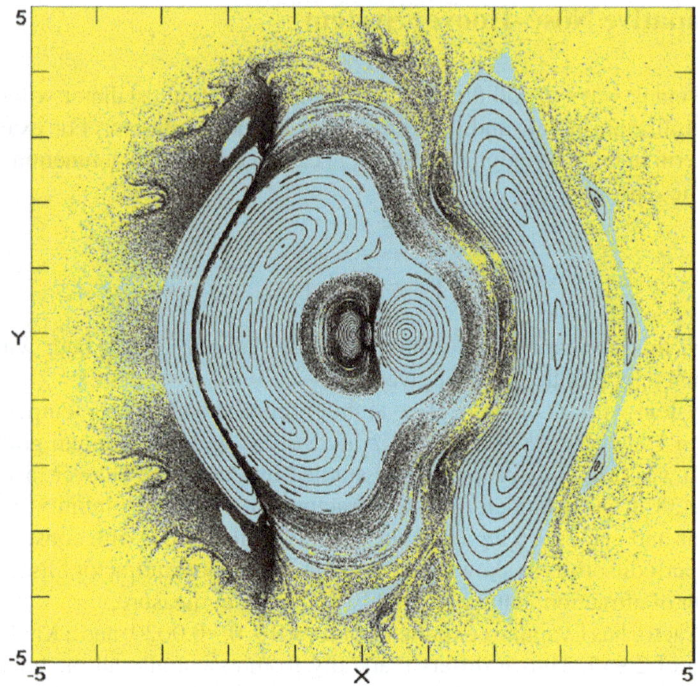

Fig. 1 Cross-section of the orbits in the $z = 0$ plane for system (2) with $f(x) = 1 + 0.38\tanh(x)$. Blue indicates the regions with conservative tori and quasiperiodic orbits, and yellow indicates the infinite basin of attraction of the dissipative hidden chaotic attractor shown in black [11]

4 Buncha System

As if the previous case were not remarkable enough, there are variants of the Nosé–Hoover system that are dissipative and ergodic with a hidden attractor that is the entirety of the three-dimensional state space. Probably the simplest and most elegant example is a reduced form of a general class of system proposed and studied by Buncha Munmuangsaen and collaborators [13] and given by

$$\begin{cases} \dot{x} = y \\ \dot{y} = -x - azy \\ \dot{z} = |y| - 1. \end{cases} \tag{3}$$

Like the Nosé–Hoover case, this system has a single bifurcation parameter a that can be put in any of the five terms, and that completely characterizes the system through a one-dimensional bifurcation diagram as shown in Fig. 3. For $a = 0$ the system is a simple conservative harmonic oscillator with an amplitude that depends only on the initial conditions.

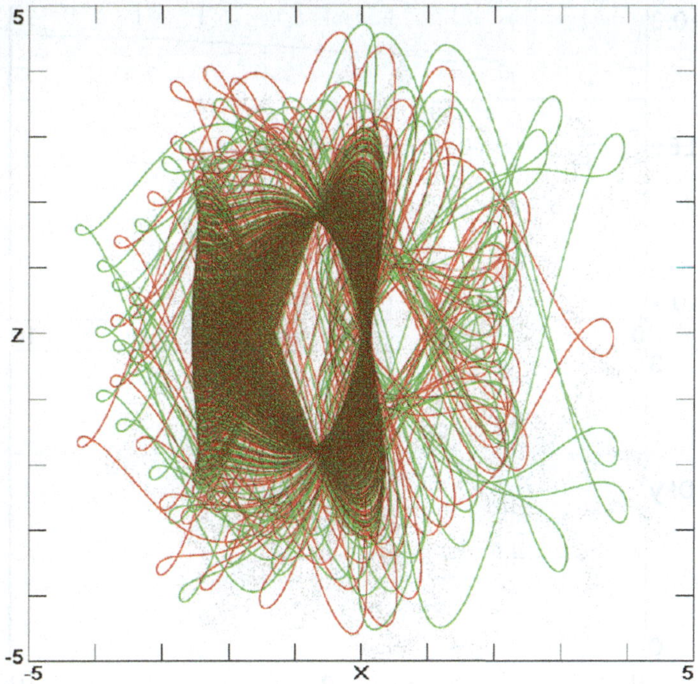

Fig. 2 An orbit on the hidden attractor (in red) and on the corresponding repellor (in green) for system (2) with $f(x) = 1 + 0.38\ \tanh(x)$

For $a > 0$, there are three distinct regions with bifurcations in the vicinity of $a = 0.9$ and $a = 2.1$. For a less than about 0.9, the dynamic is dominated by nested invariant tori with conservative quasiperiodic orbits but surrounded by a dissipative region with limit cycles and/or strange attractors. In the range of a between about 0.9 and 2.1, there is a conservative region containing nested tori that are linked by a symmetric pair of dissipative limit cycles and a long-duration chaotic transient whose orbit eventually collapses onto one of the limit cycles with a riddled basin of attraction. At $a \approx 2.0$, the limit cycles merge into one large limit cycle at $a \approx 2.07$ that gives birth to a strange attractor surrounding the tori. As a is increased further, the tori shrink and eventually vanish at $a \approx 3.07$, leaving only dissipative regions with a single strange attractor that fills all of space.

The Kaplan–Yorke dimension of the attractor continues to increase with increasing a, reaching a maximum of about 2.9924 at $a = 7$ before slowly decreasing, except for narrow periodic windows in the vicinity of $a = 2.28$, 3.78, 4.00, and 6.00, as well as other values that are unresolved in Fig. 3. In these periodic windows, there is a long-duration chaotic transient.

Although system (3) exhibits a variety of unusual behaviors, our interest here is in the regime where there is a single ergodic strange attractor that fills all of space and thus is technically "hidden" because the system has no equilibrium points. For that

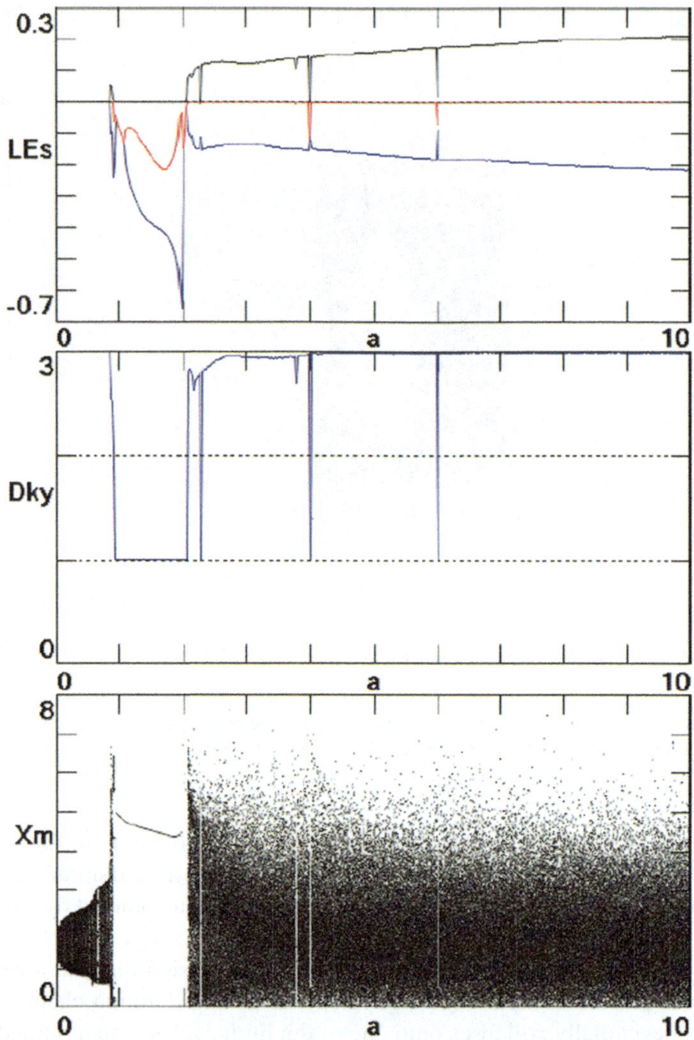

Fig. 3 Lyapunov exponents (LEs), Kaplan–Yorke dimension (Dky), and the local maxima of x (Xm) as a function of the bifurcation parameter a in (3) over the range $0 < a < 10$

purpose, we focus on the case $a = 5$ for which the Lyapunov exponents are (0.1610, 0, –0.1633), the Kaplan–Yorke dimension is 2.9858, and the orbit is as shown in Fig. 4. In this plot, the colors indicate the value of the local largest Lyapunov exponent with red positive and blue negative. While the attractor appears to be bounded, it has a fuzzy edge, and after a sufficiently long time the orbit will come arbitrarily close to every point in the three-dimensional state space.

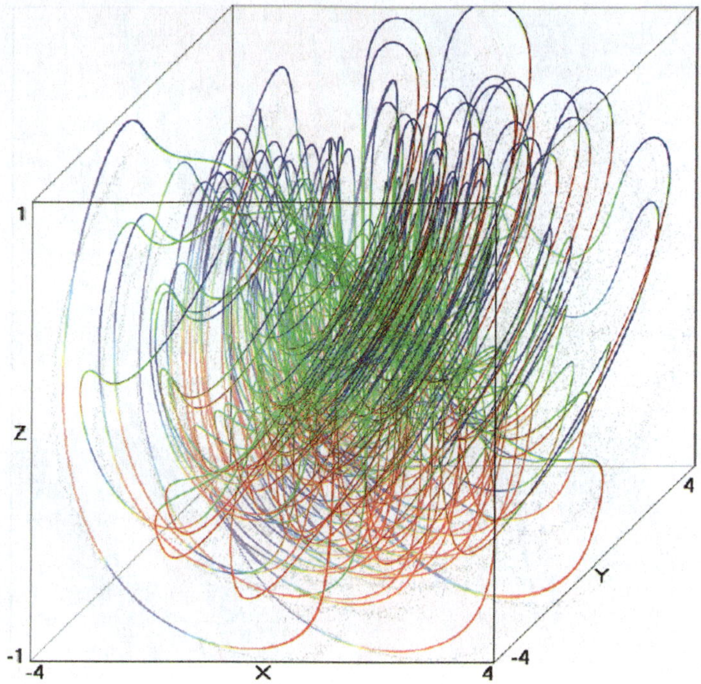

Fig. 4 An orbit on the ergodic strange attractor for system (3) with $a = 5$. The colors indicate the value of the local largest Lyapunov exponent with red positive and blue negative

To confirm that the system is ergodic with no embedded tori and quasiperiodic orbits, it suffices to examine a cross section of the flow at $z = 0$. Figure 5 shows such a plot. Other than the nullclines at $y = \pm 1$, where the orbit is tangent to the plane, there are no holes that would indicate a lack of ergodicity. A single orbit eventually visits every point in the plane, and every initial condition produces the same plot. Said differently, the attractor is globally attracting with a *Class 1a* basin of attraction [12], and the attractor fills the whole of its basin.

The local largest Lyapunov exponent has a complicated structure as evidenced by the variations in color. Furthermore, the equations are time-reversible under the transformation $(x, y, z, t) \rightarrow (x, -y, -z, -t)$ just like the previous cases. When time is reversed, the attractor becomes a repellor that looks identical to the attractor. Thus there exists a symmetric strange attractor-repellor pair that is coincident, except for a tiny offset in the z direction of $\langle z \rangle = \pm 0.0023$, and they exchange roles when time is reversed.

The attractor and repellor are both multifractal with a capacity dimension of exactly 3.0 but with a highly nonuniform measure that is far from the Gaussian that characterizes the usual conservative ergodic harmonic oscillator. The probability distribution functions for the three variables, along with the first six even moments of the distribution, are shown in Fig. 6. None of the distributions have a sharp cutoff,

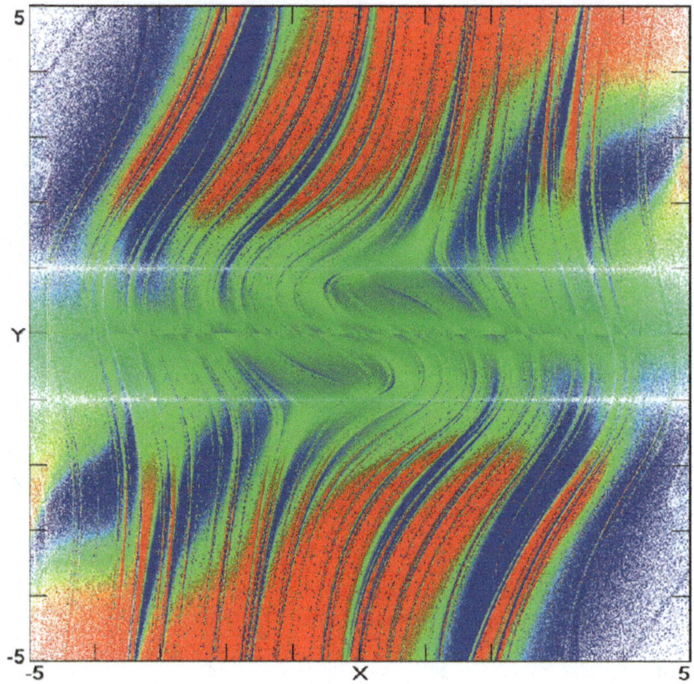

Fig. 5 Cross section of the ergodic strange attractor in the $z = 0$ plane for system (3) with $a = 5$. The colors indicate the value of the local largest Lyapunov exponent with red positive and blue negative

but rather they have long tails that extend to infinity in all directions. Hence the attractor and its basin fill the whole of the state space.

5 Signum Thermostat Dissipative System

Finally, we consider a dissipative chaotic system that is fully ergodic with a measure that more nearly approximates a Gaussian with a hidden global attractor and that is presented here for the first time. This system is a variant of the dissipative Nosé–Hoover system (2) but with the zy term replaced by $2\text{sgn}(z)y$ and is called the signum thermostat [14]. To preserve symmetry in the x probability distribution, $f(x)$ is taken as $f(x) = \exp(-\varepsilon x^2)$ to give

$$\begin{cases} \dot{x} = y \\ \dot{y} = -x - 2\text{sgn}(z)y \\ \dot{z} = y^2 - \exp(-\varepsilon x^2). \end{cases} \tag{4}$$

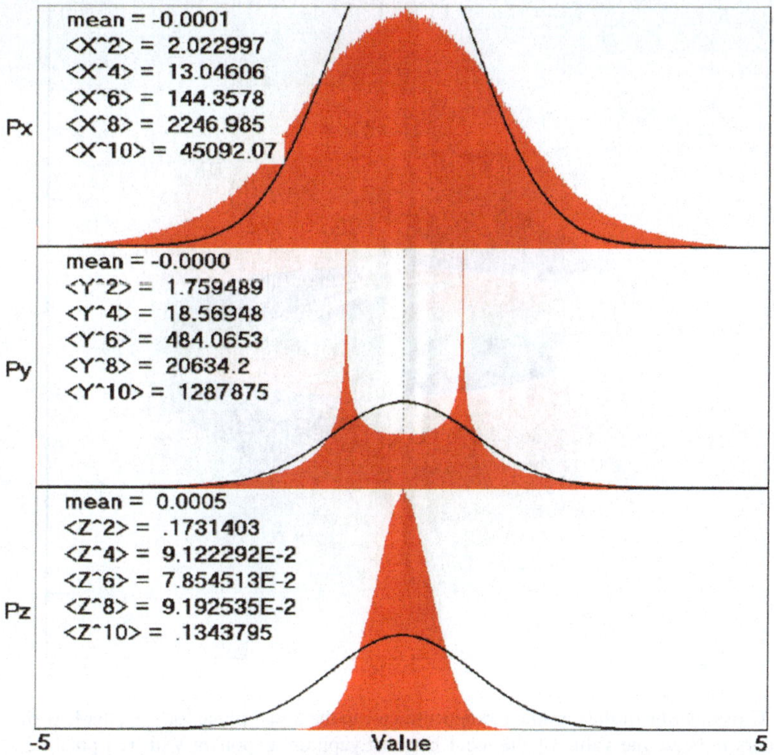

Fig. 6 Probability distribution functions of the ergodic strange attractor for system (3) with $a = 5$. The black curves show a Gaussian distribution with a variance (second moment) of 1.0

Physically, this corresponds to a harmonic oscillator in a heat bath with its highest temperature at $x = 0$ and that approaches absolute zero at $x = \pm\infty$. Since $f(x) > 0$ for all x, there is no equilibrium point, and so any attractor for the system is hidden by definition.

For $\varepsilon = 0$ (constant temperature), $\exp(-\varepsilon x^2) = 1$, and the system is nonuniformly conservative and ergodic with a chaotic sea whose probability distribution is given exactly by $P(x, y, z) = \exp(-x^2/2 - y^2/2 - 2|z|)/2\pi$. For ε small and positive, the system is dissipative and ergodic with a strange attractor whose probability distribution departs only slightly from the case with $\varepsilon = 0$.

For example, $\varepsilon = 0.1$ gives the cross section plot at $z = 0$ shown in Fig. 7. Aside from the nullclines at $y = \pm \exp(-x^2/20)$, there is no indication of quasiperiodic holes in the plot. The colors show that the local largest Lyapunov exponent has a considerable structure as is typical of these systems.

The evidence that the system is dissipative with a chaotic attractor comes from the Lyapunov exponents whose values are (0.03544, 0, –0.3636), the Kaplan–Yorke dimension whose value is 2.9746, and the time-averaged dissipation of $\langle 2\text{sgn}(z)\rangle \approx 9.2 \times 10^{-3}$. The attractor is multifractal with a capacity dimension of 3.0.

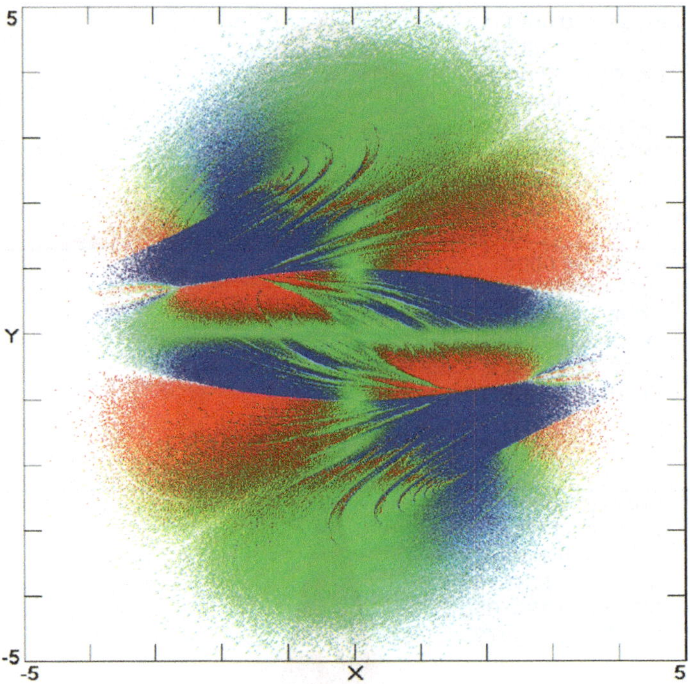

Fig. 7 Cross section of the ergodic strange attractor in the $z = 0$ plane for system (4) with $\varepsilon = 0.1$. The colors indicate the value of the local largest Lyapunov exponent with red positive and blue negative

Like the previous systems, this case is time-reversal invariant under the transformation $(x, y, z, t) \rightarrow (x, -y, -z, -t)$ with an attractor-repellor pair that fully overlap with only a tiny offset in the z-direction. The attractor has a global $Class\ 1a$ basin of attraction [12], and the attractor fills the entire basin.

Figure 8 shows the small departure of the probability distribution functions from the ones with $\varepsilon = 0$. The dissipative oscillator spends slightly less time in the vicinity of the origin where it is heated strongly as well as far from the origin where it is cooled, and relatively more time at intermediate values. Although the tails of the distributions are suppressed, they still extend to infinity in all directions so that every initial condition is on the attractor.

Smaller values of ε give distributions even closer to a Gaussian and Kaplan–Yorke dimensions that approach ever closer to the limit of 3.0. The system remains ergodic for ε up to about 3.4 except for periodic windows with long duration chaotic transients, whereupon the chaotic attractor is replaced by a globally attracting hidden limit cycle. This is not surprising since the temperature is an amplitude parameter that does not affect the dynamic in the conservative constant-temperature case with a signum thermostat. However, the probability distributions become increasingly peaked at intermediate values of x and y as ε increases.

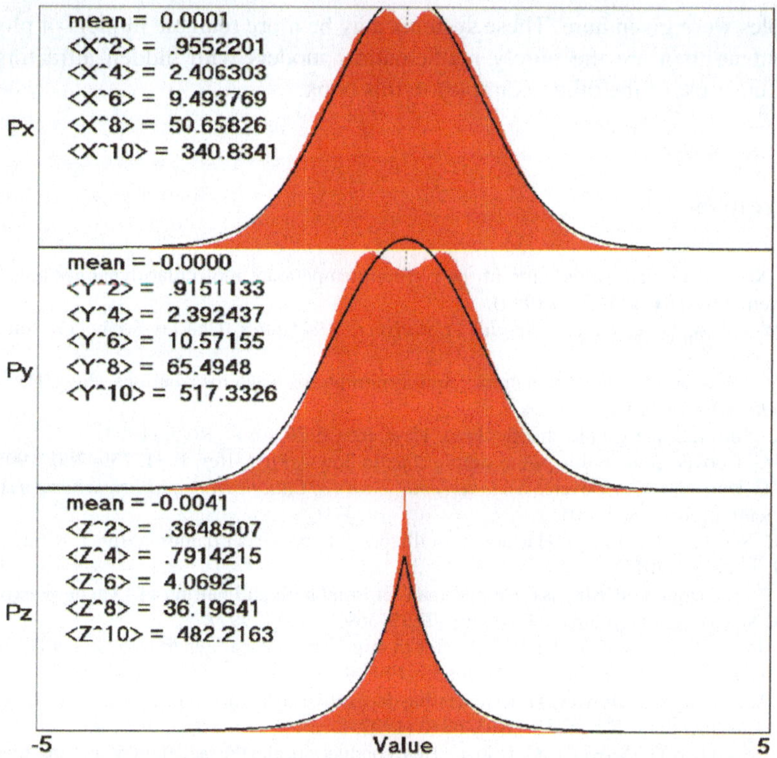

Fig. 8 Probability distribution functions of the ergodic strange attractor for system (4) with $\varepsilon = 0.1$. The black curves show the distributions for $\varepsilon = 0$

6 Summary and Conclusions

The Nosé–Hoover system is almost certainly the simplest example of a nonuniformly conservative chaotic flow without equilibria, but the chaotic sea coexists with regions of quasiperiodicity. It is the simplest example of a wide class of thermostatted oscillators, which are time-reversal invariant in accordance with Newton's laws and that exhibit aspects of thermodynamics and statistical mechanics such as a Gaussian probability distribution function. It is possible to eliminate the quasiperiodic regions and obtain systems that are fully ergodic with the orbit visiting every point in space as desired for a realistic physical model.

There are various ways to add dissipation to such systems and produce strange attractors that are hidden and that fill almost the entire state space, typically with a finite region that is occupied by tori with conservative quasiperiodic orbits. Most remarkably, it is also possible to modify the dissipative systems in such a way as to make them fully ergodic with a multifractal strange attractor that is globally attracting and fills the whole of space and yet satisfies the definition of being hidden. Two such examples were given here. These systems may be more realistic models of physical phenomena than are the purely mathematical models with hidden attractors that constitute most of the other examples in this book.

References

1. S. Nosé, A unified formulation of the constant temperature molecular dynamics methods. J. Chem. Phys. **81**, 511–519 (1984)
2. J.W. Gibbs, *Elementary Principles in Statistical Mechanics* (Charles Scribner's Sons, New York, 1902)
3. W.G. Hoover, Canonical dynamics: equilibrium phase-space distributions. Phys. Rev. A **31**, 1695–1697 (1985)
4. J.C. Sprott, Some simple chaotic flows. Phys. Rev. E **50**, R647–R650 (1994)
5. W.G. Hoover, Remark on 'some simple chaotic flows'. Phys. Rev. E **51**, 759–760 (1995)
6. W.G. Hoover, C.G. Hoover, *Simulation and Control of Chaotic Nonequilibrium Systems* (World Scientific, Singapore, 2015)
7. J.C. Sprott, W.G. Hoover, Harmonic oscillators with nonlinear damping. Int. J. Bifurc. Chaos **27**, 1730037 (2017)
8. C.P. Dettmann, G.P. Morriss, Hamiltonian reformulation and pairing of Lyapunov exponents for Nosé-Hoover dynamics. Phys. Rev. E **55**, 3693–3696 (1997)
9. J. Heidel, F. Zhang, Nonchaotic behavior in three-dimensional quadratic systems Π: the conservative case. Nonlinearity **12**, 617–633 (1999)
10. H.A. Posch, W.G. Hoover, Time-reversible dissipative attractors in three and four phase-space dimensions. Phys. Rev. E **55**, 6803–6810 (1997)
11. J.C. Sprott, W.G. Hoover, C.G. Hoover, Heat conduction, and the lack thereof, in time-reversible dynamical systems: generalized Nosé-Hoover oscillators with a temperature gradient. Phys. Rev. E **89**, 042914 (2014)
12. J.C. Sprott, A. Xiong, Classifying and quantifying basins of attraction. Chaos **25**, 083101 (2015)
13. B. Munmuangsaen, J.C. Sprott, W.J. Thio, A. Buscarino, L. Fortuna, A simple chaotic flow with a continuously adjustable attractor dimension. Int. J. Bifurc. Chaos **25**, 1530036 (2015)
14. J.C. Sprott, Ergodicity of one-dimensional oscillators with a signum thermostat. Comput. Methods Sci. Technol. **24**, 169–176 (2018)

Spontaneous Symmetry Breaking in Nonlinear Dynamic Systems

Xiong Wang

1 Introduction

The concept of symmetry dominates modern fundamental physics, both in quantum theory and in relativity. While symmetry plays a crucial role in modern physics, its dual concept "symmetry breaking" is also very important. Spontaneous Symmetry Breaking (SSB), in contrast to explicit symmetry breaking, is a spontaneous process through which a system governed by a symmetrical dynamic ends up in an asymmetrical state. It thus describes systems where the equations of motion or the Lagrangian function obey certain symmetries, but their lowest energy solutions do not exhibit that symmetry. So, the symmetry of the equations is not reflected by individual solutions, but it is reflected by the symmetrical coexistence of asymmetrical solutions. An actual measurement reflects only one solution, representing a breakdown in the symmetry of the underlying theory. "Hidden" is perhaps a better term than "broken" because the symmetry is always there in these equations.

2 SSB in Quantum Field Theory

Without spontaneous symmetry breaking (SSB), the local gauge principle requires the existence of a number of bosons as force carriers. However, some particles (the so-called W and Z bosons) would then be predicted to be massless. While in reality, they are observed to have mass. To overcome this conflict, spontaneous symmetry breaking is augmented by the Higgs mechanism to give these particles masses. It also suggests the presence of a new particle, the Higgs boson, reported as possibly

X. Wang (✉)
Institute for Advanced Study, Shenzhen University,
Shenzhen, Guangdong 518060, People's Republic of China
e-mail: wangxiong8686@szu.edu.cn

X. Wang et al. (eds.), *Chaotic Systems with Multistability and Hidden Attractors*,
Emergence, Complexity and Computation 40,
https://doi.org/10.1007/978-3-030-75821-9_26

identifiable with a boson detected in 2012. SSB occurs whenever a given field in a given Lagrangian has a nonzero vacuum expectation value. The Lagrangian appears symmetric under a symmetry group, but after randomly selecting a vacuum state, the system no longer behaves symmetrically.

2.1 Real Scalar Field Example

Consider the scalar Lagrangian given by

$$\mathcal{L} = \underbrace{\frac{1}{2}(\partial_\mu \varphi)^2}_{\text{"kinetic term"}} + \underbrace{\frac{1}{2}\mu^2\varphi^2 - \frac{\lambda}{4!}\varphi^4}_{\text{"potential term"},} \tag{1}$$

where φ is the scalar field, μ is a sort of "mass" parameter, and λ is the coupling. Observe that there is a symmetry of $\varphi \to -\varphi$ (a discrete symmetry). One can think of the potential as being

$$V(\varphi) = -\frac{1}{2}\mu^2\varphi^2 + \frac{\lambda}{4!}\varphi^4, \tag{2}$$

which has extremes when its derivative is zero. Actually, there are two, given by

$$\varphi_0 = \pm v = \pm\mu\sqrt{\frac{6}{\lambda}}, \tag{3}$$

where the constant v is the "*vacuum expectation value*".

The vacuum, that is, the lowest-energy state, is described by a randomly chosen point of these two new extremes. One can then write

$$\varphi(x) = v + \sigma(x), \tag{4}$$

and then rewrite the Lagrangian as

$$\mathcal{L} = \frac{1}{2}(\partial_\mu \sigma)^2 - \frac{1}{2}(2\mu^2)\sigma^2 - \sqrt{\frac{\lambda}{6}}\mu\sigma^3 - \frac{\lambda}{4!}\sigma^4, \tag{5}$$

where the constant terms have been dropped. It turns out that the symmetry $\varphi \to -\varphi$ is no longer identifiable.

2.2 Complex Scalar Field Example

Consider a complex scalar boson φ and $\varphi^\dagger$. The Lagrangian is

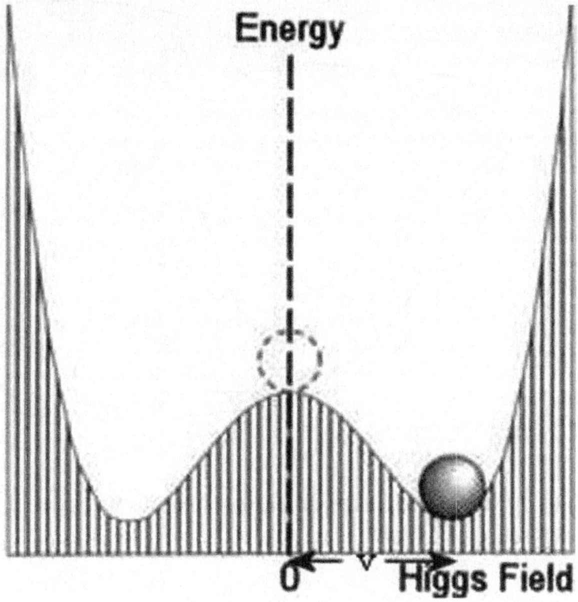

Fig. 1 Illustration of SSB in a real scalar field; figure from stick

$$\mathcal{L} = -\frac{1}{2}\partial^{\mu}\varphi^{\dagger}\partial_{\mu}\varphi - \frac{1}{2}m^2\varphi^{\dagger}\varphi\,. \tag{6}$$

Naturally, one can write this as

$$\mathcal{L} = -\frac{1}{2}\partial^{\mu}\varphi^{\dagger}\partial_{\mu}\varphi - V(\varphi^{\dagger}, \varphi), \tag{7}$$

where

$$V(\varphi^{\dagger}, \varphi) = \frac{1}{2}m^2\varphi^{\dagger}\varphi\,. \tag{8}$$

This Lagrangian has the $U(1)$ symmetry, which means that it is invariant under the transform: $\varphi \mapsto e^{i\theta}\varphi$.

In graph $V(\varphi^{\dagger}, \varphi)$, plotting V vs. $|\varphi|$, one can see a "bowl" with $V_{minimum}$ at $|\varphi|^2 = 0$. The vacuum of any theory ends up being at the lowest potential point, and therefore the vacuum of this theory is at $\varphi = 0$, as one would expect.

Now, change the potential and consider

$$V(\varphi^{\dagger}, \varphi) = \frac{1}{2}\lambda m^2(\varphi^{\dagger}\varphi - \Phi^2)^2, \tag{9}$$

where λ and Φ are real constants. Notice that the Lagrangian will still have the global $U(1)$ symmetry as before. But, by graphing V vs. $|\varphi|$, one gets Fig. 2.

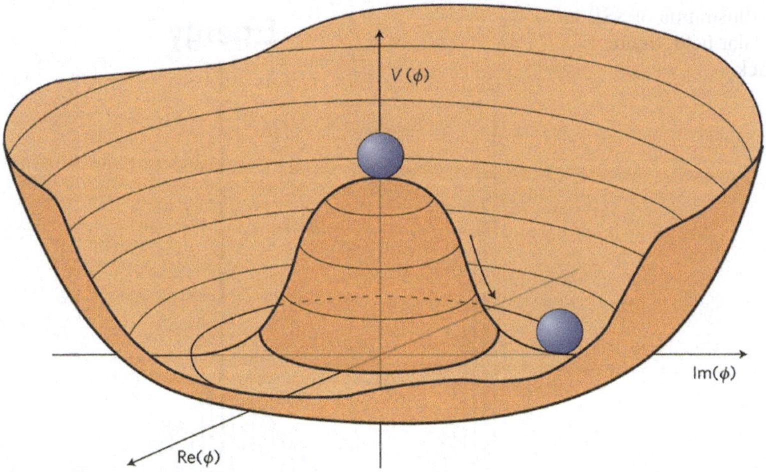

Fig. 2 Mexican hat potential function V versus Φ; figure from Eyes on a prize particle, Nature Physics 7, 2C3 (2011)

In Fig. 2, now, the vacuum $V_{minimum}$ is represented by the circle at $|\varphi| = \Phi$. In other words, there are an infinite number of vacuums in this theory. And, because the circle drawn in this figure represents a rotation through a field space, this degenerate vacuum is parameterized by $e^{i\alpha}$, the global $U(1)$. There will be a vacuum for every value of α, located at $|\varphi| = \Phi$.

In order to make sense of this theory, one must *choose* a vacuum by hand. Because the theory is completely invariant under the choice of the $U(1)$, $e^{i\alpha}$, one can choose any α and *define* that as the true vacuum. So, here, choose α to make the vacuum at $\varphi = \Phi$.

Next, it is needed to rewrite this theory in terms of the new vacuum. One can therefore expand around the constant vacuum value Φ to have a new field, as

$$\varphi \equiv \Phi + \alpha + i\beta \,, \tag{10}$$

where α and β are new real scalar fields (so, $\varphi^\dagger = \Phi + \alpha - i\beta$). One can now write out the Lagrangian as

$$\begin{aligned}\mathcal{L} &= -\frac{1}{2}\partial^\mu[\alpha - i\beta]\partial_\mu[\alpha + i\beta] - \frac{1}{2}\lambda m^2[(\Phi + \alpha - i\beta)(\Phi + \alpha + i\beta) - \Phi^2]^2 \\ &= \left[-\frac{1}{2}\partial^\mu\alpha\partial_\mu\alpha - \frac{1}{2}4\lambda m^2\Phi^2\alpha^2 - \frac{1}{2}\partial^\mu\beta\partial_\mu\beta \right] \\ &\quad -\frac{1}{2}\lambda m^2\left[4\Phi\alpha^3 + 4\Phi\alpha\beta^2 + \alpha^4 + \alpha^2\beta^2 + \beta^4 \right].\end{aligned}$$

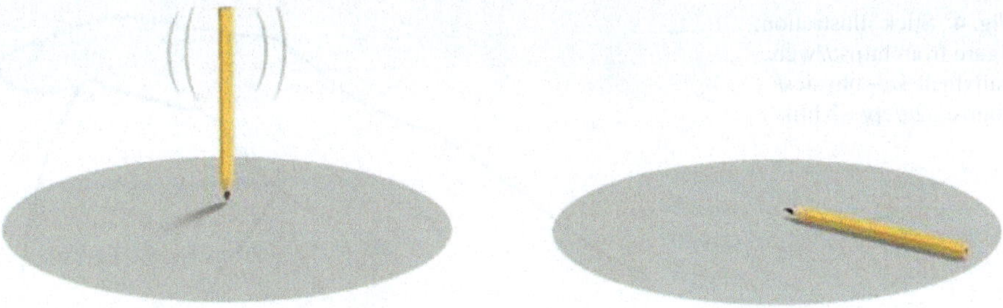

Fig. 3 Pencil illustration of SSB: Spontaneous broken symmetry. The world of this pencil is completely symmetrical. All directions are exactly equal. But this symmetry is lost when the pencil falls over. Now one direction holds. The symmetry that existed before is hidden the fallen pencil; figure from https://web.hallym.ac.kr/~physics/course/a2u/ep/ssb.htm

This is now a theory of a *massive* real scalar field α (with mass = $\sqrt{4\lambda m^2 \Phi^2}$), a *massless* real scalar field β, and five different types of interactions (one allowing three α's to interact, the second allowing one α and two β's, the third allowing four α's, the fourth allowing two α's and two β's, and the last allowing four β's to do so.) In other words, there are five different types of vertices allowed in the Feynman diagrams for this theory. Thus, the $U(1)$ symmetry is no longer manifest.

3 Other Simple Illustrations

3.1 Pencil

It is an amazing idea that the symmetry of the equations could not be reflected by individual solutions, but it is reflected by the symmetrically coexistence of asymmetrical solutions.

First, consider a pencil, standing upright at the center, which is possible but unstable. The pencil tends to fall into one stable state randomly. After it falls down, the original symmetry was broken (see Fig. 3). This is exactly the illustration of the "Mexican hat" potential of a complex field, shown in Fig. 2.

3.2 Stick

Consider another daily-life example: A linear vertical stick with a compression force applied at the top and directed along its axis. The physical description is obviously invariant for all rotations around this axis. As long as the applied force is mild enough, the stick does not bend, and the equilibrium configuration (the lowest energy con-

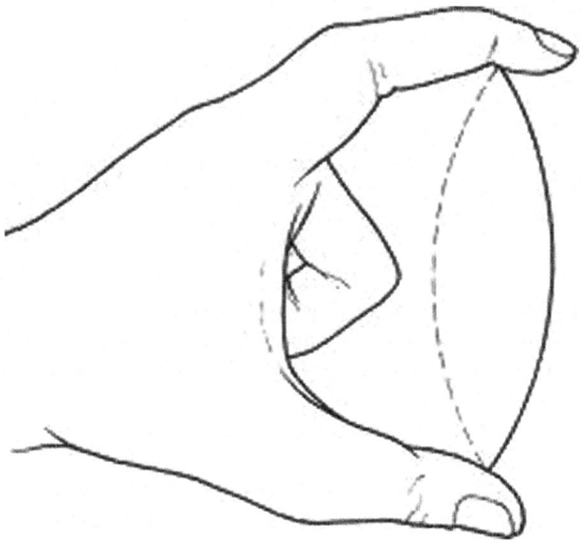

Fig. 4 Stick illustration; figure from https://web.hallym.ac.kr/~physics/course/a2u/ep/ssb.htm

figuration) is invariant under this symmetry. When the force reaches a critical value, the symmetric equilibrium configuration becomes unstable, and an infinite number of equivalent lowest energy stable states appear, which are no longer rotationally symmetric. The actual breaking of the symmetry may then easily occur due to the effect of a (however small) external asymmetric cause, and the stick bends until it reaches one of the infinite possible stable asymmetric equilibrium configurations (see Fig. 4).

3.3 Shortest Connecting Path

The above two examples are essentially the same, where all the new lowest energy solutions are asymmetric but are all related through the action of the symmetry transformation $U(1)$.

Here is yet another elegant example.

Consider the scenario of connecting four nodes located symmetrically at four vertices of a square. A simple solution one can draw by hand at once is like the one shown in Fig. 5.

This solution fully maintains the original symmetry of a foursquare, and, in fact, it is a very satisfactory solution with a total length of $2\sqrt{2} \doteq 2.828$.

However, the best solution turns out to be the one shown in Fig. 6. The total length is $1 + \sqrt{3} \doteq 2.732$, and the symmetry of the solution is less than that of the original system.

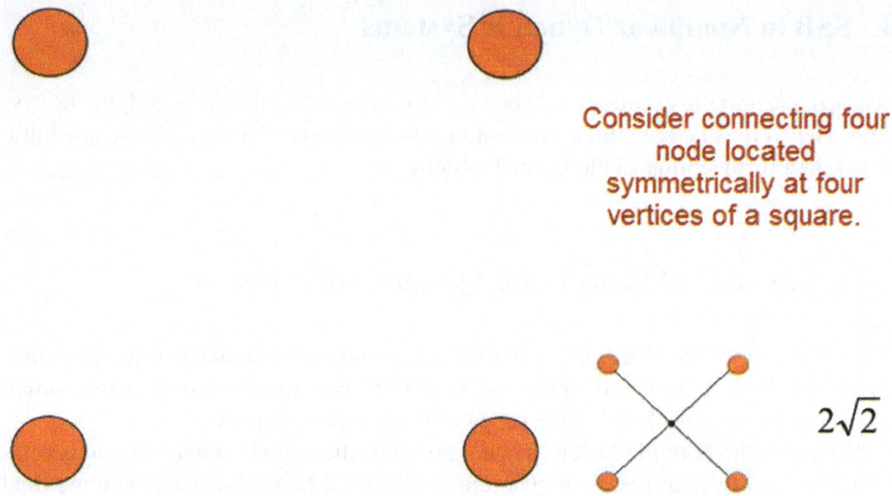

Fig. 5 Connecting four nodes located symmetrically at four vertices of a square

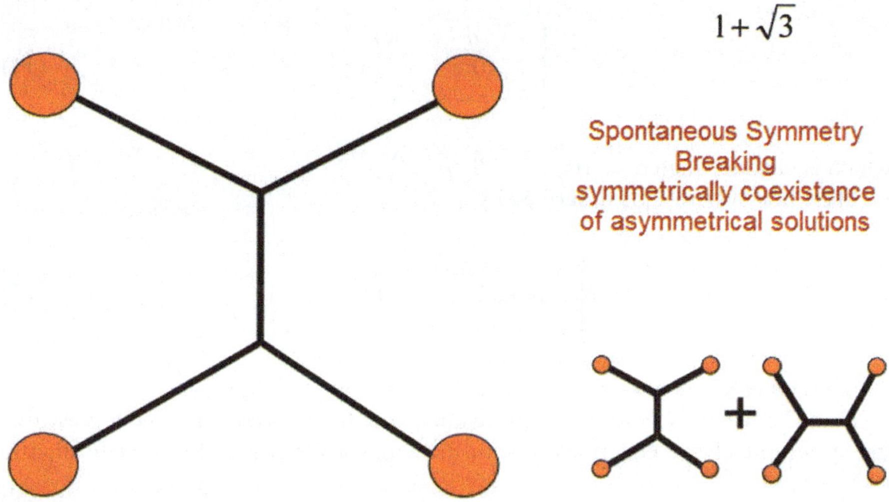

Fig. 6 Spontaneous symmetry breaking

So, the symmetry of the original problem is reflected by the symmetrical coexistence of asymmetrical solutions, each of which has less (or even no) symmetry than the original full symmetry.

4 SSB in Nonlinear Dynamic Systems

Perhaps, the best illustrations could be given by attractor(s) of nonlinear dynamic systems. Here, consider only three-dimensional autonomous systems with two quadratic terms, in a form similar to the Lorenz system.

4.1 Symmetrical System with Symmetrical Attractor

The symmetry of the algebraic equations determines the symmetry of its geometric dynamics. In fact, symmetry plays an important role in generating chaos, which somehow determines the possible shape of the resulting attractor.

For example, both the Lorenz system and the Chen system have the z-axis rotational symmetry, and they both generate a two-scroll butterfly-shaped symmetrical attractor.

The Lorenz system [1] is described by

$$\begin{cases} \dot{x} = \sigma\,(y - x) \\ \dot{y} = rx - y - xz \\ \dot{z} = -bz + xy\,, \end{cases} \tag{11}$$

which is chaotic when $\sigma = 10,\ r = 28,\ b = \frac{8}{3}$.

The Chen system[2] is described by

$$\begin{cases} \dot{x} = a(y - x) \\ \dot{y} = (c - a)x - xz + cy \\ \dot{z} = -bz + xy\,, \end{cases} \tag{12}$$

which is chaotic when $a = 35,\ b = 3,\ c = 28$.

Both Lorenz and Chen systems have z-axis rotational symmetry, in the sense that the system algebraic equations remain the same when (x, y, z) is transformed to $(-x, -y, z)$.

There are totally six possible quadratic nonlinear terms: xy, yz, xz, x^2, y^2 and z^2 in a 3D quadratic equation. Restricted by the z-axis rotational symmetry, the nonlinear terms in the second equation of the above 3D system must be either xz or yz, while in the third equation they must be either xy or z^2, or x^2, or y^2.

Based on the above observations, to maintain the z-axis rotational symmetry, the most general form of a 3D autonomous system with only linear and quadratic terms seems to be the following:

$$\begin{cases} \dot{x} = a_{11}x + a_{12}y \\ \dot{y} = a_{21}x + a_{22}y + m_1xz + m_2yz \\ \dot{z} = a_{33}z + m_3xy + m_4x^2 + m_5y^2 + m_6z^2 + c\,, \end{cases} \tag{13}$$

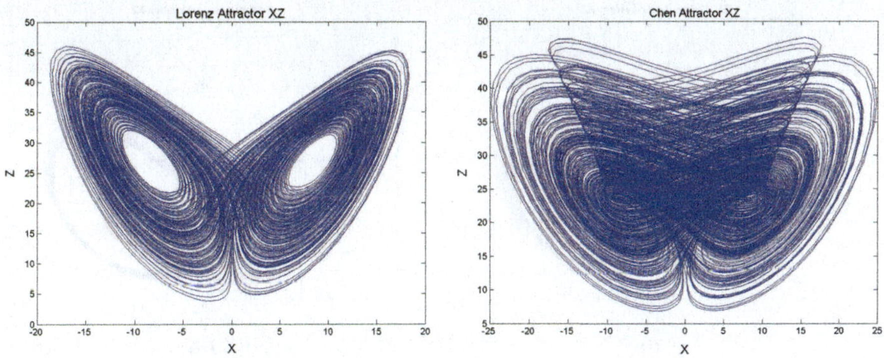

Fig. 7 Lorenz attractor and Chen attractor

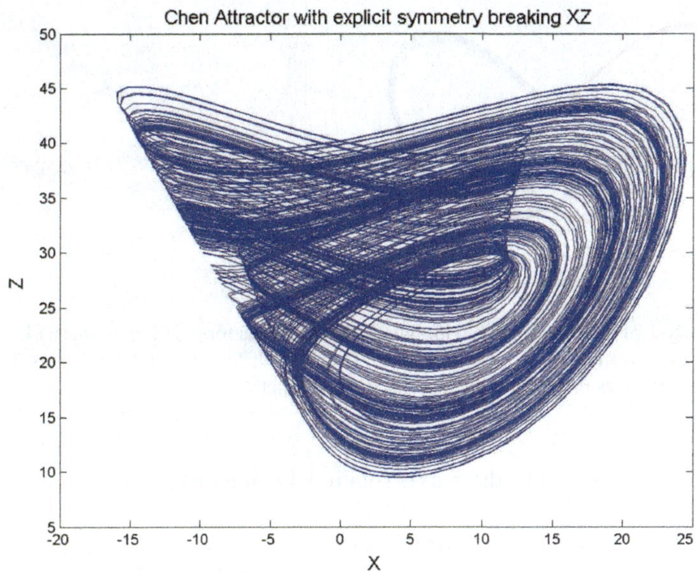

Fig. 8 Explicit symmetry breaking of the Chen system with $m = 20$

where all coefficients are real constants.

The Lorenz attractor and Chen attractor are shown in Fig. 7.

4.2 *Explicit Symmetry Breaking*

The explicit symmetry breaking can be easily realized by adding terms that break the symmetry of the algebraic equations. For example, adding a constant term m into

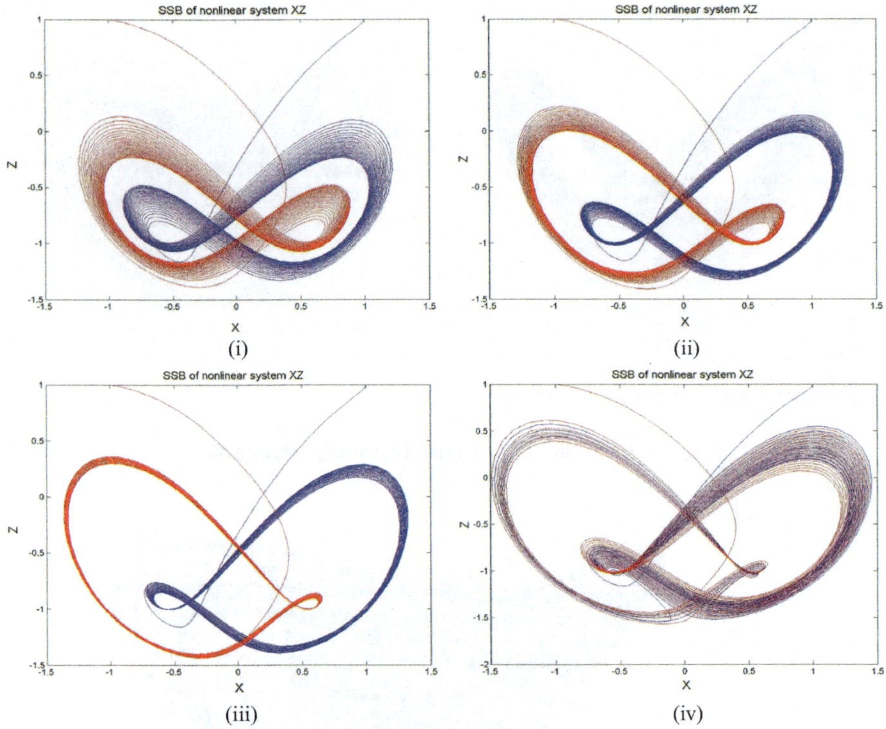

Fig. 9 Family-k of chaotic systems with two coexisting attractors: (i) $k = 0.78$, (ii) $k = 0.79$, (iii) $k = 0.8$, (iv) $k = 0.81$. The two trajectories are from two different initial conditions, for the latter case the two attractors merge into one and become symmetric

the second equation breaks the z-axis rotational symmetry:

$$\begin{cases} \dot{x} = a(y - x) \\ \dot{y} = (c - a)x - xz + cy + m \\ \dot{z} = -bz + xy\,. \end{cases} \tag{14}$$

The resulting attractor is not symmetrical anymore due to the asymmetrical dynamic equation.

4.3 SSB of Nonlinear Systems

Constrained by the algebraical symmetry of the system equations, the above systems generate symmetrical two-scroll attractors. But this is not always the case, because this statement is based on the postulation that the dynamic system has only one attractor.

A very interesting phenomenon is that nonlinear systems may have multiple coexisting attractors. One such examples is described by Ref. [3]

5 Discussion

Constrained by the algebraically symmetry of the dynamic equations, the geometric attractors generated by such a system may or may not preserve the original symmetry. It is still puzzling to me what type of systems may break the symmetry while the others preserve it, or what type of systems may have multiple attractors.

What's more, in reality, one may not know the governing equation or may not fully understand its symmetry. What we can directly observe is the physical phenomenon, where the trajectory and the attractor may be asymmetric, thus we may overlook the underlying symmetry. If the observation (attractors) cannot reflect the underlying symmetry of the fundamental physic law, then the symmetry is considered "hidden".

Thus, SSB allows the symmetry theory to describe asymmetric reality. SSB provides a way of understanding the complexity of nature without renouncing the fundamental symmetries. So, we believe or prefer symmetric to asymmetric fundamental laws.

In one word, symmetry is simple and elegant, while symmetry breaking makes the world complex and colorful.

References

1. E.N. Lorenz, Deterministic non-periodic flow. J. Atmos. Sci. **20**, 130–141 (1963)
2. G. Chen, T. Ueta, Yet another chaotic attractor. Int. J. Bifurc. Chaos **9**(07), 1465–1466 (1999)
3. X. Wang, G. Chen, A chaotic system with only one stable equilibrium. Commun. Nonlinear Sci. Numer. Simul. **17**, 1264–1272 (2012)

Multi-stability: The Source of Unity and Diversity of the World

Xiong Wang

1 Starting with Symmetry

1.1 Unity and Symmetry of Physical Laws

There are various kinds of symmetry in nature. A symmetry in the real world may be approximately satisfied. There are so many different kinds of symmetry in nature that involve different objects and different types to different degrees. What's the nature of symmetry? In essence, symmetry is to remain invariant under certain transformations.

Symmetry in mathematics is often more abstract and essential, and the basic tool for describing symmetry in mathematics is group theory. Around 1831, the young French mathematician Galois, on the basis of his predecessors' research, discovered the symmetry of the roots of algebraic equations from the well-known Weida's theorem and conceived the concept of "groups". He also creatively solved the solvability problem of algebraic equations from the viewpoint of symmetry. He systematically explained that there are no formulas for solutions of a quintic equation or equations with degrees higher than five, but there are formulas for solutions of equations with degrees four or lower. He solved two of the three great drawing problems since ancient times: "One cannot divide an angle into three equal parts at will" and "it is impossible to double a cube". These problems, which do not seem to be directly related to symmetry, are perfectly solved by Galois from the viewpoint of deep symmetry. Therefore, the study of symmetry is a shortcut to grasp the essence of a problem. Inspired by Galois's work, Lie established his theory by linking groups to differential equations. The theory of Lie group in Lie algebra provides a powerful

X. Wang (✉)
Institute for Advanced Study, Shenzhen University,
Shenzhen 518060, Guangdong, People's Republic of China
e-mail: wangxiong8686@qq.com

X. Wang et al. (eds.), *Chaotic Systems with Multistability and Hidden Attractors*,
Emergence, Complexity and Computation 40,
https://doi.org/10.1007/978-3-030-75821-9_27

tool for geometric interpretations of modern physics, and the Lie group indeed plays a very important role in particle physics.

The importance of symmetry in modern physics cannot be overemphasized. Symmetry is central to general relativity and quantum mechanics, which are the twin pillars of modern theoretical physics. "The themes of the 20th-century physics are quantization, symmetry and phase factors," Mr. Chen-Ning Yang once said. In fact, the symmetry that physics is concerned with is the symmetry of physical laws, which is a very deep kind of symmetry. That is to say, the physical laws should be kept invariant under certain transformations, which is essentially the inherent requirement of the unity of the physical laws. The direction of the development of physics is to seek a deeper understanding of nature, a more unified understanding, a greater symmetry in some sense.

It is fair to say that every major advance in modern physics, from special relativity, general relativity, quantum mechanics, quantum field theory, to gauge theory, has been shaped by the idea of transformation invariance. Dirac also pointed out that the future direction of theoretical physics is to continue to expand the invariance of transformations.

1.2 Action and Symmetry

The laws of physics, as they are called, can often be expressed in terms of the least-action principle, and the study of the symmetry of the laws becomes the study of the symmetry of the action itself.

The field theory model of a particle system based on the action principle is the basic procedure of constructing a dynamical model in quantum field theory, which is an important way to establish theory in modern physics. For an infinite degree-of-freedom system in field theory, the regular coordinates of the system are field variables, where the space coordinate $Phi(VECx, t)$ is used to identify the different degrees of freedom of the system. The action of the system is

$$S = \int d^4x \mathscr{L}\left(\varphi, \partial_\mu \varphi\right) \tag{1}$$

where $d^4x = dtd^3\mathbf{x}$ is a four-dimensional volume element of Lorentz invariant space-time; $\mathscr{L}$ is the Lagrangian of the system, which is a functional of the field quantity and its derivative, and does not depend directly on the space-time coordinates.

Based on the experience of classical mechanics and classical electromagnetic theory, it is generally expected that the equation of motion is only a second-order partial differential equation of space-time coordinates, so it is assumed that it depends only on φ and $\partial_\mu \varphi$. The Lagrangian can contain higher-order differential terms (even fractional-order terms), but the higher-order differential terms are usually not considered based on the experience of classical mechanics and classical electromagnetic theory.

The principle of field action requires that, when the field quantity have a small variation under certain conditions:

$$\delta S = 0 \tag{2}$$

we can derive the Euler–Lagrange equation. It is the equation of motion of the field φ:

$$\frac{\partial \mathscr{L}}{\partial \varphi} - \partial_\mu \frac{\partial \mathscr{L}}{\partial \left(\partial_\mu \varphi\right)} = 0 \tag{3}$$

The field theory model of relativity usually initiates from the Lagrangian of the field. A model of the field theory is a concrete assumption about the Lagrangian of the field. Knowing the Lagrangian of the field, we can get the motion equations of the field from its Euler–Lagrange equation. In quantum field theory, if we know the Lagrangian, we know the corresponding Feynman rule, and thus everything about the theory in principle.

1.3 Noether Theorem, Conserved Flows and Conserved Quantity

Noether theorem states that each symmetry of an action corresponds to a conservation law, and therefore a conservation current and a conservation quantity can be defined accordingly. The two main concepts of symmetry and conservation are closely linked. The Noether theorem, which links symmetry and conservation laws in classical physics, has since been extended to quantum mechanics. In quantum mechanics and particle physics, some new internal degrees-of-freedom have been introduced, and some new symmetries of the abstract space and the corresponding conservation laws have been realized. Especially now, the group theory is more advantageous to deal with the problems according to the symmetry of a quantum system.

In classical mechanics, the familiar correspondence is that the time translational symmetry (time translation invariance) corresponds to the conservation of energy, and the rotational symmetry (space isotropy) corresponds to the conservation of angular momentum. Time translational symmetry and conservation of energy—time translational symmetry requires that the laws of physics are the same in yesterday, today and tomorrow. If the physical laws change with time, for example, the laws of gravity change with time, it would also be possible to use the variability of gravity overtime to lift water into a reservoir when gravity is so weak that requires less work. Releasing water from a reservoir when gravity gets stronger, using hydroelectric power to generate more energy, is a first-class perpetual motion machine that literally creates energy, which goes against the energy conservation. This clearly shows the connection between the translational symmetry of time and the conservation of energy.

Noether proved an important theorem in 1918: if the action S of a field is invariant under a continuous transformation, then there exists a corresponding conservative current $j^\mu(x)$, which satisfies the continuity equation

$$\partial_\mu j^\mu(x) = 0 \tag{4}$$

In general, the invariance of $\mathscr{L}$ under some transformation is said to have some symmetry; therefore, some symmetry of the Lagrangian corresponds to a conservation law. This is a general conclusion and is not limited to symmetries associated with continuous transformations.

Integrating the continuous equation that the conservative current $j^\mu(x)$ is satisfied over the total space, we obtain

$$\int d^3\mathbf{x}\frac{\partial}{\partial t}j^0(x) + \int d^3\mathbf{x}\nabla \cdot \mathbf{j}(x) = 0 \tag{5}$$

Now, replace the second term with the surface integral at infinity by the Gauss theorem. When $\mathbf{j}(x)$ is zero at infinity, we have

$$\frac{dQ}{dt} = -q\int_\infty d\boldsymbol{\sigma} \cdot \mathbf{j}(x) = 0 \tag{6}$$

That is,

$$Q = q\int d^3\mathbf{x}j^0(x) \tag{7}$$

where q is the constant properly introduced in the definition of Q; therefore, the existence of a conserved current $j^\mu(x)$ means the existence of a conserved quantity q in the whole space whose density is proportional to $j^0(x)$.

1.4 Why Is Symmetry so Important in Physics?

Why is symmetry so important in physics? In a way, this is consistent with the fundamental goals of physics, and with our quest to understand nature, so the quest for symmetry is rooted in the genes of theoretical physics.

- Symmetry is a prerequisite for the rational definition of fundamental physical quantities. Based on the above discussion, we can define energy because of the invariance of time translation. Without these symmetries, there would be no reasonable way to define these fundamental quantities, and there would be no laws governing the physics of these quantities.
- Symmetry is a prerequisite for the universality of physical laws. What we seek is a universal law of physics, and if the laws of physics here and now cannot be applied to here and then, this will also lose its meaning. So, the nature of theoretical

physics requires that the law be as symmetric as possible, that is, as universal as possible. So, symmetry is rooted in the genes of theoretical physics.

1.5 Symmetry is Rooted in the Genes of Theoretical Physics

In some sense, if there is no symmetry, there is no reasonable definition of these basic physical quantities, so the physical laws of these quantities cannot be mentioned. What we seek is a universal law of physics, and so if the laws of physics of here and now cannot be applied to there and then, it will lose its theoretical significance. Therefore, the essence of theoretical physics requires the law to be as symmetrical as possible, that is, as universal as possible. So, symmetry is rooted in the genes of theoretical physics.

1.6 Symmetry from Outer Space-Time to Internal Space

In the past, we had learned about invariance under space-time continuous transformations, including translation invariance and Lorentz covariance. Translation invariance, which reflects the homogeneity of space and time, is the basis of the laws of energy and momentum conservation. Lorentz covariance is the invariance of rotation of a four-dimensional space-time, reflecting the isotropy of the four-dimensional space-time. It includes the invariance of 3D rotation, which is the basis of the conservation of angular momentum. In quantum field theory, the field is symmetric in the internal space. The internal space of a field is an abstract space introduced to describe the internal properties of a particle field, such as electric charge, baryon number, lepton number, isospin, flavor and color.

Modern gauge theory is an extension of the space-time symmetry to a more general local internal symmetry. This first step was taken by Chen-Ning Yang and Mills, who wanted to find the consequences of postulating the law of isospin conservation. The concept of isospin is introduced by N. Kemmerz in 1938, the isospin is assumed to be similar to the electron spin when interacting. It played an important role in the following nuclear and gauge theory. Isospin conservation is a restatement of the fact that nuclear forces are charge independent. According to Werner Heisenberg, protons and neutrons are two states of the same particle in an abstract isospin space. Since charge conservation is related to phase invariance, by analogy one would guess that strong interaction is invariant under isospin. From the philosophy of science, what is embodied in the gauge theory's thinking is that objective physical events are independent of the framework we have chosen to describe. Namely, the laws of physics have some profound inherent invariance. Isospin invariance belongs to the gauge theory category (isospin space is a type of internal space). The results of Young and Mills are significant.

In gauge theory, the internal space is "rotated" in terms of gauge group G. The writing of elements in the norm group is

$$u(\theta) = \exp\left[-i\theta^{\alpha}T^{\alpha}\right] \tag{8}$$

where T^{α} is the generator of a group G, which satisfies the communitive relation

$$\left[T^{\alpha}, T^{\beta}\right] = if^{\alpha\beta\gamma}T^{\gamma} \tag{9}$$

where $f^{\alpha\beta\gamma}$ is the structure of a group G constant; $\alpha\beta\gamma = 1, 2, \ldots, N$, with N being the dimension of the group G, which is equal to the number of generators; θ^{α} is a parameter of a group G, called a group parameter, having also N numbers. The generators of a group can be used as operators to describe the properties of field particles. But what is unsatisfactory is that, unlike the structure of space-time, the structure and symmetry of internal space are uncertain: which gauge group should be chosen is determined by the specific properties of the field particles and the law of interaction between the particles.

The symmetry of a physical system in internal space is described by the invariance of its Lagrangian under the following transformation:

Gauge transformation of the field quantity:

$$\Phi'(x) = \exp\left[-i\theta^{\alpha}T^{\alpha}\right]\Phi(x) \tag{10}$$

Gauge transformation of space-time derivative $\partial_{\mu}\Phi(x)$ of the field quantity:

$$\partial_{\mu}\Phi(x) \to \partial_{\mu}\Phi'(x) = \exp\left[-i\theta^{\alpha}T^{\alpha}\right]\partial_{\mu}\Phi(x) \tag{11}$$

The invariance of Lagrangian that can be described by

$$\mathscr{L}\left(\Phi'(x), \partial_{\mu}\Phi'(x)\right) = \mathscr{L}\left(\Phi(x), \partial_{\mu}\Phi(x)\right) \tag{12}$$

The above transformation is the global gauge transformation. Because θ^{α} is a constant independent of the space-time coordinate x, the space-time derivative of the field $\partial_{\mu}\Phi(x)$ is just like the field $\Phi(x)$, which is an important characteristic of the global gauge transformation.

1.7 From Global Symmetry to Local Symmetry

The field quantity $\Phi_{\sigma}(x)$ is transformed in the internal space rotation $\Phi_{\sigma}(x) \to \Phi'_{\sigma}(x) = \exp\left[-i\theta^{\alpha}T^{\alpha}_{\sigma\rho}\right]\Phi_{\rho}(x)$. This transformation is called the gauge transformation. When the group parameter θ^{α} is a constant independent of the space-time coordinate x, it is called the global gauge transformation. When the group parameter θ^{α} is a function of the space-time coordinate x: $\theta^{\alpha}(x)$, it is called the local gauge transformation. Global here means that the field quantities of each point of space-time do the same transformation, while local means that each point of space-time makes its own transformation.

Under the local gauge transformation, the field quantity transform is

$$\Phi(x) \to \Phi'(x) = \exp\left[-i\theta^{\alpha}(x)T^{\alpha}\right]\Phi(x) \tag{13}$$

The transformation of the derivative of the field quantity $\partial_{\mu}\Phi(x)$ is

$$\begin{aligned}\partial_{\mu}\Phi(x) \to \partial_{\mu}\Phi'(x) &= \exp\left[-i\theta^{\alpha}(x)T^{\alpha}\right]\partial_{\mu}\Phi(x) - i\partial_{\mu}\theta^{\beta}(x)T^{\beta}\exp\left[-i\theta^{\alpha}(x)T^{\alpha}\right]\Phi(x)\\ &\neq \exp\left[-i\theta^{\alpha}T^{\alpha}\right]\partial_{\mu}\Phi(x)\end{aligned} \tag{14}$$

It is different from the transformation law of the field quantity $\Phi(x)$. This is different from the global transformation. It can be seen that the field of each point of space-time is transformed according to their respective quantities in the local gauge transformation, so when the invariant Lagrangian under the global gauge transformation is extended to the local gauge transformation, the Lagrangian is no longer kept unchanged.

We hope that the Lagrangian is not only invariant under the global transformation, but also invariant under the local transformation. Since θ^{α} has nothing to do with x, but it with x is a special case, the global gauge theory is only a special case of the gauge theory, and the invariant Lagrangian under the global gauge transformation should be extended to the invariant local gauge. As mentioned earlier, the Lagrangian of global gauge invariance is not invariant under local gauge transformation because the field derivative $\partial_{\mu}\Phi(x)$ and field $\Phi(x)$ are transformed in the same way under global gauge transformation, but under the local gauge transformation, it follows different rules. So, if we replace the derivative ∂ with the covariant derivative D_{μ} and the field quantity derivative $\partial_{\mu}\Phi(x)$ with the field quantity covariant derivative $\partial_{\mu}\Phi(x)$, and require the field quantity covariant derivative $\partial_{\mu}\Phi(x)$ and the field quantity $\Phi(x)$ to be transformed in the same way, namely,

$$\begin{aligned}\Phi(x) \to \Phi'(x) = \exp\left[-i\theta^{\alpha}(x)T^{\alpha}\right]\Phi(x)\\ D_{\mu}\Phi(x) \to D'_{\mu}\Phi'(x) = \exp\left[-i\theta^{\alpha}(x)T^{\alpha}\right]D_{\mu}\Phi(x)\end{aligned} \tag{15}$$

then the Lagrangian formed by them, $\mathscr{L}\left(\Phi(x), D_{\mu}\Phi(x)\right)$, will be invariant under local gauge transformation; that is,

$$\mathscr{L}\left(\Phi'(x), D'_{\mu}\Phi'(x)\right) = \mathscr{L}\left(\Phi(x), D_{\mu}\Phi(x)\right) \tag{16}$$

1.8 Introduction of Gauge Field

Next is the selection of the covariant derivative D_μ, inspired by quantum electrodynamics, according to Yang–Mills. We introduce the gauge field potential A_μ^α and define the covariant derivative as

$$D_\mu(x) \equiv \partial_\mu + A_\mu(x), A_\mu(x) \equiv -igA_\mu^\alpha(x)T^\alpha \tag{17}$$

The gauge fields thus introduced are closely related to the gauge groups. For each generator T^α, there is a canonical potential $A_\mu^\alpha(x)$. For $\alpha = 1, 2, \ldots, N$ generators, there are n normal potentials. In addition, the gauge field is introduced by changing the derivative into covariant derivative D_μ, where ∂_μ and D_μ are vectors of four-dimensional space-time, and A_μ^α should also be vectors of four-dimensional space-time. This means that the quantum number of spins of the gauge particle corresponding to the gauge field must be 1. The quantum numbers of the spins of photons, gauge Boson, gluons, etc., are all 1, where g is the interaction constant.

By introducing the gauge field $A_\mu^\alpha(x)$ and defining the covariant derivative D_μ, the objective here is to make the covariant derivative $\partial_\mu \Phi(x)$ have the same transformation law as the field $\Phi(x)$ under the local gauge transformation. Thus, the invariant Lagrangian under the global gauge transformation is extended to the invariant Lagrangian under the local gauge transformation. This requires the transformation law of the gauge potential $A_\mu^\alpha(x)$ under the gauge transformation be determined; that is,

$$\left(\partial_\mu - igA_\mu'^\alpha(x)T^\alpha\right)u(\theta)\Phi(x) = u(\theta)\left(\partial_\mu - igA_\mu^\alpha(x)T^\alpha\right)\Phi(x) \tag{18}$$

So,

$$A_\mu'^\alpha(x)T^\alpha = u(\theta)A_\mu^\alpha(x)T^\alpha u^{-1}(\theta) + \frac{i}{g}u(\theta)\partial_\mu u^{-1}(\theta) \tag{19}$$

This is the law of transformation of the gauge potential.

1.9 Symmetry Governs Interaction: Take Electromagnetic Force for Example

Which is more essential, the equation of motion of the interaction or the symmetry?

In classical mechanics, the conservation law of mechanical quantities can be derived from Newton's equation under certain conditions. At first glance, the conservation law seems to be the result of the equation of motion. But, in essence, the law of conservation is more fundamental than the equation of motion, because it reveals some universal laws of nature, governs all processes of nature, and restricts the equations of motion in different fields.

Symmetry governs the form of action, yet physicists cannot know a priori all the symmetries involved in the world, and the symmetries that are already known are not sufficient for fully determining the form of action. Although the possible forms of action have been greatly restricted, they can still take many, many possible forms, and physicists have to take tentative methods, where the candidates for each action were examined in turn according to the physical possibilities.

Fundamental physical equations have certain symmetry requirements, such as Lorentz invariance, general covariance, gauge invariance, etc., which strongly constrain the equations and even completely determine the interactions. This is what we call the symmetric dominating interaction.

Let's take the electromagnetic force as an example to show how we can get the exact mechanical equation of the interaction from the symmetry principle, why there should be an electromagnetic field because of the symmetry principle requirement, and why electromagnetic fields interact in this way but not in that way because of the need for symmetry. This is a very good example of the power of the principle over mechanics.

The problem of mechanical systems, which boil down to the concrete form of Lagrangian, is now illustrated by the example of a single free electron, as how to construct Lagrangian. Then, equations of motion will be derived.

Take a system with only one free electron for example, and we first construct the free electron Lagrangian $\mathscr{L}$. The free electron has a mass of m, a momentum of p_μ, and a wave function of $\psi(x)$. The quantities that can be used under Lorentz invariant requirements are $\bar{\psi}(x)$, $\psi(x)$, γ_μ, m, p_μ. The relativistic invariants composed by them are

$$\mathscr{L}(x) = -\bar{\psi}(x)\left(\gamma_\mu \partial_\mu + m\right)\psi(x) \tag{20}$$

where γ_μ is a 4×4 constant matrix, and $p_\mu = -i\hbar\partial_\mu$. Suppose that this is the Lagrangian of the free electron, and by the Euler–Lagrange equation

$$\frac{\partial \mathscr{L}}{\partial \varphi_a} - \partial_\mu \frac{\partial \mathscr{L}}{\partial \partial_\mu \varphi_a} = 0 \tag{21}$$

we get

$$\left(\gamma_\mu \partial^\mu + m\right)\psi(x) = 0 \tag{22}$$

This is the Dirac free electron relativistic equation of motion. This equation is invariant under the following transformation:

$$\psi(x) \to \psi'(x) = \mathrm{e}^{\mathrm{i}\alpha}\psi(x) \tag{23}$$

where α is a constant. This transformation is called the first kind of gauge transformation, which Weyl proposed in 1929, and also called the global gauge transformation. This α is the same constant for all space-time. Under this gauge transformation, $\mathscr{L}' = \mathscr{L}$. Due to this invariance, it can be derived from Noether's theorem, as

$$\partial_\mu j^\mu = 0;\ j_\mu = -\mathrm{i}\alpha\bar{\psi}\gamma_\mu\psi \tag{24}$$

where $Q = -i\int j_\mu(x)d^3x$. Take $\alpha = e$. Then, Q is the charge, $j_\mu(x)$, $\mu = 1, 2, 3$, is the current intensity, $\partial_\mu j^\mu = 0$ is the charge-current conservation; that is, the global gauge transformation invariant leads to the charge conservation.

Now, localize the transform $\psi'(x) = \mathrm{e}^{\mathrm{i}\alpha}\psi(x)$; that is, α is not a constant but varies by x, namely, $\alpha \to \alpha(x)$, local gauge transformation $\psi'(x) = \mathrm{e}^{\mathrm{i}\alpha(x)}\psi(x)$.

It can be verified that $\mathscr{L}(x)$ is not invariant to the local gauge transformation, under the local gauge transformation

$$\mathscr{L}'(x) = \bar{\psi}'(x)[\gamma_\mu(\partial^\mu - i\partial^\mu\alpha(x)) + m]\psi'(x) \tag{25}$$

If the Lagrange density is required to be constant under local gauge transformations, an additional term must be added to $\mathscr{L}(x)$ to offset the contribution of $\alpha(x)$, namely,

$$\mathscr{L} = -\bar{\psi}(x)[\gamma_\mu(\partial^\mu - ieA^\mu(x)) + m]\psi(x) \tag{26}$$

This makes the Lagrangian invariant under the following transformation—the local gauge transformation:

$$\psi(x) \to \psi'(x) = \mathrm{e}^{\mathrm{i}\alpha}\psi(x) \tag{27}$$

$$A_\mu(x) \to A'_\mu(x) = A_\mu(x) + \frac{1}{e}\partial_\mu\alpha(x) \tag{28}$$

It makes $\mathscr{L}'(x) = \mathscr{L}(x)$.

So, the local gauge invariance requires the existence of an $A_\mu(x)$, which is the electromagnetic potential vector, also known as the gauge potential in the gauge field.

The gauge theory gives the interaction of electrons and electromagnetic fields as follows:

$$-ie(\bar{\psi}(x)\gamma_\mu\psi(x))A^\mu(x) = j_\mu A^\mu(x) \tag{29}$$

$$j_\mu = -ie\bar{\psi}(x)\gamma_\mu\psi(x) \tag{30}$$

This is the interaction between the electron current (current) and the gauge potential (electromagnetic potential).

Suppose that the electromagnetic tensor $F_{\mu\nu} = \partial_\mu A_\nu - \partial_\nu A_\mu$. The Lagrangian density of a pure electromagnetic field is

$$\mathscr{L}_A = -\frac{1}{4}F_{\mu\nu}F^{\mu\nu} \tag{31}$$

The total Lagrange density of the (electron + electromagnetic field) system is

$$\mathscr{L} = -\frac{1}{4}F_{\mu\nu}F^{\mu\nu} - \bar{\psi}\gamma_\mu(\partial^\mu + m)\psi + ie\bar{\psi}\gamma_\mu\psi A^\mu \tag{32}$$

which = interaction of electromagnetic field + free electron + current with electromagnetic field.

Here, $\mathscr{L}$ is invariant under local gauge transformation, which completely describes the electronic, electromagnetic field system. The local gauge transformation $\psi'(x) = e^{i\alpha(x)}\psi(x)$ is regarded as the element of the $U(1)$ group.

As can be seen from the above example, the relativistic symmetry principle is the starting point for constructing the Lagrange density, and the principle of least action is the starting point for deriving the dynamic equations of the system. The principle of local symmetry is the starting point of constructing perfect electronic electromagnetic force, which leads to the automatic and natural emergence of gauge fields, and becomes the medium of electronic interaction, the messenger of information.

1.10 Symmetry Governs Interaction: Take General Relativity for Example

In general, the establishment of a physical theory starts from observing data, constructing mathematical models, and then sublimating to universal mechanics. This was certainly true for Newton's theory of gravity. Firstly, Tycho Brahe and others recorded detailed observations; then Kepler and others built a phenomenological model of the movement of celestial bodies, and then Galileo and others built a model of the motion of a projectile on the earth; finally, by Newton's grand unification, the celestial and terrestrial models are unified into the mechanical equation of gravitation.

General relativity follows a different path. Unlike all other theories of physics, general relativity's theoretical development followed a path from the principle of symmetry to equations to experiments, with startling mathematical beauty and conviction. It's far simpler than any other possible scenario and has been miraculously confirmed by countless facts.

The top-down, high-rise theory of general relativity was far ahead of the experimental conditions at that time of Einstein's creation of the theory. Various predictions made by general relativity, including gravitational waves and the effects of gravity on the expansion of space-time, followed by the accumulated observations and data decades and hundreds of years after the discovery of general relativity. Theories like general relativity are almost impossible to go from data accumulation to phenomenological models to theories.

Newtonian mechanics has the symmetry of Galileo groups, special relativity has the symmetry of Poincaré groups, and general relativity has the symmetry of generalized coordinate transformation invariant groups. Electromagnetic theory and general relativity are two of the best examples demonstrating the power of the symmetry principle.

Under any coordinate transformation $x^\mu \to x'^\mu(x)$, the metric $g_{\mu\nu}(x)$ is transformed in the form of a second-order covariant tensor:

$$g_{\mu\nu}(x) \to g'_{\mu\nu}\left(x'\right) = \frac{\partial x^\rho}{\partial x'^\mu}\frac{\partial x^\sigma}{\partial x'^\nu}g_{\rho\sigma}(x) \tag{33}$$

General relativity can be regarded as a gauge theory whose gauge transformation includes a generalized coordinate transformation. Therefore, the action of gravity theory should be invariants under generalized coordinate transformation. We can write the terms that satisfy the condition in the order of increasing the dimension of energy. There is only one term of dimension 0, the cosmological term S_Λ,

$$S_\Lambda = -M_{\mathrm{P}}^2\Lambda\int \mathrm{d}^4x\sqrt{-g} \tag{34}$$

where $M_{\mathrm{P}} = (8\pi G)^{-1/2} \simeq 2.4\times 10^{18}$ GeV is the reduced Planck mass, G is the Newton gravitational constant, Λ is a cosmological constant, and $g \equiv \det g_{\mu\nu}$. The term with dimension 2 is also unique, the famous Einstein–Hilbert action:

$$S_{\mathrm{EH}} = \frac{M_{\mathrm{P}}^2}{2}\int \mathrm{d}^4x\sqrt{-g}\mathscr{R} \tag{35}$$

where the Ricci scalar $\mathscr{R} = g^{\mu\nu}\mathscr{R}_{\mu\nu}$, Ricci tensor $\mathscr{R}_{\mu\nu} = \mathscr{R}^\lambda_{\mu\lambda\nu}$, and the curvature tensor $\mathscr{R}^\lambda_{\varkappa\mu\nu}$ is defined by the affine contact $\Gamma^\lambda_{\mu\nu} = \frac{1}{2}g^{\lambda\varkappa}\left(\partial_\mu g_{\varkappa\nu} + \partial_\nu g_{\varkappa\mu} - \partial_\varkappa g_{\mu\nu}\right)$, as

$$\mathscr{R}^\lambda_{\varkappa\mu\nu} = \partial_\nu\Gamma^\lambda_{\varkappa\mu} - \partial_\mu\Gamma^\lambda_{\varkappa\nu} + \Gamma^\lambda_{\nu\sigma}\Gamma^\sigma_{\varkappa\mu} - \Gamma^\lambda_{\mu\sigma}\Gamma^\sigma_{\varkappa\nu} \tag{36}$$

In general relativity, on the other hand, the interaction between the matter field and the gravitational field obeys the principle of minimum coupling. For the Bose field, this is equivalent to replacing all the Minkowski metric $\eta_{\mu\nu}$ in the physical field action S_{matter} with the general metric $g_{\mu\nu}(x)$. The space-time integral measure d^4x is replaced by the invariant form of generalized coordinate transformation $\mathrm{d}^4x\sqrt{-g}$, and all partial derivatives ∂_μ are replaced by covariant derivative ∇_μ, which is defined by affine connection $\Gamma^\lambda_{\mu\nu}$. For the general tensor field $T^{\rho\cdots}_{\sigma\cdots}$, the covariant derivative is

$$\nabla_\mu T^{\rho\cdots}_{\sigma\cdots} = \partial_\mu T^{\rho\cdots}_{\sigma\cdots} + \Gamma^\rho_{\mu\lambda}T^{\lambda\cdots}_{\sigma\cdots} - \Gamma^\lambda_{\mu\sigma}T^{\rho\cdots}_{\lambda\cdots} + \cdots \tag{37}$$

After such a substitution, the matter field is coupled to the gravitational field by the energy tensor $T_{\mu\nu}$, which is defined as

$$T_{\mu\nu} = \frac{2}{\sqrt{-g}}\frac{\delta S_{\mathrm{matter}}}{\delta g^{\mu\nu}} \tag{38}$$

For example, for a scalar field of φ, whose interaction action is $S = \int \mathrm{d}^4x\sqrt{-g}\left[\frac{1}{2}g^{\mu\nu}\left(\partial_\mu\varphi\right)\left(\partial_\nu\varphi\right) - V(\varphi)\right]$, its momentum tensor is

$$T_{\mu\nu}[\varphi] = (\partial_\mu \varphi)(\partial_\nu \varphi) - \frac{1}{2} g_{\mu\nu} (\partial_\lambda \varphi)^2 + g_{\mu\nu} V(\varphi) \tag{39}$$

The fact that the field of matter is coupled to the gravitational field by its kinetic tensor leads to the equivalence principle. In a word, under the low energy and low speed Newton limit, the 00 component of the metric field degenerates into the Newton gravitational potential, while the 00 component of the energy tensor is the energy density of the matter field and degenerates into the inertial mass density under the Newton limit. It follows that the strength of the gravitational force felt by the place of matter is proportional to its inertial mass, which is the original form of the equivalence principle.

Finally, the field equation in general relativity can be obtained by varying the action $S_{\mathrm{GR}} \equiv S_\Lambda + S_{\mathrm{BH}} + S_{\mathrm{matter}}$ to the metric $g^{\mu\nu}$, which is $\delta S_{\mathrm{GR}}/\delta g^{\mu\nu} = 0$. The result is

$$\mathscr{R}_{\mu\nu} - \frac{1}{2} g_{\mu\nu} \mathscr{R} + \Lambda g_{\mu\nu} = -M_{\mathrm{P}}^{-2} T_{\mu\nu} \tag{40}$$

This is Einstein's field equation with a cosmological constant.

1.11 Greater Symmetry? Greater Unification?

Weyl gauge theory reveals a very important physical idea—"the conservation of charge, or the symmetry of a local $U(1)$ group, determines all electromagnetic interactions". He pointed out that "as long as the system has a symmetry of $U(1)$, there must be electromagnetic interactions between the particles in the system" and "all gauge interactions must be transferred by gauge quanta". Weyl's ideas had a "great appeal" to Chen-Ning Yang, prompting him to come up with a bold and tantalizing idea: to extend the gauge theory that Weyl discovered and proposed from the conservation of charge to the conservation of isospin.

Putting forward by Chen-Ning Yang (1922-) and Robert Mills (1927–1999) in 1954, it is now called the Yang–Mills theory. It is a so-called local "Non-Abelian gauge theory", an extension of quantum electrodynamics local "Abelian gauge theory".

Yang–Mills theory can be regarded as an extension of Weyl's gauge electromagnetic theory. But, according to the symmetry, the Yang–Mills field, like the electromagnetic field, cannot have mass. That is to say, the three gauge quanta of the Yang–Mills field should have no mass like photons, which greatly affects the practical application of the Yang–Mills field. In the 1950s, Yang–Mills gauge theory received little attention.

When Chen-Ning Yang visited Princeton, he gave a special report on his work with Mills. Wolfgang Pauli was also visiting Princeton at that time. Not long after the presentation began, as soon as Yang had written the field equation on the blackboard, Pauli asked, "what is the mass of the field?" Yang said, "we don't know", and went on. But, Pauli persisted in asking the same question again. "This is a very complex problem, and although we have studied it, we have not come to a definite conclusion", Yang answered. Pauli replied obstinately, "This is not a sufficient justification". Pauli's question was: since electromagnetic fields have no rest mass, the gauge field should also have no mass, but it must have mass in order to explain the short-range forces associated with the nucleus.

That "mass issue" has haunted Chen-Ning Yang ever since. Mr. Chen-Ning Yang, who has pursued symmetry and theoretical elegance through all his life, might not have realized that the solution required an effect of "spontaneous breaking of symmetry", which is common in nonlinear systems.

It wasn't until the 1960s and 1970s when spontaneous symmetry-breaking and the emergence of Higgs mechanism led Wilhelm Weinberg, Sheldon Glashow and Tammam Salam to develop a unified theory of electroweak interaction. The work that won the Nobel Prize in physics in 1979, 1999 and 2004, respectively, was based on the Yang–Mills field that the Yang–Mills gauge theory eventually became the basis for the unification of the strong, the weak, and the electromagnetic fields.

2 From Symmetry to Symmetry-Breaking

2.1 How the World Governed by Simple Symmetrical Laws of Physics Can Produce Rich and Complex Phenomena?

The unity of the world requires the symmetry of the laws of physics. What about the diversity of the world? Symmetry, to some extent, means indistinguishability, consistency, and repetition. For a snowflake, one only needs to know one sixth because the rest is a repeat. But our world is so rich and colorful, and so complex and diverse. Can an intrinsically symmetrical law of physics really describe this complex, multicolored world?

Here, we will combine some of the latest concepts in nonlinear dynamical systems and chaos theory to explore the question 11 how can the world governed by symmetrical laws of physics produce colorful phenomena?" We attempt to answer this question by means of the latest concept in chaos theory, "coexistence of multiple attractors". Since the world is a complex nonlinear system, multi-stability may occur in nonlinear systems, which is the source of the unity and diversity of the world. The problem of "God particle and the origin of mass", as we will see, is essentially the problem of multi-stability. Multi-stability can be said to be the unity and diversity of the world, for it is "mysterious with mysterious, the door of many wonderful".

2.2 Spontaneous Breaking of Symmetry

The direct breaking of symmetry in physical laws is something that physicists do not want to see, which would mean that the laws of physics are not perfect. In addition to the direct breaking of symmetry due to the nonlinear nature of the equations of motion, there may be a situation where the laws of physics themselves remain symmetrical, but the symmetry laws produce an asymmetry phenomenon, which is something physicists can accept.

Spontaneous symmetry-breaking is when the laws of nature that govern a physical system have some symmetry, but a physical phenomenon itself does not have such symmetry. In particle physics, it usually means that the Lagrangian of a physical system (a function that generalizes the dynamic state of the whole system) has some symmetry, while the ground state of a vacuum (the lowest energy order of the system) does not.

In fact, symmetry still exists in the underlying laws of physics, so it's actually hidden.

The simplest spontaneous breaking of symmetry. A cylindrical elastic rod will stand on the table, with the match stick and gravity, and the desktop system with the cylindrical elastic rod has the axis of rotational symmetry. If the cylinder is pressed vertically downward, it must be unstable and will always bend in one direction or another, pointing in a particular direction and breaking the previous rotational symmetry. Although this is a simple example, it shows the process of changing the equilibrium point from stable to unstable and at the same time producing numerous new equilibrium states, which is a typical bifurcation phenomenon in chaos theory.

Spontaneous symmetry-breaking in a paramagnetic ferromagnetic phase transition. With the increase of temperature, the magnetism of ferromagnet will decrease gradually. Over a certain temperature, the magnetism completely disappears. At this temperature, as long as there is no external magnetic field, the magnet cannot produce its own magnetic field, so the ferromagnetic has become paramagnetic. This transition temperature is called the Curie temperature. At Curie temperatures, magnets tend to be isotropic (except for certain special materials), where physical systems have great symmetries. Macroscopically, the material has no magnetism at this time, so there is no specific direction. When the temperature is lowered, the magnet restores its magnetism. If there is no external magnetic induction, the direction of the recovered magnetic field will be random, due to the previous state where there is no particular direction. The material restores the magnetic field, indicating that it has chosen a particular direction internally as a particular direction of the system. Symmetry is no longer maintained. This phase transition, from the state of symmetry, automatically changed to the state of asymmetry, is the spontaneous breaking of symmetry.

Crystal symmetry. The laws that describe solids are invariant throughout the Euclidean group, but solids themselves break the Euclidean group into space groups. Imagine simply that liquid water is more symmetrical, and loses some of its symmetry when it freezes or forms snowflakes. And all sorts of specific symmetries are subgroups of the three-dimensional space.

2.3 Spontaneous Symmetry-Breaking in the Condensed Matter Physics

The phase states of most matters can be understood through the view of spontaneous symmetry-breaking. For example, a crystal is formed by an arrangement of atoms in a periodic matrix, which is not invariant for all translation transformations, but only for some translation transformations with lattice vectors as intervals. The magnetic north and south poles of a magnet point in a particular direction, breaking rotational symmetry. In addition to these common examples, there are many other symmetry-breaking states, including nematic phases of liquid crystals, superfluids, and so on.

A similar Higgs mechanism is used in condensed matter to create the superconductor effect of the metal. In metal, the condensed state of the electron cooper pair spontaneously breaks the $U(1)$ symmetry of the electromagnetic force, creating a superconductor effect.

The phase states of most substances, such as crystals, magnets, and ordinary superconductors, can be understood from the Spontaneous symmetry-breaking point of view. Substances of topological phases such as fractional quantum hall effects are notable exceptions.

2.4 Spontaneous Breaking of Discrete Symmetry in Particle Physics

The vacuum of quantum mechanics is different from the vacuum of general knowledge. In quantum mechanics, a vacuum is not an empty space, and virtual particles are continually randomly generated and annihilated at any point in the space, creating mysterious quantum effects. When these quantum effects are taken into account, the lowest energy state in space is the lowest energy state of all, which has no additional energy to generate particles, otherwise known as the ground state or the vacuum state. Space in the lowest energy state is the vacuum of quantum mechanics.

Suppose that a symmetric group transformation can only transform the lowest energy state into itself, and the lowest energy state is said to be invariant under the transformation. The assumption that the Lagrangian of a physical system is invariant under a symmetric group transformation G does not mean that its lowest energy state is invariant under the transformation G. A physical system is said to have congruent symmetry if Lagrangian has the same invariance as the lowest energy state. If only the Lagrangian has invariance and the lowest energy state does not satisfy, the symmetry of the physical system is broken spontaneously, or the symmetry of the physical system is hidden, a phenomenon known as "spontaneous symmetry-breaking".

A common example can be used to explain the spontaneous symmetry-breaking phenomenon. Suppose that there is a ball on the top of the sombrero. The sphere is in a state of rotational symmetry with local maximum gravitational potential energy. This state is so unstable that a small disturbance can cause the ball to tumble to the bottom of the hat in order to reduce its potential energy. The symmetry is broken because the direction in which the ball rolls is different from the other directions. The Lagrangian of the physical system is invariant under the rotational transformation, but the lowest energy state is not invariant, hence this phenomenon is named spontaneous symmetry-breaking.

The best way to understand the meaning of these visual metaphors is through specific mathematical expressions.

First, for a real scalar field φ^4 with discrete Z_2 symmetry, the Lagrangian is

$$\begin{aligned}\mathcal{L} &= \frac{1}{2}\left(\partial_\mu\varphi\right)^2 - V(\varphi) \\ &= \frac{1}{2}\left(\partial_\mu\varphi\right)^2 - \frac{1}{2}m^2\varphi^2 - \frac{1}{4!}\lambda\varphi^4\end{aligned} \tag{41}$$

Replacing m^2 with a negative argument $-\mu^2$, we obtain

$$\begin{aligned}\mathcal{L} &= \frac{1}{2}\left(\partial_\mu\varphi\right)^2 - V(\varphi) \\ &= \frac{1}{2}\left(\partial_\mu\varphi\right)^2 + \frac{1}{2}\mu^2\varphi^2 - \frac{1}{4!}\lambda\varphi^4\end{aligned} \tag{42}$$

where μ and λ are two parameters of potential energy V, and λ describes the self-coupling of the field.

The model obviously has Z_2 symmetry, that is, $\mathcal{L}$ in the transformation

$$\varphi \to \varphi' = -\varphi \tag{43}$$

The bottom stays the same. The corresponding Hamiltonian is

$$\mathcal{H} = \int d^3x \left[\frac{1}{2}\pi^2 + \frac{1}{2}(\nabla\varphi)^2 - \frac{1}{2}\mu^2\varphi^2 + \frac{1}{4!}\lambda\varphi^4\right] \tag{44}$$

The classical configuration of the minimum energy is the field $\varphi(x) = \varphi_0$, where φ_0 makes the potential energy

$$V(\varphi) = -\frac{1}{2}\mu^2\varphi^2 + \frac{1}{4!}\lambda\varphi^4 \tag{45}$$

to take the minimum. This potential has two minima:

$$\varphi = \pm v \tag{46}$$

Among them,

$$v = \sqrt{\frac{6\mu^2}{\lambda}} \tag{47}$$

The constant v is called the vacuum expectation value (or Vev) of the field φ.

Let's look at the system near a minimum, for a positive minimum. Then, it is easy to define

$$\varphi = v + \varphi \tag{48}$$

Expressing $\mathscr{L}$ by φ, we find that the linear term of φ does not exist. Throwing away the constant term, we get the Lagrangian:

$$\mathscr{L} = \frac{1}{2}\left(\partial_\mu \varphi\right)^2 - \frac{1}{2}\left(2\mu^2\right)\varphi^2 - \sqrt{\frac{\lambda\mu^2}{6}}\varphi^3 - \frac{1}{4!}\lambda\varphi^4 \tag{49}$$

This Lagrangian describes a simple scalar field with a mass of $\sqrt{2}\mu$ and the interaction of φ^3 and φ^4. Here, the Z_2 symmetry $\varphi \to -\varphi$ is no longer obvious. The only residue lies in the relationship between the three coefficients in the upper form, which in a particular way depends only on two parameters. This is the simplest example of a spontaneous break in symmetry.

2.5 *Spontaneous Breaking of Global Continuous Symmetry in Particle Physics*

Consider the Φ^4 model of a complex scalar field with global $U(1)$ symmetry, where the Lagrangian is

$$\mathscr{L} = \left(\partial_\mu \Phi\right)^\dagger \left(\partial^\mu \Phi\right) - V(\Phi) \tag{50}$$

in which

$$V(\Phi) = -\mu^2 \Phi^\dagger \Phi + \frac{\lambda}{2}\left(\Phi^\dagger \Phi\right)^2 \tag{51}$$

where $\mu > 0$ and $\lambda > 0$ are two parameters of potential energy V, λ describes the self-interaction intensity of the field, and $\mathscr{L}$ remains invariant under the following global specification transformation:

$$\Phi \to \Phi' = \exp(i\alpha)\Phi \tag{52}$$

where α is a real constant.

The potential energy differentiates the field, so

$$\frac{\partial V}{\partial \Phi^\dagger} = -\mu^2 \Phi + \lambda\Phi\left(\Phi^\dagger \Phi\right) \tag{53}$$

It is easy to see that the point $\Phi = 0$ is at a maximum, which means that it is not a true vacuum and cannot be expanded here; the minimum is at

$$|\Phi|^2 = \frac{v^2}{2} \tag{54}$$

where

$$v = \sqrt{\frac{2\mu^2}{\lambda}} \tag{55}$$

That is,

$$|\Phi| = \frac{v}{\sqrt{2}} \tag{56}$$

This equation has infinitely many solutions corresponding to points on the circumference of a circle whose radius is $\frac{v}{\sqrt{2}}$ on the Φ complex plane, so now vacuum has infinite degeneracy, and physical vacuum is only one of them.

The phase transformation of the field Φ corresponds to a rotation of the field in the complex plane of Φ. In this rotation, the degenerate vacuum changes from one state to another. In this model of a complex scalar field, where the physical vacuum takes a particular phase, the symmetry of the field Φ is broken spontaneously. The above equation shows that the field still has a certain average value of $\frac{v}{\sqrt{2}}$ in the vacuum state, but what was measured in the experiment is only based on the excitation of the average value. We can isolate the mean and study the excitation relative to the mean:

$$\Phi = \frac{1}{\sqrt{2}}[v + \varphi + i\rho] \tag{57}$$

where

$$\begin{aligned} |\langle 0|\varphi|0\rangle| &= 0 \\ |\langle 0|\rho|0\rangle| &= 0 \end{aligned} \tag{58}$$

These two real scalar fields are physical fields that can be measured directly, so we get

$$\mathscr{L} = \frac{1}{2}\left(\partial_\mu\varphi\right)^2 + \frac{1}{2}\left(\partial_\mu\rho\right)^2 - \frac{1}{2}\lambda v^2\varphi^2 - \frac{1}{2}\lambda v\varphi\left(\varphi^2 + \rho^2\right) - \frac{1}{8}\lambda\left(\varphi^2 + \rho^2\right)^2 \tag{59}$$

As can be seen, the field φ has a mass proportional to the expected value of the vacuum $\sqrt{\lambda v^2}$, which is called the Higgs particle.

On the other hand, after separating the fields with an average value of V, we get a massless ρ field. A spontaneous break in continuous symmetry must lead to the existence of massless particles. This general conclusion is known as the Goldstone theorem. The massless field is called the Goldstone field, and the corresponding particle is called the Goldstone particle. We note that for discrete symmetry Z_2, there is no Goldstone particle, but for continuous symmetry $U(1)$, there is a Goldstone particle. In physics, the spontaneous breaking of the continuous symmetry leads to the continuous degeneracy of the ground state, and the conversion of the system between the degeneracy states does not need any exchange of energy, so the corresponding Goldstone particles cannot have mass.

2.6 Higgs Mechanism and the Origin of Mass

In the section on symmetry, it was said that the electromagnetic force is completely determined by the local symmetry principle. In order to satisfy the gauge theory, one has to set the mass of the gauge Boson to zero. On the other hand, according to the Goldstone theorem, a spontaneous break in continuous symmetry must lead to the existence of massless particles. There are two zero-mass problems, and they are magically curing each other. This is the Higgs mechanism.

We now analyze the two-zero mass problem. The massless particle of the Goldstone theorem is a global continuous symmetry. If we break the local symmetry, it will be free from the Goldstone theorem. The Yang–Mills theory has a local symmetry, but the symmetry here is not broken. So, if we let this local gauge break spontaneously, then perhaps we can save the Yang–Mills theory. We may put the two together, namely, using local symmetry to rescue the Goldstone particle and using symmetry-breaking to rescue the Yang–Mills theory, and furthermore let spontaneous symmetry-breaking get rid of the local symmetry that produces the massless vector particles. In this way, Yang–Mill's theory would be out of the woods, and since the symmetry in Yang–Mill's theory is not global but local, the Goldstone theorem would not apply to the spontaneous breaking of this symmetry, which would be the best situation for both worlds.

In the standard model, Higgs mechanism is a mass generation mechanism that allows the fundamental particles to gain mass. Why do Fermions, W-Bosons, and Z-Bosons have mass, while photons and gluons have zero mass? The Higgs mechanism would explain that. Higgs mechanism with gauge theory and spontaneous symmetry-breaking can generate the mass of particles. This is the simplest of all possible mechanisms that can generate the mass of gauge Boson while at the same time adhering to gauge theory.

According to the Higgs mechanism, Higgs field is all over the universe, and some of the fundamental particles gain mass by interacting with the Higgs field. But there is also a byproduct, the Higgs Boson. Since the Higgs field vacuum expectation is not equal to zero, the resulting spontaneous symmetry-breaking, when the continuous symmetry is spontaneously broken, and the gauge Boson gains mass, both produce the Higgs Boson with mass and a zero-mass Boson called the Goldstone Boson. By choosing an appropriate gauge, the Goldstone Boson will be offset, leaving only the Higgs Boson with mass and the gauge vector field with mass.

Fermions also gain masses by interacting with the Higgs field, but not in the same way as the W- and Z-Bosons. In gauge field theory, in order to satisfy the gauge theory, we have to set the mass of the Fermion to zero. By Yukawa coupling, Fermions also gain masses because of their spontaneous symmetry-breaking. So, in the end, the masses of all the elementary particles in the standard model come from the spontaneous breaking of the symmetry.

2.7 Electroweak Unification

In 1979, the Royal Swedish Academy of Sciences decided to award the Nobel Physics Prize to American Physicists Sheldon Glashow, Steven Weinberg and Pakistani physicist Abdus Salam in recognition of their outstanding contributions to electroweak unification. The theory of electroweak unification is one of the greatest achievements of this century. The unified theory for achieving this weak electromagnetic effect is gauge theory. The gauge theory was proposed by Chen-Ning Yang and Mills in 1954, but no substantial progress had been made ever since because of the problem in regulating particle mass. It was only in the 1960s that the spontaneous breaking of vacuum symmetry gave new life to the Yang–Mills theory. The electroweak unified theory uses the Yang–Mills theory to describe the fundamental interactions on the one hand, and the spontaneous breaking of vacuum symmetry on the other, to provide the mass of gauge particles. So, the spontaneous breaking of symmetry indeed plays an important role in history.

2.8 Chiral Symmetry-Breaking and the Origin of Quark Mass

Chiral symmetry-breaking refers to the spontaneous breaking of the chiral symmetry of the strong interaction, a form of spontaneous symmetry-breaking. The simplest example of a Chiral symmetry is the mirror symmetry of the left and the right hands. In the quantum chromodynamics, assuming that the quark has zero mass (which is the chiral limit), the Chiral symmetry holds. The quarks, however, have no real mass of zero, although the masses of the upper and lower quarks are small compared with those of the hadrons, so the chiral symmetry can be considered "approximately symmetric". Because in a quantum chromodynamics vacuum, the vacuum expectation value of antiquark-quark condensation is not equal to zero, which causes the chiral symmetry of the physical system to be spontaneously broken. This also means that the vacuum of the quantum chromodynamics will mix both chiral states of the quark, causing quarks to move through the vacuum to gain effective mass.

According to the Goldstone boson, when a continuous symmetry is spontaneously broken, a zero mass boson is produced, called a Goldstone boson. The chiral symmetry is also continuously symmetric, and its Goldstone boson is the π meson. If the chiral symmetry is completely symmetric, then the mass of the meson is zero; but, because the chiral symmetry is approximately symmetric, the π meson has a very small mass, an order of magnitude smaller than the mass of a hadron. This theory became the basis and element for the later Higgs mechanism of electroweak symmetry-breaking.

According to cosmology, chiral phase transitions occur for 10^{-6} s after the Big Bang, at the beginning of the hadronic period, as the universe continued to cool, and the temperature dropped below a critical point $KT_c \approx 173\,\text{MeV}$. The original chiral symmetry in physical systems no longer had this property, and the chiral symmetry was spontaneously broken. Until then, the quarks could not form hadron bound states and the vacuum expectation of the antiquark-quark condensation. The order parameter of the physical system was equal to zero, and the physical system obeyed the chiral symmetry. After that point, the quarks could form hadron bound states, the vacuum expectation of antiquark-quark condensation was not equal to zero, and the chiral symmetry was spontaneously broken.

3 Chaos Revolution

3.1 Stability and Chaos

We consider several possible final states of a deterministic motion when time tends to infinity, for example, in the case where the solution of a three-dimensional autonomous system. The simplest case being divergence to infinity or convergence to a point of equilibrium, or periodic and quasi-periodic motions. Each of these possibilities is simple and a little more prosaic than the variety of our real world. In fact, if the equations were nonlinear, the situations might be different.

If the equations of a three-dimensional autonomous system are nonlinear, in addition to the above possibilities, there is also a possibility that they do not diverge but are restricted to a limited region, or do not converge to any equilibrium point, or do not periodically traverse. An orbit in this situation will have a very complex motion rather than just diverging, converging or repeating. This is the situation of chaos, which is sometimes described as a periodic motion with an infinite period or as seemingly random. Deterministic chaotic motion has its own unique characteristics, as discussed in the following sections.

3.2 Unification of Initial Value Sensitivity and Initial Value Desensitization

The so-called butterfly effect, which is the sensitivity of chaotic systems to initial values, is not the same as the initial value sensitivity of divergent systems. A divergence system also has the property that a small initial value error can be enlarged as the trajectory diverges to infinity, for example two straight lines with a little difference in their slopes will further separate away after the straight lines extend indefinitely. This kind of initial value sensitivity is not the nature of chaos. In fact, chaos is bounded, which means that there is a region containing a chaotic attractor. Once its trajectory falls into this region, it will always be restricted to be within this region, and its trajectory will not go out of the region. So, on a large scale, chaotic systems are somewhat "stable".

On the one hand, chaos is very sensitive to initial values, where two trajectories that differ a little at the start will be farther apart with the evolution of time, but the trajectories will approach a chaotic attractor. In fact, for a classical system with only one chaotic attractor, the final trajectory always hovers over the attractor no matter where the initial value is, as long as it is in the so-called "basin of attraction". So, this is a kind of "insensitive" to initial values, to some extent. Because no matter what initial value the trajectory starts from the basin of attraction, its final state is pretty close to each other near the attractor. In other words, the evolution of the motion has forgotten about the initial value. These two effects constitute a dialectical unity of initial sensitivity and initial desensitization. Therefore, the "sensitivity to initial value" property of chaos is more subtle than just being a divergent system.

3.3 Unity of Essence Certainty and Appearance Randomness

Einstein once said something interesting: "Quantum mechanics is really impressive. But something inside me told me that it wasn't the right theory. This theory says a lot, but it doesn't lead us any closer to God's secrets. In any case, I am convinced that God does not play dice". What Einstein was talking about here is whether the laws of physics that describe the way the world works are random. I don't think that Einstein himself would object to being able to describe some phenomena as random, but here he referred to the question of whether the fundamental laws of nature are random.

Can physical laws be inherently random? Apart from the randomness of the collapse of the wave function in what is now called quantum mechanics, is there any intrinsic randomness in the laws of physics? The mechanical motion of the dice is completely deterministic. Can computers generate real randomness? What is really random? The so-called "computer on all the random simulation" is merely pseudo-random. In fact, in theory, all systems are deterministic, randomness is only the appearance, which is an expedient way to deal with the problem. For example, when you throw a dice, if you know exactly all the initial conditions, then all the details of the motion of the dice can be precisely determined. But since this information is physically unavailable, without caring about all the details of the motion, only the final orientation after the landing is stable. So, it's easy to look at this in a simpler way, where the orientation is random, 1/6 of the chance, so one can use a random model to describe this kind of random phenomenon, but the underlying physics is completely deterministic. To produce "real" randomness, it is necessary to create extremely highly complex physical systems.

Can an essentially deterministic equation describe a seemingly random phenomenon? Chaos gives a definite answer: as a definite differential equation, the state of its motion has some kind of "random" behavior; that is, the randomness generated spontaneously by the nonlinearity in the chaotic system.

Thus, I think the conclusion is very dialectical: **the real randomness comes from completely deterministic systems**. If we do not believe in the complete certainty of the nature of the laws of physics, we cannot believe that we use a random method to describe a random phenomenon superficially.

Without certainty, there is no randomness.

3.4 Unity of Integer-Dimensional and Fractional-Dimensional Attractors

As mentioned above, the essence of chaos is not the so-called "sensitivity to initial values". In fact, the difference between chaos and stability is essentially the difference in the structures of attractors, which is a leap from the integer dimension to the fractional dimension.

The so-called stable solution is a zero-dimensional point attractor; the so-called periodic attractor, generally refers to the stable periodic solution, is actually a one-dimensional line attractor. In a higher-dimensional phase space, a two-dimensional attractor can be generated; that is, an attractor that forms a torus on which quasi-periodic solution can be generated.

In a three-dimensional phase space, the attractor must have a dimension smaller than three, so there seems to be no possibility other than zero-dimensional point attractors, one-dimensional line attractors, and two-dimensional torus attractors. The magic of chaos is that its attractor can be fractal; that is, the attractor can have a fractional dimension, between 2 and 3, giving rise to infinite possibilities.

The geometry of integer dimension is simple. Compared to the traditional geometry for the study of the integer dimension, such as zero-dimensional points, one-dimensional lines, two-dimensional surface, three-dimensional solid, fractal geometry is much more complex. The fractal dimension indicates that the chaotic motion state has a multi-leaf and multi-layer structure, and the finer the leaf layer is, the more self-similar the structure of the infinite layer appears. Therefore, fractal geometry is also called "the geometry of nature".

What are the dimensions? The nature of the dimension is inseparable from the measurement. Here is an example of the concept of fractal dimension. A square, with its side length enlarged three times, is similar to the original, and is equivalent to nine small squares. Similarly, one side of a cube grows three times as long, similar to the original, and the equivalent of 27 small cubes. And that's where the definition of dimension comes in. In the 1970s, the French mathematician Mandelbrot explored how long the England coastline is. This problem depends on the scale used in the measurement. If kilometers are used as a unit of measurement, some twists and turns from a few meters to a few tens of meters will be ignored; if meters are used instead, the total length measured will increase, but some less than centimeters will not be reflected. Geometric objects such as coastlines, which have infinite fine structures and are self-similar, need to be characterized by fractal dimensions.

It's as if 2 and 3 are just two simple integers, but the space between 2 and 3 is filled with an infinite number of real numbers. So is the chaotic attractor, which breaks through two dimensions, and has infinite possibilities in the vast space between two and three dimensions.

Chaos, in essence, is leaping of the attractor structure from an integer dimension to a fractal dimension.

3.5 Unity of Equilibrium and Chaos: Far from Equilibrium

Is equilibrium antithetical to chaos? Is it possible for a system to be both stable and chaotic? The real world is full of all kinds of equilibria and chaos. Can equilibrium state and chaos be described in a unified way?

In statistical mechanics, it is also important to stay far from the equilibrium. Equilibrium is important, but the world we see is not drab and dead, but alive, evolving and full of life.

New discoveries in chaos science tell us that equilibrium and chaos are not opposite, but can often coexist. If we consider three-dimensional continuous autonomous systems, chaotic attractors are generally considered to appear in systems with at least one unstable saddle-focus type equilibrium point. And all the known classical chaotic systems do meet this condition. The known classical three-dimensional continuous autonomous chaotic systems (including classical Lorenz system, Chen System, Lü system, Rössler system, Sprott systems) can be used by Šil'nikov criteria to study the existence of chaos and explain the mechanism of chaos. Based on the unstable saddle focus equilibrium point of the system, the existence of an attractor is deduced from the self-intersection of a homoclinic orbit or heteroclinic orbit. Šil'nikov theorem shows that a system with a saddle-type equilibrium point can produce chaos under certain conditions, yet this is a sufficient but not a necessary condition for chaos to occur. On the other hand, the Hartman–Grobman theorem strictly states that if the hyperbolic equilibrium of a nonlinear system is stable, the dynamic behavior near the equilibrium is also asymptotically stable. But this theorem makes no provision for the dynamic behavior far away from the equilibrium. Therefore, the possibility of chaos appearing in a three-dimensional autonomous system with no or only one stable equilibrium point is not ruled out. Based on these considerations, Xiong Wang and Guanrong Chen constructed a chaotic system with only one stable equilibrium point, and the mathematical model of the system is described by

$$\begin{cases} \dot{x} = yz + a \\ \dot{y} = x^2 - y \\ \dot{z} = 1 - 4x \end{cases} \tag{60}$$

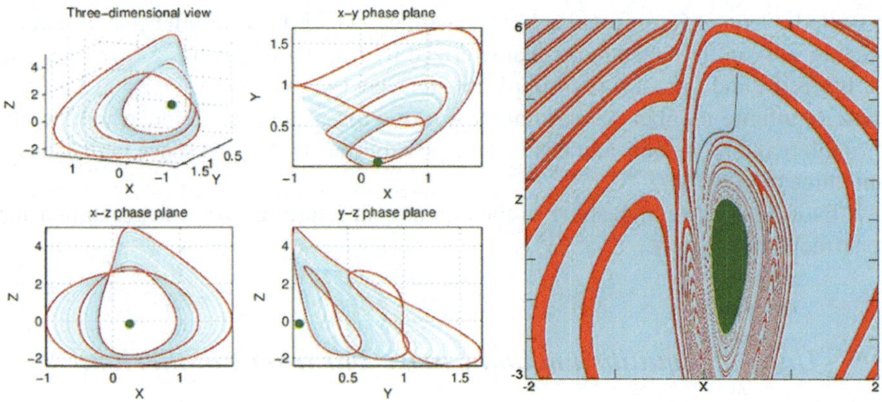

Fig. 1 There is one and only one stable equilibrium point in the chaotic system with $a = 0.01$. The left diagram shows the coexistence of three types of attractors. Green is the stable equilibrium point, red is the stable periodic orbit, and blue is the chaotic attractor. On the right, the X-Z initial value plane of the three different attractors is cut off. The green region is the region of attraction for the stable equilibrium point, the red region for the periodic orbit, and the blue region for the chaotic attractor

This result also attracted the attention of J. C. Sprott and his colleagues, and they cooperated in investigating the interesting dynamics of the system. They found that, besides the stable equilibrium point and the chaotic attractor, the system also has a periodic orbit with a small region of attraction. Three different types of attractors coexist in such a simple system, where each of these different attractors has its own complex domain of attraction. The whole phase space is thus divided by several attractors, whose respective domains of attraction are interwoven with each other, which may even have very complicated fractal boundaries. Moreover, these domains of attraction vary with the system parameters as shown in Fig. 1.

These new findings show that chaotic dynamics do not depend on the number and the stability of the local equilibrium points, and chaos is a global dynamic behavior of nonlinear systems. This kind of global chaotic dynamic behavior is also referred to as "hidden attractor". The traditional chaotic attractor is self-excited due to the existence of homoclinic and/or heteroclinic orbits at the unstable equilibrium point. If the equilibrium point is stable or there is no equilibrium point in the chaotic system, the region of attraction of the chaotic attractor does not contain any equilibrium point, so starting from any initial value may not be attracted to the chaotic attractor; or, the region of attraction of this kind of chaotic attractor is extremely small, so if we start from somewhere near the stable equilibrium point, we may not find the chaotic attractor in numerical simulations. A hidden attractor is essentially a large-scale dynamical characteristic of such a nonlinear system that does not depend on local equilibrium points. The research on this kind of newly-found chaotic system with hidden attractors has become overwhelming recently.

What is it that makes our world so vivid and colorful? I believe that it is a complex structure, such as life, arising from chaotic state far from an equilibrium, which has countless possibilities.

3.6 Coexisting Attractors

The focus of this book is about multi-stability, which traditionally refers to a dynamic system with multiple stable equilibrium points. But, what we're really talking about here is the coexistence of multiple attractors, or even the coexistence of different types of attractors.

A classical chaotic system usually has only one chaotic and a global attractor, so the symmetry of the equation determines the symmetry of the chaotic attractor. General research focuses on the analyses of the equilibrium point, power spectrum, Poincaré cross section, bifurcation diagram and Lyapunov exponent spectrum, and so on. Unlike classical chaotic systems, some newly-found chaotic systems often have many different attractors, even many different types of attractors, to coexist, where different initial values may lead to different attractors.

In this case, if the equation is symmetric, the solution of the equation, as well as the final or observable state of the solution, cannot be symmetric. That actually is a novel dynamical behavior of spontaneous symmetry-breaking, which is rare for classical systems to have, yet for the new chaotic systems it is quite common.

For classical dynamical equations, the attractors derived from the evolution of the system also maintain the symmetry. The classical Lorenz system and Chen system both satisfy the Z-rotational symmetry, for example the Lorenz system:

$$\begin{cases} \dot{x} = \sigma\,(y - x) \\ \dot{y} = rx - y - xz \\ \dot{z} = -bz + xy\,, \end{cases} \tag{61}$$

which is chaotic when $\sigma = 10,\ r = 28,\ b = \frac{8}{3}$, and the Chen system:

$$\begin{cases} \dot{x} = a(y - x) \\ \dot{y} = (c - a)x - xz + cy \\ \dot{z} = -bz + xy\,, \end{cases} \tag{62}$$

which is chaotic when $a = 35,\ b = 3,\ c = 28$.

The equations of these systems satisfy the Z-axis symmetry; that is, when (X, Y, Z) is transformed to $(-X, -Y, Z)$, the equation form remains unchanged. And each of these related systems produces an attractor with the same symmetry.

In general, the equations with the Z-axis symmetry satisfy the following general formula:

Fig. 2 One countervailing attractor spontaneously breaking down into two attractors

$$\begin{cases} \dot{x} = a_{11}x + a_{12}y \\ \dot{y} = a_{21}x + a_{22}y + m_1xz + m_2yz \\ \dot{z} = a_{33}z + m_3xy + m_4x^2 + m_5y^2 + m_6z^2 + c\,, \end{cases} \tag{63}$$

Based on this general formula, we found the following system:

$$\begin{cases} \dot{x} = -x + y \\ \dot{y} = -x - 0.1y - yz \\ \dot{z} = -0.5z + x^2 - k\,, \end{cases} \tag{64}$$

where k is a real parameter.

What is interesting about this system is that, as the k changes, the system goes from one countervailing attractor spontaneously breaking down into two attractors, both of which lose the symmetry of the original system of the equation, as shown in Fig. 2.

How to judge whether an attractor will keep symmetry from the equation itself? What is special about the critical state of spontaneous breaking of symmetry? And how to predict it? These are some new challenging problems faced by the classical chaos theory.

4 Multi-stability: A Door Open for Many Wonderful Things

4.1 Multi-stability Based on a Unified System

The premise of multi-stability is that there is a unified kinematic equation governing all the states in space, so multi-stability does not emphasize the diversity caused by different laws. What is emphasized is that the same set of physical laws can produce different observable phenomena. The dream of physicists is to build the ultimate unified equation that describes the whole world, whether it is a macroscopic universe on a very large scale, or an elementary particle world on a very small scale, or the ordinary world of everyday's life, or extreme situation such as black holes.

Our dream is **one universe, one equation**.

4.2 Vacuum and the Ground State of a System

The so-called spontaneous symmetry-breaking, in the language of quantum field theory, is that the Lagrangian of a physical system has some symmetry, but the ground state does not. In other words, the ground state of the system breaks the symmetry of the equation of motion. In quantum field theory, the ground state of the system is a vacuum state, so the spontaneous symmetry-breaking of the system shows that the symmetry of the Lagrangian is broken by the vacuum state.

The central idea here is the definition of vacuum, which is the ground state of the system. Throughout the years, people have been unremittingly exploring the vacuum, and they have gone through a circuitous, tortuous long road. At present, based on quantum field theory, the concept of "vacuum is the ground state of a quantum field system" has become a profound basic concept verified by experiments in modern physics. Quantum field theory says that the presence of a particle in space indicates that the field is in an excited state, while the absence of a particle indicates that the quantum is in the ground state. Thus, in quantum field theory, vacuum is regarded as a state without any excitation; that is, vacuum is the ground state of the quantum field system.

If the lowest ground state of the system is unique, then the vacuum state has the same symmetry as the interaction. Such a vacuum is called an ordinary vacuum or a normal vacuum. If the lowest energy ground state of a system is not one, that is, the symmetry of the actual physical state does not reflect the symmetry of the interaction. This phenomenon is called the spontaneous breaking of the vacuum symmetry. Symmetry-breaking is a new type of vacuum that is very different from the normal vacuum. This theory of spontaneous breaking of vacuum symmetry is not an idle dream without practical significance. In fact, it plays a very important role in the completion of the electroweak theory, which has been proved by a lot of experiments and is, of course, strong support for the vacuum theory. The theory of spontaneous breaking of vacuum symmetry has a profound influence not only on high energy physics, but also on astrophysics and solid state physics.

4.3 At the Heart of Higgs Mechanism Is the Concept of Multi-stability

The vacuum state of a physical system is determined by the Lagrangian. Why doesn't it have the symmetry of the Lagrangian? The secret is that many physical systems have degenerate vacuum states, and if we think of all these degenerate vacuum states as a set, it does indeed have the same symmetry as the Lagrangian. However, the actual vacuum state of a physical system is only one state in the set, which often does not have the symmetry of the whole set, which results in the break of symmetry—that is, the spontaneous break of symmetry.

This is even clearer in the language of nonlinear dynamical systems, where the equilibrium or attractor of a dynamical system is determined by a dynamical equation. Why doesn't it have the symmetry of the dynamical equation? The secret is that some dynamical systems have multiple attractors that coexist. If we think of all the attractors as a set, it still has the same symmetry as the dynamic equation, but the real observation of a trajectory and the attractor lose the symmetry of the whole dynamic equation. This may be a more direct demonstration of spontaneous symmetry-breaking.

4.4 Chaos Theory Beyond the Higgs Mechanism?

Chen-Ning Yang was not involved in the study of the Higgs mechanism, and I don't think Chen-Ning Yang's aesthetic would have been too happy with such a solution. Nor do I personally believe that the Higgs mechanism is a satisfactory answer to the question of the origin of mass. Let's compare the two theories which both reform the Lagrangian: the local gauge principle and the Higgs mechanism. The reformation of Lagrangian from local gauge principle is very clear, profound and concise. The principle does not require real natural experimental observations such as the experience of electromagnetic fields, but directly from the requirements of the principle, which deduced that there must be an electromagnetic field. This top-down high-altitude derivation reflects the power of the principle. The Higgs mechanism, on the other hand, knows from experimental experience that the weak interaction's gauge particle has mass, and artificially transforms the Lagrangian to require the introduction of an extra Higgs term. This term describes neither the particles of matter nor the particles that interact with each other, but rather, the production of mass purely for the purpose of breaking symmetry.

So, what is the origin of mass? The Higgs mechanism's answer, in a sense, is not so much an answer to a question rather than shifting it. It shifts the mass of elementary particles to the vacuum expectation of the Higgs field, the gauge coupling constant, and the Yukawa coupling constant, which we still want to understand. And, this transference, instead of becoming simpler and clearer in principle, is becoming more complex, and there is no actual reduction. The Yukawa coupling constant, for example, has an independent value for each type of fermion. Because of the existence of these parameters, the Lagrangian of the standard model contains no mass parameters, but the number of free parameters that are directly related to the mass is no less than the number of mass parameters that need to be explained a little more.

Therefore, I would like to believe that the Higgs, and the electroweak unified theory of the Higgs, is a phenomenally successful description, and that the idea of linking mass to the symmetry-breaking property of a vacuum is profound. However, they cannot be regarded as a satisfactory answer to the question of the origin of mass because they fail to realize the real reduction of the concept of mass.

My personal guess here is that there is a need for a more powerful principle than the local gauge symmetry, and a deeper understanding of vacuum. Based on the idea of multi-stability in chaos theory, perhaps we can get a more satisfying answer. At that time, the Higgs mechanism will not have to be added artificially, but rather, as a result of a more powerful principle, a more profound spontaneous breaking of the symmetry of the multi-stability effect. Thus, the Higgs mechanism is only a phenomenological approximation of the new mechanism.

5 Sources of Diversity in the World

How can the unified physical law produce all the colorful phenomena? Is an equation enough to describe such a colorful world? After the chaos revolution, we became more convinced of the problem because we knew more about how complex nonlinear equations can be. If, in the end, all physical phenomena can be expressed in one equation, it is necessary that this unified equation has a very rich connotation.

We summarize the top 10 reasons why a unified law of physics may produce all colorful phenomena and make wild guesses about the equations of the ultimate unified theory of physics that may be discovered in the future. We also make a summary of this chapter, also the end of the book.

5.1 Physical Quantity: Different Observers Can Have Different Observation Results

First, there is a wealth of variables that describe the states of motion of physical systems. The state of a physical system requires first determining the external space-time, the internal gauge symmetry, and then the physical quantities that describe the state of the system, whether it is a tensor or a spinor, a geometric quantity or an operator, or something more general.

In special relativity, for example, observers of different inertial frames of reference have different observations of the time interval and space interval between events, and even the sequence of time. But both will observe a uniform space-time interval behind it. So, a physical quantity describing the system is a covariant vector or tensor, and at this time the physical quantity already has the rich connotation. Observers with different motions can obtain very different observation results. For example, from the equations of motion of the electromagnetic field, different observers get seemingly different physical laws, where some observers see the electric side, some observers see the magnetic side.

In the future unified equation, the variable describing the motion state of the physical system itself, which has enough rich connotation, is the unified objective physical quantity that does not depend on the observer to describe the observer in all different motion states. This quantity should be some kind of generalization of the tensor.

At this level, the diversity of the observed phenomena by different observers is the first source, which explains why the same physical equation can produce very different observed phenomena.

5.2 State Space: Richness and Diversity of Motion States

Secondly, in the final equation, all the possible states of the object constitute a state space. The structure of the state space is very rich and complex. The state space itself may be nontrivial, nonlinear, and even with a complex topology. In the classic Newtonian framework of physics, a particle moves in a 3D, and the State of the particle includes the position of the particle and the velocity of the particle. The state space is equal to the real three-dimensional space, plus the velocity space. In classical Newton mechanics, all possible states of motion make up a relatively simple space.

But, when it comes to the electromagnetic field, the physical quantity describing the motion in the electromagnetic field is the electromagnetic vector field. The components of the electromagnetic vector potential have different projections to different observers, leading different observers to see different aspects of the unified electromagnetic field. Maxwell's equations are as important in electromagnetism as Newton's laws of motion is in mechanics. The theory of electromagnetism, centered at Maxwell's equations, is one of the proudest achievements in classical physics. It reveals the perfect unification of the electromagnetic force, giving physicists the belief that various interactions of matter should be unified at a higher level.

All the possible states of the electromagnetic vector potential form a function space, which is actually much more complex than the state space of Newton's mechanics. General Relativity equates gravity with the curvature of space-time, where itself is no longer just a container for physical processes but the subject of dynamics, which describes the motion of space-time in the form of a metric tensor field. The possible states of the metric field constitute a complex state space. These allow the equations to describe very complex and diverse phenomena.

5.3 Ground State: Even the Simplest Vacuum Is Rich in Content

In the space of all possible states, the simplest state, such as the ground state in a quantum field, can be non-trivial and indeed colorful.

In classical electromagnetic field theory, the field quantity satisfies the partial differential equation of space coordinates and time, so the classical field is characterized by continuity. According to the principle of quantum physics, micro-objects have the duality of particle and wave, discrete and continuous. In elementary quantum mechanics, the description of electron is quantum, and the motion of a single electron is quantized by introducing the operator corresponding to electron coordinates, momentum and their commutative relation. But its description of electromagnetic fields is still classic. Such a theory does not reflect the particle nature of the electromagnetic field and cannot accommodate photons, not to mention the generation and annihilation of photons. Thus, elementary quantum mechanics, while providing a good description of the structure of atoms and molecules, cannot directly deal with such important phenomena as spontaneous emission and absorption of light in atoms.

The picture of quantum field theory is that the whole space is filled with different fields, which permeate and interact with each other. The excited states of the field are the appearance of particles, and the different excited states are the different numbers and states of the particles. The interaction of the field can cause the change of the excited state of the field, which is represented by various reaction processes of the particles. Therefore, quantum field theory can describe spontaneous emission and absorption of light in an atom, as well as the production and annihilation of various particles in particle physics. Quantum field theory is essentially the quantum mechanics of infinite-dimensional degree-of-freedom systems. In statistical quantum physics and condensed matter physics, and some other branch of physics, the object of study is a system of infinite-dimensional degrees of freedom. In these branches, the degree of freedom of interest is often not the motion of the elementary particle, but rather, the collective motion of the system, such as fluctuations in a crystal or quantum liquid. These waves can be thought of as wave fields, and they obey the laws of quantum mechanics, so quantum field theory can also be applied to these problems.

In quantum field theory, a particle is a quantum excitation of a field, and each particle has its own field. The interactions and dynamics among particles can be described by quantum field theory. The so-called vacuum is a physical system in all possible states of some special state, that is, the lowest energy state. From the above-described physical meaning of the quantum field theory, one can see that vacuum is not without matter. The ground state field has the characteristic of zero-point vibration and quantum fluctuation of quantum mechanics. The physical effect of vacuum can be observed in the experiment when the external conditions are changed. For example, when a metal plate is placed in a vacuum, the interaction (Casimir effect) between two uncharged plates is due to the change of the zero point energy of the vacuum, and the polarization of the vacuum is due to the change of the distribution of positive and negative electrons in the vacuum under the action of the external electric field.

The stability of vacuum requires the vacuum to be an attractive subset of the state space. For different potential functions, a vacuum can be a point in the state space, such as a zero field, where the vacuum is a stable equilibrium point in the state space. However, nonlinear dynamical systems and chaos theory tell us that such subsets may have very complicated structures, or even non-trivial topological structures. If the vacuum is a special subset of the state space, then the vacuum may have some non-trivial properties such as spontaneous symmetry-breaking.

We suspect that if the vacuum itself is not a simple point or loop, but rather, some sort of "attractor" of the nontrivial topology, then it will have all sorts of weird physical effects.

5.4 Nonlinearity: Solutions of Equations Are Rich and Colorful

The development of general relativity depends to a large extent on how to find a solution to the gravitational field equation and its physical interpretation. The solution of the field equation is an important part of Einstein's theory of gravity. The field equation is a nonlinear second-order partial differential equation with space-time as the independent variable and metric as the dependent variable. Because of the nonlinear nature of the equation, it is very difficult to obtain the exact expression of the solution. It needs to choose different approximate methods in different cases. It is often assumed that the dosimeter has some specific simple forms, such as spherically symmetric form, cylindrical symmetric form, statically fixed and axially symmetric form, plane waveform, etc.

The nonlinear nature of the Einstein field equations makes general relativity very different from applying other theories of physics. For example, Maxwell's equations of electromagnetism are linearly related to the distribution of the electric field, magnetic field, electric charge and electric current (i.e., the linear superposition of the two solutions is still a solution). Another example is the Schrödinger equation in quantum mechanics, which is also linear for probabilistic wave functions. The solution of the Einstein field equations cannot be simply a linear superposition but requires a special solution generation technology. The technique of solution generation is to find some transformations and to generate new solutions or solution families by using known seed solutions. With the development of the "generation technique" of the solution, the number of exact solutions of the gravitational field equation will be increased greatly, where exact solutions are all special cases. Once encountering more concrete realistic questions or studying some meaningful questions, one has to use approximate methods.

There are two such methods that are particularly useful. They are called the post Newton approximation and the linear approximation. The first method is suitable for slow-moving particle systems, such as the solar system, which are connected together by gravity. When particles in a system of particles bound together by gravity move at low speeds, their states do not deviate much from Newton's laws. So, Newton's solution can be regarded as the zero-order approximation of the general relativity. Hence, the method is called the post Newton approximation. The theory can be used to compare general relativity and Newton's laws, and to test various gravitational effects in the celestial mechanics. The second method deals with the field in a lower-order approximation but does not assume the matter be in non-relativistic motion, so it is suitable for dealing with gravitational radiation.

In addition, there is a perturbation method, essentially seeking a new approximate solution from the strict solution known to us. Specifically, to find a reference system, the first requirement is to be very similar or close to the system under study, and the second requirement is to be able to solve it precisely. At this point, the difference between the properties of the studied system and that of the reference system is considered as a kind of perturbation, which can be approximately calculated according to the characteristics of the reference system.

Einstein believed that the final unified equation must be nonlinear, where the nonlinearity means that the equation solutions are complex and diverse, and therefore can have some solutions and linear superposition generation. Nonlinearity produces a variety of solutions that correspond to the variety of the real world.

5.5 Approximate Equations: Sources of Theoretical Diversity in Different Physics

The development of physics is that the laws of seemingly different phenomena in different fields are unified into different solutions in the same theoretical equation. The primary separated laws are only one aspect of the ultimate high-level unified law. The elementary equations are approximations of the general equations under specific conditions.

In the case of the theory of electromagnetism, Maxwell's equations were not discovered by Maxwell himself but on the basis of his predecessors' summary of the basic laws of electromagnetic phenomena. Oersted, Ampere, and some others proposed the theory that electric fields produce magnetic fields, while Faraday proposed the theory that magnetic fields produce electric fields, Faraday's law of induction. On the basis of these theories, Maxwell put forward the hypothesis of "displacement current". On this basis, Maxwell's equations was formed together, and thus electricity and magnetism reached a complete unity, forming a new theory of electromagnetic fields. Gauss's law describes how electric charges generate electric fields, Maxwell's ampere laws describe how electric currents and time-varying electric fields generate magnetic fields, Faraday's induction law describes how time-varying magnetic fields generate electric fields, and, finally, all of them together form the ultimate electromagnetic field equation.

A better example is the general relativity, which magically unified Newton's two theoretical systems: Newton's equations of motion and Newton's equations of gravity, both as approximations and corollaries of the Einstein field equation. Newton's gravity and Newton's mechanics are separate sets of physical notions. Einstein, in order to make the principle of relativity standing up for itself, suggested the Einstein field equations. What is amazing is that the equation's description of gravity naturally takes the form of Newton's gravity as the limit, and it also has incorporated Newton's equations of motion, but generalize it in curved space-time. When Einstein first built the general relativity, he thought that the fundamental equations of the general relativity had two parts: the field equation and the equations of motion. Later, Einstein and Fokker respectively proved that equations of motion could be derived from field equations, so that there was only one fundamental equation for the general relativity, the Einstein field equation. Newton's two sets of theoretical systems are thus deeply unified by Newton's true successor, Einstein.

It was thought that the kinetic equation should be attached to the principle of relativity, but it turned out that instead of this additional "mechanical equation", the "mechanics" could be derived directly from the "principle", which embodied the power of the principle. The absolute derivative of the Einstein tensor is zero, which contains abundant information about the motion of the material: for the ideal fluid, it is the equation of motion of the material field; for the ideal fluid whose pressure is zero, that is dust, the world line of dust can be obtained as the geodesic line. The world line is also a geodesic line for any object whose gravity is weak enough and small enough. So, the assumption that the world line of free particles is a geodesic is no longer an independent basic assumption, where Newton's mechanics has been absorbed, as an additional bonus to the general relativity.

In fact, the general relativity is more than an elaborated theory of gravity. The general relativity is a profound revolution in space-time and matter. The Einstein field equation is only a temporary concrete realization of the general relativity principle. In Einstein's words, "I had no doubt for a moment that this field equation formulation was only a stopgap measure in order to give the principle of relativity a preliminary, self-explanatory representation. Because it is essentially just a theory of gravitational field that has been somewhat artificially separated from the total field whose structure is not yet known". Einstein's field equation is also far more intelligent and profound than Newton's equations.

In the final unified theory, the equations of motion of physical systems must be extremely rich, and the existing mechanical equations are unified equations in different conditions under various approximations or inferences. In the process of human recognition of physical laws, various kinds of concrete phenomena have been preliminarily refined, and various equations in specific application ranges under various specific conditions have been obtained. These equations have been verified in their respective conditions and scopes of application, but these are only various parts and one-side of the underlying unified theory. Perhaps they all have some rationality from the phenomenological level. But, from the perspective of theoretical self-consistency and completeness, they are all problematic. Only by revealing the underlying unified equation can we truly establish a self-consistent and complete description of nature. Then, we can glimpse the great true beauty.

When I reach the top, I can see all mountains in a single glance.

5.6 *Initial Conditions: Sensitivity of Initial Values*

In the unified theory, the Lagrangian, or equation of motion, of a physical system must be highly nonlinear. Nonlinear equations may generate abundant dynamical behaviors, such as strange attractors, multi-stability and chaos, spontaneous breaking of symmetry, coexistence of multiple attractors, and so on. A typical property of nonlinearity is the sensitivity to initial values, unlike Newton's equations of mechanics which are rather monotone. Newton's equations of gravitation have a slightly richer set of solutions that can be a variety of conics. For the Einstein field equations of the general relativity, not to mention the wide variety of possible solutions of space-time structure, the geodesic of kinematics is also very rich. For a straight space-time, the geodesic line can only be a straight line, but for a curved space-time, the geodesic line will be very complex and colorful. Different initial conditions may produce various kinds of colorful solutions, which also make the world full of possibilities. But they are all governed by the same set of equations.

5.7 *Multi-stability: Sources of Symmetry-Breaking and Diversity of Stable Structures*

Another typical characteristic of nonlinear equations is multi-stability. Different initial values not only have very different evolutionary trajectories but also have completely different final states that may lead to different attractors. An attractor is actually a stable observable state corresponding to a stable structure, and multi-stability gives the possibility of many different stable structures under the same equation.

In the Higgs mechanism, where the vacuum state of a physical system is determined by the Lagrangian, why doesn't it have the same symmetry as the Lagrangian? The secret is that many physical systems have degenerate vacuum states. This is even clearer in the language of nonlinear dynamical systems, where the equilibrium or attractor of a dynamical system is determined by a dynamical equation. Why doesn't it have the symmetry of the dynamical equation? The secret is that some dynamical systems have multiple attractors that coexist.

We hope that in the final unified equation, the Higgs mechanism is no longer introduced externally to find the origin of mass but is required as a deep principle. An effect of spontaneously coexisting multiple attractors is due to the inherent nonlinearity of the equation.

5.8 *Attractor: Source of Diversity of Stable Structure*

Black holes are very complex solutions of general relativity. According to the experience from chaos theory, it is possible to produce more and more complex solutions such as black rings and black attractors for general relativity. In addition, these solutions are simple stable structures that could be used to describe elementary particles, or even to find some connection between elementary particles and black holes.

The final theory should not be a theory of everything which, like the current standard models, ties together all the known elementary particles. The theory is a theory, which should not contain all the details of everything, but it should give a theoretical prediction of all the possibilities. The attractor itself is rich in topological structure, which can explain all kinds of elementary particles. We may have a full new understanding of what is called an elementary particle, and these elementary particles can be described by some kind of attractors or knots and links on the attractors. But elementary particles are not fundamental; they are some special stable structures under some special energy standards.

5.9 *Key Parameters: Sources of Phase Transition and Mutation*

In chaos theory, changes in the parameters of an equation can lead to all sorts of interesting system dynamics. Bifurcations are generated by small and continuous changes in the system parameters (bifurcation parameters), resulting in sudden changes of the stability or topological structure of the system. Therefore, the final unified equation may also produce various complex dynamic characteristics under different varying parameters. For a concrete problem, we choose a parameter to describe the corresponding system characteristic, and the system will produce bifurcation within a certain parameter change, corresponding to the kind of phase transition in the system. This is similar to the fact that water can exist in the form of ice (solid phase), water (liquid phase) and steam (vapor phase), at different temperatures (parameters). We all know that at a certain temperature, several phases of water can transform into each other. Similarly, the attractors mentioned above represent some special stable structures under certain conditions, and some key parameters represent these "special conditions". Under the change of these parameters, the "bifurcation", "mutation" and "phase transition", which are common in the real world, will occur.

Under certain conditions, different vacuum phases will also transform into each other, which is the vacuum phase transition. Different types of interactions can lead to different types of vacuum states. So, for the same interaction, different types of vacuum states may also occur under different conditions. These different states of vacuum are known as different "vacuum phases" in physics.

5.10 *Emerging as a Whole: Complex Behavior Can Occur from Simple Equations*

Chaos theory tells us that, when considering the interaction of a large number of complex individuals, even if each individual is determined by a simple equation, the whole system can have very complex behavior. Because the unified equation is a complex and profound equation, even more colorful phenomena can emerge.

The so-called emergent property means that this property does not exist in any single element, and the system is expressed at a low level, but the phenomenon appears at a high level, so it is called "emergence". System science believes that the reason why the system function often appears, as "the whole is greater than the sum of the parts", is because the system has emerged a new quality, where "greater than the part" is the emerging new quality. This emergence of the system behavior is the result of a nonlinear interaction between some adaptive subjects of the system. Emergence is a transition from a low level to a high level, in the form of mutation in performance and organization of the macro system based on the evolution of the microscopic subject. In this process, new qualities can be generated from old ones. This often involves the contradiction between the so-called reductionism and holism. I have no intention to argue about the philosophy here but discuss it from a scientific point of view. In fact, both show the same thing, and simple equations can produce complex phenomena. Compared with the philosophical discussion with nothingness, it is more important to understand how to find the simple equation behind, how to explore the rich complexity that this simple equation can produce, and so on. These are all practical scientific problems.

If you are still skeptical about how simple and profound equations can produce rich and colorful results, I suggest you play with the Mandelbrot fractals. You can take a long time to zoom in on any interesting part and keep zooming in. I believe that you can see the endless magical geometry. If you have tasted Mandelbrot's fractals, I think you will be amazed that such a simple equation can produce such a colorful and endless array of fantastic fractals. The Mandelbrot set is the most bizarre and most magnificent geometric figure ever made by mankind. It was once called "the fingerprint of God". Such a set of points comes from a very simple iterative formula: $Z_{n+1} = (Z_n)^2 + C$, as shown in Fig. 3.

For nonlinear iterative formulas, all results in an infinite iteration can keep finite values with the collection of the plural C, which constitutes the Mandelbrot set.

5.11 *After the Unification of Physics, the Light of Principle Will Shine Everywhere*

It can be predicted that the final unified equation would be more nonlinear and complex, and the Einstein field equation is only a special approximation of the unified equation. Yet, even if the final equation is found, it doesn't mean the end

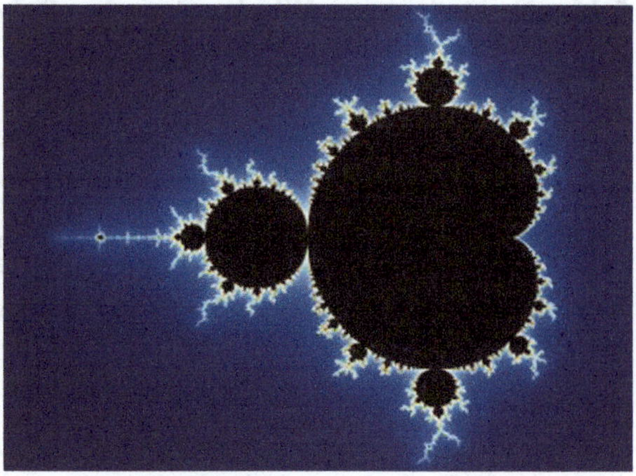

Fig. 3 God's fingerprint: Mandelbrot fractal https://en.wikipedia.org/wiki/Mandelbrot_set

Fig. 4 The symmetric equation symmetrically produces two asymmetric attractors and the spontaneous nonlinear system symmetry-breaking

of theoretical physics. The direction of research will be from the previous bottom-up approach to the future top-down development. In fact, finding a solution of the final unified equation can continue to be studied for hundreds of years, which may be aided by the enormous computing power and artificial intelligence technology generated by the development of computer technology. The future physics will start from the principal equation, combine the computer modeling and simulation, and be integrated with the massive data analysis, to solve all kinds of concrete problems and to solve all kinds of complex problems which previously thought impossible. The theoretical equations in physics and advanced mathematical theory will be more deeply involved in materials, chemistry, biology, society and many other scientific fields, which will be a greater manifestation of the power of physical equations. The light of principle will illuminate all corners of science (Fig. 4).

6 Noether Theorem

There are many quantities in nature that do not evolve over time, named conserved variables, as evidenced by a large number of experiments.

Therefore, when constructing the Lagrangian quantity that describes the physical system from the field as the starting point, it is necessary to correctly reveal the conservation law of the system.

At the beginning of the 20th century, German female mathematician Noether discovered the relationship between continuous symmetry transformation and conservation law, which makes it easy to find the conservation laws of a physical system through the interacting Lagrangian form.

Assuming that the interaction cation of the system S is invariant under the continuous transformation of the field, in the infinitesimal transformation of the parameters, one has

$$X'_{\mu} = x_{\mu} + \delta x_{\mu} \tag{65}$$

At this time, the amount of field $\varphi_a(x)$ and the Lagrangian density are $\mathscr{L}\left(\varphi(x), \partial\varphi(x)/\partial x_{\mu}\right)$, which will also have a corresponding infinitesimal change as follows:

$$\varphi'_a\left(x'\right) = \varphi_a(x) + \delta\varphi_a(x) \tag{66}$$

$$\mathscr{L}'\left(x'\right) = \mathscr{L}(x) + \delta\mathscr{L}(x) \tag{67}$$

In the above formula, in order to make the expression concise, the dependence of the Lagrangian density on the amount of field is hidden. Variation of the field quantity defined by the above formula gives

$$\begin{aligned} \delta\varphi_a(x) &= \varphi'_a\left(x'\right) - \varphi_a(x) \\ &= \varphi'_a\left(x'\right) - \varphi'_a(x) + \varphi'_a(x) - \varphi_a(x) \\ &= \delta_x\varphi'_a + \delta_\varphi\varphi_a \\ &= \frac{\partial\varphi_a}{\partial x_\mu}\delta x_\mu + \delta_\varphi\varphi_a \end{aligned} \tag{68}$$

In the above formula, the second-order small amount has been neglected. Assume that the system (the amount of action) remains invariant under the infinitesimal transformation of space-time coordinates, that is,

$$\delta S = \int_{R'} d^4x'\mathscr{L}'\left(x'\right) - \int_R d^4x\mathscr{L}(x) = 0 \tag{69}$$

An important conclusion will follow, that is Noether's theorem. In the above formula, R' and R are the integral areas of x' and x, respectively. Therefore, the change in the amount of action can be expressed as

$$\delta S = \int_{R'} d^4x' \mathscr{L}(x) + \int_{R'} d^4x' \delta \mathscr{L}(x) - \int_{R} d^4x \mathscr{L}(x) \tag{70}$$

Replacing the integral element of the above formula yields

$$D^4x' = J\left(\frac{x'}{x}\right) d^4x = \text{Det}\left(\frac{\partial x'^{\mu}}{\partial x^{\nu}}\right) d^4x = \left(1 + \frac{\partial \delta x^{\mu}}{\partial x^{\mu}}\right) d^4x \tag{71}$$

where J is the Jacobian—Jacobi Determinant. This is obtained after the second-order small amount δS is omitted.

It can be changed to

$$\delta S = \int_{R} d^4x \delta \mathscr{L}(x) + \int_{R} d^4x \mathscr{L}(x) \frac{\partial \delta x^{\mu}}{\partial x^{\mu}} \tag{72}$$

where

$$\begin{aligned} \delta \mathscr{L}(x) &= \mathscr{L}\left(\varphi_a'\left(x'\right), \partial_{\mu}' \varphi_a'\left(x'\right)\right) - \mathscr{L}\left(\varphi_a(x), \partial_{\mu} \varphi_a(x)\right) \\ &= \delta_{\varphi} \mathscr{L} + \delta_x \mathscr{L} \end{aligned} \tag{73}$$

$$\delta_x \mathscr{L} = \mathscr{L}\left(\varphi_a\left(x'\right), \partial_{\mu}' \varphi_a\left(x'\right)\right) - \mathscr{L}\left(\varphi_a(x), \partial_{\mu} \varphi_a(x)\right) = \frac{\partial \mathscr{L}}{\partial x^{\mu}} \delta x^{\mu} \tag{74}$$

$$\begin{aligned} \delta_{\varphi} \mathscr{L} &= \mathscr{L}\left(\varphi_a'\left(x'\right), \partial_{\mu}' \varphi_a'\left(x'\right)\right) - \mathscr{L}\left(\varphi_a\left(x'\right), \partial_{\mu}' \varphi_a\left(x'\right)\right) \\ &= \frac{\partial \mathscr{L}(x)}{\partial \varphi_a} \delta_{\varphi} \varphi_a(x) + \frac{\partial \mathscr{L}(x)}{\partial\left(\partial^{\mu} \varphi_a\right)} \delta_{\varphi}\left(\frac{\partial \varphi_a(x)}{\partial x_{\mu}}\right) \\ &= \left[\frac{\partial \mathscr{L}(x)}{\partial \varphi_a} \delta_{\varphi} \varphi_a(x) - \frac{\partial}{\partial x_{\mu}}\left(\frac{\partial \mathscr{L}(x)}{\partial\left(\partial^{\mu} \varphi_a\right)}\right) \delta_{\varphi} \varphi_a(x)\right] + \frac{\partial}{\partial x_{\mu}}\left[\frac{\partial \mathscr{L}(x)}{\partial\left(\partial^{\mu} \varphi_a\right)} \delta_{\varphi} \varphi_a(x)\right] \\ &= \left[\frac{\partial \mathscr{L}(x)}{\partial \varphi_a} - \frac{\partial}{\partial x_{\mu}}\left(\frac{\partial \mathscr{L}(x)}{\partial\left(\partial^{\mu} \varphi_a\right)}\right)\right] \delta_{\varphi} \varphi_a(x) + \frac{\partial}{\partial x_{\mu}}\left[\frac{\partial \mathscr{L}(x)}{\partial\left(\partial^{\mu} \varphi_a\right)} \delta_{\varphi} \varphi_a(x)\right] \end{aligned} \tag{75}$$

Note that the Lorentz indices and field indices repeated in the above expressions are summarized together. Substituting the integral gives

$$\begin{aligned} &\int_{R} d^4x \left\{\left[\frac{\partial \mathscr{L}(x)}{\partial \varphi_a} - \frac{\partial}{\partial x_{\mu}}\left(\frac{\partial \mathscr{L}(x)}{\partial\left(\partial^{\mu} \varphi_a\right)}\right)\right] \delta_{\varphi} \varphi_a(x) \right. \\ &\left. + \frac{\partial}{\partial x_{\mu}}\left[\frac{\partial \mathscr{L}(x)}{\partial\left(\partial^{\mu} \varphi_a\right)} \delta_{\varphi} \varphi_a(x) + \mathscr{L}(x) \delta x_{\mu}\right]\right\} = 0 \end{aligned}$$

The first part of the above formula should be zero according to the Lagrangian equation of the field. The second part is due to the randomicity of the volume area

R, and the integrand is always zero. Combining with $\delta\varphi_a(x)$ gives

$$\begin{aligned}
&\frac{\partial}{\partial x_\mu}\left[\frac{\partial \mathscr{L}(x)}{\partial(\partial^\mu \varphi_a)}\delta_\varphi \varphi_a(x) + \mathscr{L}(x)\delta x_\mu\right] \\
&= \frac{\partial}{\partial x_\mu}\left[\frac{\partial \mathscr{L}(x)}{\partial(\partial^\mu \varphi_a)}\left(\delta\varphi_a(x) - \frac{\partial \varphi_a}{\partial x_\nu}\delta x_\nu\right) + \mathscr{L}(x)\delta x_\mu\right] \\
&= \frac{\partial}{\partial x_\mu}\left[\frac{\partial \mathscr{L}(x)}{\partial(\partial^\mu \varphi_a)}\delta\varphi_a(x) - \left(\frac{\partial \mathscr{L}(x)}{\partial(\partial^\mu \varphi_a)}\frac{\partial \varphi_a}{\partial x_\nu} - g_{\mu\nu}\mathscr{L}(x)\right)\delta x^\nu\right] \\
&\equiv \frac{\partial J_\mu}{\partial x_\mu} = 0
\end{aligned} \tag{76}$$

The above equation shows that the system has a corresponding conserved currents with respect to a continuous symmetry transformation. This current can be determined by the Lagrangian density of the field.

According to the above formula and Gauss's theorem, one has

$$\begin{aligned}
\int d^3x \partial_\mu J^\mu &= \int d^3x \partial_0 J^0 + \int d^3x \partial_i J^i = \frac{d}{dt}\int d^3x J^0 + \int d\sigma_i J^i \\
&= \frac{d}{dt}Q + \text{ surface term}
\end{aligned} \tag{77}$$

Assuming that the field disappears quickly at infinity, the last item in the above equation, whose area is the integration component, tends to zero. From the current conserving, one obtains conservation of charge, as follows:

$$\frac{d}{dt}Q = 0 \tag{78}$$

7 Goldstone Theorem

One can write the Lagrangian as follows:

$$\mathscr{L} = \left(\partial_\mu \Phi\right)^\dagger \left(\partial^\mu \Phi\right) - V(\Phi) \tag{79}$$

where Φ is the column vector of the N-dimensional interior space. Assume that the potential energy $V(\Phi)$ has a local minimum at $\Phi = \Phi_0$, that is,

$$\left.\frac{\partial V(\Phi)}{\partial \Phi_a}\right|_{\Phi=\Phi_0} = 0, \quad \text{for } a = 1, 2, \ldots, N \tag{80}$$

Expand the field near this equilibrium point, so that the potential energy can be expanded to

$$V(\Phi) = V(\Phi_0) + \frac{1}{2}M_{ab}(\Phi - \Phi_0)^a(\Phi - \Phi_0)^b + \mathscr{O}\left((\Phi - \Phi_0)^3\right) \tag{81}$$

Because $\Phi = \Phi_0$ is a minimum, the corresponding mass matrix must be non-negative:

$$M_{ab} = \left.\frac{\partial^2 V(\Phi)}{\partial \Phi_a \partial \Phi_b}\right|_{\Phi=\Phi_0} \geq 0 \tag{82}$$

We want the Lagrangian to be invariant in the rotation of the N-dimensional interior space, which means that the Lagrangian is invariant in the following transformations:

$$\Phi \to \Phi' = U\Phi = \exp\left(i\theta_c T^c\right)\Phi \tag{83}$$

Thus, the Lagrangian is invariable. So, it follows that

$$\begin{aligned} V(\Phi) &= V\left(\Phi'\right) \\ &= V\left(\Phi_0\right) + \frac{1}{2}M_{ab}\delta\Phi^a\delta\Phi^b + \cdots \end{aligned} \tag{84}$$

where $\delta\Phi$ is Φ_0, satisfying the rotation transformation of the above condition. Consequently,

$$M_{ab}\delta\Phi^a\delta\Phi^b = 0 \tag{85}$$

if and only if $\delta\Phi^a = \delta\Phi^b = 0$; thus, $M_{ab} \neq 0$. If neither $\delta\Phi^a$ nor $\delta\Phi^b$ is zero, then there must held $M_{ab} = 0$, that is,

$$\begin{aligned} M_{ab} &= 0 \\ \delta\Phi^a &\neq 0 \\ \delta\Phi^b &\neq 0 \end{aligned} \tag{86}$$

where $\delta\Phi^a$ is nonzero. This means that the vacuum is degenerate in this dimension, the symmetry is broken spontaneously, and the rotation invariance of Φ_0^a is lost. Then, there is $M_{aa} = 0$, and the corresponding physical field mass of Φ^a is zero. That's the Goldstone theorem.

It is also shown that the dimension of the zero mass matrix, i.e., the number of Goldstone particles N_G, is equal to the dimension of symmetry-breaking in vacuum. And, the number of particles with mass, the Higgs, is normally equal to $N - N_G$.

8 Mathematical Representation of the Higgs Mechanism

The following is a detailed mathematical description of the Higgs mechanism. One can see that the definitions of vacuum and multistability play a decisive role.

(1) The Lagrange in the $U(1)$ gauge is

$$\mathscr{L}_{(1)} = \left(D_\mu\Phi\right)^\dagger \left(D^\mu\Phi\right) - \frac{1}{4}F_{\mu\nu}F^{\mu\nu} - V(\Phi) \tag{87}$$

where $D^{\mu} = (\partial^{\mu} - igA^{\mu})$, $F_{\mu\nu} = \partial_{\mu}A_{\nu} - \partial_{\nu}A_{\mu}$, $V(\Phi) = \mu^2\Phi^{\dagger}\Phi + \lambda\left(\Phi^{\dagger}\Phi\right)^2$, $\lambda > 0$, $\Phi = \frac{1}{\sqrt{2}}(\Phi_1 + i\Phi_2)$, and Φ_1, Φ_2 is a real scalar field.

The scalar field is transformed to

$$\Phi \to e^{-i\alpha(x)}\Phi \tag{88}$$

The transformation of the gauge field is

$$A_{\mu} \to A_{\mu} - \frac{1}{g}\partial_{\mu}\alpha(x) \tag{89}$$

This is the $U(1)$ local transformation.

(2) We have previously shown that $\mathscr{L}_{(1)}$ is invariant under the $U(1)$ local gauge transformation. Next, we show that the Higgs field vacuum is not zero under the $U(1)$ transformation, that is, SSB occurs.

Here, $V(\Phi)$ is analogous to the ferromagnetic case mentioned earlier: Higgs has a vacuum of $|\langle\Phi\rangle| = \sqrt{\frac{-\mu^2}{2\lambda}} \neq 0$, a vacuum broken defect, and when the angle of $\langle\Phi\rangle$ is 0, one has $\langle\Phi_1\rangle = \sqrt{\frac{-\mu^2}{\lambda}} = v$, $\langle\Phi_1\rangle = 0$. At this point, $\mathscr{L}_{(1)}$ represents a system whose local specification is break the symmetric property. So, one can change the coordinates for the new field as follows:

$$\Phi_1' \equiv \Phi_1 - v;\ \Phi_2' \equiv \Phi_2 \tag{90}$$

The purpose of this transformation is to make the vacuum zero, namely $< \Phi_1' >= 0$, $< \Phi_2' >= 0$.

Next, the effect of this violation of local gauge symmetry is examined.

Put $\Phi_{1,2}'$ into the primitive formula

$$\begin{aligned}
&(D_{\mu}\Phi)^{\dagger}(D^{\mu}\Phi) = \left[(\partial_{\mu} + igA_{\mu})\frac{1}{\sqrt{2}}(\Phi_1 - i\Phi_2)\right]\left[(\partial^{\mu} - igA^{\mu})\frac{1}{\sqrt{2}}(\Phi_1 + i\Phi_2)\right)\Big]\\
&= \frac{1}{2}\left[(\partial_{\mu} + igA_{\mu})(\Phi_1' + v - i\Phi_2')\right]\left[(\partial^{\mu} - igA^{\mu})(\Phi_1' + v + i\Phi_2')\right]\\
&= \frac{1}{2}\left[\partial_{\mu}(\Phi_1' + v - i\Phi_2') + igA_{\mu}(\Phi_1' + v - i\Phi_2')\right]\\
&\times\left[\partial^{\mu}(\Phi_1' + v + i\Phi_2') - igA^{\mu}(\Phi_1' + v + i\Phi_2')\right]\\
&= \frac{1}{2}\left(\partial_{\mu}\Phi_1' - i\partial_{\mu}\Phi_2' + igA_{\mu}\Phi_1' + igA_{\mu}v + gA_{\mu}\Phi_2'\right)\\
&\times\left(\partial^{\mu}\Phi_1' + i\partial^{\mu}\Phi_2' - igA^{\mu}\Phi_1' - igA^{\mu}v + gA^{\mu}\Phi_2'\right)\\
&= \frac{1}{2}\left[(\partial_{\mu}\Phi_1' + gA_{\mu}\Phi_2') - i(\partial_{\mu}\Phi_2' - gA_{\mu}\Phi_1') + igA_{\mu}v\right]
\end{aligned}$$

$$\times\left[\left(\partial^{\mu}\Phi_1'+gA^{\mu}\Phi_2'\right)+i\left(\partial^{\mu}\Phi_2'-gA^{\mu}\Phi_1'\right)-igA^{\mu}v\right]$$
$$=\frac{1}{2}\left[\left(\partial_{\mu}\Phi_1'+gA_{\mu}\Phi_2'\right)\left(\partial^{\mu}\Phi_1'+gA^{\mu}\Phi_2'\right)+i\left(\partial_{\mu}\Phi_1'+gA_{\mu}\Phi_2'\right)\right.$$
$$\times\left(\partial^{\mu}\Phi_2'-gA^{\mu}\Phi_1'\right)\big]-i\left(\partial_{\mu}\Phi_1'+gA_{\mu}\Phi_2'\right)gA^{\mu}v$$
$$-i\left(\partial_{\mu}\Phi_2'-gA_{\mu}\Phi_1'\right)\left(\partial^{\mu}\Phi_1'+gA^{\mu}\Phi_2'\right)$$
$$+\left(\partial_{\mu}\Phi_2'-gA_{\mu}\Phi_1'\right)\left(\partial^{\mu}\Phi_2'-gA^{\mu}\Phi_1'\right)-\left(\partial_{\mu}\Phi_2'-gA_{\mu}\Phi_1'\right)gA^{\mu}v$$
$$+igA_{\mu}v\left(\partial^{\mu}\Phi_1'+gA^{\mu}\Phi_2'\right)-gA_{\mu}v\left(\partial^{\mu}\Phi_2'-gA^{\mu}\Phi_1'\right)+g^2v^2A_{\mu}A^{\mu}$$
$$=\frac{1}{2}\left[\left(\partial_{\mu}\Phi_1'+gA_{\mu}\Phi_2'\right)^2+\left(\partial_{\mu}\Phi_2'-gA_{\mu}\Phi_1'\right)^2\right.$$
$$\left.-2gA^{\mu}v\left(\partial_{\mu}\Phi_2'-gA_{\mu}\Phi_1'\right)+g^2v^2A_{\mu}A^{\mu}\right]$$
$$=\frac{1}{2}\left(\partial_{\mu}\Phi_1'+gA_{\mu}\Phi_2'\right)^2+\frac{1}{2}\left(\partial_{\mu}\Phi_2'-gA_{\mu}\Phi_1'\right)^2$$
$$-gA^{\mu}v\left(\partial_{\mu}\Phi_2'-gA_{\mu}\Phi_1'\right)+\frac{1}{2}g^2v^2A_{\mu}A^{\mu} \tag{91}$$

At this point, the vacuum of the system is zero, which means that the system is symmetric. The last term of the above formula indicates that the gauge field $A_{\mu}A^{\mu}$ obtains mass. Although mass is obtained by the gauge field, of course, the upper formula is not physical because there is a nonphysical mixing term $-gA^{\mu}v\left(\partial_{\mu}\Phi_2'-gA_{\mu}\Phi_1'\right)$, which is a mixture of the gauge Boson and the scalar field. In order to obtain a true physical state, this term must be eliminated. One can do this by choosing the coordinates of when—Polar coordinates—they describe tiny oscillations in the shape of the vacuum, called parameterization.

(3) "Parameterization".

Instead of using the real component $\Phi_1(x)$, $\Phi_2(x)$ to indicate $\Phi(x)$, we introduce the real field $\eta(x)$ along the direction of the vacuum state vector v, and the real field $\xi(x)$ along the perpendicular direction, to the vacuum state vector v, with parameterization

$$\Phi''(x)=\frac{1}{\sqrt{2}}(v+\eta(x))e^{\frac{i(x)}{v}} \tag{92}$$

$$\left(\Phi''(x)\right)^{\dagger}=\frac{1}{\sqrt{2}}e^{\frac{-i\xi(x)}{v}}(v+\eta(x)) \tag{93}$$

It verifies that the new field vacuum value introduced in this way is 0:

$$<|\sqrt{\Phi''^{\dagger}\Phi''}|>=\frac{v}{\sqrt{2}} \tag{94}$$

Furthermore, because arg $<\Phi>=0$, so $<\xi(x)>=0$, therefore he can get

$$\begin{aligned}<|\sqrt{\Phi'^{\dagger}\Phi''}|>&=<|\sqrt{\frac{1}{2}e^{\frac{-i\varepsilon(z)}{v}}\left(v+\xi(x)\left(v+\xi(x)e^{\frac{i\zeta(z)}{v}}\right|>}\\&=\frac{1}{\sqrt{2}}<|\sqrt{v^2+2v\eta+\eta^2}|>\end{aligned} \tag{95}$$

To make it be equal to the upper form, $<\eta>=0$ is required.
Inserting $\eta(x)$ and $\xi(x)$ into $\mathscr{L}_{(1)}$ yields

$$
\begin{aligned}
\mathscr{L}_{(2)} &= \left(D_\mu \Phi''\right)^\dagger \left(D^\mu \Phi''\right) - \frac{1}{4}F_{\mu\nu}F^{\mu\nu} - V\left(\Phi''\right) \\
&= \left[\left(\partial_\mu + igA_\mu\right)\Phi''^\dagger\right]\left[\left(\partial^\mu - igA^\mu\right)\Phi''\right] - \frac{1}{4}F_{\mu\nu}F^{\mu\nu} - \left[\mu^2\Phi''^\dagger\Phi'' + \lambda\left(\Phi''^\dagger\Phi''\right)^2\right] \\
&= \left[\left(\partial_\mu + igA_\mu\right)\frac{1}{\sqrt{2}}e^{\frac{-i\xi}{v}}(v+\eta)\right]\left[\left(\partial^\mu - igA^\mu\right)\frac{1}{\sqrt{2}}(v+\eta)e^{\frac{i\xi}{v}}\right] - \frac{1}{4}F_{\mu\nu}F^{\mu\nu} \\
&- \mu^2\frac{1}{\sqrt{2}}e^{\frac{-i\xi}{v}}(v+\eta)\frac{1}{\sqrt{2}}(v+\eta)e^{\frac{i\xi}{v}} - \lambda\left[\frac{1}{\sqrt{2}}e^{\frac{-i\xi}{v}}(v+\eta)\frac{1}{\sqrt{2}}(v+\eta)e^{\frac{i\xi}{v}}\right]^2 \\
&= \frac{1}{2}\left[\partial_\mu\left(e^{\frac{-i\xi}{v}}(v+\eta)\right) + igA_\mu e^{\frac{-i\xi}{v}}(v+\eta)\right]\left[\partial^\mu\left((v+\eta)e^{\frac{i\xi}{v}}\right) - igA^\mu(v+\eta)e^{\frac{i\xi}{v}}\right] \\
&- \frac{1}{4}F_{\mu\nu}F^{\mu\nu} - \frac{1}{2}\mu^2(v+\eta)^2 - \frac{1}{4}\lambda(v+\eta)^4 \\
&= \frac{1}{2}\left[\partial_\mu\left(e^{\frac{-i\xi}{v}}v\right) + \partial_\mu\left(e^{\frac{-i\xi}{v}}\eta\right) igA_\mu(v+\eta)e^{\frac{-i\xi}{v}}\right] \\
&\times\left[\partial^\mu\left(ve^{\frac{i\xi}{v}}\right) + \partial^\mu\left(\eta e^{\frac{i\xi}{v}}\right) - igA^\mu(v+\eta)e^{\frac{i\xi}{v}}\right] \\
&- \frac{1}{4}F_{\mu\nu}F^{\mu\nu} - \frac{1}{2}\mu^2\left(v^2+2v\eta+\eta^2\right) - \frac{1}{4}\lambda\left(v^4+4v^3\eta+6v^2\eta^2+4v\eta^3+\eta^4\right) \\
&= \tfrac{1}{2}\left[v\left(-\tfrac{i}{v}\right)e^{-\frac{i\xi}{v}}\partial_\mu\xi + \eta\left(-\tfrac{\xi}{v}\right)e^{-\frac{i\xi}{v}}\partial_\mu\xi + e^{-\frac{i\xi}{v}}\partial_\mu\xi + igA_\mu(v+\eta)e^{-\frac{i\xi}{v}}\right] \\
&\times\left[v\left(\tfrac{i}{v}\right)e^{\frac{i}{v}}\partial^\mu\xi + \eta\left(\tfrac{i}{v}\right)e^{\frac{i\xi}{v}}\partial^\mu\xi + e^{\frac{i\xi}{v}}\partial^\mu\xi - igA^\mu(v+\eta)e^{\frac{\xi}{v}}\right] \\
&- \tfrac{1}{4}F_{\mu\nu}F^{\mu\nu} + \left(\tfrac{1}{2}\lambda v^2v^2 + \tfrac{1}{2}\lambda v^2 2v\eta + \tfrac{1}{2}\lambda v^2\eta^2\right) \\
&- \left(\tfrac{1}{4}\lambda v^4 + \tfrac{1}{4}\lambda 4v^3\eta + \tfrac{1}{4}\lambda 6v^2\eta^2 + \tfrac{1}{4}\lambda 4v\eta^3 + \tfrac{1}{4}\lambda\eta^4\right) \\
&= \tfrac{1}{2}\left[-i\partial_\mu\xi - i\tfrac{\eta}{v}\partial_\mu\xi + \partial_\mu\eta + igA_\mu(v+\eta)\right]e^{-\frac{i\xi}{v}} \\
&e^{\frac{i\xi}{v}}\left[i\partial^\mu\xi + i\tfrac{\eta}{v}\partial^\mu\xi + \partial^\mu\eta - igA^\mu(v+\eta)\right] \\
&- \tfrac{1}{4}F_{\mu\nu}F^{\mu\nu} + \left(\tfrac{1}{2}\lambda v^4 + \lambda v^3\eta + \tfrac{1}{2}\lambda v^2\eta^2\right) \\
&- \left(\tfrac{1}{4}\lambda v^4 + \lambda v^3\eta + \tfrac{3}{2}\lambda v^2\eta^2 + \lambda v\eta^3 + \tfrac{1}{4}\lambda\eta^4\right) \\
&= \frac{1}{2}\left[\partial_\mu\xi\partial^\mu\xi + \frac{\eta}{v}\partial_\mu\xi\partial^\mu\xi - i\partial_\mu\xi\partial^\mu\eta - \partial_\mu\xi gA^\mu(v+\eta)\right. \\
&+ \frac{\eta}{v}\partial_\mu\xi\partial^\mu\xi + \frac{\eta^2}{v^2}\partial_\mu\xi\partial^\mu\xi - i\frac{\eta}{v}\partial_\mu\xi\partial^\mu\eta - \frac{\eta}{v}\partial_\mu\xi gA^\mu(v+\eta) \\
&+ i\partial_\mu\eta\partial^\mu\xi + i\frac{\eta}{v}\partial_\mu\eta\partial^\mu\xi + \partial_\mu\eta\partial^\mu\eta - i\partial_\mu\eta gA^\mu(v+\eta) \\
&- gA_\mu(v+\eta)\partial^\mu\xi - gA_\mu(v+\eta)\frac{\eta}{v}\partial^\mu\xi + igA_\mu(v+\eta)\partial^\mu\eta - \frac{1}{4}F_{\mu\nu}F^{\mu\nu} \\
&\left.+ gA_\mu(v+\eta)gA^\mu(v+\eta) + \frac{1}{4}\lambda v^4 - \lambda v^2\eta^2 - \lambda v\eta^3 - \frac{1}{4}\lambda\eta^4\right] \\
&= \frac{1}{2}\left[\partial_\mu\xi\partial^\mu\xi + 2\frac{\eta}{v}\partial_\mu\xi\partial^\mu\xi - 2gA^\mu(v+\eta)\partial_\mu\xi + \frac{\eta^2}{v^2}\partial_\mu\xi\partial^\mu\xi\right. \\
&\left.-2\frac{\eta}{v}gA^\mu(v+\eta)\partial_\mu\xi + \partial_\mu\eta\partial^\mu\eta + g^2(v+\eta)^2A_\mu A^\mu\right] \\
&- \frac{1}{4}F_{\mu\nu}F^{\mu\nu} - \frac{1}{4}\mu^2v^2 + \mu^2\eta^2 - \lambda v\eta^3 - \frac{1}{4}\lambda\eta^4
\end{aligned}
$$

$$
\begin{aligned}
&= \frac{1}{2}\left[\left(1+2\frac{\eta}{v}+\frac{\eta^2}{v^2}\right)\partial_\mu\xi\partial^\mu\xi-\left(1+\frac{\eta}{v}\right)2gA^\mu(v+\eta)\partial_\mu\xi+\partial_\mu\eta\partial^\mu\eta\right.\\
&\left.+g^2(v+\eta)^2A_\mu A^\mu\right]-\frac{1}{4}F_{\mu\nu}F^{\mu\nu}-\frac{1}{4}\mu^2v^2+\mu^2\eta^2-\lambda v\eta^3-\frac{1}{4}\lambda\eta^4\\
&= \frac{1}{2}\left[\left(1+\frac{\eta}{v}\right)^2\partial_\mu\xi\partial^\mu\xi-2g\frac{1}{v}(v+\eta)^2A^\mu\partial_\mu\xi+\partial_\mu\eta\partial^\mu\eta\right.\\
&\left.+g^2(v+\eta)^2A_\mu A^\mu\right]-\frac{1}{4}F_{\mu\nu}F^{\mu\nu}-\frac{1}{4}\mu^2v^2+\mu^2\eta^2-\lambda v\eta^3-\frac{1}{4}\lambda\eta^4\\
&\approx \frac{1}{2}\left[\partial_\mu\eta\partial^\mu\eta+\partial_\mu\xi\partial^\mu\xi-2gvA^\mu\partial_\mu\xi+g^2v^2A_\mu A^\mu\right]\\
&-\frac{1}{4}F_{\mu\nu}F^{\mu\nu}+\mu^2\eta^2-\lambda v\eta^3-\frac{1}{4}\lambda\eta^4
\end{aligned}
\tag{96}
$$

At this point, the vacuum of a system described by $\mathscr{L}_{(2)}$ is symmetric, but its motion is no longer symmetric. According to the final expression of $\mathscr{L}_{(2)}$, the η field has real mass, $m_\eta=\sqrt{-2\mu^2}$, and the gauge field A_μ also has mass, $m_{A_\mu}=gv$, but the ξ field still has no mass. From the following derivation, it is shown to be Goldstone boson. In the expression of $\mathscr{L}_{(2)}$, there is a direct coupling between ξ and A_μ, which we do not want to have. That is the unitary specification to be introduced next.

(4) 'Unitary normalization'

Make a unitary normalization

$$
\Phi'''(x)=e^{\frac{-i\xi(x)}{v}}\eta''(x)=\frac{1}{\sqrt{2}}(v+\eta(x)) \tag{97}
$$

$$
\left(\Phi'''(x)\right)^\dagger=\left(\eta''(x)\right)^\dagger e^{\frac{i\xi(x)}{v}}=\frac{1}{\sqrt{2}}(v+\eta(x)) \tag{98}
$$

$$
A_\mu(x)=A_\mu(x)-\frac{1}{gv}\partial_\mu\xi(x)\equiv B_\mu(x) \tag{99}
$$

Replacing the above unitary gauge with the corresponding physical quantity in $\mathscr{L}_{(1)}$ leads to

$$
\begin{aligned}
\mathscr{L}_{(3)} &= \left[\left(\partial_\mu+igB_\mu\right)\left(\Phi'''\right)^\dagger\right]\left[\left(\partial^\mu-igB^\mu\right)\Phi'''\right]\\
&-\frac{1}{4}F'_{\mu\nu}F'^{\mu\nu}-\left[\mu^2\left(\Phi'''\right)^\dagger\Phi'''+\lambda\left(\left(\Phi'''\right)^\dagger\Phi'''\right)^2\right]\\
&=\left[\left(\partial_\mu+igB_\mu\right)\frac{1}{\sqrt{2}}(v+\eta)\right]\left[\left(\partial^\mu-igB^\mu\right)\frac{1}{\sqrt{2}}(v+\eta)\right]-\frac{1}{4}F'_{\mu\nu}F'^{\mu\nu}\\
&-(\mu)^2\frac{1}{\sqrt{2}}(v+\eta)\frac{1}{\sqrt{2}}(v+\eta)-\lambda\left[\frac{1}{\sqrt{2}}(v+\eta)\frac{1}{\sqrt{2}}(v+\eta)\right]^2
\end{aligned}
$$

$$
\begin{aligned}
&= \frac{1}{2}\left[\partial_\mu \eta + igB_\mu(v+\eta)\right]\left[\partial^\mu \eta - igB^\mu(v+\eta)\right] - \frac{1}{4}F'_{\mu\nu}F'^{\mu\nu} \\
&- \frac{1}{2}\mu^2(v+\eta)^2 - \frac{1}{4}\lambda(v+\eta)^2 \\
&= \frac{1}{2}\left[\partial_\mu \eta \partial^\mu \eta - i\partial_\mu \eta g B^\mu(v+\eta) + igB_\mu(v+\eta)\partial^\mu \eta\right. \\
&\left.+g^2(v+\eta)^2 B_\mu B^\mu\right] - \frac{1}{4}F'_{\mu\nu}F^{\mu\nu} + \mu^2\eta^2 - \lambda v\eta^3 - \frac{1}{4}\lambda\eta^4 \\
&= \frac{1}{2}\left(\partial_\mu \eta\right)^2 + \mu^2\eta^2 - \frac{1}{4}\left(\partial_\mu B_\nu - \partial_\nu B_\mu\right)^2 + \frac{1}{2}g^2v^2 B_\mu B^\mu \\
&+ \frac{1}{2}g^2\eta(2v+\eta)B_\mu B^\mu - \lambda v\eta^3 - \frac{1}{4}\lambda\eta^4
\end{aligned} \tag{100}
$$

where $F'_{\mu\nu} = \partial_\mu B_\nu - \partial_\nu B_\mu$.

As can be seen from $\mathscr{L}_{(3)}$, it has a physical meaning. It describes a spin of 1 with a mass of gauge Boson B_μ and with a mass of $m_{B_\mu} = gv$, and a spin of 0 with a mass of scalar Boson η and $m_\eta = \sqrt{2}|\mu|$. The Goldstone Boson field ξ in the middle disappears, where η is the Higgs Boson.

The above very simple example provides a complete introduction to the specific content of the Higgs mechanism.

Now, we summarize the general properties of the Higgs mechanism:

(1) the symmetry of the interaction is conserved;

(2) the massless gauge theory satisfies renormalization;

(3) the total number of degrees of freedom of polarization is conserved. For example, in the $U(1)$ gauge transformation mentioned earlier, the total polarization degree of freedom before SSB $= 4 = (2A_\mu)$+(2Φ), while after SSB the total polarization degree of freedom is $= 4 = (3B_\mu)$+(1η);

(4) non-physical field In the case of $U(1)$ is $\xi(x)$;

(5) the number of gauge Boson that acquires mass, the number of symmetry-breaking of the possible number of GBs, and also in the case of $U(1)$, is 1;

(6) a possible combination of GBs and the gauge Boson to acquire mass, and in the case of $U(1)$, this combination is the mixed term $gvA^\mu\partial_\mu\Phi'_2$;

7) the Higgs mechanism itself does not predict the existence of the Higgs, which appears only when the property of conservation of the degree of freedom of polarization is required.

Author Index

X. Wang et al. (eds.), *Chaotic Systems with Multistability and Hidden Attractors*, Emergence, Complexity and Computation 40,
https://doi.org/10.1007/978-3-030-75821-9

Subject Index

X. Wang et al. (eds.), *Chaotic Systems with Multistability and Hidden Attractors*, Emergence, Complexity and Computation 40,
https://doi.org/10.1007/978-3-030-75821-9

The manufacturer's authorised representative in the EU is Springer Nature Customer Service Centre GmbH, Europaplatz 3, 69115 Heidelberg, Germany. If you have any concerns regarding our products, please contact ProductSafety@springernature.com

Printed and bound by CPI Group (UK) Ltd, Croydon, CR0 4YY
28/09/2026
02224065-0019